Brain Structure and Its Origins

Brain Structure and Its Origins

in Development and in Evolution of Behavior and the Mind

Gerald E. Schneider

The MIT Press
Cambridge, Massachusetts
London, England

© 2014 Massachusetts Institute of Technology

All rights reserved. No part of this book may be reproduced in any form by any electronic or mechanical means (including photocopying, recording, or information storage and retrieval) without permission in writing from the publisher.

MIT Press books may be purchased at special quantity discounts for business or sales promotional use. For information, please email special_sales@mitpress.mit.edu.

This book was set in Syntax LT Std and Times Ten LT Std by Toppan Best-set Premedia Limited, Hong Kong. Printed and bound in the United States of America.

Library of Congress Cataloging-in-Publication Data

Schneider, Gerald E.
 Brain structure and its origins : in development and in evolution of behavior and the mind / Gerald E. Schneider.
   pages cm
 Includes bibliographical references and index.
 ISBN 978-0-262-02673-4 (hardcover : alk. paper) 1. Brain. 2. Brain—Evolution. I. Title.
 QP376.S348 2014
 612.8′2—cd23
                          2013023059

10 9 8 7 6 5 4 3 2 1

This book is dedicated to the memory of my mother and father and my youngest sister, Linda Jean. Their enthusiastic support is always with me.

Contents

Detailed contents xi
Preface xxiii

PART I: INTRODUCTORY ORIENTATION 1

1   Getting Ready for a Brain Structure Primer 3
2   Methods for Mapping Pathways and Interconnections That Enable the Integrative Activity of the CNS 29

PART II: THE CENTRAL NERVOUS SYSTEM, FROM INITIAL STEPS TO ADVANCED CHORDATES 53

3   Evolution of Multicellular Organisms with Neuron-Based Coordination 55
4   Expansions of the Neuronal Apparatus of Success 67

PART III: INTRODUCTION TO CONNECTION PATTERNS AND SPECIALIZATIONS IN THE EVOLVING CNS 87

5   The Ancestors of Mammals: Sketch of a Pre-mammalian Brain 89
6   Some Specializations Involving Head Receptors and Brain Expansions 107
7   The Components of the Forebrain Including the Specialty of the Mammals: The Neocortex 117

PART IV: DEVELOPMENT AND DIFFERENTIATION: SPINAL LEVEL 137

8   The Neural Tube Forms in the Embryo, and CNS Development Begins 139
9   The Lower Levels of Background Support: Spinal Cord and the Innervation of the Viscera 153
    Intermission: The Ventricular System, the Meninges, and the Glial Cells 175

PART V: DIFFERENTIATION OF THE BRAIN VESICLES   179

10   Hindbrain Organization, Specializations, and Distortions   181
11   Why a Midbrain? Notes on Evolution, Structure, and Functions   205
12   Picturing the Forebrain with a Focus on Mammals   217
13   Growth of the Great Networks of Nervous Systems   235

PART VI: A BRIEF STUDY OF MOTOR SYSTEMS   263

14   Overview of Motor System Structure   265
15   Descending Pathways and Evolution   283
16   The Temporal Patterns of Movements   299

PART VII: BRAIN STATES   309

17   Widespread Changes in Brain State   311

PART VIII: SENSORY SYSTEMS   323

18   Taste   325
19   Olfaction   333
20   Visual Systems: Origins and Functions   355
21   Visual Systems: The Retinal Projections   371
22   The Visual Endbrain Structures   393
23   Auditory Systems   417

PART IX: THE FOREBRAIN AND ITS ADAPTIVE PRIZES: A SNAPSHOT   447

24   Forebrain Origins: From Primitive Appendage to Modern Dominance   449

PART X: THE HYPOTHALAMUS AND LIMBIC SYSTEM   465

25   Regulating the Internal Milieu and the Basic Instincts   467
26   Core Pathways of the Limbic System, with Memory for Meaningful Places   483
27   Hormones and the Shaping of Brain Structures   501
28   The Medial Pallium Becomes the Hippocampus   513
29   The Limbic Striatum and Its Outpost in the Temporal Lobe   537

PART XI: CORPUS STRIATUM 559

30   The Major Subpallial Structure of the Endbrain   561
31   Lost Dopamine Axons: Consequences and Remedies   583
     Intermission: Neurogenesis in Mature Brains   589

PART XII: THE CROWN OF THE MAMMALIAN CNS: THE NEOCORTEX   593

32   Structural Origins of Object Cognition, Place Cognition, Dexterity, and Planning   595
33   Basic Neocortical Organization: Cells, Modules, and Connections   617
34   Structural Change in Development and in Maturity   645

Figure Credits   669
Index   679

Detailed Contents

Preface   xxiii

PART I: INTRODUCTORY ORIENTATION   1

1 Getting Ready for a Brain Structure Primer   3
   The Nature of This Book   3
   Brain Space: Specifying Directions and Major Regions   5
   CNS Tissue   9
   Primitive Cellular Mechanisms   12
   Irritability and Conduction   13
   The Specialized Membrane of the Axon   15
   Specializations for Irritability   19
   Movement   19
   Secretion   20
   Synaptic Morphology and Types   21
   Related Cellular Dynamics   23
   Parallel Channels of Information Flow and Integrative Activity   25
   Endogenous Activity   25
   Readings   27

2 Methods for Mapping Pathways and Interconnections That Enable the Integrative Activity of the CNS   29
   Histology and Brain Architecture   29
   What Is Connected to What?   35
   Tracing a Pathway from Sensory Input to the Response (S to R)   38
   Pathways Seen as Bundles and Ribbons of Fibers   41
   Experimental Studies Needed to Trace Connections in the Brain   41
   Experimental Neuroanatomy: Tract-Tracing Methods   42
   Readings   51

PART II: THE CENTRAL NERVOUS SYSTEM, FROM INITIAL STEPS TO ADVANCED CHORDATES 53

3 Evolution of Multicellular Organisms with Neuron-Based Coordination 55
   Basic Survival Skills 55
   The First Neurons and Nerve Networks 56
   The Nervous System in Basic Concept 58
   Some Terminology for Brain Talk 59
   Introducing the Simplest Chordates 60
   Natural Selection and the Logic of Evolution 63
   The Behavioral Demands That Changed the Neural Tube 63
   Readings 65

4 Expansions of the Neuronal Apparatus of Success 67
   Why We Can Talk about Broad Aspects of Brain Evolution with Some Confidence 67
   Three Enlargements Formed the Primitive Brain 67
   Expansions of the Hindbrain 69
   Early Forebrain Expansion Served Olfactory Functions 72
   Midbrain Expansions, Distance Receptors, and Decussations 75
   A Second Expansion of the Forebrain with Invasion of Non-olfactory Inputs 77
   Concurrent Evolutionary Changes: Motor Control and Plasticity 79
   Concurrent Evolutionary Changes: Motivational States 80
   A Third Expansion of the Forebrain: Anticipating Events and Planning Actions 81
   Brain Sizes Relative to Body Weight Compared for Present-Day Animals 81
   Where We Stand in the Path to Our Goal 83
   Readings 84

PART III: INTRODUCTION TO CONNECTION PATTERNS AND SPECIALIZATIONS IN THE EVOLVING CNS 87

5 The Ancestors of Mammals: Sketch of a Pre-mammalian Brain 89
   The "Old Chassis" for the Building of Further Additions 90
   Subdividing the CNS 94
   The Basic Setup: Major Types of Neurons 96
   Sensory Channels of Conduction: Local Reflex, and Not so Local 97
   Old Ribbons to the Brain 101
   Ribbons to the Little Brain: The Cerebellum 103
   Readings 106

6 Some Specializations Involving Head Receptors and Brain Expansions 107
   Electroreception and the Cerebellum 107
   Infrared Sensors in Pit Vipers 109
   Echolocation and the Auditory System 110
   The Visual Systems of Primates 111
   Whisker Fields and Barrel Fields 111
   Other Behavioral Specializations with Corresponding Enlargements of Brain Parts 112
   Readings 116

7 The Components of the Forebrain Including the Specialty of the Mammals: The Neocortex 117

Forebrain Removal Experiments 117
Cat Behavior after Forebrain Removal 117
Rat Behavior Is Less Drastically Affected by Forebrain Removal 118
Pigeons after Forebrain Removal, Sparing the Optic Tracts 119
Forebrain Function: A Conclusion 120
The Problem of Species Differences in Brain Lesion Effects 120
The Phenomena of Diaschisis: An Answer to the Problem 120
What Can We Conclude about Interpreting Effects of Brain Lesions? 123
Role of the Corpus Striatum 123
The Rostral End of the Brainstem Located between the Hemispheres ('Tweenbrain) 124
Limbic System of the Forebrain 125
Olfactory System Origins of Major Functions of the Limbic Endbrain 126
Overview of the Roles of Corpus Striatum and Limbic Forebrain 128
Neocortex, the Grand Innovation of the Mammals: Appearance of New Pathways 128
But What Does It Do? Why Did Neocortex Evolve? 134
Taking Stock 134
Readings 135

PART IV: DEVELOPMENT AND DIFFERENTIATION: SPINAL LEVEL 137

8 The Neural Tube Forms in the Embryo, and CNS Development Begins 139

Before There Is Any CNS 139
The Onset of a Nervous System 140
Molecules from the Notochord Induce the Formation of the Nervous System 141
The Neural Crest Gives Rise to the PNS 142
Cell Proliferation in the Early Neural Tube 143
Diversity in Neuronal Migration 145
Differentiation of the Neurons Begins 149
Notes 150
Readings 150

9 The Lower Levels of Background Support: Spinal Cord and the Innervation of the Viscera 153

Major Features of Cord Structure 153
Questions from Comparing Different Species 156
The Local Reflex Channel and the Older Lemniscal Channels 158
The Mammalian Highway for Ascending Somatosensory Information 161
Cerebellar Channel 162
The Pathways of Regulation within the Spinal Cord Itself 163
The Pathways of Influence and Control from the Brain 164
A Reminder 165
Maintaining Stability of the Internal Environment: The Autonomic Nervous System 166
Three Major Divisions of the Motor System 166
A Sketch of the Autonomic Nervous System 167
Chemical Mediation at Autonomic Nervous System Synapses 171

xiv Detailed Contents

        The Enteric Nervous System   171
        Levels of Control of the Internal Environment   173
        Readings   173

Intermission: The Ventricular System, the Meninges, and the Glial Cells   175
        Readings   178

PART V: DIFFERENTIATION OF THE BRAIN VESICLES   179

10   Hindbrain Organization, Specializations, and Distortions   181
        A Glamorized Spinal Cord   181
        Vital Functions of the Hindbrain   183
        Routine Maintenance Services   184
        Hindbrain Participation in Mammalian Higher Functions   184
        The Isodendritic Core of the Brainstem   185
        Segmentation of the Hindbrain   186
        Columns and Cranial Nerves   188
        The Adult Hindbrain: Cell Groups and Axons of Passage   192
        Somatosensory Inputs from the Face   193
        The Evolution of Crossed Projections   194
        Hindbrain Sensory Channels in Mammals   197
        Hindbrain Specializations and Mosaic Evolution   199
        Readings   204

11   Why a Midbrain? Notes on Evolution, Structure, and Functions   205
        Primitive Vision   206
        Primitive Olfaction   206
        A Structural Consequence of the Priority of Escape Behavior for Survival   207
        The Midbrain Correlation Centers   208
        Outputs of Midbrain for Motor Control   210
        Mosaic Evolution of Midbrain   212
        Long Axons Passing through the Midbrain   212
        Readings   216

12   Picturing the Forebrain with a Focus on Mammals   217
        Pictures of Ancestral and Modern Endbrain   218
        Words for Forebrain Parts   220
        Major Structural and Functional Subdivisions of the 'Tweenbrain   220
        Major Parts of the Telencephalon of Mammals   222
        Origins and Course of Two Major Pathways of the Forebrain   222
        The Neocortex Is Involved in Both Major Systems   224
        Interim Review of Neuroanatomy   226
        Segmentation of the Forebrain   227
        Notes on Neocortical Origins   227
        Readings   232

13 Growth of the Great Networks of Nervous Systems   235
The Axonal Growth Cone   235
Signals That Shape the Development of Neuronal Circuits   239
Four Types of Chemical Guidance   244
Two Modes of Axon Growth   247
Formation of Maps in the Brain   247
Plasticity in Brain Maps   249
More Plasticity in the CNS: Collateral Sprouting   252
Modulation of Competitive Axonal Growth Vigor   255
Rules of Sprouting Apply to Development, with Implications for Evolutionary Change   256
Plasticity in the Small Interneurons of the Adult Brain   256
Structural Regression during Development and Its Purposes   257
Axon Loss in the Damaged CNS: Is Regeneration Possible?   258
Readings   259

PART VI: A BRIEF STUDY OF MOTOR SYSTEMS   263

14 Overview of Motor System Structure   265
A Functional Starting Point for the Study: Three Major Types of Movement Critical for Survival   265
Midbrain Control of the Three Types of General-Purpose Movement   266
The Midbrain Was the Connecting Link between the Primitive Forebrain and Motor Systems   267
Head Receptors and Locomotor Approach and Avoidance   268
Initiation of Foraging by Activity Intrinsic to the Brain   268
The Motor System Hierarchy   269
Locomotor Pattern Generation and Its Adjustments by Vestibular and Cerebellar Systems   269
Orienting of Head and Body   272
Grasping: The Third Major Type of Movement Controlled by the Midbrain   275
Comparative Anatomy of the Red Nucleus and Its Projection to the Spinal Cord   275
A Structural Approach to Understanding Motor Control: Begin with the Motor Neurons   277
The Spatial Arrangements of Somatic Motor Neurons in the Spinal Cord   278
Readings   281

15 Descending Pathways and Evolution   283
Axons Descending from Brain to Spinal Cord: Functional Groupings   283
Functions of the Descending Pathways: The Corticospinal Tract   286
Functions of the Descending Pathways: The Medial Hindbrain Tracks   287
Functions of the Descending Pathways: The Lateral Brainstem Tracks   287
A Conclusion with Application to Humans   289
The Brain Disconnected from the Motor Pattern Generators   289
Importance of the Corticospinal Tract for Innate and Learned Movements That Require Special Dexterity   290
The Nature of the Spinal Motor Pattern Generators   291
Motor Cortex in Phylogeny   291
Corticospinal Projections in Phylogeny   293
The Highest Levels of Motor Control   296
Readings   296

16 The Temporal Patterns of Movements   299
Three Types of Mechanism   299
Explaining Movement Dynamics in Terms of S-R Circuits   300
Central Programs Rather than Reflex Chaining   302
Many Fixed Action Patterns Are Centrally Generated   302
Reverberating Circuits within the Brain and Spinal Cord   303
Endogenous Activity of Single CNS Neurons   304
The Endogenous Clock   304
How It All Works at the Circuit Level   305
The Circuits Are Not Always Fixed   305
How Adequate Are These Concepts?   306
Readings   306

PART VII: BRAIN STATES   309

17 Widespread Changes in Brain State   311
Brain States Influenced by Widely Projecting Axon Systems   311
Cholinergic Systems   312
The Monoamine-Containing Systems   313
Serotonin, Another Monoamine Neurotransmitter Influencing Behavioral State   315
Diencephalic Origins of Other Widely Projecting Axon Systems   316
How Many Different Brain States?   318
Readings   320

PART VIII: SENSORY SYSTEMS   323

18 Taste   325
Pre-chordate Taste and Other Chemoreceptor Systems   325
Olfaction or Taste?   326
Visceral and Taste Inputs to the Hindbrain   326
Innervation of the Tongue   327
Distribution of Mammalian Taste Receptors   327
From Tongue to Telencephalon   328
Purposes of Taste: Routes to Motor Control   330
Readings   331

19 Olfaction   333
Sections through the Forebrain of Vertebrates   333
Olfactory Bulb Projections in Primitive Vertebrates   336
Variations in Relative Size of Olfactory Systems   338
Olfactory Bulb Projections in Mammals   339
Human and Small Mammalian Brains   341
Neuronal Organization as Depicted by Ramón y Cajal   341
The Axons of the Lateral Olfactory Tract   344
Overview   346

Spatial Organization of the Primary Sensory Neurons   347
Beyond the Mitral Cells   349
Ongoing Plasticity in the Olfactory Bulb by Cell Turnover   350
Olfaction and Behavior   351
Readings   352

20   Visual Systems: Origins and Functions   355
Origins of Vision, 1: Light Detection   356
Origins of Vision, 2: Image Formation   358
  Predator Avoidance and Escape: A Hypothesis Concerning Evolution and the Origins of Crossed Projections   359
  Orienting Toward or Around Visually Detected Objects and Other Responses   361
  The Midbrain Tectum and Orienting Toward Novel Objects, Food, or Potential Mates or Rivals   363
  Identifying Animals, Objects and Textures   364
  The Invasion of the Endbrain by Visual Pathways: Likely Evolutionary Steps in Pre-mammalian and Mammalian Ancestors   365
More about the Third Role of Visual Images   366
Expansions and Specializations in the Visual System   367
Readings   369

21   Visual Systems: The Retinal Projections   371
Two Views of the Optic Tract and Its Terminations   371
Distortions in Large Primates   374
How the Optic Tract Looks in the Brain of an Adult Animal   375
Looking at the Exposed Brain from Above   380
The Embryonic Optic Tract   381
Midbrain Tectum: Species Differences   384
Lamination of the Midbrain Tectum   385
Topographic Organization of the Retinal Projection to the Midbrain Surface   385
Notes   391
Readings   392

22   The Visual Endbrain Structures   393
Multiple Routes from Retina to the Endbrain   393
The Visual System's Two Major Routes to the Endbrain in Phylogeny   395
The Route through the Lateral Geniculate Body   397
Early Myelination of the Optic Radiations   399
The Brain and Neocortex in Human Development and in Phylogeny   399
Evolutionary Multiplications of Cortical Representations of the Retina   404
Comparing Species: Evidence for Older and Newer Visual Cortical Areas   407
And Where Do We Go from Here?   407
Transcortical Pathways from Visual Cortex   409
Three Visual Pathways and Their Functions   411
The Third Major Transcortical Pathway   412
Comments on Transcortical Interconnections   414
Readings   415

23 Auditory Systems 417
Embryonic Placodes Give Rise to Auditory and Vestibular Nerves 417
The Auditory Pathway (in Brief) 418
Some Special Functions of the Auditory System: Antipredator Behaviors 418
Aversiveness of Noise and the Role of the Limbic System 419
Learned Fear: Importance of the Forebrain 420
Predation Had Other Requirements: Identifying and Localizing Prey Animals 421
Transformations of Sound Vibrations in Middle Ear and Cochlea 423
Evolution of Jaw Bones of Ancestral Reptiles into the Ossicles of the Middle Ear in Mammals 424
Mammalian Ear Structures and Dynamics 425
Initial Coding of Information 426
The Flow of Auditory Information in the CNS 428
Terminations of the Auditory Nerve 428
Thalamic Projections Carry Auditory Information to the Limbic System and to Neocortex 430
Review: Multiple Routes Carrying Auditory Information to the Forebrain 430
Distinct Pathways for Two Major Functions: Orienting and Identification 431
Location Specificity Arises in Cells of the Hindbrain's Trapezoid Body 431
How Direction Can Be Derived from Precise Time-of-Arrival at the Two Ears 432
A Second Hindbrain Mechanism for Sound Localization 434
How Did This Spatial Localization Apparatus of the Hindbrain Evolve? 434
Judging a Sound's Direction in the Vertical Axis 434
Location Information Reaches the Midbrain's Superior Colliculus 434
Location Information Also Reaches the Endbrain 435
The Second Function of Ascending Auditory Pathways: Pattern Identification 435
Temporal Pattern Selectivity: Examples 438
Functionally Distinct Auditory Pathways in the Neocortex 440
Auditory System Specializations 442
Readings 445

PART IX: THE FOREBRAIN AND ITS ADAPTIVE PRIZES: A SNAPSHOT 447

24 Forebrain Origins: From Primitive Appendage to Modern Dominance 449
Life without a Forebrain 449
In Search of Ideas about Origins: Primitive Forebrains 450
Brain Expansions in the Vertebrates 452
More about the Early Forebrain 456
The Scene Is Set for the Early Expansion of the Striatum 456
Another Kind of Plasticity in the Primitive Endbrain 457
Parallel Evolution of Pallium and Subpallium 459
Visualizing the Early Striatum and Pallium 460
Next Steps into the Forebrain 463
Readings 463

PART X: THE HYPOTHALAMUS AND LIMBIC SYSTEM 465

25 Regulating the Internal Milieu and the Basic Instincts 467
Nature of the Hypothalamus and Affiliated Structures: The Limbic System 467
Approaching the Limbic System from Below: Two Kinds of Arousal from the Midbrain 468
Autonomic and Endocrine Functions of the Hypothalamus 470
Homeostatic Regulation of the Internal Milieu 472
Regulation of Cyclic and Episodic Behaviors 473
The Center of Motivational State Control 474
Distinguishing between Appetitive and Consummatory Behaviors Involved in a Drive 474
Computational Neuroethology 475
Hunger, Feeding, and Brain Circuits 476
Electrically Elicited Drive States 476
Connections with Other Systems 478
Drive and Reward Involve Distinct Axon Populations in the Medial Forebrain Bundle 479
Readings 479

26 Core Pathways of the Limbic System, with Memory for Meaningful Places 483
Cell Groups in the Hypothalamus 483
Feedback from Visceral Afferents and from Blood Chemistry 484
Limbic System Interconnections within the Forebrain: The Circuit of Papez 486
Why the Revival of Interest in This Circuit? 490
Visualizing the Circuit of Papez 490
What Is the Functional Significance of the Return Pathway? 492
Place Memories in the Neuronal Pathways of Feeling and Emotion 493
Review of Structures in the Limbic System 493
A Variety of Ways the Hypothalamus Sends Its Influences to the Neocortex 494
Mental State and the Hypothalamus 497
Influence of the Hypothalamus on the Brainstem and Spinal Cord 497
Review 498
Functional Specificity in the Limbic Midbrain 498
Readings 499

27 Hormones and the Shaping of Brain Structures 501
Sex Differences in the Human CNS: Evidence from Pathologies 502
What Determines an Individual's Sexual Orientation? 502
Hormone Peaks and Brain Differentiation 503
More Sex Differences in the CNS 504
Sexual Dimorphism Underlying Singing in the Canary and Other Songbirds 506
"A Brain for All Seasons" 507
How Many More Sex Differences in the Brain Will Be Found? 508
Back to the Anatomy of the Limbic System 510
Readings 511

28 The Medial Pallium Becomes the Hippocampus 513
Evidence of an Internal Map: Place Cells 513
Head-Direction Cells and the Circuit of Papez 515
Thoughts about the Origins of the Medial Pallium 517
The Expansion of the Pallium, with a Focus on the Medial Pallium 518
Cells and Circuits of Ammon's Horn (*Cornu Ammonis*) 520
Further Notes on Hippocampal Anatomy 520
Major Connections between the Mammalian Hippocampus and Neocortex 523
Local Circuits within the Hippocampus 527
Memory and the Hippocampus 527
Subcortical Projections to and from the Hippocampus 528
Anatomical Plasticity in the Hippocampus 529
Brain States and Memory Consolidation during Sleep 531
Readings 533

29 The Limbic Striatum and Its Outpost in the Temporal Lobe 537
A Brain System of Forebrain and Midbrain Underlying Motivation, Emotion, and Autonomic Regulation 537
What Is the Amygdala? 539
The Stria Terminalis: A Major Output of the Amygdala 542
Sensory Pathways to the Amygdala 542
Amygdala Connections with Systems of Higher Cognitive Functions 545
Behavioral Change Caused by Artificial Activation of the Amygdala 545
Behavioral Change after Amygdala Lesions 547
How Sounds That Warn of Danger Come to Cause Fear and Avoidance 547
The More Rostrally Located Ventral Striatum, in the "Basal Forebrain" 548
Disordered Functions of the Limbic System 551
Other Behavioral Difficulties in People Are Tied to Amygdala Functions 554
Central Roles of Basal Forebrain and Amygdala for Learned Affects 556
Readings 557

PART XI: CORPUS STRIATUM 559

30 The Major Subpallial Structure of the Endbrain 561
Beginnings: A Link between Olfactory Inputs and Motor Control 561
Non-olfactory Inputs Invade the Striatum 562
Visualizing the Striatum and Other Major Structures in the Cerebral Hemisphere 564
Major Outputs of the Mammalian Endbrain: The Position of the Striatum 564
Simplified Pictures of Dorsal Striatal Connections 568
Diaschisis after Loss of Inhibitory Connections 571
The Striatal Satellites Add Complexity to the System 573
Neocortex Dominates the Inputs to Dorsal Striatum of Large Mammals 574
Topography in the Ventral Striatum 575
Complex Chemoarchitecture of the Striatum 578
Compartmental Organization within the Striatal System 578
Finding the Corpus Striatum in Large Mammalian Brains 580
Overview 580
Readings 582

31 Lost Dopamine Axons: Consequences and Remedies 583
　Fetal Nigral Tissue Transplantation Initiated as a Therapy for Parkinson's Disease 583
　Initial Promise of Transplant Treatments 585
　Deep Brain Stimulation or Selective Surgical Lesions in Advanced Parkinson's Disease 585
　Additional Alternative Treatments 586
　Readings 588

Intermission: Neurogenesis in Mature Brains 589
　Readings 590

PART XII: THE CROWN OF THE MAMMALIAN CNS: THE NEOCORTEX 593

32 Structural Origins of Object Cognition, Place Cognition, Dexterity, and Planning 595
　The Primitive Olfactory Functions of Endbrain: Identification of Object and Place 595
　The Advantages and Consequences of Non-olfactory Inputs to the Endbrain 597
　In Mammals, Why Did Neocortex Evolve Rather than Wulst? 597
　The Orderly Architecture of Neocortex Illustrated by the Human Brain 598
　Basic Sensorimotor Functions of Neocortex 601
　Functional Regions Based on the Early Evolution of Place Cognition and Object Cognition 601
　Sensory Cortex That Led to the Evolution of Fine Motor Control 603
　Anticipation as a Major Innovation of the Neocortex 603
　Origins of Planning 604
　Neocortex as Anticipator and Planner: Illustrative Evidence 604
　Endbrain Evolution: Suggested Major Steps Reviewed and Extended 605
　Expansions of Neocortical Areas 609
　Evolutionary Changes in Thalamocortical Axon Trajectories 611
　Readings 614

33 Basic Neocortical Organization: Cells, Modules, and Connections 617
　Two Major Types of Neocortical Neuron 617
　The Major Morphological Characteristics of the Neocortex 617
　Modules of the Neocortex: Areas and Columns 619
　Cortical Types and Variations 623
　Connections of the Neocortical Modules 624
　Convergence of Inputs, Great and Small 626
　Expansions of the Neocortex Reviewed 627
　Axons to and from Neocortex 627
　Ascending Pathways to Neocortex Go Mainly through the Thalamus: The Basic Pattern 628
　Illustrations of Thalamocortical Connections: Human Brain 629
　Illustrations of Thalamocortical Connections: Brain of Small Rodent or Insectivore 633
　Transcortical Association Fibers 634
　Comparative Neocortical Anatomy from a Moscow Perspective 639
　Studies of Development and Plasticity Lend Support to Classifications of Major Neocortical Types 640
　Review of Ideas on Thalamic Evolution 642
　Readings 643

34 Structural Change in Development and in Maturity   645

Neuronal Proliferation   645
Neocortical Expansion in Development and in Evolution   646
When Cortical Cells Are Born and How They Migrate   650
How Do Distinct Cortical Areas Become Specified in Development?   652
The Role of Activity in the Development of Neocortical Connections   654
Apparent Shaping of Thalamocortical Projections by Abnormal Activity   656
How Does Activity Affect Axons and Their Connections?   656
Altered Neocortical Structure after Early Blindness   659
Alteration of Receptive Fields by Repetitive Stimulation in *Adult* Somatosensory Cortex   660
Review: Where Are the Engrams?   663
Is Everything Plastic?   664
Conclusions about the Neocortex   664
    Posterior (Postcentral) Portions of the Hemisphere   664
    Rostral (Frontal) Portions of the Hemisphere   665
    Brain State Changes Can Alter the Processes of Both Posterior and Frontal Portions of the Neocortex   665
The Study of Brain Structure and Its Origins Lays the Groundwork for Understanding the Underpinnings of Behavior and Mental Abilities   666
Readings   666

Figure Credits   669
Index   679

Preface

The first draft of this book was based on classes I have taught in the MIT Brain and Cognitive Sciences Department for many years, including classes for undergraduates. The book is also based on an introduction to neuroanatomy I teach to beginning graduate students. Finding no available textbook that fitted what I wanted for these classes led me to start writing the book. In the writing, I have assumed that readers have had some exposure to biology or to anatomy and physiology. This book can be useful to advanced undergraduates, to graduate students studying psychology, biology, zoology, or neuroscience, and to interested readers in other fields. More generally, the book is for any persons who wants to gain greater familiarity with the nervous system and how it works. It will be useful also for scientists, bioengineers, and medical personnel who have plunged into work on details of a particular portion of the brain and want to fill out their picture of how that part fits into the whole system.

The book is not primarily about evolution. It is about the structure and functional significance of the central nervous system (CNS), especially the major structures and their interconnections in the brains of vertebrates with a special focus on mammals. CNS structure and its variations among vertebrates can be learned and understood more easily if one knows something about *why* it is so. The "why" can be answered in terms of functions, in terms of evolution, or in terms of development. In this book, I consider each of these three aspects of why the brain is put together as it is, because they make the structure more comprehensible as well as more interesting and memorable.

Brain evolution is presented in broad outline, based primarily on comparative neuroanatomical findings. The great importance of functional adaptations in evolution as described by Charles Darwin is assumed. Darwinian theory as updated in modern times, together with comparative neuroanatomical findings, are used to guide the formulation of suggested explanations of some key steps in early chordate CNS evolution.

The major goal is to help the student of brain anatomy and function gain an outline that she or he can use in further explorations of this field. It is a field that can seem formidably intimidating at first but fascinating and rewarding to the explorers who persist in going through the territory and learning the rudiments of the language.

I am indebted to many teachers, especially to Hans-Lukas Teuber, Norman Geschwind, and Walle J. H. Nauta when I was a graduate and postdoctoral student at MIT. I also owe much to teachers and advisors who preceded them in my education, especially to Joseph Spradley and Arthur Holmes at Wheaton College (Illinois), from whom I learned that studies of physics and philosophy prepared me for intellectual adventures outside those specialties. The book would not have been completed without the unfailing assistance and encouragement of my wife, Aiping Liang Schneider. I also acknowledge the help of Jeffrey Meldman of the MIT Sloan School, who read an early draft of the manuscript and helped me to clarify the ideas and to improve the English and the logic. The MIT students who have been reading drafts of the chapters as they prepared for my classes have also given me encouragement, as have the teaching assistants who assure me that the book is worthwhile for them as well as for the students. Two teaching assistants have given special help during final stages of editing: Andrew Bolton and Beverly Cope. I also thank my faculty colleagues and colleagues at other institutions, and other friends for their help and support. George Adelman at MIT has promoted this project and given help from its onset. Finally, I am grateful to the various members of my family for their encouragement and humor during the whole process.

INTRODUCTORY ORIENTATION

1 Getting Ready for a Brain Structure Primer

It is not difficult for a student to become fascinated with the brain. Even for scientists who have been investigating it for decades, it is an organ of endless intrigue and interest. We do not have to focus only on the brains inside our own heads. We can still be interested and often amazed by the brain of a small animal like a mouse, a hamster, or a songbird. A very small amount of damage in such a brain may not be lethal, but if that damage is in certain places, the animal will not be able to move certain parts of its body, or make any movements at all, or it will not be able to find its food, or it will lose its normal hunger—or eat too much. Damage in another part of this amazing structure will make the creature forget things it has learned or render it nearly incapable of remembering where it has hidden its food or even its nest. Or it can lose one of its keen sensory abilities.

How can we explain such things as these? To do so, we have to know something about the cells that make up the structure of the nervous system and how those cells are connected to the sense organs and to the muscles and to each other. How the nerve cells, or neurons, are connected to each other in the brain and spinal cord—the central nervous system—is not a simple subject to get into. The goal of this book is to get you into it deeply enough so that you will have in mind an outline of the neuroanatomy of vertebrates, and especially of mammals. Reaching that goal in a way that will help you make some sense of it along the way, and remember the outline for later studies, will be facilitated by discussions of how our complex brains have originated in evolutionary history and in individual development. We will be helped by examinations of brain structures in various animals other than our own species. Because adaptive function is the purpose of brain structure as it is shaped by the processes of evolution, we will be paying attention to nervous system functions all along the way.

The Nature of This Book

This book is for readers who want to learn about how the brain is put together and why. It is not like commonly encountered neuroanatomy textbooks. There are many such books that are filled with the names and descriptions of a very large number of structures and even greater numbers of connections among those structures, so a beginner can get lost before he or she ever acquires much of an outline of the central nervous system. In writing this primer, I have tried to keep in mind the reader who is beginning to learn about the brain, perhaps after encountering it in an introductory biology or psychology class. In certain respects it resembles an introductory behavioral neurology text except that unlike such a book, it deals more with comparative anatomy and brain evolution, and probably a little more with development.

Concerning those seemingly endless numbers of connections between the many parts of the central nervous system (the brain and spinal cord): This book does include quite a few, but far fewer than actually exist and fewer than are mentioned in many neuroanatomy books. So why do I include some and not others? What is the basis for my selections? The answer is not difficult. Connections are described if they are major ones—the connections that appear to be most important to explain behavioral and other functions. Some of the smaller connections are also described not because they are dominant connections in mammals but because they provide clues to an evolutionary past and were likely of crucial importance in early ancestors. (As for all those other connections, the reader can always search the Web for more information.)

As we begin a journey into the world of the central nervous system, what is most needed? It seems to me that we need a sketch that makes sense, a basic outline or map. The kind of "sense" I mean is of two types. First, it should make sense in explaining how the brain and spinal cord control physiological and behavioral functions. But an engineer can design a number of alternative systems that could control the same functions. So a second kind of sense is also needed in the brain sketches we will draw: This is the sense provided by ideas about the progression of evolutionary changes that led to modern brain structure.

A few words to the neuroanatomy student: Most students find that learning neuroanatomy makes demands on them that are different from the demands of other academic subjects. For most people, it is not realistic to expect really to learn or remember something in brain anatomy after the first encounter. Neuroanatomy is usually learned by repeated exposure and study, with many reviews and extensions of the knowledge already gained. This fact has guided the writing of this book. At the beginning, many readers will find it to be a review with a slightly different slant. What comes next is more novel, and after that it becomes still more novel.

If at first some of it seems a bit confusing, this is not cause for worry. The early chapters provide only a first run through the brain. There will be more such runs, each of them with a different thrust, and each time you will remember more and it will make more sense.

Learning neuroanatomy is like learning both a new language and a map of a new world. New words will generally be defined the first time they are used, and they are also defined in a glossary that can be found online at http://mitpress.mit.edu/brainstructure. Definitions for most of the words can also be found using Internet searches.

To get started, we have to know something about how to read diagrams and pictures of the brain and its parts, and that means knowing some special terminology. We also have to know something about neurons, the cells whose activities make all behavior of organisms possible. In introducing that topic, we will include some thoughts on neuron origins. After everyone knows at least some very basic physiology of nerve cells, we will review some of the methods used by neuroscientists to study the structures of brains.

Brain Space: Specifying Directions and Major Regions

Anatomists describe brain structure with the help of drawings, diagrams, and photographs. We can think of these illustrations as maps. Directions in these maps are not those of geographic maps, but of the body with respect to head and tail, back and belly, and left and right of the midline (figure 1.1). Rostral means toward the head end of the body (toward the rostrum), and caudal means toward the opposite end of the body (toward the tail). Toward the back is called dorsal; toward the belly is called ventral. In most animals, rostral-caudal corresponds to anterior-posterior. But in the upright human being in the regions below the head, anterior-posterior corresponds to ventral-dorsal. The other directions we need in locating something in the body or brain are given in terms of distances from the midline, to the right or to the left. If we move away from the midline, we are moving laterally, and if we move toward the midline, we are moving medially. Hence, we have a coordinate system for specifying positions in the brain as well as the rest of the body. We specify distances and give directions as rostral or caudal to something, and as dorsal or ventral to something, and as a specific distance left or right of the midline.

Many photographs and drawings we see in the study of brain structure are of slices cut through the brain in different ways (figure 1.2). A cross section, cut at approximately right angles to the long axis of the brain, is called a transverse or frontal section. A section cut parallel to the midline (as if the animal had walked forward into a whirling vertical saw blade) is called a parasagittal or sagittal section. A midsagittal section divides the brain exactly in half, and the left and right sides are fairly symmetric. A horizontal section is cut in a plane more or less at right angles to the other two. Other planes may be referred to as oblique sections, which are less common, but they can be useful for study of specific structures.

Now let's get down to the business of naming major distinct subdivisions of the central nervous system, which includes both the brain and the spinal cord. These major subdivisions will be discussed repeatedly in the book, from various points of view, so that soon you will remember them (if you do not know them already). Look carefully at figure 1.3. This drawing depicts an embryonic mammalian central nervous system (CNS), such as

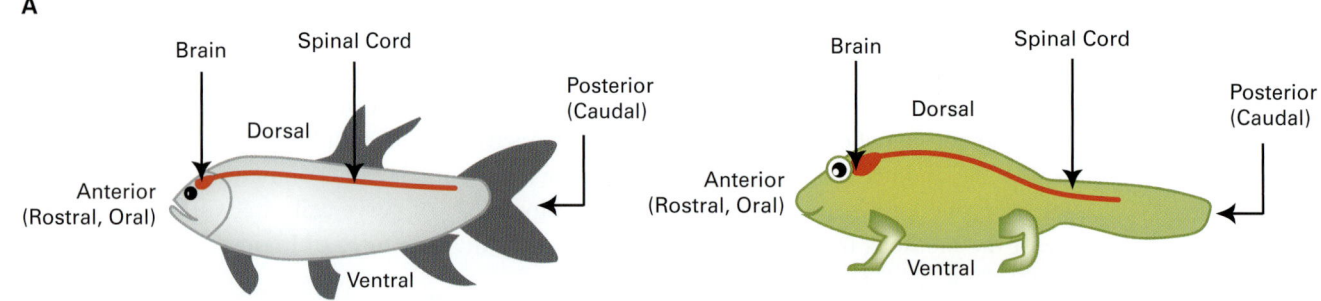

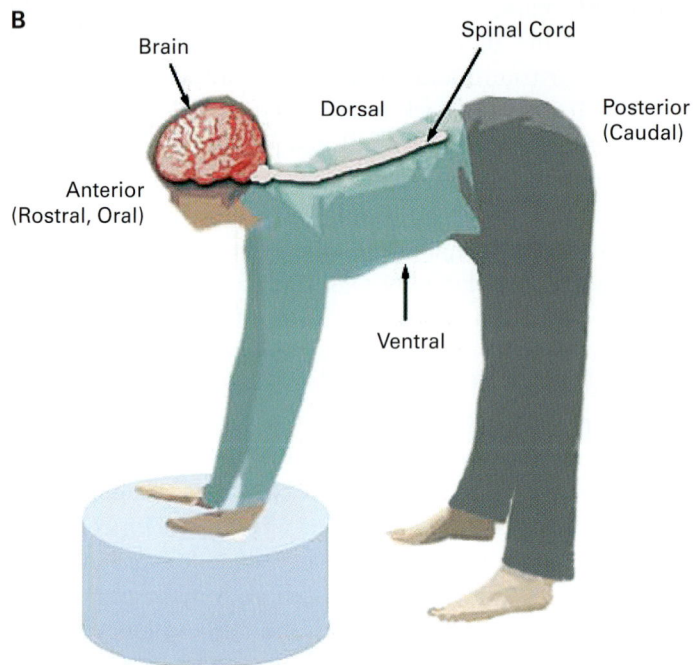

Figure 1.1
(A) Directions in the body and CNS of a fish and an amphibian are named. (B) Directions in the body and CNS of a human in the quadruped position are named. These terms can be applied to any tetrapod animal. Panel (A) based on Butler and Hodos (2005).

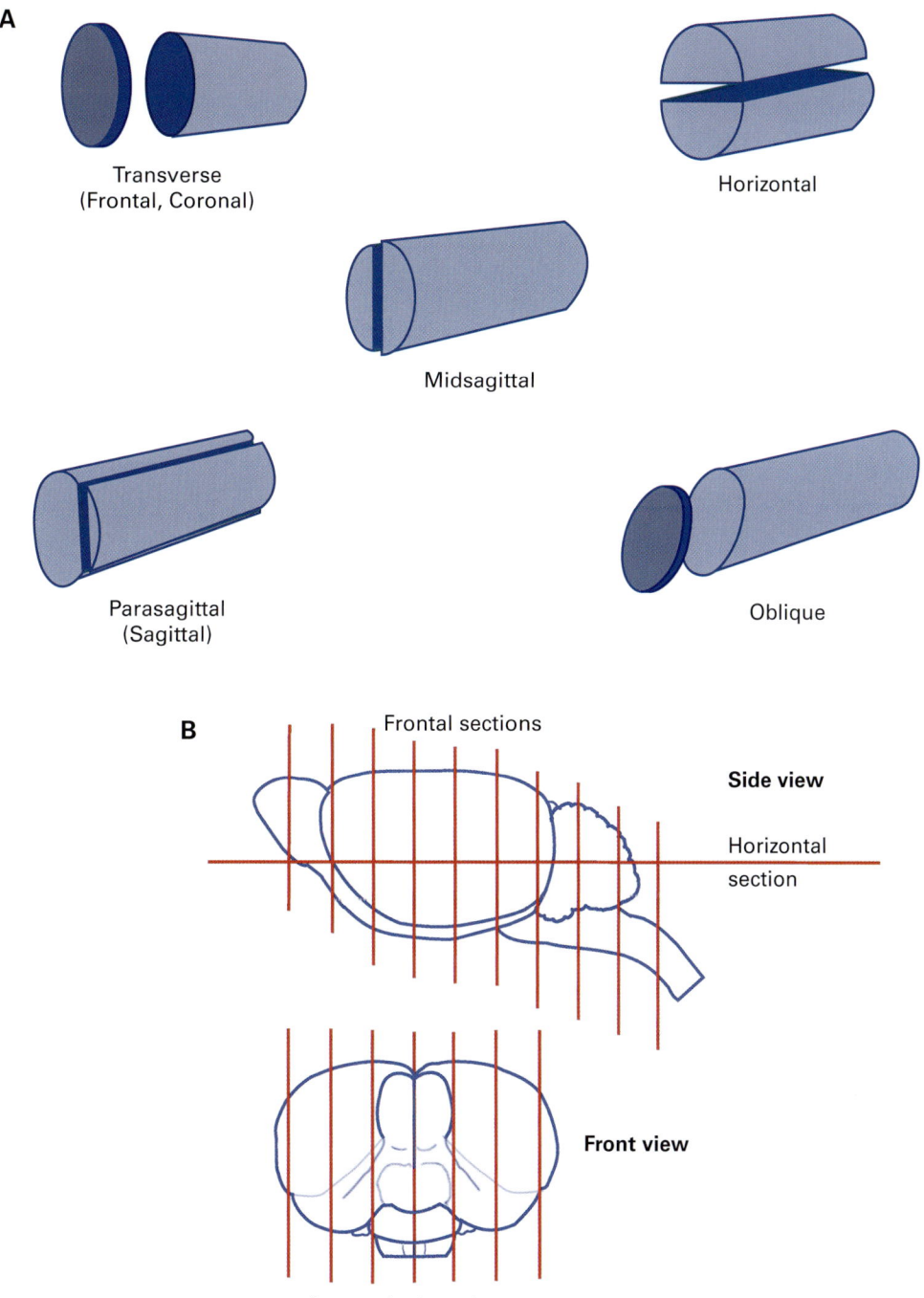

Figure 1.2
(A) The meanings of terms for different planes of section are illustrated using a cylinder as a first approximation of the shape of a brain. (B) In sketches of a side view and a front view of the brain of a small rodent like the hamster or mouse, the three most common planes of section used in neuroanatomical studies are indicated. Positions of multiple cuts are shown for frontal and parasagittal planes, and only one position is shown for the horizontal plane. Panel (A) based on Butler and Hodos (2005).

8   Chapter 1

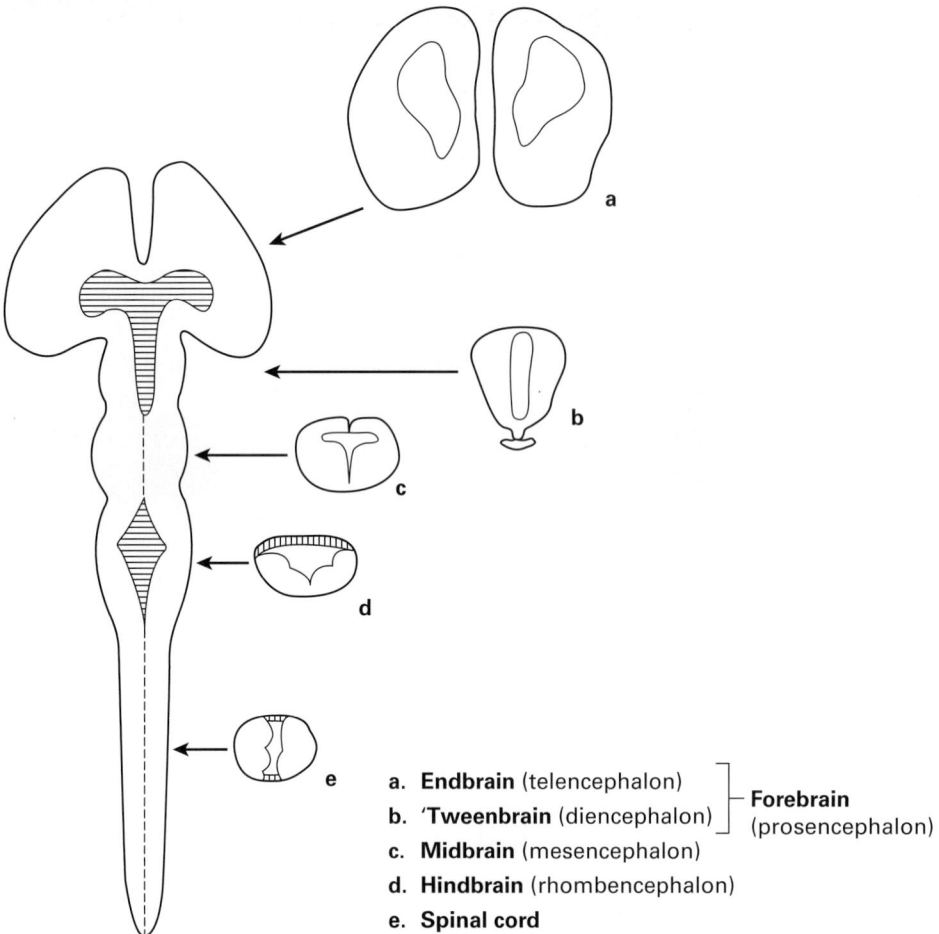

a. **Endbrain** (telencephalon) ⎤
b. **'Tweenbrain** (diencephalon) ⎦ **Forebrain** (prosencephalon)
c. **Midbrain** (mesencephalon)
d. **Hindbrain** (rhombencephalon)
e. **Spinal cord**

Figure 1.3
Simplified drawings of the thickening embryonic neural tube of an advanced vertebrate. At the left is a view from the dorsal side of the developing CNS that has been straightened out and has had the cerebral hemispheres of the endbrain pushed apart. Thin walls of the centrally located, fluid-filled ventricular space are indicated by horizontal striations. To the right are frontal sections taken at the levels indicated. The endbrain section represents a slightly later stage of development.

that of a human brain toward the end of the first trimester of a pregnancy. The CNS of another mammal would look very similar. The caudal-most portion is the spinal cord. The frontal section shows that the cord is a tube, actually a fluid-filled tube, with thickened walls. Rostral to the spinal cord is the hindbrain, where the cord structure is modified a little and the neural tube is wider. The hindbrain is also known as the rhombencephalon because of the rhombus shape you can see when looking at it from above. Encephalon means "in the head" in Greek. Proceeding rostrally, we come next to the midbrain, or mesencephalon. Then we reach what in higher vertebrates is the largest major subdivision, the forebrain (prosencephalon), which includes the 'tweenbrain (diencephalon) and the endbrain (telencephalon). The largest parts of the endbrain appear to have ballooned out from the 'tweenbrain, forming two outpouchings, one on each side. These are often called the cerebral hemispheres. In the figure, the two hemispheres are separated as in earliest development, although in the actual brain at this stage they have become pushed together above and in front of the diencephalon.

This description is full of a lot of strange names! Names for CNS structures were often given by early anatomists using the Latin or Greek languages, but now the English translations are frequently used as well. We will sometimes but not always use names both in the old languages (the classical Greek or Latin) and in English, but will use what you need to know in order to follow the readings. I should also tell you that many structures in the brain have several synonymous names, and I will try not to confuse you by switching from one name to another without defining the terms.

So now you have just begun to learn a new language, the language of neuroanatomy. If you encounter a word you cannot remember, please refer to the glossary or check it on the Internet.

CNS Tissue

The embryonic central nervous system depicted in figure 1.3 develops into the adult brain and spinal cord (figures 1.4 and 1.5) by an extended process of cell multiplication and cell growth, and also cell death. We will look more closely at these phenomena, but first let's ask a very basic question about what we are dealing with: What is the nature of this *central nervous system* that is of such critical importance in the functioning of our bodies, underlying and making possible the actions, urges and feelings, and thoughts and memories that constitute our lives?

An experimental psychologist in England (Richard Gregory) has remarked, "One of the difficulties in understanding the brain is that it is like nothing so much as a lump of porridge." Indeed, if you open the skull of an animal just after death, the brain inside does seem like that!

But just what is the real nature of this brain and the spinal cord—this CNS? It is a tissue, and it is a system made up of specific small elements. What kind of tissue is it? It

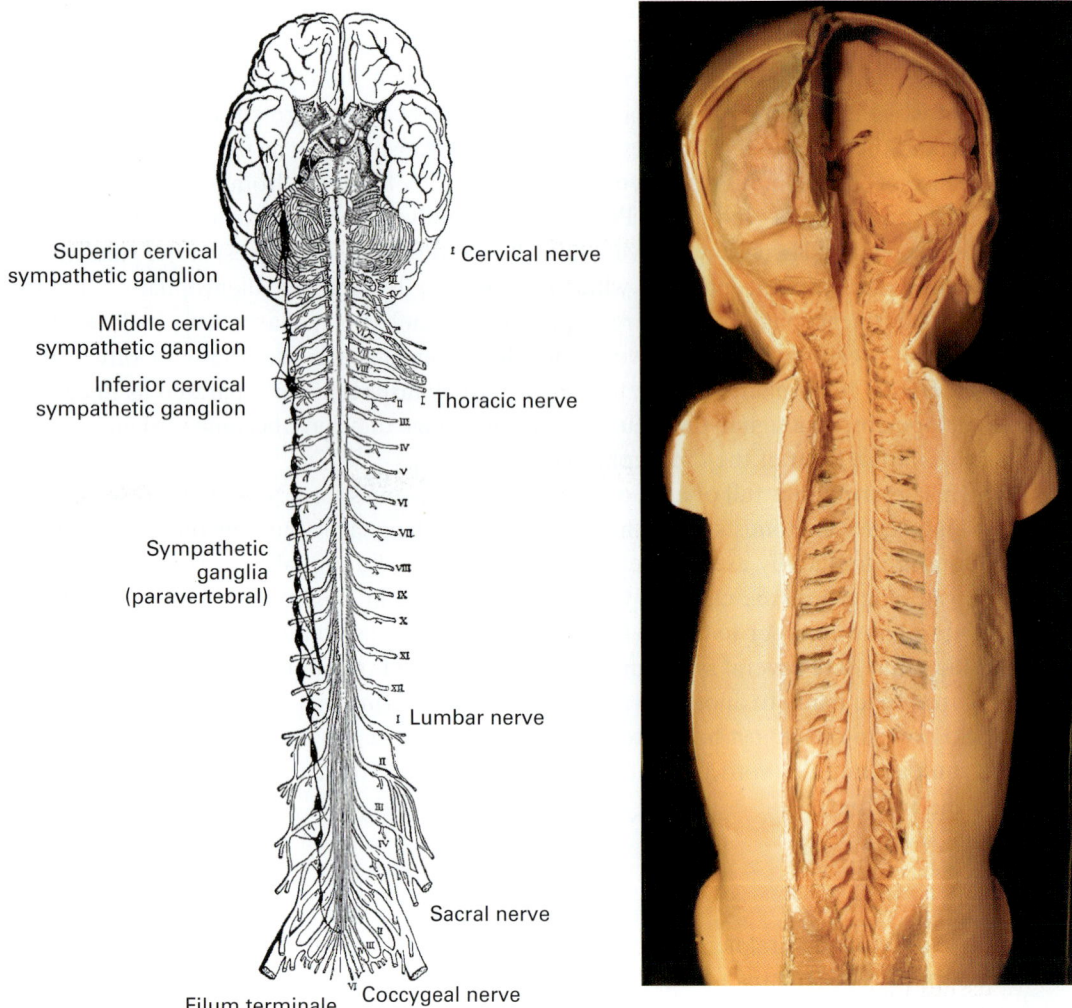

Figure 1.4
(Left) Drawing of ventral view of adult human CNS with stumps of spinal and cranial nerves. (Right) Photograph of a dissected body of a young human, opened to reveal the brain and spinal cord with proximal portions of the spinal nerves. Left panel from Herrick (1922); right panel courtesy of Robert Kretz (University of Fribourg).

Getting Ready for a Brain Structure Primer

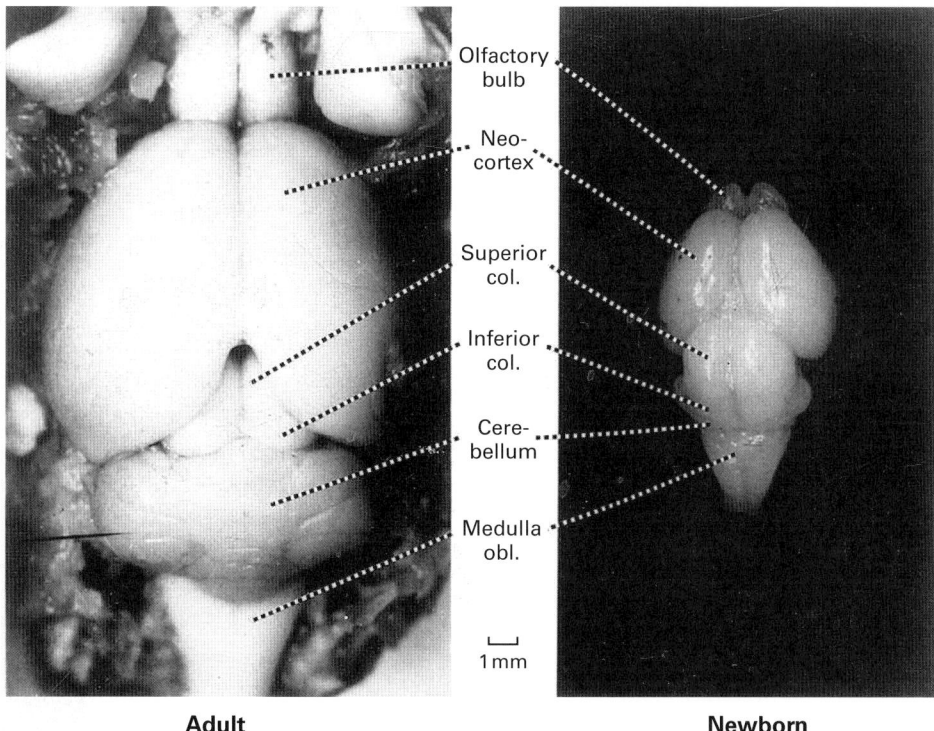

Figure 1.5
(Left) Dorsal view of the brain of an adult hamster (similar to that of rat or mouse). The skull top has been removed, but the brain has not been lifted out of the skull. (Right) Similar view of the brain of a newborn hamster. col, colliculus; obl, oblongata.

develops from the surface layer of the early embryo, the primitive ectoderm, the same layer that gives rise to the skin. So it is an ectodermal tissue, which is different from the tissues that make up the muscles and skeleton (the mesodermal tissues) and the tissues that make up the internal organs (the endodermal tissues). The neural ectoderm, like other tissues of the body, is composed of cells. Those cells have been seen and studied by scientists exploring nervous system structures using microscopes, usually after the cells have been made more visible by staining techniques or by other methods of marking specific components.

The CNS is an unusual tissue in the way that its cells are interconnected, forming a complicated communication system handling vast amounts of information. It also functions as a gland, as some of its cells secrete substances into the bloodstream—the neural hormones.

Table 1.1
Primitive cellular mechanisms present in one-celled organisms and retained in the evolution of neurons

- Irritability and conduction
- Specializations of membrane for irritability
- Movement
- Secretion
- Parallel channels of information flow; integrative activity
- Endogenous activity

Primitive Cellular Mechanisms

What is the functional architecture of the brain? Soon we will illustrate this, beginning at a very basic level. (For a quick preview, see figure 3.2.) But before we start to focus on the structural organization of the CNS, it is important to review some very basic properties of its cells. These properties can be called the primitive cellular mechanisms, as they derive from properties found in one-celled organisms, the Protozoa, and are retained in the evolution of neurons. Thus, single-celled organisms and neurons alike are responsive to various kinds of irritation (stimulation), and these responses spread to other parts of the cells. The cells do interesting things, on their own and in response to things they encounter in their environments: They respond to and integrate various kinds of inputs representing information arriving through multiple channels; they move; and they secrete chemicals (table 1.1).

But if Protozoa do all these things, why do organisms need neurons? Being a single cell is very limiting at least in situations where small size can result in less competitiveness. Adaptive advantages of cellular specializations led eventually to the evolution of multicellular organisms. In multicellular organisms, different cells evolved different specializations, including specializations for irritability and conduction, movement, secretion, integration, and endogenous activity. All these primitive cellular mechanisms are found in neurons except that movement, using contractile proteins, is not characteristic of neurons except during developmental stages of the cells.

Consider, first, specializations for irritability. Protozoa, of course, respond to stimulation, including mechanical stimulation caused by being touched or by bumping into things. In sponges and other metazoan animals, there are surface cells specialized for being responsive to contact or to chemicals produced by prey animals. In cnidarians like hydra, such specialized cells are called primary sensory neurons, distinct from other neurons that respond to the activity of the primary sensory cells. With the evolution of bilaterally symmetric animals with forward locomotion, specialized head receptors brought advantages that ensured their continuance in more advanced creatures. With these receptors, there was an evolution of an enlargement of the nearby nervous system they connected to—the brain—enabling processing and routing of the information. We will return to these topics soon.

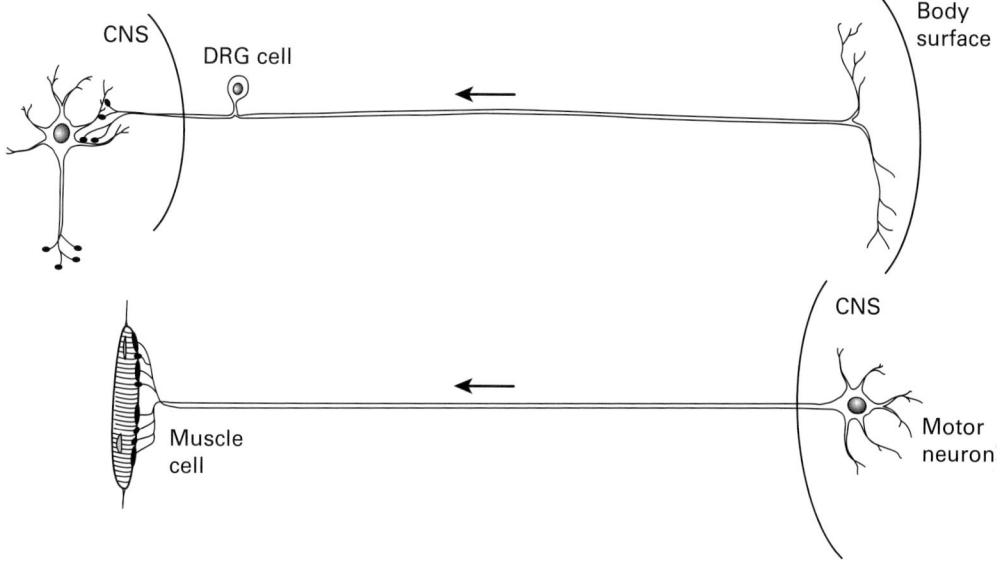

Figure 1.6
Drawings of two neurons with axons located mostly outside the central nervous system. The top drawing shows a *primary sensory* nerve cell with the cell body located in a dorsal root ganglion near the spinal cord. The dendritic region is located at the far right near the skin surface. Nerve impulses travel along the axon from this region toward the CNS (spinal cord), where the axonal endings contact neurons. One such neuron, called a secondary sensory neuron, is shown. In the bottom drawing, a motor neuron of the spinal cord or brainstem is illustrated. Its axon carries nerve impulses to endings on striated muscle cells. In these drawings, the axons are shown without a myelin sheath, which appears on all but the smallest axons during development. The myelin sheath is formed by glial cells and results in a large increase in the speed of conduction of action potentials.

Irritability and Conduction

For illustrating the properties of irritability of the cell membrane and conduction, consider two neurons in vertebrates, pictured in the drawings of figure 1.6. One is a dorsal-root ganglion cell (DRG cell), so-called because its cell body is located in a clump of cells, a ganglion, located just outside the central nervous system within a *root*—a bundle of nerve fibers, or axons—attached to the spinal cord. The cell is very elongated. One end is near the body surface, where it appears as many little branches that respond to pressures or to stretching. The response is a partial depolarization of the electrical potential difference across the cell's membrane. The little branches—the dendrites—form the receptive portion of the neuron. They converge on a single long process, the axon. The axon runs together with others, in a bundle of fibers carrying sensory information, all the way to the CNS. When the electrical depolarization in the dendrites is great enough at the point where it reaches the beginning of the axon, a sudden change occurs—an *action potential* is initiated. This action potential is a complete depolarization of the cell membrane of the axon (in fact, this depolarization is more than complete, as it overshoots the zero point by a small amount before the membrane recovers), which in this specialized portion, the conductive portion, of the cell, moves down the axon without any diminution, all the way to the terminals at the other end of the cell. In this case, those terminals are located within the spinal cord. The action potential, or spike potential, reaches the axon endings, which contact the surfaces of other neurons where they cause small changes in membrane potentials. Such a contact is called a *synapse* (from a Greek word meaning "a joining together or touching"), so named by the English physiologist Charles Scott Sherrington, who could not see them but who was able to describe their properties by their functional effects in the spinal cord. The axonal end branches, the telodendria, where the synaptic endings are located at tiny enlargements (end buttons, or boutons, from the French name *boutons termineaux*), are thus the transmissive portion of the neuron.

Now, consider the neuron illustrated in the lower part of the figure. This one is a motor neuron, defined as a neuron with an axon that exits the CNS and contacts a cell like a muscle or gland cell. In this motor neuron, we can see the same three functional portions of the cell (receptive, conductive, and transmissive). The receptive portion includes the cell body and dendrites, all within the CNS. The axon begins at the cell body and transmits action potentials over a long distance to a muscle cell, where specialized synapses are located. Action potentials in this case cause secretion of a chemical (a neurotransmitter) that causes contraction of the muscle cell.

Note that the position of the cell body is very different in the two neurons named in figure 1.6. The shape of the dorsal root ganglion cell of mammals, known as a *pseudo-unipolar* shape, is not found in all animals. This was illustrated (figure 1.7) by the neuroanatomist Ramón y Cajal, a key person in the history of the field and a fabulously productive scientist (figure 1.8). His illustration depicts primary somatosensory neurons in a series of animals: an earthworm (at the top), a mollusk, a lower fish, and an amphibian or reptile or bird or mammal (at the bottom). In the earthworm, the cell body is located within the surface epithelium. Such a location occurs for mammals only in the case of the primary olfactory neurons, where the cell bodies are found in the nasal epithelium. In the mollusk and in the lower fish, the primary sensory neurons are bipolar in shape, which is the shape of neurons in the mammalian auditory and vestibular systems.

Let's review and add a little bit to the descriptions of the major parts and activities of neurons (table 1.2). The cell body, also called the soma, usually has multiple processes extending from it. Except for the axon, these are the dendrites. The entire cell has an electrically polarized membrane, with the inside negative with respect to the outside due to an uneven distribution of ions. In the resting membrane there are more sodium ions on the outside, whereas on the inside there are large, negatively charged ions that cannot cross the membrane and also more potassium ions. The electrical potential difference across this membrane is about 70 millivolts, negative on the inside with respect to the

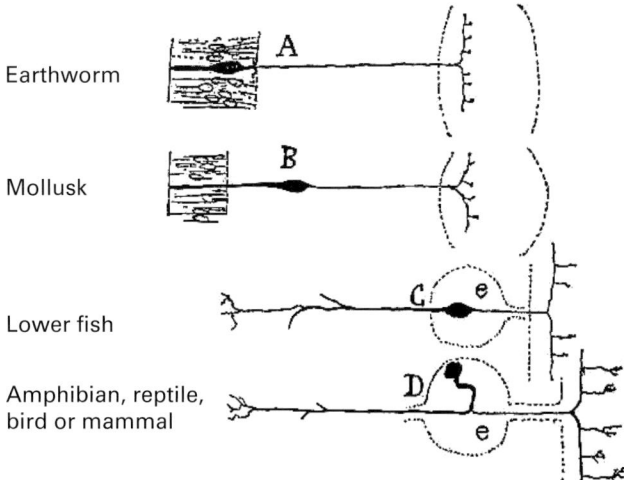

Figure 1.7
Primary somatosensory neurons are illustrated for widely different species of animals. (A) An earthworm, at the top, has the cell body of the primary sensory neurons in the epithelium. (B) The second illustration is for a mollusk; the cell body is below the epithelium. (C) Third from the top is a primary sensory neuron of a lower fish; the cell body of the bipolar neuron is much closer to the spinal cord. (D) At the bottom is the typical appearance of a dorsal root ganglion cell of the type found in amphibians, reptiles, birds, and mammals; the shape of this neuron is called *pseudo-unipolar*. (e) Dorsal root ganglion. Adapted from Ramón y Cajal (1989).

outside (figure 1.9). The membrane is relatively impermeable to the sodium ions (also to potassium) in the resting state. When stimulated mechanically even slightly or when contacting certain chemicals, the membrane depolarizes a little bit. Ion channels in the membrane are opened a small amount by the stimulation. The depolarization in the dendrites and cell body is graded—greater with greater stimulation—and this effect diminishes with distance from the point of stimulation. We can say that the change is conducted, very rapidly, away from the site of stimulation in a decremental fashion. The farther from the point of stimulation, the smaller the change is. This depolarization is called excitatory because it moves the membrane potential at the starting point of the axon closer to a critical value, or threshold, where an action potential will be triggered.

The Specialized Membrane of the Axon

The axonal membrane is very different. Once a threshold level of depolarization is reached, an action potential is triggered. This could occur only after the evolution of voltage-gated ion channels; these are found even in jellyfish axons. Sodium channels open, and there is a local implosion of sodium ions, moving positive charges into the cell, so many that the cell at that region becomes briefly positive on the inside with respect to the extracellular fluid (figure 1.10). Very quickly, potassium channels open also, and

Figure 1.8
Photograph of the pioneering neuroanatomist Santiago Ramón y Cajal by his microscope while drawing. From the Cajal Institute, Madrid.

Table 1.2
Names for major parts and activities of neurons

- Cell body (soma) and its branches (dendrites)
    Membrane potential
    The cell's irritability: membrane depolarization when stimulated (*excitation*)
    Graded spread (conduction) of membrane potential change away from the point of stimulation
- Axon and its end arborization (telodendria) with *synaptic* contacts on other neurons or on muscle or gland cells
    The axonal membrane is specialized for conduction of *action potentials*.
    Action potentials are conducted in a nondecremental fashion.
    Action potentials are found even in jellyfish. This specialization depended on the evolution of voltage-gated ion channels.

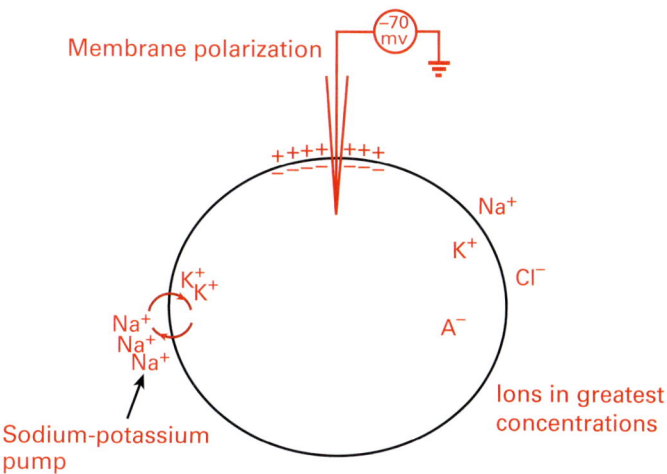

Figure 1.9
A cartoon that illustrates the distribution of major ions inside and outside the resting neuronal membrane. The membrane is impermeable to negative anions within the cell (A⁻). The large number of positive sodium ions (Na⁺) on the outside greatly outnumber negative ions there, and there is a much greater density of these sodium ions on the outside compared with the inside of the cell. The membrane is relatively impermeable to sodium ions, but an active pumping mechanism is constantly moving sodium ions out of the cell and bringing potassium ions (K⁺) in. Three sodium ions are moved out when two potassium ions are moved in.

potassium ions move outward, causing the membrane potential to begin to move back to the resting state. The voltage-gated ion channels close again. The membrane recovers its resting potential after a short time. Many action potentials would cause so much sodium and potassium to move across the membrane that it would lose its polarization. However, there are proteins that act as active pumps, moving sodium ions out and potassium ions back into the cell. (This goes on continually, but a leakage of these ions back across the membrane increases as relative concentration differences increase, so normally, at rest, an equilibrium state is reached.)

The action potential in each piece of membrane triggers the spike-like change in the adjacent membrane, and the spike travels down the axon to the endings (figure 1.10). The conduction is nondecremental: Once triggered, the spike travels down the axon without diminution. It will travel in both directions if triggered artificially somewhere in the middle of the axon's length. This is what occurs when you strike your "funny bone" in the elbow. The mechanical stimulation of the ulnar nerve triggers action potentials in many axons whose dendritic regions are in the hand and forearm, hence you feel a burning sensation in those regions although they were not actually stimulated.

During development in the jawed vertebrates (all vertebrates except hagfishes and lampreys), a myelin sheath develops on all but the smallest axons. Myelin is formed by the membranes of glial cells (oligodendrocytes in the CNS and Schwann cells in the

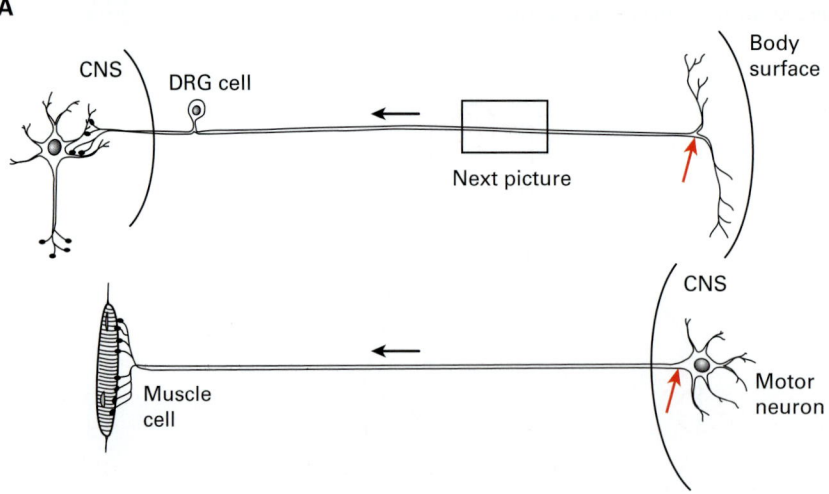

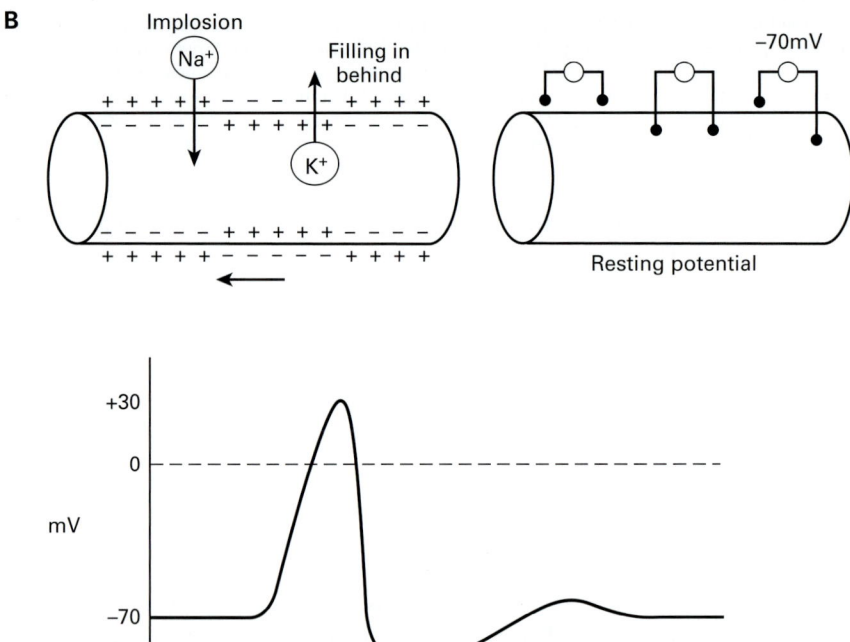

Figure 1.10
(A) The two neurons of figure 1.6 are shown in order to illustrate the position of the axon hillock in these cells (red arrows) and to indicate a small section of the axon that is shown magnified in panel (B). (B) The membrane potentials along the active portion of an axon during an action potential are indicated. At the upper left, the potential distribution is shown at an instant of time. At the arrival of the action potential, the resting potential changes rapidly because of voltage-gated ion channels. There is an implosion of sodium ions as the membrane becomes permeable to these ions, momentarily reversing the potential (to positive on the inside). The membrane potential rapidly reverses again, led by the movement of potassium ions outward as the membrane becomes permeable to these ions. At the upper right, the recording of the resting potential by tiny electrodes is illustrated. The graph in the lower part of the panel shows the recording of an action potential by a microelectrode placed at one position inside the axon. Time is represented on the abscissa. Alternatively, the graph represents a snapshot distribution of potentials along the active portion of the axon.

peripheral nervous system [PNS]). It blocks the flow of ions across the membrane except at periodically placed nodes where it is absent. The result is a great increase in speed of the action potential, as the depolarization spreads decrementally very rapidly from one node to another, triggering an action potential at each node. Thus, an action potential jumps from one node to the next, so the conduction is saltatory (jumping).

Specializations for Irritability

The membranes of all neurons respond to stimulation, but some of them have portions of the membrane specialized for responding to particular kinds of change. Most neurons within the CNS have specialized regions where other neurons contact them and stimulate them by release of chemicals called neurotransmitters. But the dendritic parts of some neurons have evolved responsiveness to changes originating outside the organism—they are receptor neurons. Recall the examples shown in figure 1.7. There are also specialized receptor cells that are epithelial in origin but are not classified as neurons and have no axons. They respond to changes by depolarization, which then causes stimulation of primary sensory neurons. Some receptor cells, as well as receptor neurons, have evolved sensitivities to various chemicals (e.g., receptors in the olfactory and taste systems or receptor cells for detecting carbon dioxide levels in the blood). Other cells have specialized for detecting light energy at certain portions of the spectrum (e.g., retinal receptor cells). There are receptive neurons (sometimes together with specialized accessory cells) that are especially sensitive to pressure or to stretch of skin or muscle. Other mechanoreceptors have specialized for detecting sound vibrations. There are receptors that can detect warming or cooling. In some animals, there are also receptors for detection of electrical potential changes in water, or infrared radiation from body heat, or magnetic fields.

Altogether, there is a remarkable range of sensory abilities representing specializations for irritability of cells. As is well known, there is great variability among species in the degree of sensitivity of these detectors (e.g., the exquisite odor sensitivities of dogs and the night vision and hearing of owls). Later, we will study the neural circuits that have evolved for filtering and analyzing the large amounts of information detected by these various receptive cells.

Movement

Organisms move about, independently or in groups of fellow creatures. Even single-celled creatures move themselves. This requires energy transfer to contractile proteins. The distribution of such proteins varies. Amoebas can move by control of cytoplasmic flow; paramecia can locomote by the moving of cilia that protrude from the cell membrane. Multicellular organisms have evolved specialized contractile cells—the muscle cells—in order to locomote or produce contractions for other purposes.

Growing neurons contain an abundance of contractile proteins, especially actin, in order to extend their long processes and to migrate during development. Such movements are much more restricted in mature nerve cells, but some of the other cells of the CNS, the glial cells, move in the accomplishment of particular functions. The neurons are not specialized for moving themselves except during the early developmental period. The cells that retain and specialize in the ability to contract are, of course, the muscle cells, which accomplish this by the interaction of two proteins, actin and myosin, which obtain the necessary energy for this function from adenosine triphosphate (ATP) molecules.

Secretion

An output mechanism in Protozoa, used for poison attacks on prey, is secretion from the cell into the surrounding fluid environment. It is used this way by specific cells in sponges as well. Poisons are made and secreted by specialized cells in many animal and plant species for protection from predators or for predation. In mammals, secretions have many other functions, but are used especially for communication (e.g., by the specialized secretory cells of endocrine glands). There are a few neuron types in vertebrates that have similar functions, secreting substances into the bloodstream.

Sponge cells also use secretion to communicate with adjacent cells in the same organism. It is this activity that was either retained or evolved again in neurons, for it is the major means of transmission of information at synaptic contacts. Before this was known for sure, it was only one hypothesis in the early years of the twentieth century, when it rivaled a competing hypothesis that neural transmission was purely electrical in nature. The discovery that provided the first strong evidence for the chemical nature of synaptic transmission is a fascinating story.

Otto Loewi was one of the early neuroscientists working on this problem. One night he had a dream: In his dream, he saw how chemical transmission at the synapse could be demonstrated. In the morning (so the story goes), he struggled to read a note he had made that night about the dream and to recall what he had dreamed but he could not read the note and could not remember! Fortunately, he had the dream again, and this time he took no chances. He got up and went directly to his laboratory and conducted the experiment he had seen in his dream.

He placed two frog hearts in fluid in separate Petri dishes. The hearts had the two nerves known as the accelerator nerve and the decelerator nerve still attached. He already knew that electrical stimulation of the accelerator nerve caused the heart to speed up, and stimulation of the decelerator nerve caused it to slow down. His next step was to stimulate electrically, for some time, one of these nerves. Thus, when he stimulated the accelerator nerve of one of the hearts for a while, the heart beat more rapidly than the other heart. Now, he used a pipette to move fluid from the stimulated heart's dish to the other dish, and behold! The unstimulated heart now beat faster. The stimulation of that second heart had to be through the fluid transferred by pipette. Thus, it had to be chemical. Loewi called the active chemical released by the accelerator nerve *Acceleransstoff* in German.

Loewi repeated the experiment stimulating the other attached nerve, which caused the heart to slow down. In this case, the fluid transfer resulted in a slowing of the second heart, so he named the active substance from the decelerator nerve, which is a branch of the vagus nerve, *Vagusstoff*. Later, it was discovered by others that the accelerator nerve released noradrenaline, and the vagus nerve released acetylcholine. Loewi received a Nobel Prize for his discovery.

It was established in subsequent years that electrical transmission also occurs, but the contacts between cells where this happens are much less common than the sites of chemical transmission. Neither type of synapse can be seen with a light microscope. Synapses were not actually identified anatomically until pictures were taken with an electron microscope in the 1950s. In those pictures, chemical synapses can be identified by thickenings in the membranes of both neurons (the presynaptic and postsynaptic neurons) and by a collection of tiny vesicles on the presynaptic side (figure 1.11). These vesicles are known to contain neurotransmitter molecules. When the membrane undergoes a rapid depolarization, as when an action potential arrives, some vesicles bind to the membrane and open to the space outside the cell (the synaptic cleft), releasing the neurotransmitter molecules, which bind to specific membrane receptors in the postsynaptic membrane. Different neurotransmitter and receptor combinations cause different effects on the postsynaptic membrane: A small depolarization is called an excitatory postsynaptic potential (EPSP) because it moves the membrane closer to the threshold for triggering an action potential (at the beginning of the neuron's axon, the axon hillock). A small hyperpolarization is called an inhibitory postsynaptic potential (IPSP) because it moves the membrane away from the threshold (figure 1.12).

It has been discovered that a variety of peptides are also secreted along with the neurotransmitter at many synapses. These peptides are called neuromodulators. They can be found in other vesicles mixed in with those containing neurotransmitter at chemical synapses.

Synaptic Morphology and Types

Electrical synapses can also be identified in pictures taken with an electron microscope. Known as *gap junctions*, they have no synaptic vesicles, and the synaptic cleft is much narrower. They can be identified already in sponges. In adult vertebrates, chemical synapses appear to be much more common than electrical synapses, judging from results reported by electron microscopists. However, in embryonic stages, electrical synapses may be much more numerous.

The earlier studies of synapses by neuroanatomists using electron microscopes showed that dendrites of neurons were commonly studded with synapses, with the dendritic

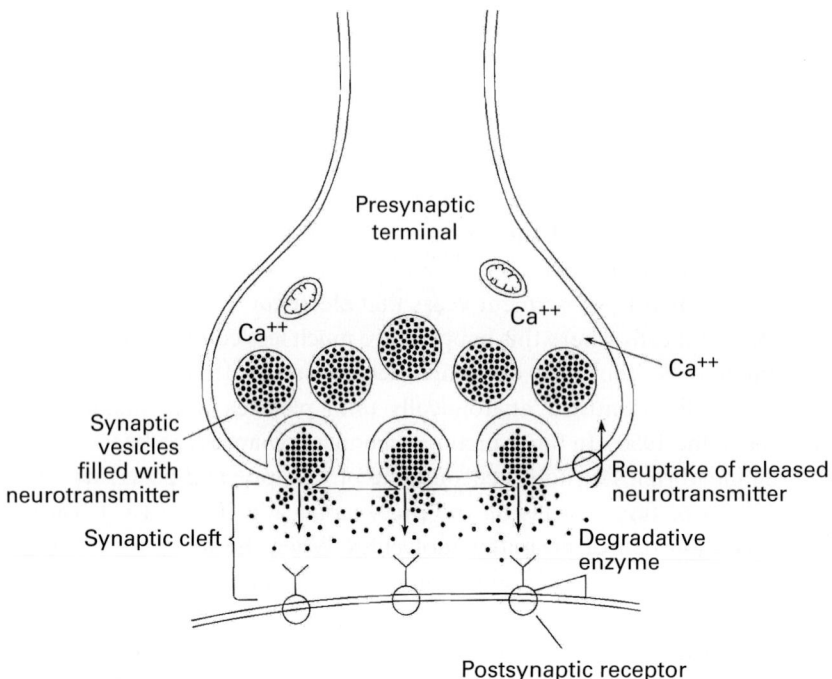

Figure 1.11
Cartoon illustrating an enlargement of a chemical synapse, with notes on the movements of molecules and synaptic vesicles with the arrival of an action potential, resulting in alterations in the potential across the postsynaptic membrane. From Becker, Breedlove, Crews, and McCarthy (2002).

membrane on the postsynaptic side and various axonal endings on the presynaptic side. The synapses of axons were often formed by the membrane of enlarged endings known from the Golgi studies of Ramón y Cajal (see chapter 2) and others. These enlargements are the end buttons, or *boutons* from the French. A single bouton can form a number of different synapses. In addition to the axo-dendritic synapses, there were many axo-somatic synapses on the cell bodies of neurons. (The cell body is the soma.)

Soon, close inspection of the electron microscope pictures revealed that additional types of synapses are also found (figure 1.13), some of them fairly commonly in certain brain regions. Thus, in addition to axo-dendritic (on dendritic shafts or on dendritic protrusions called spines) and axo-somatic synapses, there are axo-axonal synapses. More surprisingly, there are also dendro-dendritic and dendro-axonal synapses. Physiologists have found functional correlates of axo-axonal contacts. If activity at an axonal ending causes a depolarization of another axon's ending, then any action potential arriving at that second axon's ending will be reduced in size (because of the reduced membrane potential). The consequence is a reduced release of neurotransmitter, so the effect is called

23   Getting Ready for a Brain Structure Primer

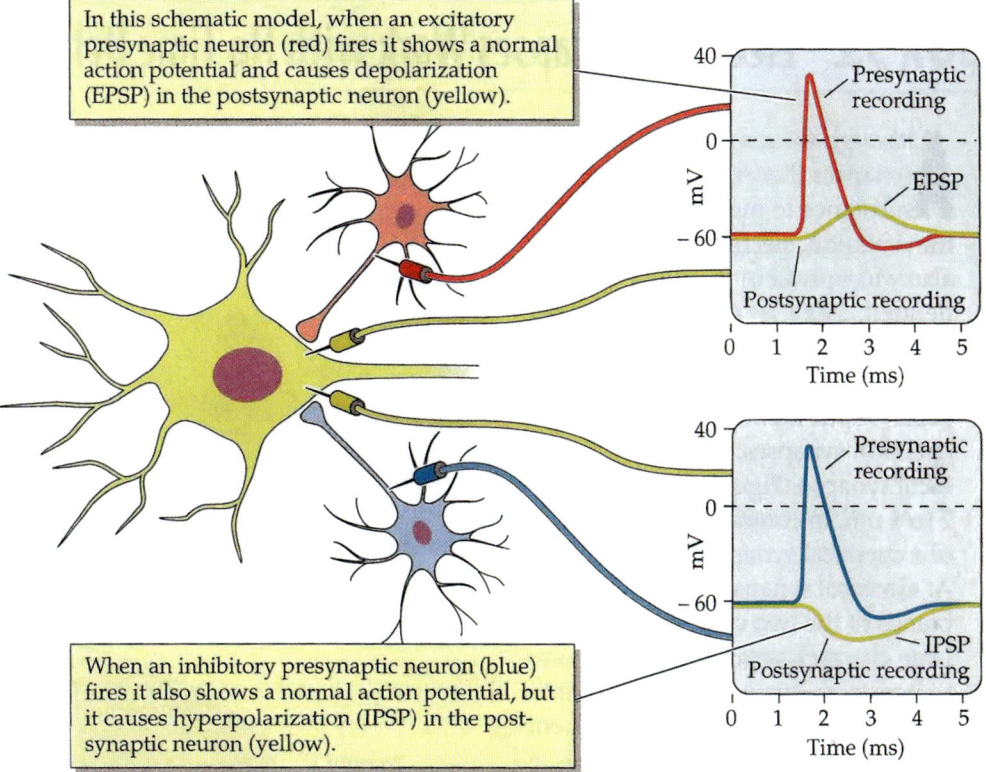

Figure 1.12
Cartoon illustrating the recording of an EPSP and an IPSP in neurons, and the action potentials that caused these synaptic potentials. Adapted from Breedlove, Rosenweig, and Watson (2007).

presynaptic inhibition. Similarly, an axo-axonal contact can cause *presynaptic facilitation* if it causes hyperpolarization.

These additional types of synapses can be found in more complex synaptic arrangements. For example, in a reciprocal synapse, an axo-dendritic contact can be adjacent to a dendro-axonal contact, with vesicles indicating that the influences go in opposite (reciprocal) directions, indicating a kind of feedback arrangement.

Related Cellular Dynamics

When a cell releases a chemical into the surrounding intercellular fluid, as at a synapse, the process is called exocytosis. Cells also take up substances from the surrounding fluid. This is called endocytosis. Endocytosis occurs around synaptic endings when the membrane of an axon takes up a neurotransmitter it has just released (called reuptake) or it

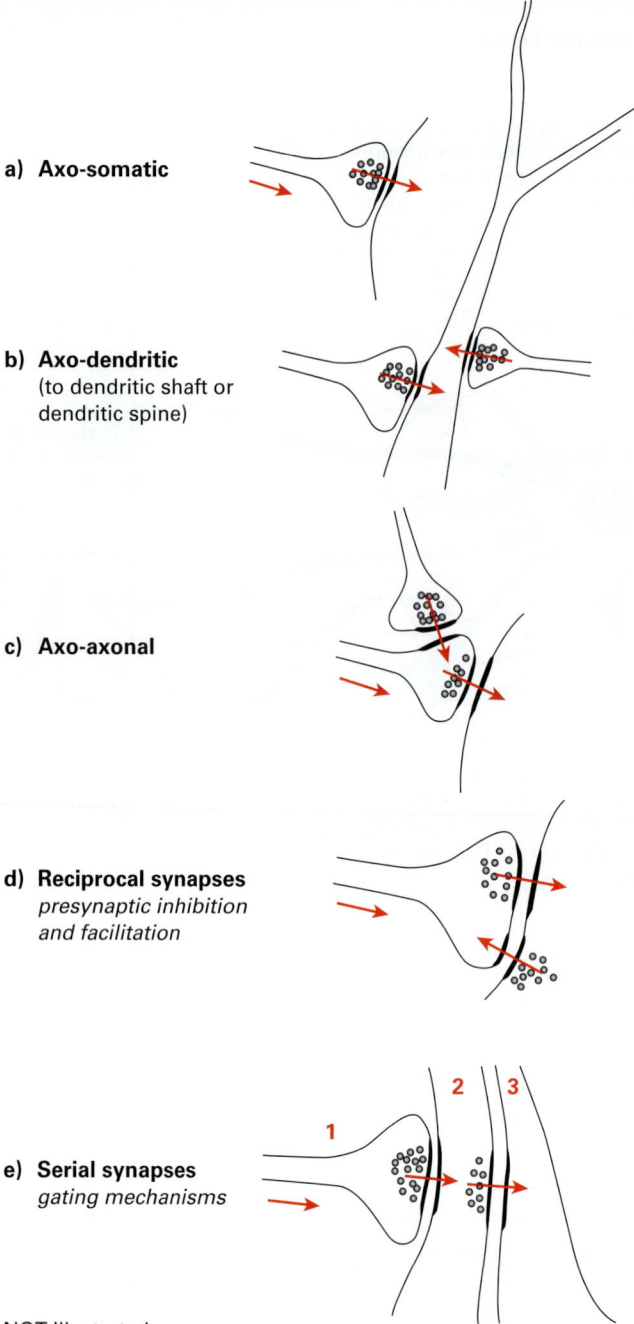

Figure 1.13
Illustration of sections, seen at electron microscopic magnifications, through synapses found in varied structural arrangements within the CNS. Each arrangement indicates specific functional effects. Synapses were first seen, anatomically, via electron microscopy in the 1950s. (a) Axo-somatic synapse, between an axon ending and a cell body. (b) The most common arrangement: Axo-dendritic synapses. (Both types can be excitatory or inhibitory.) (c) Axo-axonal synapses are also found. They can increase or decrease the effects of action potentials arriving in the axon contacted (see text). (d) A reciprocal synapse, indicating a contact that works in both directions (mentioned in the text). (e) A serial synapse: The membrane potential in the middle process (2) modulates the effects of the initial synapse (by cell 1) on the third cell contacted (3).

takes up the breakdown products of the neurotransmitter if the neurotransmitter has been broken down by enzymatic action. For example, acetylcholine is broken down, and thus its action at a synapse is halted, by acetylcholinesterase.

Substances taken up at an axonal ending—not only neurotransmitters but other substances as well—are often transported up the axon to the cell body. This process is called retrograde transport, moving opposite to the direction of information flow in the axon. Retrograde transport is an active process involving the protein dynein.

Anterograde transport is another form of intracellular transport of molecules and organelles. This movement involves the protein kinesin. It is used by the neuron to move molecules from the site of synthesis in the cell body to the endings of its axon. Membrane is moved in this fashion in the form of vesicles, especially during development when the axon is adding to its length by adding membrane at its endings.

In the next chapter, we will see how endocytosis and intracellular transport, both anterograde and retrograde, are used by neuroanatomists to trace axon pathways and connections.

Parallel Channels of Information Flow and Integrative Activity

Every multicellular organism has a special need, different from the needs of single-celled organisms: Information in the form of various kinds of stimulation can come at it from various directions, hitting different cells on different sides of the animal. Thus, there is a special need for integrative action: How does one end of the animal influence the other end? How does one side coordinate with the other side? (How does the left hand know what the right hand is doing?) With multiple inputs and multiple possible outputs, how can conflicts be avoided or at least reduced?

This problem, if it was to be solved without every cell being similar and in contact with every adjacent cell as in the sponge, made the evolution of nervous systems inevitable. The need for integrative connections increased as the size of the central nervous system expanded. The adaptive value of integrative communicative activity resulted in the evolution of interconnections among multiple subsystems of the nervous system of animals. It is the nature and organization of these subsystems and their connections that is the topic of this book. How such connections can be studied is the topic of the next chapter. How they function is the topic of behavioral and of physiological studies, including studies using neurochemical and electrophysiological methods.

Endogenous Activity

The last primitive cellular mechanism discussed here is the one most neglected and often overlooked in such a survey. One reason may be that the topic is often introduced with little regard to evolution. Yet, endogenous activity occurs in all animals including Protozoa and Cnidaria. Another reason for the neglect is the simplicity of the reflex model for thinking in anatomical terms about the neurology of behavior. Reflexes begin with a stimulus; endogenous activity is generated without any external stimulus. When a scientist believes that all behavior can be explained in terms of a stimulus–response (S-R) model, then he is not comfortable dealing with actions where there is no stimulus from the outside world. The absence of a stimulus does not mean a lack of a cause, but for a long time the causes of endogenous activities were unknown. It is only with modern cell biological approaches that endogenously generated activities of organisms are beginning to be understood at the cellular level.

If the S-R model can explain how the activity of the nervous system underlies the actions of an organism, then it should be the easiest to apply to the behavior of animals with the simplest nervous systems. For example, consider the behavior of the little cnidarian, hydra (figure 1.14). When it is foraging for food, to move about this little animal shows a peculiar somersaulting behavior. What is the stimulus that elicits such behavior? In an interesting study, hydra were placed in a very uniform, homogeneous aquarium environment, eliminating all stimuli to the greatest degree possible. Even under these conditions, the hydra initiated periodically its characteristic foraging locomotion. The action sequence appears to be generated from within. It could arise from some internal change in non-neural tissue that signals hunger or it could arise from within some cells of the nervous system.

How could nerve cells activate themselves without a stimulus? It has been discovered that some neurons do just that.

In the 1960s and early 1970s, Felix Strumwasser was recording from single neurons of sea slugs of the genus *Aplysia*. He found neurons that showed rhythmically oscillating membrane potential changes even when the cell was isolated from neuronal inputs. Such an endogenously generated rhythmic potential could cause rhythmic bursts of action potentials. Figure 1.15 illustrates such a finding by sketching the results obtained from a recording from an identifiable large secretory neuron of the abdominal ganglion. The period of the rhythm was 40 seconds. The membrane potential changes continued to oscillate even if the action potentials were blocked with tetrodotoxin (obtained from the

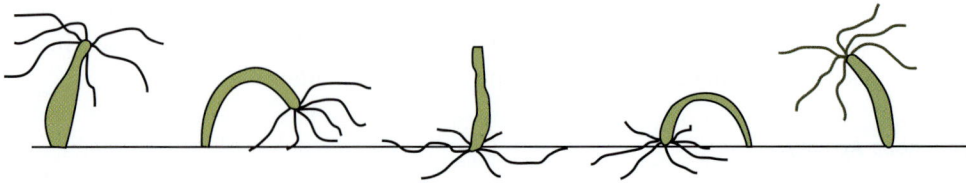

Figure 1.14
Illustration of locomotor behavior in hydra, which can be internally generated as discussed in the text. Adapted from Keeton (1976).

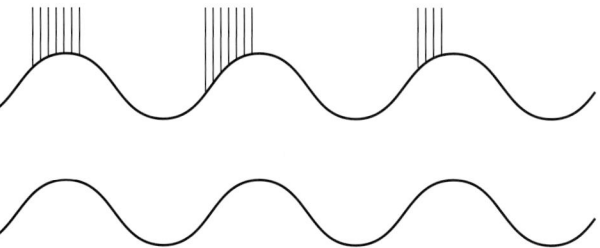

Figure 1.15
Membrane potentials of the type recorded in *Aplysia* abdominal ganglion neurons with rhythmically changing slow potentials that appear to have endogenous causes. There is evidence for molecular mechanisms within the membrane of some neurons that can cause such oscillations, as they change cyclically. The top sketch shows action potentials occurring with each upswing of depolarization. The bottom sketch shows a consequence of tetrodotoxin (TTX), which blocks action potentials but leaves the changing slow potentials intact.

Japanese puffer fish). However, it did not persist if the membrane's sodium pump was blocked with another substance, ouabain.

Strumwasser and his co-workers continued recording from such an *Aplysia* neuron for a very long time and claimed that there were changes of approximately a day in periodicity—a circadian rhythm. The phase of this rhythm could be changed by light, indicating that the cell could act as a biological clock entrained by the day–night cycle.

It is well known that the daily activity rhythm of vertebrate animals depends on internal, endogenous clocks known as biological clocks. In the absence of exposure to daily changes of light and dark, such activity rhythms persist with a circadian rhythmicity: The length of each period is approximately but not exactly 24 hours. Mice showing such a "free-running" circadian activity rhythm were found in one experiment to be affected by the addition of heavy water—D_2O—to their drinking water. The rhythm slowed down with more D_2O, indicating an effect of the inertia of the water molecules!

There have been many electrophysiological and molecular studies of endogenous activity in neurons since the early work. Rhythmically changing potentials have been found to be generated by molecules located at specific places in the cell membrane, designated as pacemaker loci. Because a cell may have different pacemakers generating different rhythms, such a cell could, in theory, generate a rhythmic waveform of any complexity (according to the mathematics of Fourier analysis), as the various simple waveforms would summate at the axon hillock.

Readings

Good reviews of various sensory receptors can be found in texts such as the following:

Møller, A. R. (2003). *Sensory systems: Anatomy and physiology.* Boston: Academic Press.

Shepherd, G. M. (1988). *Neurobiology*. New York: Oxford University Press.

Squire, L., Berg, D., Bloom, F. E., du Lac, S., Ghosh, A., & Spitzer, N. C., eds. (2013). *Fundamental neuroscience*. Boston: Academic Press.

Concerning hydra (figure 1.14):

Lenhoff, H. M. (1961). Activation of the feeding reflex in *Hydra littoralis* I. Role played by reduced glutathione, and quantitative assay of the feeding reflex. *Journal of General Physiology, 45,* 331–344.

See various papers published in *American Zoologist* in 1965.

The quotation from Richard Gregory is from the following source:

Gregory, R.L. (1978) *Eye and brain*, 3rd edition. New York: McGraw Hill, World University Library. See chapter 4, p. 42.

2 Methods for Mapping Pathways and Interconnections That Enable the Integrative Activity of the CNS

As the central nervous system—the brain and spinal cord—evolved and its size and complexity increased, its parts had to become interconnected. The integration of the various activities of an animal does not happen by magic. It happens because of the many connections among the neurons of its CNS. These interconnect the multiple subsystems that influence and control the multiple functions of the organism. This is not to say that conflicts never occur; they certainly do, but great numbers of axonal connections have evolved to reduce the conflicts. Also, such interconnections can be increased with the efforts of learning.

How are we able to discover the nature of these connections? If we open a skull, the brain it encloses looks "like nothing so much as a lump of porridge," recalling R. L. Gregory's remark cited in the previous chapter; or perhaps it is more like a cream-colored gelatin pudding permeated with a spidery red network (the blood vessels). We don't see the individual neurons, most of which are too small and not visually distinct. To obtain data for making sense of this lump of porridge, we must apply neuroanatomical (neuromorphological) techniques.

We can make much more sense of it when we use multiple methods to study the same brains. For example, in addition to neuroanatomical methods we can use neurophysiology for electrical stimulation and recording. We can use the methods of neurochemistry and neuropharmacology. We can also conduct behavioral studies in conjunction with brain studies.

In recent years, various imaging methods have also been used. These imaging methods have the advantage of being useful for study of the large brains of humans, cetaceans, and other animals without a need to cut them up. However, these methods have important limitations for the study of pathways and connections in the CNS.

Histology and Brain Architecture

Neuroanatomists study the component parts of the brain and its pathways and interconnections mainly by experimental work with animals using various specialized techniques.

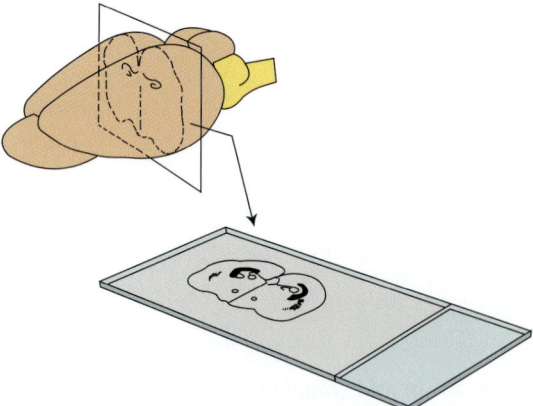

Figure 2.1
Illustration of the sectioning of a rodent brain into thin frontal (transverse) sections and mounting them on glass slides for staining and viewing through a microscope. To section a brain, it is usually hardened in a fixative and then frozen and/or embedded in a material that holds it together during the cutting, mounting, and staining processes. Adapted from Breedlove, Rosenzweig, and Watson (2007).

The initial steps of these techniques are usually the same or very similar, often after an experimental procedure involving surgery. We will say more about such procedures soon. The initial steps begin with methods of fixation of the tissue, usually with formaldehyde or glutaraldehyde. These fixatives also cause the brain to become firm. The fixed brain is then embedded in some material that holds it together or it is frozen, and then it is cut into thin sections using a *microtome* for precise control of a very sharp blade (figure 2.1). Techniques for visualizing nerve cells or axons often involve staining procedures, which are usually carried out on the sections rather than before the cutting of those sections. The sections are mounted on glass slides, either before or after the staining procedures.

Before studies of connections can make much sense, the anatomist has to be able to visualize the distinct subdivisions of the central nervous system—the separable groups of neurons or fibers. He does this by using techniques that reveal the nerve cells, often just the cell bodies, or the axons, using dyes to bind components of the tissue selectively. Distinguishing the different cell groups of the brain using such techniques is called working out the **cytoarchitecture** of the brain. When an anatomist uses the fiber groups and their arrangements in his analysis, he is using **fiber architecture**—the spatial distribution of the axons that form the connections among neurons.

In figures 2.2 and 2.3 we see examples of brain sections stained selectively for cell bodies. The staining method uses a basic dye to bind to structures of the perikaryal region called the Nissl substance, after the anatomist Franz Nissl. In figure 2.2, a section from the brain of a rat reveals some cytoarchitectural divisions, with neocortex above displaying

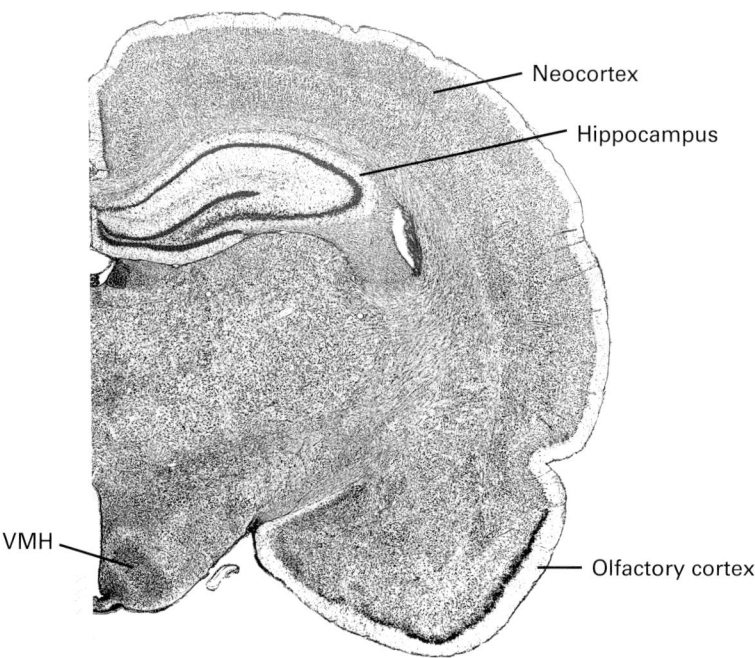

Figure 2.2
Part of a frontal section through a rat brain stained for cell bodies with a Nissl stain. Cytoarchitectural details can be seen. Note the distinct layering of neurons in the neocortex at the top of the picture and the single cell layer of the hippocampus below the cortex. Note also the cell groups that stand out because of their staining intensity or cell spacing or size in the diencephalon below the cortex. VMH, ventromedial nucleus of the hypothalamus. From Swanson (2002).

its characteristic layering, and the diencephalon ('tweenbrain) below revealing cell groupings, some distinct and some not so distinct. In figure 2.3, we can see the distinct cell layers of visual cortical areas in a human brain.

In figures 2.4A and 2.5, we see examples of fiber architecture obtained by staining brain sections with a chemical that binds to myelin, making the myelinated fibers appear black. The first shows a frontal section of a human midbrain. Its most obvious fiber bundles are the cerebral peduncles at the ventrolateral surface, carrying large numbers of axons from the neocortex on their way to the hindbrain and spinal cord. Figure 2.4B illustrates a very different and more sensitive method for revealing myelin in the CNS. In this figure, frontal sections of the superficial part of the midbrain of a young hamster are shown. Immunohistochemistry has been used to mark the membrane of the glial cells that form myelin, using an antibody to a protein in the membrane of oligodendrocytes.

Figure 2.5 shows a horizontally cut section from the brain of a 7-week-old human, revealing three pathways that are early to acquire myelin. These are fibers carrying

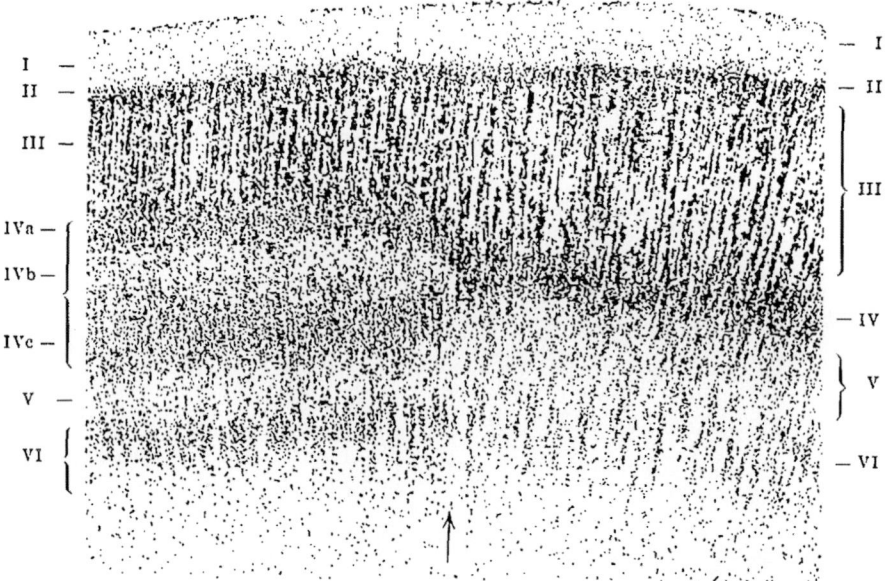

Figure 2.3
Cytoarchitectural differences between two neocortical areas in the human brain, sectioned and stained for cell bodies as in figure 2.2. These cortical areas are in the occipital region and are involved in the processing of visual information. At the left is area 17, the primary visual cortex. At the right the less complex lamination of area 18 is evident. From Brodmann (1999).

somatosensory, auditory, and visual information from the 'tweenbrain to sensory areas of the neocortex. Also very obvious are fibers of the cerebellum, the so-called cerebellar white matter. The name *white matter* comes from the fact that in unstained tissue, the fibers appear more white than adjacent tissue where cell bodies are more abundant.

There are methods used by neuroanatomists for subdividing the central nervous system besides methods for visualizing cell bodies and axons. Other methods are used to reveal the brain's **chemoarchitecture**. Chemical substances that play specific functional roles are concentrated in some regions more than in others. By stains that allow us to visualize these substances, we can distinguish some brain regions from others. For example, in figure 2.6 we can see specific regions darkened by a marker for acetylcholinesterase—an enzyme that breaks down the neurotransmitter acetylcholine. Using such techniques, structural patterns not revealed by cell or fiber arrangements may be revealed. (See also figure 2.7).

Other methods of mapping the brain are increasingly being used. One method, mentioned already in describing figure 2.4B, uses the ability of antibodies to bind selectively to specific protein molecules. Thus, the techniques of immunohistochemistry allow a neuroscientist to visualize the distribution of a specific kind of receptor (figure 2.8A).

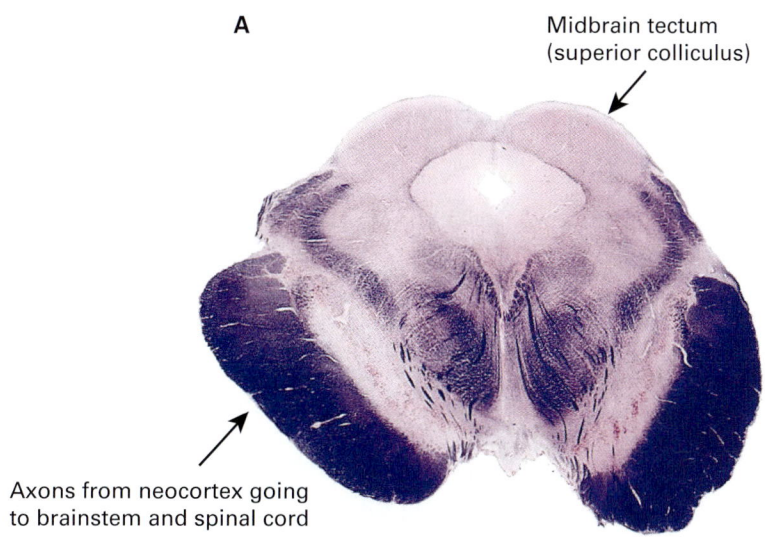

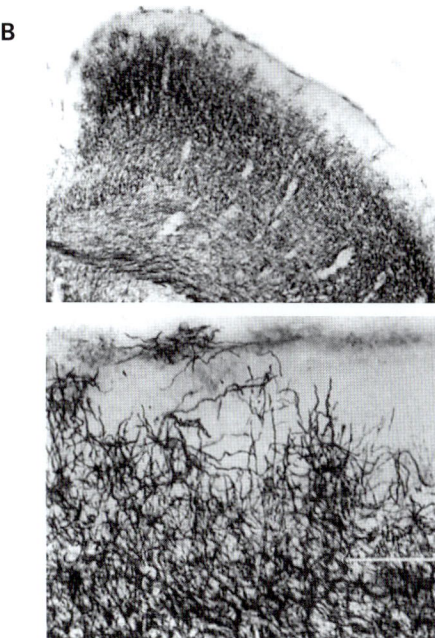

Figure 2.4
(A) Transverse section of a human midbrain stained for myelinated fibers with iron hematoxylin, which appears dark. Cell bodies have been lightly stained a pink color with eosin. The very large bundles of fibers at the bottom on either side are the cerebral peduncles, which consist of fibers from the neocortex to more caudal structures. (B) In frontal sections of the 21-day-old Syrian hamster's superior colliculus, immunohistochemistry made use of an antibody that binds to the membrane of oligodendrocytes, the myelin-forming glial cell in the CNS. The dorsal surface of the right midbrain is seen at the top of each photograph. Oligodendrocytes can be seen, many of them forming myelin. Individual glial cells are visible, especially in the higher-power photograph at the bottom. Panel (A) from Nolte and Angevine (2007) (these images can also be found on the authors' Creighton University Web site); panel (B) from Jhaveri, Erzurumlu, Friedman, and Schneider (1992).

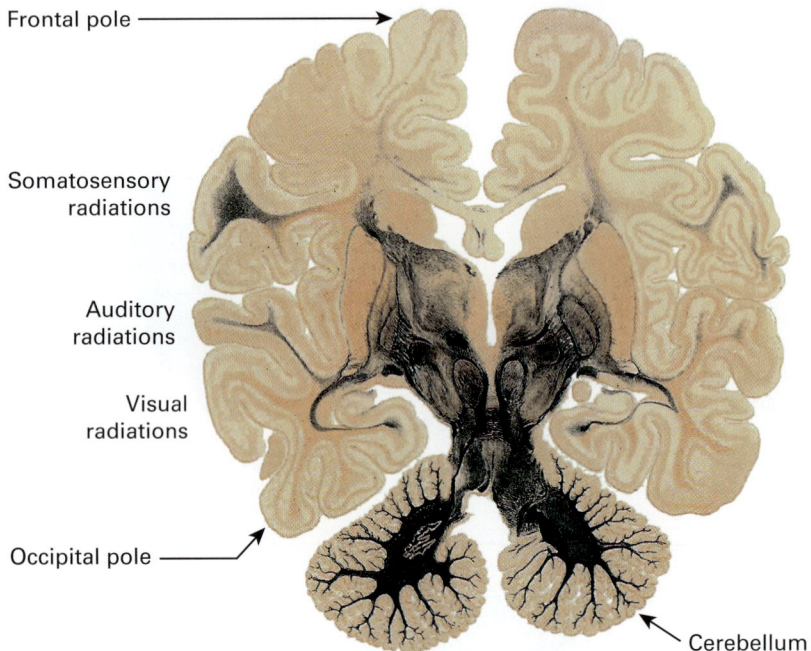

Figure 2.5
Horizontal section of the brain of a 7-week-old human fetus stained for myelin. Very little myelin has developed in the neocortex at this stage. One can see thalamocortical radiations of the somatosensory, auditory, and visual systems because they myelinate relatively early. At this fetal stage, brainstem and cerebellar fibers have much more myelin than in most of the cortex. Adapted from Flechsig (1920).

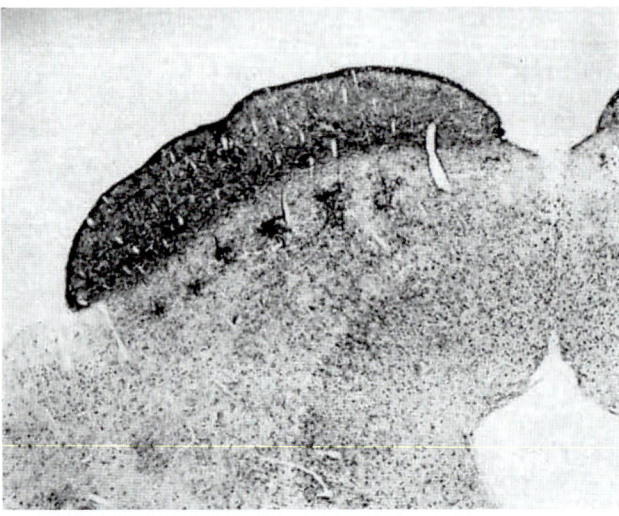

Figure 2.6
An example of the brain's chemoarchitecture: a frontal section of the rat midbrain that has been stained for the enzyme acetylcholinesterase. One can see layers and patches within the superior colliculus. From Graybiel (1978).

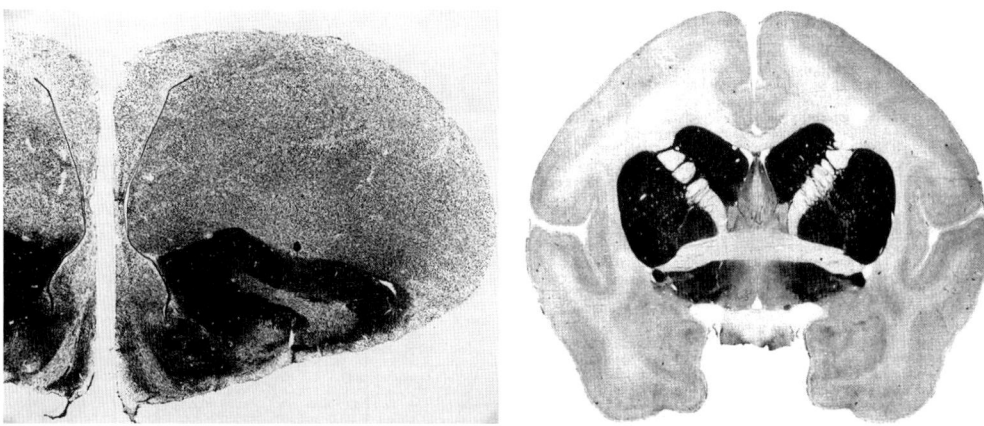

Figure 2.7
Acetylcholinesterase (AChE) histochemistry has been a useful tool for comparative neuroanatomists. The enzyme is found at high levels in the corpus striatum. A comparison of the staining patterns in the telencephalon of (A) the pigeon and (B) the squirrel monkey has provided evidence that the bird striatum is much more restricted than it was once thought. The heavy AChE staining does not include a large dorsal territory in the bird's endbrain that does not look like cortex but has also been found not to be like the mammalian striatum. The dorsal area, which does not contain large amounts of AChE, has been found to resemble mammalian cortical structures in its thalamic connections. Left panel from Karten and Dubbeldam (1973); right panel from Maclean (1972).

Large catalogs of brain maps are being assembled to store the results of studies of gene expression patterns in particular species (figure 2.8B).

What Is Connected to What?

With some knowledge of various subdivisions of the brain—its architecture—we can name the subdivisions and parts and begin to have a language for these organs of the mind. However, it is only when we can map the brain's pathways for information transmission and specify its synaptic interconnections that we can begin to understand something of how it serves as the organizer and controller of behavior. We cannot discover these connections using only the methods of staining for normal cell bodies or fibers described thus far.

As descriptive neuroanatomy was developing, there was one method of staining of nerve cells that could yield information about connections. This was the method invented by Camillo Golgi and used most extensively and comprehensively by the great Spanish neuroanatomist Santiago Ramón y Cajal in the latter part of the nineteenth century and the early part of the twentieth century. The stain is remarkable in its ability to reveal, in optimal cases, the entire extent of a neuron including all of its long extensions, the dendrites and axons. It stains only a small proportion of the total population of neurons,

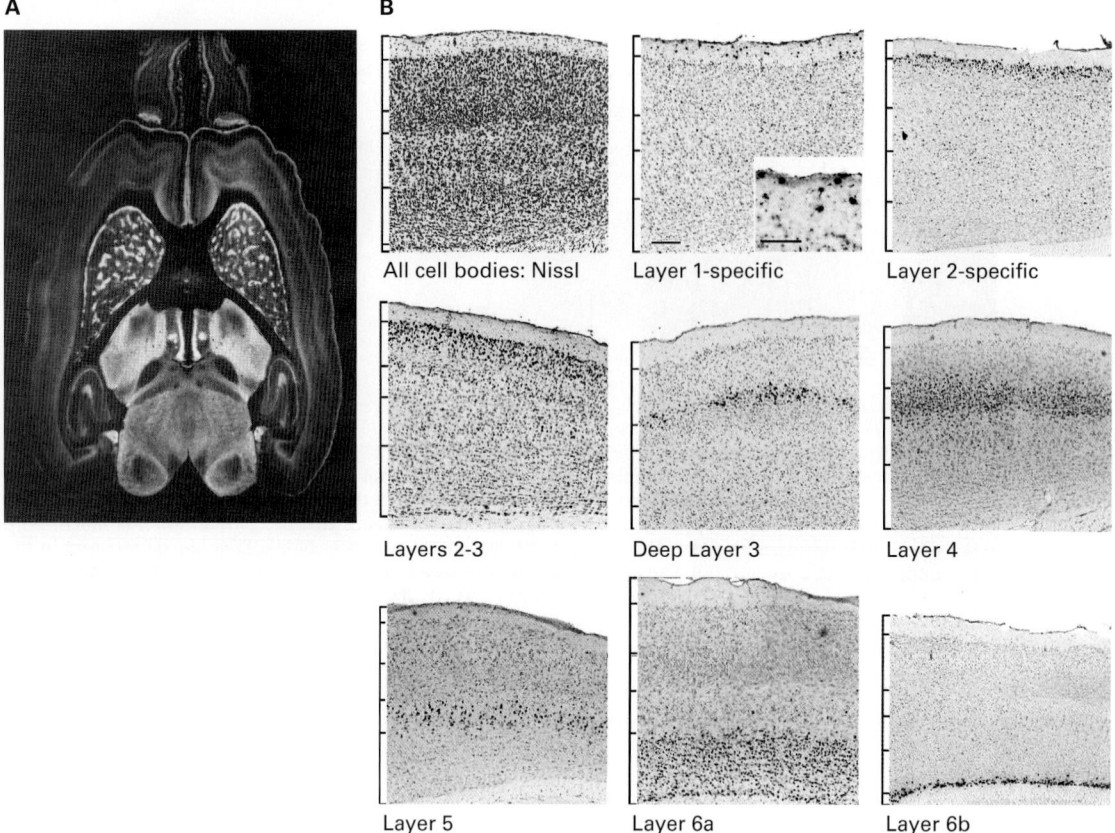

Figure 2.8
(A) Example of immunohistochemistry: opiate receptor localization in the rat brain, shown in a horizontal section; anterior is up. Opiate receptors were labeled in the live animal by injection of tritium-labeled naloxone. Sections of the brain were coated with photographic emulsion and exposed (in the dark) for 15 weeks. A striking pattern of selective labeling of brain structures became visible when the emulsion was developed. Parts of the thalamus (in the 'tweenbrain; i.e., the diencephalon) are most heavily labeled, but many other structures also show a selective pattern of labeling. (B) Expression patterns of layer-specific genes in the mouse neocortex, from a paper with 108 authors. Panel (A) from Herkenhan and Pert (1982); panel (B) from Lein et al. (2007).

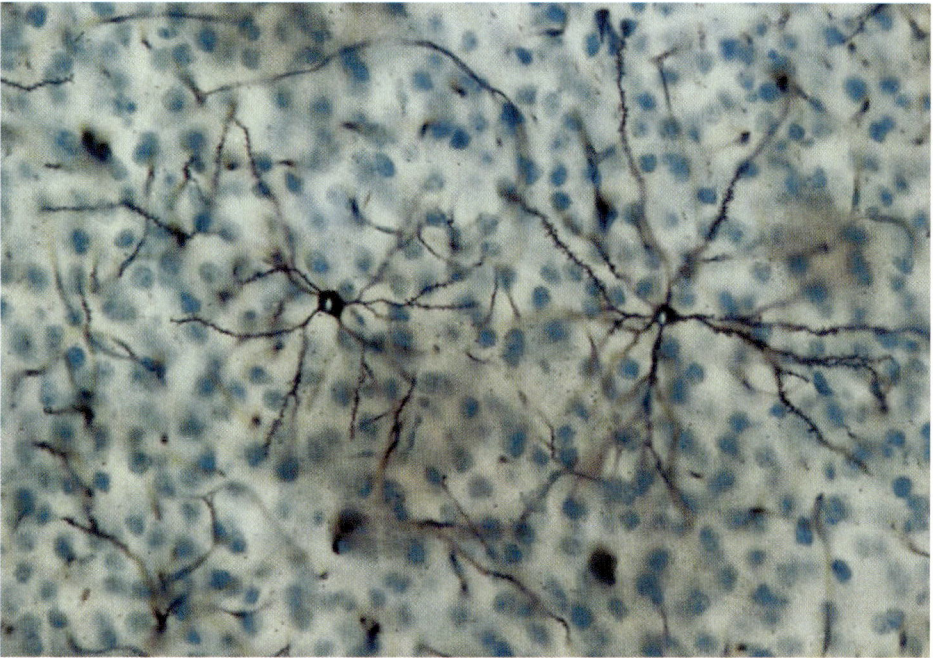

Figure 2.9
Stellate-shaped neurons in the somatosensory cortex of a rat stained with the Golgi method, with other cells in the thick section counterstained with a Nissl method for cell bodies. As is typical with the method, the Golgi technique stains only a few of the neurons. The earliest version of the method was developed by Camillo Golgi and was soon put to major use by Santiago Ramón y Cajal. The method is capable of revealing the cell body, the dendrites, and often the axon as well. In this photograph, the depth of focus is much less than the thickness of the section or the depth of the neuron, as can be seen by following dendrites in the photograph that become blurred. These dendrites can be brought into sharp focus by adjusting the plane of focus. From Glaser and Van der Loos (1981).

which is most fortunate because if it were to stain all of them, the result would be nearly total darkness because of the great density of cells. This property is illustrated in figure 2.9. In this picture, the Golgi-stained neurons are seen filled with a black substance, while the remaining cells are revealed only by their cell bodies stained with a Nissl method.

Cajal used the Golgi method, and variations of his own invention, in a monumental, long and comprehensive series of studies that covered all the major parts of the vertebrate nervous system, with a concentration on mammals. In figure 2.10, we see a drawing he made that reproduces what he saw in thick sections of the spinal cord of a newborn rat. In his sections, he saw many axons coming into the cord from peripheral nerves and ramifying in terminal end arbors of various configurations. In the figure 2.11, we see a Cajal drawing of large neurons in the ventral spinal cord of a fetal cat. He has drawn some of these neurons with their long branching dendrites in black, and in red he has

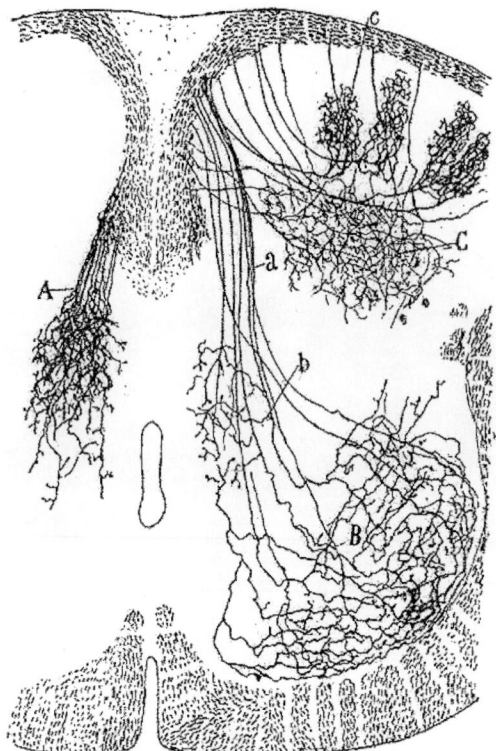

Figure 2.10
Some axons in the spinal cord of a newborn rat, drawn by Ramón y Cajal. These axons are all collaterals of dorsal root fibers that originated in skin, joint, and muscle tissue. As is the standard for such drawings, dorsal is at the top. A, axon collaterals in the intermediate nucleus; B, axonal arborizations in the motor nucleus (ventral horn); C, extended ramifications in the head of the dorsal horn; a, sensory-motor bundle; b, collateral to the intermediate nucleus, c, deep collaterals in the substantia gelatinosa. From Ramón y Cajal (1995).

included the portions of their axons that had branches in the same regions. We know that some of the neurons are motor neurons because their axons send a main branch out of the cord.

Cajal published large numbers of such drawings, which he apparently produced from memory after periods of intense microscopic study of sections of the Golgi-stained brains. His visual memory was amazingly accurate, as attested by the more painstaking tracings of similar material by other anatomists.

Tracing a Pathway from Sensory Input to the Response (S to R)

After many studies of the spinal cord using this neuroanatomical technique, Ramón y Cajal was able to follow a complete pathway from axons carrying sensory input into the

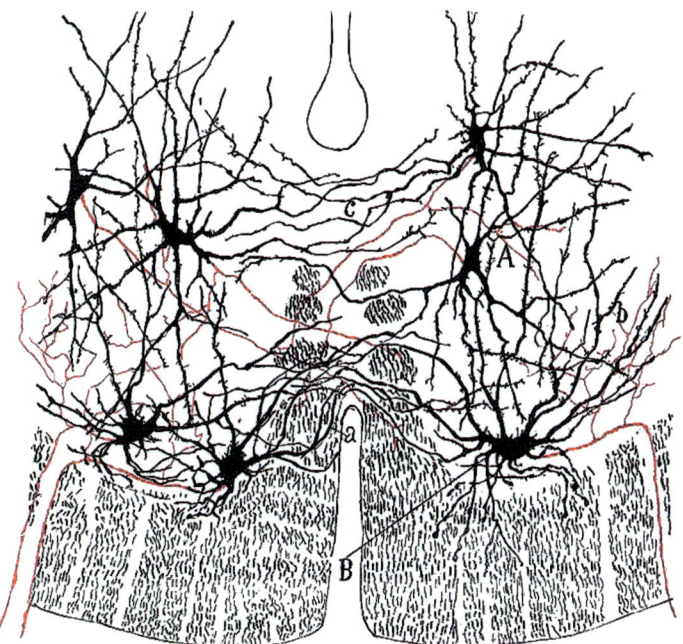

Figure 2.11
Neurons of the spinal cord of a fetal cat stained with the Golgi method and drawn by Ramón y Cajal. Cell bodies and dendrites are in black and axons in red. In the ventromedial spinal cord, it is common for some axons and dendrites to cross the midline. A, commissural neuron; B, motor neuron in the medial nucleus; a, commissural dendrites; b, dorsal dendrites; c, commissure that is formed by the dendrites of spinal neurons. From Ramón y Cajal (1995).

spinal cord to motor neurons with fibers going to muscle cells. His interpretive summary of what he saw is shown in figure 2.12. He had found specific evidence, for the first time, for the stimulus–response (S-R) model—explaining all behavior in terms of connections through the nervous system between a stimulus (S), which initiates the behavior, and a response (R). This was the reflex model of how behavior is produced and controlled, the model that originated with René Descartes in the seventeenth century and was championed by philosophers (e.g., LaMettrie) in the following century and boosted in the nineteenth century by the Russian reflexologist Ivan Sechenov and his famous student Ivan Pavlov. It was Pavlov who provided the evidence that made the reflex model much more comprehensive, the evidence that the S-R connections of a reflex could be altered by learning (conditioning).

In the first half of the twentieth century, the pioneering physiological psychologist Karl Lashley was assuming the S-R model when, as a young student, he had the opportunity to view slides of a frog brain. The thought arose in his mind that if he could use this kind of material to see all the connections in the entire brain and spinal cord of the frog, it

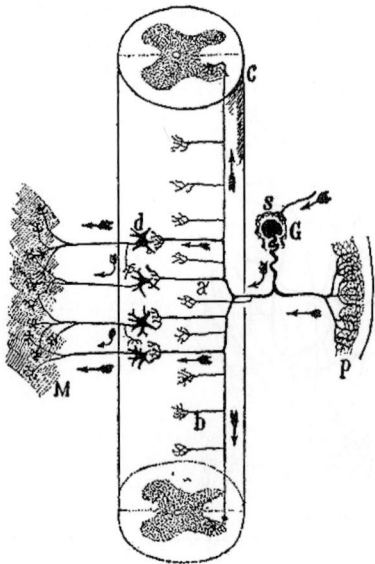

Figure 2.12
Diagram of a local reflex pathway in the spinal cord, seen for the first time in its entirety by Ramón y Cajal. His summary drawing was included in his autobiography. Using the Golgi method, Cajal could discern local circuits well, but he could not always be very certain about the origins or the destinations of long axons. The monosynaptic reflex pathway in this picture is now known to originate in stretch receptors of a muscle, and the motor neurons synapse with striated muscle fibers of the same muscle. G, spinal dorsal root ganglion; M, muscles; P, sensory endings, believed by Cajal to be in skin, now known to be in muscle spindle organs; S, pericellular sympathetic arborization; a, reflexomotor collaterals; b, short collaterals; d, motor neurons. Arrows indicate direction of impulse flow (Cajal called it current flow). From Ramón y Cajal (1995).

would be possible to explain the frog's behavior (according to the assumptions he was making).

Later, Lashley argued against the adequacy of S-R theory for explaining temporal order in rapid sequences of behavior. Also, the accumulating evidence for endogenous activity in the nervous system does not fit (recall the last section in chapter 1), and it became obvious to ethologists and other behaviorists that motivational systems can initiate behavior independent of external stimulus triggers. Nevertheless, the S-R model remains a common assumption among neuroscientists including neuroanatomists.

Even if reflex models are inadequate as complete explanations of all behavior, particularly of what we call cognitive processes, nevertheless it seems quite clear that continuously active reflexes are a major part of the background of behavior. Furthermore, brain pathways and their connections are important not only for reflex functions but for many other functions as well. How do we know about these connections, and how can more be learned about them?

Pathways Seen as Bundles and Ribbons of Fibers

Early anatomists used dissection techniques, which were made much easier with the development of fixatives and their improvement. In brain dissections, one can see clearly the major blood vessels and many of their ramifications. The major fiber tracts can be discerned as white bundles and ribbons, because the myelin that covers the larger axons reflects light more than does the tissue containing mostly neuron cell bodies. Thus, we can distinguish white matter and gray matter. We have already mentioned a few of the staining methods that have enabled neuroanatomists to visualize cell bodies or fibers.

For a few pathways, dissection and such staining methods can be used to follow large groups of axons from origins to a likely region of termination, such as from the eye to the roof of the midbrain. However, through the first half of the twentieth century, only with the Golgi method could an anatomist see axonal endings. That method is capricious in what is stained in any one brain, and it is most useful only for the short, local connections. It is usually impossible to follow an axon continuously for long distances. The brain has such large numbers of axons that, unless only a very few are stained, an anatomist loses track of single fibers in the maze of others.

Experimental Studies Needed to Trace Connections in the Brain

Consider a historical example of the problem. When it became clear that vision in mammals depends heavily on the cerebral hemispheres, it was important to know how information gets from the eye to the visual parts of the hemispheres. Effects of brain damage showed that the information must reach the occipital regions. In the latter part of the nineteenth century, it was believed that the pathway went from the eye via the optic tracts to the roof of the midbrain—the optic tectum—and from there the pathway proceeded rostrally into the hemispheres. Gradually, evidence accumulated from experimental studies of patterns of degeneration after brain lesions and destroyed this belief. Each optic tract, on its way to the midbrain, passed over and through a cell group in the 'tweenbrain called the lateral geniculate body, and it was discovered that cells in the lateral geniculate degenerated after lesions of the occipital cortex of the overlying hemisphere, a process called **retrograde degeneration**. Studies of this kind of degeneration showed that the cells underwent atrophy and death after their axonal endings were destroyed. From this and other evidence, it became clear that the major pathway from eye to hemisphere in many mammals, including humans, goes from retina to lateral geniculate body to the visual areas of the posterior hemisphere (the visual cortex).

This example illustrates how a particular experimental method can be used to trace brain pathways and connections. Electrophysiologists developed another experimental method for establishing the reality of a neuronal connection. The great British physiologist Charles Scott Sherrington, and scientists who followed him, used ***antidromic stimulation and recording*** for this purpose. Antidromic means "against the stream," in this case, opposite to the normal direction of axonal conduction. If you stimulate the axonal end arbors, you can record action potentials after a very short delay in the region of the cell bodies of those axons, and your recordings will show a particular waveform that is not like a postsynaptic potential. If you record in any other places in the CNS, you will not see this waveform.

Experimental Neuroanatomy: Tract-Tracing Methods

The methods of retrograde degeneration and antidromic electrical stimulation and recording were not suitable for finding axonal pathways where the scientist did not already have a good idea where to look. It was not practical to search in unknown territories of the central nervous system—as these territories were vast indeed. Needed were experimental neuroanatomical techniques for tracing axons from their cell bodies to their endings. The first techniques that could be used for this kind of tracing employed the changes that occur in axons that are degenerating. Brain damage that destroys neuronal cell bodies, or that severs axons, causes the axons of those cells, or the portion of the axons distal to the point of damage, to undergo degenerative changes that are called **anterograde degeneration**. For example, the myelin sheath slowly breaks up and degenerates, and as it does so, its staining properties change. The Marchi method makes use of these changes. It enables one selectively to visualize the degenerating myelin. It was used, for example, by Karl Lashley to trace the axons from the retina of the rat to the major areas of optic tract termination in the 'tweenbrain and midbrain.

The neuroanatomist Walle J. H. Nauta (figure 2.13), trained at Utrecht in the Netherlands, admired the Marchi method but realized a very great limitation imposed by the fact that most axons lose their myelin before they reach their points of termination. Some small axons, in fact, have no myelin at all throughout their length. He experimented with histological methods using silver nitrate, which were capable of staining with silver the unmyelinated axons. He made modifications of a method used by Bielschowsky, and while he was working in Switzerland as a neuroanatomy instructor at a medical school, he co-authored with P. A. Gygax a paper published in 1951 on a technique they had developed, which included the suppression of the staining of normal axons so the degenerating axons stood out in sharp contrast to the background tissue. This launched the modern era of discovering what is connected to what in the brain and spinal cord, using experimental studies of animals. Nauta emigrated to the United States and worked first at the Walter Reed Army Hospital training students and doing groundbreaking studies of connections in the brains of monkeys, cats, and rats. In 1964, he came to the Massachusetts Institute of Technology, becoming the first neuroanatomy professor appointed to the faculty of a department of psychology. (He was hired by the well-known

Figure 2.13
Walle J. H. Nauta (1916–1994) and P. A. Gygax developed a silver staining method that helped bring experimental neuroanatomy into the modern age. Born in 1916 in Indonesia, Nauta earned the MD and PhD degrees at the University of Utrecht. He taught at the University of Leiden (1946–1947) where he continued his basic research, and then emigrated to the United States where he did research at the Walter Reed Army Institute of Research (1951–1964) and taught at the University of Maryland (1955–1964). He came to MIT in 1964 where he conducted research with students and research associates in the Department of Brain and Cognitive Sciences until his retirement in 1986; he was named an Institute Professor in 1973. Photograph courtesy of Haring Nauta.

neuropsychologist Hans-Lukas Teuber in a move that presaged the development of modern neuroscience.)

At MIT, Walle Nauta attracted young scientists who soon developed modifications of his famous silver staining method, most notably Robert Fink and Lennart Heimer (1967). With these modifications, the method became more sensitive and capable of staining very clearly the terminal enlargements of axons in the early stages of degeneration.

This author was a graduate student at MIT when Nauta came there. After consulting with him on neuroanatomical issues encountered in my dissertation research, I joined Nauta's laboratory and worked there for two productive postdoctoral years. The new silver stain is illustrated in figure 2.14. Regions of termination of the optic tract are revealed in an adult hamster with a neonatal brain lesion. With this particular experiment, we discovered that an anomalous pathway had developed. Axons that were still developing at the time of the lesion of one side of the midbrain surface found their way across the dorsal midline and terminated in the surface layers of the midbrain roof on the opposite side. The sensitive new axon tracing methods enabled this kind of discovery.

Meanwhile, cell biological studies of neurons led to evidence for major intracellular movement of substances by active transport mechanisms. For example, if an amino acid like leucine or proline was labeled with tritium, its fate could be followed after injection

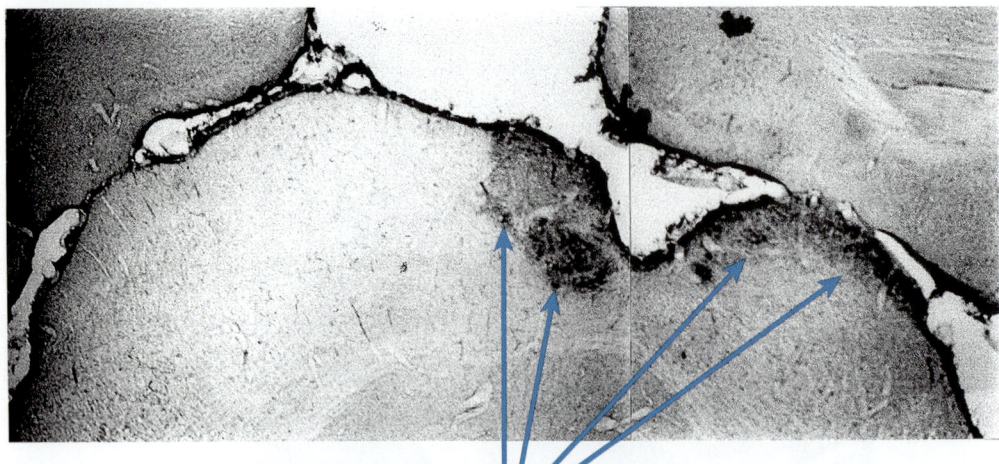

Silver-stained degenerating axon terminals (boutons)

Figure 2.14
Frontal section of the dorsal midbrain of a Syrian hamster with a unilateral lesion of the midbrain surface on the first postnatal day. Axons from the left eye are degenerating because of a left eye removal and are stained with a modified Nauta silver stain for degenerating axons. The photograph shows the axons from the left retina coursing over the surface of the damaged right side of the midbrain, where there are also many terminations. Axons also cross the midline abnormally and terminate in the medial part of the undamaged left superior colliculus of the adult animal.

of the labeled amino acid into the region of neuronal cell bodies. The amino acid is taken into the cells, incorporated into proteins being synthesized there, and then transported by active axonal transport down the axon to the axonal endings, where it accumulates. If the brain is obtained, fixed and sectioned, then coated in total darkness with photographic emulsion and kept in darkness, the emulsion will be exposed by the radioactive emissions so that later the photographic emulsion can be developed and the locations of the radioactivity can be seen. Thus, the method of **autoradiography using labeled amino acids** became a tract-tracing method with some clear advantages. For example, the labeled amino acids could not be incorporated into proteins within axons. This happened only in cell bodies, so injections into white matter did not result in labeling of the endings of the injected axons. This was an advantage over degeneration methods, as any damage to axons of passage resulted in anterograde degeneration, and this could lead to false interpretations of data obtained in tract-tracing experiments.

Soon it was also discovered that various proteins injected into cellular regions of the brain were transported by the axons, not only to the terminal endings but often also in the opposite direction, up the axons from endings to the cell bodies. The enzyme **horseradish peroxidase (HRP)** was the first to be used extensively for tract tracing in both directions, from neuronal cell bodies to the axonal endings of those cells, and from terminal end-arbors by retrograde transport to the cell bodies of origin of those axons. In figure 2.15, we can see HRP-labeled axons of the optic tract terminating in the lateral geniculate nuclei of the 'tweenbrain and in adjacent cell groups.

To visualize the HRP injected into an animal's brain, the brain sections must be processed by special histochemical techniques. Alternatively, one can use an antibody to HRP to bind to the enzyme in the sections. If that antibody is linked to another molecule that can be visualized, then the HRP can be localized in a sensitive manner. Such a procedure is called **immunohistochemistry**. It is useful for localizing other tracer molecules as well. For example, a portion of the cholera toxin molecule, subunit B (abbreviated CT-B), can be a very sensitive label of axonal pathways (figure 2.16). When visualized by immunohistochemistry, one can see the entire extent of the labeled axons, including their terminal endings, appearing as in a Golgi stain (see also figure 20.1).

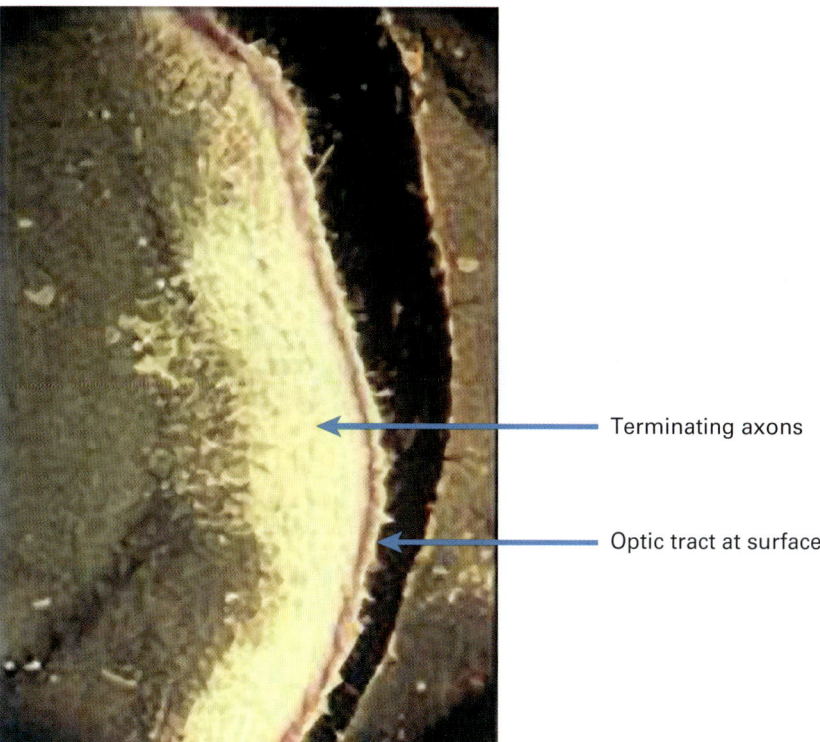

Figure 2.15
Frontal section (dorsal is up) showing one side of the diencephalon of a hamster pup, demonstrating anterograde transport of horseradish peroxidase (HRP) from the retina. The HRP appears bright because of the dark-field microscopic technique used for the photograph. Labeled axons can be seen coursing over and through the lateral geniculate nuclei where they are forming terminal arbors.

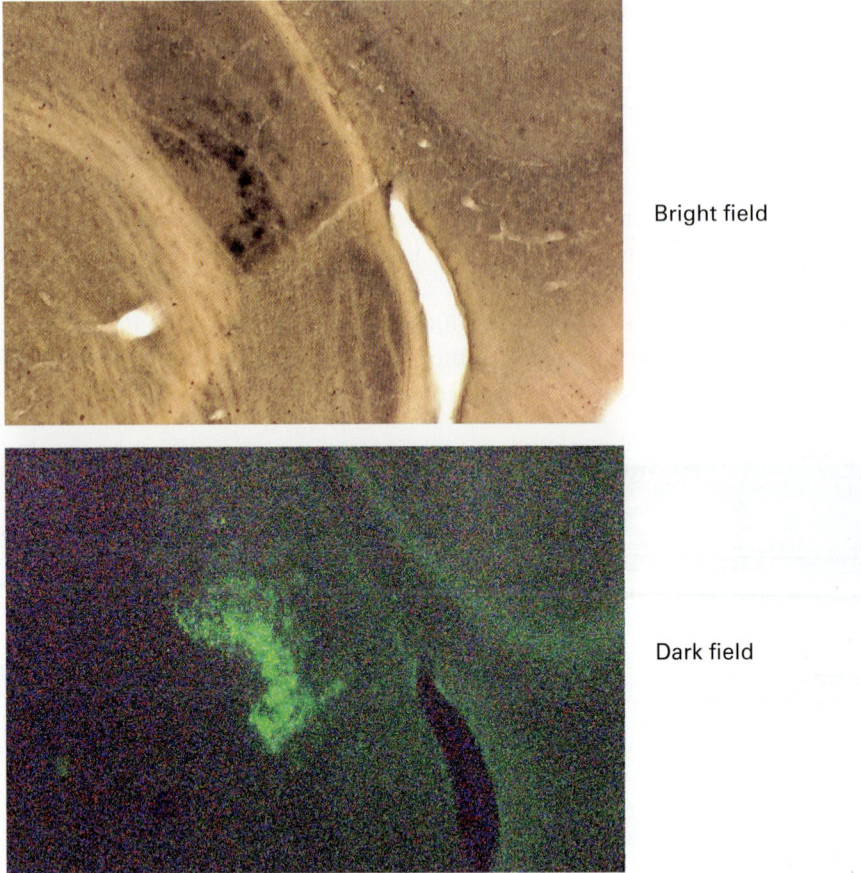

Bright field

Dark field

Figure 2.16
A frontal section through the lateral geniculate body of a hamster, a site of termination of fibers from the retina. Some of the retinal fibers have been labeled by cholera toxin, subunit B, injected into a part of the retina of the opposite eye and later visualized by immunohistochemistry. In the bottom photograph, the labeled axons appear brighter than the background with dark-field microscopy. The same axons are seen in the top photograph with light-field microscopy. In the light-field photograph, unlabeled axon bundles are seen as lighter than the background. Part of the hippocampal formation, free of labeled axons, can be seen at the right.

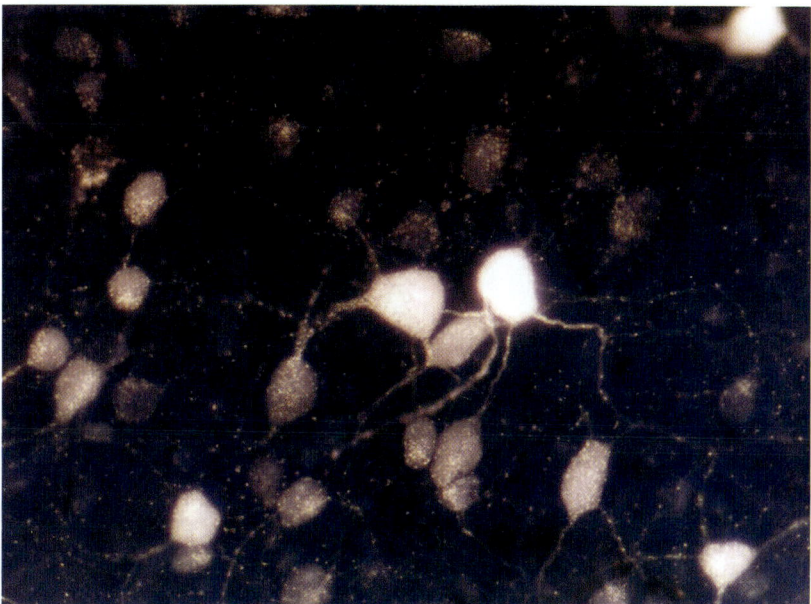

Figure 2.17
Photograph of part of a flat-mounted retina of a hamster viewed with fluorescence microscopy. A tracer substance called fluorogold was injected into the superior colliculus on the opposite side; this substance has been transported retrogradely down some of the optic-tract axons to retinal ganglion cells—which you can see glowing in the photograph.

It is possible to do sophisticated neuroanatomical studies now without any special histochemical procedures: One can use one or more of a number of **fluorescent molecules**. When such a molecule is injected into a region of cell bodies or axonal endings, it may be taken up by the cells or axons and transported in the anterograde or retrograde direction. Some fluorescent molecules are more useful for anterograde tracing of axons, others for retrograde labeling of cell bodies from their axon endings. Figure 2.17 shows cells in the retina of a hamster where many of the neurons are showing a bright fluorescence when illuminated by a particular wavelength of light. The molecule injected into a midbrain termination zone of the axons of these cells was fluorogold. The next picture (figure 2.18) shows retinal cells with their nuclei fluorescing because of the accumulation of a molecule called nuclear yellow. A cell in the center of the picture is labeled outside the nucleus with HRP. This neuron has a branching axon: The nuclear yellow was taken up by one branch when it was injected in the optic tract on the opposite side of the brain. The HRP was taken up by the other branch at an injection site in the optic tract on the other side.

Such double-labeling methods to find evidence of axonal branching patterns can make use of two different fluorescent molecules if they fluoresce under different wavelengths

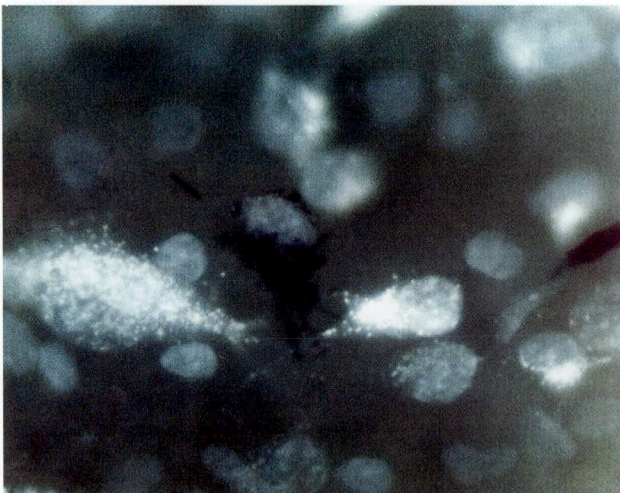

Figure 2.18
Photograph of part of a flat-mounted retina of a hamster showing two separate labeling substances. Nuclear yellow had been injected into a cut through the opposite (contralateral) optic tract, and HRP had been injected into a cut through the ipsilateral optic tract. Note that one retinal ganglion cell has been double labeled. This neuron, therefore, appears to have an axon that sends branches into both contralateral and ipsilateral optic tracts. A dark-field photograph showing the nuclear yellow fluorescence has been superimposed on a light-field photograph showing the HRP.

of light. Figure 2.19 displays two photographs of an identical patch of retina. The picture on the right shows a number of cell bodies labeled with fluorogold. The left-hand photograph shows that two of these cells are also marked with fluorescent beads that have their fluorescence stimulated by a different wavelength. The tiny beads were injected in one optic-tract terminal area, and the fluorogold was injected in another optic-tract terminal area. The result demonstrates that some retinal cells send terminal branches into both of these areas.

Figure 2.20 illustrates a technique developed more recently than the previously mentioned ones: it does not use histological methods at all, but involves a type of magnetic resonance imaging of living brains. Called *diffusion tensor imaging*, it is being used to follow major axon pathways in the human brain. It is based on the use of magnetic resonance imaging methods to detect the axis of movement of water in the brain tissue. Water diffuses much more readily up and down the axons than it does across the axons. Computer analysis enables the creation of images of axon groups, following the axons from any one region to other regions connected to it. The direction of axonal information flow cannot be distinguished with the method, and the resolution of the images does not extend to the cellular level, but the pictures nevertheless are providing new information about major pathways in human brains. They are proving to be particularly useful in picturing

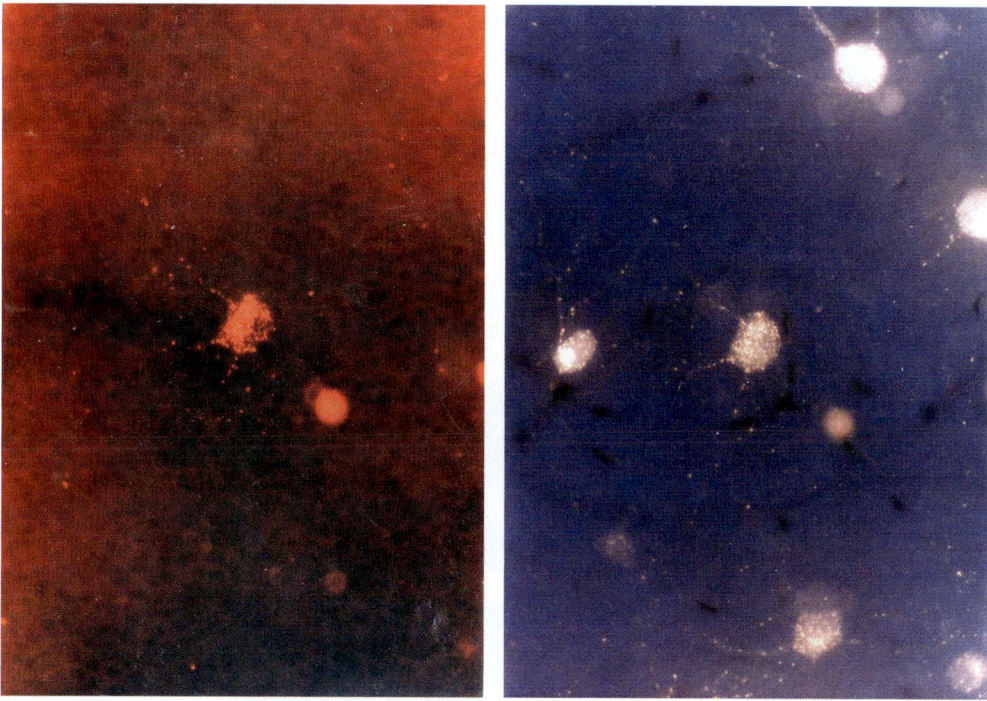

Figure 2.19
Two fluorescence photographs of the same portion of a flat-mounted retina, one using a wavelength that reveals fluorogold (right), the other using a wavelength that reveals tiny fluorescent beads (left). The fluorogold had been injected into the midbrain's superior colliculus, whereas the beads had been injected into the ventral nucleus of the lateral geniculate body on the same side of the brain. The co-localization results reveal, in this area of the retina, two retinal ganglion cells that appear to send branches into both structures.

pathologies that affect axonal pathways (e.g., tumors that distort axon trajectories or strokes that destroy axon pathways). The figure shows the patterns of axons coursing into and out of the neocortex. They pass through the corpus striatum into the cerebral peduncles of the 'tweenbrain and midbrain and continue through the hindbrain into the spinal cord.

Since about 2005, neuroanatomical studies of brain connections have acquired a new glamour because of the increasing application of computational methods in this field. Even in the popular press, one reads about a new field of *connectomics* and efforts to construct a *connectome* for the brain of humans and for the brains of other species. This has led to an increased awareness of the importance of brain connections for understanding the organization and control of behavior and cognitive processes. The study of brain connections is, of course, not new. However, there are new methods being brought into the field. Many of these methods are based on advances in cell and molecular biology.

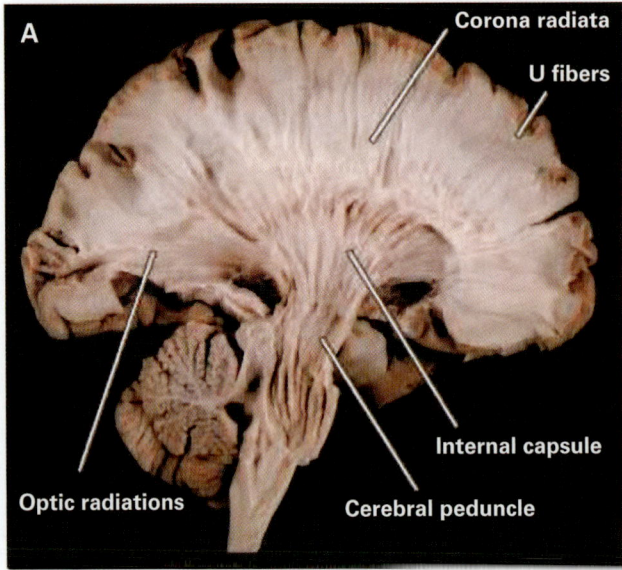
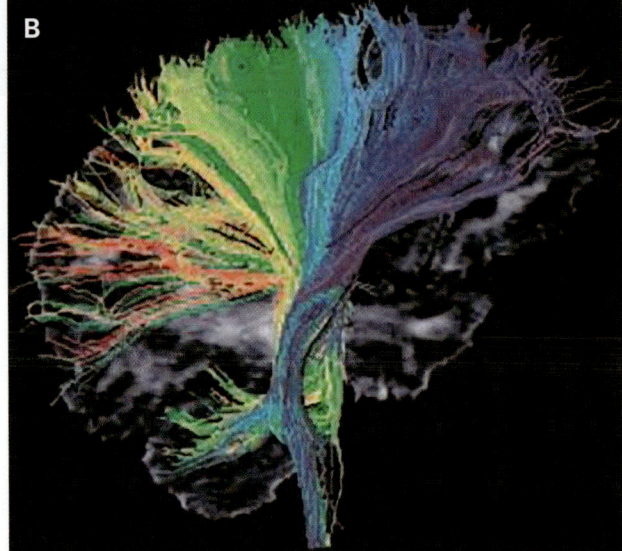

Figure 2.20
(A) This dissection of a fixed human brain shows fiber bundles coursing from the neocortex, down through the corona radiata and through the internal capsule into the cerebral peduncle toward the spinal cord. (B) In the brain of an alive human, diffusion tensor imaging has been used to follow fibers from the corona radiata to the cerebral peduncle. Fibers originating (or ending) in various cortical regions are separated by color coding. Note also fibers to and from the rostral hindbrain passing into the cerebellum. From studies of animal brains, we know that these fibers do not originate in the cerebral cortex. From Jellison, Field, Medow, Lazar, Salamat, and Alexander (2004).

However, investigators have also been developing computer methods to store and analyze large amounts of neuroanatomical data from various sources. In addition, there are attempts to apply computational methods to speed the acquisition of data on neuronal connections. Early work of this nature was done at MIT and reported in 1972 by Cyrus Levinthal and Randle Ware, who used computers to aid three-dimensional reconstruction of brain elements from electron microscope tissue—an enormously laborious task without computer aid. Only much more recently have others been attempting such tasks on small pieces of brain tissue.

Advances in the field of neuroanatomy have been characterized by surges that have resulted from the development of more sensitive or easier-to-use techniques. This is very likely to continue in the future, especially with the influx of new methods from the areas mentioned earlier. Of course, the learning of multiple methods, no matter how new, will not lead to major conceptual advances unless they are applied in the right way by creative scientists who are not only broadly educated in the neurosciences and other fields, but also are not satisfied with the current state of our concepts of brain structure and function.

Readings

Scientific reports describing specific neuroanatomical techniques can be found in various professional journals, including journals that specialize in methodology. An older one is *Stain Technology*, published 1925–1990 and continuing as *Biotechnic & Histochemistry*; a more recently established one is the *Journal of Neuroscience Methods*, published since 1979.

The belief in the latter part of the nineteenth century that the primary visual pathway from eyes to the cerebral hemispheres went by way of the roof of the midbrain—the optic tectum—was expressed by D. Ferrier (1876), *The Functions of the Brain* (London: Smith, Elder).

Much of the work of Ramón y Cajal is available in English translation:

Ramón y Cajal, S. (1995). *Histology of the nervous system of man and vertebrates* (2 vol.). (Translated from French by Neely Swanson and Larry Swanson.) New York: Oxford University Press.

Santiago Ramón y Cajal's interesting autobiography is also available in English:

Ramón y Cajal, S. (1966). *Recollections of my life.* (English translation by E. Horne Craigie and Juan Cano.) Cambridge, MA: The MIT Press.

The history of neuroscience, once summarized within works on the history of psychology, physiology, and neurology, has been the major focus of some recent books:

Finger, S. (1994). *Origins of neuroscience: A history of explorations into brain function.* New York: Oxford University Press.

Gross, C. G. (1998). *Brain, vision, memory: Tales in the history of neuroscience.* Cambridge MA: The MIT Press.

An interesting account of the twentieth century development of tract-tracing techniques and a summary of various more recent axonal tracing methods:

Köbbert, C., Apps, R., Bechmann, I., Lanciego, J. L., Mey, J., & Thanos, S. (2000). Current concepts of neuroanatomical tracing. *Progress in Neurobiology, 62,* 327–351.

Nauta, W. J. H. (1993). Some early travails of tracing axonal pathways in the brain. *Journal of Neuroscience, 13,* 1337–1345.

An early application of computational methods to neuroanatomy:

Levinthal, C., & Ware, R. (1972). Three dimensional reconstruction from serial sections. *Nature, 236,* 207–210.

Recent developments of new imaging methods and applications of molecular biology have been summarized:

Benjamin, R. A., & Ehlers, M. D. (2009). Molecular genetics and imaging technologies for circuit-based neuroanatomy. *Nature, 461,* 900–907.

II THE CENTRAL NERVOUS SYSTEM, FROM INITIAL STEPS TO ADVANCED CHORDATES

3 Evolution of Multicellular Organisms with Neuron-Based Coordination

The important point is the advent, shadowy though it is, of the great intermediate net: a barrier of intermediate neurons that interposes itself between sensory neurons and motor neurons quite early in evolution.

—W. J. H. Nauta and M. Feirtag, 1986

Despite all the activities that single-celled organisms were capable of, their small size severely restricted the environments they could thrive in. Although the initial evolutionary steps toward multicellular organisms are unclear, the elaboration of these creatures, with their greater body size and their capacity to evolve cellular specializations, seems to have been unstoppable. To begin to understand, at least in sparse outline, how and why this happened, we will look first at the basics of behaviors enabling survival of an animal. Adaptive changes in such behaviors are basic to evolution. Then we will review anatomical discoveries in studies of the simplest nervous systems. In these nervous systems, we can discern the earliest stages that must have characterized nervous system evolution. This will give us a simple model for describing elements of nervous systems, and we can then introduce some terms to be used throughout this book.

It was the appearance of the first chordate animals—the most primitive members of the phylum to which our own species, and all vertebrates, belong—that set the stage for the long series of evolutionary elaborations of the central nervous system that led to the brains of advanced mammals and birds.

Basic Survival Skills

To survive, every animal from amoeba to human being must be able to perform certain basic actions (at least during active stages of its life). Most fundamental, in my view, are (1) locomotor approach and avoidance movements, (2) orienting (turning) toward or away from something, and (3) foraging behavior patterns and exploration of places and objects. The third is made up largely of the first two plus the impelling drives of various kinds of motivational states. Each evolutionary advance had to incorporate these basic multipurpose action patterns, as they are needed for various goal-directed activities. The actions take place on a background of essential, ongoing maintenance activity that includes respiration, balance and posture, temperature regulation, and so forth. In multicellular organisms, all of these activities require nervous system control and integration.

The First Neurons and Nerve Networks

Even in the sponges, although they do not have a true nervous system, one can find cells that resemble neurons of more advanced creatures. Sponges have contractile cells that can respond to stimulation without intervention by neurons. These cells can conduct electrical potential changes in a *neuroid* or *myoid* manner (i.e., resembling conduction in neuronal or muscle cells). Such cells are critical for the basic actions of the sponge that enable it to survive.

But it is in the Ctenophora and in Cnidaria like jellyfish and hydra (previously grouped together in the phylum Coelenterata) that we find the beginnings of specializations for intercellular communication that characterize the cells of a nervous system. The earliest stages in the evolution of nervous systems were proposed on the basis of anatomical studies of various species within these phyla by George Parker of Harvard University, published in 1919. Parker's studies were reviewed and updated by George Mackie in 1970 (figure 3.1A).

In Mackie's first stage of evolution, the surface cells not only respond to external stimulation, but they are also contractile and are connected electrically to each other by gap junctions—sites of very close membrane apposition. This situation resembles intercellular interactions seen in sponges. In a second stage, some cells lose their connection to the surface and remain contractile, still connected electrically to the surface myoepithelium. In a third stage, protoneurons evolve: They are a kind of sensorimotor neuron, connected to the surface and to underlying contractile cells (myocytes). All cells are still electrically coupled. In Mackie's fourth stage, there appear true neurons, not only sensory neurons but also motor neurons connected to them. These motor neurons connect to the contractile cells. Furthermore, chemical synapses evolve, adding specificity to the connections, although there is still electrical coupling between many of the epithelial cells at the surface as well as among the contractile cells.

In Nauta's view, these stages do not go far enough in depicting the evolution of a central nervous system. Even in cnidarians like hydra, there is evidence for the appearance of some intermediate neurons, interposed between sensory neurons and motor neurons. This is depicted in figure 3.1B. It is this kind of interposed neuron that has proliferated the most in the evolution of central nervous systems. They have evolved as a great intermediate net, underlying so much of the integrative activity of the nervous system.

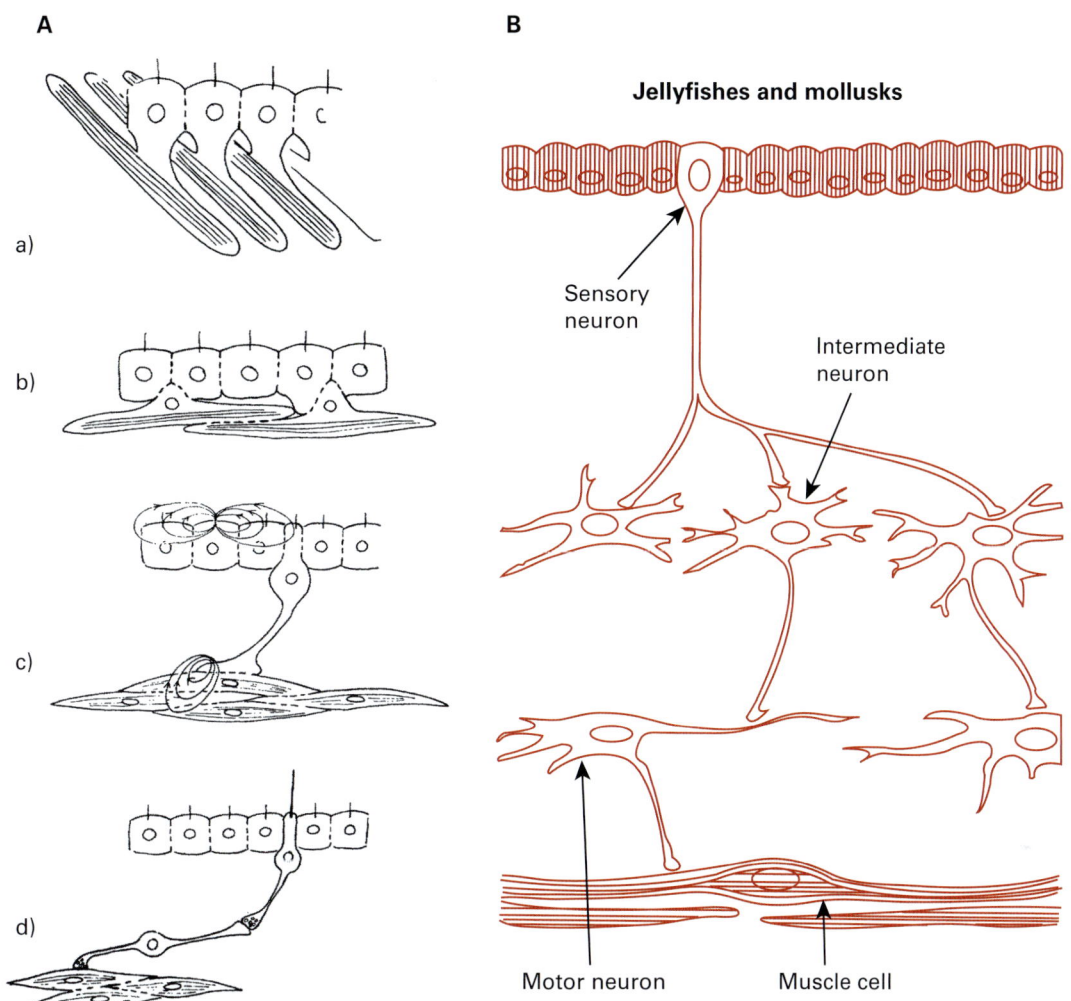

Figure 3.1
(A) George Mackie (1970) envisioned the following earliest stages in the evolution of conducting tissues into a nervous system, based on a review of various studies of ctenophores and cnidarians. (a) A primordial myoepithelium: surface cells that respond to stimuli and are also contractile. (b) Protomyocytes (contractile cells) start to sink into the interior from the surface epithelium. (c) Protoneurons evolve, carrying excitation from the exterior to the myocytes. All cells are still electrically coupled (dashed lines). (d) Neurosensory cells and neurons evolve. Chemical synapses evolve, while many epithelial and muscles cells remain electrically coupled. From Mackie (1970). (B) Nauta emphasized a further step that was critical in the evolution of central nervous systems. Even in Cnidaria like hydra, a third neuron can be found in the chain of conduction, in between the sensory neuron in the epithelium and the neuron connected to muscle cells. The intermediate cell is part of a network of neurons that evolves into the central nervous system of more advanced animals. Based on Nauta and Feirtag (1986).

58   Chapter 3

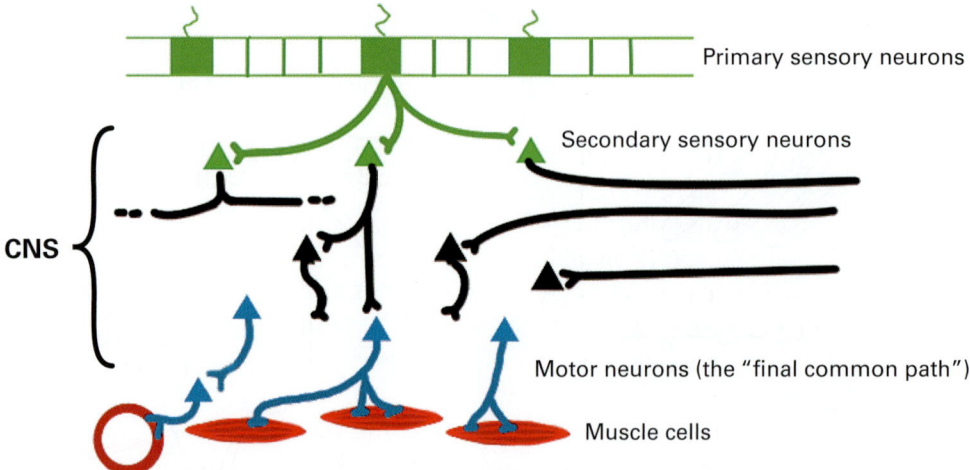

In black: cells of "the great intermediate net"

Figure 3.2
Schematic nervous system, including peripheral nervous system (PNS) and central nervous system (CNS). (All neuronal elements—cell bodies and axons—that are not in the CNS are included in the PNS.) Primary sensory neurons in the surface layer of the body send their axons to secondary sensory neurons within the CNS. Axons of secondary sensory neurons contact interneurons (of the great intermediate net) or they connect to motor neurons directly. The motor neurons are defined by axons that leave the CNS and contact striated muscle cells or cells of a peripheral ganglion that connects to smooth muscle cells. Note the longer CNS pathways that could represent axons to and from a brain.

The Nervous System in Basic Concept

This leads to a generalized conception of the nervous system (figure 3.2). We call the neurons of the epithelium that respond to sensory stimulation the *primary sensory neurons*. Axons of these neurons enter the *central nervous system* (abbreviated CNS) where they form synaptic contacts with *secondary sensory neurons*. These secondary sensory neurons have axons of various lengths. Long axons extend to distant parts of the great intermediate network, whereas shorter axons make connections with local *interneurons* or they connect directly with *motor neurons*. Connections with motor neurons can trigger movements. When motor neuron activation results from inputs at the same level of the central nervous system, we call the responsible circuit a local reflex pathway. Those same motor neurons and the interneurons that connect to them also receive inputs from more distant parts of the great intermediate network of the central nervous system.

The motor neurons have axons that leave the central nervous system, and many of them connect with muscle cells. Axons of other motor neurons also leave the CNS, but these axons synapse with neurons in *ganglia* (singular *ganglion*) located closer to the contractile cells (of smooth muscle) or the gland cells in the viscera that they innervate. The connections involving these ganglia of the peripheral nervous system (defined as all parts of the nervous system outside the CNS) are designated as connections of the autonomic nervous system (also called the visceral or vegetative nervous system) in vertebrates.

The motor neurons of the CNS constitute the "final common path" for nervous system control of actions, as all movements of the organism depend on the activation of these neurons. They are the output cells; their axons are the only routes out.

Looking at the simplified picture of the nervous system shown in figure 3.2, we can easily conceive of the idea that all behavior is initiated by the sensory inputs, in accord with the stimulus–response (S-R) model described in the previous chapter. Even if we realize that there is much animal behavior, and certainly human behavior, that cannot be fully explained in such a way, it is easy to think of simpler creatures as reflex machines. However, there is clear evidence that behavior can also be initiated by endogenous nervous system activity even in the small coelenterate hydra (mentioned in chapter 1). These little animals are active when kept in an extremely uniform and unchanging environment. The locomotor behavior pattern illustrated in figure 1.14 occurs periodically in such an environment, apparently triggered by endogenous activity of neurons connected to the locomotor apparatus of their interneuronal network. Motivational states of this nature, which are periodically changing, are a common feature of animal behavior and represent an important caveat to reflex explanations of behavior.

Some Terminology for Brain Talk

At this point, we are ready to review some important terms used for describing the structure of nervous systems and also to define a few additional terms we will be using. Below, these terms are listed together with short definitions.

- *Primary sensory neuron* This is the first neuron in a sensory pathway. The cell body is located outside the CNS, in the peripheral nervous system (PNS).
- *Secondary sensory neuron* This is the type of neuron within the CNS that receives synaptic inputs directly from primary sensory neurons.
- *Interneuron* A neuron of the great intermediate net of the CNS. All neurons in the CNS can be called interneurons except, by definition, the secondary sensory neurons and the motor neurons. However, many anatomists use the term in a more specific way to exclude the cells of origin of longer axon pathways. These anatomists use the term to designate interneurons with short axons, terminating locally.
- *Motor neuron* Cell with an axon that leaves the CNS, travels through a peripheral nerve, and has endings on effector cells—usually muscle cells. Some axons that leave the CNS are from *preganglionic motor neurons*, as they terminate on *ganglion cells* in clumps of neurons outside the CNS; these ganglion cells have axons that terminate on smooth muscle or gland cells.

- *Ganglia (singular: ganglion)* Groups of neurons located outside the CNS. The term is sometimes also used for a few specific cell groups within the CNS.
- *Cell group,* or nucleus, *within the CNS* A group of neurons in the CNS that can be characterized as different from surrounding regions by its appearance in stained sections and/or by its functions.
- *Nerve* A bundle of axons located outside the CNS (in the PNS).
- *Tract or fasciculus (plural: fasciculi)* A group of axons located next to each other in the CNS; usually they have the same region of origin and they also share destinations. In the spinal cord, a group of such tracts located together are called a column (namely, the dorsal column, lateral column, ventral column). Some smaller tracts are also called columns. These groups of axons may also be referred to as bundles.
- *Neural tube* The entire central nervous system in animals of the phylum Chordata comes from this embryonic structure, initially a fluid-filled tube with walls that are onecell thick. The CNS remains a tube in its basic structure, but the lateral walls become thicker and thicker as cells proliferate and differentiate.
- *Notochord* The cartilaginous rod-like structure located near the dorsal surface of the body of all embryonic chordates. The neural tube forms just dorsal to this structure; in vertebrates, the vertebrae of the spinal column form around it.

Introducing the Simplest Chordates

To begin a characterization of the evolution of central nervous systems focusing on the vertebrates, it makes sense to start with a tiny creature that is sometimes called the simplest living member of our phylum, the Chordata. This worm-like invertebrate chordate, a cephalochordate, has characteristics that suggest similarities to what the earliest chordates must have been like (figure 3.3). It is *Branchiostoma*, often called amphioxus because of its body shape; amphioxus means "sharp at both ends."

I have sketched the body plan of amphioxus in a simplified way in figures 3.4 and 3.5 to show the location of the nervous system. A dorsal nerve cord is found above the cartilaginous notochord, which stiffens the soft body. The dorsal nerve cord is the central nervous system in the form of a neural tube, located beneath the surface epithelium along the animal's back. The peripheral nerves distribute sensory fibers mostly near the body surface and motor fibers at a deeper position. These nerves enter and exit the tube of the CNS (the dorsal nerve cord) following, in the main, the "law of roots" (the Bell-Magendie law) that holds for all vertebrates: Sensory input fibers enter as dorsal roots and axons of motor neurons exit as ventral roots. (At each level along the axis of the spinal cord of vertebrates, there is a pair of dorsal roots [right and left] and a pair of ventral roots, but in amphioxus there is only a single dorsal root and a single ventral root on the opposite

Evolution of Multicellular Organisms with Neuron-Based Coordination

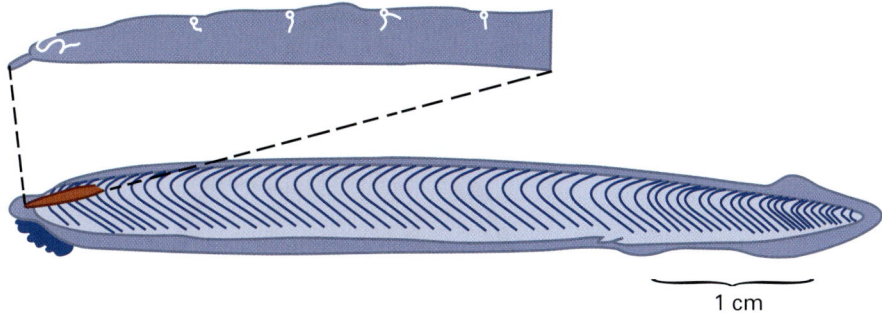

Figure 3.3
Branchiostoma (amphioxus) is a tiny present-day chordate (an invertebrate cephalochordate) that has characteristics that suggest similarities to what the earliest chordates must have been like. It is sometimes called the simplest living chordate. The brain region of the CNS, at the rostral end of a dorsally located CNS, does not show any marked enlargement when compared with the more caudal CNS. Based on Striedter (2005).

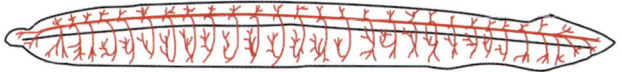

Figure 3.4
Simplified drawing of the body plan of amphioxus, "the simplest living chordate," based on a published reconstruction. The meaning of its name, "sharp at both ends," is clear. A dorsal nerve cord (shown in red) is found above a cartilaginous rod called the notochord (shown in black), a structure that characterizes all members of the chordate phylum. The dorsal nerve cord is the central nervous system, which has the form of a neural tube, derived from the surface layer of the embryo. Sensory nerves from the periphery connect to it dorsally. Processes of contractile cells reach neuronal connections within the ventral part of the nerve cord (see figure 3.5).

side at each anterior-to-posterior level, and the side where a root is found alternates as one passes from one level to another.)

Recent studies of amphioxus have used molecular markers that discriminate between different segments of the CNS in vertebrates. Because of conservation of these proteins in evolution, the markers are also specific for different regions of the CNS of the primitive chordate. The findings indicate that most of the neural tube in amphioxus corresponds to the brainstem and spinal cord of vertebrates. The three *primary brain vesicles* are present in amphioxus (i.e., the hindbrain, midbrain, and forebrain), but there are no cerebral hemispheres—which in vertebrates balloon out from the sides of the forebrain and constitute the largest part of the brain in all mammals. The amphioxus forebrain mostly corresponds to 'tweenbrain (diencephalon) of vertebrates, with regions corresponding to the pineal gland (epiphysis) and the pituitary (hypophysis). This forebrain has two visual inputs: One of these probably corresponds, by its position, to the pineal eye (parietal eye) in vertebrates, whereas the other is from an unpaired light-sensitive organ, the frontal eye spot, which has a cell arrangement that seems to be related to the arrangement of cells

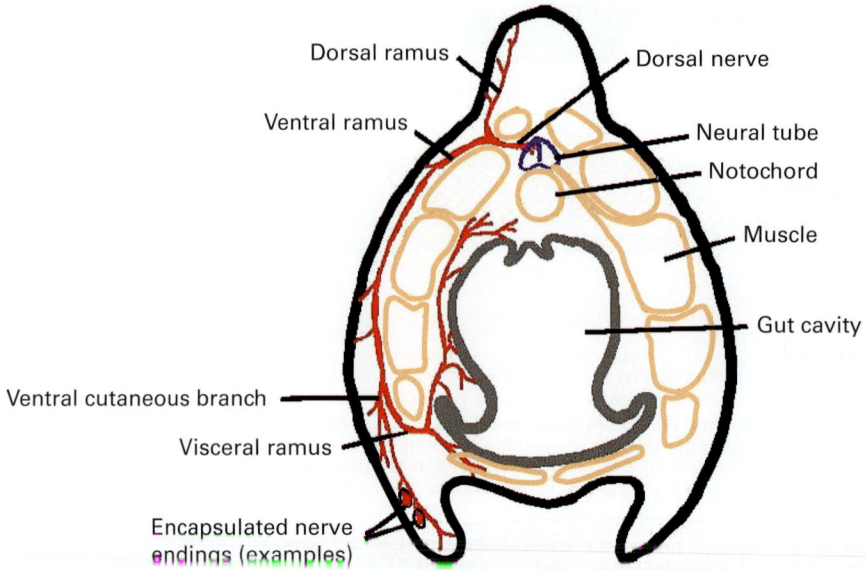

Figure 3.5
Sketch of a transverse section through the body of amphioxus. Along the body axis, the peripheral nerves attach to the CNS in dorsal attachments (dorsal nerves) that are mostly sensory and visceral motor. The ventral attachments are apparently the processes of muscle cells that abut motor neurons in the ventral part of the neural tube. At a single rostrocaudal level, one finds a dorsal nerve on one side and a ventral attachment on the other. Dorsal nerves and ventral attachments alternate sides along the body axis. Based on Wicht and Lacalli (2005).

in the vertebrate eye. This creature has little that can be called an endbrain. As already mentioned, there are no hemispheres, which are the largest part of the endbrain in vertebrates, but there are two nerves that may correspond to the endbrain nerves in vertebrates, the terminal nerve and the better known olfactory nerve. (The terminal nerve, often ignored in mammals, innervates the nasal septum in vertebrates.) However, the homologies are not certain.

Thus, in this little chordate we have a neural tube that has not differentiated or expanded very much at its head end. It may seem difficult to relate this primitive state to the vertebrate neural tube. And yet we know that from such primitive beginnings, over many millennia, the CNS of advanced vertebrates including our own species evolved by a progressive elaboration of the neural tube. Survival of the primitive chordates depended on details of their behavioral abilities, some of which had the highest priority for survival. Adaptation to new or altered environments and new predators and new food sources or prey depended on behavioral changes that came with step-by-step evolutionary changes in the neural tube. In the words of H. Chandler Elliott (1969), "Every brain system grows logically from the tube."

Natural Selection and the Logic of Evolution

Before I start to sketch for you some likely major steps in the brain's evolution, it will help to explain what most scientists mean by evolution and the rules it follows. Evolution of animals means the changes that occur, in populations, because of underlying genetic changes. These changes include changes in behavior resulting from changes in the central nervous system. The genetic changes occur, for the most part, by natural selection, as proposed originally by Charles Darwin and supported by large amounts of evidence accumulated by him and in large numbers of subsequent investigations. The genes that survive in one generation and are passed on most frequently to the next generation are the genes in the animals with the best functional adaptations—animals that are able to leave more descendants.

The process of natural selection is a process of selection among genetic variations within a species. The variations within a species are enhanced by sexual reproduction, involving the chance resorting of genes. Additional variations are introduced into a population by gene mutations. Because certain genotypes result in more surviving offspring than others, some genes increase in frequency within the population while others decrease or disappear.

Keep in mind that most genetic variations are small, resulting from differences in one or a very few genes (and consequently the resulting proteins). A mutation generally affects only a single gene, although one genetic change can alter the expression pattern of many others. As we proceed in this book, we will discuss relatively large changes in the central nervous system much more than the series of small steps by which those changes must have occurred.

The evolution of chordate animals has included successive elaborations of the basic plan of the neural tube in its simplest form as illustrated, at least in a suggestive way, by amphioxus. The consequences of these elaborations in CNS structure were the functional/behavioral changes that promoted survival and reproduction, and hence the perpetuation and proliferation of the underlying genes.

When we draw the main outlines of brain evolution, we are sketching the course of adaptive changes that led to present-day ability repertoires. Our artistry is guided by comparisons of extant species—their nervous systems and their behaviors—and by the development of both brain and behavior, and by paleontology.

We undertake this composition with the goal of introducing you to brain structure—neuroanatomy—in a way that helps you to understand not only what it is but also why it is so. Knowing why will inspire you to remember what is there.

The Behavioral Demands That Changed the Neural Tube

Action patterns that enabled survival and reproduction of individual animals had the highest priority (see the earlier section "Basic Survival Skills"). To accomplish these critical things, ongoing background support was essential. To make all this possible, interfaces with the outside world were necessary on both the sensory side and the motor side.

We will review more specifically those behavioral demands. Then we will survey major evolutionary changes that have occurred in the CNS in order to enable the behavioral abilities and changes in those abilities.

Animals need both approach and avoidance actions. These actions are controlled by motivational systems as well as by specific sensory inputs. The actions include antipredatory behaviors, feeding and drinking behaviors, and reproductive behavior. It was also important that goal hierarchies be part of the controlling mechanisms (e.g., fleeing from a predator usually gets priority over eating and mating).

When generating these action patterns, an animal also has to maintain stability of the internal environment (e.g., by changes in respiration, blood pressure, and heart rate). Changes in endocrine systems also helped prepare the body for action, so CNS connections evolved for regulation of the endocrine systems that was more rapid than simple secretions into the general circulation would allow.

No matter what patterns of behavior an animal is engaged in, it also has to maintain stability in space by adjustments in posture and balance. These ongoing background support functions involve many reflexes. We can say that we and other animals are always wearing a "mantle of reflexes."

Interfaces with the outside world have evolved in major ways with changing demands of the environment including other animals competing for food and shelter and other resources. Sensory detection and analysis in the forward-locomoting chordates has depended heavily on head receptors and the brain structures of the CNS pathways connected with those receptors (for specialized touch, taste, and hearing; for changes in head position and movement; for vision and olfaction). The neural apparatus for sensory processing has become more and more sophisticated and, in some species, highly specialized. Motor control and coordination has also evolved in major ways with changes in body size, shape, and appendages, and with changing demands of the environment. All of this enabled better and more precise control of orienting toward or away from sources of input, and it improved the processes of exploring, foraging, and seeking behavior and of escape behavior as well. We know that, eventually, the integrative systems of the CNS evolved what we call cognitive abilities for anticipating events (what is about to be sensed) and for planning actions (preparing for actions in the immediate and sometimes the more distant future, for achieving specific goals). These abilities appear to have evolved from apparatus for object identification, of objects both generic and individual and both living and nonliving, and from apparatus for control of attention to goals of motivational states, with prioritizing of those motives and goals, and from apparatus for fine motor control and coordination.

All of these functions, as they changed and improved, entailed progressive changes in the neural tube with the addition of, or changes in, sensory analyzing mechanisms, motor control systems, structures enabling correlation between the sensory and motor sides, structures containing the mechanisms of complex central programs, multiple levels of motivational control, and mechanisms for anticipating and planning, both short term and long term.

Readings

Allman, J. (2000). *Evolving brains.* New York: Scientific American Library.

Butler, A. B., & Hodos, W. (2005). *Comparative vertebrate neuroanatomy: Evolution and adaptation*, 2nd ed. New York: Wiley-Interscience. [See sections in chapter 31 on "Brain Evolution within Chordates" and "The Origin of Vertebrates."]

Elliott, H. C. (1969). *The shape of intelligence: The evolution of the human brain.* New York: Scribner.

Mackie, G. O. (1970). Neuroid conduction and the evolution of conducting tissues. *The Quarterly Review of Biology*, 45, 319–332. See also: Mackie, G. O. (1990). The elementary nervous system revisited. *American Zoologist, 30*, 907–920. Mackie has revised and updated the work of George Parker: Parker, G. H. (1919). *The elementary nervous system.* Philadelphia: J.B. Lippincott Co. (Available in Google Books.)

Nauta, W. J. H., & Feirtag, M. (1986). *Fundamental neuroanatomy.* New York: Freeman.

Wicht, H., & Lacalli, T. C. (2005). The nervous system of amphioxus: Structure, development, and evolutionary significance. *Canadian Journal of Zoology, 83*, 122–150.

For an update on the cephalochordate amphioxus, see the following review:

Fritzsch, B., & Glover, J. C. (2007). Evolution of the deuterostome central nervous system: An intercalation of developmental patterning processes with cellular specification processes. In Kaas, J. H., ed. *Evolution of nervous systems*, vol. 2, pp. 1–24. Oxford: Academic Press.

On hydra behavior: see *American Zoologist*, 1965, for a series of reports. Others can be found by Web searching using Google Scholar.

4 Expansions of the Neuronal Apparatus of Success

Every brain system grows logically from the tube.
—H. Chandler Elliott, 1969

In the previous chapter, we considered why the CNS has evolved the way it did and what the evolutionary changes have accomplished for organisms. Now we want to see how this is expressed in the basic organization of the brain. We will continue to focus mainly on chordates, from primitive worm-like animals all the way to humans, whales, and birds. The following descriptions of early CNS anatomy are educated speculation based on comparisons of a wide range of species and on knowledge of evolution of animal behavior and structure.

Why We Can Talk about Broad Aspects of Brain Evolution with Some Confidence

It is true that there are many details concerning evolution, especially brain evolution, that are not known for sure. However, the beginnings are very likely, from what we know of the simplest living metazoans and the simplest creatures with forward locomotion and bilateral symmetry. The results of evolution are written in the book of existing species and in paleontological records of extinct species. It is abundantly clear that currently living species have changed, in evolution, in greatly varying ways and amounts, especially the latter. There are some strong correlations between the relative size of structures and the functional demands, and examples of this will be given below. Laying it all out, with help from cladistics and DNA analyses of genetic similarities and differences, we are able to construct the likely major trends.

Three Enlargements Formed the Primitive Brain

Let's now start with the neural tube of a (surmised) very early chordate (figure 4.1, left), a worm-like animal similar to amphioxus, which was described briefly in the previous chapter. This neural tube, the dorsally located CNS of the animal, had many attached

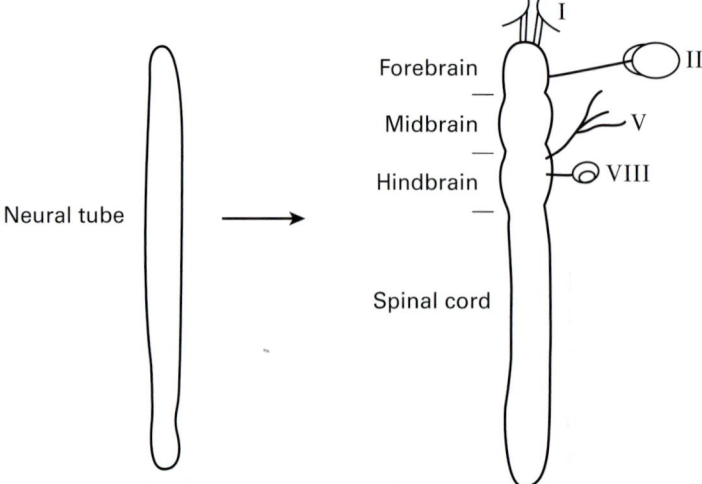

Figure 4.1
Evolution of chordate brain, 1 (neural tube, forward locomotion, and head receptors): Selective advantages of special sensory receptors at the head end of forward-locomoting animals led to the evolution of expansions at the rostral end of the neural tube. These expansions above the spinal cord were the primary brain vesicles: hindbrain, midbrain, and forebrain. Examples of inputs to the hindbrain are somatosensory inputs from the head region via what became cranial nerve V, and vestibular inputs via part of cranial nerve VIII. Examples of inputs directly to the forebrain are inputs from light receptors and from olfactory receptors; these led to the formation of cranial nerves II and I. The midbrain evolved into a crucial link between forebrain and hindbrain for control of orienting and escape movements.

peripheral nerves, entering and exiting the CNS at multiple distinct levels along its longitudinal axis. The dorsal roots carried sensory information from the body surface, allowing the animal to respond to touch and certain chemical and temperature changes.

At the anterior end were some other specializations, including input from light receptors. With forward locomotion—basic to the earliest evolution of chordates—even small evolutionary increases in certain types of information carried into the CNS from the region the animal was approaching promoted the chances of survival. Also, information about the orientation of the head end of the body was important for control of movement. These factors led to changes, both quantitative and qualitative, in sensory inputs at the head end and in the neuronal apparatus for processing this information and sending it to motor control mechanisms in the CNS. The corresponding sensory nerves increased in size.

The changes caused some expansion of the brain region that controlled orienting to, or fleeing from, objects arising from somatosensory inputs encountered by the forward-moving animal (figure 4.1, right). These inputs came from the head region into the hindbrain. Chemosensory inputs evolved as a gustatory system (taste system) of the hindbrain, of special importance for food selection and rejection, and for use in motivating search behavior by memories of good and bad foods. Along with new sensory abilities, motor control of the turning of the head and control of the mouth via midbrain and hindbrain circuits were added to lower-level (spinal) control of the body.

Hindbrain circuits also evolved for using vestibular inputs to maintain stability of body orientation. Such circuits also evolved for analyzing and responding to vibrations or electrical currents in the water.

Stability of the internal milieu was equally important, and with locomotion through varied environments, some CNS control of visceral mechanisms controlling respiration, circulation, maintenance of body temperature, and so forth, was advantageous. As noted in chapter 3, the central control enabled changes that were much more rapid than the effects of substances secreted into the general circulation or of local controls. Therefore, mechanisms already present in the lower parts of the body, with local (spinal and peripheral) control, began to be influenced by circuitry at the anterior end, which was evolving into a brain. This circuitry evolved in each of the brain vesicles, the three primitive enlargements called the hindbrain, midbrain, and forebrain. Central control by endocrine mechanisms came to depend mainly on structures at the base of the caudal forebrain—which became the hypothalamus and the pituitary organ. (In chapter 3, we noted that there are indications of these structures in amphioxus.)

At the anterior extreme of the neural tube, specializations for detecting olfactory stimuli and visual stimuli appeared and began to expand, causing the forebrain to increase in size (figure 4.1, right). (In the case of visual input, at the outset only the presence and amount of light was signaled—see chapter 20.) In addition to circuitry for sensory analysis, motor apparatus for directing these receptors, by turning of the head end of the body, increased as well.

Expansions of the Hindbrain

Expansion of the hindbrain is illustrated in the sketch shown in figure 4.2. Inputs that evolved into cranial nerves V (somatosensory, face area) and VIII (vestibular and auditory) are depicted. Inputs from taste organs entered through cranial nerves VII, IX, and X and also contributed strongly to the expansion of the hindbrain. Another major influence in some animals came through the lateral line nerves from mechanoreceptors and from electroreceptors distributed in a row along the length of the body.

Along with the increases in sensory information coming into the hindbrain, programs for various action patterns triggered by these inputs also evolved in the local reticular formation (the mixture of neurons and fibers that constitutes the core of the brainstem), with changes in genetically determined neural circuitry. Such action patterns included orienting and feeding patterns triggered and guided by touch and taste or by water vibrations and pressure changes or electric field disturbances. Other action patterns triggered by the same senses or by inputs reaching the midbrain and forebrain were movements of

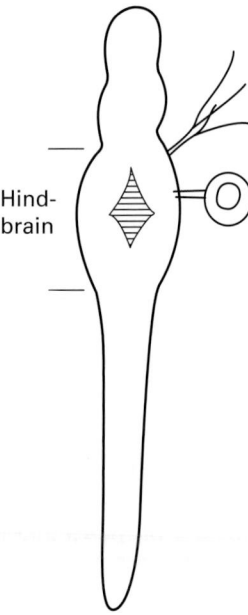

Figure 4.2
Evolution of chordate brain, 2 (expansion of hindbrain): Expansions of the hindbrain occurred with evolution of sensory analyzers for various inputs from face and head (somatosensory, gustatory, vestibular, auditory) or for other inputs entering the hindbrain via cranial nerves (e.g., from electroreceptors). On the motor side, the hindbrain networks evolved programs for control of action patterns triggered by specific sensory patterns.

escape or aggression and patterns of mating and parental behavior. With the evolution of social mammals, there was a large increase in patterns of emotional expression produced by contractions of muscles in the face, controlled by axons that exit the hindbrain through cranial nerve VII in present-day mammals.

That functional adaptations require expansions in the CNS can be illustrated from comparative anatomy, using illustrations of brains of fish from C. L. Herrick. First, look at figure 4.3, which shows a dorsal view of the brain of a freshwater mooneye. Note the size and shape of the hindbrain, the region between the midbrain and spinal cord (including cerebellum and medulla oblongata). Next, look at figure 4.4, showing a dorsal view of the brain of a buffalofish. Fish of this species receive and process taste input from a specialized palatal organ. The sensory information from this organ goes to a huge *vagal lobe*. A branch of the vagus nerve innervates the taste organ located in the palate. The secondary sensory cells receiving taste input from this organ are so numerous that they cause a large bulge that protrudes from the dorsal hindbrain behind the cerebellum. This apparatus, including mechanisms for both sensory analysis and associated motor control, plays a critical role in filtering of murky waters for small food particles at and near the bottom of a river.

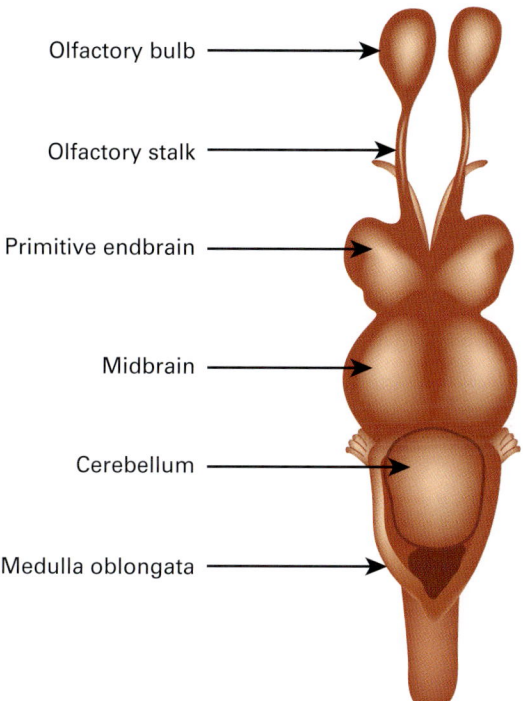

Figure 4.3
Dorsal view of brain of the freshwater mooneye (*Hyodon tergisus*), based on an illustration by C. L. Herrick. The brain shows effects of evolution of an olfactory apparatus attached to a slightly enlarged endbrain, a midbrain that is enlarged more than the endbrain, and a hindbrain that is also enlarged compared with the spinal cord caudal to it. The hindbrain includes the medulla oblongata (the slightly enlarged rostral elongation of the spinal cord) and a cerebellar region at the rostral end. Stumps of the fifth cranial nerves can be seen attached, on both sides, to the rostral end of the hindbrain. Note: the 'tweenbrain (diencephalon) in this and in figures 4.4 and 4.5 is mostly hidden by the endbrain. Based on Herrick (1962).

Next, compare these two brains to the brain of a catfish depicted in figure 4.5. The vagal lobe is, again, enlarged, but less than in the buffalofish. A *facial lobe* that is enlarged relative to its size in many other species is also evident. The facial lobe, receiving input through the sensory part of the seventh cranial nerve, receives and processes taste inputs in the catfish from all over the body surface, rather than just from the face as in many species (including the mammals). The next illustration, shown in figure 4.6, originally published by Herrick in 1903, demonstrates this by depicting the surprising distribution of the seventh cranial nerve, known as the facial nerve, in another species of catfish, which has a brain with a shape similar to the one shown in the previous figure. Unlike the facial sensory nerve in other species, this nerve in the catfish innervates taste receptors in the skin of the entire body.

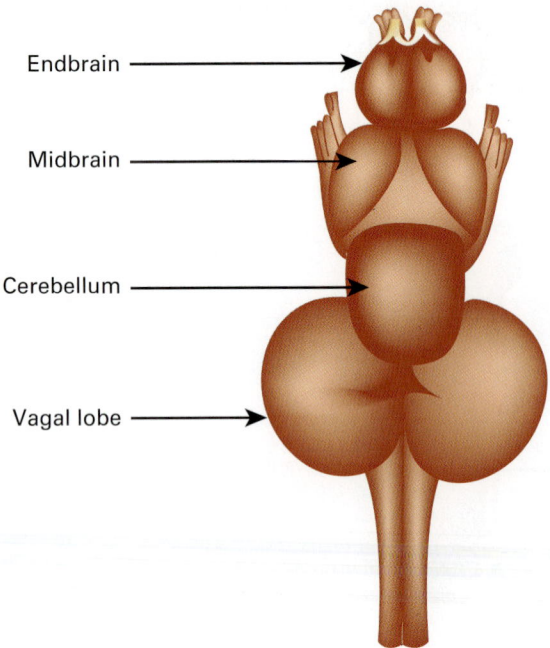

Figure 4.4
Dorsal view of brain of a buffalofish (*Carpiodes tumidus*). The most remarkable feature is a great expansion of the hindbrain. The cerebellum is similar in relative size to that in the mooneye, but the evolution of a specialized palatal organ for analysis of taste inputs during feeding has been accompanied by a large expansion of the group of secondary sensory neurons that receive taste inputs from the throat via the tenth cranial nerve—the vagus nerve—hence the structure is called the vagal lobe. Based on Herrick (1962).

From this series of illustrations, it is obvious that the evolution of a specialized sense organ that is heavily innervated can be accompanied by correspondingly enlarged central nervous system structures that process the information.

Early Forebrain Expansion Served Olfactory Functions

The expansion of the hindbrain illustrated in figure 4.2 omits another expansion that was very likely happening concurrently in evolution—the expansion of the forebrain. Forebrain expansion almost certainly occurred initially because of the adaptive value of the olfactory sense for approach and avoidance functions (figure 4.7). Approach and avoidance functions are critical in feeding, mating, predator avoidance, and predation—so an ocean creature, or a little amphibian or reptile, with better olfaction must have had an edge in the competition to survive and pass on its genes. Serving these functions were links to brain systems controlling locomotion. The most important links were very likely in the forerunner of the corpus striatum of the endbrain. The striatum projected output

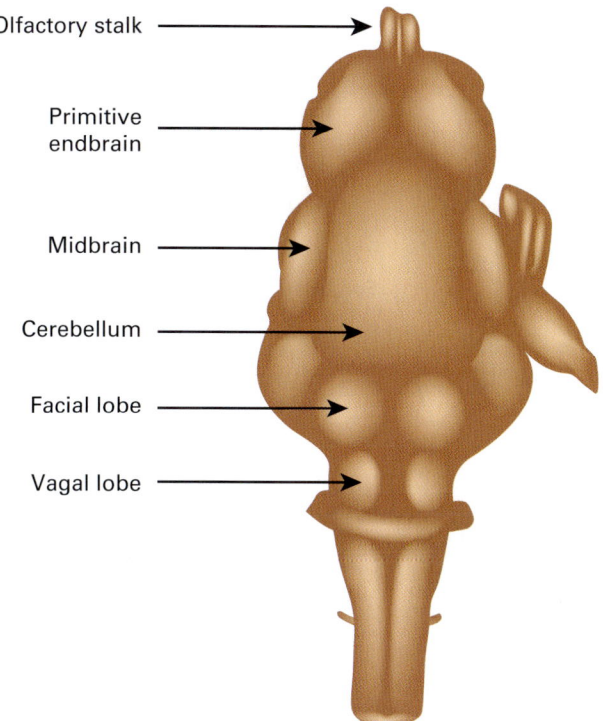

Figure 4.5
Brain of a catfish (*Pilodictis olivaris*). The endbrain and midbrain are wider than in the brains shown in figures 4.3 and 4.4. In the hindbrain, in addition to a greatly enlarged cerebellum, the vagal lobe is enlarged although much less than in the buffalofish. An enlarged *facial lobe* is also evident. The latter is the site of secondary sensory neurons receiving taste inputs via the seventh cranial nerve (the facial nerve). The seventh cranial nerve in this animal receives taste inputs from all over the body surface, not just from the tongue and oral cavity as in mammals. Based on Herrick (1962).

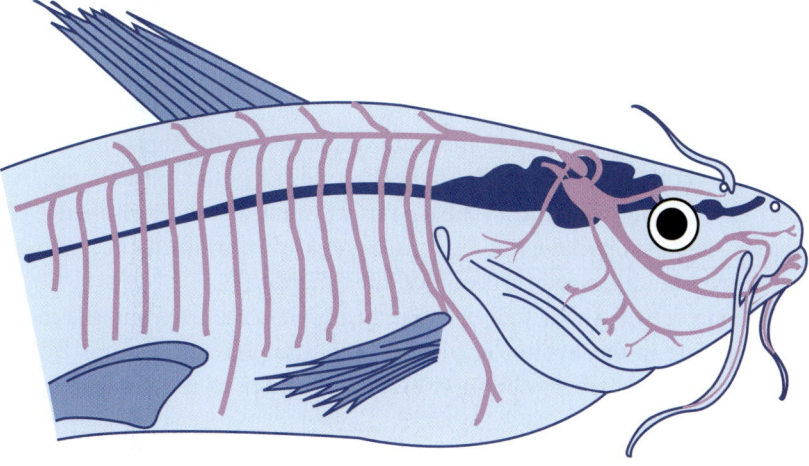

Figure 4.6
The small catfish (*Amiurus melas*): The illustration depicts the position of the central nervous system and the distribution of the seventh cranial nerve, which innervates taste receptors in the skin of the entire body. Based on Herrick (1962).

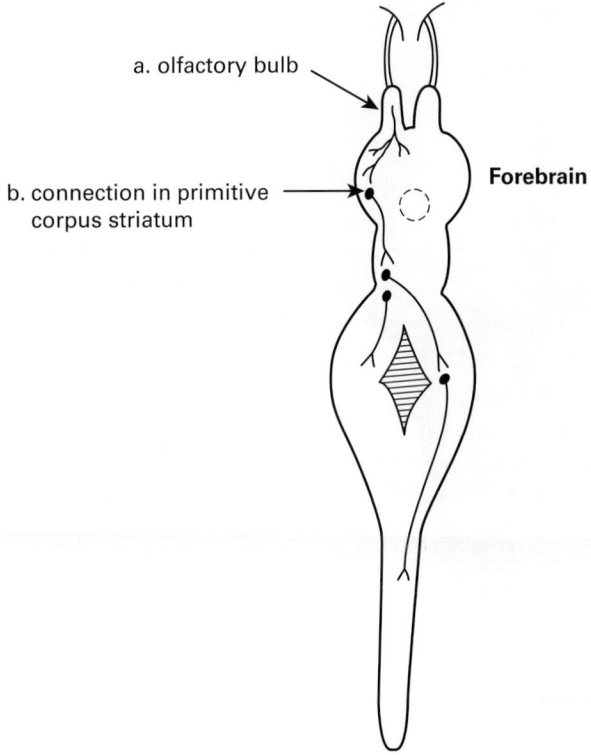

Figure 4.7
Evolution of chordate brain, 3 (first expansion of forebrain): Concurrently with expansions of the hindbrain, there appears to have been an expansion of the forebrain, largely because of the adaptive value of the olfactory sense for approach and avoidance functions (feeding, mating behavior, predatory avoidance). On the output side, links to locomotion were the most critical. Most likely, the primitive striatum received projections from the olfactory system and had caudal projections that reached 'tweenbrain and midbrain structures involved in locomotor initiation.

axons, directly or indirectly, to the midbrain (indicated in the figure). Many axons of this pathway went via a connection in the hypothalamus or in the epithalamus. A most useful property of the connections within the corpus striatum was their plasticity: They could be strengthened or weakened depending on feedback from consequences of actions. (These connections will be discussed further in a later chapter.)

The midbrain served as the link between forebrain structures and the motor system. It was only very late in the evolution of species that longer connections from an enlarged forebrain began to bypass midbrain structures in order to reach more caudally located neurons of motor control.

Midbrain Expansions, Distance Receptors, and Decussations

We have just mentioned the midbrain as a locus of locomotor control. This region is in the caudal midbrain and is not the midbrain tectum—the roof of the midbrain, which is so prominent in many present-day vertebrate brains. The tectum evolved for other reasons as a very important center of sensorimotor integration, with input from multiple sensory analyzers and outputs to the motor apparatus. One of these outputs controlled escape movements; another controlled turning movements of the head in response to visual and auditory inputs as well as somatosensory inputs. The importance of orienting and escape behavior resulted in the expansion of the midbrain tectum early in the evolution of vertebrates. This is depicted in figure 4.8, which demonstrates connections from the eye to the midbrain tectum and pathways from the tectum controlling orienting and escape movements. You can see the midbrain tectum also in figure 4.3, showing the brain of the mooneye: note the prominent cerebellum located just caudal to the midbrain. The medial cerebellar structures (omitted in figure 4.8) are involved in the coordination of head and body orientation and other aspects of maintaining stability in space, using vestibular, visual, and tactile information.

The evolution of the distance-receptor senses of vision and audition brought clear advantages over olfaction in control of escape and orienting behavior, advantages of speed and sensory acuity, and also advantages for early warning and anticipation of events. On the output side, the tectum gives rise to two major descending pathways, an ipsilateral one (passing to structures on the same side) and a contralaterally projecting one. The links to these output systems can be modulated by motivational states, including those triggered by the olfactory sense.

Figure 4.8 depicts the projection from a laterally placed eye to the opposite side of the midbrain tectum (often called the optic lobe, and in mammals the superior colliculus [*colliculus* meaning "little hill"]). It also illustrates the two major output pathways.

Note that neurons and pathways are shown on one side of the brain, but actually there are matching neurons and pathways on the other side. This is done to keep the drawing simpler, a practice that will be followed in many of the figures in this book.

Why sensory pathways decussate (i.e., cross to the opposite side) is a frequently asked question for which there is no generally accepted answer. Not only in the visual system, but in the somatosensory and auditory systems as well, one side of the external world is represented on the opposite side of the midbrain and the forebrain. The question of why the decussations that cause this kind of representation have evolved will be discussed in chapter 10 and reviewed further in chapter 20.

My hypothesis on the evolution of crossed projections can be summarized as follows. Very early in the evolution of the midbrain and forebrain, before the hemispheres appeared, visual inputs from lateral eyes projected bilaterally but then in evolution they became mostly crossed. This resulted in later evolution of decussations of nonvisual

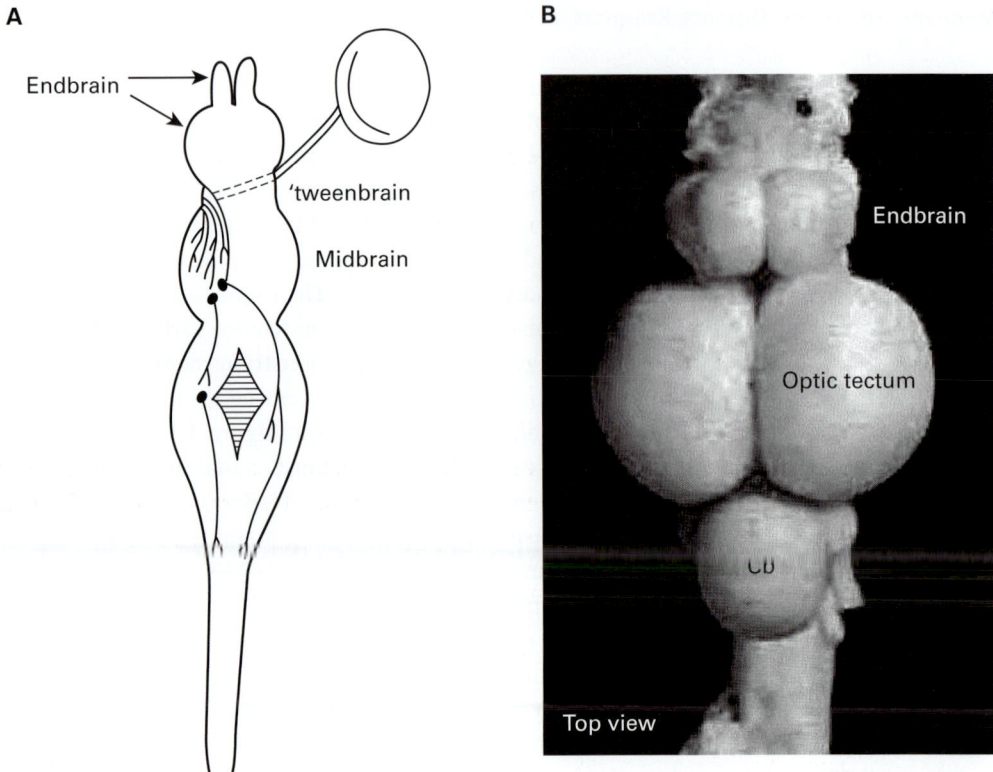

Figure 4.8
(A) Evolution of chordate brain, 4 (expansion of midbrain): The midbrain also expanded in some of the early chordates with evolution of the visual sense including image formation, and the auditory sense with auditory localization. These distance senses had advantages over olfaction for speed and sensory acuity, for early warning, and for anticipation of events. On the motor side, besides the rapid activation of flight movements, the midbrain tectum controlled turning of head and eyes. There was also modulation of these movements by motivational states. Note: Neurons and pathways depicted on one side of the brain are generally the same on the other side. Drawing them on only one side is done to make the drawing simpler. The cerebellum is omitted in the drawing. (B) Top view of the brain of a teleost fish, the great barracuda (*Sphyraena barracuda*), which, like most predatory fish, has a large optic tectum (at the roof of the midbrain) receiving projections from the eyes, and a relatively small endbrain. In this brain, the olfactory bulbs have been torn away, but some tissue in front of the endbrain, including olfactory stalks, has not been removed. Cb, cerebellum. Panel (b) from Schroeder (1980).

pathways as well. The original crossing of axons from the eyes must have evolved because it was more adaptive—supporting the better survival of the organism—because crossed pathways were able to reach crucial output mechanisms most quickly. These mechanisms must have controlled rapid escape/avoidance movements—movements most critical for staying alive when living near predators. In chapter 10, when we study somatosensory connections in the hindbrain, we will present additional reasons and details. The ideas will be reviewed also when we study the visual system (in chapter 20).

A Second Expansion of the Forebrain with Invasion of Non-olfactory Inputs

We have described a first expansion of the forebrain already, largely attributable to the influence of olfactory inputs at the rostral-most end of the neural tube (figures 4.1 and 4.7). A second expansion occurred later with the invasion of more and more axons carrying non-olfactory sensory information. The expansion that resulted led to a ballooning of the endbrain out of the remaining, more caudal, forebrain (figure 4.9). This produced

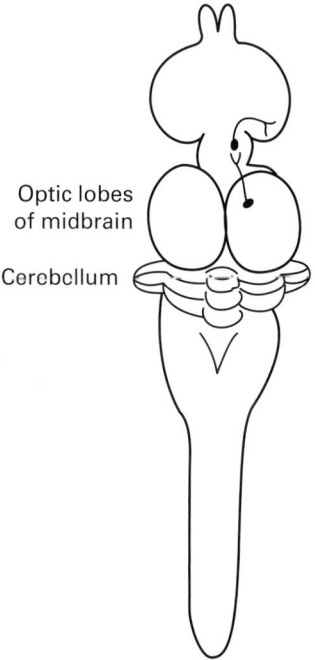

Figure 4.9
Evolution of chordate brain, 5a (second expansion of forebrain): Non-olfactory systems invade the 'tweenbrain and endbrain. These systems took advantage of the plasticity of links in the striatum. Thus, thalamic axons carrying visual, somatosensory, and auditory information reached the corpus striatum and the pallium (cortex). These connections resulted in pronounced expansions of the endbrain. Also in the expanding cerebellum, the original vestibular information was joined by information from the other modalities.

Box 4.1

For many years, a number of comparative neuroanatomists have been rejecting this kind of hypothesis concerning the expansion of the endbrain. The reasons are embedded in the history of the field. Before the advent of the Nauta methods for the experimental tracing of axonal projections, the early comparative neurologists (e.g., C. J. Herrick) had believed that olfaction completely dominated the endbrain of fishes, amphibians, and even some reptiles. In the 1970s and 1980s, the use of silver staining of degenerating axons made new experiments possible. The olfactory bulb was found to project only to restricted portions of the endbrain of the fishes and amphibians used in new studies. This did not fit the claims of Herrick and others. The endbrains of many fishes, amphibians, and reptiles were found to be surprisingly similar in basic organization to the brains of mammals. The experimental studies, though restricted to a small number of modern species, led investigators to conclude that there was a pervasive conservation of the major connections throughout the vertebrates.

There is no doubt a remarkable conservation in vertebrate brain organization and connections. However, I believe that the rejection of the invasion hypothesis was premature. The baby was thrown out with the bath water.

The idea that the endbrain appeared in evolution with all the major senses dividing up the pallium or the subpallium from the beginning is highly implausible. Instead, we have to think of plausible smaller steps between the earliest chordates, without cerebral hemispheres, and the jawed vertebrates. To help us, we have few extant species to study for guidance. Representatives of many important steps along the course of progressive parcellation of the endbrain have no doubt become extinct.

Whatever connections are found in extant vertebrates, they had to evolve from the earliest chordates that did not even possess cerebral hemispheres. When we review the olfactory system in chapter 19, we will note some findings in fishes that may be more closely related to the earliest vertebrates than those used in the studies just mentioned. The more recent findings are compatible with the views presented in this book.

the cerebral hemispheres, with the expansion of the corpus striatum ventrally and the expansion of the dorsal pallium (often called a dorsal cortex) above it. In birds, also in many reptiles, much of the expansion occurred below the ventricle, in the derivatives of the *dorsal ventricular ridge* (box 4.1).

We can guess a few of the details of how this happened. With an increasing importance of the midbrain roof for processing visual and auditory information in coordination with somatosensory inputs and for directing escape actions and orienting and feeding movements, this structure expanded. Its outputs became increasingly dominant, and the axons of those outputs became more widely distributed. Some of them terminated in the forebrain, where they could influence motivational states and endocrine control. Some neurons of the caudal forebrain receiving midbrain tectal inputs—in the diencephalic region—often responding to multiple modalities including visual and auditory, made connections in the corpus striatum, which was dominated by olfaction and linked not only to the hypothalamus but also to the locomotor control mechanisms of the caudal midbrain (separate from the tectum). Thus, non-olfactory inputs reached the striatum, in the ventral wall of the rostral neural tube, where they could take advantage of plastic links to motor control. The striatal region receiving non-olfactory inputs became increasingly segregated in the dorsal striatum, and this region expanded because of the importance of its functions, especially the plasticity of its circuits. Animals with enlargement of this region survived better and therefore passed on their genes more successfully. Thus, the ventral wall of the rostral neural tube became thicker with the addition of more neurons.

Some rostrally projecting sensory system axons also reached the dorsal pallium (cortex)—the wall of the neural tube above the ventricle in each cerebral hemisphere of the endbrain. Initially multisensory in nature, as was the case for the dorsal striatum, these endbrain regions became more and more segregated from olfactory dominance, with differing requirements of sensory analysis. Over many millennia, portions of these regions, both below and above the ventricle, not only became segregated from olfactory areas, but in addition the different non-olfactory senses became separated from each other, reducing the amount of multisensory convergence in these regions. The processes of segregation, or parcellation, of brain structures may have occurred, at least initially, by a kind of competition. Competition between certain populations of axons has been observed in studies of brain development—both normal development and development of abnormal connections after early brain damage.

This story omits much more than it includes—for instance, it says nothing yet about outputs of the dorsal cortex. Investigation of outputs of the dorsal cortex in modern reptiles has found projections that resemble those of pallial regions near the hippocampus of mammals, that is, the connections do not resemble those of mammalian neocortex.

At this point, we will add only that in primitive proto-mammals (or mammal-like reptiles), the dorsal cortex probably began to expand and evolve functional segregations. In addition, neuronal migrations into it formed additional layers of cells. These were momentous evolutionary moments, as they led to an evolution of the neocortex, a region that enlarged more and more as mammals spread and their endbrains evolved.

The story omits mention of sensory pathways that terminated in neither striatum nor in the pallium, but in a region called the dorsal ventricular ridge. This structure seems to have evolved from the lateral and ventral pallium of the embryo. In members of the three major animal groups we call birds, reptiles, and mammals, it matures very differently as it enlarges. The enlargement was greatest in some groups of birds, whereas in mammals it was not this region but rather the neocortex that enlarged the most.

Concurrent Evolutionary Changes: Motor Control and Plasticity

We have been stressing the role of sensory pathways and sensory processing in their influences on expansions and innovations in the brain over evolutionary time. It does appear that the demands of sensory analysis on neuron number and space outstripped the demands of motor control. However, the motor system was evolving all the while. All sensory systems are intimately linked with systems that control actions (systems for motor control). Evolutionary change occurs because some genetic variations are expressed in animals that survive and reproduce in greater numbers (and hence are more adaptive), so the responsible genes increase in a population of animals. With increased sensory acuity came increased accuracy in orienting movements. With increased object identification came more actions to manipulate and use the identified objects. Changed motor control abilities also influenced the evolution of sensory systems. More speed and more complex manipulation made more demands on ability to anticipate and predict sensory changes (see discussion later on a third expansion of the forebrain). It also increased the demands on maintaining stability in space and on rapid adjustments in blood flow to stabilize the internal environment. Adaptive changes are responses to such demands.

With an increase in numbers and complexity of instinctive movements (the *fixed action patterns* of the ethologists, which are largely "fixed" in the genes), there were corresponding evolutionary changes in hindbrain and spinal cord circuits that controlled them, as well as in connections with motivational control systems, centered on caudal forebrain and parts of the midbrain, and with sensory systems. Elaborations that became particularly prominent in mammals were structures for reward-driven learning and habit formation. These structures appeared early in forebrain evolution, in the olfactory-dominated corpus striatum, linked to motivational systems of the 'tweenbrain and midbrain. Sensory inputs to these structures for learning expanded with the invasion of the forebrain by non-olfactory pathways and the resulting expansions of corpus striatum and, even more, the overlying cortex in mammals.

For reward-driven learning and habit formation, there also had to be an early evolution of pathways that signaled events worth repeating, the reward pathways. For example, pathways from the midbrain reached the primitive striatal area, conveying such signals (pathways that remain in modern mammals). One pathway, with little doubt, signaled tastes that resulted from approach and intake of good foods.

Concurrent Evolutionary Changes: Motivational States

Another concurrent evolutionary development was the elaboration of systems for modulating other brain systems in response to visceral and social needs. These were the structures in control of cyclic and episodic changes in motivational state, present in the most primitive animals but elaborated with expansions of the forebrain. With these expansions, there was an elaboration of more complex goal hierarchies and more subtle communication of needs and desires among social animals.

The structures that achieved these functions were initially, in the early evolution of the brain vesicles, in the hindbrain and midbrain systems controlling visceral functions. These came to be dominated by the caudal forebrain, the portion we call the 'tweenbrain or diencephalon, especially the hypothalamus ventrally and the epithalamus dorsally. With the expansion of the more rostral forebrain, the part we call the endbrain, there was increasing influence over these functions by rostral extensions of the hypothalamus in the basal forebrain and corpus striatum and in derivatives of the pallium. In mammals, these structures include portions of the cerebral hemispheres that we have come to call the *limbic system* because of their location at the margin—the limbus—of the cortex.

A Third Expansion of the Forebrain: Anticipating Events and Planning Actions

Another major expansion of the forebrain has occurred in more recent evolution of mammals, an expansion mostly of the neocortex and of 'tweenbrain structures closely connected with it. On the sensory side there has evolved an ability to form images that simulate objects and events, facilitating an anticipation of what is likely to appear next in the sensory world. On the motor side, there has evolved more and more sophisticated planning of and preparing for actions. These are nonreflex functions involving memory and internal representations of the external world. They are cognitive functions.

These abilities have evolved in mammals with the evolution of structures most prominently centered in the neocortex of the endbrain. The portions that expanded the most, in the mammals with larger bodies and brains, were the so-called association areas of the cortex. But also, the parts of the corpus striatum and the cerebellum closely connected to those areas of neocortex have expanded concurrently. These expansions are sketched in figure 4.10.

The adaptive advantages of fine, dexterous movements, especially with the evolution of distal appendages with capacity for manipulation, resulted eventually in evolution of motor cortex together with an expansion of the cerebellar hemispheres. This has occurred throughout mammalian evolution, but has become most pronounced in the large primates and other animals with great dexterity, as in raccoons. Some birds also have highly developed motor control of the digits of the feet for grasping and manipulation, and have also evolved regions of the endbrain for motor control. This has been investigated less than in mammals, but it is known that as in mammals, these regions are closely associated with somatosensory regions, and they have some direct projections to the spinal cord.

Brain Sizes Relative to Body Weight Compared for Present-Day Animals

The major expansions of the brain that we have been describing have been presented as relative expansions. However, absolute changes in brain size and weight have occurred

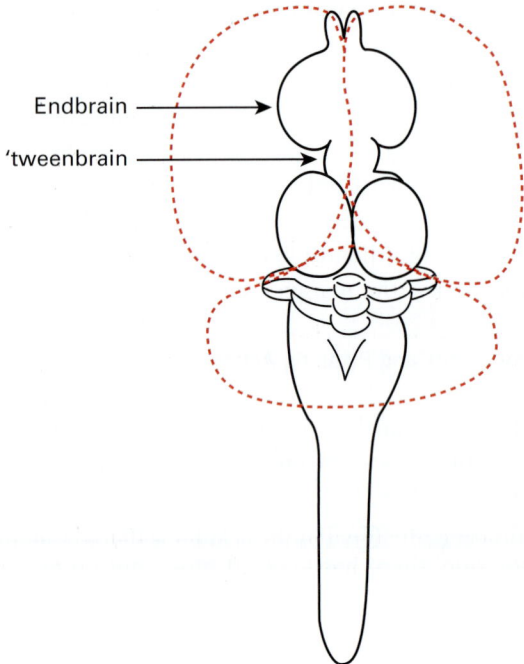

Figure 4.10
Evolution of chordate brain, 5b (third expansion of forebrain): Greater expansion of the endbrain. In mammals this was dominated by the expanding area of the neocortex (dotted red lines at top). In birds, this expansion was dominated by the *Wulst* and the derivatives of the *dorsal ventricular ridge* of the embryo. Correlated with this was an expansion of the cerebellar hemispheres (dotted red line below the hemispheres) and also the dorsal striatum. In the sketch, a pathway from the midbrain tectum to the thalamus, and hence to the endbrain, is depicted. Ascending pathways to the thalamus bypassing the midbrain also evolved. On the motor side, the advantages of control of fine movements, especially with the evolution of distal appendages with capacity for manipulation, resulted in mammals in the evolution of motor cortex as well as the cerebellar hemispheres.

as well. The size of the brain has been measured, usually as brain weight, in a large number of different species in a few studies. The results of these measures show that the major determinant of brain size is the size of the body. The plots that are most meaningful show brain weights plotted against body weights on log-log scales. Figure 4.11 shows such a plot of data from a large collection of brains from animals of widely different species. Such a plot shows *relative* brain sizes as distances above or below a diagonal line. The human brain shows the greatest relative size although it is much smaller than the brain of animals with much larger bodies. Figure 4.12 shows a similar plot for mammals alone. The data plotted in this way show that different major groups of animals can be represented by different diagonal lines, and that such lines reveal differing relative brain sizes for different major groups.

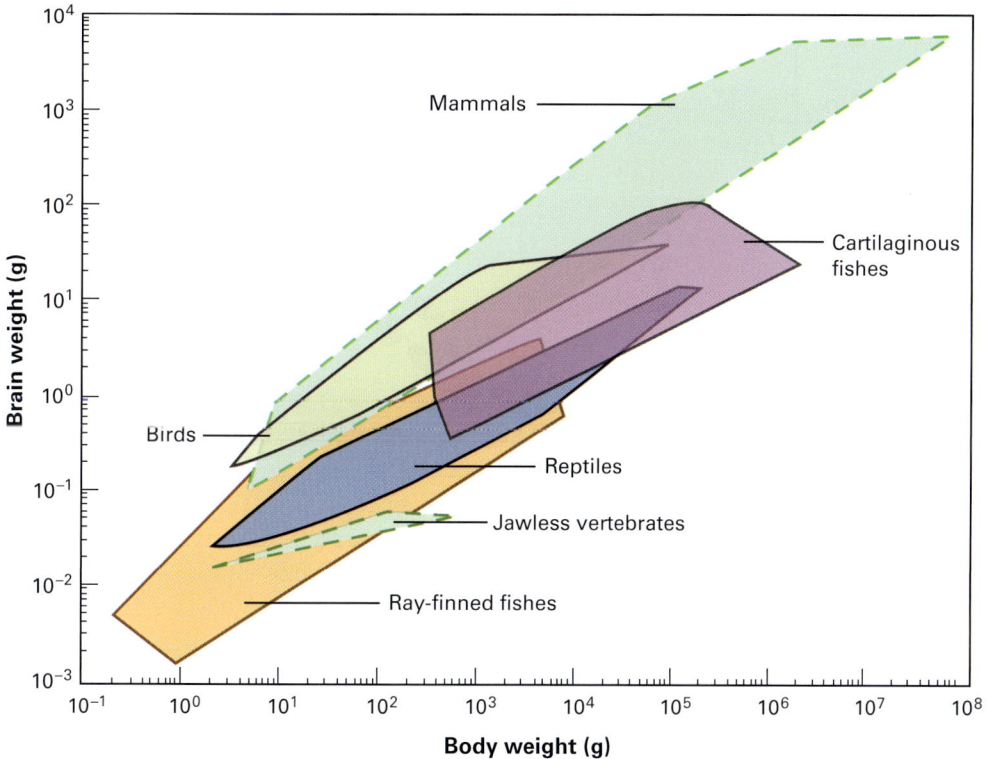

Figure 4.11
Vertebrate brain–body weight scaling. Numbers are graphed in logarithmic coordinates. Illustrated in green is the total envelope of points for the brain weight and body weight of a large number of mammals. Another envelope, in yellow-green color, encompasses the brain versus body-weight points for a large number of birds. Additional envelopes are drawn for cartilaginous fishes, for ray-finned fishes, for reptiles, and for jawless vertebrates (fish). Based on Striedter (2005).

Where We Stand in the Path to Our Goal

Now we have an outline of the major regions of the CNS of vertebrates, especially the mammals. We have sketched some major steps in brain evolution. However, only a little about connections within the brain has been included thus far. In the next chapter, it will be the major connections and their organization that will become the focus.

Don't worry if it all seems a bit confusing. This is only our first run through the brain. There will be more such runs, each of them with a different thrust, and each time you will remember more and it will make more sense.

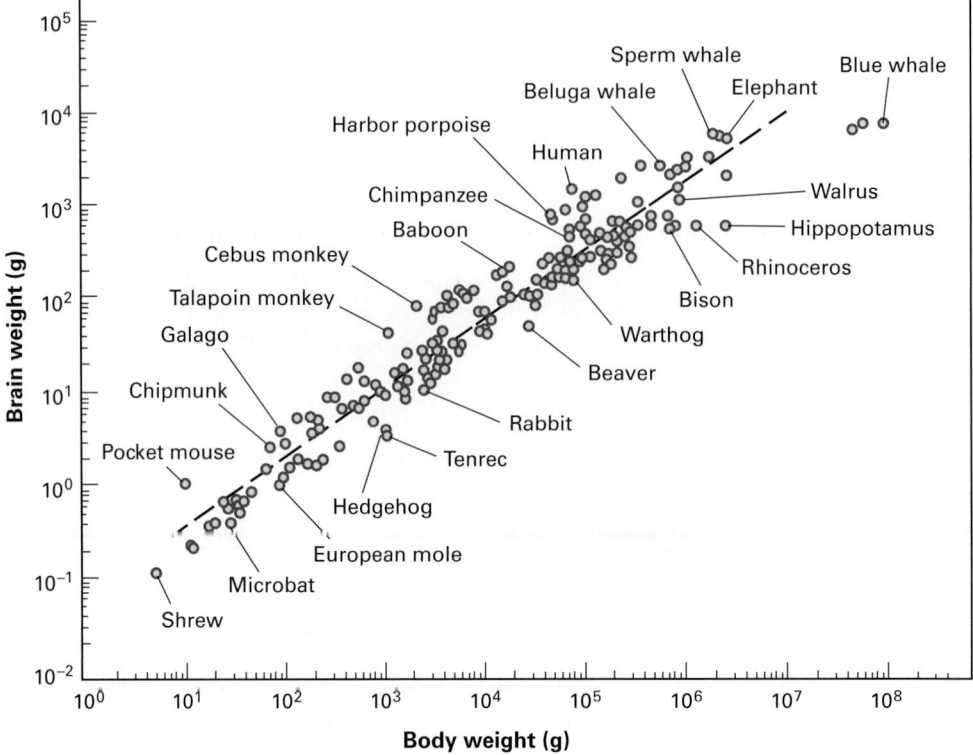

Figure 4.12
Placental mammal brain weights versus body weights plotted in logarithmic coordinates for 213 separate species. Based on Striedter (2005).

Readings

The opening quotation is from Elliott, H. C. (1969). *The shape of intelligence: The evolution of the human brain.* New York: Scribner.

Sources of much information from the fields of comparative neuroanatomy and brain evolution:

Butler, A. B., & Hodos, W. (2005). *Comparative vertebrate neuroanatomy: Evolution and adaptation*, 2nd ed. New York: Wiley-Interscience.

Striedter, G. F. (2005). *Principles of brain evolution.* Sunderland MA: Sinauer Associates. This book contains references to studies of brain size in a large number of species, including reports by Stephan et al. and by Jerison.

Concerning phenomena of competition between populations of axons during brain development and after brain damage, see the book-length review by Ray Lund and two papers by the author of this book:

>Lund, R. D. (1978). *Development and plasticity of the brain: An introduction.* New York: Oxford University Press.

>Schneider, G. E. (1973). Early lesions of superior colliculus: Factors affecting the formation of abnormal retinal projections. *Brain, Behavior and Evolution, 8,* 73–109.

>Schneider, G. E. (1979). Is it really better to have your brain lesion early? A revision of the "Kennard principle." *Neuropsychologia, 17,* 557–583.

The regulation of body temperature is carried out at many levels of the CNS, from spinal levels to the forebrain. This was made clear for mammals by the investigations of Evelyn Satinoff, which she reviewed in the following paper published in 1978:

>Satinoff, E. (1978). Neural organization and evolution of thermal regulation in mammals. *Science, 201,* 16–22.

An important idea in this field was developed by the comparative neuroanatomist Sven Ebbesson, who wrote a paper in 1980 on the "parcellation theory":

>Ebbesson, S. O. E. (1980). The parcellation theory and its relation to interspecific variability in brain organization, evolutionary and ontogenetic development, and neuronal plasticity. *Cell and Tissue Research, 213,* 179–212.

III INTRODUCTION TO CONNECTION PATTERNS AND SPECIALIZATIONS IN THE EVOLVING CNS

5 The Ancestors of Mammals: Sketch of a Pre-mammalian Brain

Observations of body shapes of the embryos of widely different vertebrate species revealed something remarkable to Karl Ernst von Baer in 1828. Embryos at early stages of development could be very difficult to distinguish, even when they are of species as different as a mammal, a bird, and a reptile. This was used by Charles Darwin as a finding that was in accord with evolutionary change in animal forms (figure 5.1). A contemporary of Darwin, the German biologist and philosopher Ernst Haeckel, and others of the time propounded the idea that "ontogeny recapitulates phylogeny"—an idea that went far beyond the von Baer observations. His drawings of embryos of a small collection of vertebrates were not very accurate, but the idea behind the arrangement he put together has seemed so compelling to many that his illustrations have been copied and republished very often (figure 5.2). The illustrations show the appearance of embryos at three stages of development in a collection of animals including a fish, salamander, tortoise, chick, and several mammals ending with human. At an early stage, the embryos look fairly similar, whereas at later stages they look more and more distinct. The fish and salamander begin to look very different at an earlier stage than the others. Recent restudy of this issue has included a wider range of species and more detailed anatomical comparisons, and results have shown that all vertebrates actually do not go through an early stage of looking nearly identical. Details make them distinguishable throughout embryogenesis—much more distinguishable that the Haeckel illustrations imply (see caption of figure 5.2).

These kinds of developmental sketches hardly prove anything for certain about phylogeny, but they may lead us to expect some similarities in the CNS of all vertebrates—and these similarities have indeed been found. We can use information from neuroanatomical studies of amphibians and from skulls of some extinct animals to construct a schema of what major subdivisions and connections in pre-mammalian brains must have been like. This will give us an outline of brain organization that prepares us for understanding the mammalian brain, as mammalian brains evolved by additions to what preceded them much more than by any subtractions or radical transformations. Remember that gene-based changes generally have to occur in small steps.

Karl Ernst von Baer

Charles Darwin

Figure 5.1
Likenesses of Karl Ernst von Baer (1792–1876) on the left and Charles Darwin (1809–1882) on the right, when Darwin was 45 years old and preparing to publish *On the Origin of Species by Means of Natural Selection*. Von Baer was one of the founders of embryology. He discovered the notochord and the blastula, among other things. He noticed that embryos of different species at early stages could be difficult to distinguish, even embryos of species as different as a mammal, a bird, and a reptile. This observation was cited by Darwin (1859) but attributed to Agassiz.

The "Old Chassis" for the Building of Further Additions

So let's imagine our ancestors of the Triassic and early Jurassic period, little creatures coexisting with the earliest dinosaurs and their relatives and no doubt hard pressed to avoid being eaten by them. We will sketch a brain that could represent that of an early cynodont (figure 5.3). Some of these mammal-like reptiles (therapsids, a group that had separated from the line leading to reptiles and birds) evolved skull characteristics that were like the mammals that came later. The brain case of some of them indicates a shape and size of a brain that resembles that of some small mammals of today. The brain sketch (figure 5.4B) is based on a classroom drawing used in neuroanatomy lectures by W. J. H.

The Ancestors of Mammals: Sketch of a Pre-mammalian Brain

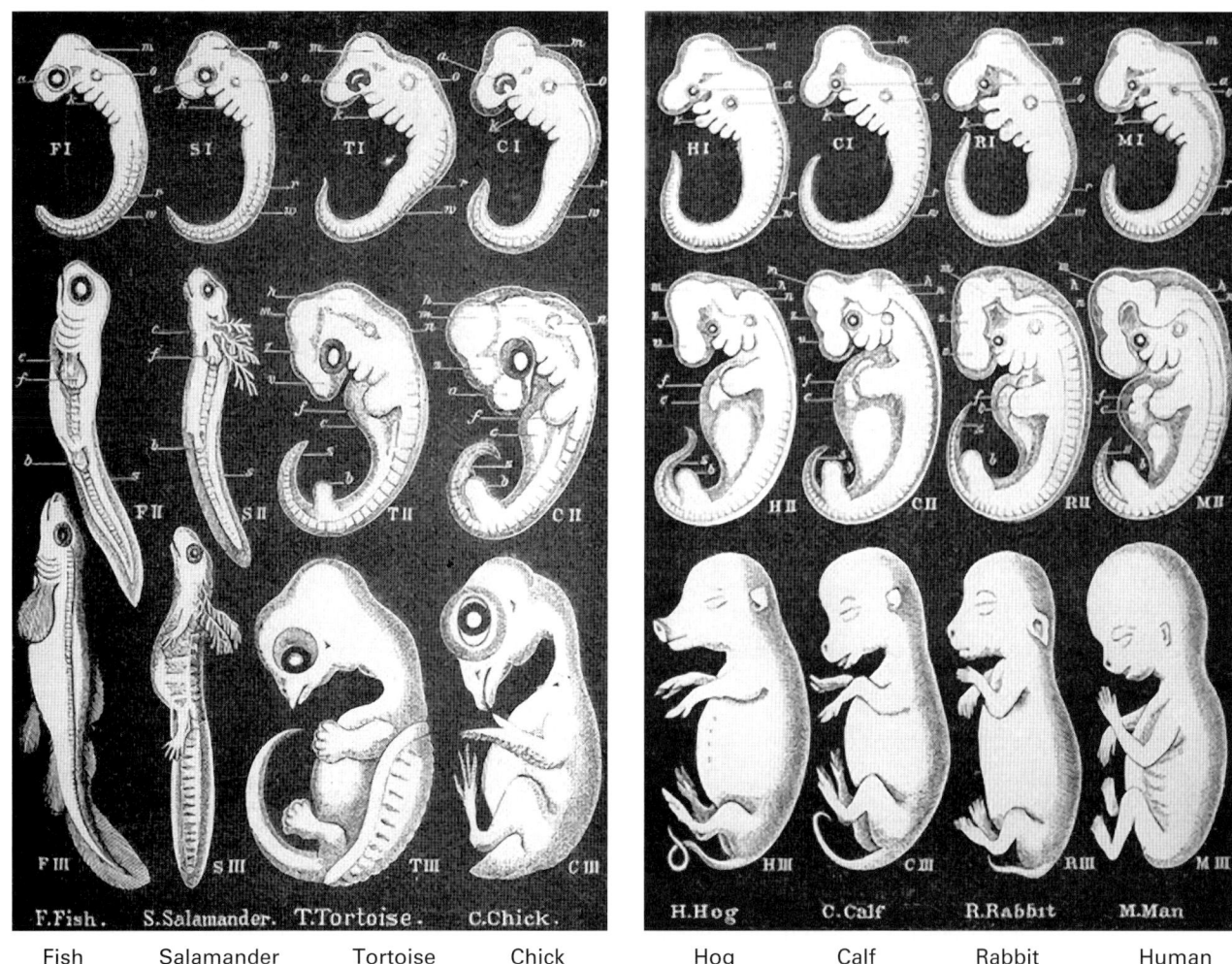

Figure 5.2
A frequently reprinted illustration by Ernst Haeckel (1879), who promoted Darwin's theory in Germany. The drawings were used by Haeckel in support of the idea that "ontogeny recapitulates phylogeny" and have been reproduced many times up to the present. However, the drawings were not accurately done, especially for the fish and amphibian at early stages. Even for the other groups, the trend toward more similarity in shape at earlier stages, first noticed earlier by von Baer, is at best very rough and cannot be taken as a precise rule. In the words of Charles Darwin, at early stages one finds "the embryos of different species within the same class, generally, but not universally, resembling each other" (*On the Origin of Species by Means of Natural Selection*, chapter 13, section on embryology). From Haeckel (1897).

Figure 5.3
Cynodonts ("dog-teeth") are extinct proto-mammals, or mammal-like reptiles, that existed from the late Permian age through the Triassic, the Jurassic, and into the mid-Cretaceous period. They evolved increasingly small body sizes, and their skulls became more and more like small mammals of today. It is easy to find, on the Web, reconstructions that scientists and other enthusiasts have made from skull and skeletal remains to suggest their appearance. Pictures of Procynosuchus and Cynognathus courtesy of Nobumichi Tamura.

Nauta at MIT (neuroanatomy professor at MIT from 1964 to 1986). After describing this sketch in some detail, we will illustrate major mammalian additions to the ancestral brain.

You can think of the ancestral brain sketches as representing the "old chassis" upon which new parts were built in evolution. Little was really discarded after the ancestral cynodonts. (You still have that ancestral brain in the core of your human brain!) Note also that "old" is a relative term. The amphibian brain is very advanced in comparison to that of more primitive chordates, and this was certainly true of the pre-mammalian cynodonts. Later, we will discuss in more detail some of the evolutionary transitions that gave rise to major changes in the CNS.

The Ancestors of Mammals: Sketch of a Pre-mammalian Brain

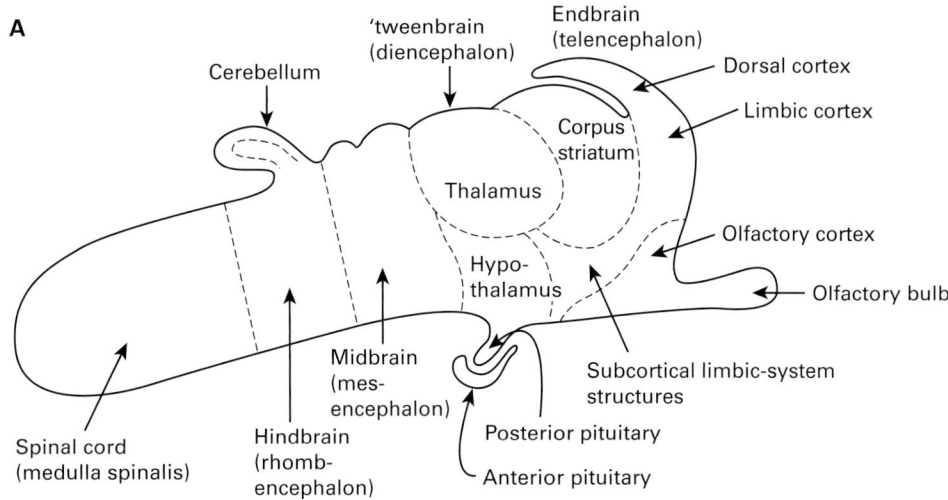

Figure 5.4
Diagrammatic sketch of the hypothetical CNS of an animal believed to have preceded the advent of mammals in chordate evolution. The relative size of the spinal cord and brainstem have been enlarged for illustrative purposes. One can imagine this to characterize the brain and spinal cord of an early cynodont. (A) Major subdivisions are indicated; (B) a few representative neuronal pathways and connections are sketched, including primary sensory neurons of the olfactory, auditory, and somatosensory systems. Secondary sensory neurons and interneurons are indicated by open triangles, and motor neurons by filled triangles. An alpha motor neuron connected to striated muscle cells is indicated, as is a preganglionic motor neuron of the autonomic nervous system, which connects with a peripheral ganglion cell that has an axon projecting to smooth muscle or gland tissue.

We begin our sketch by showing the basic subdivisions of the brain and some major neuron types (figure 5.4). Then we will define distinct sensory channels of conduction, thus organizing the information that is pouring into the CNS all the time. Next, an overview of cephalic structures will be presented, and finally we will sketch the major mammalian addition, the neocortex, and some major changes in pathways that occurred with that addition.

Subdividing the CNS

English names of brain subdivisions are used in figure 5.4A, but the names derived from Latin and Greek are also given. The forebrain is shown with its two major divisions, the 'tweenbrain (diencephalon) and the endbrain (telencephalon). Each of these subdivisions is subdivided. The 'tweenbrain of modern mammals has four major subdivisions, each of them divided into various cell groups, but for simplicity, just two diencephalic subdivisions (plus the pituitary) are shown in the figure—the thalamus and hypothalamus—the divisions most frequently discussed because of their functional importance.

In the sketches and photos of the endbrain in chapter 4, you probably remember the olfactory bulb at the rostral end. We have also mentioned previously the major subcortical component of the endbrain, the corpus striatum. In figure 5.4, above the striatum are shown the limbic cortex and the dorsal cortex (or dorsal pallium). In addition to the olfactory cortex (behind the olfactory bulbs), most, if not all, of the pallium in pre-mammalian brains corresponded to non-neocortical areas of mammals that are portions of the limbic system—the group of structures most closely connected with the motivational and visceral regulatory mechanisms of the hypothalamus. Included was the dorsal cortex, which in amphibians and reptiles receives non-olfactory inputs from the 'tweenbrain. So does the medial pallium with which it is closely connected; the medial pallium in these animals is the homolog of the hippocampus.

The sketch of the ancestral brain (figure 5.4) shows spinal cord and brainstem regions that are unrealistically enlarged in order to give space for sketching some pathways and connections. A more realistic drawing is shown in figure 5.5. In this drawing, we see a dorsal view of the thickening neural tube of an embryonic mammal, believed to resemble the more mature neural tube of an adult proto-mammal. In the drawing, the cerebellum is not shown for simplicity's sake; at this embryonic stage in mammals, it is forming in the roof of the rostral hindbrain. Also shown are frontal sections representing five different levels of the CNS. The roof plate of the neural tube, where the wall of the tube remains only one cell thick, is indicated with horizontal striations. The dorsal view reveals the origin of the name for the hindbrain: the rhombencephalon, for the widened roof plate there has a rhombic shape. In the endbrain, the two hemispheres have been pushed apart for greater clarity, showing the thin roof plate that is continuous with the roof plate of the 'tweenbrain. You can imagine those hemispheres pushed up and toward each other

The Ancestors of Mammals: Sketch of a Pre-mammalian Brain

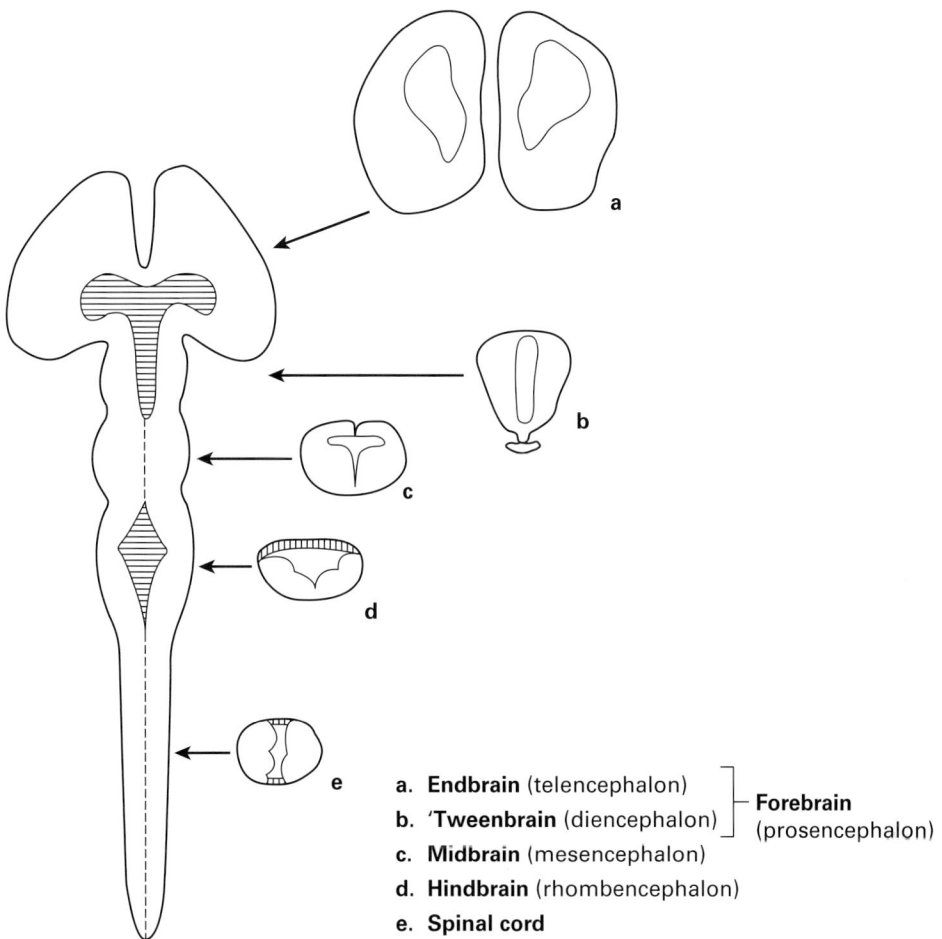

a. **Endbrain** (telencephalon)
b. **'Tweenbrain** (diencephalon) ⎤ **Forebrain**
c. **Midbrain** (mesencephalon)   ⎦ (prosencephalon)
d. **Hindbrain** (rhombencephalon)
e. **Spinal cord**

Figure 5.5
Simplified drawings of the thickening embryonic neural tube of a proto-mammal. At the left is a view from the dorsal side of the developing CNS that has been straightened out and has had the cerebral hemispheres of the endbrain pushed apart. Thin walls of the centrally located, fluid-filled ventricular space are indicated by horizontal striations. To the right are frontal sections taken at the levels indicated. The endbrain section represents a slightly later stage of development. The cerebellum has been omitted.

until they come against each other at the midline, thus fitting the endbrain more compactly into the skull.

The Basic Setup: Major Types of Neurons

Now look back at figure 5.4B and find the three primary sensory neurons depicted there as examples. Three different shapes of such neurons are shown (cf. figure 1.7): You find a dorsal root ganglion cell with its axon entering the dorsal part of the spinal cord through a spinal nerve. The cell body is located near, but not within, the CNS. The *pseudo-unipolar* shape of the cell is typical for primary sensory neurons of the spinal dorsal roots of quadrupeds and birds. A neuron of this type was illustrated in chapter 1 in the discussion of axonal conduction. More rostrally in figure 5.4B you see a bipolar cell of the auditory or vestibular nerve (cranial nerve VIII) that enters the dorsal part of the hindbrain. Most rostrally, an olfactory nerve fiber is depicted (part of cranial nerve I) entering an olfactory bulb. It originates from a primary sensory neuron in the olfactory epithelium of the nasal passages.

Now, find the secondary sensory neurons depicted in the sketch. These are the cells that are contacted directly by the axons of primary sensory neurons, at synapses formed on their dendrites and cell bodies. The secondary sensory neurons receiving input from the dorsal root axons are found mostly in the dorsal horn of the spinal cord, which is the spinal gray matter nearest the back of the animal. In the case of the cranial nerves entering the hindbrain, sensory fibers terminate on cells grouped together in *nuclei*. In the case of the example fiber of the eighth cranial nerve, synaptic contacts are made with neurons of the cochlear nuclei or the vestibular nuclei, located under the lateral part of the cerebellum. For the last example in the brain sketch, we see one of the thin axons of the olfactory nerve contacting a secondary sensory neuron of the olfactory bulb (i.e., the axonal ending synapses on the olfactory bulb neuron). As we will study later (chapter 19), the axonal endings form a terminal bush-like arbor that is entangled with a dendritic arbor of a neuron called a mitral cell. Within the intimate tangle, many synaptic contacts are made.

You can locate the motor neurons depicted in the sketch of figure 5.4B by finding the darkened cells. They are located in the ventral part of the CNS of the spinal cord and in groups of motor neurons within the hindbrain and midbrain. (Not shown is a different kind of effector neuron located in the hypothalamus that secretes a hormone into the bloodstream.)

Before we add to this picture of motor neurons, it is now easy to find the interneurons of the great intermediate net in the sketch. They are all the remaining neurons—all the neurons of the CNS except for the secondary sensory neurons and the motor neurons. Only a small number of examples are included in the picture, for use in illustrating some of the pathways within this ancient brain. The interneurons vary greatly in their axonal lengths and in the shapes of their cells and dendritic arbors (not illustrated). The local interneurons, or short-axon interneurons, are the cells most likely to be named interneurons in neuroanatomy parlance. Three local interneurons are shown connecting to motor neurons in the brain sketch.

Returning briefly to the motor neurons, you should notice that two types of these neurons are illustrated in the sketch. Each of them has an axon that exits the CNS ventrally. The axon of one type synapses directly with cells of striated muscles. The axon of the other type comes from a preganglionic motor neuron of the autonomic nervous system, the system of neurons that innervates glands and smooth muscles of the viscera. The axon from the CNS ends on a ganglionic motor neuron—a neuron located within one of the autonomic ganglia—that has an axon innervating a visceral end-organ, either a smooth muscle or a gland cell. Innervation of the viscera will be discussed further in a later chapter.

Sensory Channels of Conduction: Local Reflex, and Not so Local

We are presenting this nervous system that had evolved in animals like the early cynodonts, prior to the appearance of mammals, as a system of communicating cells. To begin to understand how this system worked, let's discuss what happens to the sensory inputs after they enter this system through a sensory nerve. We will focus now on the somatosensory inputs to the spinal cord. Sensory inputs can be organized by grouping them into specific channels. The first sensory channel is the local reflex channel. For dorsal root inputs, we can define a local reflex channel as a pathway limited to a single segment of the spinal cord. This is illustrated in figure 5.6. The dorsal root input activates secondary

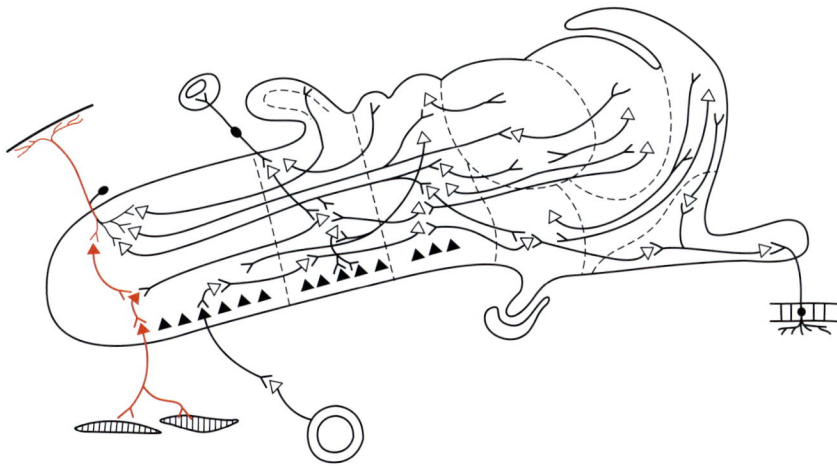

Figure 5.6
The sketch of a pre-mammalian brain (see figure 5.4), with a local reflex channel highlighted in red. The reflex includes three synapses, with input and output at the same level of the CNS.

sensory neurons, some of which project their axons to interneurons that, in turn, connect to motor neurons with axons that leave the CNS through the ventral root at the same spinal level. These motor neurons send their axons to striated muscle cells.

What might have been the function of such a local reflex channel? A likely function for the polysynaptic reflex just indicated is the withdrawal of a limb or part of a limb. Thus, the motor neurons are those that connect with flexor muscles. The withdrawal reflex, triggered by inputs that are potentially harmful, has been shown in experimental work by physiologists to involve at least two synaptic connections in the cord.

It is probably rare for the spinal activity of a reflex like this to be confined to a single level. The dorsal root axons generally branch upon entry into the cord, and the end arbors are distributed over several spinal segments, albeit they are most dense at the level of entry. If the stimulus is not very intense, the resulting activation of flexor reflex activity would be limited to the local level, if it reaches above threshold for firing motor neurons at all. If it is more intense, it can activate spinal interneurons at multiple levels and excite motor neurons of more flexor muscles connected to a larger range of spinal segments. This "spread of effect" is a result of spatial and temporal summation in the activation of neurons by multiple simultaneous or near-simultaneous excitatory inputs. In this way, many, probably most, reflexes activated by dorsal root inputs are intersegmental reflexes. The intersegmental reflexes always involve connections between different levels of the spinal cord, connections made by interneurons. The axons that interconnect different levels of the cord are part of the *propriospinal* system. The term refers to pathways that never leave the spinal cord. If the axons are a little longer and connect to levels above the cord, in the brainstem, then we call the pathways suprasegmental.

We can go further in defining "local" for the local reflex channel of sensory conduction. In the periphery, the area of skin innervated by a single pair (left and right) of dorsal roots is called a *dermatome*. A map of the dermatomes in the human is illustrated in figure 5.7, and a human CNS with spinal nerves attached is shown in figure 5.8. Such maps can readily be found in medical textbooks. The dermatomes of humans have been mapped in two major ways. One way is to outline the hypersensitive regions resulting from irritation of single spinal roots (e.g., from a herniated intervertebral disk). The other way is to chart the skin areas where sensibility remains after severance of adjacent spinal nerves or dorsal roots. Results of such studies have shown that single spinal nerves (or dorsal roots) innervate an area of the body surface (on the same side of the body as the nerve or root) that overlaps that of the immediately adjacent spinal nerves, but the innervation is densest in the area we call the dermatome for that nerve.

The term *myotome* is the corresponding term for the somatic motor system. It is used for the striated muscles innervated by a single ventral root. Thus, if a reflex is strictly a local reflex, it will involve stimulation of a single dermatome and contraction of muscles in a single corresponding myotome. This may be a rare occurrence.

The Ancestors of Mammals: Sketch of a Pre-mammalian Brain

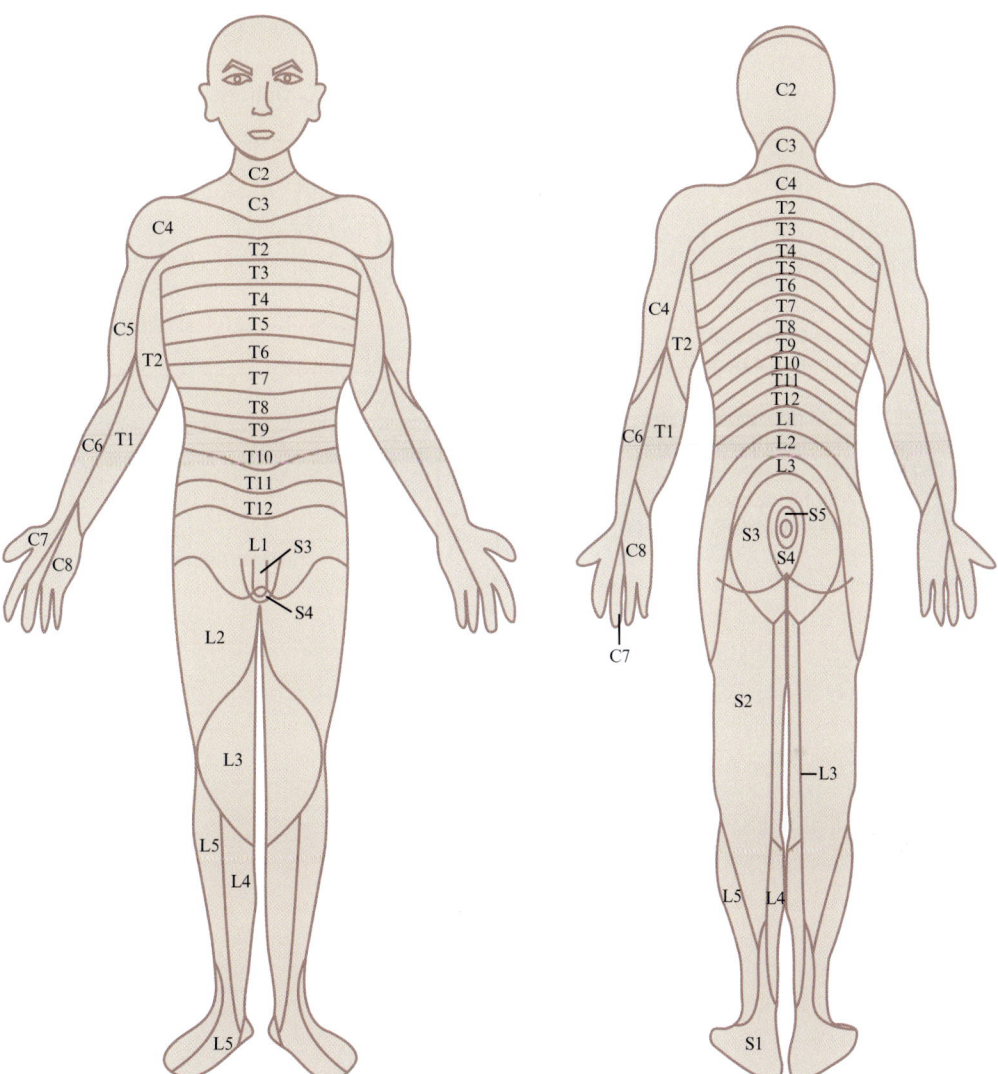

Figure 5.7
Human dermatome map. Each dermatome represents the area of the body surface that receives the densest innervation by a single pair of spinal nerves. The face and top of the head are innervated by the fifth cranial nerve. C, cervical; T, thoracic; L, lumbar; S, sacral.

100 Chapter 5

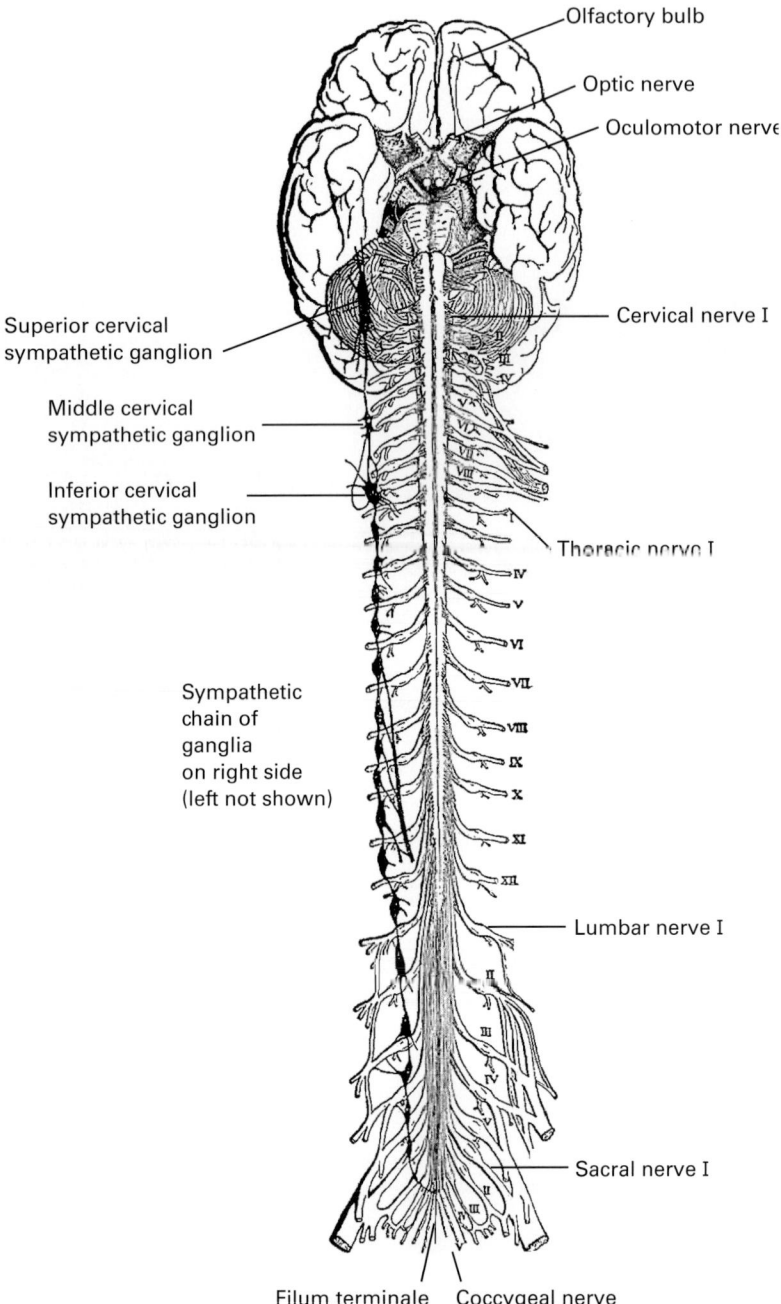

Figure 5.8
Ventral view of brain and spinal cord of an adult human, dissected free of the remainder of the body. The proximal portions of the spinal nerves are attached. Enlargements caused by dorsal root ganglia are difficult to see. From Herrick (1922).

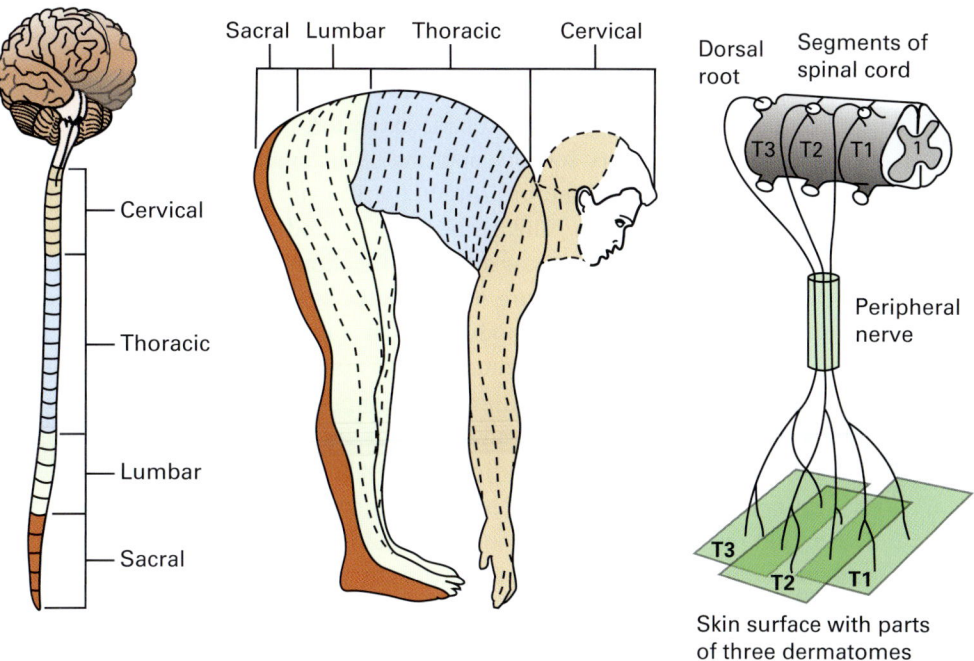

Figure 5.9
Human dermatome map shown with the human body in a quadruped position. Compare with figure 5.7: In the quadruped position, the map in the limbs is easier to understand. Note also that the face and forehead are not innervated by spinal nerves. (They are innervated by the fifth cranial nerve). Based on Breedlove, Rosenzweig, and Watson (2007).

The complexity of the pattern of dermatomes in the limbs of a human can be understood more readily if the dermatome map is shown for a human seen from the side with the body arranged in a quadruped position (figure 5.9). Note also that the dermatomes defined by the spinal nerves do not include the face and anterior part of the top of the head. That head region is innervated not by a spinal nerve but by the fifth cranial nerve, usually called the trigeminal nerve because of its three major branches (ophthalmic, maxillary, and mandibular). We will discuss the trigeminal nerve and its central connections when we consider the hindbrain in more detail.

Old Ribbons to the Brain

The next sensory channels of conduction to be looked at are called lemniscal channels. A lemniscus in the central nervous system is a band of fibers running together over some distance, usually to a common destination. We will describe, for the pre-mammalian brain, two lemniscal pathways. One carries somatosensory information rostrally from the spinal cord into the brainstem. The second carries such sensory information to the cerebellum—the spinocerebellar pathways.

The first and more ancient of these pathways is composed of spinoreticular axons. The spinoreticular pathway is like a rostral extension of the propriospinal system of axons. Originating in secondary sensory neurons of the spinal gray matter, these axons project bilaterally to the core of the brainstem, the so-called reticular formation. Most of them stay on the ipsilateral side. This is the least studied of the somatosensory pathways, but we can make a reasonable guess about its functions from what is known about functions of the reticular formation where it terminates. It is very likely that it serves as input to neurons involved in sensory modulation of autonomic activity (e.g., heart rate, blood pressure, breathing rate and volume). It also must contribute to the control of autonomic and defensive behavioral responses to pain inputs, and probably contributes more generally to the perception of pain and what that entails. Other likely functions include temperature regulation, and in addition it probably serves on the input side for control of sexual behavior patterns. These axons terminate, with widely branching axons, at hindbrain and midbrain levels, and some of them extend into the more ancient portions of the 'tweenbrain—see the ascending axons in the schematic picture of figure 5.10.

If the spinoreticular pathway, ascending mostly on the ipsilateral side, is really the most ancient lemniscal pathway from the spinal cord in the evolution of vertebrates, then studies of living animals most closely resembling the most primitive vertebrates should provide some supporting evidence. A cladogram of vertebrates shows three animals or

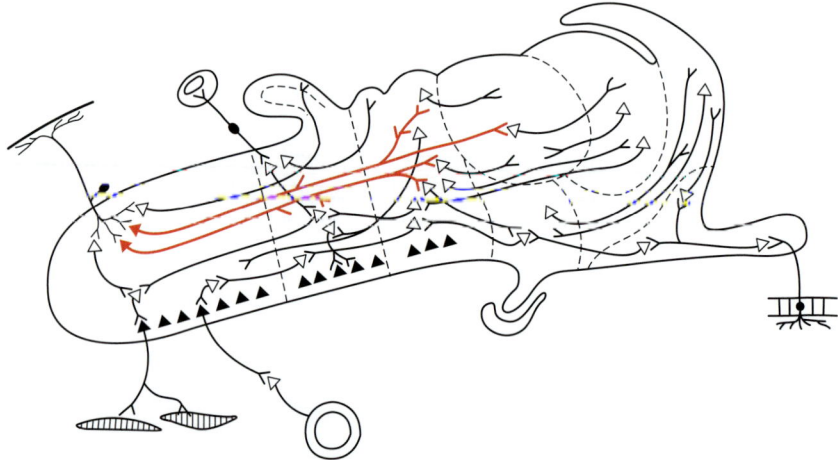

Figure 5.10
On the sketch of a pre-mammalian brain, axons that represent the spinoreticular or spinothalamic pathways are highlighted. Ascending spinoreticular axons distribute to both sides of the brain but are mostly ipsilateral. Spinothalamic tract axons decussate near their cells of origin (not shown); some of them reach as far as the thalamus.

groups of animals that have descended with the least change from the ancestors of all vertebrates (figure 5.11). The most ancient group (i.e., the one whose lineage branched off from other animals the earliest) is represented by amphioxus, the little worm-like non-vertebrate chordate we considered earlier. The two most primitive vertebrates, the jawless vertebrates, are the hagfish and the sea lamprey. Studies of amphioxus have indicated inputs from the body surface that enter the CNS and contact neurons that distribute outputs to both sides. This resembles, at least in this one respect, the axons of the spinoreticular pathways that have been studied in higher vertebrates. Hagfishes and sea lampreys, the most primitive living vertebrates, have spinoreticular pathways while they lack other ascending lemniscal pathways.

The other ascending somatosensory pathway of the pre-mammalian CNS is called the spinothalamic tract. This pathway is often referred to as the paleolemniscus—the old lemniscus. Although it is not older than the spinoreticular pathways, the data indicate that it did originate before the advent of mammals. It is found also in the reptiles and birds where it has been traced. The spinothalamic tract probably evolved as a specialization of a portion of the widely branching spinoreticular axons.

This pathway, unlike the spinoreticular, is a decussating pathway: The axons of the secondary sensory neurons in the dorsal horn of the spinal cord that compose the spinothalamic tract cross to the other side of the cord before ascending toward the brain. Despite the name spinothalamic, most of the axons of this pathway do not reach the thalamus, but terminate in the hindbrain and the midbrain, in the reticular formation and in the deep layers of the midbrain tectum. Two axons that ascend from the spinal cord into the brain in figures 5.4B and figure 5.10 could be spinothalamic tract axons if they decussate in the cord. (Otherwise, they would be spinoreticular axons.) In those sketches, a thalamic neuron receiving this somatosensory input is shown projecting to the corpus striatum. The spinothalamic tract is drawn on a dorsal view of the developing embryonic neural tube of a mammal in figure 5.12. In this figure, a thalamic somatosensory neuron is shown projecting to a termination zone in the cortex.

Remember that we are not representing the earliest chordates here, but rather the proto-mammalian or early mammalian brain. There was a very long period of evolution between those early chordates and the advent of mammals.

Interesting questions for research into somatosensory pathways can be suggested. What can vertebrate animals do in the absence of somatosensory pathways to the forebrain? or in the absence of all *crossed* ascending pathways?

Ribbons to the Little Brain: The Cerebellum

The second type of lemniscal channel from the spinal cord takes information to the cerebellum. The word *cerebellum* is a diminutive form of the Latin word for brain, *cerebrum*. The cerebellar channel from the spinal cord is made up of the spinocerebellar tracts,

104   Chapter 5

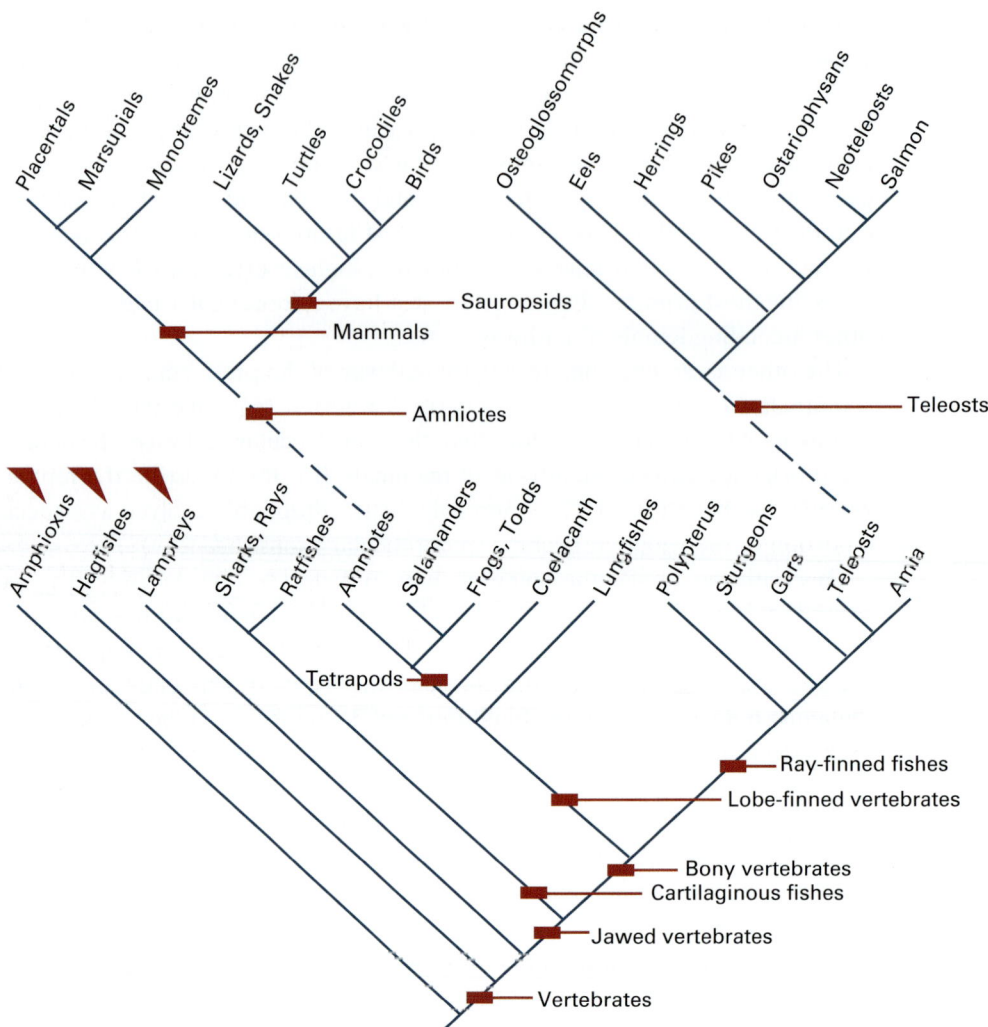

Figure 5.11
A cladogram of the vertebrates, based on multiple sources. In this particular kind of diagram, time is represented on the vertical dimension, so ancestral animals have locations lower in the tree structure. Living species are at the upper end of the branches. In this cladogram, extinct species are not shown, nor are lines of descent that were terminated by extinctions without leading to surviving lines. A number of groups are given their common names; cephalochordates are represented by one species (amphioxus). Three groups that retain close similarities to the ancestors of all vertebrates or all jawed vertebrates are specially marked. The cladogram includes only a subset of all vertebrate lineages. Based on Striedter (2005), who compiled information from multiple sources.

The Ancestors of Mammals: Sketch of a Pre-mammalian Brain

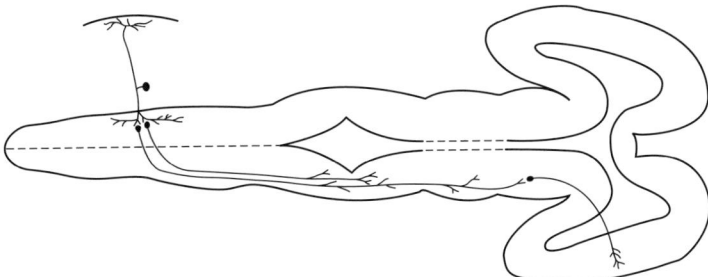

Figure 5.12
Spinothalamic tract axons are sketched in this dorsal view drawing of an embryonic mammalian CNS. Note the decussation in the spinal cord. Axons terminate in the hindbrain and midbrain; some of them reach the 'tween-brain. Here, termination of an axon in the thalamus is illustrated on a cell that projects its axon to the neocortex. Spinothalamic pathways are found in reptiles, birds, and cartilaginous fish as well as in mammals.

including one that is uncrossed and another that crosses to the opposite side in the spinal cord before its ascent. Both tracts terminate in the cerebellar cortex. Such a pathway is easy for you to find in the sketch of the pre-mammalian brain shown in figure 5.4B.

A suggestion about why the cerebellum and cerebellar channels of sensory systems evolved is as follows. A special problem existed with the evolution of multiple sensory pathways for control of the same outputs; namely, the problem of timing. Different conduction delays and summation times in different pathways controlling the same movement resulted in a considerable problem of coordination. This problem increased with increasing body size (changing during development) and the increased role of head receptors in addition to body receptors in control of the same movements. It increased further with need for precise coordination of larger numbers of muscles, especially the distal muscles of the limbs. The cerebellum appears to have evolved to deal with this problem: It can make adjustments of relative timing that can be varied according to feedback received. Its connections are plastic; they can adjust the relative timing of the cerebellar outputs. These outputs go directly to the spinal cord, and in mammals they also go by rapidly conducting pathways to the major structures providing higher control of movement patterns.

The basic cell types and circuitry of the cerebellum are highly conserved in vertebrate phylogeny. Layering becomes more regular, and in tetrapods new inhibitory cell types appear. Also, the deep nuclei of the cerebellum, connected with the cortex, appear in tetrapods as the output structures.

However, the size and form of the cerebellum vary considerably across species. In the weakly electric fishes, there was a huge expansion of the cerebellar cortex, especially in the hyperfolded cerebellum of mormyrid fishes. In these fishes, it may be the great demands for analysis of precise timing of input signals that made this structure's circuitry so useful.

In land animals with increased locomotor abilities, and most prominently in those with fine manipulation ability, there evolved a dramatic lateral expansion and increase in surface area of the cerebellum. There was a considerable growth of the cerebellar hemispheres and the lateral deep nuclei that receive the outputs of the hemispheres.

As the cerebral hemispheres expanded in evolution, so did the cerebellum. With the evolution of higher mammals, all parts of the vastly expanded neocortex, including large areas devoted to what we refer to as cognitive functions, appear to have connections with the cerebellum via an expanded structure that provides input to it, the pons (bridge) at the base of the rostral hindbrain. Coordinated timing of the multiple representations of the external world used to anticipate changes in objects and to predict or plan actions can be as important as the actual performance. Thus, the cerebellum may play the same kind of role in our inner world as it does for our actions in the external world.

Readings

Allman, J. (2000). *Evolving brains*. New York: Scientific American Library. Chapter 5 includes a summary of evidence indicating that a cynodont (a mammal-like reptile) was ancestral to mammals.

Northcutt, R. G. (1984). Evolution of the vertebrate central nervous system: Patterns and processes. *American Zoologist, 24*, 701–716. Includes review of data that show that the most primitive living vertebrates (hagfishes and sea lampreys) have spinoreticular pathways but not other ascending lemniscal pathways.

Richardson, M. K., et al. (1997). There is no highly conserved embryonic stage in the vertebrates: Implications for current theories of evolution and development. *Anatomy and Embryology, 196*, 91–106.

General sources for topics in this chapter include books cited in previous chapters (by Butler and Hodos, by Nauta and Feirtag, and by Striedter), plus the following book:

Brodal, P. (2004). *The central nervous system: Structure and function*, 3rd ed. New York: Oxford University Press.

Cerebellar structure and function in mormyrid fishes:

Bell, C. C., & Maler, L. (2005). Central neuroanatomy of electrosensory systems in fish. In Bullock, T. H., Hopkins, C. D., & Fay, R. R., eds. *Electroreception*. Springer Handbook of Auditory Research, vol. 21. New York: Springer.

6 Some Specializations Involving Head Receptors and Brain Expansions

We are ready to add details to the brain sketch by focusing more on the forebrain and its functions. We will pay special attention to the parts that changed the most with the evolution of mammals. An enjoyable introduction to all of that will be an interlude: A brief look in this chapter at some specializations in particular species. These specializations involve several non-olfactory receptors: electroreceptors, receptors for infrared radiation, receptors for audition and for vision, and tactile receptors of the head and the hand. Studies of the brain, in the species investigated, have revealed that the specializations have resulted in some marked expansions in parts of the brain. Some of these specialized brain structures are being studied intensively in neuroscience laboratories. Some of them involve senses that humans and other mammals do not possess. Others are specializations of primates including our own species.

Electroreception and the Cerebellum

The first specialization is one we have already mentioned, involving a sense not found in mammals. Electroreception is a most useful sensory ability in some fishes. The receptors are located along the sides of the body, in the lateral line. They are innervated by cranial nerves that are not found in many species—the lateral line nerves, up to six of them. In many species with electrosensory ability, the receptors detect the weak bioelectric fields generated by other animals. In some species, electroreceptive ability is active: These animals have the ability to generate pulses of electric current around their bodies. With the receptor system, they can detect distortions in the resulting electric field caused by the presence of objects or animals near them. The usefulness of this ability is easy to understand when we consider the murky waters where some of these fish dwell and have to feed and avoid predators. A few of them, like the electric eels and the torpedo fish, can emit an electric discharge sufficiently intense to stun their prey. In most, the electroreception ability is for detection of objects and animals and for localization of them.

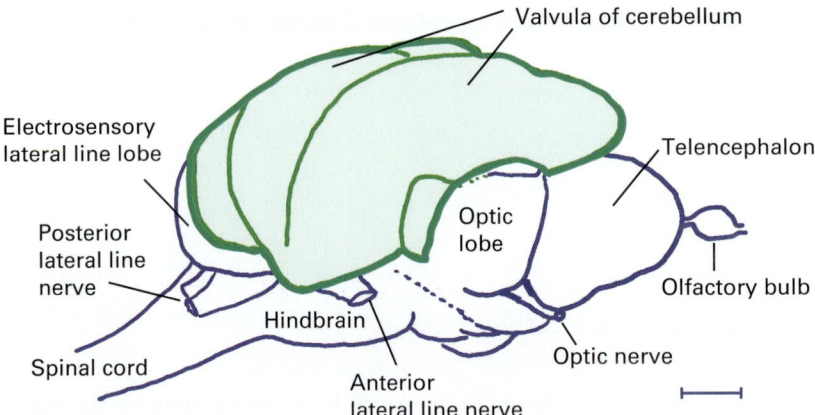

Figure 6.1
Example of mosaic evolution in teleost fish: The cerebellum—depicted in green in the drawing—especially the rostral lobe (the valvula), is enlarged in mormyrids disproportionately. The mormyrids also have very large electrosensory lateral line lobes. There is strong evidence that the cerebellum participates in the decoding of sensory signals of electroreception. The signals come from the lateral line by way of lateral line nerves (cranial nerves) to the secondary sensory neurons of the lateral line lobes of the hindbrain. The drawing depicts a lateral view of the brain of *Brienomyrus brachyistius*. Scale bar, 1 mm. Based on Friedman and Hopkins (1998).

The lateral line nerves enter the hindbrain and terminate on dorsally located secondary sensory neurons. The group of neurons receiving this input is large enough to distort the shape of the hindbrain in the fish with electroreception abilities. The electrosensory lateral line lobes—the location of secondary sensory neurons that receive inputs from the primary sensory neurons carrying inputs from electroreceptors—are particularly large in mormyrid fish. The neurons of the lateral line lobes project to other structures, including the cerebellum, where anatomists have found an even more spectacular enlargement in the brains of these fish. Part of the cerebellum called the valvula is so large that it extends over nearly the entire brain including the endbrain (figure 6.1). This structure is involved in the analysis of the electrosensory inputs, an analysis in which details of relative timing of signals and time delays are particularly important. This system has been the subject of extensive anatomical and electrophysiological studies.

Electroreception is also found in certain mammals, the monotremes, especially in the duck-billed platypus. The bill in a platypus is sensitive to both touch (pressure changes) and electric fields. These inputs come into the rostral hindbrain, not through lateral line nerves but through the fifth cranial nerve. Changes in electric fields in the water caused by a moving animal reach the platypus earlier than vibrations in the water. In the somatosensory neocortex, comparisons of these two types of input and their timing are used to compute location information useful for prey capture. The relevant areas of neocortex are enlarged in the platypus.

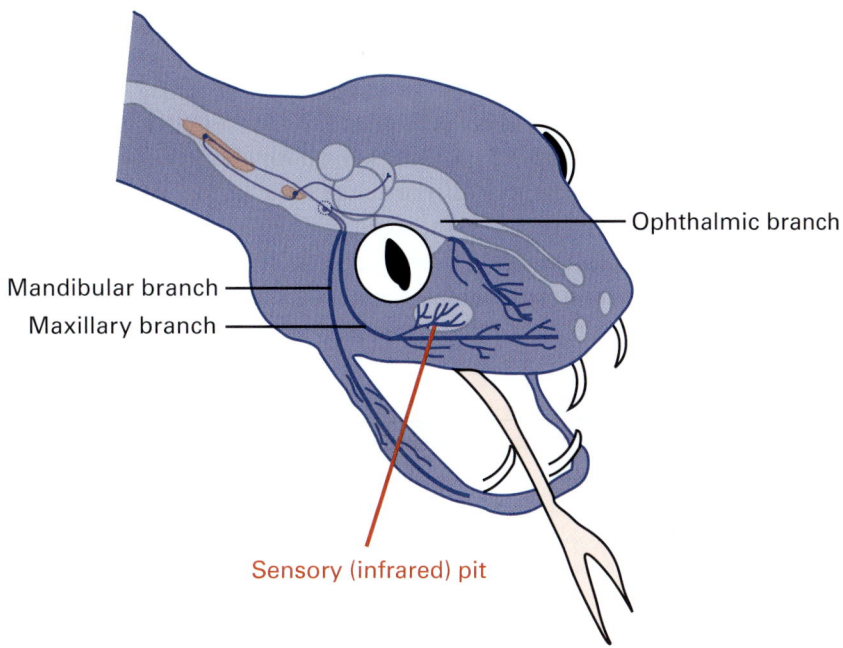

Figure 6.2
A pit viper has infrared receptors at the bottom of a depression on its face. Directional detection of infrared radiation allows the snake to localize sources of heat, like the bodies of living mammals. The figure shows the position of the CNS within the head and neck region and the three branches of cranial nerve V (the trigeminal nerve). The sensory pit is innervated by the middle (maxillary) branch of this nerve. The fibers of the nerve contact secondary sensory neurons in the trigeminal nucleus, a cell group that extends from the rostral hindbrain caudally into the most rostral portion of the spinal cord. The information from the pit organ reaches the midbrain tectum on the opposite side, triggering orienting movements. Based on Butler and Hodos (2005).

Infrared Sensors in Pit Vipers

Other species have evolved specialized head receptors not found in mammals. In vertebrates, the surface of the forward part of the head is innervated by a cranial nerve rather than by spinal nerves as in the rest of the body (see figures 5.9 and 10.17). In the snakes called the *pit vipers*, a portion of the middle branch of this cranial nerve, the fifth (the trigeminal nerve), innervates a small region in front of the eye on both sides of the face (figure 6.2). The dense innervation here is responsive to infrared radiation, as is caused by body heat. The area of this innervation is recessed to form a small pit, making the infrared detection sensitive to direction so the snake can localize the source of body heat. Thus, a rattlesnake can locate its prey independent of vision or sound. Control of the orienting movements of the snake involves the midbrain tectum, mentioned in chapter 4 because of its importance in the evolution of spatial orienting in response to visual, auditory, and somatosensory inputs (see figure 4.8).

Echolocation and the Auditory System

Rather than pulses of electric current, some mammals have evolved the ability to emit pulses of sound and to detect the reflections of the sound pulses in order to discern the nature and location of nearby objects. Best known for this echolocation ability are bats and dolphins. Correlated with this ability are enlargements in the central auditory systems of these animals. If we look down on the exposed brain of an echolocating bat or on the exposed brainstem of a dolphin (after removal of the large hemispheres), we notice that the midbrain surface structures serving the auditory sense are relatively much larger than in other mammals (figure 6.3). Electrophysiological studies have been carried out in bats,

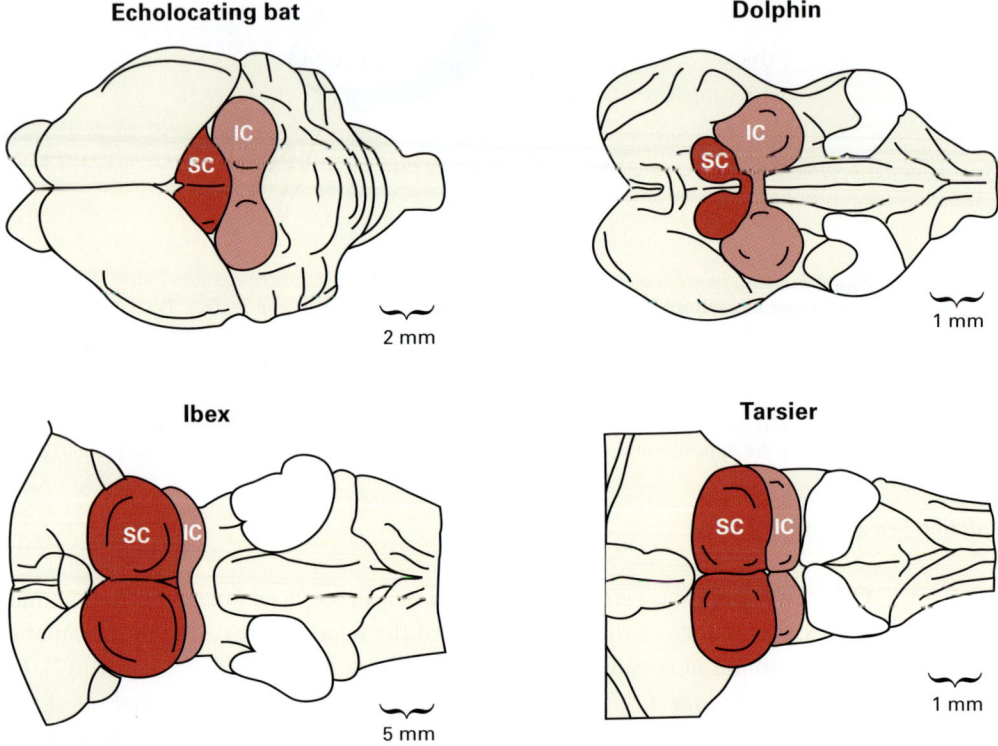

Figure 6.3
The midbrain roof viewed from the dorsal side in four mammals. Except for the brain shown at the upper left, the cerebral hemispheres have been removed to expose the midbrain. At the surface of the mammalian midbrain are the superior and inferior colliculi (*colliculi* means "little hills"). The superior colliculi receive input directly from the retina; the inferior colliculi are major structures of the auditory pathways. These dorsal views showing the midbrain surface in four mammals illustrate mosaic evolution. The two animals with echolocation abilities, a bat and a dolphin, show relatively enlarged inferior colliculi. The ibex (a wild goat) and the tarsier (a small primate of Southeast Asia) both depend on vision for detecting and avoiding predators, and the tarsier also uses vision for detection and manual capture of insects and other small prey animals. Based on Striedter (2005).

and the investigators have also reported enlarged auditory areas concerned with spatial location in the neocortex.

The Visual Systems of Primates

Humans and other primates have developed remarkable visual systems—the brain structures that process information originating in the retinae. The primates have high visual acuity and the ability to use vision to learn to recognize a very large number of objects and to discern subtle visual cues from other members of the species, and also to navigate accurately the spatial world using vision alone. With the evolution of these visual abilities was an underlying evolution of brain structures, particularly the areas of the neocortex that process visual inputs. These areas not only expanded in size but also proliferated in number, occupying not only the occipital lobe and adjacent parts of the parietal cortex but also the whole ventral portion of the temporal neocortex. See, for example, the map of the visual areas in the owl monkey neocortex illustrated in figure 6.4. In this monkey, the visual areas appear to occupy more than a third of the neocortical mantle.

Whisker Fields and Barrel Fields

Small rodents like mice and rats do not have a visual brain approaching that of primates, but their brain shows another specialization that is just as remarkable. Each of these

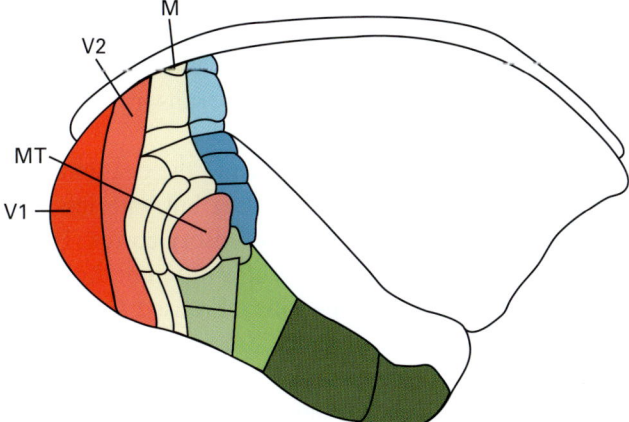

Figure 6.4
This lateral view of the right cerebral hemisphere of an owl monkey shows many distinct visual areas. The primary visual cortex is area V1. The other areas are visual association areas. In addition to their inputs from the thalamus, these areas receive visual input from V1 and/or from each other. The owl monkey depends heavily on high-acuity vision, and a relatively large proportion of its neocortex is devoted mainly to vision. M, visual area M; MT, middle temporal visual area; V2, second visual area. Based on Allman (2000).

animals has a special ability to discriminate textures and objects in the space very close to the animal's head, within the field of its facial vibrissae. In the somatosensory areas of the neocortex, inputs to each whisker are represented in the activity of a special set of neurons packed within a barrel-like arrangement. This has led neuroscientists to designate as the *barrel field* the neocortical region receiving inputs from the vibrissae, in a portion of the face area innervated by the trigeminal nerve. In figure 6.5, the barrels can be seen in tangentially cut sections of a rat neocortex stained for cell bodies with a Nissl method.

After the discovery of this specialized structural pattern in the neocortex of rat and mouse, a close inspection of the cell groups in the hindbrain receiving trigeminal nerve inputs and of the thalamic region to which these hindbrain cells project (the cells with axons that project to the neocortical barrel fields) revealed clearly separated clusters of neurons—one for each of the vibrissae.

As you may expect from this description, visual and auditory regions of the neocortex of these rodents are smaller than the somatosensory regions. In total area of cortex, only the olfactory sense has a larger representation, but the olfactory cortex in the ventral parts of each hemisphere is not neocortical in its structure, but rather a type of cortex present long before the advent of mammals.

Other Behavioral Specializations with Corresponding Enlargements of Brain Parts

The hand in raccoons, some monkeys, apes, and, especially in humans has evolved very high tactile acuity and also fine motor control. In the brains of these species, neurophysiological and neuroanatomical studies have found expansions of the regions where there are neurons involved in analyzing somatic inputs from the hands and in moving the hands and its parts. Particularly striking are expansions of somatic sensory and motor neocortex and of the cerebellar hemispheres.

In figure 6.6, we see pictures comparing the raccoon and the coatimundi, a species of similar size that also uses hands (front paws) for feeding but with less dexterity. Also shown are the outlines of the hand area of the somatosensory cortex. The hand area of the raccoon is not only enlarged but has different gyri for different digits.

The spider monkey shows a very different specialization used for some of the same functions as the hand in the raccoon and in primates. This monkey has a prehensile tail with expanded representations of that tail in sensory and motor areas of neocortex.

In this context, we should include two more functions in primates that have without doubt led to marked expansions of the neocortex. The first is a function greatly developed in many primate groups, and probably the most in humans. It is the elaborate social organization abilities of these groups, with the considerable planning and problem-solving abilities that are part of it. The portion of the brain that has expanded the most with the evolution of these abilities encompasses the prefrontal neocortical areas. Along with the

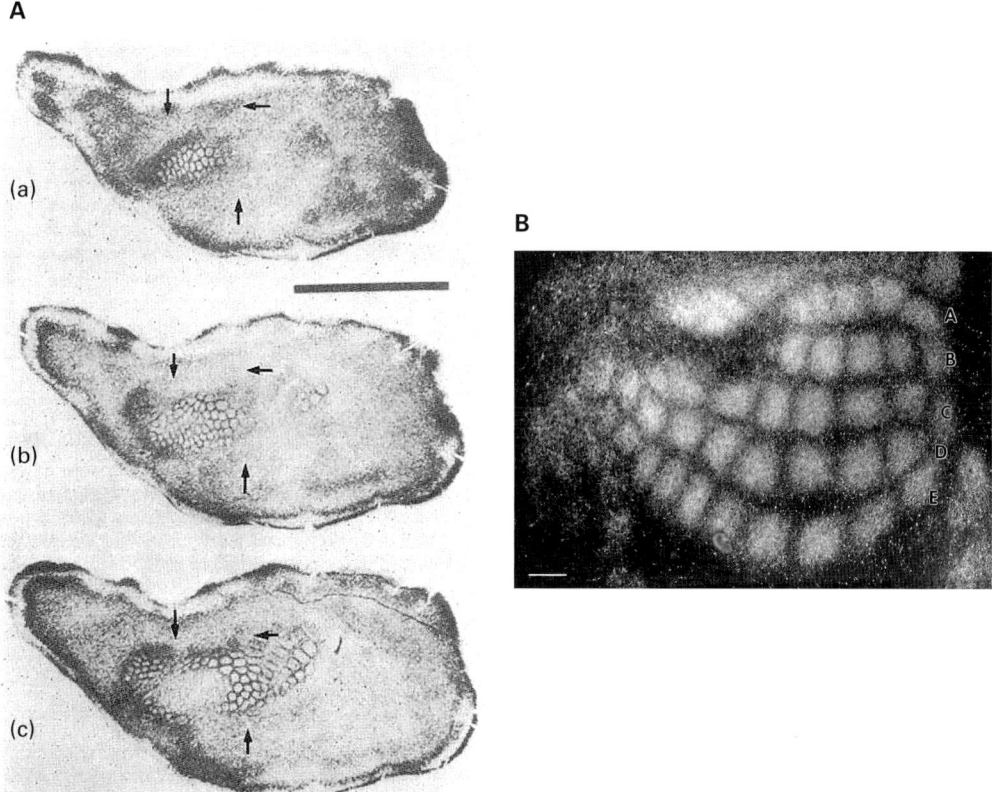

Figure 6.5
The *barrel field* representing the mystacial vibrissae in the somatosensory neocortex of rodents. (A) Three successive tangential sections of mouse hemisphere, stained for cell bodies with a Nissl method. The barrel fields are prominent where the section passes through layer 4 of the neocortex. Each barrel represents one whisker; the neurons are activated by movement of that whisker. Arrows mark positions that are exactly in line if the three sections are placed over each other. Section (a) is the most superficial, (b) is slightly deeper, and (c) is deepest. Sections are 50 μm thick, stained with methylene blue. Scale bar, 2 mm. (B) Taken with a fluorescence microscope, the photograph is from a section cut parallel to the surface of the somatosensory neocortex of a 4-day-old rat at the depth of layer 4. It shows distinct groups of thalamocortical axons, each group representing a single whisker in the barrel field. Each group of axons is surrounded by neurons, arranged like the walls of a barrel. The spatial arrangement has a topographic order matching the arrangement of the mystacial vibrissae of the face, labeled with capital letters A–E. Scale bar, 200 μm. Panel (A) from Woolsey and Van der Loos (1970); panel (B) photograph courtesy of S. Jhaveri.

Figure 6.6
The hand area of the somatosensory neocortex in two species of mammal: the raccoon on the left and the coatimundi on the right. Different regions of the body surface are marked in the drawings, and the corresponding parts of the cortex are marked similarly on the hemisphere surfaces. The raccoon has much greater sensory acuity in the hand and a correspondingly larger hand area in the neocortex. In that area of cortex, there is a separate gyrus for each digit only in the raccoon. Adapted from Allman (2000).

prefrontal functions, supporting functions have also evolved with specializations of the posterior areas of neocortex (e.g., the ability to discriminate differences in faces and to remember them).

The abilities that are probably the most unique in humans are the language abilities. These abilities have evolved with expansion of particular portions of multimodal association cortex of parietal, temporal, and frontal lobes. In addition, hemispheric specialization has also evolved, with the language functions most heavily dependent on the left hemisphere in most humans.

The human brain is not alone in having a very large prefrontal cortex. Figure 6.7 illustrates the brain of an echidna—the spiny anteater—a monotreme of Australia, which has a surprisingly large amount of its neocortex occupied by prefrontal areas. There are few studies of the behavior of echidna, so we do not have much information about behavioral specializations that could account for the expansion of the prefrontal areas. We do know that it has electrosensory receptors in its snout, but these are probably represented in somatosensory cortex as in another monotreme with electroreception, the duck-billed platypus. We can speculate that the prefrontal cortex is involved in working memory and planning for actions of the immediate future in the catching of fast-moving insects with its specialized tongue.

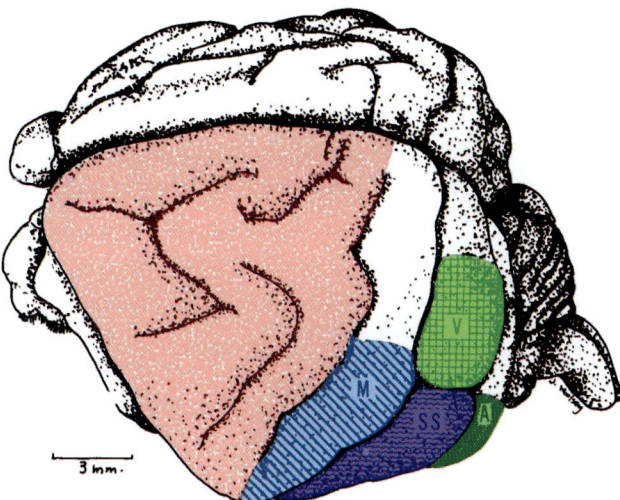

Figure 6.7
The brain of an Australian monotreme, a spiny anteater—the echidna—is depicted, with estimates of the positions of the visual, auditory, and somatosensory cortical areas outlined in light green, darker green, and dark blue, respectively. The motor cortex is shown in lighter blue. The very large prefrontal cortex, shown to have projections from the mediodorsal nucleus of the thalamus as in placental mammals, is shown in pink. A, auditory cortex; M, motor cortex; SS, somatosensory cortex; V, visual cortex. Adapted from Ulinksi (1984).

Readings

> Newman, E. A., & Hartline, P. H. (1982). The infrared "vision" of snakes. *Scientific American*, *246*, 116–127.

Electrosensory systems: See the volume on electroreception listed in chapter 5; also, see references in Striedter (2005), listed previously.

Studies of bat echolocation and the CNS:

> Suga, N. (2009). Echolocation II: Neurophysiology. In Squire, L. R., ed. *Encyclopedia of neuroscience*, pp. 801–812. New York: Elsevier.

> Covey, E. (2005). Neurobiological specializations in echolocating bats. *The Anatomical Record Part A, 287A,* 1103–1116.

Somatosensory system of rodents, showing "barrels" in neocortex: The original anatomical study is the 1970 paper listed below. The 2011 paper is an example of more recent papers, which can be found using Google Scholar:

> Li, H., & Crair, M. C. (2011). How do barrels form in somatosensory cortex? *Annals of the New York Academy of Sciences, 1225*(1), 119–129.

> Woolsey, T. A., & Van Der Loos, H. (1970). The structural organization of layer IV in the somatosensory region (SI) of mouse cerebral cortex. *Brain Research, 17,* 205–242.

Visual system of primates: introduced by Allman (2000), listed at the end of chapter 3.

An interesting historical review with excellent figures is the following paper:

> Zeigler, H. P. (2011). Wally Welker and neurobehavioral evolution: an appreciation and bibliography. *Annals of the New York Academy of Sciences, 1225*(1), 1–13.

7 The Components of the Forebrain Including the Specialty of the Mammals: The Neocortex

In the ancestral brain we have sketched, the brain of pre-mammalian animals (figure 5.4), the forebrain had evolved first under the influence of olfactory inputs and later under the influences of other senses, but the neocortex had not yet evolved from its predecessor, the simpler dorsal pallium (cortex) of the endbrain. The neocortex was the greatest innovation of the mammalian brain, and in the large mammals it has become the largest part of the entire brain, especially in humans. With this change, new pathways evolved, and soon we will be describing them. But first things first. The forebrain and each of its parts evolved because of the functions served. So our first question is: What functions? Why a forebrain? And later: Why a neocortex?

Forebrain Removal Experiments

To try to get at some answers to these questions, we will review some experiments with animals where very drastic damage to the brain was carried out. The most drastic procedure was surgical removal of the entire forebrain, or the complete surgical separation of the forebrain from more caudal brain structures. The question becomes: What can an animal do *without* a forebrain? We will review and interpret experiments on chronic decerebrate cats, rats, and pigeons. (*Decerebrate* here means "without a cerebrum"; i.e., without most of the forebrain. *Chronic* means that the animals lived a long time after the lesion was made.) Then, we will try to separate major contributions of 'tweenbrain (diencephalon), corpus striatum, limbic system (cortical and subcortical components), and neocortex.

We will see that there are large differences in decerebration effects in the three species. We will consider why this is the case, proposing both qualitative and quantitative factors.

Cat Behavior after Forebrain Removal

First, we look at cats studied in laboratories after this drastic neurosurgery. After finishing experiments published in 1958, Bard and Macht concluded that the decerebrate cat was "a purely reflex animal." The animal "fails to do any act that requires performance of a series of reflexes in a proper sequence." It is no doubt incorrect to consider normal behavior to be such a "series of reflexes," but let's consider what such a cat can actually do (according to Bard and Macht and also two later studies published in 1965). Also we will ask: Is it able to learn?

First, we notice that cats with the forebrain removed or disconnected are both anosmic and blind because the two cranial nerves carrying olfactory and visual inputs into the brain, the first and second, have been destroyed or are no longer connected to the brain remaining below the damage.

We discover that the cats fail to eat spontaneously even if long deprived of food, but they show licking and chewing responses to food placed in the mouth. To keep the cats alive, the experimenters must force-feed them. They show no hunger motivation even after a long recovery period.

They have other major problems as well. They fail to groom themselves, and they show no spontaneous sexual or other social behavior. However, sexual postures and movements can be elicited by genital stimulation.

In contrast, these drastically altered cats show good standing and sitting. They can right themselves and walk although they generally do not initiate walking on their own. Their standing and locomotion is not fully normal, as they display abnormal posture and gait, and they fail to reposition limbs that have been displaced.

The cats show rage responses (e.g., to tail pinching), but these responses have been called *sham rage* because the cats fail to bite or strike out and the rage behavior disappears very quickly after removal of the eliciting stimulus. Thus, they are not showing anything like a normal cat's rage mood; it appears to be the result of innate responses to intense stimulation.

Autonomic responses are seen, such as piloerection (fluffing up of body hairs) and other thermoregulatory responses, but only in response to extreme temperatures unless the hypothalamic portion of the 'tweenbrain is left attached to the more caudal brainstem. (In the Bard and Macht experiments, a hypothalamic island was usually spared because of need for crucial visceral regulation; e.g., water balance. The island of tissue was attached to the posterior pituitary.)

The decerebrate cats showed minimal learning. In experiments on learning abilities, conditioned eyeblink or respiratory changes could be obtained, but there was a rapid loss of the conditioned responses.

Rat Behavior Is Less Drastically Affected by Forebrain Removal

In 1964, Woods published results of his studies of chronic decerebration in rats. Comparing these results with the cat studies, it is clear that, despite some similarities, the rats were not as extremely deficient.

The effects soon after the surgery were the most similar to those reported for the cat studies, but there was more rapid recovery of righting reflexes and locomotion. The rats recovered more eating and drinking responses, but they showed no seeking of food and hence, as for the cats, the brain damage was fatal without forced feeding.

Grooming occurred in the decerebrate rats, and thus they showed a "series of reflexes," more accurately described by ethologists as fixed action patterns (instinctive movement patterns, fixed in the sense that the basic patterns are fixed in the genes). The rats were reported to show typical rodent defensive behavior, including vocalizations, escape attempts, clawing, and biting. Also seen were some orienting responses to sounds, indicating that some auditory localization in space was retained.

Why did these rats show more recovery than cats with equivalent brain damage? Is it because of less "encephalization of function"—an evolutionary shift in functional controls to the forebrain in cats compared with rats? Or was it a difference in *diaschisis* effects; that is, in disruptive effects of the lesion on the functioning of distant parts of the nervous system that have lost some of their connections? We will describe this kind of effect after reviewing decerebration effects in pigeons.

Pigeons after Forebrain Removal, Sparing the Optic Tracts

Prior to any of the decerebration experiments on cats and rats reviewed earlier, Visser and Rademacher in 1935 and 1937 had published related experiments with pigeons. They had removed the forebrain in some of the birds but had spared the optic tract pathway to the midbrain tectum.

The pigeons were reported to show recovery that apparently was faster than in the rats later studied by Woods. They showed a basic repertoire of "unlearned" reactions: Thrown into the air, they flew, avoided vertical sticks, and landed on horizontal sticks. However, they also landed on the backs of dogs and cats! They lost many more complicated behavior patterns that seemed to the investigators to depend to some degree on learning.

It is too extreme to conclude that the decerebrate pigeons lost all learned behavior and retained unlearned behavior, although in a quantitative sense this may seem to be true. Other experiments have provided evidence that even spinal animals (with spinal cord severed so it is no longer connected with the brain) can acquire conditioned leg withdrawal. Other types of conditioning have been found to occur without forebrain (e.g., conditioned eyeblink responses).

The forebrain, among its functions, appears to be important in linking together, by learning, species-typical action patterns. Much of this linking of fixed action patterns (instinctive movements) occurs normally during development.

Forebrain Function: A Conclusion

My mentor in neuroanatomy, Walle Nauta, concluded from experiments like those just reviewed that without the forebrain, vertebrate animals can maintain their stability in space, at least at a crude level. They can also maintain the stability of the internal milieu fairly well, but this is best if the hypothalamus is left intact. However, without a forebrain, an animal has very poor "stability in time." By this, Nauta meant that the behavior of such an animal is determined by current sensory inputs. It has little or no motivation-initiated behavior and little or no long-term memory. It appears to react only to the here and now and is not affected by either past or future. We have to add that a mammal without forebrain also has major losses in sensory and motor acuity.

As we proceed, we will be augmenting this picture of forebrain functions, particularly of endbrain functions.

The Problem of Species Differences in Brain Lesion Effects

If instead of removing both the neocortex and corpus striatum, an experimenter removes only the neocortex, how much difference would it make in the behavioral effects? The first, more drastic lesion, is similar to the decerebrations we have been reviewing. Compared with decerebrates, cats with sparing of the corpus striatum have shown more behavior sequences (e.g., grooming, spontaneous eating, and mating behavior). But remember that the decerebrate rats showed grooming even without the sparing of the striatum. Why the species difference?

Even with sparing of the remainder of the forebrain, if we remove only neocortex in various species, we will find clear species differences. In general, it appears that the greater the amount of neocortex we remove, the longer lasting the behavioral defects and the more likely that some of the effects will be permanent. (The defects meant here are the most noticeable deficiencies in motor and sensorimotor abilities.) Why the species differences? Does quantity of brain tissue itself matter?

It is worthwhile to pause to consider this issue, a very basic one in neurology.

The Phenomena of Diaschisis: An Answer to the Problem

The word *diaschisis* means "a splitting in two," and in neurology it refers to a schism in the brain caused by damage. More specifically, it refers to quantitative effects of lesions, effects that are greater with larger lesions. The effects are a depression of function of parts of the brain that have lost many excitatory connections because of the lesion. It can be defined most simply as deafferentation depression. *Deafferentation* means "loss of inputs," and the depression results from loss of so many excitatory connections that the cells are dominated by inhibition, or there is no longer sufficient spatial summation for them to reach threshold for firing action potentials very often.

A good example of diaschisis effects is the phenomenon of *spinal shock*. After spinal cord transection, there is an immediate loss of spinal reflexes dependent on inputs below the site of transection, even of local, segmental reflexes. The circuits for these reflexes are intact, and yet they fail to function. The reason is that too many descending connections have been lost, so the neurons in the reflex circuit have become physiologically depressed, unable to reach threshold very often with the inputs that remain. However, after some time, the spinal reflexes recover, and this recovery happens faster for animals with fewer descending connections that have been interrupted by the lesion. Thus, a frog—with fewer descending connections from brain to spinal cord—recovers much faster than a rat, and the rat recovers much faster than a cat. A human is even slower to recover spinal reflex functions after spinal transection.

Let's consider in more detail what is happening and review what underlies the recovery of function when it occurs. Sadly, it does not always occur.

In figure 7.1, neuronal connections in two different species are diagrammed. In each diagram, at the top is a brain region like the cortex, and below is a subcortical region, like the spinal cord, which receives connections from the brain region and also receives local

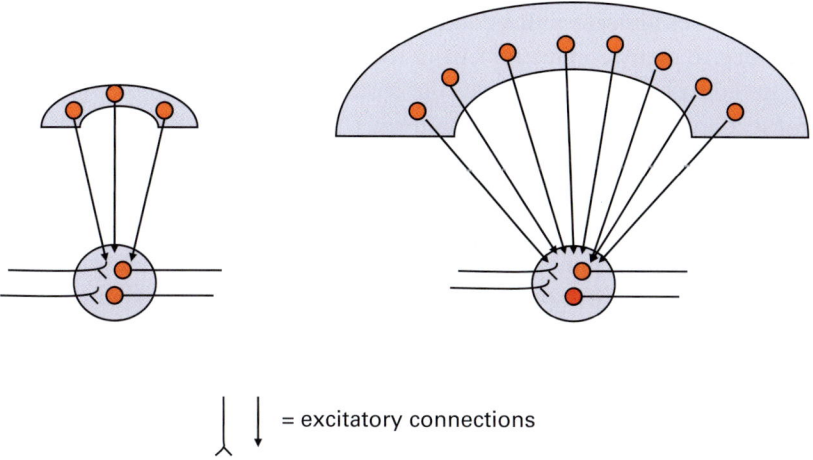

Figure 7.1
The diagram at the left is a simplified schematic of corticospinal connections in an animal with a small cortex. The right-hand diagram shows the more numerous corticospinal connections in an animal with a much larger cortex. In the two diagrams, the upper region represents the neocortex and the lower, round region represents the spinal cord. (It could also represent the brainstem.) Output neurons of the cortex project to neurons of the brainstem and spinal cord, many of which participate in reflex pathways. Reflexes are indicated by sensory inputs coming in at the left and motor neurons projecting out at the right. Each arrowhead or fiber ending in the diagram represents a large number of synapses. Interneurons that receive inputs from the cortex or from sensory inputs are omitted.

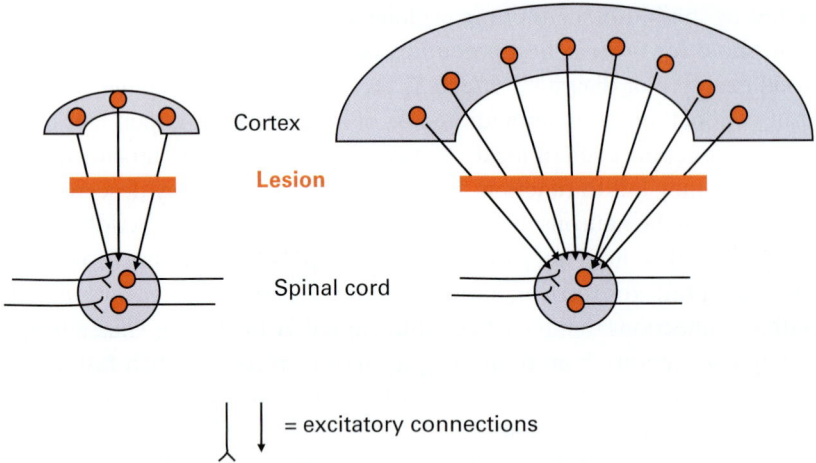

Figure 7.2
The red line in the diagrams represents a lesion that destroys the corticospinal connections in each of the two animals represented.

inputs. Excitatory connections are indicated. The connections from the brain are more numerous in the animal with the larger brain, diagrammed on the right. We can imagine that these pictures indicate reflex connections in the spinal cord, involving spinal neurons that also receive many connections from the brain.

Next, we add a lesion that removes the cortical connections to the spinal cord. It could be a transection of the corticospinal pathway (figure 7.2). The result is a loss of excitatory connections to the spinal neurons, and this loss is proportionately greater in the animal with the larger brain, shown on the right (figure 7.3). The resulting deafferentation depression is therefore much greater in that animal, the animal that has lost the greater number of connections. It is much more difficult to stimulate the spinal neurons enough for them to reach threshold for firing action potentials. This is the phenomenon of *corticospinal diaschisis* as described and analyzed by von Monakow in 1914.

How, then, do the lower circuits ever recover so that the spinal reflexes return? Two very different phenomena of CNS plasticity have been found to underlie such recovery. One is called collateral sprouting, and the other is called denervation supersensitivity. When collateral sprouting occurs, intact axons sprout new terminals that tend to replace the degenerating terminals of the transected axons. In denervation supersensitivity, the denervated neurons compensate for the loss of balance in their inputs by increasing the number of receptors for the neurotransmitters of the degenerating axons. Thereby, these cells respond more strongly to the intact axons that remain.

When these recovery processes are happening in humans with spinal injury, they often go too far. The overcompensation results in hypersensitive reflexes that interfere

The Components of the Forebrain

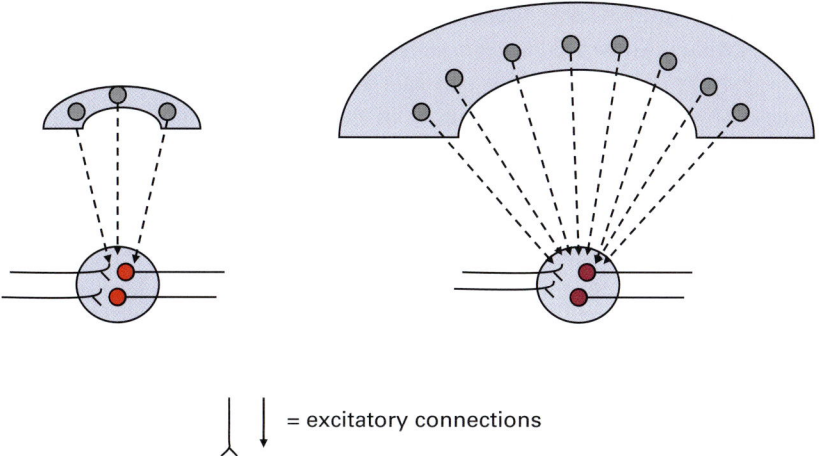

Figure 7.3
The diaschisis caused by the lesions in the two species is represented. With the loss of many excitatory corticospinal connections, the functional pathways at the spinal level become depressed. Even simple reflexes fail to function. This denervation depression is greater in the animal with the larger neocortex because many more excitatory connections have been removed. This phenomenon was called corticospinal diaschisis by von Monokow in 1914.

with normal function. They have what is known as reflex spasticity, which requires treatment.

What Can We Conclude about Interpreting Effects of Brain Lesions?

The diaschisis phenomena show that brain lesions cannot be considered as mere subtractions of functions of the removed tissue. Quantitative effects of lesions and the resulting losses of connections have to be considered. Without understanding the neuroanatomy, including the relative sizes of the pathways destroyed, we cannot comprehend some of the consequences of brain damage. The anatomical changes have physiological consequences of the kind we have reviewed, and these consequences may affect the functioning of the whole organism.

Now we return to an overview of forebrain components.

Role of the Corpus Striatum

The corpus striatum, as we have seen in our sketch of an ancestral brain, is the largest subcortical structure of the forebrain. In the long history of the evolving forebrain, it has served as an important link between sensory analyzers and the motor control systems of the brain.

Experimental studies have found that the striatum of mammals is crucial for habit formation—what is known as *procedural learning* or *implicit learning*. This was introduced briefly in chapter 4. Later, we will consider further how this originated long before the appearance of mammals, in the evolution of the early chordates.

The study by Bard and Macht described earlier indicated that one aspect of habit formation that the corpus striatum may be important for is the (learned) linking of components of instinctive behavior patterns during development, resulting in performance of these movements in a particular order. Paul Leyhausen, an ethologist, described the acquisition of these kinds of links in his behavioral studies of kittens and cats. The greater the number of distinct innate behavior patterns that make up an adult behavior pattern in an animal, the greater the importance of this type of learning. (This may be the situation for our own species more than most.)

Anatomical connections of the striatum fit these functions, and these will be described later. Another issue, important in human medical care, is the explanation for human pathologies of the striatum and its connections.

The Rostral End of the Brainstem Located between the Hemispheres ('Tweenbrain)

Components of the 'tweenbrain (diencephalon) seem to be present even before there were cerebral hemispheres. Such components have been identified in amphioxus, the tiny worm-like chordate that we have discussed (see figures 3.3 and 3.4).

Information from light sensors entered the neural tube here, even in the early chordates. In figure 7.4, these inputs are added to the sketch of the pre-mammalian brain. The optic inputs play a special role in organizing the daily rhythm of activity, which is so different during day and night, and also varies in many animals when light levels vary during the daytime. Hypothalamic and epithalamic afferents from the lateral eyes and from the dorsally located parietal eye (pineal eye) entrain the endogenous circadian activity rhythm in vertebrates. This rhythm, under the control of the biological clock, and these visual inputs, modulate the behavior of the entire system. This kind of widespread modulation is a special role of these portions of the 'tweenbrain, particularly of the hypothalamus.

Other system-wide modulations occur via the bloodstream, through secretions of the endocrine system's diencephalic component, the pituitary organ, part of which is an extension of the hypothalamus. Widespread modulation of the CNS itself is accomplished by the hypothalamus (called by Sherrington "the head ganglion of the autonomic nervous system"), which controls major motivational states in addition to sleep and waking.

Hypothalamic gating functions were without doubt very important in the evolution of the other diencephalic components (subthalamus and thalamus). Specific evidence for such hypothalamic gating or modulation of neuronal activity in the thalamus and subthalamus has been found. First, it is indicated by studies of axonal projections. Second, it has been demonstrated by electrophysiological studies of the biting attack behavior of

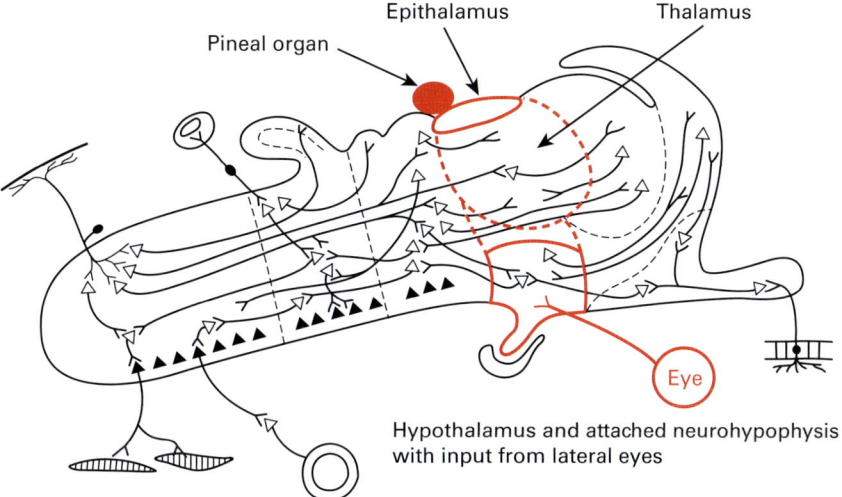

Figure 7.4
In this simplified sketch of the pre-mammalian brain (a copy of figure 5.4), note the *diencephalon*, or 'tween-brain, outlined in red dashed and solid lines (see definitions on figure 7.5). From dorsal to ventral, it is comprised of the epithalamus, thalamus, subthalamus, and hypothalamus. The neural portion of the pituitary, or neurohypophysis, is attached to the hypothalamus. The retina of each lateral eye is an outgrowth of the diencephalon; it sends axons to the hypothalamus and to other places in the diencephalon and midbrain. Dorsally, the epithalamus also receives input from light detectors. In many vertebrates, this input is direct, from a single eye on top of the head—the parietal eye or pineal eye. Much of this input goes to the nearby pineal organ. In vertebrates without a parietal eye, light affects the pineal organ by a more indirect pathway.

cats. In studies at Yale University in the 1960s and 1970s, John Flynn and his students used electrical stimulation of the lateral hypothalamus to cause an altered state: the predator mood. Such stimulation also causes changes in neuronal firing patterns in thalamus and neocortex (as well as elsewhere) (see also chapters 17 and 25).

In the evolution of sensory pathways ascending to the endbrain, this hypothalamic function may have been a major reason why a connection to thalamus, rather than more direct pathways, was almost always retained.

Limbic System of the Forebrain

When we discuss the hypothalamus and various parts of the forebrain most closely connected with it, we are discussing the *limbic system*. The word *limbic* means "at the border or the fringe." The cortical (pallial) components of the limbic endbrain in mammals are located at the fringes of the cortical expanse. We know the importance of the hypothalamus in the control of autonomic functions and in control of motivational states. The endbrain components modulate these functions and are associated with feelings and their emotional expression.

Studies of neuronal connections (hodology) can give us more ideas about limbic system origins and functions. In the sketch of an ancestral pre-mammalian brain (figure 7.5), we see the close connections with the olfactory system. In fact, an older name for the limbic endbrain structures was the rhinencephalon, or nose brain. Connection patterns also indicate that with the limbic system, we must include the basal forebrain structures that are positioned as a forward extension of the hypothalamus. Also grouped with these structures is the ventral portion of the corpus striatum. In the evolution of the endbrain, the ventral striatum was, in all likelihood, the oldest part of the striatum.

Omitted from figure 7.5 is a pathway that also must have evolved very early, conveying input from taste systems that signaled results of approach and intake of good food. Pathways that signal reward, positive or negative, have been the subject of many laboratory studies of learning in animals. Axons that have probably served this function since very early in chordate evolution reach the ventral striatum from the midbrain and hypothalamus. Less studied are the neural pathways that carry taste information to this system (see chapter 18).

Olfactory System Origins of Major Functions of the Limbic Endbrain

An ancient function of olfaction was most probably in categorizing things encountered as attractive or abhorrent, good or bad, by remembering the odors of those things. It added an initial valence to object identification. This function has continued, aided by learning through associative links in the limbic endbrain, particularly in the subcortical structure known as the amygdala. The amygdala is an outpost of the corpus striatum (that has joined with pallial components) located in the posterolateral hemisphere. As you might expect for a structure that is important in object identification, the amygdala receives inputs from neocortical representations of the other sensory systems (visual, somatosensory, auditory) as well as inputs directly from the olfactory bulbs.

Another function of olfaction that influenced evolution of the limbic endbrain in a major way was the distinguishing of good places from bad places in the local environment, and the retention of such place identities. This involved retention of a spatial map of the local environment, with memory of good or bad places as distinct from the identity of any specific objects in those spaces. Place learning evolved with a special dependence on the limbic endbrain structures that have expanded most in the posterior hemispheres, near the visual and auditory analyzers of the neocortex. These limbic endbrain structures are called the hippocampal formation in mammals. It evolved from the medial pallium of ancestral species. The hippocampus was named for its appearance in dissections of the human brain, where its shape reminded anatomists of a seahorse.

The Components of the Forebrain

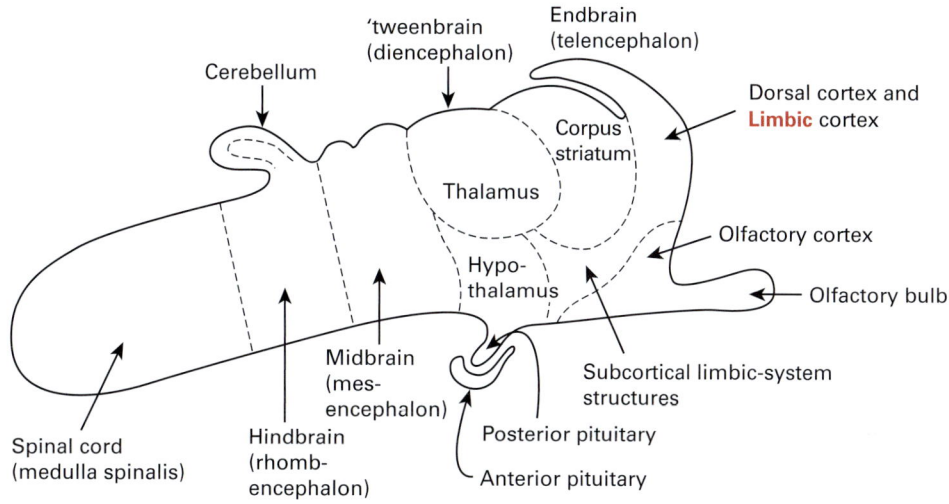

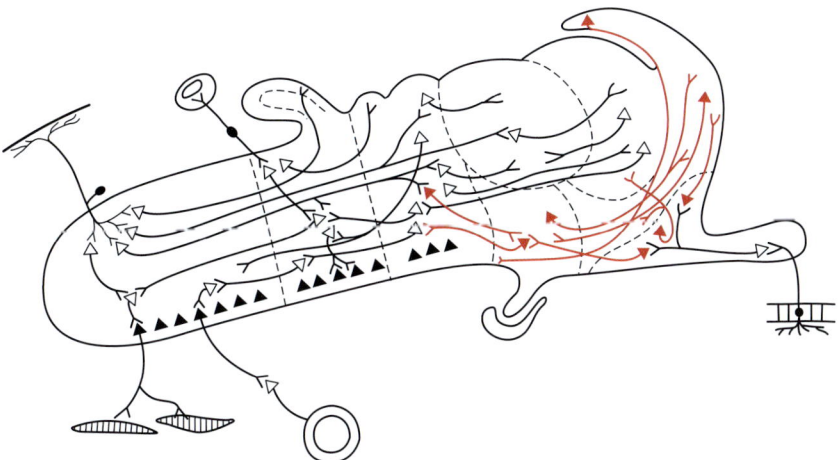

Figure 7.5
In this diagrammatic sketch of a pre-mammalian brain, representative connections of what became the *limbic system* in mammals are shown in red. Note the ties to the olfactory system and the close connections to and from the hypothalamus.

Overview of the Roles of Corpus Striatum and Limbic Forebrain

Before we finish with this coverage of our pre-mammalian heritage, I will summarize briefly the roles of these two major components of the forebrain from an evolutionary perspective.

The corpus striatum can be viewed as an ancient link between olfaction and the fundamental action patterns of vertebrates. Other sensory systems reached this anterior part of the brain because of the special adaptive value of plastic connections to structures enabling action, so the striatum became a central structure for habit formation (implicit learning). This included the learned linking of innate movement patterns for more rapid achievement of specific goals.

The limbic forebrain has evolved with some expansions in mammals because of particular functions crucial for survival and reproduction. On the one hand, there was the very ancient importance of olfactory input analysis for the ability to remember what is attractive and what is repelling and to distinguish and remember good and bad places. On the other hand, there was the crucial survival value of centralized control of visceral functions and of basic motivations, with an increasing importance of positive and negative signals (rewards and punishments) as feedback for learning to reach goals and to rank diverse goals.

In general, these forebrain components have exerted higher control of actions—what to do and when to do it, throughout vertebrate evolution, and with increasing roles for learning.

But higher vertebrates developed even more sophisticated higher control with the expansion of the telencephalic non-olfactory cortex; the part of this cortex that expanded the most is the neocortex. It is to this topic that we now turn.

Neocortex, the Grand Innovation of the Mammals: Appearance of New Pathways

At this point, we start with a schematic sketch of the mammalian brain. In figure 7.6, you can see the sketch of an ancestral pre-mammalian brain with a simple outline of what became superimposed on it. In the figure 7.7, you see the outlines of the schematic mammalian brain, in side view, with its major subdivisions; in the lower part of the figure is a top view of the embryonic mammalian brain with the spinothalamic tract sketched in. (To review the spinothalamic tract, see chapter 5.)

In the early mammals, as the neocortex evolved its unique six-layered structure and expanded in area far beyond the size of the dorsal cortex of the ancestral forms, its connections increased as well. Not only did the spinothalamic tract increase in numbers of direct connections with the thalamus, but also another route enlarged, a more rapidly conducting pathway in mammals. Anatomists have sometimes called this pathway the *neolemniscus*—the new ribbon of fibers from the spinal cord. However, a pathway like it,

The Components of the Forebrain

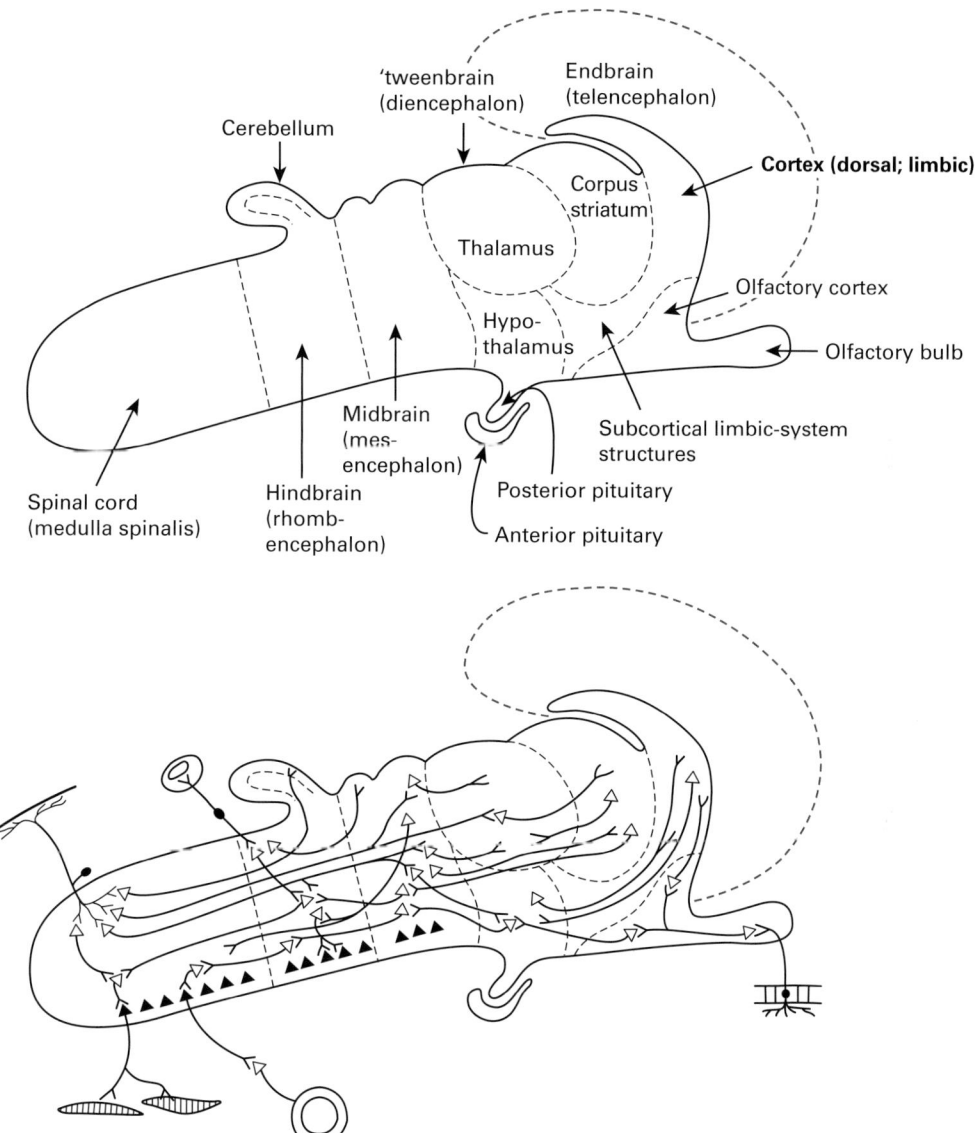

Figure 7.6
The addition of neocortex to the ancestral pre-mammalian brain by expansion and differentiation of part of the dorsal cortex is indicated with dashed lines on the illustration.

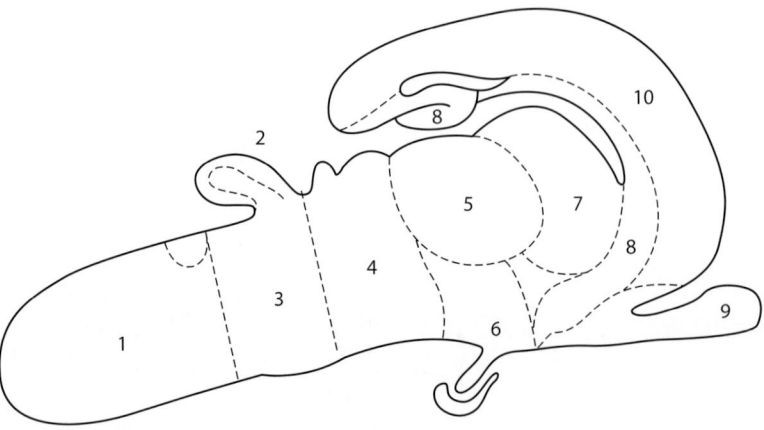

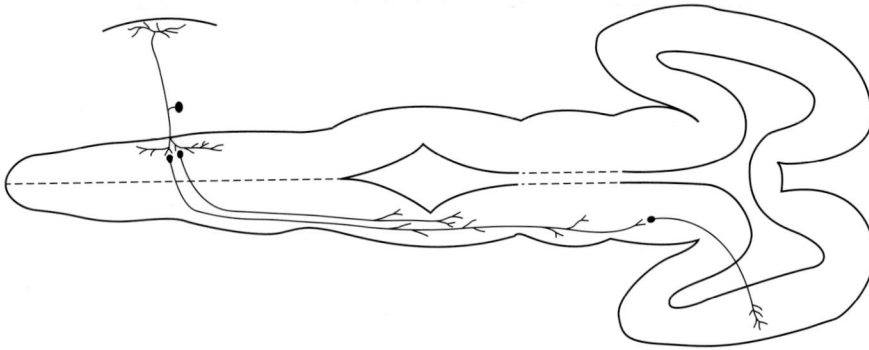

Figure 7.7
Diagrams of the mammalian brain. (Top) Schematic side view with numbering of 10 major brain subdivisions. (Bottom) Dorsal view of embryonic brain showing sketch of the spinothalamic tract. 1, spinal cord; 2, cerebellum; 3, remainder of hindbrain; 4, midbrain; 5, thalamus (and epithalamus); 6, hypothalamus (and subthalamus); 7, corpus striatum; 8, limbic endbrain structures; 9, olfactory bulb; 10, neocortex.

at least a small one, has been found not only in all amniotes but also in anuran amphibians. In this pathway, the primary sensory axons ascend toward the brain in the dorsal columns without a synapse until they reach the rostral-most end of the spinal cord (figure 7.8). There they terminate in the nuclei of the dorsal columns. These nuclei include a more medial cell group for the axons from the more caudal parts of the body (including the legs and feet) and a more lateral one for the axons from the more rostral parts (including the arms and hands). (Note: Many neuroanatomy texts designate the location of these cell groups as the caudal-most part of the dorsal hindbrain.)

From the neurons of the dorsal column nuclei, axons ascend all the way to the 'tweenbrain in a bundle termed the medial lemniscus, which terminates in the thalamus. The thalamus is the "inner room" or chamber that serves as the portal to the neocortex. Axons of the thalamic cells that receive somatosensory inputs from the spinothalamic and medial lemniscus axons project to the somatosensory areas of the neocortex (figure 7.8A). The dorsal column–medial lemniscus pathway is most important for discrimination of fine touch and pressure and for high degrees of spatial resolution—abilities that increased greatly in many mammals.

This "neo"-lemniscus, consisting of the dorsal column–medial lemniscus axons, is a more rapid route from body surface to neocortex than the spinoreticular or the spinothalamic tracts. There are some collateral projections of the ascending axons, for example to the deep layers of the midbrain tectum (superior colliculus) at least in some species.

The neocortex of modern mammals gives rise to descending axons that enable the cortical output, unlike the output of the older components of the endbrain, to bypass the midbrain and go directly to the hindbrain and to the spinal cord. Axons from output neurons of the somatosensory neocortex (pyramidal cells of the fifth layer) can be traced all the way to the dorsal regions of the spinal cord where secondary sensory cells are located (figure 7.8B). This is part of the corticospinal pathway or tract. It is called the pyramidal tract where it is found at the ventral surface of the hindbrain because of the shape of its cross section. On its caudally directed route to the cord, this pathway has various collateral connections (see figure 7.8). The somatosensory area known as the *motor cortex* has even more direct control of the motor system via descending axons from its large pyramidal cells of layer 5 directly to some motor neurons of the ventral cord and, more frequently, to the pool of spinal interneurons that project to the motor neurons. This pathway is an important part of the corticospinal tract. Again, on the way to the cord, there are collateral connections at all levels of the CNS.

Birds likewise have evolved structures of the forebrain that receive somatosensory projections, including a somatosensory Wulst—part of the hyperpallium. They also have a motor Wulst that projects axons directly to the spinal cord. The Wulst also has a visual area that is particularly large in raptors. The Wulst is considered to be a homolog of the mammalian neocortex.

132   Chapter 7

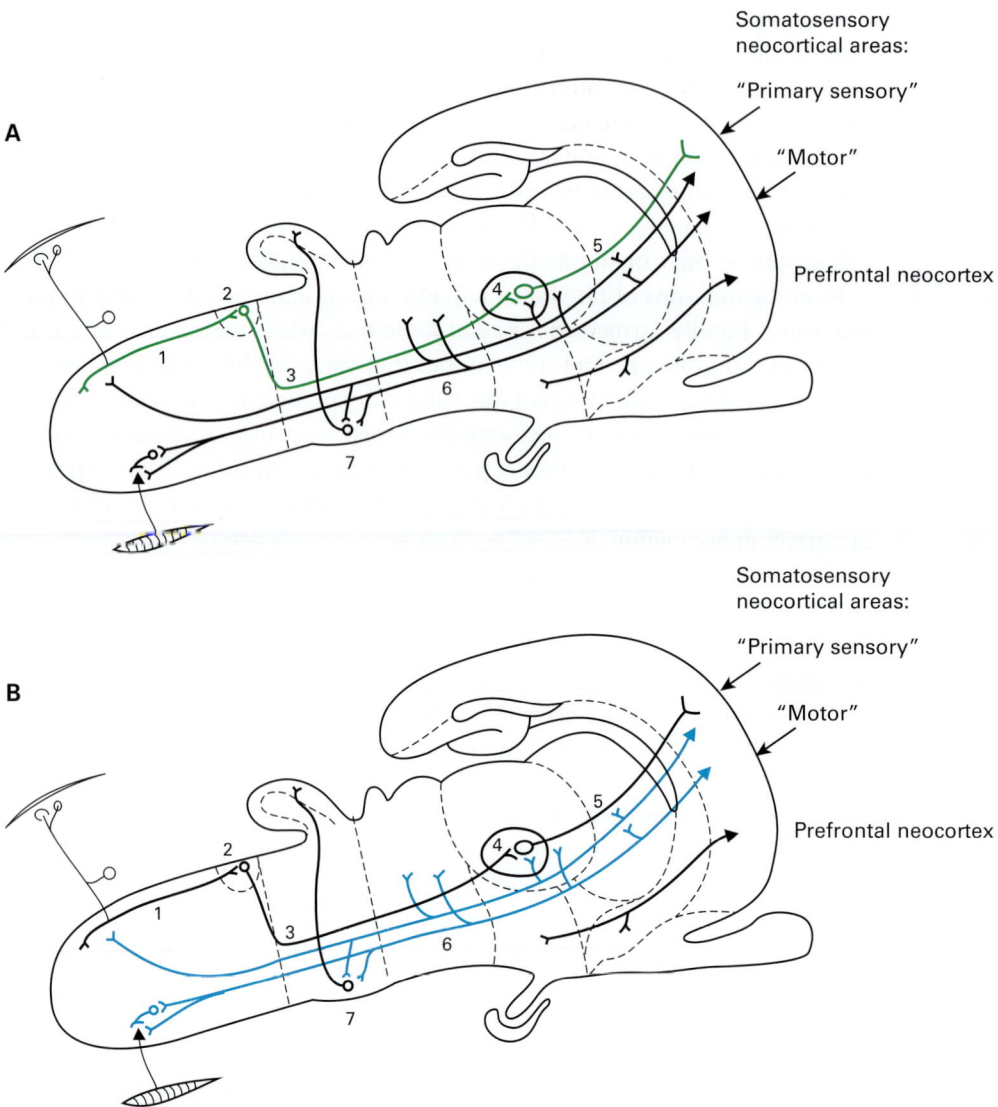

Figure 7.8
(A) In this diagrammatic sketch of the mammalian brain, cell bodies of representative neurons of the dorsal column–medial lemniscus pathway and its connection to the neocortex are shown in color. Students are encouraged to trace the entire route from the body surface through two synapses to the somatosensory cortex. 1, dorsal column; 2, nuclei of the dorsal column (nuc. Gracilis and nuc. cuneatus); 3, medial lemniscus; 4, ventrobasal nucleus of thalamus (nuc. ventralis posterior); 5, thalamocortical axons in the *internal capsule*; 6, corticofugal axons, including corticospinal components (called *pyramidal tract* in hindbrain below the pons); 7, pons. (B) In this diagrammatic sketch of the mammalian brain, cell bodies of representative corticospinal neurons are shown in color. The more caudal cell represents a layer-5 neuron of the somatosensory neocortex; the more rostral one represents a layer-5 neuron of the motor cortex. Numbers: same as in panel (A). Remember that in premammalian brains (see figure 7.4), which had a much smaller endbrain, the only routes between the forebrain and motor neurons required a link in the midbrain.

Because of these connections of neocortex to all levels of the CNS, this cortex, as it enlarged, exerted more and more influence on the behavior of the animal. In fact, it became more and more dominant. All of its regions project to the corpus striatum, to the thalamus and the subthalamus, to various parts of the midbrain, and to the cerebellum via cells of the pontine gray matter. Nothing escapes its influence, although its influence on some functions is weaker and mostly indirect, especially the autonomic nervous system functions.

The descriptions of the neolemniscus and the corticospinal pathways omit an aspect that we can see when we trace the same routes on a dorsal view sketch of the CNS (figure 7.9). Each of these pathways decussates at the caudal end of the hindbrain. It will be helpful to the reader to identify each portion of each axonal pathway as they are labeled

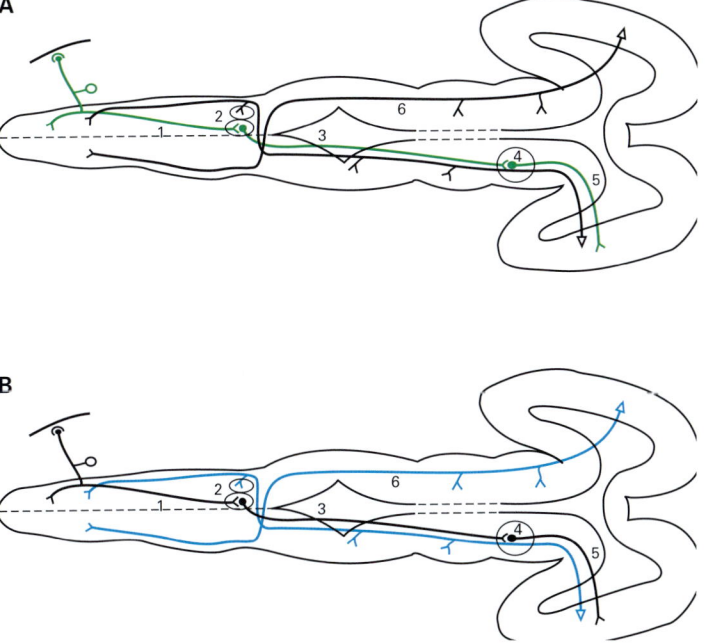

Figure 7.9
(A) In this dorsal view of the embryonic mammalian brain, cell bodies of representative neurons of the dorsal column–medial lemniscus pathway and its connection to the somatosensory neocortex are shown in color. Students are encouraged to trace the entire route from the body surface through two synapses to the somatosensory cortex. Note the decussation of the medial lemniscus at the caudal end of the hindbrain. (These crossing axons are called the internal arcuate fibers.) Numbers: same as in figure 7.8A. (B) In this dorsal view of the embryonic mammalian brain, representative pyramidal cells of the neocortex that send their axons to the spinal cord are shown with the cell bodies in color. The cell in the right hemisphere is in the somatosensory cortex. The cell in the left hemisphere is in the motor cortex. Note the decussation of the pyramidal tract at the caudal end of the hindbrain. Students are encouraged to trace the entire route from neocortex to the terminations in the spinal cord. No synapses occur before the terminations in the cord.

in figures 7.8 and 7.9, and to learn, in either view of the tracts, where the decussations occur.

But What Does It Do? Why Did Neocortex Evolve?

Precision of perception and motor control: In evolution, there was an increasing specialization of thalamic cell groups and the corresponding neocortical areas. Most of these cell groups that were connected to sensory pathways, and their neocortical areas, became dominated by a single sensory modality. Thus, early in mammalian evolution, the so-called primary sensory cortical areas evolved, adding to the animals' sensory and motor acuity and control. This evolution made possible more and more detailed perception of the sensory world of objects, and more and more accurate and delicate movements. The increased precision affected fixed action patterns (instinctive actions) as well as learned actions. Perceptual separation of objects from the background of stimuli became better. Experimental studies of effects of damage to neocortical areas or to pathways leading from them, such as the corticospinal tract, have supported these claims.

Here are a few examples of experimental results: Ablation of primary visual cortex in the tree shrew reduces visual acuity and also disturbs discrimination of embedded patterns. In hamsters, ablation of this cortical region spares orienting abilities, but animals more easily lose track of a moving stimulus. Pyramidal tract lesions in hamsters disrupt seed-shelling ability, which is an innate ability in this animal that requires great dexterity.

More significantly, in my view, neocortical expansion is associated with an increasing *ability of mammals to anticipate stimuli* and an increasing *ability to plan actions* in advance. This was described in chapter 4 in the context of an evolutionary expansion of the forebrain (the third such expansion, depicted in figure 4.10). Anticipation depends on abilities to form images of objects and scenes, using an internal model of the external world that is built up in the memory, a kind of simulation. Such imaging depends on posterior neocortex. Planning abilities also use the internal model; these abilities depend more on the frontal neocortical areas.

Taking Stock

Let's take stock of where we are at this point in learning the anatomy of the central nervous system and where we will go next. We have a basic sketch of the entire system, albeit somewhat rudimentary. Next, we will get more involved in learning about the basic structural divisions we have seen in this sketch. We will be aided by studies of CNS evolution and by studies of development in mammals, including humans.

Readings

Chronic decerebration experiments with cats and rats:

> Bard, P., & Macht, M. B. (1958). The behaviour of chronically decerebrate cats. In Wolstenholme, G. E. W., & O'Connor, C. M., eds. *Neurological basis of behavior: In commemoration of Sir Charles Sherrington, 1857–1952* (Ciba Foundation Symposium). London: Churchill.
>
> Emmers, R., Chun, R. W., & Wang, G. H. (1965). Behavior and reflexes of chronic thalamic cats. *Archives Italiennes de Biologie, 103*, 178–193.
>
> Markel, E., & Adam, G. (1965). Elementary temporary connection in the mesencephalic cat. *Acta physiologica Academiae Scientiarum Hungaricae, 26*, 81–87.
>
> Woods, J. W. (1964). Behavior of chronic decerebrate rats. *Journal of Neurophysiology 27*, 635–644.

Studies of the effects on visually guided behavior of forebrain ablation in pigeons were published in German by J. A. Visser and G. G. J. Rademacher in 1935 and 1937 in the journal *Archives néerlandaises de physiologie de l'homme et des animaux*, volumes 20 and 22 (cited by G. E Schneider, PhD thesis, MIT, 1966). It is easier to find more recent studies of effects of smaller lesions.

The phenomenon of *corticospinal diaschisis* was described and analyzed by von Monakow in 1914:

> Von Monakow, C. (1914). *'Diaschisis,' localization in the cerebrum and functional impairment by cortical loci*. Wiesbaden: J. F. Bergmann. Pages 26–34 were translated from German by G. Harris in Pribram, K. H., ed. (1969). *Brain and behavior 1: Mood, states and mind*. New York: Penguin Books.

See also: Feeney, D. M., & Baron, J. C. (1986). Diaschisis. *Stroke, 17*, 817–830.

Recovery from brain injury is a very complex issue involving more than recovery from diaschisis effects. For interesting reviews of the issues, see the following papers:

> Sabel, B. A. (1999). Restoration of vision I: Neurobiological mechanisms of restoration and plasticity after brain damage — A review. *Restorative Neurology and Neuroscience, 15*, 177–200.
>
> Seitz, R. J., Azari, N. P., Knorr, U., Binkofski, F., Herzog, H., & Freund, H.-J. (1999). The role of diaschisis in stroke recovery. *Stroke, 30*, 1844–1850.

Evidence for diaschisis effects in the human neocortex have been found in recent years in studies of brain-damaged patients using functional magnetic resonance imaging (fMRI).

Paul Leyhausen described the acquisition of links between different fixed motor patterns (instinctive movements) in his behavioral studies of kittens and cats:

> Leyhausen, P. (1965). On the function of the relative hierarchy of moods (as exemplified by the phylogenetic and ontogenetic development of prey-catching in carnivores. English translation in Lorenz, K., & Leyhausen, P. (1973). *Motivation of human and animal behavior: An ethological view,* pp. 144–241. New York: Van Nostrand Reinhold.

"Anticipation depends on abilities to form images of objects and scenes, using an internal model of the external world that is built up in the memory, a kind of simulation": This statement near the end of the chapter is supported by articles in the special issue of *Brain Research* published in March 2007.

Neuropsychological and neurophysiologic studies of the role of prefrontal cortical areas in planning, choice of actions, and the guiding of attention to reach goals is discussed in the following review:

> Miller, E. K., & Cohen, J. D. (2001). An integrative theory of prefrontal cortical function. *Annual Review of Neuroscience, 24,* 167–202.

IV DEVELOPMENT AND DIFFERENTIATION: SPINAL LEVEL

8 The Neural Tube Forms in the Embryo, and CNS Development Begins

The central nervous system of chordates evolved as a fairly integrated single system, and similarly its various parts develop together although maturation is earlier, sometimes much earlier, in some regions than in others. In the development of a mammal, the processes that form the neural tube begin in the cervical cord region, only slightly ahead of the caudally and rostrally adjacent regions. Lagging behind the rest are the most caudal and most rostral ends of the tube. It is noteworthy that within the brain, structures that are the last to mature tend to be those that appeared most recently in evolution.

Now we will have a look at how the CNS first forms in the embryo, focusing on the spinal level. Then we will survey the outcome by inspecting the structure of the adult spinal cord. Before moving on to the hindbrain, we will sketch nervous system structures that innervate the smooth muscles and glands of the viscera.

Before There Is Any CNS

The chordate embryo is at the *neurula* stage when the neural tube starts to form. Before it reaches this stage, it must pass through several steps. The initial fertilized egg cell divides, and those cells divide, and this continues. A *morula* has formed—a clump of cells resembling a raspberry. The ball of cells develops a fluid-filled interior as the embryo becomes a *blastula*, before a remarkable change begins to transform it. This change is called gastrulation, involving an invagination that changes the ball into a donut shape (figure 8.1). The wall of the "hole in the donut" will develop into the alimentary tract.

The patterned changes of gastrulation mold the basic form of the embryo. These and later transformations are brought about by basic cellular activities. These activities have been summarized by Lewis Wolpert as the following four cellular events: contractions, changes in adhesion (via changing expression of cell adhesion molecules, or CAMs), cell movement, and growth. As we sketch the progressive changes that form the neural tube, we will see examples of these event-causing processes. Already in gastrulation we see them. In figure 8.1, note the invagination beginning in the fifth drawing, together with a

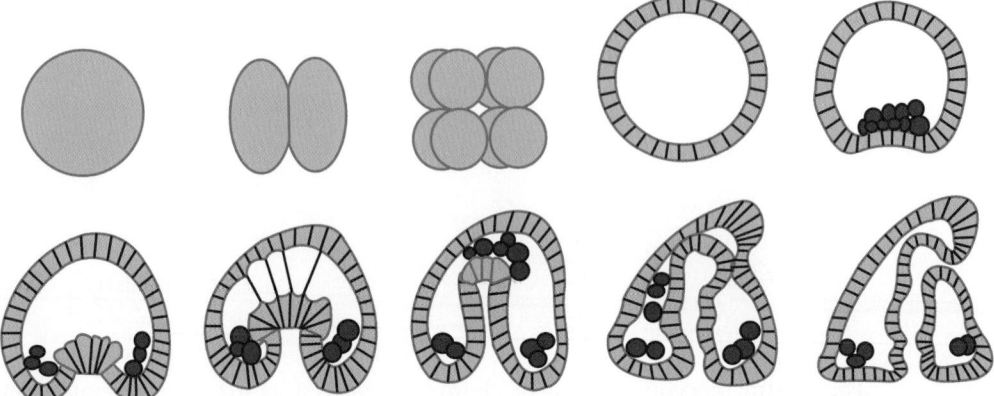

Figure 8.1
Pictures from embryogenesis: From fertilized egg (upper left) to morula (upper row, third drawing) to blastula (upper row, fourth and fifth drawings) to gastrula (last drawing, lower right). Note the role of filopodia (lower row, second drawing) in gastrulation, as described in the text. Based on Wolpert (1991).

movement of some cells into the interior of the blastula. This invagination appears to be caused by contractions that are greater at the outside edge of the cells in one region than at the inside. Then, from the invaginated cells, there is a growth of slender processes, or filopodia, extending across the blastula to the cells of the opposite side. The tips of the filopodia adhere to the cells there, and then they contract, pulling the indentation of the surface farther within, resulting soon in a meeting of the two walls of the embryo. With further changes in adhesion, an opening appears in the side opposite to the original invagination, and the hole in the donut is completed.

One can easily imagine similar events resulting in other movements of the embryo's cells (e.g., as cells of the future skeleton find their way along the inner wall of the embryo) (figure 8.2).

The Onset of a Nervous System

The entire vertebrate nervous system forms from the ectoderm—the outermost layer of the embryo that also forms the skin. What happens at the spinal level can be described with the help of the drawings shown in figure 8.3. The drawings depict transverse sections of the dorsal surface region of the developing embryo. Neurulation—the formation of the neural tube (which becomes the CNS) and the peripheral nervous system (PNS)— begins with a thickening of the ectodermal cells along the midline, above the underlying notochord. The notochord is the rod-like cartilage that forms along the back of the embryo from the mesoderm, the middle layer of cells that will give rise to bones and muscle. The thickened region of the ectoderm is called the neural plate. Soon a groove in the neural plate forms at the midline, as the plate begins to invaginate toward the

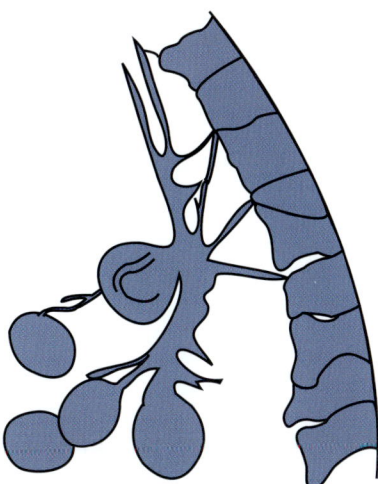

Figure 8.2
Cells that are precursors of the skeleton find their way along the inner wall of the embryo. Filopodial extensions extend to the inner surface of the wall where they may adhere. Contractions of the filopodia pull the skeletal precursors toward their destinations. Based on Wolpert (1991).

notochord. This invagination is not unlike the process of gastrulation, but there are striking differences. Not only does it occur along most of the back of the embryo, but also it results in the formation of a tube of ectodermal cells as this region is pinched off. The pinching off leaves the neural tube located below the embryonic skin and above the notochord. Some cells do not become part of the tube nor do they remain with the surface cells (figure 8.3; see also figure 8.6 later in text). They are the neural crest cells that become the cells of the PNS. Neural crest cells at the spinal level also migrate farther away from the neural tube to become the core of the adrenal gland and melanocytes of the skin. They also become Schwann cells—the myelin-forming cells of the peripheral nerves.

This phenomenon of closure of the neural tube begins in the middle portions of the tube, the part that will become the cervical spinal cord (figure 8.4). This begins in human embryos at about 21 days postconception. Then the neural tube "zips up" in both rostral and caudal directions, so that 3 days later the tube remains unclosed only at the caudal end and in the midbrain and forebrain regions (figure 8.5).

Students with Internet connections can find helpful animations of neurulation with neural tube closure.

Molecules from the Notochord Induce the Formation of the Nervous System

That the region of the notochord actually causes the nervous system to develop by its proximity to the overlying ectoderm is a concept developed by Hans Spemann, with evidence from experiments done by Hilde Proescholdt for her doctoral dissertation.[1]

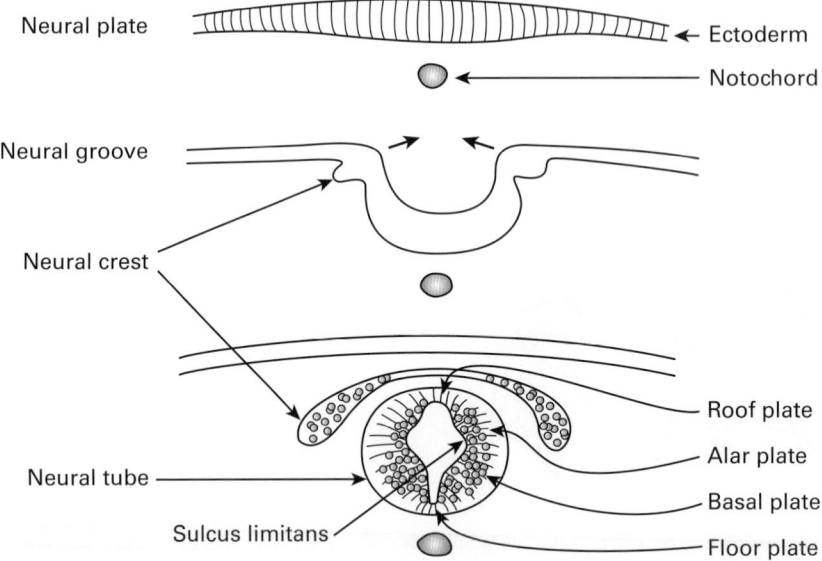

Figure 8.3
Formation and closure of the neural tube are illustrated in sketches of sections made at right angles to the surface ectoderm of the embryo. The ectoderm overlying the notochord thickens, forming the neural plate. The thickened plate invaginates, forming a groove along the future back of the embryo. The groove forms a tube that loses its continuity with the surface, forming the neural tube. Cells at the edge of the neural groove, the neural crest cells, remain outside the tube. As the walls of the neural tube begin to thicken, the major parts of the tube are identifiable as indicated.

Late in the twentieth century, specific inducing molecules were discovered. The protein sonic hedgehog (SHH) diffuses from the notochord, inducing neurulation, and later functioning as a *ventralizing factor*, influencing the differentiation of cells in the ventral portion of the spinal cord (basal plate cells, including motor neurons; see figure 8.3).

Molecular influences on the dorsal horn region of the cord have also been discovered. Dorsalizing factors are secreted by the ectoderm adjacent to the neural plate. Bone morphogenetic proteins BMP-4 and BMP-7 play such a role here. (They play other roles elsewhere in the embryo, as the name implies.)

The Neural Crest Gives Rise to the PNS

The neural crest cells were mentioned earlier and are shown in figure 8.6. These cells not only form the dorsal root ganglia, the primary sensory neurons that carry somatic sensory information into the cord, but also some of them migrate farther in the ventral direction and form the ganglia of the autonomic nervous system. The position of paravertebral ganglia of the sympathetic nervous system is shown in the figure, as is the position of unpaired prevertebral ganglia. Some neural crest cells migrate farther, forming the gland

The Neural Tube Forms in the Embryo

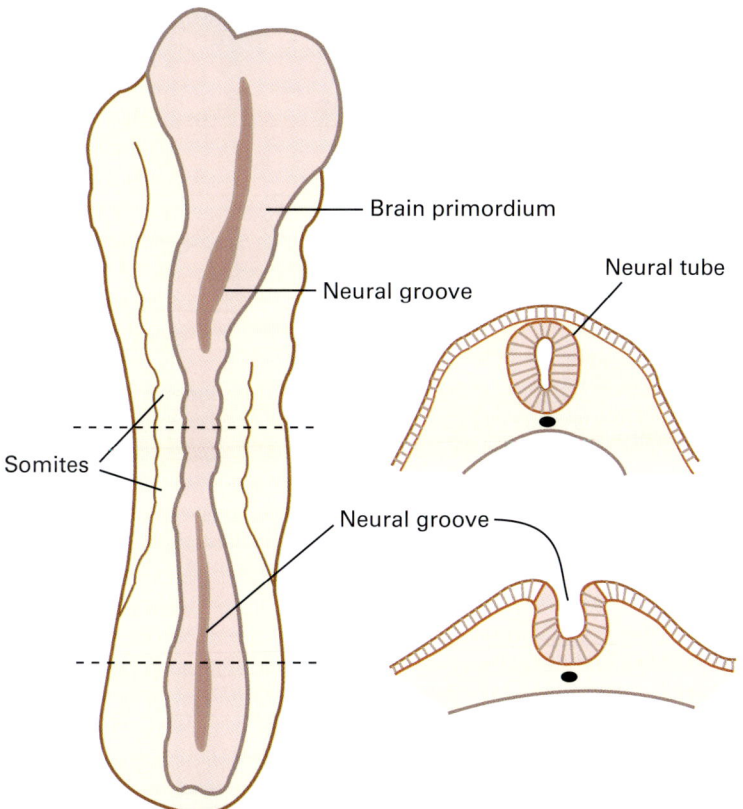

Figure 8.4
In this illustration of the entire embryo, the closure of the neural tube has begun in the cervical (neck) region. The neural groove remains visible at more rostral and more caudal levels. At the right, sections taken at the levels indicated are shown (see figure 8.3). Segmentation of the embryo is indicated by the appearance of somites in the region of tube closure. Based on Brodal (2004).

cells of the adrenal medulla. Crest cells that migrate in a dorsolateral direction distribute throughout the skin and form the melanocytes.

The cells that form the myelin sheath around the axons of dorsal root ganglia cells, the Schwann cells, also migrate from the neural crest. Within the head region, these cells also form other tissues. Some of them migrate caudally to form ganglia of the parasympathetic nervous system (see chapter 9).

Cell Proliferation in the Early Neural Tube

Meanwhile, the walls of the one-cell-thick neural tube are thickening because of continued cell proliferation. Mitoses occur when the cell nuclei are adjacent to the ventricle. Thus, the

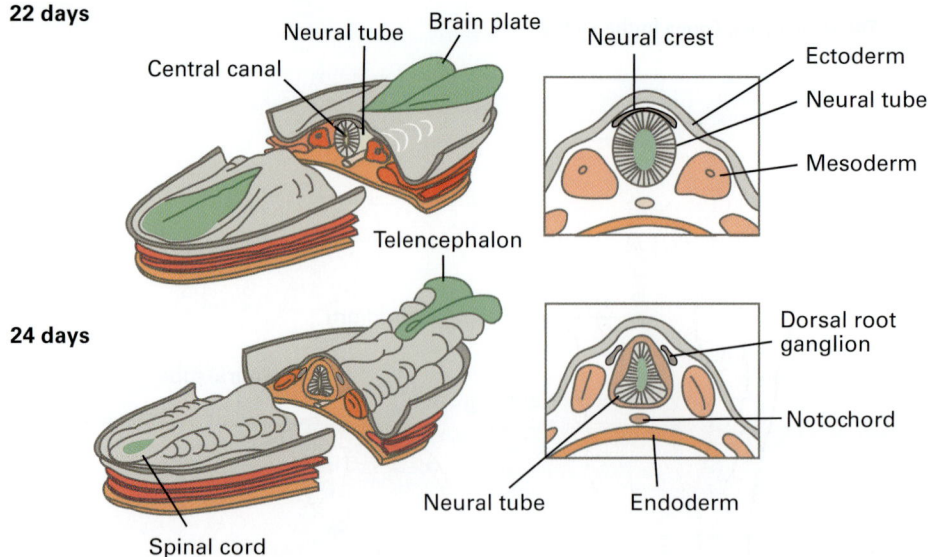

Figure 8.5
At 22 days postconception, the human neural tube has closed slightly further than in figure 8.4. At 24 days, the closure of the neural tube is nearly complete caudally, whereas at the rostral end it remains open at the level of the midbrain and forebrain. Based on Breedlove, Rosenweig, and Watson (2007).

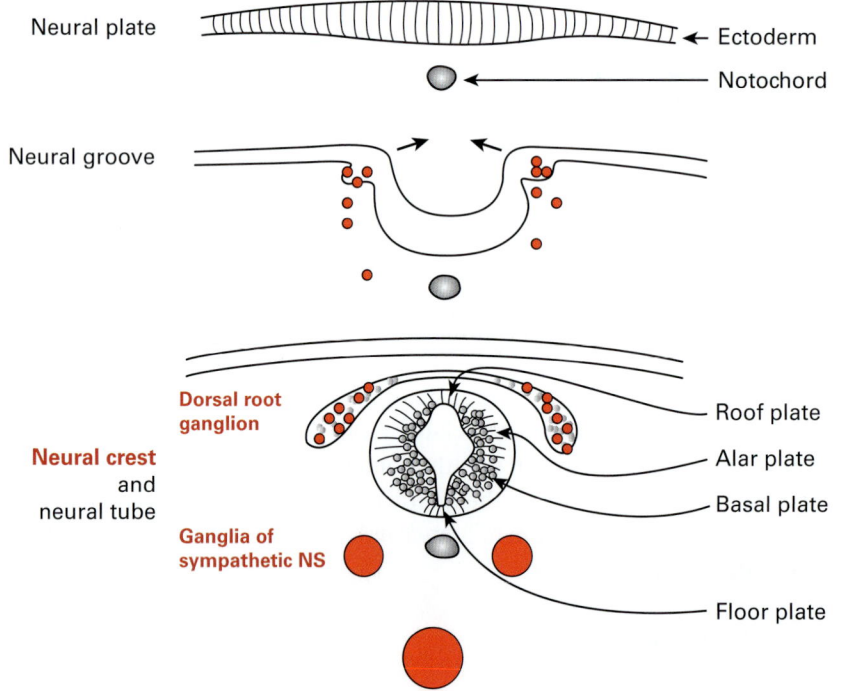

Figure 8.6
Illustration of neural tube closure as in figure 8.3, with more information about the fate of neural crest cells, shown in red. These cells not only form the dorsal root ganglion cells but they also migrate more ventrally to form the paravertebral and prevertebral ganglia of the sympathetic nervous system. Not indicated are the crest cells that move into the ectoderm and become melanocytes (pigment cells of the skin).

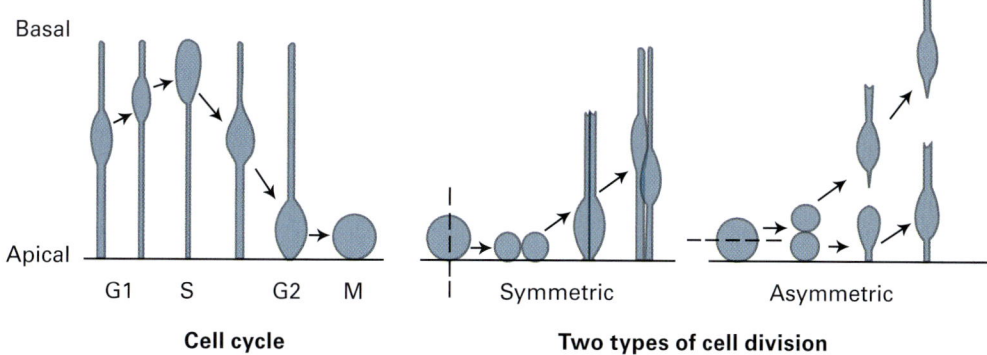

Figure 8.7
Cell proliferation is illustrated in diagrammatic depictions of cells as they would appear in sections cut at a right angle to the ventricular surface. In each part of the figure, changes in the cells are shown from a state pictured at the left, with later steps shown to the right (arrows indicate time progression). (Left) The cell cycle is illustrated. The cell body moves away from the ventricular surface (G1), then it replicates its DNA (synthesis, S); the cell body moves back toward the ventricular surface (G2), where it begins the process of mitosis (M). (Center) When mitosis is symmetric, both daughter cells remain attached to the ventricular surface and reenter the cell cycle: They extend primitive processes, along which the cell body moves in the G1 phase. (Right) When mitosis is asymmetric with respect to the ventricular surface, the daughter cell located farther from the ventricle becomes postmitotic and migrates away, while the cell next to the ventricle reenters the mitotic cycle. Based on McConnell (1995).

ventricular layer is called the *matrix layer* (the mother layer) of the developing spinal cord. At first there is only symmetric cell division, in which the two daughter cells both remain in a proliferative state. Then some of them start to divide asymmetrically, where one daughter cell becomes postmitotic and migrates away from the ventricular layer (figure 8.7). This migration, at least in the earliest period, is actually a movement of the cell nucleus through the elongated cell toward the pial surface—a nuclear translocation.

For a long time, the cells maintain contact with both the outer, pial surface and the ventricular surface, and thus it can be said that the neural tube is still one cell thick although the cell bodies are located at varying distances from the ventricle. It is called a *pseudostratified epithelium*. The cell nucleus moves within the elongated cell, first during the steps of cell division (figure 8.7, left). Later, when a cell becomes postmitotic, the nucleus also moves within the elongated cell, but farther, to its position in the more mature cord. Cajal's drawings of neurons in the developing CNS show many elongated cells (e.g., see figures 8.8 and 8.9). Many of them in the young mammals he was studying were, with little doubt, migrating by nuclear translocation.

Diversity in Neuronal Migration

There are other types, or modes, of migration of embryonic neurons, but in every mode the basic mechanism may be the same: A process is extended from the cell—it may

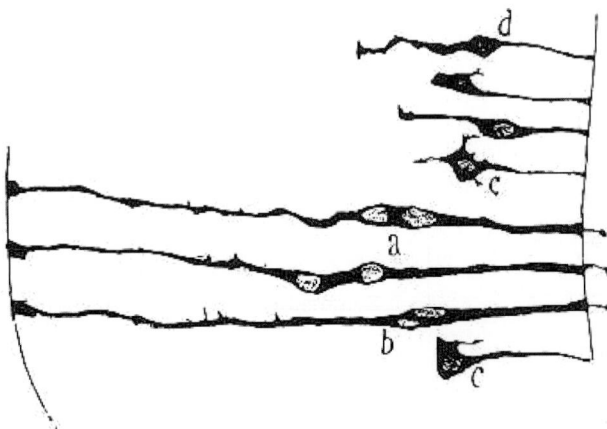

Figure 8.8
During the early stages of the thickening of the neural tube, the walls of the tube remain a neuroepithelium, the ectodermal tissue of the neural plate. All cells abut the ventricular surface, and many also reach the pial surface. The neural tube is still one cell thick, until some cells lose their attachment to the ventricular surface. At this stage, illustrated by Ramón y Cajal from a Golgi-stained preparation, it is not clear which cells will become neurons and which may be destined to become glial cells. a, b, binucleated cells; c, cells that appear to be derived from a cell division next to the ventricle; d, cell with an elongating outer process. From Ramón y Cajal (1995).

be very long or it may be short—and the nucleus moves along this process (it *translocates*); the trailing cell process may or may not be pulled back into the cell body right away. The guidance of the growing extension along which the nucleus moves involves various molecular and mechanical factors. When the neurons are already elongated, attached to the pial surface and extending to or near the ventricular surface, the migration is simply a nuclear translocation. In some locations in the CNS, as in the embryonic neocortex, where the migration distances are longer, the process grows along the surface of an elongated glial cell already in place, a radial glial cell. Thus, the embryonic neuron moves along this surface to its destination. In other locations, the migrating neurons are guided by other substrate factors that provide the necessary adhesive properties.

These three modes or types of cell migration and their similarity in cellular mechanisms have not always been appreciated. Thus, when evidence for radial glial guidance in embryonic neocortex was reported (by Rakic in 1972), it was presented as a theory that superseded the nuclear translocation propounded earlier (by Morest in 1970). Other types of cell movement were neglected as a controversy developed. The phenomenon of radial glial guidance was strongly supported by careful anatomical reconstructions from electron microscopic images and eventually from direct observations of cell movements in explant cultures. Then, convincing evidence for the reality of nuclear translocation without glial cell guidance in the developing chick midbrain was found and presented at a meeting of

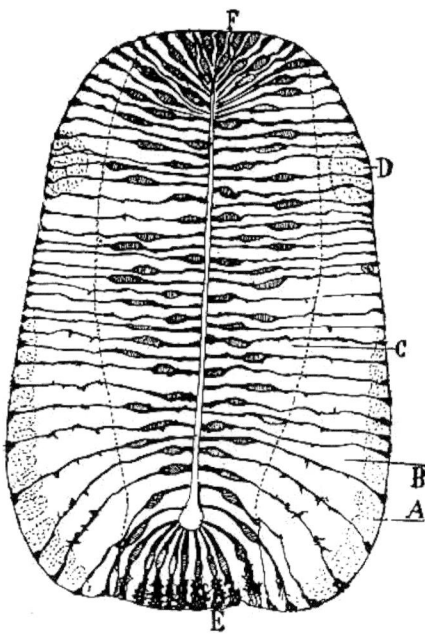

Figure 8.9
In this Cajal drawing of a chick spinal cord at 3 days of incubation, all the cells are attached to both inner and outer limiting membranes (ventricular and pial surfaces), and therefore the walls of the neural tube have the appearance of a neuroepithelium. However, other, more differentiated cells are present. They are either not stained or were not drawn by Cajal. Figure 8.10 indicates that some of the cells of this type may be future neurons. A, region of the ventral funiculus; D, the oval bundle of His; E, neuroepithelial cask or wedge (floor plate); F, neuroepithelium of the dorsal median sulcus (roof plate). From Ramón y Cajal (1995).

neuroanatomists called the Cajal Club and later published (by Domesick and Morest in 1977), so neuroscientists realized that there was more than one way for neurons to migrate.

The strongest evidence came from studies of Golgi-stained brains of chicks at various stages of development in the egg. The investigators focused on one cell type in the chick's optic tectum, the "shepherd's crook cell," easily identified because of the consistent bend near the beginning of the axon. In figure 8.10, the drawing at the far right depicts a maturing shepherd's crook cell on day 13 of incubation, showing the initial part of its axon near the cell body, not far below the tectal surface. The drawings to the left show the same cell type, identified by the appearance of the axon, at successively earlier stages. The axon begins elongating from the primitive process that extends toward the pia at a stage when the cell body is still near the ventricular surface: see the drawing at the far left depicting the tectum at day 7 or 8 of incubation. During a 6-day period, the bulge indicating the position of the cell body translocates toward the pial surface, and the cell retracts much of its primitive process on the ventricular side.

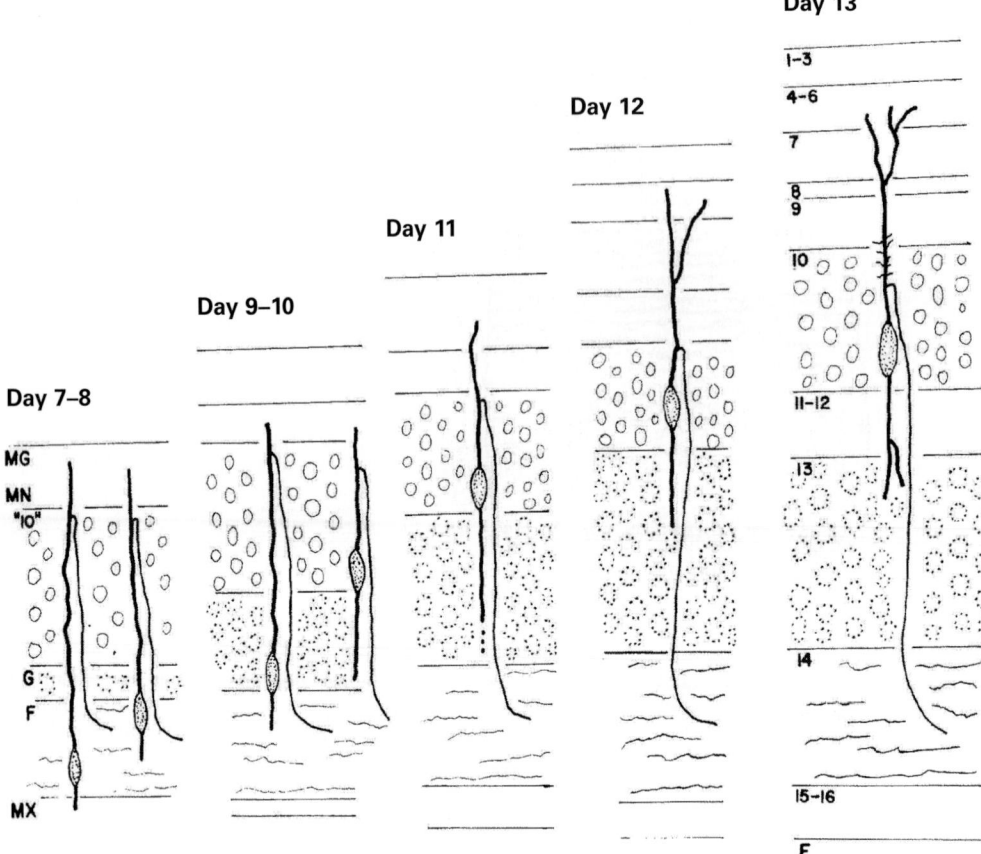

Figure 8.10
Summary drawings based on Golgi stained neurons of the developing chick midbrain in frontal sections from a study by V. Domesick and D. K. Morest (1977). At the far right is a "shepherd's crook cell" in the oldest embryo. This is a neuron that can be identified by an axon that has a peculiar bend (crook) near the cell body. In this drawing, the same kind of neuron is identified at progressively earlier stages of development, using the shepherd's crook as the property that makes this cell type identifiable. When the axon begins elongating beyond the crook shape, the cell body is near the ventricle—it has not yet moved very far up the primitive process toward the midbrain surface. At later stages, as the axon continues to elongate, the cell body is closer to the point of axon origin. When it gets closer to its mature location, the primitive process extending toward the ventricle has detached from the ventricular surface and appears to be retracting. This study provided strong evidence for migration by nuclear translocation and showed that elongated cells of the neuroepithelium may be future neurons. E, ependymal zone; F, fiber layer; G, ganglion cell layer; MG, marginal zone; MN, mantle zone; MX, matrix zone; "10," presumptive layer 10. From Domesick and Morest (1977).

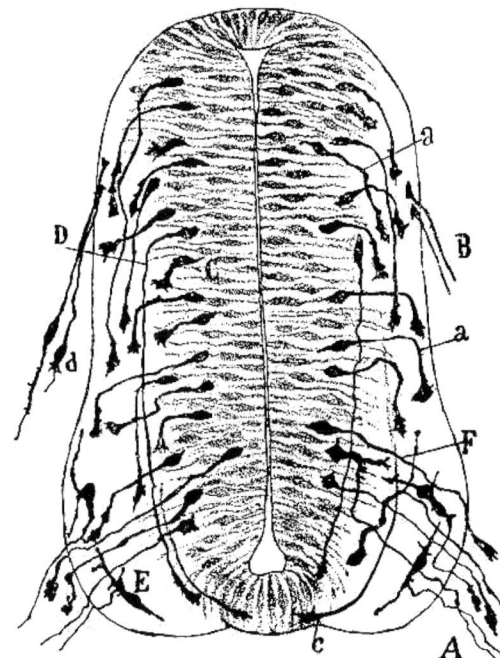

Figure 8.11
Drawing of a Golgi-stained chick spinal cord on day 3 of incubation. This is the same age as in figure 8.9, but here Cajal has drawn cells showing early stages of differentiation into neurons. Note the growth cones at the tips of elongating axons. The axons growing away from the cord ventrally are axons of motor neurons. A, ventral root; B, dorsal root; C, young neuroblast; D, commissural neuroblast; E, motor neurons that have already formed dendrites; F, motor neuron armed with a growth cone; a, neuroblast with an inner process; c, commissural growth cone. From Ramón y Cajal (1995).

Differentiation of the Neurons Begins

The series of drawings of a developing shepherd's crook cell illustrates how a neuron can begin to differentiate some time before its migration is complete. The same thing can be found in studies of the developing spinal cord. In figure 8.9, we saw the radially arranged cells of the day-3 chick spinal cord. In figure 8.11, Cajal has drawn a spinal cord from another chick of the same age, but different cells have been stained—cells with actively growing axons. We see the growth cones—structures that Cajal recognized as the tips of actively elongating axons—on axons of motor neurons exiting the ventral part of the cord, also on axons extending from dorsal root ganglion cells outside the cord, and on axons of interneurons within the cord, including axons starting across the midline in the floor plate.

We will return to axonal growth and other aspects of neuronal differentiation later (see box 8.1 for a review of neuronal development terms and events). First we will inspect the

> **Box 8.1**
>
> **Review**
>
> **Some neuronal development terms to be familiar with:**
>
> - Ectoderm (vs. mesoderm) (students should also look up the term *endoderm*—the third embryonic layer)
> - Ventricular layer (= matrix layer)[2]
> - Modes of migration
> - Radial glia (sometimes referred to as radial astrocytes)
> - Sulcus limitans, separating alar and basal plates
> - Neural crest
> - Dorsal and ventral roots and rootlets
>
> **Major events in development of the CNS:**
>
> 1. Neurulation and formation of neural tube
> 2. Proliferation
> 3. Migration
> 4. Differentiation, with growth of axons and dendrites

adult spinal cord, and then the development and structure of the brain vesicles—hindbrain, midbrain, and forebrain—with notes on their evolution.

Notes

1. Hans Spemann won a Nobel Prize for this in 1935. Sadly, Hilde Proescholdt, who had become Mrs. Hilde Mangold, had died at the age of 26 in a kitchen accident. She never received the credit she deserved for her important discovery.

2. Note on neuroembryological terms: When neurons migrate away from the ventricular layer, they accumulate in the intermediate layer (also called the mantle layer). Between the intermediate layer and the pia, there is a layer of fibers known as the marginal layer or zonal layer with few or no neurons. The cells that remain at the ventricular surface in the mature animal, lining the ventricles, are called ependymal cells; the lining is known as the ependyma (meaning "an upper garment" in Greek).

Readings

For drawings of Santiago Ramón y Cajal, see books listed in chapter 2.

Basic cellular activities summarized as four cellular events:

> Wolpert, L. (1991). *The triumph of the embryo*. New York: Oxford University Press.

Discovery of identity of molecules involved in inducing the CNS, as well as ventralizing and dorsalizing factors: Reviewed in more recent editions of the above book, in which Wolpert has been joined by additional authors.

> Wolpert, L., Jessell, T., Lawrence, P., Meyerowitz, E., Robertson, E., & Smith, J. (2007). *Principles of development*, 3rd ed. New York: Oxford University Press.

Migration by nuclear translocation:

> Domesick, V. B., & Morest, D. K. (1977). Migration and differentiation of shepherd's crook cells in the optic tectum of the chick embryo. *Neuroscience, 2* 477–491.
>
> Morest, D. K. (1970). A study of neurogenesis in the forebrain of opossum pouch young. *Anatomy and Embryology, 130*, 265–305.

Migration by radial glial guidance in embryonic neocortex:

> Rakic, P. (1972). Mode of cell migration to the superficial layers of fetal monkey neocortex. *Journal of Comparative Neurology, 145*, 61–83.

Growth cones will be a topic we encounter again in chapter 13. For a review of cellular mechanisms of growth cones, see the following article:

> Lowery, L. A., & Van Vactor, D. (2009). The trip of the tip: Understanding the growth cone machinery. *Nature Reviews Molecular Cell Biology, 10*, 332–343.

9 The Lower Levels of Background Support: Spinal Cord and the Innervation of the Viscera

As development of the neural tube continues, migrating neurons reach their destinations and differentiate into their adult shapes with growth of dendrites and axons. Some interesting findings of basic research on axonal growth will be reviewed in chapter 13. In this chapter, we will picture the mature CNS below the head region, extending from the cervical spinal cord of the neck at the rostral (upper) end, through thoracic and lumbar levels to the sacral and coccygeal spinal cord of the tail region. As the spinal cord grows and matures, so does the peripheral nervous system, invading every region of the enlarging organism.

In figure 9.1, we illustrate the adult spinal cord of a human. Later, we will check the anatomy of several important sensory channels. We will also take note of some major descending pathways of brain control over spinal activities. Also, we will indicate how the spinal nervous system is organized locally, independent of these higher influences.

Major Features of Cord Structure

In the shape of the mature spinal cord, can you see what has happened to the embryonic cord? We examined a section of the embryonic cord, sketched in figure 8.6. What has happened in the adult to the fluid-filled ventricle, and where are the roof plate and the floor plate? If you inspect figure 9.1 closely, you may notice a relatively tiny ventricle, the central canal of the cord, in the very center of each section (see also figures 9.2 and 9.3, and the drawings of the cord later in the chapter). It is not really smaller than in the embryo; it is just that the size of the rest of the cord has increased so much. The roof plate above the ventricle and the floor plate below it are no longer one cell thick, as they contain axons and dendrites that bridge the two sides. With cell proliferation and growth of axons and dendrites, the lateral walls of the neural tube have grown so thick that a ballooning out has occurred on each side, leaving a deep groove at the midline.

Inspecting figure 9.4, we see a hemisection of the lumbar level of the rat spinal cord. The first obvious feature is the contrast between the dark tissue of the outer parts of each half of the cord and the lighter, butterfly wing–shaped region this darker tissue surrounds.

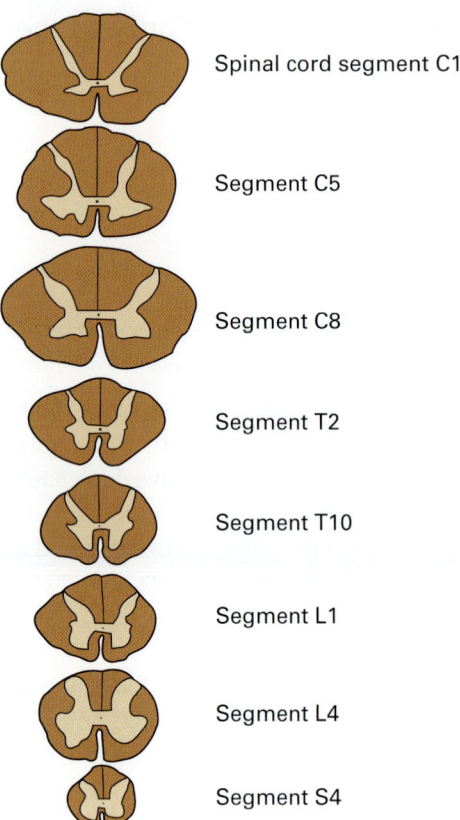

Figure 9.1
The human spinal cord is illustrated in simple charts of sections through different levels from the first cervical segment (C1) to the fourth sacral segment (S4). Spinal gray matter is shown in lighter brown; the columns of myelinated axons are shown in darker brown. The lateral horn is visible at thoracic and upper lumbar levels. This is the position of preganglionic motor neurons of the sympathetic nervous system. Based on Truex and Carpenter (1964).

The method of projecting an unstained section has made the myelinated axons appear dark so the spinal columns of axons are very visible: the dorsal columns, lateral columns, and ventral columns. This is the *white matter* of the cord, so called because the myelinated axons appear to be white in the unstained brain. In the unstained brain, regions where cell bodies are denser appear as slightly darker tissue and are called gray matter.

The cell bodies of the spinal gray matter are revealed by the Nissl stain, illustrated for the cat cervical cord in figure 9.2. In that section, the large cells located ventrally (in the lower part of the section) are motor neurons with axons that innervate the upper limb (see later for a further discussion of this figure).

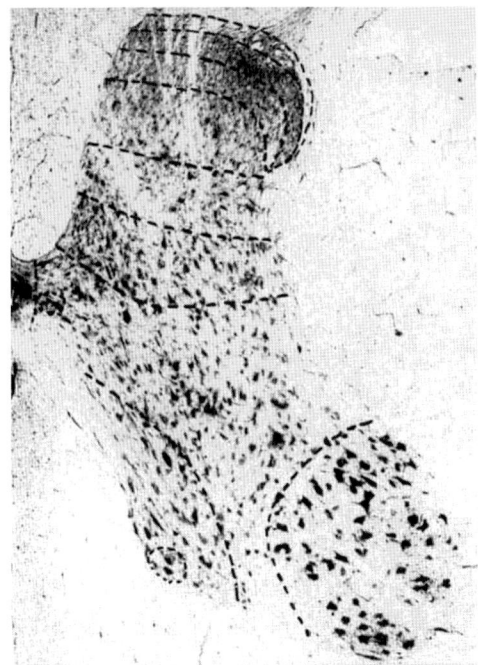

Figure 9.2
Cat spinal cord, seventh cervical segment as seen in a Nissl-stained frontal section in a 1954 publication by Bror Rexed, whose outlines of layers of the spinal gray matter have gained wide acceptance. The layers are indicated here. From Rexed (1954).

Looking at the small tracings of multiple levels of the cord in figure 9.1, you can see how the amount of white matter changes from rostral to caudal levels. The reasons for this are easy to figure out: For the descending pathways from the brain, more and more axons are leaving the white matter to terminate in the gray matter as they descend further and further. For the ascending pathways, more and more axons are being added to the white matter as more and more rostral levels are reached.

Next, look carefully at the shape and size of the gray matter at different cord levels. The ventral and dorsal horns are larger in two regions, one at cervical levels and the other at lumbar levels. The thickness of the cord is greater in these two regions in quadruped animals because of the greater number of neurons connecting with the limbs. There are many more motor neurons for the greater mass of muscle tissue and more secondary sensory neurons receiving input from the larger numbers of primary sensory axons coming from the more densely innervated surfaces of the distal appendages (paws or hands, feet). Think for a moment about the huge size of the rear legs of some of the large dinosaurs,

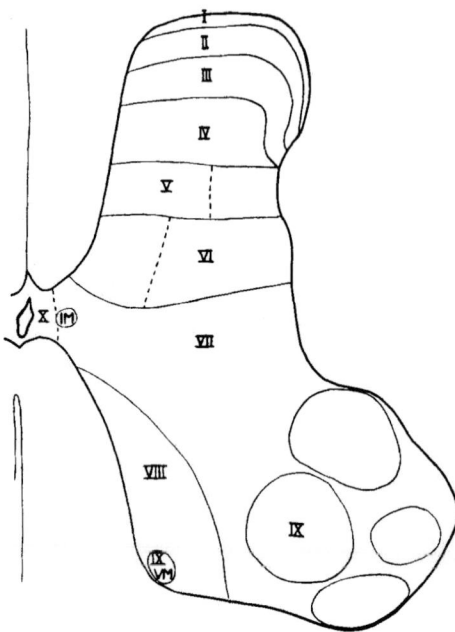

Figure 9.3
In a chart of the same section illustrated in the photograph of figure 9.2, the layers of Rexed, based on cytoarchitecture, are drawn. Motor neurons are found in layers VIII and IX. From Rexed (1954).

especially *Brontosaurus*. In *Brontosaurus* (now officially called *Apatosaurus*), studies of bones by paleontologists show that the lumbar enlargement was larger than the brain!

One additional feature should be remembered from these pictures: the presence of a small *lateral horn* in thoracic and upper lumbar cord levels. This small protrusion of gray matter into the lateral columns is caused by the *preganglionic motor neurons* of the sympathetic nervous system. They are found from the first thoracic level down to the second or third lumbar level. We will go into more details of the anatomy of the autonomic nervous system later.

Questions from Comparing Different Species

The abundant myelin seen in the spinal cord of mammals is not found in all animals, not even in all chordates. In fact, there is no myelin at all in the jawless vertebrates—the hagfish and lampreys—or in most invertebrates. So, why myelin? This wrapping of axons by the membrane of the peripheral glial cells, the Schwann cells, evolved in early vertebrates as an important adaptation—it gave the central communication system composed of these neurons more speed. Without myelin, the rate of conduction of action potentials

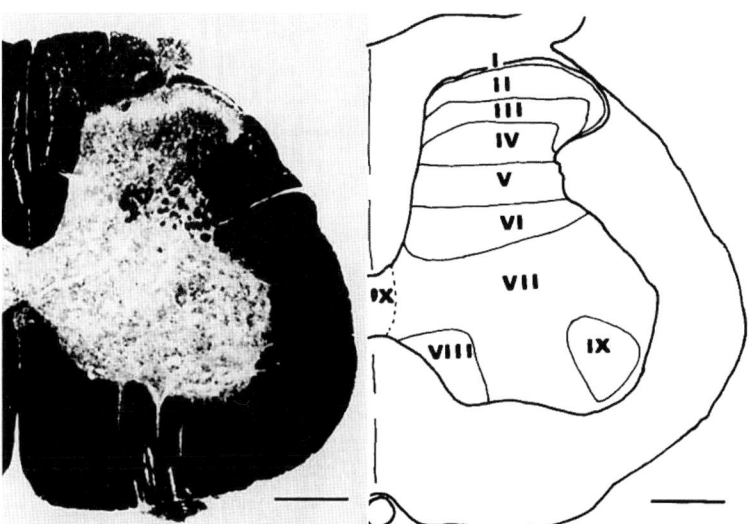

Figure 9.4
(Left) Lumbar spinal cord of rat in an unstained coronal section (L4 level). The myelinated axons in the 50-μm-thick section appear as very dark, similar to what would be seen in a myelin stained section. (Right) Chart of the same level of the cord, outlining the cytoarchitectural layers as specified in Rexed's 1952 paper. Scale bar, 380 μm (corrected estimate). From Gibson et al. (1981).

along axons increases as the axon increases in diameter. There are obvious limits to how large axons can become; without some other means of increasing conduction rate, the possible size variations in any species was constrained in evolution. Giant axons did evolve in some animals, like the squid, but only for control of the most crucial survival action, the escape response. But with myelin around much of the axon, ion flow across the membrane became restricted to small "nodes" of bare membrane. The result is that the action potential jumps from node to node (the conduction is *saltatory*), and current spread in the internodal regions is very fast.

Most of the long axons in the vertebrate CNS are myelinated (except when the diameter is small, near their terminals), as are all but the smallest axons of the peripheral nervous system. In a section like the one shown in figure 9.4, myelin is found around the long axons of dorsal, lateral, and ventral columns. Only the small tract just outside the dorsal tip of the dorsal horn has little or no myelin. Within the gray matter regions, terminating axons tend to lose their myelin in the end arbors close to the synaptic terminals.

How does the spinal cord of humans differ from spinal cords of other mammals? Think about what may be most unique in our species. We have great sensory acuity and probably finer motor control in our hands than other mammals. The sensory acuity requires a high density of innervation of our fingers, and correspondingly large numbers of axons carrying the information to the cervical enlargement of the spinal cord, and from the cord to the brain via the dorsal columns. The fine motor control requires large numbers of motor neurons with each one innervating very small numbers of muscle fibers. It also requires large numbers of corticospinal axons for control of those motor neurons. For these reasons, the human cervical spinal cord has larger white matter tracts, and the cervical enlargement is relatively great.

Figure 9.2 illustrates the results of cytoarchitectural studies of the cat spinal cord at the cervical enlargement. The neurons appear to be organized into distinct layers. This organization was described in 1952 by the Swedish neuroanatomist Bror Rexed [pronounced with two syllables: rex'-ed]. Rexed's layers were found to correspond well to the functional as well as the anatomical organization of the spinal cord in many species and have been widely adopted. In the ventral horn, the motor neurons that innervate different muscle groups of the limbs can be divided into separate regions, all labeled layer 9 in Rexed's terminology. The motor neurons that innervate more proximal muscles are in layer 8. The layers and motor neuron groupings can be seen in myelin stained sections as well (figure 9.4).

The Local Reflex Channel and the Older Lemniscal Channels

The sensory channels were introduced in chapter 5 when we sketched an ancestral pre-mammalian brain: a local reflex channel (and also intersegmental reflex channels), the old lemniscal channels that are followed by somatosensory information to the brain (spinoreticular and spinothalamic tracts), and the cerebellar channels. Then, in chapter 7 we described the ascending somatosensory channel that has become greatly enlarged in mammals—the dorsal column–medial lemniscus pathway. We can trace these sensory channels in the mammalian spinal cord.

To begin doing this, we look at the dorsal root fiber inputs (figure 9.5). Fibers of different sizes and different peripheral sources have somewhat different termination patterns. For example, the largest axons, originating in the muscle spindle organs, which detect muscle stretch, terminate in several places. Some collaterals end on cord interneurons, especially in layer 6 of Rexed where some of the secondary sensory neurons have axons that connect with the cerebellum. Some of the large dorsal-root axons connect directly with large motor neurons (alpha motor neurons), completing a monosynaptic muscle-to-muscle reflex arc. This is the pathway of the stretch reflex. Activity of this reflex maintains muscle tone or tension, which determines the spring-like properties of muscles. (The sensitivity of the muscle spindles to muscle stretch is adjusted continuously by tiny muscle fibers within them; these fibers are innervated by smaller motor neurons, the gamma motor neurons.) The muscles containing the most spindles are those of fine motor control, as in the fingers.

The Lower Levels of Background Support

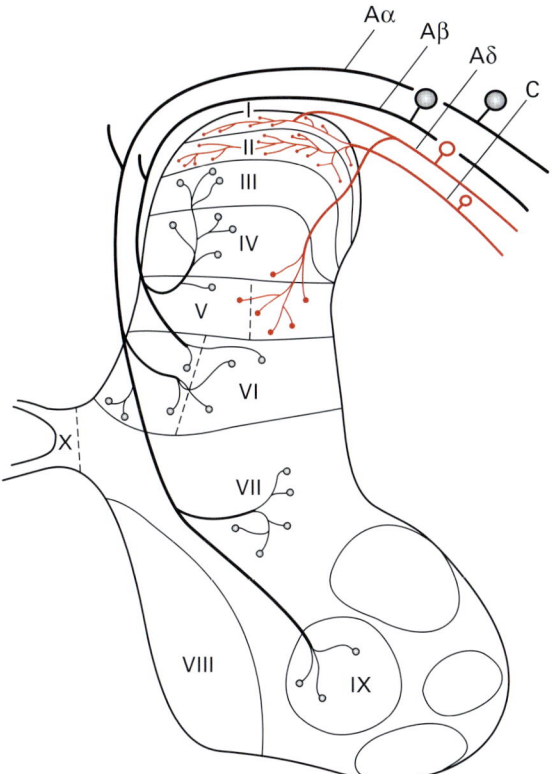

Figure 9.5
Chart of the gray matter as seen in a spinal hemisection. The layers of Rexed are indicated. The terminations of four major types of dorsal root fibers are illustrated by simplified sketches of four representative axons of different sizes and different origins. Based on Brodal (2004).

Another local reflex pathway begins with connections of skin afferents to cord interneurons. These interneurons, responding to intense pressure or other potentially harmful stimuli, connect in turn to flexor motor neurons. Activation causes withdrawal of the limb from the contact that is initiating the reflex. Such a local polysynaptic reflex arc is illustrated in schematic fashion in figure 9.6. Some of the same interneurons involved in withdrawal reflexes connect with interneurons that inhibit the neurons that connect to extensor muscles opposing the flexors being stimulated (not illustrated). This is reciprocal inhibition. Other interneurons send axons to the opposite side of the cord and excite extension of the opposite limb, preventing the animal from falling. This is the crossed extensor reflex.

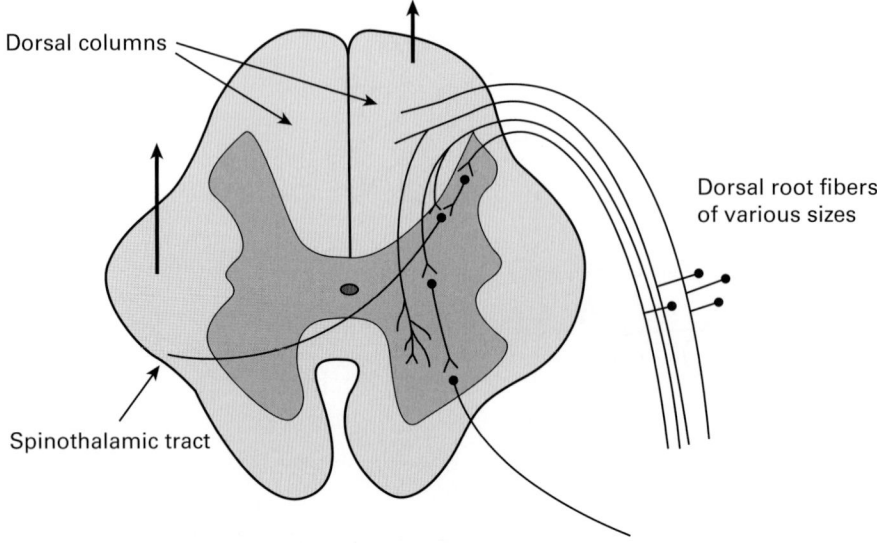

Figure 9.6
In this drawing of a frontal section of the adult spinal cord, both reflex and lemniscal channels of conduction are sketched. Smaller fibers of a dorsal root are shown synapsing on neurons of the dorsal horn. One of these neurons has a short axon that terminates on ventral horn motor neurons (like the one illustrated). This neuron could innervate a flexor muscle, completing a two-synapse pathway of a withdrawal reflex. Another neuron in the dorsal horn sends its axon across the midline to the lateral column on the opposite side, joining the spinothalamic tract. Other primary sensory axons of the dorsal root, mostly larger ones, send branches into the dorsal column where they ascend to a dorsal column nucleus.

Figure 9.6 also includes sketches of the origins of two lemniscal pathways for carrying somatosensory information to the brain. (Remember that these drawings are extremely simplified. See figures 2.10 and 2.11 for two of Cajal's drawings of spinal axons and cells with their dendritic arbors.) Axons forming the spinoreticular tract (figure 9.7) pass into the white matter of the lateral and ventral columns, most of them remaining on the ipsilateral side on their route to destinations in the hindbrain and midbrain reticular formation, with a few of them extending into the older parts of the 'tweenbrain (hypothalamus, subthalamus, parts of the dorsal thalamus that project to corpus striatum). Collaterals of these axons may terminate also at other levels of the spinal cord.

We also see the origins of the spinothalamic tract in figures 9.6 and 9.7: Neurons in the dorsal horn send their axons through the floor plate to the anterolateral white matter where they begin their ascent toward the brain. These axons reach the hindbrain and midbrain reticular formation, the deep layers of the superior colliculus, and the caudal diencephalon. In the diencephalon, some of them terminate in the ventral posterior nucleus, which projects to the somatosensory areas of the neocortex.

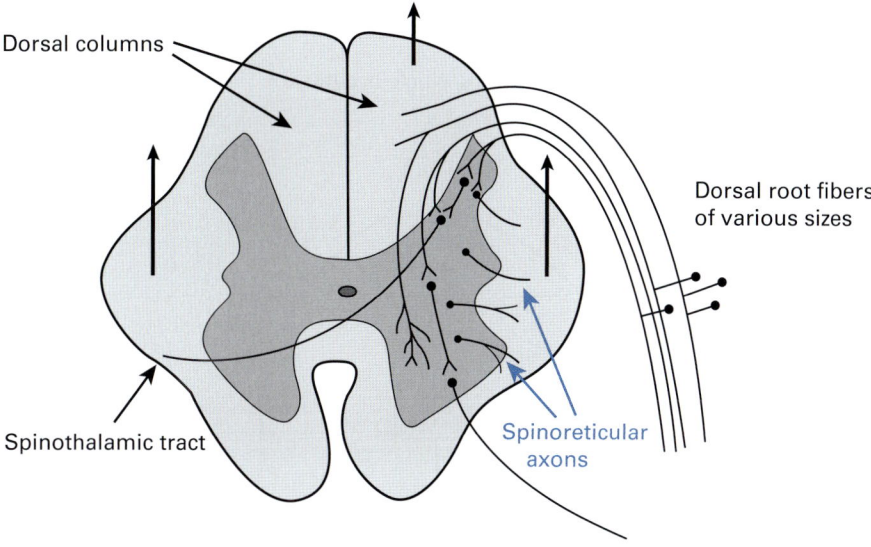

Figure 9.7
The drawing of a frontal section of the adult spinal cord in figure 9.6 is repeated with the addition of cells of origin of spinoreticular tract axons. Such axons are part of the oldest lemniscal pathway, ascending to the brain mostly on the ipsilateral side. Many of the cells of origin receive connections from dorsal root axons. Many axons of this nature are *propriospinal*—part of an interneuronal network interconnecting different levels of the spinal cord, including some axons that terminate on the opposite side of the midline. Those that carry somatosensory information to the brain terminate on neurons of the reticular formation of the hindbrain and midbrain; the longest axons reach the 'tweenbrain.

The Mammalian Highway for Ascending Somatosensory Information

Finally, in the same figures, note the axons entering the dorsal columns. These are primary sensory axons that ascend toward the brain—the first part of the lemniscal channel that has expanded so much in the evolution of mammals. Some collateral branches terminate more locally in the cord, but the ascending dorsal column axons continue far rostrally, terminating in cell groups at the very rostral end of the spinal cord and caudal-most part of the hindbrain. These cell groups are the dorsal column nuclei, two on either side.

As we move rostrally from the caudal spinal cord, we find that the axons being added to the dorsal columns at each level are added to the lateral side. The relative positions of the axons are maintained all the way up to the dorsal column nuclei and in their pattern of termination there. Thus, at cervical levels, the medial portion of the dorsal columns contains axons from the caudal body (tail region, hind limbs, and lower trunk), and the lateral portion contains axons from the rostral body (trunk, forelimbs, and neck). The result is that in the dorsal column nuclei, there is a topographic representation of the body, with the tail represented medially and the forelimbs and neck represented

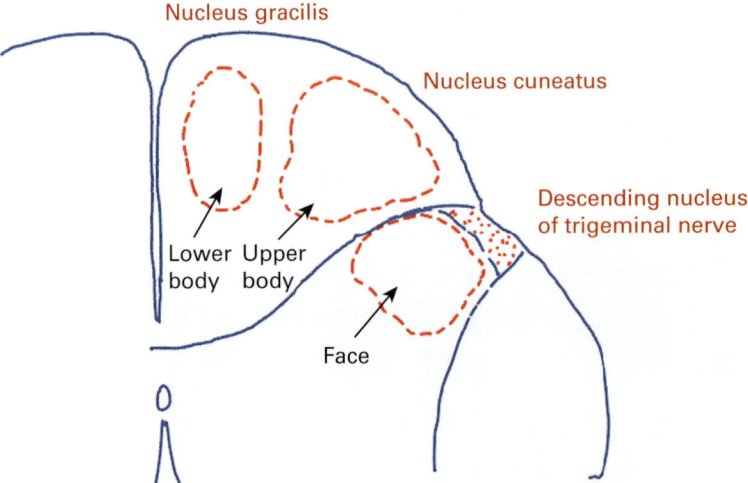

Figure 9.8
A partial sketch of the dorsal part of the spinal cord at its rostral-most end, showing the outlines of three cell groups that receive inputs from the body surface. The most medial cell group (nucleus gracilis) receives dorsal column inputs from the lower body, and the cell group lateral to it (nucleus cuneatus) receives dorsal column inputs from the upper body except for most of the head. Lateral to the position of the dorsal columns, in the dorsal horn of the spinal gray matter, neurons receive inputs from the face via axons from the fifth cranial nerve (the trigeminal nerve). Small axons of the descending tract of primary sensory axons from the trigeminal nerve travel caudally in the tract of Lissauer (stippled). The cell group is the descending nucleus of the trigeminal nerve (fifth cranial nerve).

laterally. The medial cell group, with cells activated from the lower body, is called nucleus gracilis because of its slender shape—longer in its rostrocaudal extent and narrower. The more lateral nucleus is called nucleus cuneatus (wedge shaped).

In the cervical cord, it may surprise you to learn that the representation of the body includes the face region, although inputs from that region enter the hindbrain and not the spinal cord. Although many of the primary sensory axons of the fifth cranial nerve from the face do terminate in hindbrain cell groups, some of them descend, terminating as far caudally as the cervical spinal cord (figure 9.8). The axons that descend to the dorsal horn of the cervical cord are found in Lissauer's zone just outside the dorsal horn. These axons carry information about pain and temperature from the face.

Cerebellar Channel

We don't want to forget another sensory channel, one that is probably as ancient or more ancient than the spinothalamic tract (sometimes called the old lemniscus, or paleolemniscus, the only more ancient lemniscus being the spinoreticular). This is the cerebellar channel. Some axons to the cerebellum from the cord traverse a crossed pathway, with decussations like those of the spinothalamic tract. Others follow an ipsilateral route. The

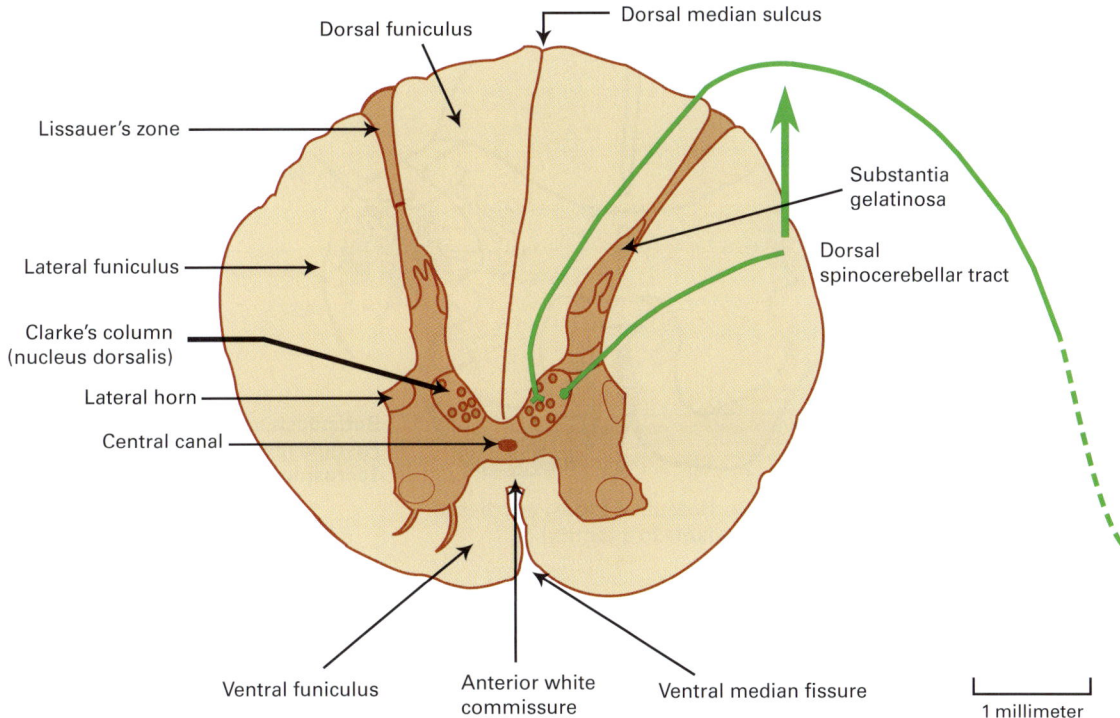

Figure 9.9
Chart of a section of the human spinal cord at the fifth thoracic segment, showing the position of Clarke's column, which receives proprioceptive inputs from dorsal root fibers. These inputs carry information on joint movements of the lower limbs and the trunk. The axons of Clarke's column neurons send axons into the lateral column on the ipsilateral side, where they ascend to the cerebellum as the dorsal spinocerebellar tract. Other portions of the spinal cross-section are noted on the drawing. Based on Nauta and Feirtag (1986).

ipsilateral axons follow the dorsal spinocerebellar tract, which originates in the secondary sensory cell group called Clarke's column, named after the English anatomist Jacob A. L. Clarke (1817–1880), or nucleus dorsalis if you prefer the Latin, found in the medial part of the spinal gray as shown in figure 9.9.

The Pathways of Regulation within the Spinal Cord Itself

On the left side in figure 9.10, propriospinal axons are sketched. These are the axons intrinsic to the spinal cord itself—they originate in the cord and terminate in the cord and control innate patterns of behavior including the patterns of walking and other locomotor gaits. Control of such patterns requires connections that pass from one level of the cord to another level or to several other levels. The axons are found adjacent to the spinal gray matter. They include, of course, the pathways underlying every intersegmental reflex.

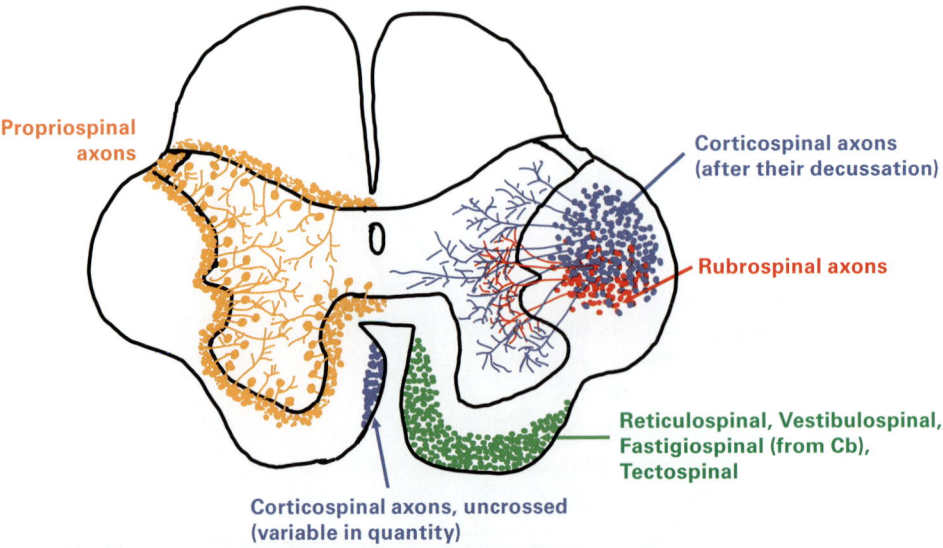

Figure 9.10
Sketch of some descending and intrinsic axons of the adult spinal cord. Left side: Propriospinal axons have their sources and their terminations within the spinal cord. Intersegmental axons are found in the white matter close to the spinal gray matter. The position of an uncrossed component of the corticospinal tract is also illustrated, without showing the distribution of its axons in the ventromedial gray matter. Propriospinal axons that cross to the contralateral side are also omitted from the figure. Right side: The positions and terminations of the corticospinal axons in primates are illustrated, as is the position of the rubrospinal tract from the midbrain. In addition, the position of descending axons that control axial muscle groups is shown (reticulospinal, vestibulospinal, tectospinal tracts, and fastigiospinal tract from medial cerebellum). Cb, cerebellum.

The interconnections within the spinal cord, in order to have control over various fixed motor patterns (the motor components of fixed action patterns, which also include non-spinal regulation by motivational states), are necessarily quite complex. Yet, their formation does not require learning. This circuitry is programmed in the genes—programs that are very similar across large numbers of species. You don't have to learn to walk. You begin to walk when the circuits of the spinal cord are sufficiently developed, together with descending influences from the brainstem (primarily hindbrain and midbrain). However, learning certainly does help to shape these movement patterns. Some of this plasticity depends on changes in the cerebellum. Additional shaping of movement details depends on changes in pathways within, and coming from, the endbrain's corpus striatum and neocortex.

The Pathways of Influence and Control from the Brain

On the right-hand side of figure 9.10, you can see where major descending pathways can be found in a section of the human spinal cord. It is similar for many other species, with some exceptions.

Various fixed motor patterns (see earlier) depend on reticulospinal axons that descend from the reticular formation of the hindbrain and midbrain to spinal cord interneurons at various levels. Such axons are found mainly in the ventral columns and often are distributed to both sides. Instinctive behaviors controlled by these pathways—acting on intricate networks of spinal neurons—include feeding and drinking patterns, courting and mating behavior, and emotional displays of social life. Reticulospinal pathways are also critical for vital functions like the control of breathing and heart rate, functions with many influences from the limbic forebrain. Locomotor patterns, though controlled at the spinal level, are initiated and halted by the activity of pathways from more rostral levels, particularly from locomotor centers in the hypothalamus and in the caudal midbrain reticular formation.

Stability in space depends not only on spinal reflexes initiated by pressures on the soles of the feet and from receptors in joints, but also on modulating inputs from descending pathways. These come from the vestibular system, including a direct pathway from vestibular nuclei, the vestibulospinal tract. They also come from the visual system (e.g., from the optic tectum via the tectospinal tract to the cervical cord) and from less direct pathways through the medial brainstem reticular formation.

Most movements are modulated by the cerebellum. For example, the movements of balance and postural maintenance are influenced by a direct descending pathway from the cerebellum, the fastigiospinal tract, and by cerebellar influences on other systems. Inputs to the cerebellum, originally in evolution probably from the vestibular system, come also from other sensory systems, especially the somatosensory and visual systems.

Limb control, especially control of individual hands and feet, is strongly influenced by axons descending in the rubrospinal tract, originating in the red nucleus of the midbrain. Such limb control, as well as control of many other movements, has become dominated in many mammals, especially in higher primates, by the somatic sensory and motor areas of neocortex via the descending axons of corticospinal pathways. These descending axons from neocortex enable fine control of individual digits via direct connections to motor neurons of the spinal enlargements. We will discuss motor control systems further in chapters 14–16.

The corticospinal tract, except for a small uncrossed component, is located in the lateral columns of primates. However, in the rat and in some other animal species, this tract is found in the deepest part of the dorsal columns.

A Reminder

You are reading many strange new words as you begin to acquire a language for discussing brain structure. There is no reason to worry about memorizing every new term as it comes at you. Many teachers, and the author of this book, are well aware of the difficulties of gaining an acquaintance with this language and with the intricacies of brain organization. In this book we have surveyed the entire central nervous system when thinking about brain evolution, and then we have gone through the whole system again when trying to reconstruct an ancestral pre-mammalian brain and the transformations that occurred with the early evolution of mammals. When studying development we see some of this another time, and then when looking at specific systems and regions of the mature mammalian brain we will see it all again. Each time through it, you may encounter some additional terms or concepts. When you see new terms, just try to understand the concepts and visualize the locations involved without worrying about your memory for the words. Each time you encounter the same structures and words you will remember a little more, until eventually the more important aspects of brain structure will become part of your mental model of this inner world.

Maintaining Stability of the Internal Environment: The Autonomic Nervous System

Several times we have mentioned and illustrated the preganglionic motor neurons of the sympathetic nervous system, located in the lateral horn of the spinal cord at thoracic and upper lumbar levels. These neurons connect to ganglia located adjacent to the CNS. The neurons of these ganglia connect, throughout the body and head, to smooth muscles of hollow organs including blood vessels, and to gland tissue. The sympathetic nervous system axons are not the only sources of innervation of these tissues. The same organs generally have a dual innervation. The other source of innervation is from the parasympathetic nervous system, with preganglionic motor neurons located in the hindbrain or midbrain or in the sacral part of the spinal cord. For the parasympathetic system, the peripheral ganglia are located immediately adjacent to the organs being targeted. Thus, the two divisions of the autonomic nervous system (ANS) have contrasting patterns of end-organ innervation, illustrated very simply in figure 9.11.

The separation of peripheral ganglia that give rise to functionally contrasting types of visceral innervation is not a feature of all vertebrates. It is not seen in fish, and we see only the beginnings of something like a separate parasympathetic system in amphibians. Reptiles are in more respects similar to mammals, but not in every detail.

Three Major Divisions of the Motor System

At this point, we can step back and see three major divisions of the motor system of vertebrates including the mammals. They are characterized by contrasting arrangements of motor neurons (figure 9.12).

One system is what we usually think of when we hear the term *motor system*. Axons of motor neurons form synaptic connections with striated muscle cells; the connections are formed at terminal enlargements of the axons called motor endplates. At the specialized synapses at the endplates on striated muscles of mammals, the release of the

The Lower Levels of Background Support

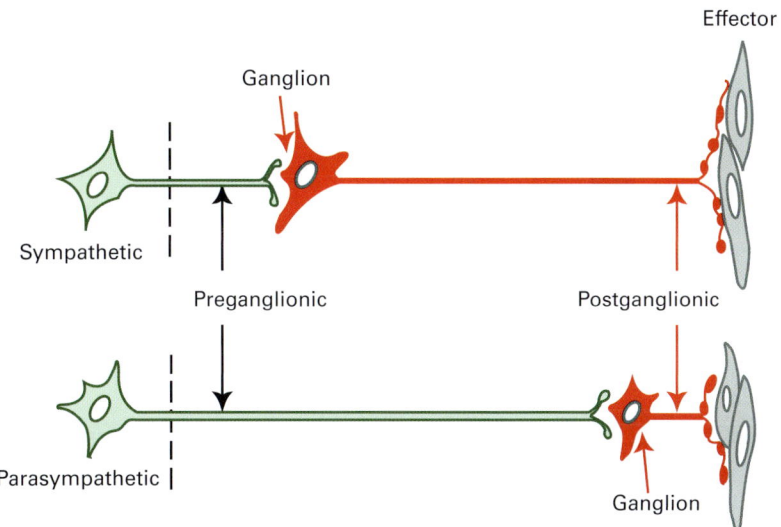

Figure 9.11
Schematic illustration of the structural arrangements within the autonomic nervous system. In the sympathetic system, the axon of the preganglionic motor neuron is relatively short, and the axon of the sympathetic ganglion cell is long. In the parasympathetic system, the preganglionic axon is long, whereas the ganglion cell axon is short. The peripheral ganglia of the sympathetic system are located near the spinal vertebrae. The peripheral ganglia of the parasympathetic system are located very close to the target organs. In some cases, the ganglionic neurons are located within the tissues of the organ. The neurotransmitters of the postganglionic neurons are different: norepinephrine for the sympathetic (with the exception of the innervation of sweat glands) and acetylcholine for the parasympathetic. Based on Brodal (2004), and others.

neurotransmitter acetylcholine causes endplate potentials in the muscle cells that trigger muscle contraction.

The second system is the autonomic nervous system, where a preganglionic motor neuron synapses with a peripheral ganglion cell. Acetylcholine is released at the synapse. The ganglion cell sends its axon to smooth muscle or gland cells, where it releases a neurotransmitter in paracrine fashion—it is a release into the extracellular space where diffusion of the substance reaches a number of nearby cells rather than just one as in a conventional synapse.

The third system has motor neurons that operate in endocrine fashion. These are neuroendocrine cells that, when activated, release their neurotransmitter into the bloodstream and thereby reach end organs where they have an effect.

A Sketch of the Autonomic Nervous System

Using a simplified picture of the embryonic mammalian neural tube, we can show the locations of the preganglionic motor neurons of the sympathetic and parasympathetic

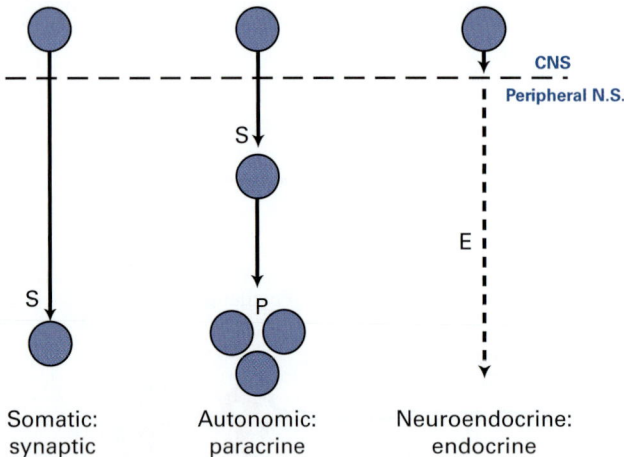

Figure 9.12
Schematic picture of the arrangement of motor neurons in the three major divisions of the motor system, not including the enteric nervous system (see text). The neurotransmitters used at the end organs of the autonomic nervous system are released in paracrine fashion (P), affecting more than one cell. This kind of neurotransmitter release is in contrast to the very local effects at synapses (S) and the very widespread effects in neuroendocrine and endocrine systems (E) in which effects are exerted through the bloodstream. Based on Swanson (2003).

divisions and their patterns of peripheral innervation of a few illustrative organs (figure 9.13). The same patterns are repeated for nearly all visceral organs.

Using the figure, consider the reflex action of the iris. When a brighter light enters the eye, the pupils quickly adjust by becoming smaller. Although the pupillary constriction response can occur in some animals by purely local events in the eye, the main reflex is a parasympathetic one. Information on light intensity passes down the optic nerve and reaches neurons in the epithalamic region called the pretectal area, and from there it goes to the cells of the parasympathetic nervous system in the midbrain, in a cell group named for its discoverers in the nineteenth century, the nucleus of Edinger–Westphal. Axons from the Edinger–Westphal (E-W) nucleus leave the CNS at the base of the midbrain as part of the third cranial nerve and reach the ciliary ganglion behind the eyeball. Neurons of the ciliary ganglion innervate the iris, where their activation causes the pupil to constrict.

Pupillary dilation, in contrast, is not caused only by an absence of constriction when the light is dim; that is only one factor. The pupils dilate also as a response to fear or anger or to a lesser extent even as a response to intense interest (which may be sexual but not necessarily). The dilation is caused by actions of the sympathetic nervous system. Preganglionic neurons in the upper thoracic spinal cord, if triggered by inputs (e.g., from the midbrain or hypothalamus) send action potentials along axons connecting with ganglion cells of the sympathetic chain of ganglia, particularly with neurons of the most rostral ganglion, the superior cervical. Axons from that ganglion reach various tissues in

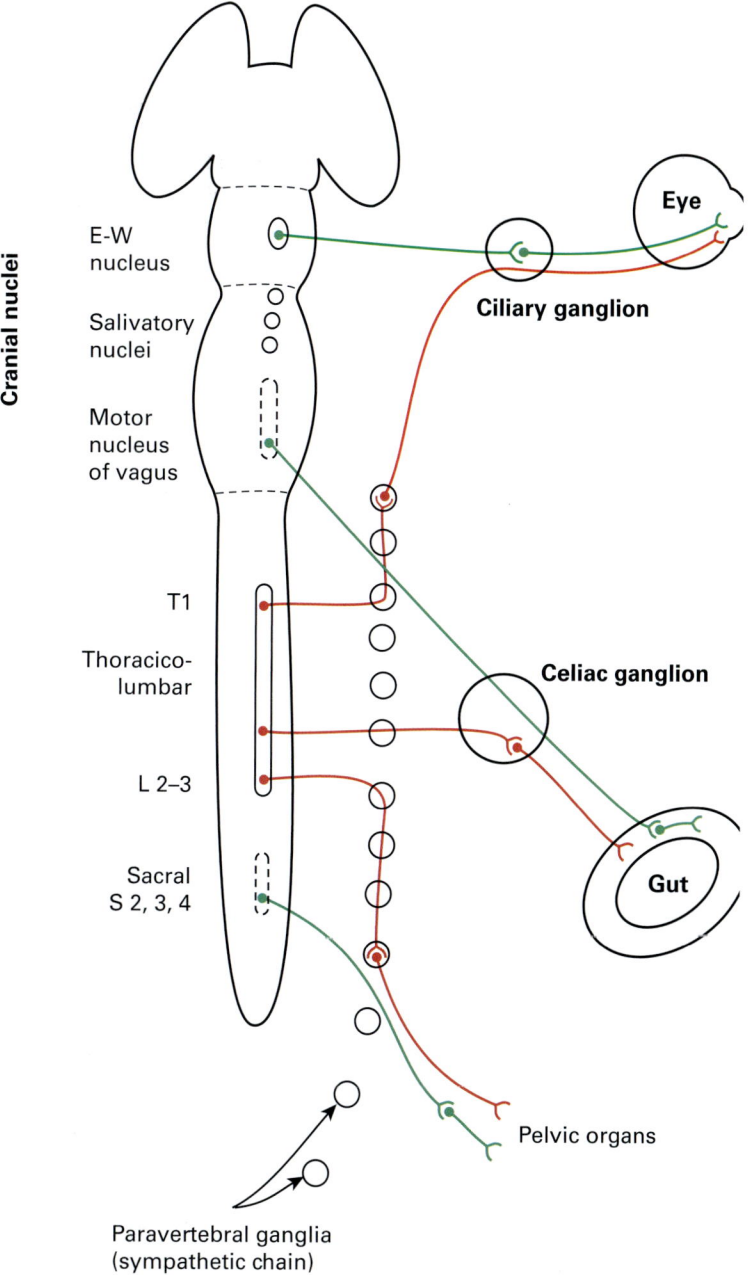

Figure 9.13
Dorsal view (partial schematic view) of an embryonic mammalian CNS with the locations of preganglionic and ganglionic neurons of the autonomic nervous systems sketched in. The pathways for innervation of the iris of the eye and of the smooth muscles of the intestinal tract are illustrated. Innervation patterns for other organs follow a similar pattern. A glance at the locations of the preganglionic motor neurons shows why the sympathetic system is called the thoracico-lumbar system, and the parasympathetic system is called the cranio-sacral system. E-W, Edinger–Westphal.

the head including the iris, where their action causes pupillary dilation. The pupils may also dilate because of circulating epinephrine (adrenaline) secreted by the medulla of the adrenal glands.

Now look at the autonomic innervation of part of the intestinal tract (the gut) near the stomach. Again we see the same kind of dual innervation pattern (figure 9.13). This time, preganglionic motor neurons of the parasympathetic nervous system are located in the hindbrain, in the dorsal motor nucleus of the vagus nerve (the tenth cranial nerve). These axons are very long, in this case reaching from the hindbrain to the intestines below the stomach where they terminate in paracrine fashion on parasympathetic ganglion cells located in the tissues surrounding the intestines. Activation of these cells stimulates intestinal motility and secretions.

The sympathetic nervous system also innervates the same region of the intestines. The preganglionic motor neurons are located in the midthoracic cord, with axons reaching, for example, a prevertebral ganglion like the celiac. Neurons in that ganglion send axons to the walls of the gut, where their activity inhibits motility, as in the states of fear and flight or of aggression. This allows a shift of maximum oxygen and glucose utilization to the muscles rather than the digestive system.

Because the parasympathetic system's preganglionic motor neurons are located in the brain (midbrain and hindbrain, as indicated in figure 9.13) and in the sacral spinal cord, this system is often called the cranio-sacral system. The sympathetic preganglionic motor neurons, in contrast, are all located within the cord, in the lateral columns of the thoracic and lumbar portions, which led to this system getting the name thoracolumbar system.

Functions of sympathetic and of parasympathetic innervation of various peripheral organs are summarized in table 9.1. We can summarize the sympathetic nervous system

Table 9.1
Important functions of some autonomic pathways

Gland or Muscle Tissue	Sympathetic Functions	Parasympathetic Functions
Iris	Dilates pupil (mydriasis)	Constricts pupil (miosis)
Lacrimal gland	Little effect on secretion	Stimulates secretion
Salivary glands	Secretion reduced; less watery	Secretion increased; watery
Sweat glands	Stimulates secretion	Little effect
Lungs, bronchi	Dilates the lumen	Constricts
Heart	Speeds heart rate; increased ventricular contraction	Slows heart rate
Stomach, intestines	Inhibits motility and secretions	Stimulates motility and secretions
Anal sphincters	Constricts except with very intense activation	Relaxes
Sex organs	Orgastic contraction of ductus deferens, seminal vesicle, prostatic or uterine muscles; vasoconstriction	Vasodilation, engorgement of erectile tissue
Urinary bladder	Relaxes wall of bladder; constricts internal sphincter; inhibits emptying	Contracts bladder, relaxes sphincter, promotes emptying
Adrenal medulla	Stimulates secretion	Little or no effect
Blood vessels, skin of trunk and extremities	Constricts	—

functions as those of "fight or flight"; when it is active in one region, it tends to be active everywhere. By contrast, the parasympathetic system works more locally, promoting the normal functions of the visceral organs. It tends to be more active during periods of the organism's inactivity and rest.

Chemical Mediation at Autonomic Nervous System Synapses

In the previous chapter, the formation of the sympathetic ganglia from the neural crest at thoracic and upper lumbar spinal levels was described: see figure 8.6. In figure 9.14, we see a schematic adult thoracic spinal cord section that includes sympathetic and dorsal root ganglia. As indicated in the figure, axons from the lateral horn, the preganglionic sympathetic system's axons, come out of the cord through a ventral root, then leave the root to synapse with cells of paravertebral or prevertebral ganglia. The axons of paravertebral ganglia rejoin the ventral root to become part of a spinal nerve, thus following the peripheral system of nerves to the periphery. The sympathetic axons reach the smooth muscle of blood vessels, the erector pili muscles (enabling body hairs to erect), and the sweat glands of the skin.

The preganglionic motor neurons of both sympathetic and parasympathetic systems use acetylcholine as their neurotransmitter. However, the two systems differ in neurotransmitters used by the postganglionic motor neurons that innervate the destination organs. (Recall from chapter 1 the discovery of chemical transmission at synapses by Otto Loewi.) Cells of the parasympathetic ganglia found near their innervated organs use acetylcholine here as well, but cells of the sympathetic ganglia use norepinephrine (noradrenaline), a molecule similar in structure to the epinephrine (adrenaline) secreted by the adrenal medulla. An exception is that the sympathetic innervation of sweat glands uses acetylcholine as the neurotransmitter.

At this point, you should be able to make a fairly accurate sketch of the autonomic innervation of the heart, showing the accelerator and the decelerator nerves (the nerves that release chemical neurotransmitters as first demonstrated by Otto Loewi (see chapter 1). You can add a heart to the sketch in figure 9.13. The accelerator nerves consist of axons from the paravertebral ganglia, from the second to about the fifth. The fibers that terminate in the heart release norepinephrine when activated. The preganglionic motor neurons are located in the lateral horn of the upper thoracic cord. The decelerator nerve is a branch of the vagus nerve, the name for the tenth cranial nerve, which comes from the hindbrain. Fibers of this nerve terminate on cells of the cardiac ganglion, which has a pericardial location.

The Enteric Nervous System

An interesting advance in the study of innervation of the viscera was made in the latter part of the twentieth century. The discovery was that the peripheral ganglia innervated

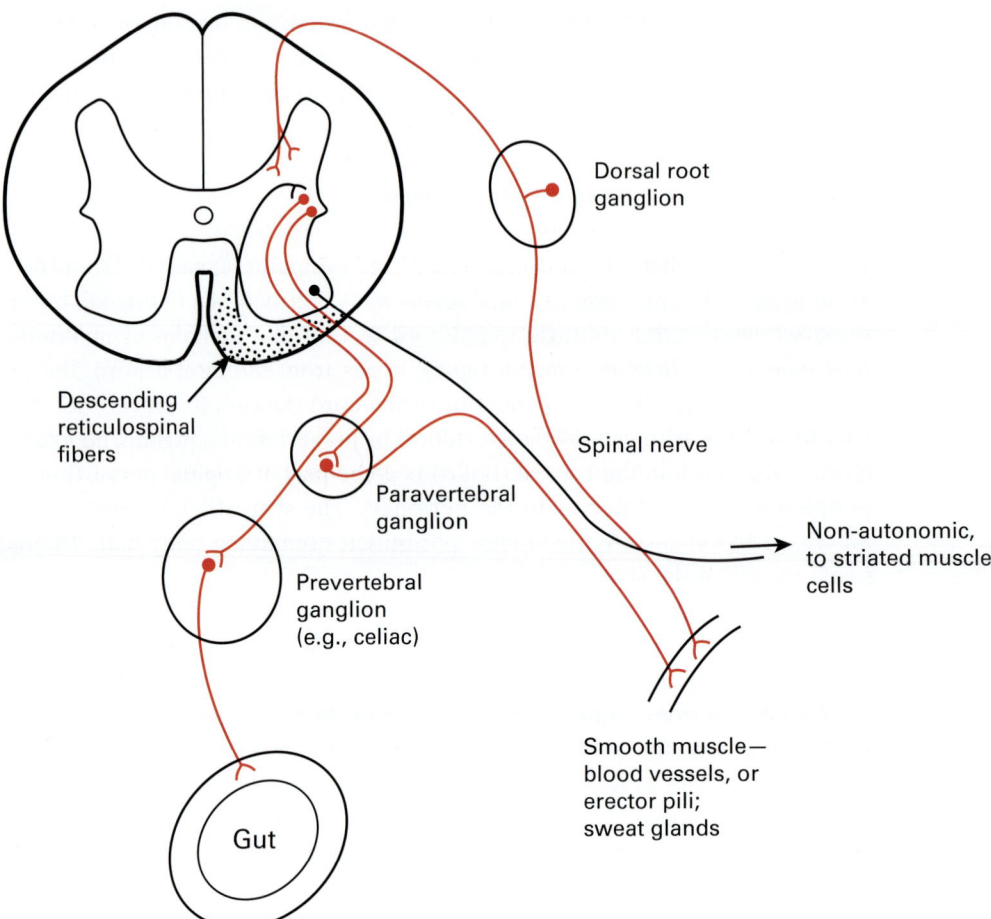

Figure 9.14
A section of the mammalian spinal cord in the thoracic region is drawn, with the addition of cells and axons of the sympathetic nervous system. The innervation patterns for the paravertebral and prevertebral ganglia are illustrated in relation to a spinal nerve.

by the parasympathetic system are not just passive links in a control line starting in the CNS. Rather, neurons on or near the walls of the intestine form an interconnected network, or group of plexi, that functions even without connections to the CNS. It is like a brain in the gut, one that, it has been estimated, may contain as many neurons as the entire spinal cord (although estimates vary). This is the enteric nervous system, modulated but not fully controlled by influences from the CNS.

In the wall of the intestine, anatomists have described at least two major plexi—an outer plexus (myenteric) and an inner plexus (submucosal)—consisting of a large number of neurons and glial cells in an organized and interconnected network. Various neurotransmitters are used by these neurons, not just acetylcholine as once was the belief when the connections of the vagus nerve were seen as more paramount.

A similar anatomical complexity may exist for heart innervation, as there is evidence that the cardiac ganglion may have functioning interconnections, with endogenous activity independent of the brain. Thus, the heart may also have its own brain.

Levels of Control of the Internal Environment

The existence of the enteric nervous system indicates that, at the very lowest level of control of the internal environment of the alimentary tract, there is considerable autonomy. There is evidence that within the central nervous system, there is some autonomy at the lowest levels, but each higher level adds more refinement. A function that has been studied at all levels is temperature regulation. For this function, there are spinal mechanisms of control, and hindbrain mechanisms that improve temperature regulation, midbrain mechanisms that improve it further, and finally the precise regulatory controls of the hypothalamus of the diencephalon. There is no doubt that a similar type of hierarchy of control exists for other functional systems as well.

It is to the neuroanatomy of brain levels above the spinal cord that we now turn.

Readings

Hartline, D. K., & Colman, D. R. (2007). Rapid conduction and the evolution of giant axons and myelinated fibers. *Current Biology, 17,* R29–R35.

Many neuroanatomy texts include a description of spinal cord structure, for example:

Brodal, P. (2004). *The central nervous system: Structure and function,* 3rd ed. New York: Oxford University Press.

Nauta, W. J. H. & Feirtag, M. (1986). *Fundamental neuroanatomy.* New York: Freeman.

The spinal cord was a focus of many physiological studies by prominent early neuroscientists, paramount among whom was Charles Scott Sherrington. His article on the spinal cord in *The Encyclopedia Britannica* was very well written, and with revisions has long been present in these volumes. Another good description of spinal cord studies is the following book:

> Creed, R. S., Denny-Brown, D., Eccles, J. C., Liddell, E. G. T. & Sherrington, C. S. (1932). *Reflex activity of the spinal cord.* Oxford: Clarendon Press.

CNS regulation of body temperature: See reviews by Evelyn Satinoff; for example, the paper noted in chapter 4:

> Satinoff, E. (1978). Neural organization and evolution of thermal regulation in mammals. *Science, 201,* 16–22.

There are many specialized articles, and books as well, with information about sensory pathways of the spinal cord. There is special medical interest in the phenomena of pain perception and its underlying CNS mechanisms. An example is the following book:

> Willis, W. D. Jr., & Coggeshall, R. E. (2004). *Sensory mechanisms of the spinal cord. Volume 1, Primary afferent neurons and the spinal dorsal horn,* 3rd ed. New York: Plenum.

Intermission

The Ventricular System, the Meninges, and the Glial Cells

With the closure of the embryonic neural tube, fluid that surrounded the embryo becomes enclosed inside the tube. This is the central canal of the spinal cord and the ventricles of the brain (see figure 8.3). Thus, a clear fluid bathes the entire embryonic CNS from both the outside and the inside. From this fluid, the early developing cells obtain such essentials as oxygen, glucose, and various ions. But once the neural tube forms and the CNS becomes enclosed within the skin-covered body, the fluid both outside and inside the neural tube loses direct continuity with the amniotic fluid. Nevertheless, the CNS retains a fluid surround: the cerebrospinal fluid (CSF), which fills the ventricles and also surrounds the CNS within its coverings, the meninges.

This fluid is important for the mature CNS. The CNS obtains nutrients not only from the blood but also through the CSF. Cells in specialized regions next to the ventricles regulate the body's fluid balance. The CSF also functions as a communication medium: Chemicals secreted into it move through it and reach other regions near the ventricles. Where is this important fluid made, and how does it reach every level of the CNS?

The cells that line the ventricle are the ependymal cells. Remember that when the walls of the neural tube were only one cell thick, with every cell reaching from the ventricular surface to the outer surface (the pial surface), the entire CNS was a neural ependyma. In maturity, there are still cells lining the ventricle, and they are called the ependyma. Regions of the ependyma become specialized for making and secreting cerebrospinal fluid. These regions are called the choroid plexus. The cells proliferate in these regions, still attached to the ependymal layer, and they form seaweed-like strands that float in the CSF in the ventricles inside the cerebral hemispheres—the lateral ventricles. Choroid plexus is also found at the dorsal end of the diencephalic ventricle—the third ventricle—and also in the widened hindbrain ventricle—the fourth ventricle.

In animals with large cerebral hemispheres, most of the CSF is made in the choroid plexus of the lateral ventricles (figure 9a.1). From there it flows caudally through the third ventricle and the midbrain ventricle—which becomes a narrow channel through the

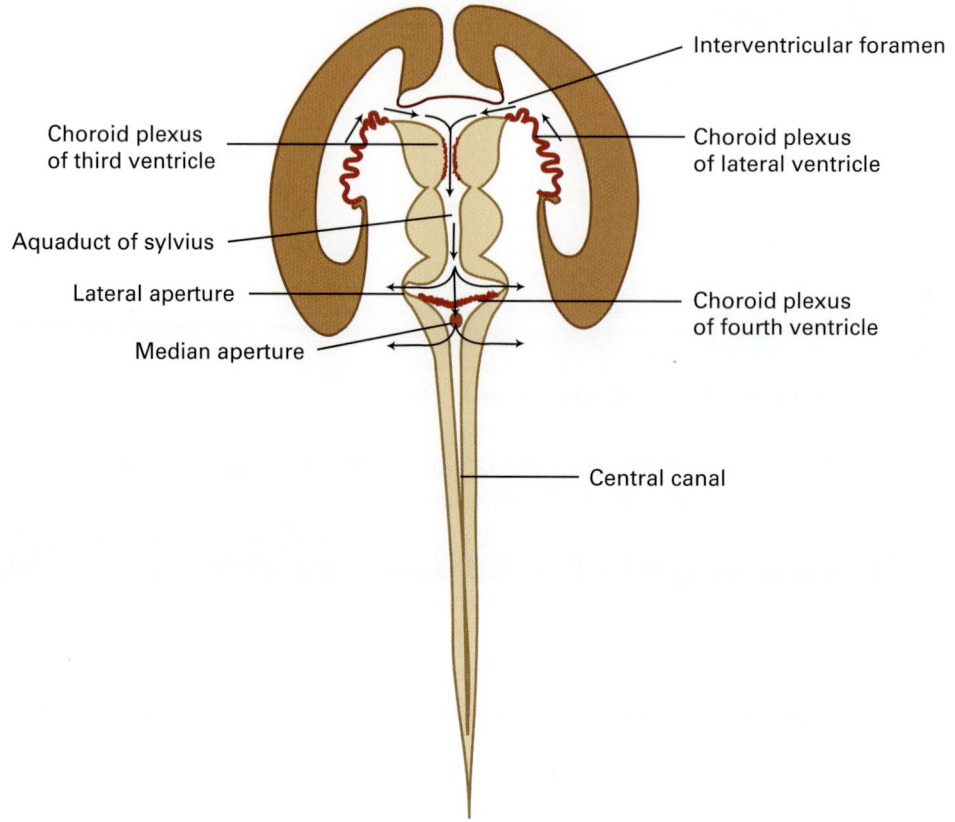

Figure 9a.1
Simplified horizontal section of the embryonic mammalian brain, adjusted to show the ventricular system and the flow of cerebrospinal fluid. The section is at a level below the cerebellum, illustrating the apertures (foramina) in the roof plate where the fluid flows out of the neural tube into the subarachnoid space surrounding the CNS. The two lateral apertures are the foramina of Luschka, and the median aperture is the foramen of Magendie (named after their discoverers). The choroid plexus (indicated in red color) consists of specialized ependymal cells that make cerebrospinal fluid. Based on Le Gros Clark (1976).

central gray area, the cerebral aqueduct or Aqueduct of Sylvius—and into the fourth ventricle. It is from the fourth ventricle that it can flow through passages into the fluid spaces surrounding the brain, within the layers of meningeal membranes. An opening is called a foramen (plural *foramina*). The foramina of Luschke are the lateral apertures, at the lateral-most margins of the roof plate. The foramen of Magendie is a median aperture at the caudal end of the widened roof plate (in the obex region).

The cerebrospinal fluid that flows outside the central nervous system is found in the *subarachnoid space* (figure 9a.2). The arachnoid membrane is located just under the outermost, and toughest, of the meningeal layers. In humans, this outer covering of the CNS,

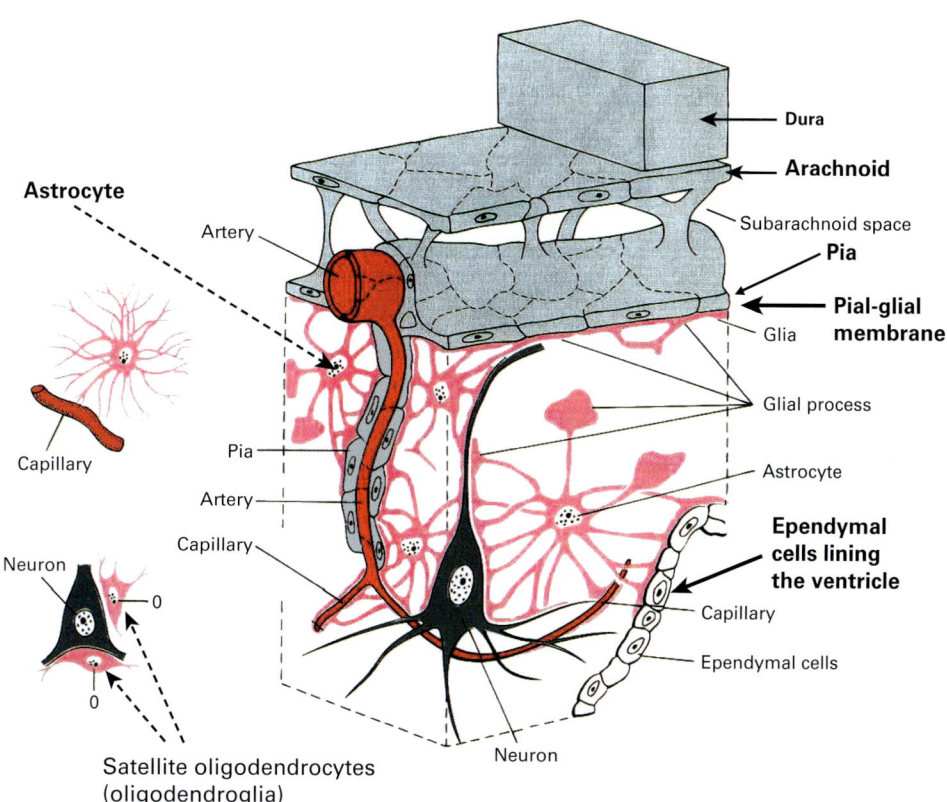

Figure 9a.2
The cellular arrangements of the three meningeal layers and glial cells in relation to blood vessels and neurons. The ependymal cells that line the ventricle are also illustrated. Based on Brodal (1998).

the *dura mater* (hard or tough mother), is like a canvas sheet. The arachnoid (spider-like) membrane gets its name from the web of fine strands that connect it with the underlying *pia mater* (soft or tender mother). The pial cells cover the surface of the CNS and also surround larger blood vessels. Abutting the pial cells everywhere are the end feet of glial cells called astrocytes, because of their starburst shapes. The astrocytic processes also surround capillaries, so the blood vessels never directly touch nerve cells. Figure 9a.2 shows these spatial relationships.

Glial cells, not only astrocytes but also the *satellite oligodendrocytes*, abut nearly all of the surfaces of neurons of mature vertebrates where they are not in contact with other neurons (e.g., at synapses).

These relationships have been seen in great detail in pictures taken with transmission electron microscopy. Reproductions of such pictures can be found in many neuroanatomy textbooks.

Now we will return to neuronal interrelationships within the CNS, as we review the caudal-most of the three main vesicles (enlargements) of the developing brain: the hindbrain in its differentiated state.

Readings

See chapter 10 in the book by Nauta and Feirtag (cited in chapters 3 and 9 of this volume); see also chapter 1 in the book by Brodal (cited in Chapter 9 of this volume) and other texts.

Ultrastructural photos (taken with an electron microscope) of central nervous system tissues:

> Peters, A., Palay, S. L., & Webster, H. (1991). *The fine structure of the nervous system: Neurons and their supporting cells.* New York: Oxford University Press.

V DIFFERENTIATION OF THE BRAIN VESICLES

10   Hindbrain Organization, Specializations, and Distortions

We can easily imagine that as the primitive chordates evolved forward locomotion, the importance of sensory inputs at the rostral end of the small creatures increased in importance, and the recipient structures of the neural tube increased in size and information processing power (see chapters 3 and 4). Three somewhat distinct enlargements characterize the brain in the earliest stages of development, and with little doubt in early stages of evolution. These three major divisions of the end of the neural tube in the head (the encephalon) we have met already. They are the brain vesicles—swellings of the rostral end of the fluid-filled neural tube—and are called the hindbrain, the midbrain, and the forebrain. (See figure 1.3 and note the alternative terms from the classical Greek.) The forebrain becomes the largest part. It includes the 'tweenbrain, located between two developmental outpouchings of the neural tube that are called the cerebral hemispheres and that become the major components of the endbrain. The endbrain also includes, as you may recall from previous illustrations, a forward extension of the ventral 'tweenbrain, known as the basal forebrain, which reaches the olfactory bulbs.

If we look at the mammalian brain in its early embryonic stages, we see three prominent flexures (figure 10.1). Viewed from the side, the neural tube bends, concave ventrally, in the region where cervical spinal cord becomes caudal hindbrain. This is the cervical flexure. Rostrally, in the midbrain region, another bend is seen, the mesencephalic flexure. Soon another bend appears, in the opposite direction, at the level of the rostral hindbrain—the pontine flexure. With the formation of the pontine flexure, the roof plate of the neural tube becomes widened. The wider roof plate when seen from the dorsal side has a rhombus shape, which led to the Greek name for the hindbrain—*rhombencephalon*. Another useful landmark is the point where the widening of the roof plate ends caudally, in the caudal hindbrain (see figure 5.5). That point is called the obex.

A Glamorized Spinal Cord

As we follow the embryonic CNS rostrally from the spinal cord into the hindbrain, we continue to see the alar and basal plate regions, but the roof plate widens as we come

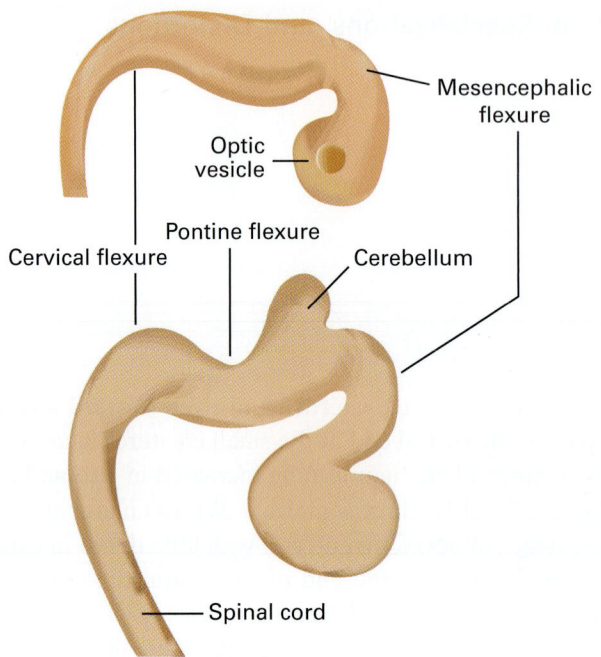

Figure 10.1
The flexures of the developing neural tube at its rostral end, viewed from the right side. The lower picture shows a slightly later stage of development. Based on Nauta and Feirtag (1986).

to the region of a wider fourth ventricle (see figure 1.3, level d). The change results in a caudal brainstem that is a little wider than the spinal cord, with the alar plate located mostly lateral to the basal plate. We no longer see regularly spaced dorsal and ventral roots; in fact, the Bell–Magendie law (the "law of roots") no longer holds. When we leave the spinal cord and enter the hindbrain, some cranial nerves are mixed nerves— containing both sensory and motor components, whereas others are like a dorsal root or like a ventral root. Secondary sensory neurons are found in specific clusters (the sensory nuclei) in the alar plate region, and motor neurons are also found grouped together in the motor nuclei, in the basal plate (figure 10.2). The structural arrangement of the caudal hindbrain can be understood as a simple modification of spinal cord organization. Because of this, the hindbrain could be called the forecord, and in fact the Latin name for the caudal hindbrain, the *medulla oblongata*, means the prolongation of the spinal cord into the brain. (Another name for the spinal cord is the *medulla spinalis*, or marrow of the spinal column.) Later, we will see how the development of the cerebellum causes an additional modification, or distortion, in this structural arrangement.

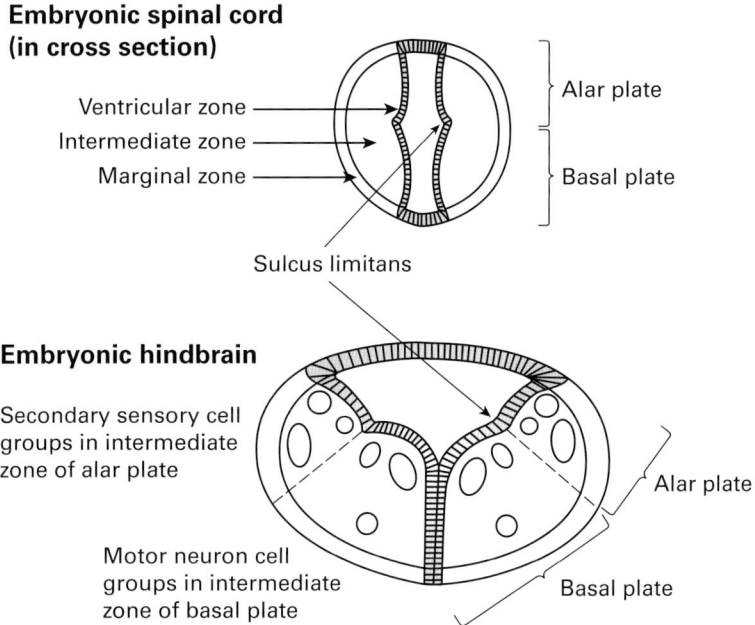

Figure 10.2
Comparison of the embryonic spinal cord and embryonic hindbrain. In the spinal cord as the walls of the neural tube thicken with cell proliferation, the roof plate and floor plate remain one cell thick (see figure 8.6). A sulcus (groove) separates the dorsal horn, where most of the secondary sensory neurons are located, and the ventral horn, where motor neurons are located. The groove is called the sulcus limitans (limiting sulcus) separating the alar and basal plates. It can be followed rostrally through the hindbrain and midbrain. In the hindbrain, the roof plate becomes wider, as shown, with the formation of the pontine flexure (figure 10.1). This results in the alar plate lying mostly lateral to the basal plate. The motor neurons are found in specific groupings related to different functions. Similarly, secondary sensory neuron groups are formed in the alar plate region.

Vital Functions of the Hindbrain

As the hindbrain evolved, it became crucial for functions that went beyond its roles in sensorimotor control of the head region. These were functions for which centralized control of the entire body had important advantages. This brain region is known as a center of vital functions critical to life, the most vital being its role in the control of breathing. Air breathing evolved long after chordates first appeared, so the question naturally arises, why was its control so centralized? The first reason is that the structures important in breathing develop from the *branchial arches* of the embryonic body and are innervated from the hindbrain. The term *branchial* refers to the gills, so important in respiration in fishes, but the branchial arches form other structures as well. It is not difficult to think of other reasons that made centralized control of breathing adaptive. Adjustments in respiration are part of many fixed action patterns controlled by hindbrain mechanisms: These include swallowing, vomiting, grooming, and righting responses. Increased respiration is part of the preparation for any actions that involve large increases in energy output as in attacking or in fleeing or any other rapid locomotion.

Closely associated with the control of breathing are adjustments in heart rate and blood flow. Hindbrain mechanisms added central modulation of heart functions governed locally and by the spinal cord.

Routine Maintenance Services

The great systems neuroanatomist Walle J. H. Nauta (see chapter 2) in his lectures used to call the hindbrain the support services department of the brain, constantly active but little noticed as it makes its contributions to the stability of the internal environment and to stability in space. Its neuronal circuits provide a higher level of control over various functions of the spinal cord, and they add controls of face and head movements. We have already mentioned the fixed action patterns (instinctive movements) organized by circuits of the hindbrain reticular formation, and more can be added (e.g., eyeblink, grooming patterns, and various emotional expressions such as smiling and frowning).

Stability in space was greatly improved by ongoing adjustments in posture and stance controlled by the vestibular system of the hindbrain, which has direct connections with the cord. The cerebellum of the rostral hindbrain was probably dominated by the vestibular system early in its evolution, and with inputs from the other sensory systems this "little brain" became the adjuster of relative timing of movements controlled by diverse inputs (see chapter 5 for a hypothesis about cerebellar evolution).

Another critical role of hindbrain mechanisms involves nonspecific widespread modulation of other brain areas. I am referring to the hindbrain's role in the animal's going to sleep or waking up, and in the accompanying changes in the arousal level of the entire system. To play such a role, some neurons of the hindbrain reticular formation have axons that are very widely branching, with multiple regions of termination (see the example in figure 10.3). In sleep-waking functions, the midbrain and 'tweenbrain also play major roles.

Hindbrain Participation in Mammalian Higher Functions

The functions of human speech depend on hindbrain controls of the tongue, lips, and breath. Emotional displays, especially in facial expressions, depend on contractions of facial musculature controlled by the hindbrain through its cranial nerves. Motor neurons controlling eye movements, a crucial component of many cognitive and social activities, are found in hindbrain and midbrain. Thus, in many higher functions, the hindbrain fulfills its reputation as a basic service provider, serving functions controlled by more rostral

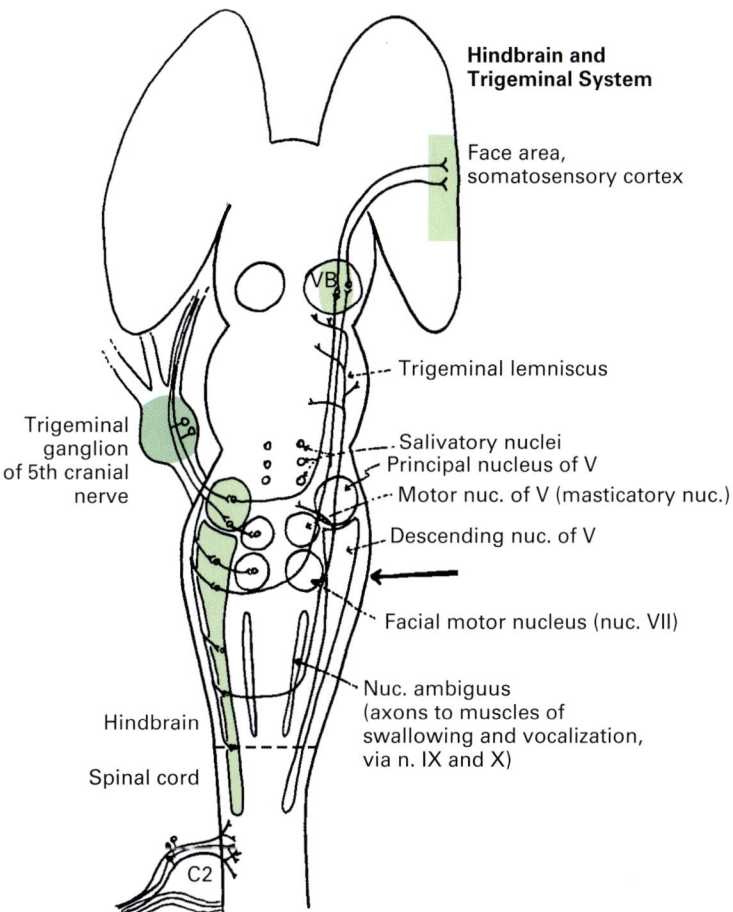

Figure 10.17
The somatosensory pathways that begin with the trigeminal nerve and ascend from brainstem nuclei to the forebrain are depicted for mammals on a dorsal view sketch of the embryonic brain. Sensory channels of the trigeminal system are like those from the spinal cord, including extensive ipsilateral and bilateral connections with neurons of the reticular formation of the hindbrain and midbrain (not included in the drawing). The primary sensory neurons are in the trigeminal ganglion, and the axons of these ganglion cells synapse in the principal nucleus and the descending nucleus of the fifth nerve. Many axons from the secondary sensory neurons decussate and ascend in the trigeminal lemniscus as far as the medial part of the ventral posterior nucleus of the thalamus (ventrobasal nucleus, VB). Axons from that nucleus project to the somatosensory neocortex. The heavy dashed line marks the border between the hindbrain and spinal cord. The arrow indicates level of figure 10.16.

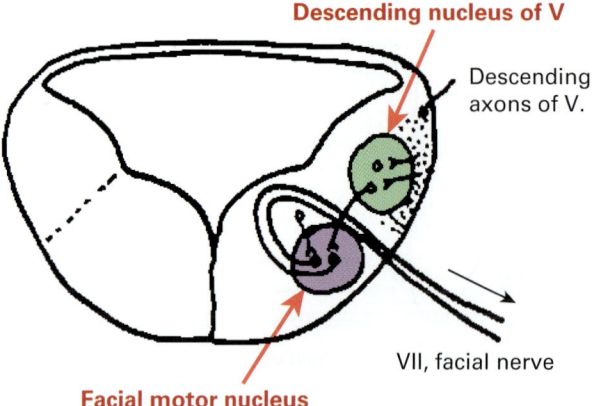

Figure 10.16
The pathway for an eyeblink reflex is depicted in this drawing of a hindbrain section at the level of the facial motor nucleus. Primary sensory axons from the trigeminal nerve (cranial nerve V) carry inputs from the cornea and the skin near the eye. These axons descend to the descending nucleus of the fifth cranial nerve where they form synapses. Axons of the postsynaptic cells extend to the facial motor nucleus (and to interneurons near it). The axons from the facial motor neurons follow an arching pathway before they exit from the side of the hindbrain in the facial nerve (seventh cranial nerve), extending to the muscles of the eyelids that control the eyeblink movement.

is similar to what we have discussed for the spinal cord. In the figure, trigeminoreticular axons are omitted; only the crossed pathway reaching the thalamus is depicted. This pathway, the trigeminal lemniscus, contains axons corresponding to both spinothalamic and medial lemniscal pathways from the spinal cord.

The figure is not as complex as it may appear at first. The elongated secondary sensory cell group of the trigeminal nerve may make it seem a little complicated. When we depict CNS pathways in any diagram, we have to simplify, simplify, simplify. If we add too many structures and too many connections, such a diagram can quickly come to resemble a bowl of spaghetti! This is not because everything is connected to everything else—far from it—but because there are so many different cell groups, each with many specific connections, some more extensive and some less.

Hindbrain Specializations and Mosaic Evolution

This is a good place for a break with a more relaxing topic before we proceed into the midbrain. You may remember some of the brain specializations pictured in chapters 4 and 6. We return to that topic here. A number of remarkable sensory and other specializations have evolved together with corresponding neuronal apparatus in hindbrain cell groups connected with them. In some cases, portions of the hindbrain have become so enlarged that the basic organization appears to become distorted. When certain

Embryonic hindbrain

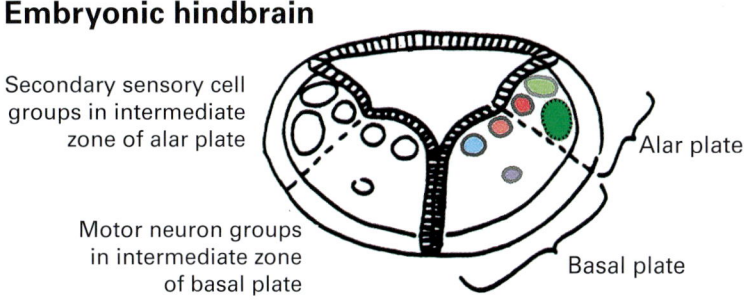

Adult hindbrain, principal cell columns and fiber tracts

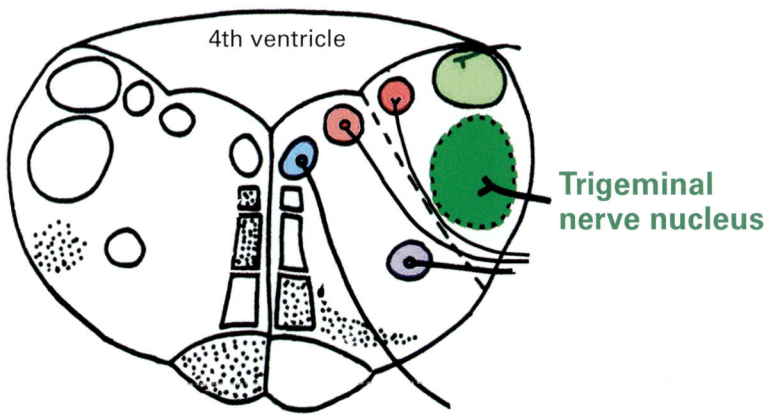

Figure 10.15
Sketches of mammalian embryonic and adult caudal hindbrain sections, from figures 10.6 and 10.9. The secondary sensory cell group of the trigeminal nerve (cranial nerve V) is indicated in green.

interneurons in or near the descending nucleus of the trigeminal nerve and interneurons connecting with motor neurons in the seventh nerve nucleus or directly with the motor neurons.

We could show the same pathway in a sketch of a more panoramic view of the trigeminal system by using an outline of the embryonic neural tube (figure 10.17). In the sketch, we show that axons of the trigeminal nerve extend caudally in the descending tract of the fifth nerve to synapse in the descending nucleus. The portion of the nucleus located within the cervical spinal cord is where inputs from the face triggered by pain-causing stimuli connect. From neurons at all levels of the brainstem trigeminal nuclei, one can trace reflex channels, cerebellar channels, and ascending lemniscal channels of conduction. The system

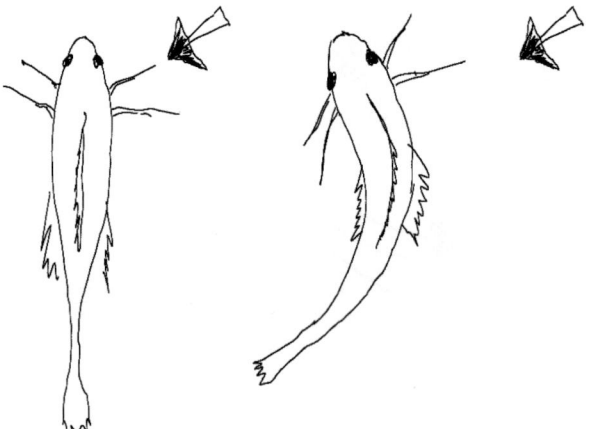

Figure 10.14
Primitive escape response. Drawing of a hypothetical primitive chordate responding to a sudden somatosensory or visual stimulus caused by something approaching it from the right side. Rapid pathways to motor neurons on the left side cause a left-sided contraction that twists the animal away from the stimulus. (The turn would be accompanied or followed by rapid forward locomotion.)

Consequently, with further evolution, the **visual** inputs, from lateral eye mainly to contralateral midbrain, became more organized for better stimulus localization. This projection was useful for orienting toward some objects as well as for escaping from others. This led to a descending pathway that crossed the midline: the *tectospinal* tract, for turning head and eyes toward a stimulus. The **somatosensory** inputs to the midbrain tectum evolved into a predominantly crossed pathway in order to match the visual projection. They responded to the same areas of space around the head and were useful for triggering the same responses.

Thus, the right side of the space around the head came to be represented on the left side of the midbrain, and subsequently the forebrain as well. The other side developed in mirror-image fashion.

Hindbrain Sensory Channels in Mammals

Sensory pathways throughout the vertebrates share many features, so we can talk about mammals and not be far off the mark for other vertebrates. If we look at a mid or caudal hindbrain section in a mammal, both embryonic and adult, it is easy to specify the secondary sensory cells receiving trigeminal nerve input (figure 10.15) as they are usually the largest cell group in the alar plate region, with input from what in many mammals is the largest cranial nerve. If our section is at the level of the facial motor nucleus, we can sketch in a pathway for the eyeblink reflex (figure 10.16), with synaptic connections between

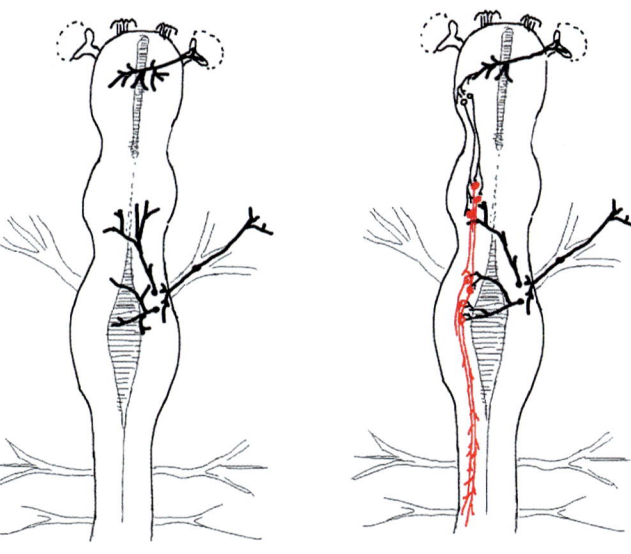

1) Bilateral optic and somatosensory projections

2) Contralateral projections result in faster contralateral body flexion

Figure 10.13
Dorsal views of brain of hypothetical ancient chordates at two evolutionary stages. (Left) A bilateral distribution of some trigeminoreticular axons is illustrated (see figure 10.12). In the same brain, the primitive input from the lateral eyes to both sides of the midline in the hypothalamus is illustrated. (Right) Later in evolution, many secondary sensory neurons receiving input from the face project to the contralateral side, synapsing with neurons that have descending axons that trigger escape movements. Likewise, axons from each eye project beyond the hypothalamus to the contralateral side to neurons with axons that conduct rapidly to escape mechanisms. See figure 10.14 for a depiction of the behavioral consequence of the activation of these descending pathways by visual detection of sudden changes or tactile detection of novel or intense stimuli.

However, the animals possessing this sensory system were faced with the life-or-death problem of avoiding predators, and any advance warnings of predator attack led to greater survival and reproduction. Such warnings could be indicated by optical changes caused by moving shadows or sudden changes in light intensity, as well as by changes in water pressure or direct contacts detected at the skin. The quicker the escape response, initiated by contraction of muscles on the opposite side of the body, the better for the animal (figure 10.14).

This escape response initiation, if it was to occur with the shortest possible latency, required connections that crossed the midline (figure 10.13, right side). Therefore, a crossed projection from the lateral eye expanded in evolution to form a rapidly conducting route to spinal motor neurons on the opposite side. As shown in the figure, somatosensory pathways no doubt reached the same brainstem neurons that connected with these spinal motor neurons.

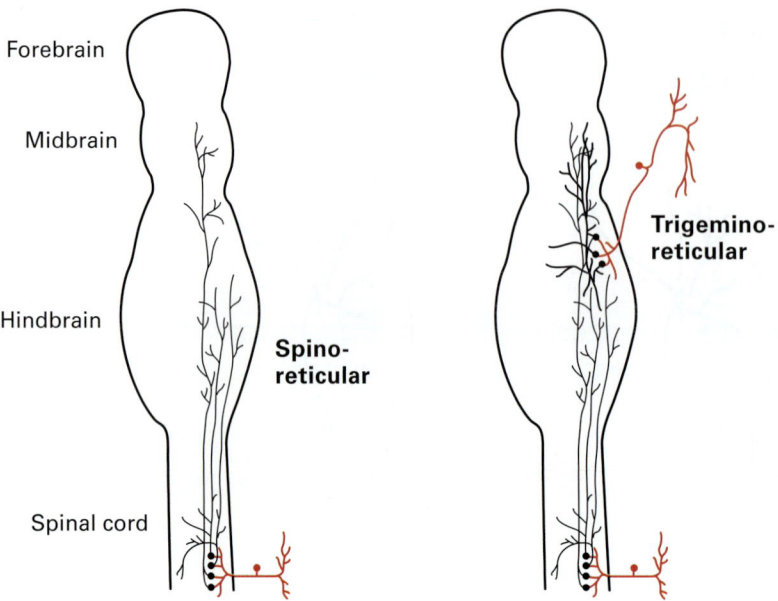

Figure 10.12
Dorsal views of hypothetical ancient chordate brain, with sketches of the oldest lemniscal pathways carrying somatosensory information from the body surface into the brain. On the left, spinoreticular axons are depicted. On the right, axons carrying inputs from the head are sketched—the trigeminoreticular axons. As in modern vertebrates, most of the secondary sensory axons of these pathways remain on the ipsilateral side, but some axons reach the contralateral side.

systems were ipsilateral or bilateral (see chapter 4 where the basic idea was introduced and chapter 5 where the spinoreticular axons were discussed).

In figure 10.12, you see a sketch of the ancient spinoreticular pathway and a duplicate sketch where trigeminoreticular axons have been added. These added axons are simply the cranial equivalent of those arising more caudally. Together, these axons form a lemniscal pathway in mammals, bilateral, inherited from ancient chordates, carrying sensory information from the face to the brainstem reticular formation, with the longest axons reaching the older parts of the 'tweenbrain.

How did such projections, predominantly uncrossed, lead to predominantly crossed pathways that became so dominant in vertebrates? The hypothesis is illustrated in the next figures.

In figure 10.13, on the left is illustrated the earliest, bilateral projections of neurons connected to the primitive lateral eyes and to somatosensory receptors of the face. This kind of optic projection was adequate for conveying information on the presence or absence of light and the intensity of light and for influencing the daily rhythm of activity.

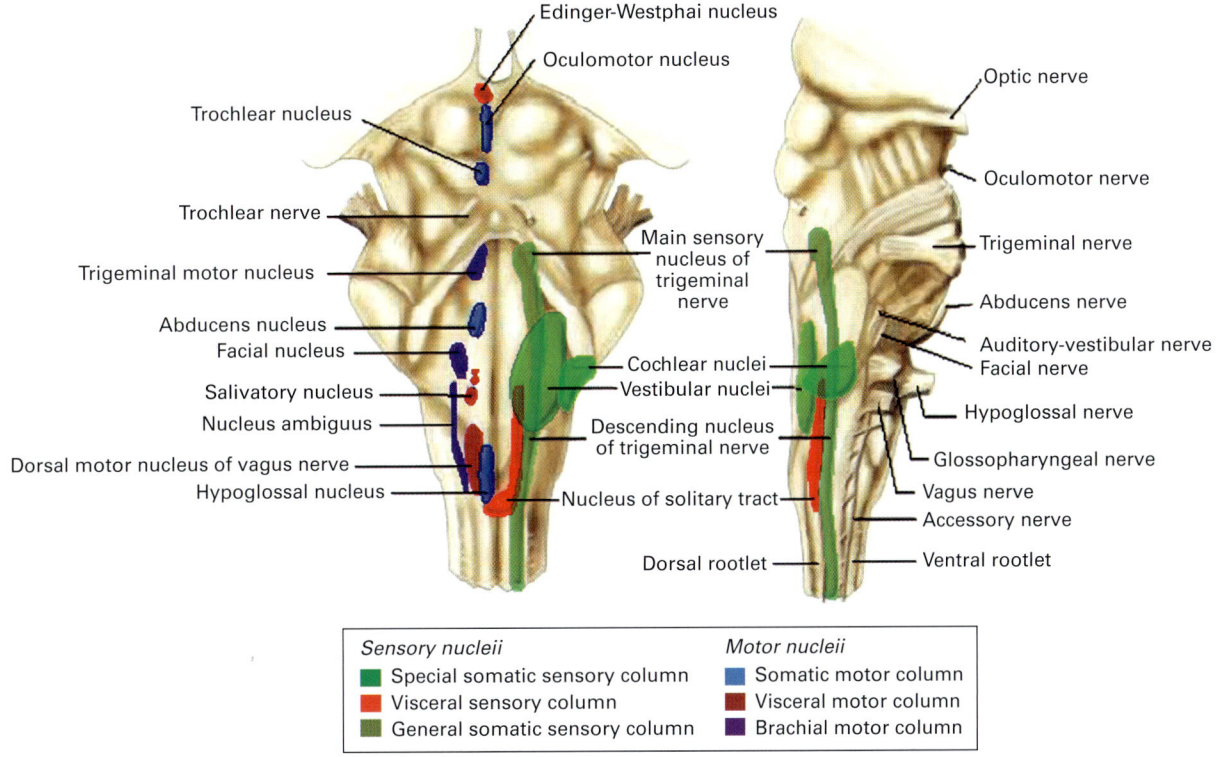

Figure 10.11
Drawings of a dorsal view of the human brainstem, from caudal thalamus rostrally to the rostral end of the spinal cord. The cerebellum has been removed (the cerebellar peduncles have been sectioned), as have the cerebral hemispheres. Imagining the brain to be transparent, in the dorsal view the positions of the motor neuron cell groups are shown on the left side of the brainstem, and the positions of the secondary sensory cell groups are shown on the right. In a side view, the positions of the secondary sensory cell groups are shown. Based on Nauta and Feirtag (1986).

terms, a latecomer. It was no doubt long preceded by trigeminoreticular pathways that were primarily ipsilateral. Then we have to ask: Why do the axons decussate? That is, why did decussating pathways to the forebrain and midbrain evolve as the dominant pathways?

The Evolution of Crossed Projections

Here I will present my hypothesis that the decussation of somatosensory axons to the midbrain and forebrain evolved with or after the evolution of the crossed retinal projection to brainstem structures like the midbrain tectum. The earliest projections in these

Hindbrain Organization, Specializations, and Distortions

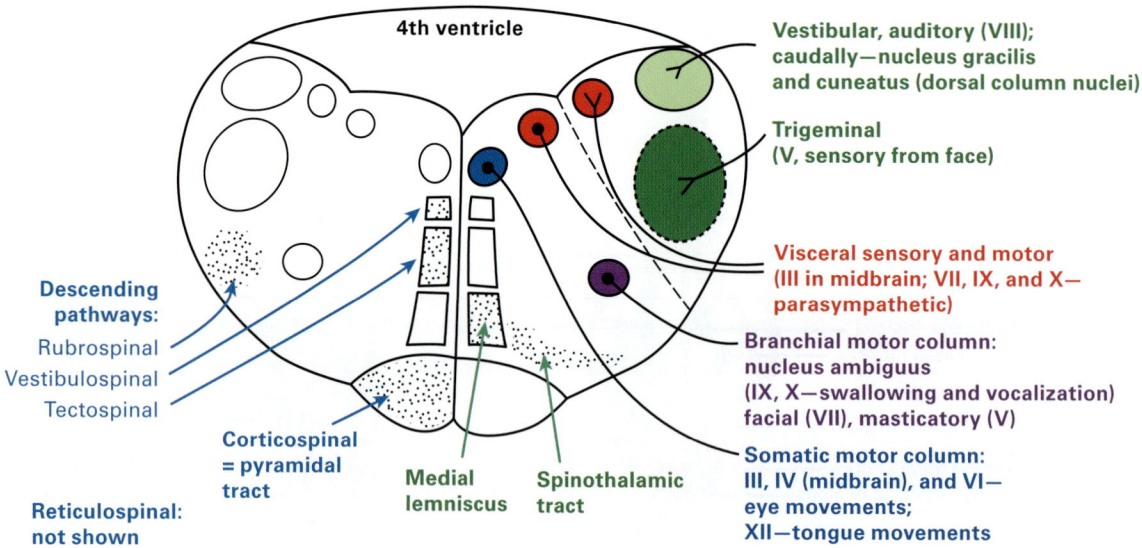

Figure 10.10
Drawing of frontal section of mammalian adult caudal hindbrain, with a schema of principal cell columns and fiber tracts sketched in. On the right side, the secondary sensory cell groups and motor neuron groups are indicated together with the cranial nerves carrying their inputs and outputs. On the left and bottom, major groups of fibers of passage are labeled. Visceral sensory and motor neurons and fibers are depicted in red.

Somatosensory Inputs from the Face

Next, we look at the sensory pathways of the CNS that start with the fifth cranial nerve, the trigeminal nerve. We can follow these axons first to their terminations within groups of secondary sensory neurons in the hindbrain: the brainstem trigeminal complex (trigeminal nuclei). This complex extends caudally into the cervical spinal cord. The functions of the connections of these secondary sensory neurons were important factors influencing the evolution not only of the hindbrain but also of more rostral structures.

We can start by tracing reflex pathways for functions like the eyeblink (see later). We can postulate connections from a trigeminal nucleus (containing secondary sensory neurons) to the facial motor nucleus (seventh nerve nucleus), which contains motor neurons that cause eyeblink. Such connections can with little doubt be found. More difficult to find or understand would be the inputs from the reticular formation responsible for the periodic, spontaneous blinking of the eyes.

Next, we can proceed, as in a typical neuroanatomy lecture, by following the fibers of the trigeminal lemniscus: the axons of trigeminal nucleus neurons, which cross the midline and ascend to the ventral-posterior nucleus of the thalamus (the medial part), which projects to the somatosensory neocortex. However, this pathway is, in evolutionary

Table 10.1
The cranial nerves

	Functions	Origins/Innervation
I. Olfactory	Sensory, special	Olfactory epithelium
II. Optic	Sensory, special	Retinal ganglion cells
III. Oculomotor	Motor, somatic Autonomic	Oculomotor nucleus of midbrain
IV. Trochlear	Motor, somatic	Trochlear nerve nucleus of midbrain
V. Trigeminal	Sensory, somatic (general) Motor, branchial	Trigeminal ganglion Masticatory nucleus
VI. Abducens	Motor, somatic	Abducens nucleus of hindbrain (innervates the lateral rectus muscle of eye)
VII. Facial	Motor, branchial Motor, visceral Sensory, visceral Sensory, somatic	Facial motor nucleus Superior salivatory nucleus Gustatory n. (nuc. solitary tract) (skin of outer ear)
VIII. Auditory–vestibular	Sensory, special	Inner ear: cochlea, vestibular canals
IX. Glossopharyngeal	Sensory, visceral Autonomic, parasympathetic Motor, branchial Sensory, somatic	Taste buds, posterior two-thirds of tongue Inferior salivatory nucleus Nuc. ambiguus (to pharynx) Trigeminal ganglion (skin, outer ear)
X. Vagus	Sensory, visceral Autonomic, parasympathetic Motor, branchial Sensory, somatic	Taste buds, epiglottis, esophagus Dorsal mot. nuc. of vagus (to thoracic and abdominal viscera) Nuc. ambiguus (to larynx, pharynx) (skin of outer ear)
XI. Spinal accessory	Motor, branchial	Accessory nuc. (spinal cord)
XII. Hypoglossal	Motor, somatic	Hypoglossal nuc. (tongue)

n, nerve; mot, motor; nuc., nucleus.

The Adult Hindbrain: Cell Groups and Axons of Passage

At this point we are ready to look at the hindbrain of an adult mammal. The caudal hindbrain is easiest to understand because it is more like the embryonic form than is the rostral hindbrain containing the cerebellum and cell groups that project to the cerebellum. Figure 10.10 shows a sketch of a frontal section with principal cell columns and fiber tracts identified. You can see the locations of secondary sensory cell groups in the alar plate and motor neuron cell groups in the basal plate. You can also see the locations of groups of axons passing through the hindbrain, ascending or descending. The neuroanatomical names add to the apparent complexity of this picture, but it will seem simpler to you after you have studied the brain more and more. This figure is presented for reference. You should understand it, but there is no pressing need to memorize it.

It would be more interesting to look at these structures in a three-dimensional model of the brainstem, if the model were transparent enough so we could see the sensory and motor neuron groups. Such models can be created and viewed in three-dimensional computer animations. Or we can simply look at pictures of such a model; for example, for the human brainstem as shown in figure 10.11.

Columns in spinal cord

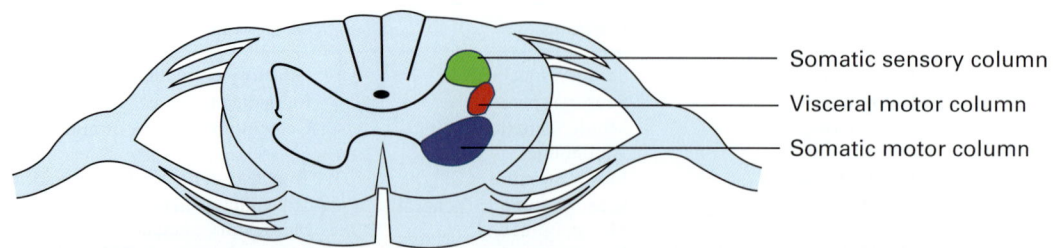

Columns in hindbrain

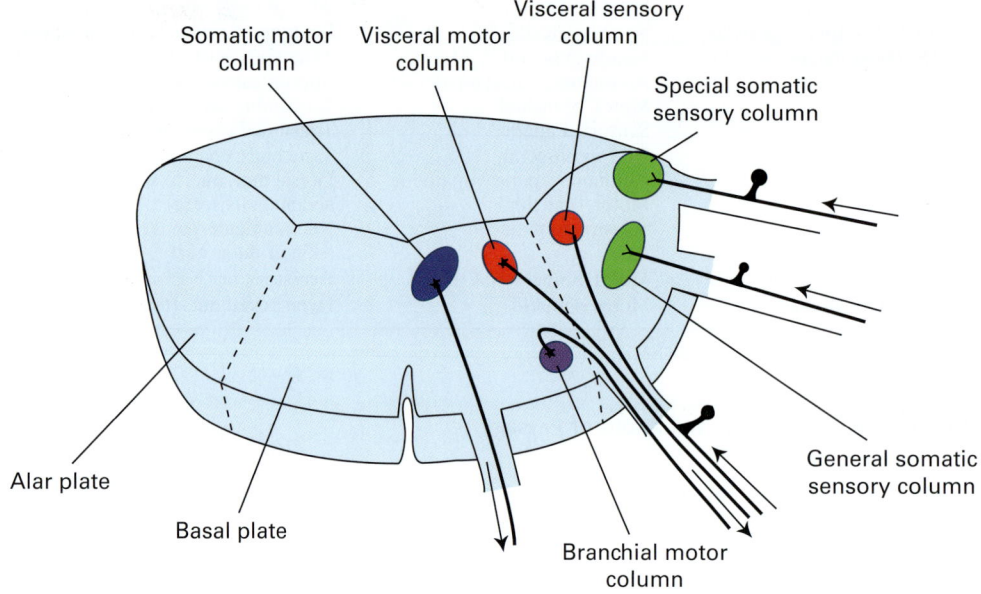

Figure 10.9
Comparison of spinal cord and hindbrain in the positions of secondary sensory neuron columns and the positions of motor neuron columns. The nature of spinal and cranial nerves is also illustrated. In the hindbrain, there is a separation of the branchial and somatic motor columns and a separation of the somatic sensory column for the head from the other sensory columns. Based on Nauta and Feirtag (1986).

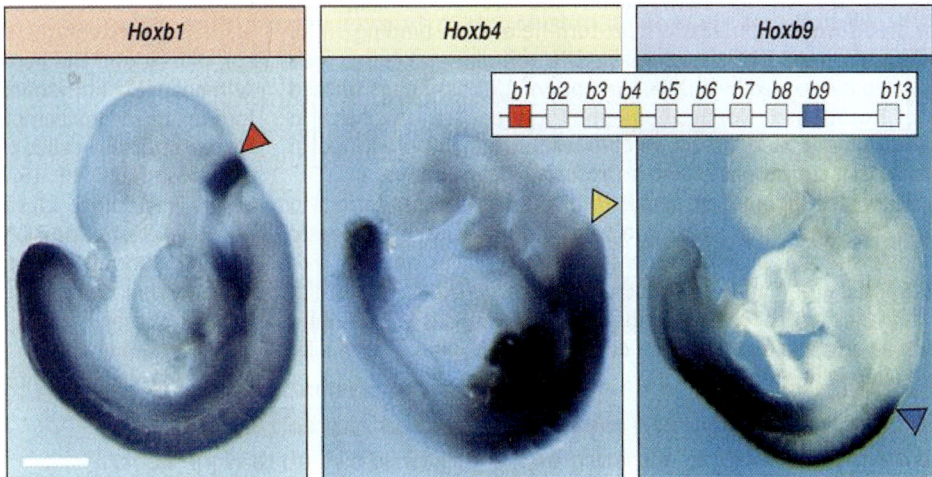

Figure 10.8
Photographs of mouse embryos at embryonic day 9.5, immunostained using antibodies specific for the protein products of the indicated hox genes. The position of rostral-most expression is different for the three proteins, in an arrangement that corresponds to the positions of the genes on the chromosome (inset). Note that these genes are expressed in both neural and non-neural tissues. From Wolpert, Jessell, Meyerowitz, Robertson, and Smith (2007). Photographs by Alex P. Gould.

When comparative anatomists study the brains of various vertebrate species, as well as the developing brain of mammals, they can distinguish at least 25 separate cranial nerves. Some of these nerves are transiently distinguishable during ontogeny or they are of little or no known function in mammals, and in some of them two nerves have combined to form a single nerve. Thus, the facial (VII), glossopharyngeal (IX) and vagus (X) each contain two distinct parts. Some cranial nerves are found only in specific groups of animals as the six lateral line nerves.

In humans and other mammals, we traditionally identify 12 cranial nerves, numbered from the most rostral to the most caudal. Table 10.1 lists these nerves. For a helpful mnemonic for remembering the cranial nerves, see the short section inserted between chapters 17 and 18.

Note that cranial nerves I and II are forebrain nerves, and the second nerve is actually a tract of the CNS, as its fibers are axons from the retina, which is part of the CNS. Cranial nerves III and IV are midbrain nerves, and all the rest are hindbrain nerves.

The primary sensory neurons of the cranial nerves do not all originate from the neural crest of the head region. Some of them originate in specialized epithelial cells of *sensory placodes*. These include the primary sensory neurons of the olfactory epithelium, originating in the nasal placode. Placodes also contribute to the fifth, seventh, eighth, ninth, and tenth cranial nerves.

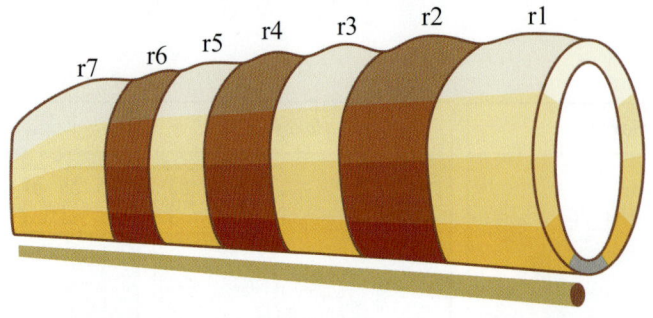

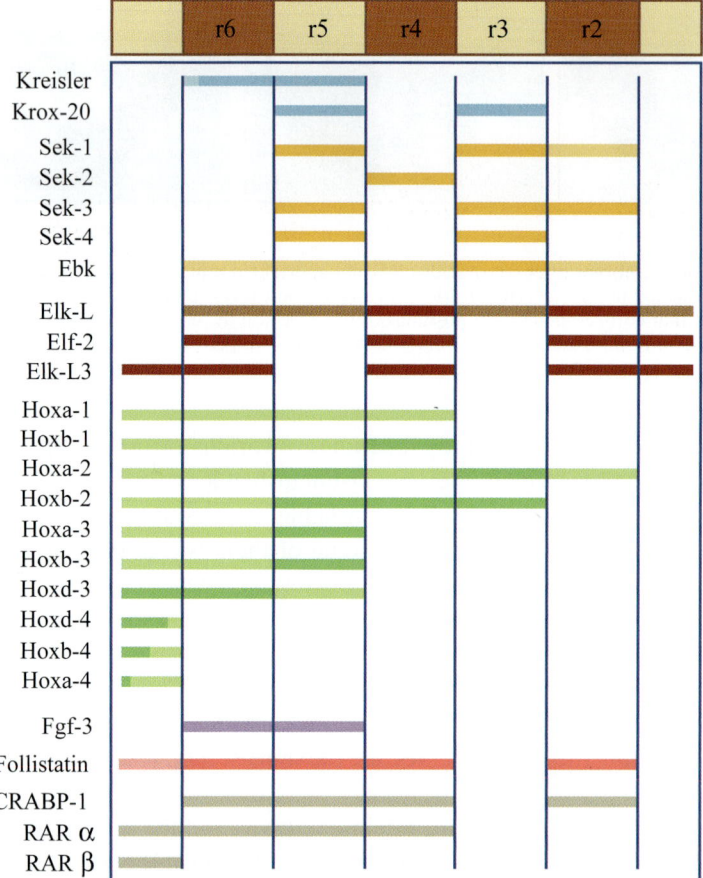

Figure by MIT OpenCourseWare.

Figure 10.7
The upper drawing is a simplified drawing of the embryonic segmentation of the hindbrain. These *rhombomeres* can be seen transiently in a surface view as a series of bulges. The lower drawing illustrates expression patterns of transcription factors in the tissue of these hindbrain segments. Many of these genes are expressed in the segments on either side of another segment that shows much less or no expression. When pictured in this way, the expression patterns remind the viewer of a bar code. From Lumsden and Krumlauf (1996).

188   Chapter 10

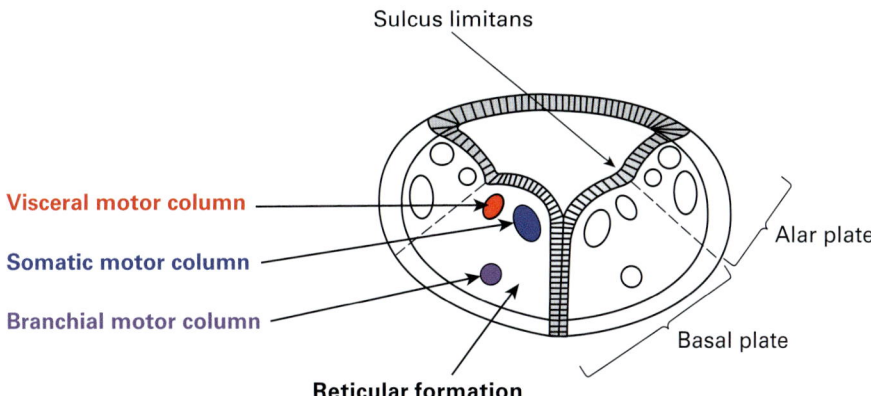

Figure 10.6
Drawing of a frontal section of the embryonic mammalian caudal hindbrain. In the alar plate, the positions of secondary sensory cell groups are indicated. In the basal plate, the positions of motor neuron groups are shown. In a cell-body stain, these groups usually stand out from the neurons of the reticular formation. Most lateral is the visceral motor column, consisting of preganglionic motor neurons of the parasympathetic nervous system—the dorsal motor nucleus of the vagus nerve. More medially located is the somatic motor column, the position of motor neurons that at rostral hindbrain levels innervate eye muscles (the abducens); more caudally, somatic motor neurons innervate the tongue muscles. Located more ventrally are the motor neuron groups of the branchial motor column. These neurons innervate the muscles that develop in the region of the embryonic branchial arches. Rostrally are the neurons controlling jaw muscles; caudal to these are the neurons that innervate facial muscles. More caudal still are the neurons that innervate the muscles of swallowing and vocalization.

hox genes are expressed differentially at various levels of both CNS and non-CNS tissues of the body.

Beyond the anterior-to-posterior specification of levels of the body and the nervous system are many questions about how hindbrain functions are specified genetically. We know that many fixed action patterns (innate behavior patterns) and many unlearned reflexes are organized in the hindbrain reticular formation, but we know how these patterns are specified only in general terms, from our knowledge of how some organized maps are determined (see chapter 13).

Columns and Cranial Nerves

The secondary sensory neurons and the motor neurons are found in longitudinal columns in both spinal cord and hindbrain, but in the hindbrain these columns are not continuous as in the cord. In figure 10.9, the positions of these columns are illustrated in schematic frontal slabs of cord and hindbrain (embryonic human or adult rodent). The figure includes the approximate positions of cranial and spinal nerves or nerve roots that enter and exit the CNS.

Hindbrain Organization, Specializations, and Distortions

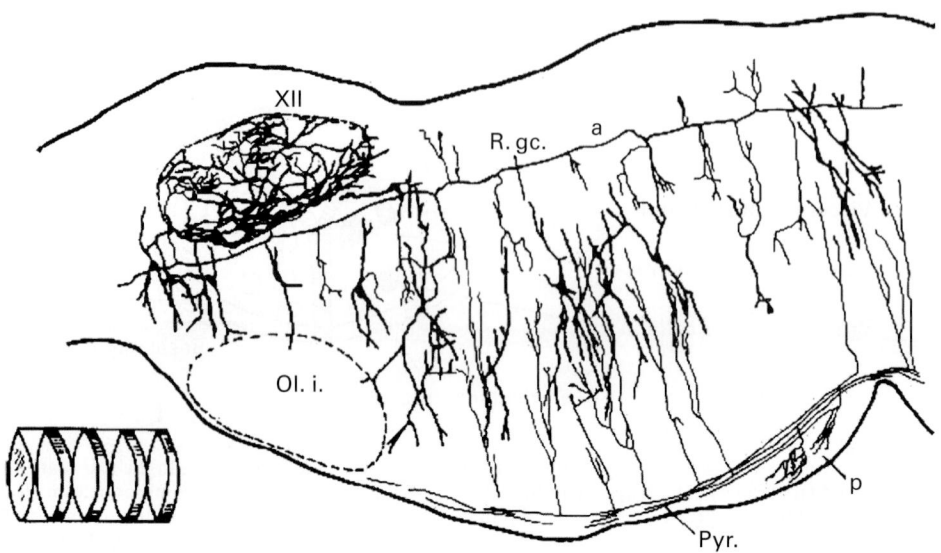

Figure 10.5
Golgi stain, parasagittal section of the hindbrain of a young rat (traced by Scheibel and Scheibel, 1958). Note the dendritic orientation of reticular formation neurons and the similar orientation of collaterals of pyramidal tract (Pyr.) axons and of the axon (a) of a reticular formation neuron (part of nucleus labeled R. gc., reticularis gigantocellularis). This provides evidence for a series of segments formed by the distribution of dendrites and axons. Anatomists refer to these as neuropil segments. Note the contrasting cellular architecture of the hypoglossal nucleus (the motor neurons of cranial nerve XII, labeled XII) and of the axonal arbors in the pons (P). Ol. i., inferior olive. From Scheibel and Scheibel (1958).

regular segmental-like divisions as at spinal levels. Motor neurons of the so-called branchial motor column (figure 10.6) innervate the branchial arch tissues, including the jaw muscles (motor component of the fifth cranial nerve) and facial muscles (facial motor, part of the seventh cranial nerve). In addition, the branchial arch tissues include the pharynx and larynx, innervated by the ninth and tenth cranial nerves and important for swallowing and vocalization.

Although the cranial nerves do not correspond to brainstem segmentation, some kind of hindbrain segmentation does become apparent at specific stages of the developing embryo. These segments, seven or eight in number (depending on details of their definition), are called the *rhombomeres*. They appear in surface views of the hindbrain as a series of bumps. There have been debates about the functions of this segmentation, but debates about their reality have been quieted by gene expression data (figure 10.7). The genes encode for transcription factors. Many of these contain a similar sequence of amino acids and are known to be involved in segmentation of the body of animals from fruit flies to mammals along the anterior to posterior axis. They are known as the homeobox genes, usually called hox genes in vertebrates. In figure 10.8, you can see evidence that

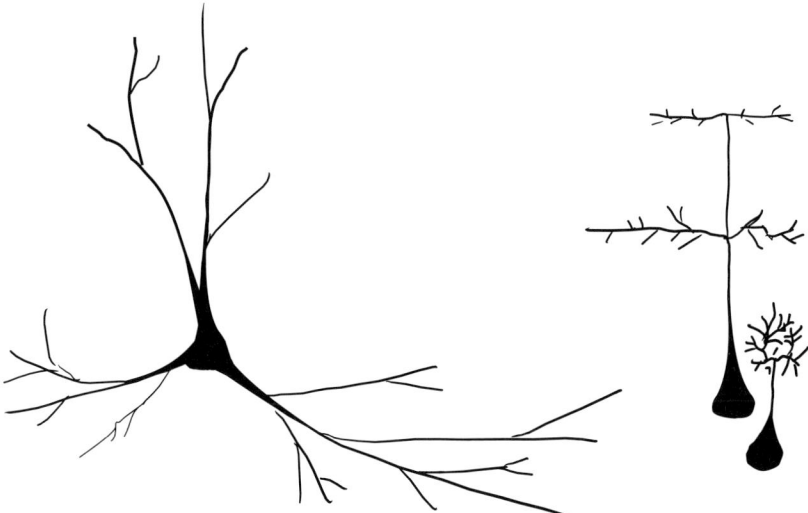

Figure 10.4
Two major types of neurons in the brainstem were described by Ramon-Moliner and Nauta and are sketched here. They are distinguished by the shapes of their dendritic trees and designated as isodendritic (left) and idiodendritic (right). The prefix *iso-* means "the same"; the dendrites are very similar to each other so they do not make the cell a recognizable type separable from others. This type of neuron, similar in shape to many motor neurons, is found throughout the reticular formation at the core of the midbrain and hindbrain. The prefix *idio-* means "distinctive"; the dendrites make the cells distinctive and recognizable types. Such neurons are found in many secondary sensory cell groups as well as in other locations.

mostly in planes at right angles to the long axis of the brainstem, so they form a series of disk-shaped neuropil segments. The collaterals of pyramidal tract axons are distributed in similar planes, raising the possibility of a relationship to some kind of functional modules. This arrangement contrasts with the cellular arrangements in specific sensory or motor nuclei (e.g., see the motor neurons of the hypoglossal nucleus in figure 10.5).

Segmentation of the Hindbrain

At spinal levels, the embryonic mesoderm becomes distinctly segmented as the *somites* appear on either side of the body near the cord. The spinal nerves are constrained, as they grow to the periphery, to extend in between adjacent somites, and thereby a kind of segmentation is imposed on spinal levels of the cord. (Differences in gene expression have also been discovered.) Other kinds of segmentation appear in the head region of the embryo, including the somitomeres and visceral arches. The visceral arches include the branchial arches, from which come the gill arches of fishes. Much of this mesodermal tissue becomes innervated by the hindbrain cranial nerves, but these nerves do not form

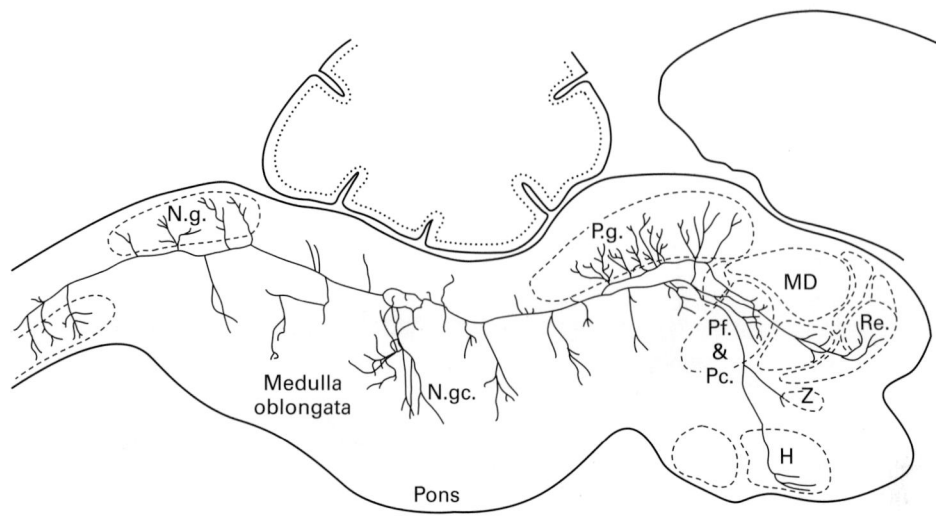

Figure 10.3
In a thick parasagittal section of a 2-day-old rat brainstem that had been processed with the rapid Golgi stain, Scheibel and Scheibel (1958) traced a large neuron of the hindbrain reticular formation. The axon, as shown, has ascending and descending branches, each with a widespread distribution of terminations. The descending axon reaches into the rostral spinal cord and also terminates in a dorsal column nucleus. The ascending axon has terminal branches in the midbrain's periventricular region (the central gray, or periaqueductal gray), in several medially located cell groups in the thalamus, and in the hypothalamus. MD, mediodorsal nucleus of thalamus; N.g., nucleus gracilis; N.gc., nucleus reticularis gigantocellularis; Pf. & Pc., parafasciular and paracentralis nuclei; P.g., periaqueductal gray; Re, nucleus reunions (a thalamic midline nucleus); Z, zona incerta (in subthalamus). From Scheibel and Scheibel (1958).

levels of the CNS. Comparatively recently evolved, these functions make use of the much more primitive output mechanisms of the hindbrain and midbrain. They do this by means of many direct connections with axons that originate in the forebrain.

The Isodendritic Core of the Brainstem

The neurons of the reticular formation of the hindbrain and midbrain have what has been called an *isodendritic* morphology (figure 10.4, left): The dendrites, like those of motor neurons, are few and without the peculiarities of shape that characterize many more specialized neurons of the CNS. Some of these neurons have widely branched axons that appear equally unspecialized, as illustrated in figure 10.3. On the right side of figure 10.4, two neurons with more specialized dendritic shapes are sketched. Such neurons have been termed *idiodendritic* because of their distinctively shaped dendrites.

In the isodendritic core of the brainstem, although the dendrites of reticular formation neurons do not have specialized shapes, a pattern can nevertheless be seen in the way the cell bodies and dendrites are arranged (figure 10.5). The dendrites are arranged cell groups enlarge proportionately much more than others, we refer to it as mosaic evolution.

Thus, in chapter 4 we saw the huge vagal lobe of the freshwater buffalofish, with the specialized palatal organ containing taste receptors. In chapter 6, we discussed the incredibly expanded cerebellar valvula of the mormyrid fish, which use electroreception to locate objects in the murky waters around them. We also saw a little of the evidence for the expansion of the visual systems of primates, and of portions of the somatosensory system representations of the vibrissae of rats and mice, and of the hands in raccoons.

The cerebellum is very large in mammals, especially in humans. During the developmental period, there are cell migrations from the alar plate of the rostral hindbrain that cause major "distortions" of hindbrain anatomy in large mammals (figures 10.18 and 10.19). As indicated in the drawings, the rather massive growth of both cerebellum and pons occurs with the migration of neuroblasts from the so-called rhombic lip region of the rostral hindbrain's alar plate. The cerebellum grows as large numbers of neuroblasts migrate into the roof plate and then continue to proliferate. The growth of the pons, and also other structures that project to the cerebellum like the inferior olive, results from cell migrations from the same rhombic lip region. Thus, these pre-cerebellar cell groups are alar plate derivatives despite their adult locations within what was originally the basal plate region.

The large size of the pontine gray matter corresponds to the large size of the major source of inputs to it, namely the neocortex, and to the large size of the target of the axons from the pontine cells, the cerebellar cortex. Recall figure 7.8, which shows the corticopontine connection in a schematic mammalian brain. The photograph of figure 10.20 is of a frontal section through the human rostral hindbrain. The cerebellum is so huge that most of it has been cut off in order to prepare the section more easily. The very large pons and cerebellum cause a distortion, in proportionate sizes of the components, of the basic hindbrain plan.

202   Chapter 10

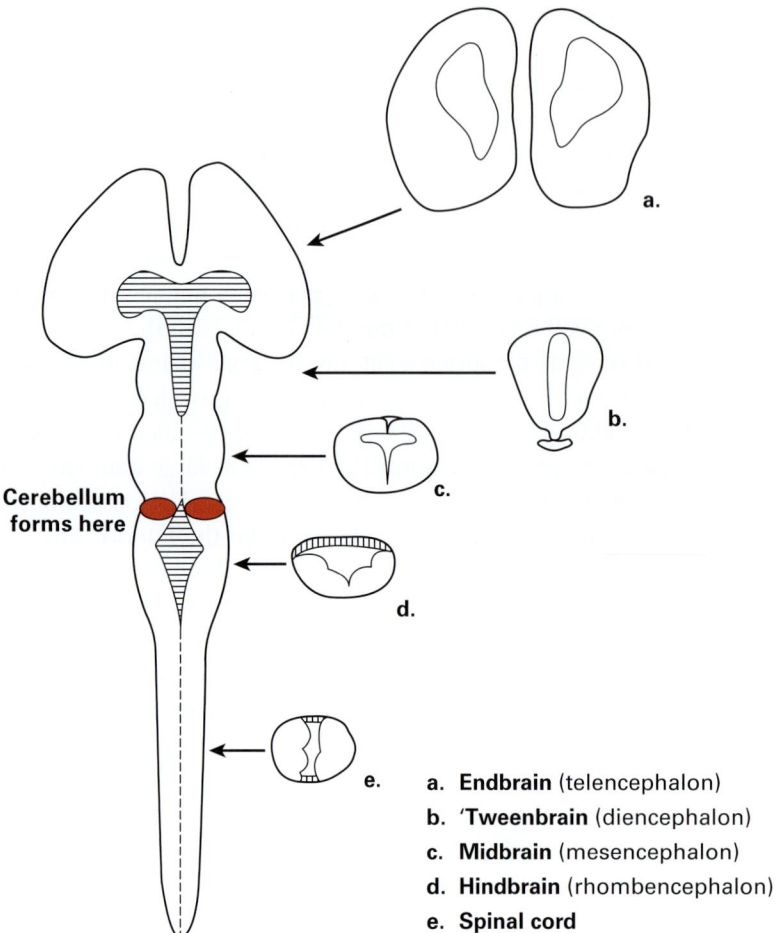

Figure 10.18
In this sketch of the embryonic mammalian brain and spinal cord, the position of the rhombic lip is shown—the embryonic proliferative zone that develops into the cerebellum.

Hindbrain Organization, Specializations, and Distortions

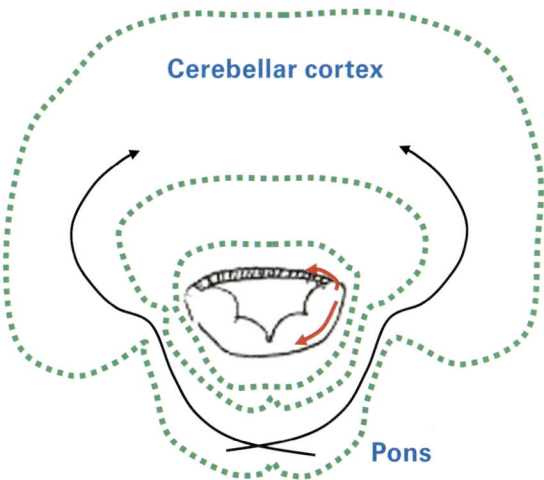

Figure 10.19
A sketch of the embryonic hindbrain is shown together with the pathways of cells that migrate from the proliferative zone of the rhombic lip. Some of the ventrally migrating cells form the gray matter of the pons. Cells that migrate into the roof plate continue to proliferate as they form the cerebellum. With dotted lines, successive stages of the resulting great enlargement of the rostral hindbrain are indicated. This great amount of growth causes a major distortion of the original outline of the hindbrain at the pontine level. In addition, the course of axons from the neurons of the pontine gray matter to the cerebellar cortex on the opposite side is indicated. Red arrows: migration of neuroblasts. Black arrows: course of axons from pontine gray to cerebellar cortex.

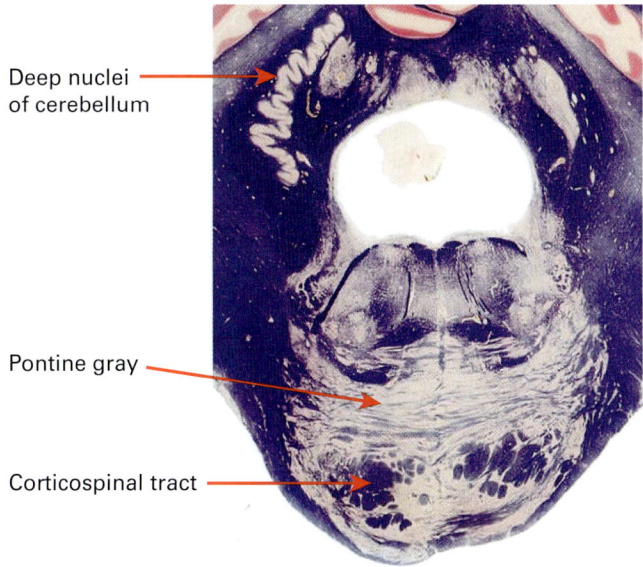

Figure 10.20
Photograph of part of a frontal section of the rostral human adult hindbrain stained for myelin (blue-black). The original shape of the hindbrain is hidden in the center of the picture, with a large pons ventrally and a very large cerebellum dorsally. The deep nuclei of the cerebellum can be seen, but most of the cerebellar cortex has been cut off. From Nolte and Angevine (1995).

Readings

The hindbrain is introduced and described in many neuroanatomy texts, including those cited earlier (books by Walle J. H. Nauta and Michael Feirtag, and by Per Brodal). Another book with a particularly interesting description:

>Swanson, L. W. (2011). *Brain architecture: Understanding the basic plan,* 2nd ed. New York: Oxford University Press.

Comparative anatomists have distinguished at least 25 separate cranial nerves: See Butler and Hodos (2005, book cited in chapter 3), their chapters 9–12.

Listings of cranial nerves in humans and other mammals can be found in many anatomy textbooks. See, for example, Nolte, J. (2002). *The human brain: An introduction to its functional anatomy*, 5th ed. St. Louis: Mosby.

An interesting paper on the developmental patterning of the hindbrain and other levels of the CNS:

>Lumsden, A., & Krumlauf, R. (1996). Patterning the vertebrate neuraxis. *Science, 274,* 1109–1115.

11 Why a Midbrain? Notes on Evolution, Structure, and Functions

Although the primitive chordate amphioxus, living in sandy ocean floors near the shore in warm regions, does possess midbrain and forebrain components of a CNS (see figure 3.3), we know next to nothing about connections and functions of these components. Gene expression studies have provided the strongest evidence for separate midbrain, 'tweenbrain, and endbrain regions, although the endbrain does not include cerebral hemispheres (see chapter 3).

All the chordates, therefore, contain the same brain vesicles (hindbrain, midbrain, forebrain), but in the vertebrates the endbrain has expanded in the cerebral hemispheres.

It is likely that very early in chordate evolution, the endbrain was the recipient of olfactory inputs, and the 'tweenbrain region was a controller of body state by secretions into the circulation. This body-state control changed with a daily rhythm that was synchronized with the day–night cycle by direct visual inputs sensitive at least to different levels of illumination. (We will explain this later.) It also changed with status of blood components and temperature that were detected by 'tweenbrain and/or other sensors. But what about the midbrain?

We know enough about midbrain functions to suggest why it evolved along with the forebrain. The midbrain, together with those primitive components of the forebrain, was a kind of rostral extension of the hindbrain that enabled visual and olfactory influences on fixed action patterns (FAPs; like locomotion, orienting, and foraging patterns) and that added more control of movements by system-wide needs—the motivational states. Thus, the midbrain received inputs from visual and olfactory processors in 'tweenbrain and endbrain, as well as sensory inputs from more caudal structures, including visceral inputs. Also, it was strongly interconnected with the hypothalamus of the caudoventral forebrain.

Thus, we can suggest the roles of 'tweenbrain and endbrain in relation to the midbrain in primitive chordates in a simple hypothesis. Olfactory and visual inputs to the forebrain structures influenced movement via connections to the midbrain. In addition, motivational states, often reflecting the state of the internal environment detected by monitoring blood components (e.g., glucose) and temperature, also influenced movements by using connections from the hypothalamus to the midbrain, as well as by causing changes in the endocrine system via the attached pituitary organs.

Primitive Vision

There is little doubt that an early role of optic input to the 'tweenbrain was the control of daily cycles of activity, with entrainment of the endogenous clock by the day–night cycle and determination of locomotor activity levels by ambient light levels. Both the pineal eye and the lateral eyes have such a role. The pineal eye has disappeared in many vertebrates, but the pineal gland retains its role in controlling the daily rhythm of melatonin secretion, important for an animal's going to sleep. For this function it receives indirect visual input originating in the lateral eyes. The lateral eyes retain the important role of timing the daily rhythm of overall activity levels. Also, breeding status and breeding seasons are controlled by the length of the daylight period in many animals.

The retina develops as an outpouching of the neural tube in the hypothalamic region of the 'tweenbrain, and it projects bilaterally to neurons located close to the region where the developmental outpouching toward the surface of the head occurred. These neurons are found near the midline just above the optic chiasm in modern mammals—the suprachiasmatic nuclei (see chapter 20). Many other, much sparser, connections of the retina to the hypothalamus have also been reported. It is very likely that the retinofugal axons that extend further from the optic stalk to more caudal parts of the neural tube evolved later, including axons to the subthalamus, dorsal thalamus, epithalamus (pretectal area), and midbrain roof (tectum).

The suprachiasmatic nucleus influences many activities of the organism that depend on time of day, especially the actions of going to sleep influenced by anterior hypothalamus, which receives axons directly from the nucleus. Various cyclic motivational states and behaviors under hypothalamic control are also influenced by this system (foraging and feeding, drinking, nesting, etc.). Hypothalamic structures and functions will be considered further in chapters 25 and 26.

Primitive Olfaction

Olfaction was and remains an important controller of behavioral state. In many vertebrates, it plays a critical role in the detecting of sexual and individual identity of other members of the same species. These functions strongly influenced the evolution of an outpost of the ventral striatal region, the amygdala (see chapter 29)—a structure that also includes pallial components.

Olfaction was a "distance sense"—for detecting stimuli originating at some distance from the body—and thus became important early in chordate evolution for advanced warnings as well as for guidance toward food and social opportunities. With rudimentary plasticity of connections (learning), it could guide adaptive habits associated with a "good place" or a "bad place," and this must have influenced the evolution of the medial pallium—which became the hippocampal cortical region.

Besides this spatial function, olfaction was also important, in conjunction with taste inputs, for discriminating what was "good to consume" and what was "bad to consume." In this also, learning was important, so the plasticity of the striatal links to approach and avoidance controllers, which could change with feedback from the taste system and from previously learned associations, was important for survival.

Connections from the forebrain to more caudal structures provided the means by which olfactory inputs could control an animal's movements. The major links were in the midbrain. By means of axonal connections to the midbrain, directly or through connections in the 'tweenbrain, olfactory input could cause locomotion toward or away from the source. It could also result in escape from predatory threat or it could sensitize the animal to orient toward food or mild novelty. For these functions, the relevant structures were a midbrain locomotor area (MLA) and the midbrain tectum.

A Structural Consequence of the Priority of Escape Behavior for Survival

Optic inputs were very useful for triggering rapid escape from approaching predators or potential predators, especially when those inputs could give information about location. Here the hypothesis developed in the previous chapter is summarized and extended as follows.

In very early chordates—as in pre-chordate ancestors—escape mechanisms had evolved in the somatosensory system and were present in the hindbrain (see figure 10.13). The most rapid route from the visual world seen by a lateral eye to the mechanisms for turning away required a crossed projection to the midbrain. This was in order to engage an already existing descending uncrossed pathway for escape behavior. This was the impetus for an evolutionary progression from bilateral optic projections to contralateral projections (see figure 10.13).

Later, with evolution of optical image formation and some topographic organization of the retinal projection, the orienting mechanisms of the tectum evolved. These required a re-crossing of the midline for greatest efficiency of connections, hence there evolved the crossed tectofugal pathways for orienting toward objects that we find in modern vertebrates.

It makes sense that the evolution of a crossed visual representation in the midbrain led to the crossed topographic representations of the outer world in the midbrain and forebrain for somatosensory and auditory inputs as well. The somatosensory and auditory inputs were most useful when they matched the visual. Related matching phenomena have been found in experiments even in modern mammals and birds: There is strong evidence that changes in the visuotopic map in the midbrain, during development or as a result of an experimental distortion, results in a shift in auditory and somatosensory representations in order to match the visual representation.

The Midbrain Correlation Centers

We have begun to answer the evolutionary question "Why a midbrain?" It evolved as a major controller of various instinctive movements and the motivational states that modulate them (the fixed action patterns). With the adaptive advantages of speed in the triggering of escape responses, visual inputs from the evolving retina to the 'tweenbrain and midbrain became crossed, and the importance of these responses and of coordination with other modalities resulted in the auditory and somatosensory inputs from the hindbrain and spinal cord to the midbrain becoming crossed as well, matching the visual. These three inputs interacted with motivational modulators in the midbrain tectum, where sophisticated circuits for input analysis connected with outputs that controlled major multipurpose movements of locomotion, orienting and grasping.

It is likely that all this first evolved when major functions of the forebrain were control of the internal milieu and motivational states, and its major sensorimotor function involved only olfaction. The advent of optic inputs that could distinguish at least right and left led to evolution of non-olfactory forebrain sensorimotor functions as well as midbrain sensorimotor functions, but when this began it is likely that the midbrain tectum was the dominant sensorimotor interface.

Now we can describe in a little more detail some of the midbrain's specific structures. Figure 11.1 shows the position and shape of the midbrain in the embryonic mammalian neural tube. The midbrain was discussed briefly in chapter 4 (see figure 4.8). Already mentioned in that chapter were two important entities that served multipurpose movements: the MLA of the caudal midbrain reticular formation and the midbrain tectum, called the superior colliculus (SC) in mammals (rostral or uppermost "little hill" of the dorsal midbrain surface). The MLA is an important controller of approach and avoidance movements, and it receives strong connections from a hypothalamic locomotor region, and also receives a direct projection from the ventral striatum/pallidum. The midbrain tectum is often called the optic tectum in vertebrates because of the retinal projections to its surface layers. Its deeper layers receive auditory and somatosensory inputs in mammals, and each of these layers receives strong neocortical projections in addition. Its outputs elicit two major functions: escape movements via an ipsilateral descending projection to the reticular formation of the midbrain and hindbrain, and orienting toward an important or novel stimulus via a crossed descending projection (the tectospinal tract). The origin of the tectospinal tract is illustrated by a sketch of one neuron of origin in a deep tectal layer in figure 11.2.

Figure 11.2 also shows a central region surrounding a small fluid-filled ventricle. This region is the central gray area (CGA). Below it near the midline at the base of the

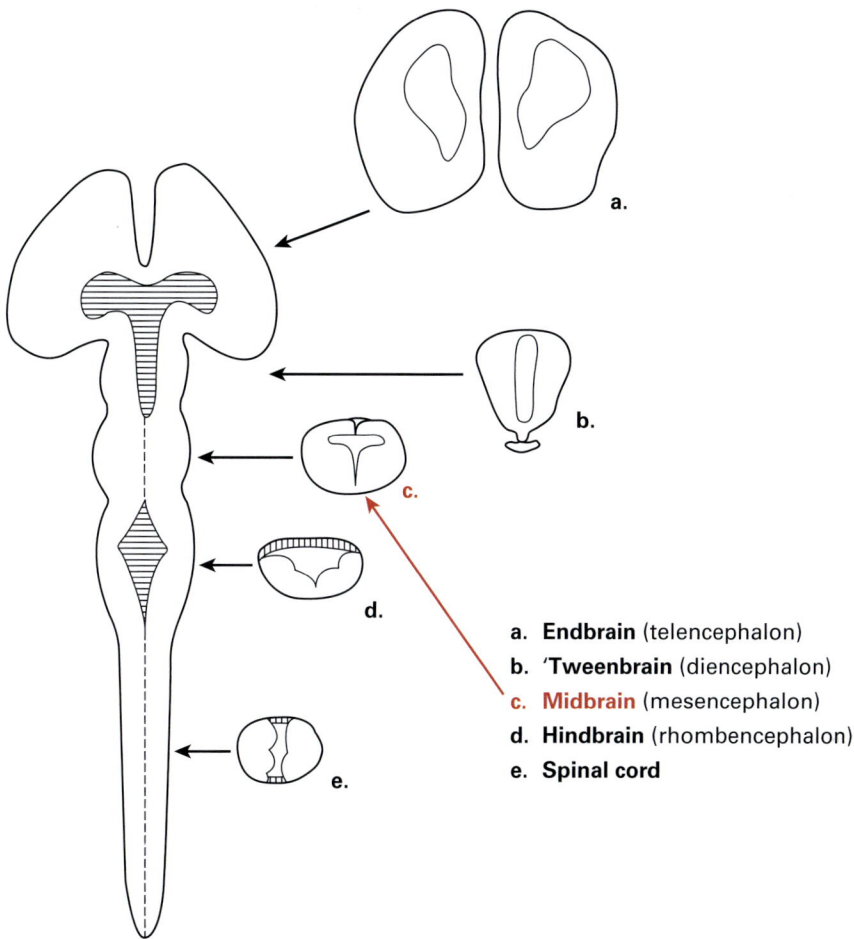

Figure 11.1
Schematic sketch of a dorsal view of an embryonic mammalian central nervous system (omitting the olfactory bulbs and the cerebellum), pointing out the position of the midbrain and its shape in a frontal section.

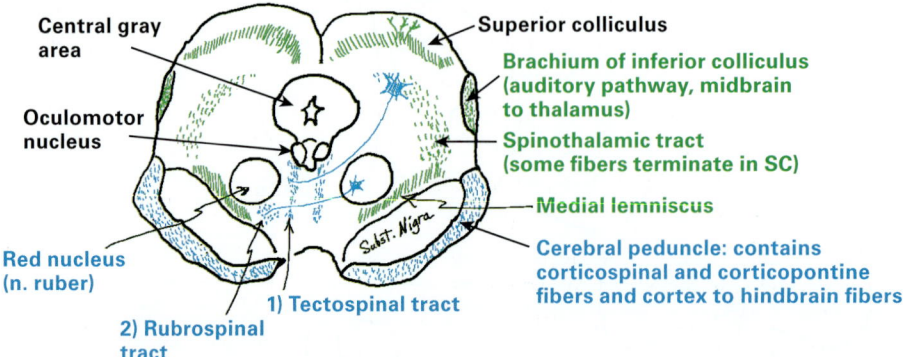

Figure 11.2
Drawing of a frontal section of the midbrain of a small mammal like the rat, mouse, or hamster. Depicted in blue are examples of output axons for the functions of (i) turning of the head and eyes and (ii) grasping with the limbs. These axons originate in the midbrain and project to hindbrain and spinal cord. Also in blue are the axons of the cerebral peduncles, containing mostly axons from the neocortex to the pontine gray and to the hindbrain and spinal cord. In green are axons of sensory systems in, and passing through, the midbrain (omitting visceral sensory axons). The central gray area, surrounding the cerebral aqueduct (the ventricle, called the aqueduct of Sylvius in the midbrain), is also labeled.

midbrain is the ventral tegmental area (VTA), which is depicted in figure 11.4 later in this chapter (also in figure 26.8). These so-called limbic midbrain areas receive strong visceral inputs and also nonvisceral inputs. Activation of cells in these regions has strongly rewarding or punishing effects. The pleasure and pain that accompanies their activation (from VTA and CGA, respectively) is often accompanied by emotional expressions.

Just caudal to the optic tectum is another part of the midbrain roof, the auditory tectum or inferior colliculus (IC) in mammals. As is the case for the superior colliculus, the auditory "little hill" is dual, one on each side of the midline. Ventral to the IC in the caudolateral midbrain are the nuclei of the lateral lemniscus. The lateral lemniscus, as we shall see in chapter 23, is a major channel carrying auditory information from the hindbrain auditory structures to the midbrain. These midbrain auditory cell groups process auditory information and relay it to intermediate layers of the SC and to the forebrain.

In figure 11.2, you can see the outline of a round cell group located in the tegmentum (below the SC and CGA), known as the red nucleus (*nucleus ruber* in Latin) because of a pinkish color noticed in dissections of human brain. This cell group is important for control of the limbs in animals without much neocortex and may play other roles in grasping movements.

Outputs of Midbrain for Motor Control

Embedded in the above descriptions and in the various scientific papers on the midbrain's anatomy, physiology, and functions, we can discern the origins of three different kinds of descending pathways for control of three distinct kinds of multipurpose movements. In the midbrain of large numbers of vertebrates, we find the origins of (1) locomotor commands from the midbrain locomotor area; (2) orienting movement commands via the tectospinal tract, a crossed pathway from deep tectal layers to medial hindbrain reticular formation and to cervical spinal cord; and (3) limb movement commands via the rubrospinal tract, from the red nucleus to the lateral hindbrain reticular formation on the opposite side and to interneurons of the spinal enlargements, controlling distal muscles. By these three pathways, the midbrain controls three types of body movements critical for survival. The first pathway's activity elicits locomotion for escape/avoidance or approach. Approach behavior can be for exploring or foraging or for other kinds of seeking behavior. The second pathway's activation causes turning of the head and/or eyes toward a stimulus, with associated postural adjustments. The third pathway is important for limb movements used for grasping or for exploring, especially in animals without a large corticospinal tract. Animals who do not perform grasping with a limb may grasp with the mouth at the end of an orienting movement, with the jaw opening controlled, not by the red nucleus, but by inputs to the motor nucleus of the fifth cranial nerve.

The midbrain locomotor area has been defined by electrophysiological studies. It is found in the caudal midbrain tegmentum (in part of the reticular formation below the caudal tectum) and is of ancient origin, probably crucial for approach and avoidance in animals without a large neocortex. Many of its neurons are connected to hindbrain reticular formation neurons that give rise to reticulospinal pathways. It receives some inputs from the primitive corpus striatum, originally (it is likely) for olfactory control of locomotor functions. Such control probably made use of pathways with synapses in the hypothalamic locomotor region and in the epithalamus (habenular nuclei) that were larger than the more direct pathways from striatum.

The turning away from something or rapidly escaping from it, and the turning toward something novel or something that promises a reward, are basic movements controlled by outputs from the midbrain tectum (superior colliculus of mammals). Orienting toward a stimulus is controlled by the tectospinal tract (figure 11.2). Not shown is the origin of the uncrossed tectal output pathway that triggers escape movements.

The orienting movements are strongly influenced by corpus striatum via the so-called *substantia nigra* in the ventral midbrain. (The term means "black substance" for the dark pigment found there in human brain dissections.) The nigra, seen just below the fibers of the medial lemniscus in figure 11.2, receives strong projections from the striatum and projects to the midbrain tectum.

The grasping of objects using a limb involves movements in which the red nucleus is involved and important. (It is probably important for other limb movements as well, but its importance for control of movements diminishes in evolution as the neocortex expands.) The output pathway from the midbrain that has this function is illustrated in figure 11.2, where you can see the origins of the rubrospinal tract from the large-celled portion of the red nucleus.

This account oversimplifies, probably fairly greatly, the motor functions of the midbrain. Activity of the midbrain locomotor area and the adjacent regions probably elicits many variations in body motions. The function of the red nucleus has not been specified for limbless animals. The central gray area, perhaps the most primitive core of the midbrain, contains regions not only involved in pain and autonomic functions but also capable of eliciting sexual behaviors, predatory attack or defensive attack behaviors, and other functions resembling those known to be under forebrain control, especially by hypothalamic activities.

Nevertheless, the three types of multipurpose movement emphasized in this treatment of the midbrain are those with the most direct, long-axon links to lower motor mechanisms. They were especially important when the midbrain was a necessary link between forebrain and the motor system in pre-mammalian vertebrates.

Mosaic Evolution of Midbrain

The species differences in midbrain structures have resulted in size distortions that require an exercise in topology to understand. This is another example of *mosaic evolution*, as opposed to *concerted evolution* in which different brain structures increase proportionately in size. We saw examples of mosaic evolution of hindbrain sensory structures in chapters 4 and 10.

In figure 11.3 are drawings of fiber-stained sections of the midbrain in three different species: a small rodent like a rat or hamster, a human, and a tree shrew. The Malayan tree shrew is a small, very visual insectivorous mammal with visuomotor abilities and a midbrain that resemble the abilities and midbrain of squirrels. The huge optic tectum in tree shrews and squirrels can be compared to the optic lobes of birds and many teleost fish. Note the much smaller equivalent structure, the superior colliculus, in rats and in humans. Humans and other primates are very visual animals in comparison with the rat, but our visual systems have a greatly expanded neocortical component that is dominant over the midbrain tectum.

Note also the relatively huge fiber bundle at the ventral edge of the midbrain in humans. This contains the corticospinal and corticopontine fibers, which increase in number with increases in size of the neocortex.

Long Axons Passing through the Midbrain

From spinal cord and from hindbrain come several major ascending pathways that have to pass through the midbrain on their way to the forebrain. Where are they in the midbrain? Also, there are major descending pathways we have already described—pathways

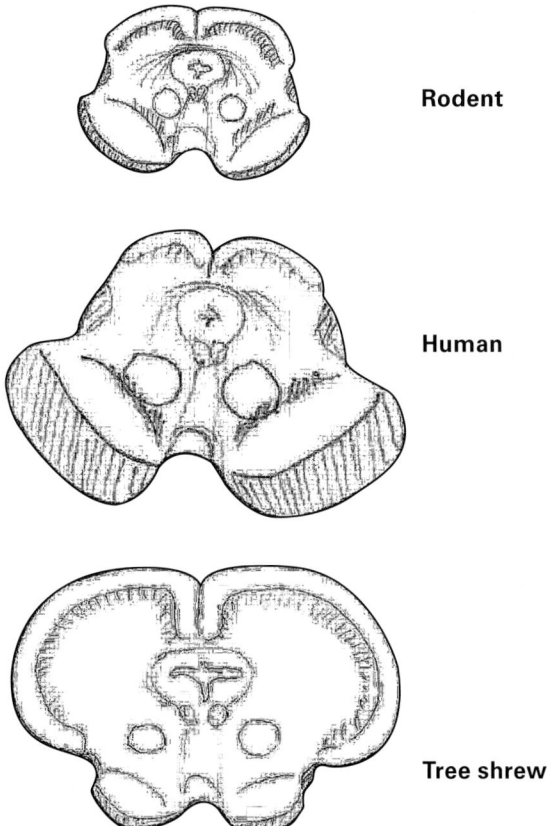

Figure 11.3
Drawings of frontal sections of the midbrain of three different species (not drawn to the same scale). The sections are illustrative of the midbrain of a small rodent at the top, a human in the middle, and a tree shrew or squirrel at the bottom. At the dorsal surface, fibers from the retina terminate in the superficial layers of the superior colliculus (midbrain tectum). In these three species, there are large differences in the relative sizes of the superior colliculi dorsally and of the cerebral peduncles—fibers from the neocortex—ventrally.

that evolved in mammals along with the innovation of a neocortex and its expansion. One of these was just mentioned: the corticospinal and corticopontine fibers. They are found in the cerebral peduncles at the base of the midbrain on either side. The word *peduncle* means "little foot" in Latin and is often used in botany to designate a stalk, as a stalk holding up a flower. The "flower" in this case is the blossoming neocortex.

You can find these axons and others indicated in the frontal section of the midbrain shown in figure 11.2. The section in the figure is drawn from sections of hamster and rat brains; for the human, see figure 11.3. Except for the relative sizes of cell groups and fiber bundles, the structures are very similar in these different mammals.

Not shown in the figures are ascending visceral sensory axons found in the CGA and axons that form reciprocal connections between the CGA and the hypothalamus (the *dorsal longitudinal fasciculus*).

Ascending somatosensory pathways include the spinothalamic tract axons. Some axons following this pathway have terminations in the dorsal midbrain, in the deeper layers of SC, thus forming a spinotectal projection. The medial lemniscus can be found, together with axons of the trigeminal lemniscus, located just dorsal to the substantia nigra and ventral to the red nucleus. These lemniscal axon bundles extend laterally and dorsally in the midbrain so they become contiguous with the spinothalamic axons. The fibers carrying auditory information rostrally from the IC are found lateral to the somatosensory pathways, at the lateral margin of the midbrain between the lateral edge of the SC and the lateral-most fibers of the cerebral peduncle (figure 11.2).

Other axon groups have been mentioned previously, such as those coursing from the vestibular nuclei to the oculomotor nuclei (the *medial longitudinal fasciculus*), except for a major bundle of axons extending rostrally from the cerebellum. In midbrain sections, these axons cross the midline (the decussation of the brachium conjunctivum) and travel through and below the red nucleus, mixing somewhat with the somatosensory system's axons. Some terminate in the red nucleus and in the subthalamic region rostral to it, but their major destination in mammals is the ventral nucleus of the thalamus, rostral to the ventroposterior nucleus where the somatosensory axons terminate. They end mainly in the ventrolateral thalamic nucleus.

This description includes many details that will probably roll off you at this point in your learning about brain structure. More interesting is the functional organization of the midbrain. We have already described it as the major link between the olfactory and visual sensory inputs to the forebrain and the neural control of actions critical for survival. However, it was not just a sensorimotor system, but was also another kind of instigator of action: It was a center of motivational control, a maintainer of the stability of the internal milieu as well as of the body in space. It could change the probability that the sensorimotor links would lead to actual behavior. This was the limbic core of the midbrain—the central gray area and the ventral tegmental area.

We can clearly separate these limbic system structures from the structures that constitute the apparatus of the somatic systems. By *somatic* we mean connected to the somatic sensory and motor systems. We can broaden this definition to include the distance senses of vision and audition (and electroreception) as well. By *limbic* we mean connected to the hypothalamus and autonomic nervous system and the closely associated limbic endbrain structures found at the fringe (the limbus) of the hemispheres.

The locations of these two functionally distinct regions of the midbrain are shown in figure 11.4. These functional divisions continue rostrally into the 'tweenbrain as indicated in the figure. This carries us directly into the subject of the next chapter—the forebrain.

Why a Midbrain?

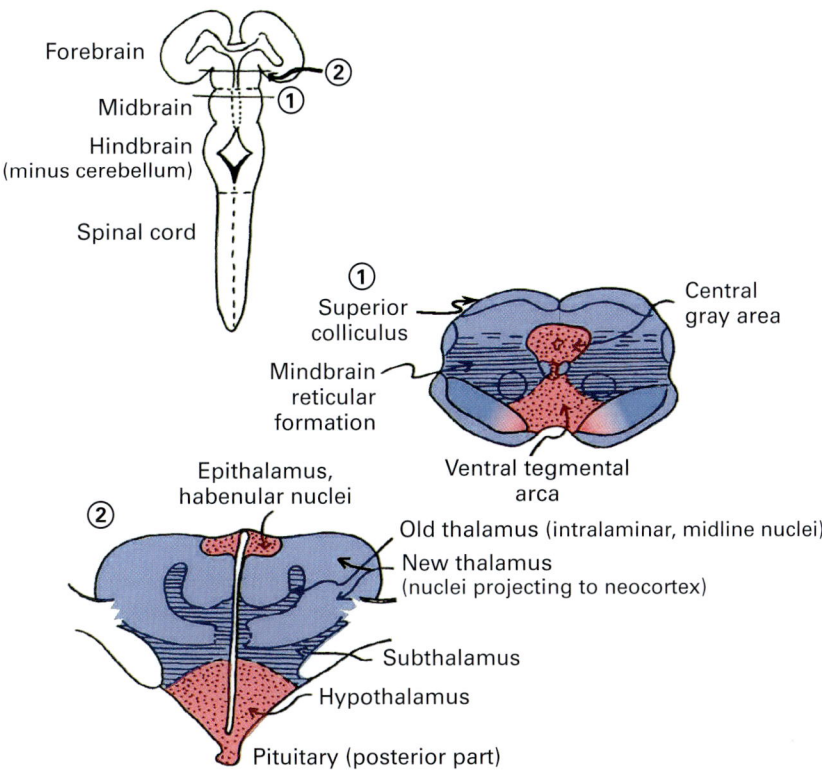

Figure 11.4
Somatic regions (blue) and limbic regions (red) of the midbrain of a small mammal can be followed into the 'tweenbrain as shown in simplified drawings of frontal sections. The levels of the two sections are indicated on a dorsal view of an embryonic CNS at the top. The midbrain reticular formation together with the red nucleus (horizontal striations) is continuous with (and closely connected to) the subthalamus and the midline and intralaminar nuclei of the thalamus. The central gray area and the ventral tegmental area of the midbrain, marked by stipple and red color in section 1, are continuous with (and closely connected to) the hypothalamus and epithalamus in section 2. The red and stippled regions of the midbrain were called the "limbic midbrain areas" by W. J. H. Nauta. (Limbic regions of the endbrain are also closely connected with the hypothalamus.) Based on lectures by W. J. H. Nauta at MIT in his class "Outline of neuroanatomy" in the late 1960s.

Readings

Many, much sparser, connections of the retina to the hypothalamus other than the projection to the suprachiasmatic nuclei have also been reported. See, for example, the following paper on Syrian hamsters:

> Ling, C. Y., Schneider, G. E. & Jhaveri, S. (1998). Target-specific morphology of retinal axon arbors in the adult hamster. *Visual Neuroscience, 15,* 559–579.

Many experimental studies have supported the role of the midbrain tectum (superior colliculus of mammals) in orienting movements (turning of head and eyes). For example, see the author's paper in 1969:

> Schneider, G. E. (1969). Two visual systems: Brain mechanisms for localization and discrimination are dissociated by tectal and cortical lesions. *Science, 163,* 895–902.

Functions of tectal output pathways include escape movements, which result from activation of an ipsilateral descending projection to the reticular formation of the midbrain and hindbrain:

> Dean, P., Redgrave, P. & Westby, G. W. M. (1989). Event or emergency? Two response systems in the mammalian superior colliculus. *Trends in Neurosciences, 12,* 137–147.

> Merker, B. H. (1980). *The sentinel hypothesis: A role for the mammalian superior colliculus.* PhD thesis, Massachusetts Institute of Technology.

For other functions of the midbrain, see chapters 14 and 15.

12 Picturing the Forebrain with a Focus on Mammals

In previous chapters (chapters 4 and 7) we have glimpsed the early evolution of the forebrain and have entertained an idea about why each side of the forebrain (and midbrain) came to represent the opposite side of the external world and the body surface (chapters 10 and 11).

Because of the forward locomotion by ancient bilaterally symmetric organisms, head receptors were extremely important. At the rostral end of these animals, specialized olfactory receptor neurons evolved in the surface epithelium. The connections of these cells to the underlying neural tube in the earliest chordates were undoubtedly a major factor in the expansion of the forebrain vesicle at the rostral end of the CNS (see figures 4.1, 4.7). Circuitry for analyzing and routing the olfactory information led to expansion of the earliest cortex of the endbrain. This cortex formed connections with neural structures—the precursors of ventral striatum and hypothalamus—that could control basic movements of approach and avoidance via axonal pathways to the midbrain. Modifiability of the connections in the primitive striatum enabled animals to alter their responses according to the good or bad consequences of previous actions.

Not only olfactory detectors but also light detectors influenced neurons of the forebrain and midbrain. Light detectors connected directly to the caudal forebrain. Because day and night required very different behaviors optimal for survival, the optic inputs were as important as the olfactory in modulation of other activities of the CNS, and more important for determining a daily rhythm of activity. This was accomplished not only by widespread modulatory influences on the CNS, but also by changes in many body tissues through hormonal secretions from the early pituitary, which was connected with the ventral forebrain.

In a frontal section through the thickening neural tube of a mammalian embryo, at the rostral end, parts of the ventricular layer of proliferating cells are particularly thick (figure 12.1), as it must have been in the ancient chordates with expanding olfactory systems. Other vertebrate embryos are similar. These proliferating cells form the cortex at the surface of the maturing neural tube and also the striatal and basal forebrain structures. Note in the figure the descriptive terms that have been used to designate certain portions

218   Chapter 12

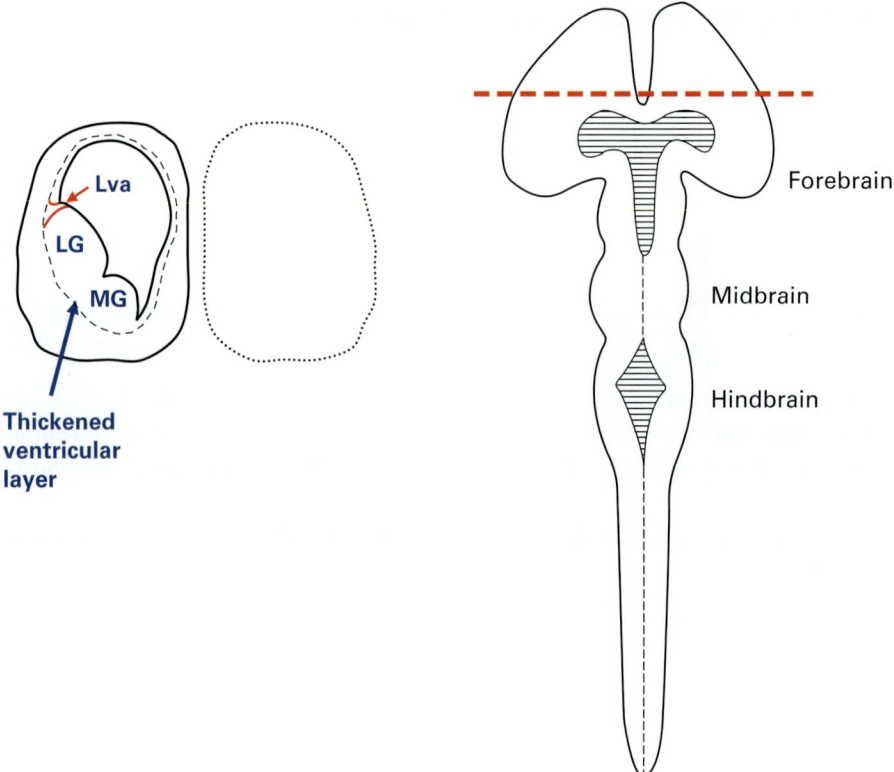

Figure 12.1
(Right) Sketch of the embryonic CNS of a mammal seen in dorsal view, with the level of the section at the left indicated by a dashed line. The hemispheres are pushed away from each other, and the anterior-posterior axis is straightened, eliminating the flexures, for instructive purposes. (Left) Sketch of a frontal section at the rostral end of the thickening neural tube of a mammalian embryonic CNS. The ventricular layer of proliferating cells is much thicker in the region below the ventricle. Lva, lateral ventricular angle region; LG, lateral ganglionic eminence; MG, medial ganglionic eminence.

of the embryonic ventricular layer: the lateral ventricular angle region is a source of cells that migrate to the lateral and ventral pallium, with contributions also to lateral parts of the dorsal pallium. (Much of the dorsal pallium becomes the neocortex in mammals.) The lateral and the medial ganglionic eminence give rise to the corpus striatum and to the adjacent pallidum; these proliferative zones also contribute short-axon cells to the neocortex.

Pictures of Ancestral and Modern Endbrain

Recent data from studies of expression patterns of regulatory genes, like the hox genes, in various species have provided important information relevant to endbrain evolution

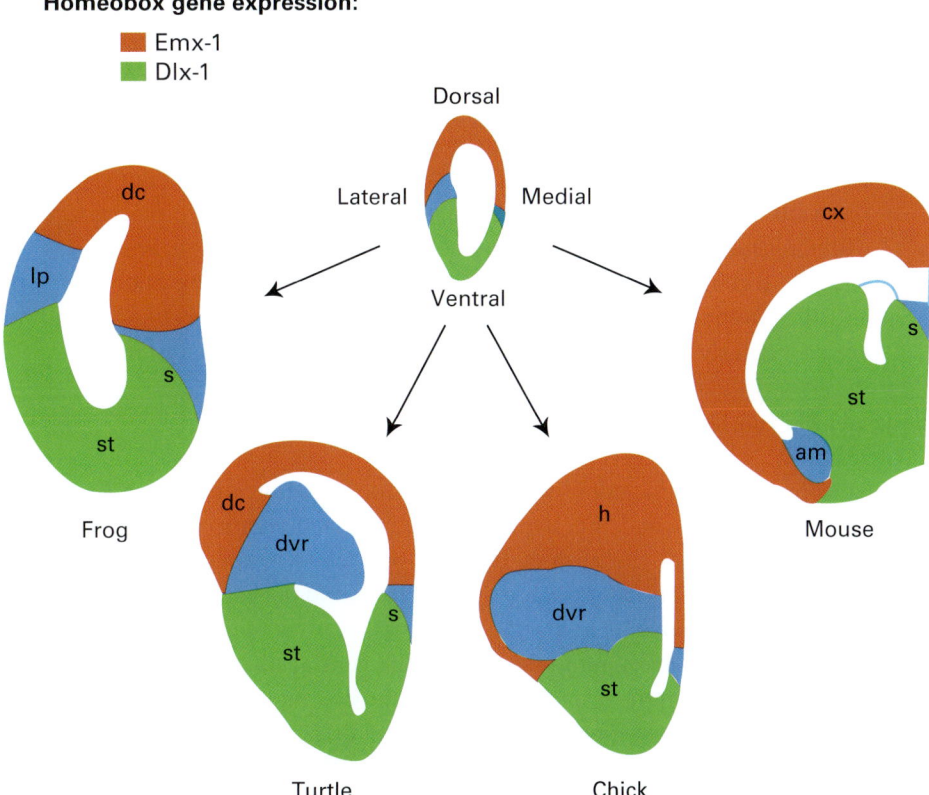

Figure 12.2
Evolution of the telencephalon is postulated on the basis of expression patterns of regulatory genes (homeobox genes Emx-1, Dlx-1) during development of four species representing amphibian, reptile, bird, and mammal. am, amygdala and claustrum; cx, neocortex; dc, dorsal cortex; dvr, dorsal ventricular ridge; h, hyperpallium and arcopallium; lp, lateral pallium; s, septum; st, striatum. Based on Fernandez, Pieau, Repérant, Boncinelli, and Wassef (1998).

and phylogeny. Prior to these studies, cross-species comparisons were made using primarily morphological data (cytoarchitecture, connection patterns). Figure 12.2 shows results of such a study that led the authors to sketch the hypothetical endbrain of an ancestor common to a modern frog, turtle, chick, and mouse and to show the corresponding endbrain parts in each of those species.

We will return to these pictures of the endbrain at the end of this chapter with some additional notes concerning forebrain evolution. But first, it will be useful if you get more familiar with some major features of forebrain structure. Pausing for a brief review of major concepts in brain structure will remind you to review these before we begin a study of major functional systems of the brain. We will embark on that study after we conclude part V of the book with a look at a unique aspect of neuronal differentiation, the growth of axons.

Words for Forebrain Parts

In figure 12.3, you see an illustration of the embryonic mammalian CNS at a stage when the walls of the neural tube are thickening to the point where the mature shapes can be distinguished. You also see sections of each major subdivision. Focus now on the rostral end: the forebrain (prosencephalon), which includes both the endbrain (telencephalon) and the 'tweenbrain (diencephalon). The Greek terms *diencephalon* and *telencephalon* are particularly common in neuroanatomical discussions and writings.

Major Structural and Functional Subdivisions of the 'Tweenbrain

The diencephalon includes four major divisions in adult vertebrates. The most well known are the thalamus and the hypothalamus, already mentioned frequently in this book. (*Thalamus* is from the ancient Greek word for an "inner room" or "chamber." *Hypo* means "below.") Caudo-dorsal to the thalamus and rostral to the midbrain is the epithalamus (meaning above or upon the thalamus). Ventral (and in the embryo, rostral) to the thalamus is the subthalamus, the embryonic ventral thalamus.

With the evolution of neocortex in mammals and the great increases in its size, the thalamus became the largest structure of the diencephalon. This is because it is such an important source of input to the neocortex. This inner chamber is the antechamber through which sensory and processed premotor messages must pass in order to reach the crown of the CNS (the neocortex).

Figure 11.4 shows both the midbrain and the 'tweenbrain divided into two major functional regions, those related to somatic sensory and motor systems and those related to the limbic system. In the diencephalon, the thalamus and subthalamus are primarily related to somatic functions, whereas the hypothalamus and epithalamus are primarily related to limbic system functions.

One way to understand this distinction between two functional regions is to look at their major inputs. The thalamus and subthalamus (the dorsal thalamus and ventral thalamus of the embryo) receive somatic inputs from the lemniscal pathways, including those we have described that carry somatosensory information from spinal cord and hindbrain. They also receive connections from the midbrain tectum and tegmentum—the somatic parts of the midbrain. The hypothalamus and epithalamus receive various signals from the visceral organs that arrive primarily via polysynaptic pathways. Connections from the midbrain come mainly from the central gray area and the ventral tegmental area. Connections from the endbrain come from the structures we group together in the limbic

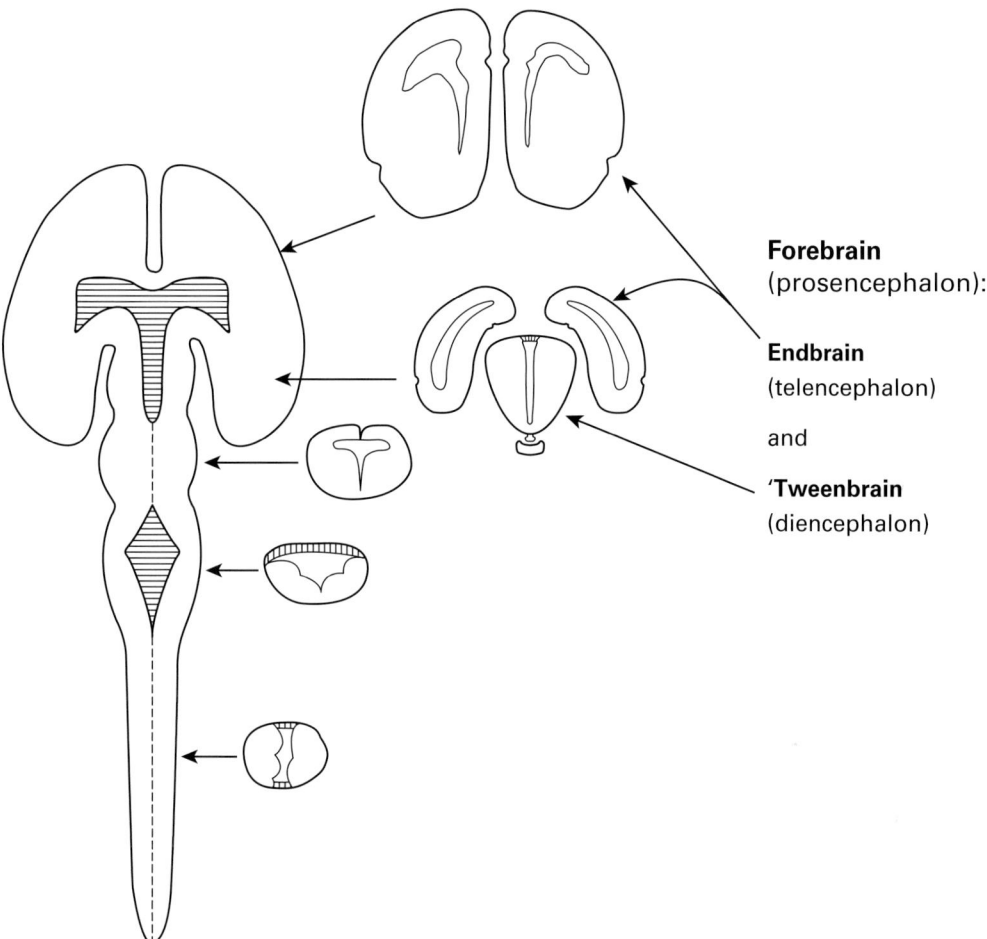

Figure 12.3
Sketch of embryonic mammalian CNS, dorsal view with flexures straightened out and the cerebellum removed. The two major forebrain components are pointed out.

endbrain, all with activity involved in control or modulation of inherited motivations and autonomic nervous system functions.

An important connection between the limbic and somatic portions of the 'tweenbrain is from hypothalamus to thalamus, a connection that is both direct (to midline thalamic cells with widespread intrathalamic connections) and indirect via the midbrain. This connection is the substrate for a kind of gating or modulation of activity passing through the thalamus to the neocortex.

There are some inputs that go to both somatic and limbic regions. Thus, pathways carrying taste information from the hindbrain reach the dorsal thalamus (ventroposteromedial nucleus, VPM) and thence the neocortex, and taste pathways also reach the hypothalamus and ventral tegmental area. The first pathways are concerned with identification of substances touching the tongue or throat; the second pathways are concerned with innate positive or negative reactions to these substances. Somatosensory inputs perceived as pain also reach both regions, in addition to triggering reflex responses that activate flexor muscles. Pain pathways to the thalamus, mostly to the older portions of the thalamus, activate the neocortex and the corpus striatum. The pain pathways reaching the central gray area (of the midbrain) and the hypothalamus cause sympathetic arousal as well as negative reward.

Major Parts of the Telencephalon of Mammals

The endbrain includes the corpus striatum, the olfactory and limbic endbrain structures (often grouped together), and the neocortex (or, prior to mammals and in many non-mammals, the dorsal cortex). This group of structures can be divided into two major functional territories, the same division as for the midbrain and 'tweenbrain (somatic and limbic).

Most of the neocortex is related to somatic functions, broadly speaking, but parts of this cortex are also tied to limbic structures—the cortical and subcortical structures most closely connected to hypothalamus. The corpus striatum can be similarly parcellated. The dorsal striatum, sometimes called the neostriatum, is dominated by sensory and motor functions of somatic systems. The ventral striatum structures are part of the limbic system, originally dominated by olfactory inputs, and closely connected with hypothalamus and epithalamus. The ventral striatum includes the olfactory tubercle, with superficial layers receiving projections of the olfactory bulb and deeper layers projecting caudally as far as the midbrain.

Origins and Course of Two Major Pathways of the Forebrain

A neuroanatomical validation of the division of the endbrain into two domains is the existence of two distinct pathways between endbrain and more caudal structures; namely, the medial and the lateral forebrain bundles (figures 12.4 and 12.5).

Picturing the Forebrain with a Focus on Mammals

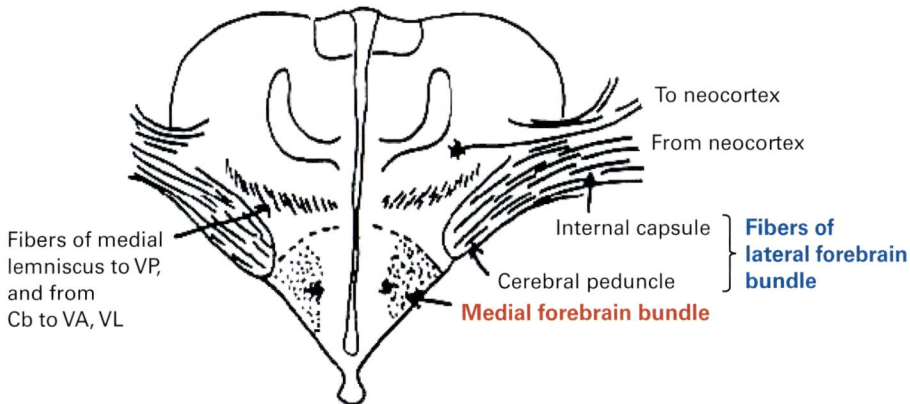

Figure 12.4
Sketch of a frontal section of the mammalian 'tweenbrain (diencephalon) showing the positions of the medial forebrain bundle (MFB) and the lateral forebrain bundle (LFB). These axon groups come from limbic system regions (MFB) and somatic regions (LFB). Cb, cerebellum; VA, ventral anterior nucleus of thalamus; VL, ventral lateral nucleus of thalamus; VP, ventral posterior nucleus of thalamus.

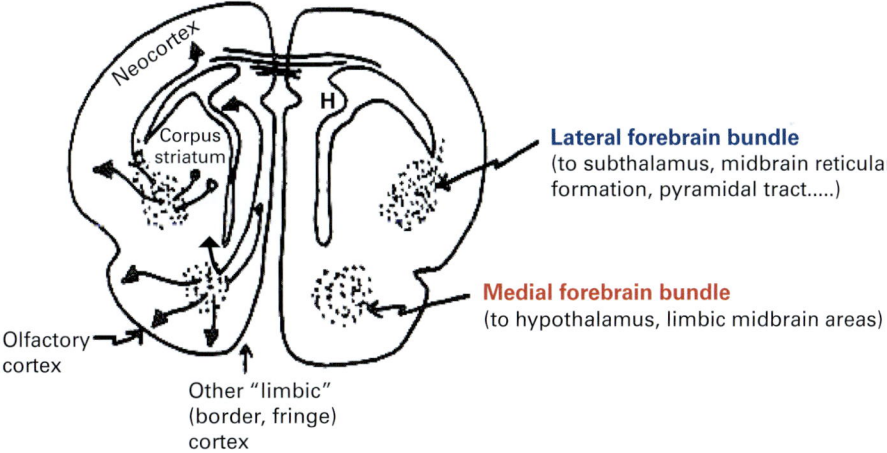

Figure 12.5
Sketch of a frontal section of the endbrain (telencephalon) of a small mammal illustrating the origins of the medial and lateral forebrain bundles. The location of most of the hippocampal region (H) is generally caudal to this level.

The medial forebrain bundle (MFB) contains axons projecting from the olfactory and the limbic cortical regions of the endbrain, and from subcortical limbic system structures (amygdala, ventral striatum, and basal forebrain), to the hypothalamus—with axons coursing through the lateral hypothalamic area and to the limbic midbrain areas (ventral tegmental area and central gray area). Axons linking these structures also course in the opposite direction; the MFB is a two-way street. (In additional to axons that course through the lateral hypothalamus, some axons of the medial system course over the dorsal surface of the 'tweenbrain, in a bundle called the stria medularis; many of these axons terminate in the habenular nuclei of the epithalamus.)

The lateral forebrain bundle contains axons projecting to and from the corpus striatum (mostly dorsal striatum) and to and from the neocortex. Axons passing rostrally through the lateral forebrain bundle are mostly thalamocortical axons. Caudal to the thalamus, most fibers in this bundle are going caudalward. Figure 12.6 shows how this bundle can be followed caudally to the spinal cord as the corticospinal tract. Axons from the corpus striatum do not project beyond the midbrain, but axons from the neocortex terminate at all levels. These axons from the neocortex are designated by different terms at different levels: First there is the *neocortical white matter*, deep to the cortical cell layers. Axons of the white matter enter the *internal capsule* coursing through the corpus striatum (with many terminations there). These axons become the *cerebral peduncle* as they course over the lateral edge of the diencephalon (where many of them also terminate). The peduncle axons continue caudally over the ventral surface of the midbrain, and enter the pons, where a large group of them terminate on cells in the pontine gray matter and the remaining axons emerge from the caudal edge of the pons as the *pyramidal tract*. After most of the pyramidal tract axons decussate at the caudal end of the hindbrain, they are simply known as the *corticospinal tract* fibers.

The Neocortex Is Involved in Both Major Systems

Figure 12.7 presents a schematic summary of some major endbrain connections with more caudal levels, emphasizing output connections. Most inputs to the thalamus are not included in the diagram. The figure shows how the neocortex is well connected with both somatic and limbic structures.

The figure leaves out a great many details. For example, in the diagram, the ventral striatum is lumped together with limbic structures. The ventral striatum is critical in habit formation and is probably the most primitive part of the corpus striatum in evolution. Also not indicated are the reward and punishment mechanisms, for which ascending projections (e.g., from taste and pain systems) play a special role in the learning of habits.

225   Picturing the Forebrain with a Focus on Mammals

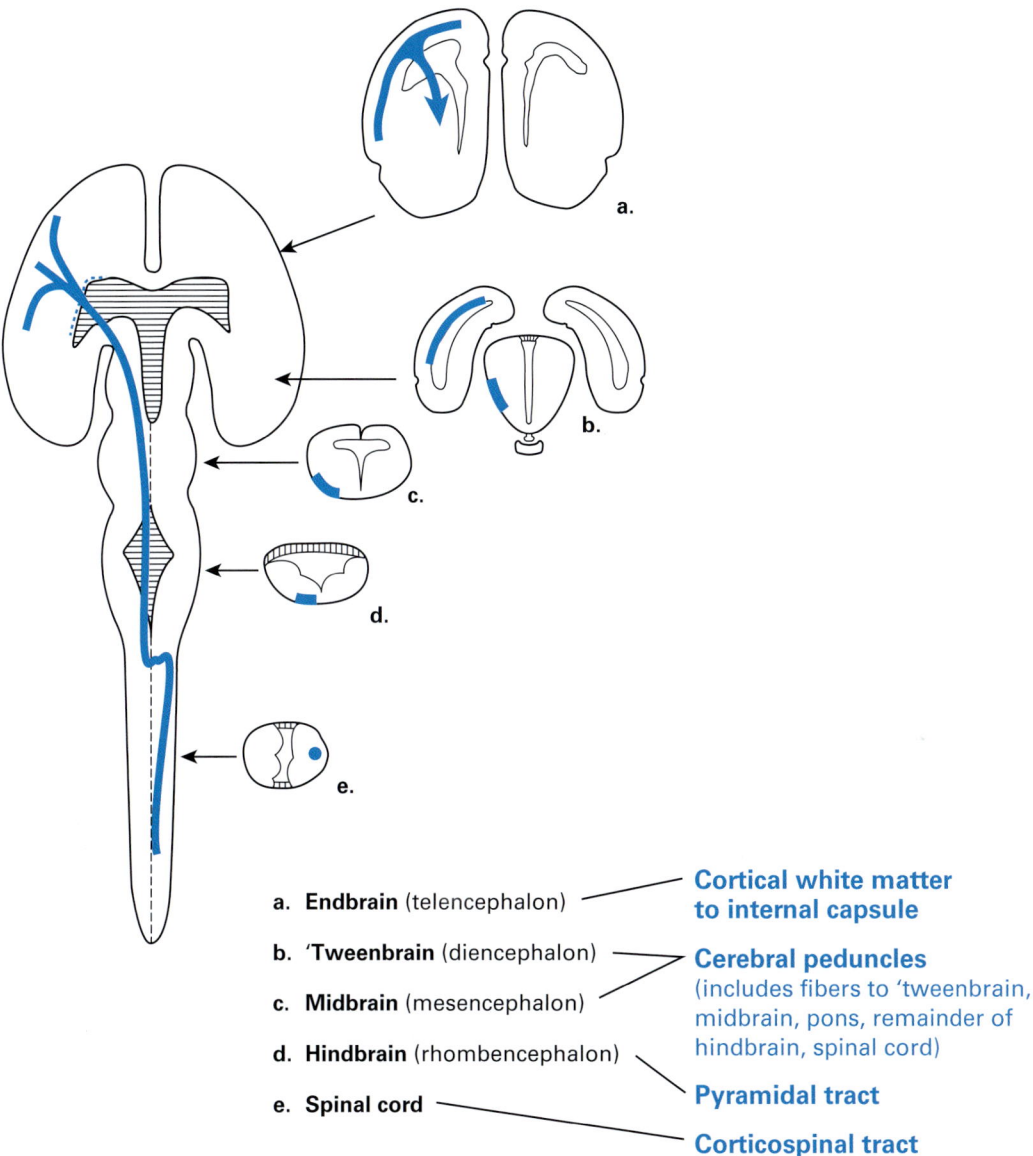

a. **Endbrain** (telencephalon) — **Cortical white matter to internal capsule**

b. **'Tweenbrain** (diencephalon) — **Cerebral peduncles** (includes fibers to 'tweenbrain, midbrain, pons, remainder of hindbrain, spinal cord)

c. **Midbrain** (mesencephalon)

d. **Hindbrain** (rhombencephalon) — **Pyramidal tract**

e. **Spinal cord** — **Corticospinal tract**

Figure 12.6
The position and course of the lateral forebrain bundle is illustrated on a sketch of the embryonic mammalian brain (dorsal view with flexures straightened out). The locations of these fibers and names used for them at the different anterior-posterior levels are shown.

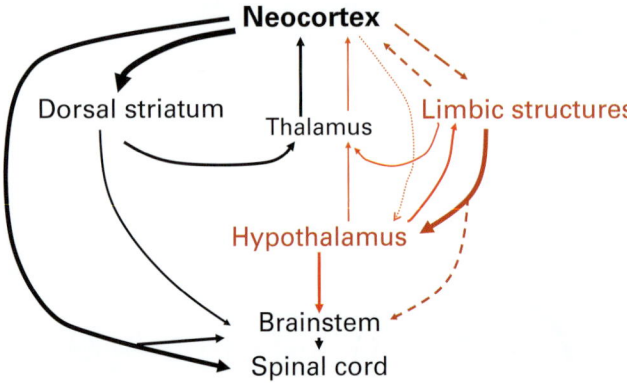

Figure 12.7
Some major connections of the endbrain are illustrated in a simplified diagram.

Interim Review of Neuroanatomy

The following list summarizes many of the more important points covered from chapter 1 to the current chapter. It will be helpful for what follows if students review the meaning of each of these terms and phrases.

- Subdivisions of CNS; definitions of cell types
  Shapes of the neural tube at various levels
- Sensory channels of conduction; dermatomes
- Diaschisis: lesion-produced deafferentation causing a functional depression of neurons
- Evolution of neocortex with major ascending and descending pathways to it and from it
- Spinal cord structure; differences between levels
- Propriospinal system
- Autonomic nervous system: its components
- Hindbrain organization; distortions of the basic plan
- Cranial nerves: the fifth cranial nerve (trigeminal nerve)
- Midbrain: tectum and tegmentum; species differences; outputs for three major types of movement
- Diencephalon: two major and two additional subdivisions (functional/structural)
- Telencephalon: the endbrain (cerebral hemispheres and basal forebrain); origins of two major pathways for descending axons (both contain some ascending axons also)

- Some major axonal pathways in mammals:

    Spinoreticular, trigeminoreticular tracts (mostly ipsilateral)

    Spinothalamic tract; longest axons to ventrobasal nucleus of thalamus (VB; also called the ventral posterior nucleus, medial and lateral parts: VPM and VPL)

    Dorsal columns of spinal cord, connecting to the medial lemniscus pathway, which goes to the ventrobasal nucleus of thalamus

    Corticospinal and corticopontine pathways

Segmentation of the Forebrain

The embryonic mammalian brain, when pictured at the stage when the hemispheres are just beginning to balloon out from the prosencephalic brain vesicle, has a curved axis as illustrated in figure 12.8. Structural studies had indicated some kind of segmentation above the rhombomeres (see chapter 10); for example, a segmentation of the diencephalon into at least four caudal to rostral divisions (neuromeres) that correspond to dorsal to ventral divisions in the mature brain. Gene expression data have provided stronger evidence for neuromeric models of the brain that include the endbrain (figure 12.8). Note in the illustrations that because of the curvature of the axis of the embryonic brain, a section that is parallel to a frontal section through the midbrain would show the diencephalic neuromeres (the prosomeres) arranged so the caudal-most segment is most dorsal and the rostral-most segment is most ventral. Thus, in the embryo, the epithalamus arises from the most caudal of these segments, and the hypothalamus arises from the most rostral ones. Neuromeric segmentation in the endbrain has been more difficult to specify. Gene expression data indicate the existence of neuromeres also in the non-amniotes lamprey and zebrafish.

Notes on Neocortical Origins

The uniqueness of the neocortex in mammals has long intrigued scientists and others interested in brain origins. Where did this magnificent multilayered cortex of our brains come from? When axonal tracing methods became sensitive enough to do reliable experimental studies of pathways and their connections in various living species, some fascinating new ideas arose. Such neuroanatomical studies gave evidence that noncortical structures in reptiles and birds are related to neocortex of mammals. (The techniques used for tracing axonal pathways and their connections in modern studies were, initially, silver staining methods developed by Nauta and his collaborators. Later, even more sensitive techniques were developed, as introduced in chapter 2.)

Each sensory system connects to thalamic cell groups, which project mainly to the neocortex of mammals. Axon tracing studies carried out in birds and reptiles showed that the

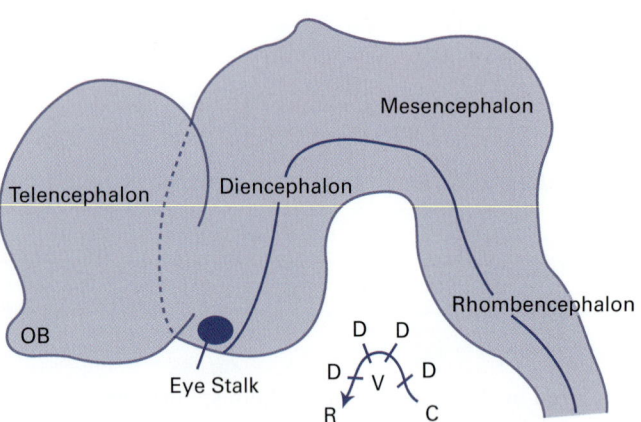
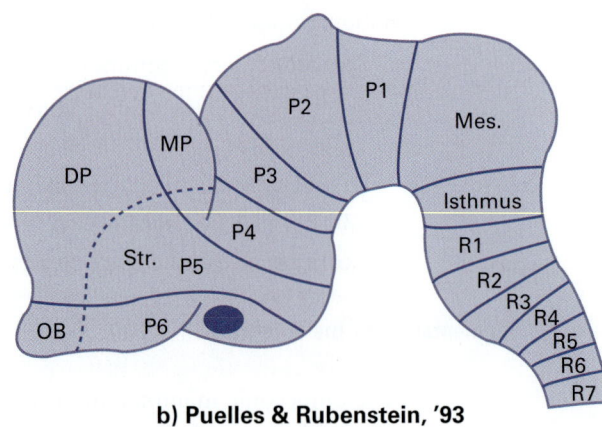
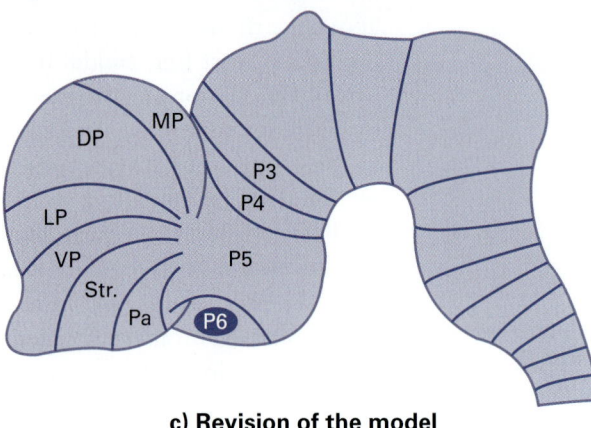

Figure 12.8
Neuromeric models of the embryonic mammalian brain, illustrated on views from the left side, rostral to the left. Gene expression data formed the basis for the model, which has been revised as shown from its original statement in 1993. Gene expression data indicate the existence of neuromeres also in the non-amniotes lamprey and zebrafish. Original model: Puelles and Rubenstein (1993). DP, dorsal pallium; LP, lateral pallium; Mes, mesomere; MP, medial pallium; OB, olfactory bulb; P, prosomere; Pa, pallidum; R, rhombomere; Str, striatum; VP, ventral pallium. Orientation line: C, caudal; D, dorsal; R, rostral; V, ventral. Based on Striedter (2005).

thalamus contains unimodal cell groups that project to structures that, at the time of the initial discoveries, appeared to be part of the corpus striatum. Such projections carrying visual or auditory information could be found in endbrain structures that were called the ectostriatum and the hyperstriatum of pigeons and owls, for example. Later, these regions have been renamed in order to better reflect their origins and homologies.

The data led to a novel proposal by a young comparative neuroanatomist, Harvey Karten, in the late 1960s when he was with Walle Nauta at MIT. He was doing experimental studies of visual and auditory pathways to the forebrain in pigeons. He suggested that some embryonic cells in mammals migrate from a striatal location to populate the neocortex where they differentiate (e.g., into cells of layer 4 that receive sensory projections from the thalamus) (figure 12.9). Thus, many neocortical cells in mammals could be

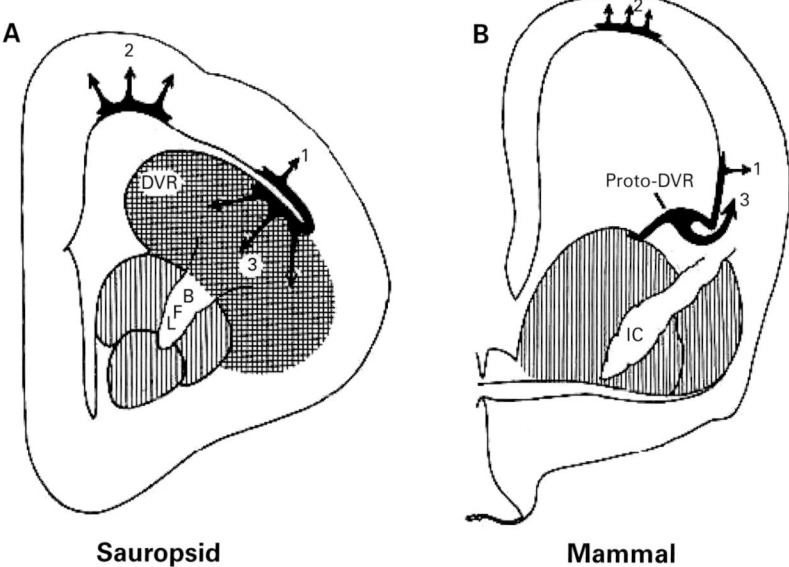

Figure 12.9
Sketches of frontal sections of embryonic sauropsid (reptilian or avian) and mammalian endbrain, illustrating migrations of cells from the ventricular layer as proposed in an original hypothesis of Harvey Karten in the 1960s, published by Karten and Nauta in 1970, and later updated slightly. (A) In sauropsids, embryonic neurons migrate from the ventricular layer into the dorsal cortex (reptiles) or hyperpallium (birds); see arrows *1* and *2*. Below the ventricle, postmitotic neurons migrate into the dorsal ventricular ridge (DVR), called the nidopallium in mature birds; see arrow *3*. (B) In mammals, the postmitotic neurons migrate from the ventricular layer into the overlying neocortex (arrows 1 and 2); cells from the nearby part of the lateral ganglionic eminence, called "proto-DVR" in this figure, migrate into upper layers of the lateral portions of the neocortex (arrow 3). Such migrations from the lateral ventricular angle (see figure 12.1) have been seen in Golgi studies in the laboratory of F. Valverde in Madrid (published in 1996); details remain to be clarified, and also it is not known whether such cell migrations contribute to the posterior, visual areas of neocortex (occipital and temporal areas). It is known that cells from the lateral ventricular angle region, which appears from genetic data to be homologous to the DVR of sauropsids, also migrate to the olfactory cortex and to the amygdala. From Reiner, Yamamoto, and Karten (2005).

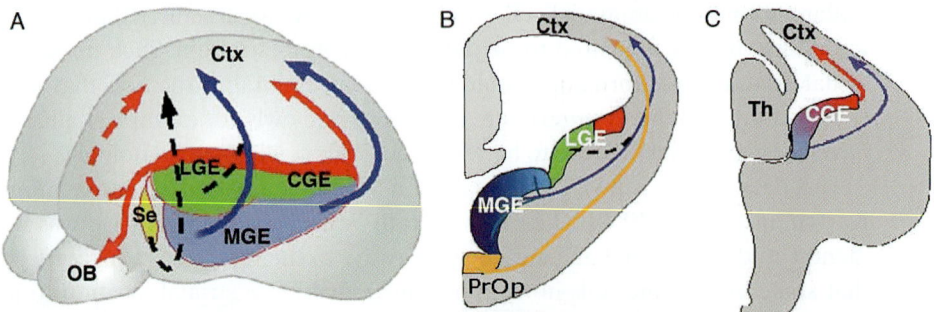

Figure 12.10
Sources and paths of migration in developing mouse brains of cells that become inhibitory interneurons, using GABA as neurotransmitter. These short-axon neurons come from the layer of proliferating cells in subcortical locations in a three-dimensional picture of the brain (A), and in coronal sections at a midrostral level (B) and a midcaudal level (C). Blue arrows depict migrations of cells that also express parvalbumin or somatostatin, from the medial ganglionic eminence (MGE) and the ventral part of the caudal ganglionic eminence (CGE). Red areas and arrow represent the origin and routes for calretinin-expressing inhibitory interneurons. The orange arrow depicts migration of neuropeptide Y–expressing inhibitory interneurons from the preoptic area (PrOp); others may come from the adjacent septal area (Se). Ctx, neocortex; LGE, lateral ganglionic eminence; Th, thalamus. Adapted from Welagen and Anderson (2011).

directly homologous to subcortical cells in birds. Later studies have given at least partial support for the idea: some embryonic neurons do migrate from the region of proliferating cells that has been referred to as the lateral ventricular angle (see figure 12.1) into more lateral parts of the neocortex. Other cells from this region—a region called the *dorsal ventricular ridge* in reptiles and birds—migrate more ventrally to form the olfactory cortex and amygdala region.

Later studies have revealed other important migrations to neocortex from the subcortical proliferative zones, mostly from the areas located more medially and more caudally, known as the medial ganglionic eminence (see figure 12.1) and the caudal ganglionic eminence. These cells become the short-axon inhibitory interneurons of the neocortex—using γ-aminobutyric acid (GABA) as the neurotransmitter. See figure 12.10 for a summary of some of these results. Note the inclusion of a more recent discovery that inhibitory neurons of the neocortex also come from the preoptic region of the hypothalamus.

Developmental studies of avian brains have provided evidence that also in these brains, inhibitory interneurons migrate from striatal positions to pallial derivatives—the dorsal ventricular ridge and the more superficially located cortex called the Wulst.

Figure 12.11 updates these results for humans and other primates. Karten's hypothesis has certainly resulted in a broadening of ideas on brain evolution and development and has led to new experiments on cell migrations during brain development (see also chapter 32).

Gene expression studies have given important clarifications of relationships of forebrain components in widely different species and have strongly suggested some likely

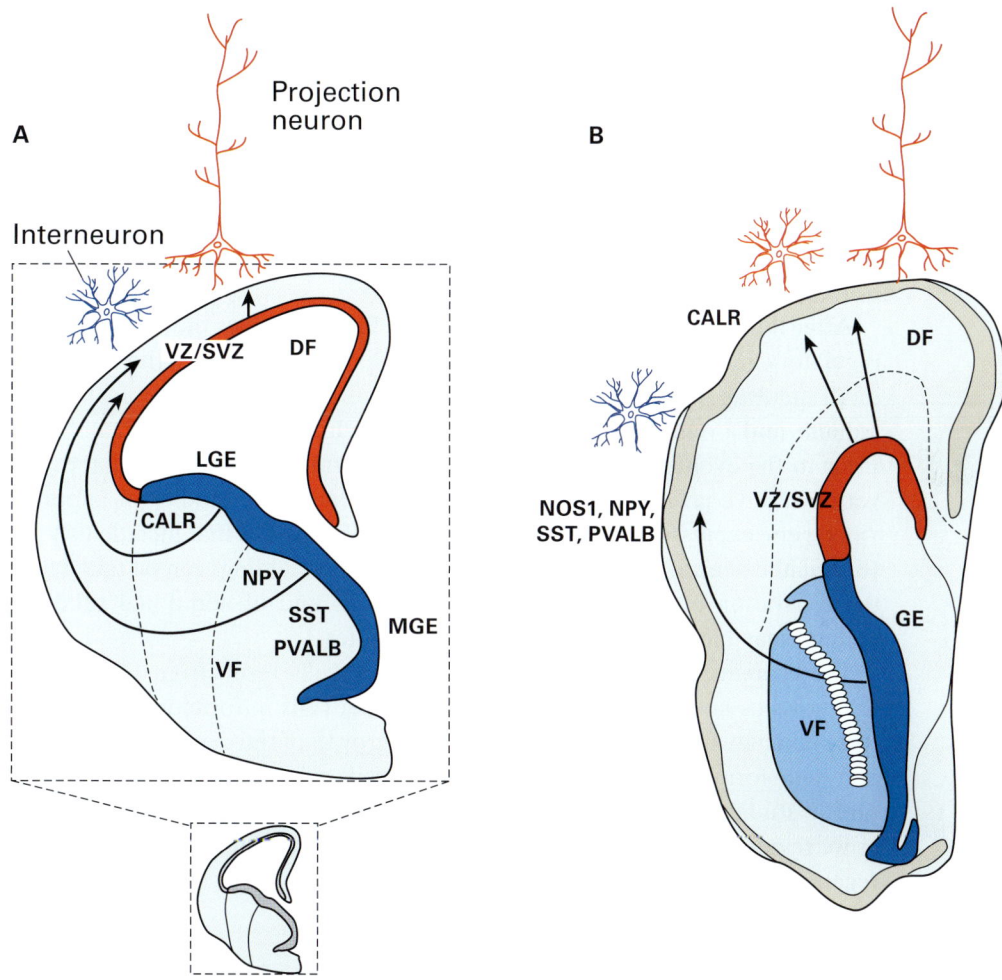

Figure 12.11
Drawings of frontal sections of the mouse brain (A) and the human brain (B) published by Pasco Rakic in a review paper in 2009 (not drawn to same scale). The figure summarizes findings on the embryonic sources and migrations of pyramidal cells and short-axon interneurons of various types defined by chemical markers. Not illustrated is another possible source: the sub-pial granular layer seen in embryonic humans and other large-brained mammals (see chapter 32). DF, dorsal forebrain; VF, ventral forebrain; VZ/SVZ, ventricular zone/subventricular zone; GE, ganglionic eminence; LGE, lateral ganglionic eminence; MGE, medial ganglionic eminence; CALR, calretinin; NOS1, nitric oxide synthase 1; NPY, neuropeptide Y; PVALB, parvalbumin; SST, somatostatin. From Rakic (2009).

evolutionary relationships. Look back at figure 12.2 for an early example of these important results. The regions that receive projections of auditory and visual cell groups in the bird and turtle thalamus that are in receipt of sensory information from the midbrain tectum correspond to the dorsal ventricular ridge in this figure. You can see that according to the gene expression data, the dorsal ventricular ridge does not correspond to the mammalian corpus striatum, but rather to portions of the amygdala. (The gene data group the amygdala with an adjacent subcortical structure, the claustrum. It should also be noted that gene expression data are complex and that the interpretations of evolutionary relationships are not free of controversy.)

After the work represented in figure 12.2, further studies of the expression patterns of regulatory genes in a number of species have led investigators to the view that the pallium of all quadrupeds has four major divisions: a medial pallium, a dorsal pallium, a lateral pallium, and a ventral pallium. Ventrally, all quadrupeds have striatal and pallidal regions. Prior to the evolution of quadrupeds, the cerebral hemispheres had appeared, as in the very primitive jawless vertebrates like the sea lampreys. Studies of lampreys have provided gene expression evidence for endbrain subdivisions that include two pallial regions (dorsal and ventral) and one subpallial region. Such findings can be taken to indicate that the separation of striatal and pallidal divisions and of additional subdivisions of the pallium may have arisen in the jawed vertebrates.

The proteins expressed by developing brain cells result from cellular differentiation that governs how those cells interact with other cells in forming the multicellular structures and networks of the mature brain. The growth of the dendrites and axons of neurons and their formation of contacts with other cells are crucial to this maturation. Next we turn to studies of this major aspect of nervous system differentiation at the cellular level, as we review the growth of axons and also the closely related topics of their plasticity and regeneration.

Readings

> Striedter, G. F. (2005). *Principles of brain evolution.* Sunderland, MA: Sinauer Associates.

Segmentation of the diencephalon into at least four caudal to rostral divisions (neuromeres) that correspond to dorsal to ventral divisions in the mature brain was described by a developmental neuroanatomical study:

> Coggeshall, R. E. (1964). A study of diencephalic development in the albino rat. *Journal of Comparative Neurology, 122,* 241–269.

More recently, descriptions of neuromeric divisions of the CNS have focused on molecular methods; for example, see the following paper:

> Puelles, L., & Rubenstein, J. L. R. (2003). Forebrain gene expression domains and the evolving prosomeric model. *Trends in Neuroscience, 26,* 469–476.

See also the note on the work of Sven Ebbesson at the end of chapter 4, and his paper cited there.

Studies of endbrain development and evolution:

> Butler, A. B., Reiner, A., & Karten, H. J. (2011). Evolution of the amniote pallium and the origins of mammalian neocortex. *Annals of the New York Academy of Sciences, 1225,* 14–27.
>
> Cobos, I. Puelles, L., & Martinez, S. (2001). The avian telencephalic subpallium originates inhibitory neurons that invade tangentially the pallium (dorsal ventricular ridge and cortical areas). *Developmental Biology, 239,* 30–45.
>
> De Carlos, J. A., López-Mascaraque, L., & Valverde, F. (1996). Dynamics of cell migration from the lateral ganglionic eminence in the rat. *Journal of Neuroscience, 16,* 6146–6156.
>
> Jarvis, E. D., et al. (2005). Avian brains and a new understanding of vertebrate brain evolution. *Nature Reviews, Neuroscience, 6,* 151–159.
>
> Karten, H. J. (1991). Homology and evolutionary origins of the "neocortex." *Brain, Behavior and Evolution, 38,* 264–272.
>
> Karten, H. J. (1997). Evolutionary developmental biology meets the brain: The origins of mammalian cortex. *PNAS, 94,* 2800–2804.
>
> Medina, L., & Abellán, A. (2009). Development and evolution of the pallium. *Seminars in Cell & Developmental Biology, 20,* 698–711.
>
> Miyoshi, G., Hjerling-Leffler, J., Karayannis, T., Sousa, V. H., Butt, S. J. B., Battiste, J., & Fishell, G. (2010). Genetic fate mapping reveals that the caudal ganglionic eminence produces a large and diverse population of superficial cortical interneurons. *Journal of Neuroscience, 30,* 1582–1594.
>
> Molnar, Z., & Butler, A. B. (2002). The corticostriatal junction: A crucial region for forebrain development and evolution. *BioEssays, 24,* 530–541.
>
> Nauta, W. J. H., & Karten, H. J. (1970). A general profile of the vertebrate brain with sidelights on the ancestry of the cerebral cortex. In Schmitt, F. O., ed. *The neurosciences: Second study program.* New York: Rockefeller University Press.
>
> Rakic, P. (2009). Evolution of the neocortex: A perspective from developmental biology. *Nature Reviews, Neuroscience, 10,* 724–735.
>
> Welagen, J., & Anderson, S. (2011). Origins of neocortical interneurons in mice. *Developmental Neurobiology, 71,* 10–17.

13 Growth of the Great Networks of Nervous Systems

We discussed development of the nervous system previously, especially in our introduction to spinal cord development (chapter 8). After the earliest embryonic stages of morula formation and gastrulation, the central nervous system begins to differentiate from the primitive ectodermal layer with the formation of the neural plate. The neural plate invaginates to form the neural tube—induced by the underlying notochord in all chordate species. The walls of the neural tube thicken with the proliferation of CNS cells. These cell bodies of the neuroectoderm, when they leave the mitotic cycle, migrate from their birthplaces to their various destinations. During, or more usually after, that migration, the neurons began to differentiate by growing extensions—axons, dendrites, and dendritic spines. Many of these cellular processes are not permanent, as they are sculpted by loss of branches and also by loss of some cells. The cells and processes that remain undergo maturation as they form their end arbors with enlargements of synaptic regions and development of membrane specializations at synaptic contacts. The resulting networks of interconnected neurons are not always stable. Some of the synaptic connections of some of the neurons show plasticity depending on their activity and on the activity of neighboring connections. This plasticity is more widespread and often greater during the developmental period than after attainment of maturity, but it also occurs in older brains.

The Axonal Growth Cone

Ramón y Cajal was able to characterize major morphological features of axon growth from Golgi-stained sections of the CNS of animals at various stages of early development (figure 13.1). He was able to reconstruct the dynamics of axonal extension (e.g., for the axon of a motor neuron or of an interneuron of the spinal cord or brain stem) from its earliest appearance with the formation of a motile *growth cone* at its tip, to its formation of terminal branches (end arbor) and in many cases some collateral branches.

The first verification of Cajal's descriptions was made after the methods of tissue culture were invented and applied to nervous system tissue. This technical advance in neuroembryology is attributed to Ross Granville Harrison (reports in 1907, 1910), who

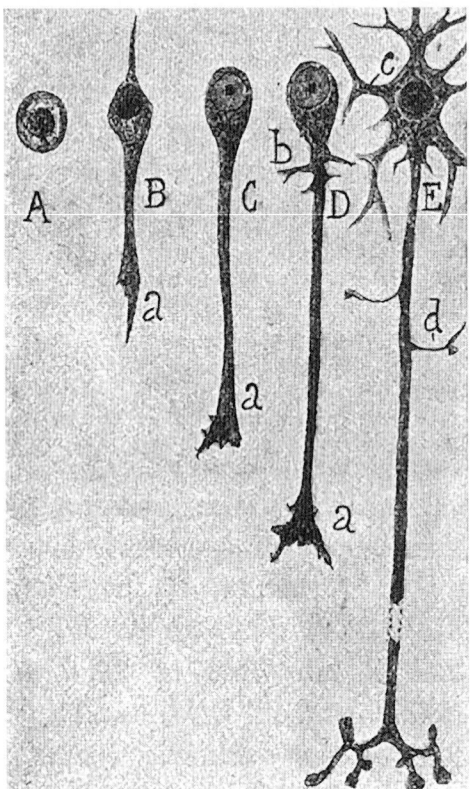

Figure 13.1
Drawings by Ramón y Cajal showing neuron development with growth of axon and dendrites as seen in Golgi-stained brains of animals at various ages. A, Neuroblast before differentiation. B, Initiation of axonal growth. The active growth tip is labeled by the letter *a*. C and D, The axon continues to elongate, and the active growth tip's cone shape is evident. Collateral processes (b) near the cell body may be transient filopodia or they may be axonal collaterals; they may also become dendrites (c), which have become evident in E. Note also in E: axonal collaterals (d) and growth of an axonal end arbor. From Ramón y Cajal (1989).

first reported the observation of a living growth cone. Later, Speidel was able to observe living growth cones of axons growing in the transparent tail of tadpoles. Cajal had described the cone of growth, and named it, accurately. Thereafter, in recalling his studies of neuronal development, he penned a poetic and memorable description based only on his pictures of Golgi-stained growing axons and dendrites "frozen" in action. Cajal was describing the elaborate growth of neuronal processes in the cerebellum. He did not know about the complex intracellular molecular changes going on within the cells—indeed, even today we are only beginning to understand those changes. His observations led him to marvel at what he had seen, and this led him to express the following question: "What mysterious forces precede the appearance of the processes, promote their growth and ramification, stimulate the corresponding migration of the cells and fibres in predetermined directions, as if in obedience to a skillfully arranged architectural plan, and finally establish those protoplasmic kisses, the intercellular articulations, which seem to constitute the final ecstasy of an epic love story?"

The beautiful drawings of Golgi-stained neuronal axons and dendrites by Ramón y Cajal and other neuroanatomists remind the viewer of the structure of trees. As with trees, the new growth occurs mainly at the tips of the stems and branches. New axonal membrane is made in the cell body and transported down the axon to the growth cones, where it is added to the cell membrane when the vesicles fuse with the membrane of the growth cone. This has been inferred from electron microscope photographs. Also, vesicles moving down axons can be observed in tissue culture; such vesicles can contain various proteins needed for function of the growing, and later of the mature, axon. The fact of membrane addition at the axonal tips has been verified by experiments in which carmine particles were added to cultures of rat sympathetic ganglion neurons growing axons: the particles adhere to cell membranes. The particles sticking to axonal membranes do not move farther from the cell body, and it is observed that particle-free membrane appears at the axonal ending as the axon extends. The new membrane is added at the tip and is particle free.

Figure 13.2A depicts an axonal growth cone as seen in a tissue culture dish. You can see the trunk of the growing axon expanding at its end, so the trunk is the apex of a cone shape. The edges are usually not smooth because there are a number of filamentous extensions. These are the filopodia (plural of filopodium, a thread-like foot). The filopodia are very motile, constantly extending and retracting. The movements are possible because of the contractile threads of proteins in their core containing actin. The neuronal cell membrane of a filopodium contains cell adhesion molecules. If these bind to the substrate, the contraction of the filopodium pulls on the growth cone.

In figure 13.2B, an axon of a chick dorsal root ganglion growing in tissue culture is photographed at three times over a period of 14 minutes. An arrow marks a constant position so the rate of elongation can be calculated. Figure 13.2C shows photos of axons growing in the hamster optic tract on embryonic day 13. These axons have been filled with a fluorescent marker, di-I. Despite the lower magnification, one can see the similarities of the growth cones to those in tissue culture.

A model of a growth cone is depicted schematically in figure 13.3. It can be conceptualized as a fan-shaped thin membranous extension of the axon tip containing a meshwork of contractile proteins (actin). The thin membranous portions are called lamellipodia. The actin-containing filopodial extensions are distributed around the perimeter of the growth cone. The direction of axonal extension tends to be in the direction of the vector sum of the tensions exerted by the filopodia that are adhering to the substrate sufficiently strongly. Meanwhile, as the growth cone moves along substrate surfaces, the addition of new membrane from vesicles transported from the neuronal soma (cell body) makes axonal extension possible without stretching of the existing length of axon.

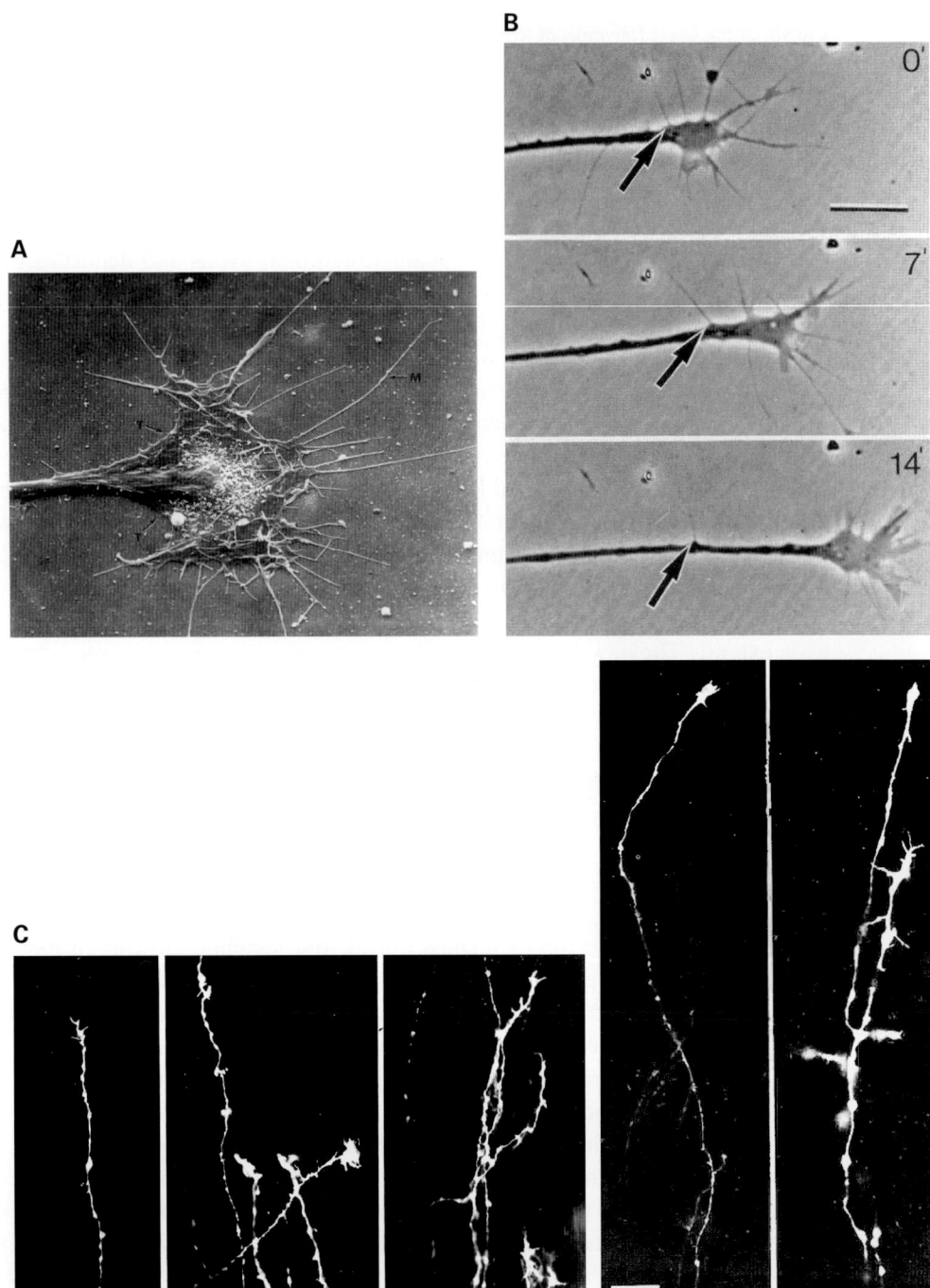

Figure 13.2
(A) Growth cone of an axon as seen in a photograph taken with a scanning electron microscope. (B) Axon of dorsal root ganglion cell from a chick embryo growing in tissue culture for 14 minutes. Scale bar, 10 μm. The axon is elongating at almost 85 μm per hour (2 mm per day). (C) Growth cones of optic tract in embryonic day 13 hamster embryo, labeled with the fluorescent dye di-I. Scale bar, 30 μm. Panel (A) from Wessells and Nuttall (1978); panel (B) from Bray and Chapman (1985); panel (C) from studies of S. Jhaveri, R Erzurumlu, and G. E. Schneider.

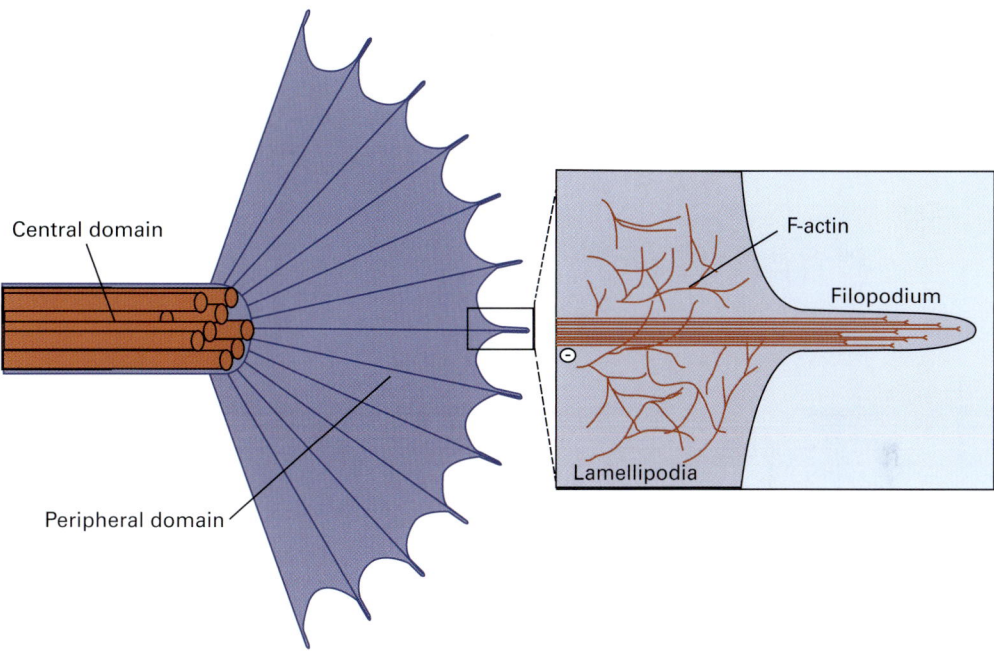

Figure 13.3
Model of an axonal growth cone and its filopodia. Inset shows a model of an actin-filled filopodium and the contrasting distribution of actin filaments in the lamellipodium. Based on Lin, Thompson, and Forscher (1994).

You can get a good mental picture of the dynamics of axonal growth if you see a video clip of a living growth cone in tissue culture with the action speeded up by time-lapse photography through a light microscope. (You can find examples on the Web by doing an Internet search for "growth cone video.") Reports of studies of dendritic growth and dendritic growth cones can also be found; specific molecular and morphological differences between dendritic and axonal growth cones have been reported.

Signals That Shape the Development of Neuronal Circuits

Axonal growth can be influenced in major ways by specific proteins secreted by cells of the body, in some cases by neurons. Such proteins are called growth factors. The first such factor to be discovered was simply called the nerve growth factor, or NGF, by Rita Levi-Montalcini and Victor Hamburger, who found it in extracts of mouse salivary glands. Figure 13.4 illustrates the dramatic trophic effects that can be exerted by NGF on axonal growth from a peripheral ganglion. The figure shows low-magnification photographs of dorsal root ganglia in tissue culture. Few axons grow out of the ganglion when NGF has not been added to the culture medium. When it has been added, large numbers of axons

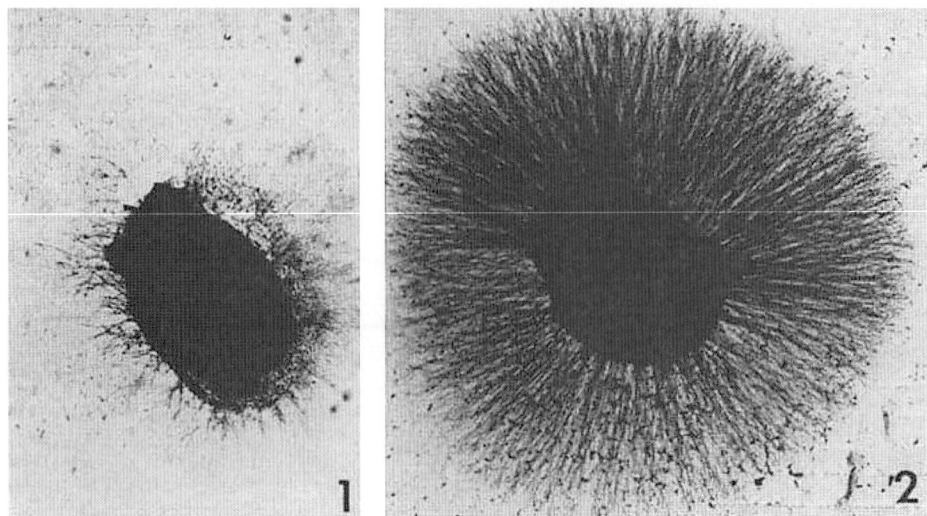

Figure 13.4
Using cultures of explants of dorsal root ganglia or sympathetic ganglia, Rita Levi-Montalcini, working with Victor Hamburger, developed an assay for the presence of a growth factor they discovered, the nerve growth factor (NGF), the first in a family of growth factors. These two photographs illustrate the method of the assay. They show the outgrowth of axons from a dorsal root ganglion in culture: 1, without NGF; 2, with NGF added to the culture medium. From Levi-Montalcini (1964).

grow out in a radiating pattern. Both dorsal root ganglia and ganglia of the sympathetic nervous system respond in this fashion to NGF.

Nerve growth factor was the first to be discovered in a family of neurotrophic molecules now called the neurotrophins. Additional families of growth factors have also been discovered, each growth factor having some unique properties. They tend to influence different groups of neurons in the CNS or PNS.

Growth factors are only one part of the story of axons growing and finding their way in the developing nervous system. An intriguing experiment on this issue was designed to investigate how axons of primary sensory neurons can find their way from the periphery of the developing grasshopper leg to a central ganglion. They do not simply grow straight up the leg toward the ganglion, but follow a somewhat crooked trajectory that is consistent from one grasshopper to another. Microscopic studies of the developing leg reveal non-neural cells in consistent locations along the axon trajectory (figure 13.5). Could the axons be using these cells as guideposts? To test that idea, the investigators performed laser ablations of one of the cells at a stage before the arrival of the elongating fibers. The change disrupted the trajectory of the growing axons. Apparently, the filopodia of the growth cone of the pioneering axon (the first one to grow) are long enough to reach from one of the peripheral cells to the next, so the axon can "follow the dots"—the

Growth of the Great Networks of Nervous Systems

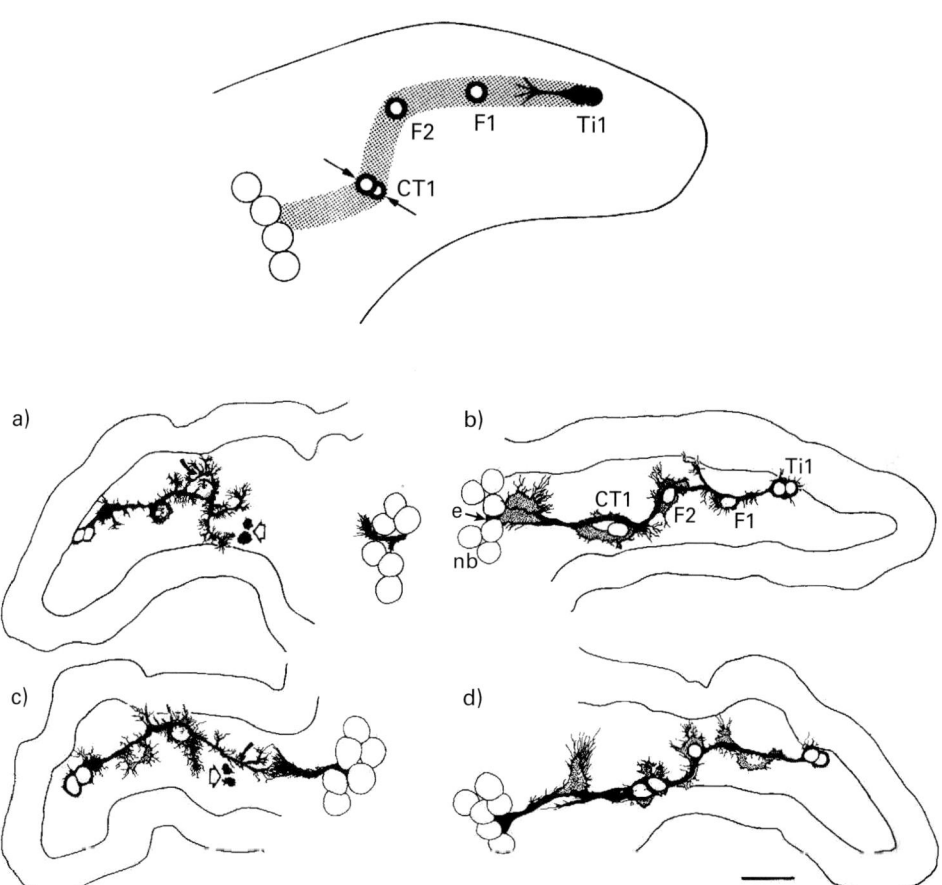

Figure 13.5
(Upper) Hypothesis concerning the guidance of axonal growth from the periphery of the embryonic grasshopper leg to the central ganglion by following a series of distinctive *guidepost cells* (F1, F2, CT1) before reaching the target ganglion (represented by four open circles at left). This idea predicts that ablation of cell CT1 (arrows) will result in an abnormal trajectory of axonal growth. The sketch shows the normally expected trajectory in gray. (Lower) Evidence for guidepost cells: effects of ablating cell CT1 on the left side (a, c). Control limbs are shown on the right (b, d). Scale bar, 100 μm. nb, central neuroblasts; e, efferent axons emerging from central ganglion. From Bentley and Caudy (1983).

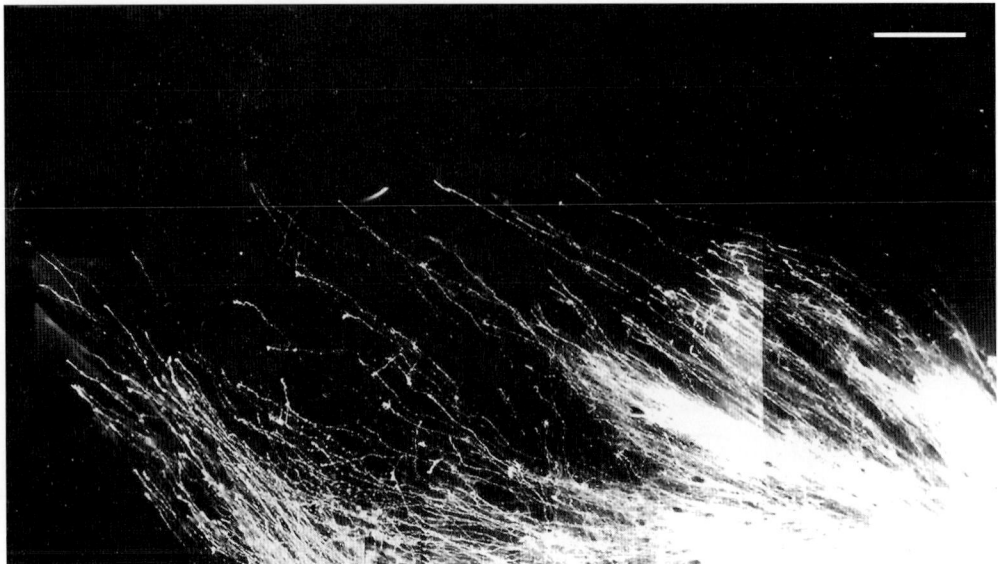

Figure 13.6
Axons of the growing optic tract in an embryonic hamster, labeled with the fluorescent dye di-I. The pioneer axons are spaced out over the width of the tract. Later-growing axons are seen growing in between the pioneers, apparently spacing themselves from the axons already there rather than growing along them. Scale bar, 140 μm.

dots being the guidepost cells. Later-growing axons from the leg periphery follow the first one up the leg to the ganglion.

It is important to point out here that this scenario in insect development is not always what happens during axon growth. In the developing mammalian optic tract, the later-growing axons do not grow along the surfaces of the earlier growing pioneers. Instead, they space themselves between them (figure 13.6).

Long before the specific molecules involved, or even the cell types involved, began to be discovered, experiments provided strong evidence that tissues of the embryonic CNS contain cues, or signals, that guide axons growing through specific regions. An oft-cited experiment was published in 1965 by Hibbard (figure 13.7). He did experiments using transplants of pieces of the amphibian hindbrain. The hindbrain of these little animals contains giant neurons that trigger escape movements when activated. A giant neuron on each side of the brain gives rise to an axon that decussates and turns caudally, extending into the spinal cord. A section of hindbrain containing a pair of these cells, the Mauthner cells, can be transplanted from one animal into another. Hibbard placed such transplants just rostral to the Mauthner cells of the host animal. He put them either in a normal orientation or he turned the transplant 180 degrees. He wanted to find out whether the axons would grow in a direction determined by their tissue of origin or by the host

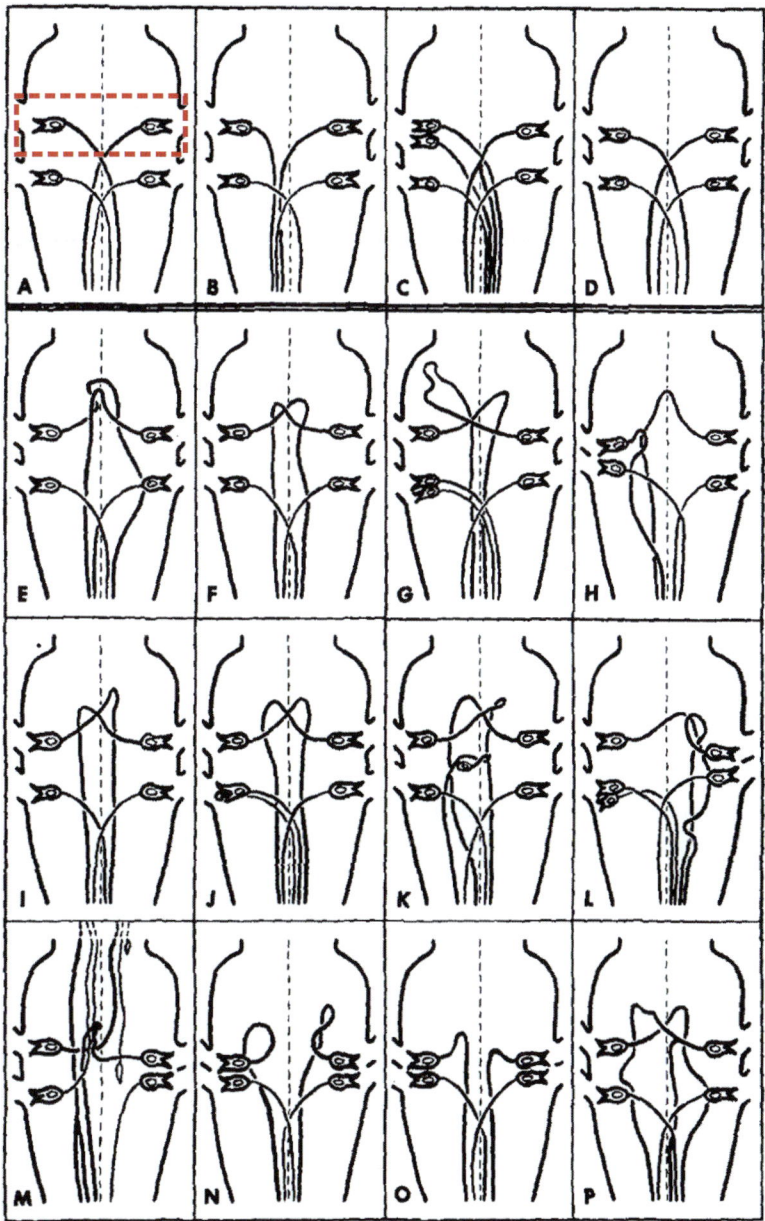

Figure 13.7
Illustration of results obtained by Hibbard (1965) on the growth of Mauthner cell axons in the amphibian hindbrain: In his paradigm, he removed a section of the hindbrain that contains the Mauthner cell pair, and then transplanted this segment into another animal just rostral to the normal location, either in the normal orientation (A–D) or a reversed orientation (E–P). The transplanted segment is indicated in the upper left drawing. These results indicate that cues within the tissue provide initial guidance of axons growing through it. From Hibbard (1965).

hindbrain tissue. The answer was clear. Axons from the reversed transplants grew across the midline and turned in a direction that was rostral in the host animal—so they followed cues from their tissue of origin (the donor tissue). However, when they grew into the more rostral host tissue, they turned around and grew caudally, showing their response to signals from the host brain. Then, even when they reentered their tissue of origin, they continued growing caudally in the host, toward the spinal cord. Thus, the directional cues within the donor tissue or the responsiveness of the growing axons to those cues had changed.

When they reviewed the state of knowledge of neuronal development in 1985, Dale Purves and Jeff Lichtman saw four mechanisms for guiding axon outgrowth. They listed these mechanisms as stereotropism, galvanotropism, tropism based on differential adhesion, and chemotropism based on diffusible substances or on membrane contact. More recent studies have not changed this picture but have supplemented it with many detailed discoveries. Experimental work since that time has been focused on chemical guidance, adding new details and uncovering many of the specific molecules involved.

Stereotropism is a term that refers to effects of substrate surfaces on the directions of axonal growth. Axons do not grow through fluid spaces; they grow on surfaces, in the brain as well as in tissue culture dishes. Not only is the relative adhesivity of a surface important, but also the orientation and curvature can affect the direction of growth. Growth cones do not penetrate other neurons or the glial cells they meet; they grow around them, a directional effect independent of any attraction or repulsion they may have. Axons grow through a veritable forest of other cells and cell extensions. Some axons tend to grow along the surface of the CNS, along the endfeet of radially oriented embryonic neurons and/or glia. Later, cells may migrate through these axons, obscuring their original surface position, as in the case of the mammalian optic tract within the midbrain tectum (superior colliculus).

Galvanotropism is a term that refers to directional effects of electric currents within the nervous system on growing axons. The evidence for such effects in normal development is very limited. That there are effects of even very small electric currents on growth cones has been clear in some tissue culture experiments.

Four Types of Chemical Guidance

Experimental studies have led to clear distinctions among four types of chemical guidance (figure 13.8). These types are basically a classification scheme: Chemical guidance occurs by either attraction or repulsion (or inhibition) of growing axons. Each of these two major types of effect can occur because of either diffusing chemicals or effects of direct cell–cell contact.

The attraction effects of cell-to-cell adhesion result from cell adhesion molecules such as the neuronal cell adhesion molecule, N-CAM, which can be found in the membranes

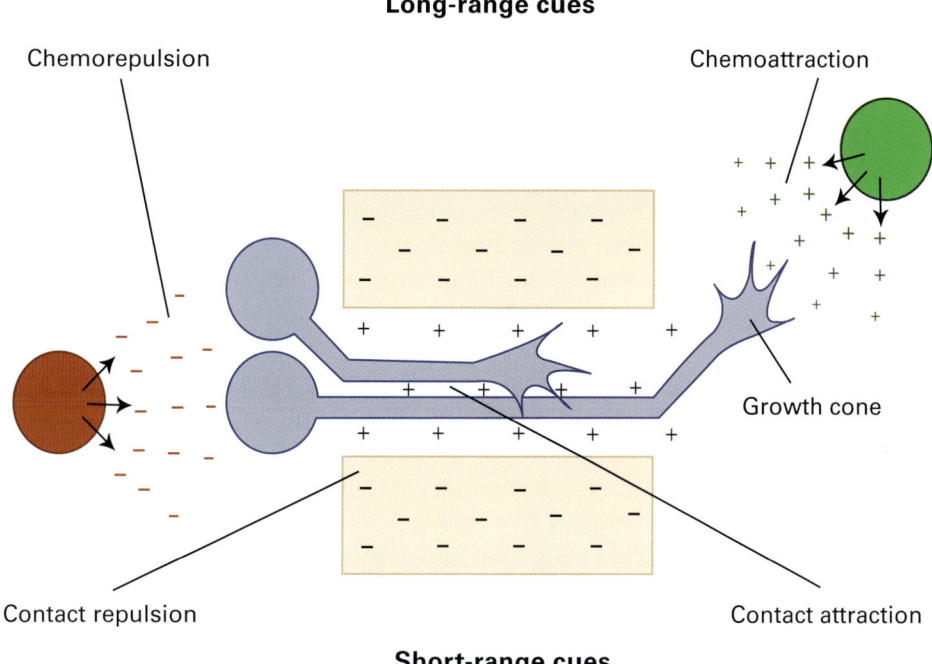

Figure 13.8
Schematic illustration of four types of chemical guidance of growing axons. A pair of axons is shown growing through a channel containing contact attraction molecules on its surfaces and avoiding surfaces on either side because they contain contact repulsion molecules. At the left, the initial growth of the axon is away from a source of molecules causing chemorepulsion. At the right, an emerging axon has turned toward a source of molecules causing chemoattraction. Based on Tessier-Lavigen and Goodman (1996).

of growing axons while they are elongating in close proximity to each other and forming tight bundles, or fasciculi. (A fasciculus is one bundle of axons.) Growing axons may also adhere to extracellular matrix (ECM) molecules like the laminins and cadherins or to molecules in the cell walls of other neurons, like the semaphorins. Diffusing molecules can also attract axons toward greater concentrations, if the growth cones contain responsive molecules in their membranes. Diffusible molecules with such effects include NGF and other neurotrophins, and the netrins, and molecules of other families of growth factors.

Inhibition of axon growth and outright barriers to axon growth include midline barriers. For example, contact repulsion is shown by some axons to certain proteoglycans secreted by midline radial glia. A membrane protein that inhibits the growth of many axon types is found in the membranes of oligodendrocytes. This protein was given the descriptive name Nogo by its discoverers.

Secreted and certain transmembrane proteins can also cause inhibition of axon growth when the axons contain specific receptor molecules. These proteins include the semaphorins and the ephrins. Surprisingly, when growing axons are in certain states, as at certain stages of development, neurotrophins and other growth factors can also cause inhibition of growth.

Although *chemoaffinity* had long before been postulated as an important determinant of specific directional growth and connection formation in the nervous system (even by Ramón y Cajal, and championed by Roger Sperry in a 1963 review paper), it was first seen in live tissue culture by Jonathan Horton at the Harvard Medical School when he was experimenting with NGF. While observing axons from dorsal root ganglia growing on the substrate of the culture dish, he placed a glass pipette containing a solution of NGF into the fluid media. He saw the axons turn toward the source of the NGF diffusing from the pipette. When he moved the pipette, the axons changed directions, always growing toward it.

This directional effect can be considered to be distinct from the effects of NGF on cell survival and amount of axon growth, which are nondirectional in nature. We can make a clear distinction between *tropic* and *trophic* effects: *tropic* means influencing the direction of growth. There are two types of *trophic* effects: survival promoting and growth promoting. Survival is promoted by the prevention or reduction of cell death. Growth-promoting effects result in more vigorous growth or growth of a greater number of processes. With more growth vigor, the axon is more able to compete with other axons.

NGF has both trophic and tropic effects on neurons of dorsal root ganglia and sympathetic ganglia. An axon of a sympathetic or dorsal root ganglion will grow toward the source of NGF by growing up a gradient of this molecule. Its cell body may not survive unless the ending takes up NGF, and the axonal growth vigor is greater with more NGF.

Studies of spinal cord development have suggested ways in which the presence of growth-promoting and growth-repelling molecules can orchestrate the formation of specific connections. For example, semaphorin III is produced in the ventral half of the embryonic spinal cord. Its presence there may repel axons of temperature and pain sensory neurons while allowing the axons of the large sensory fibers that respond to muscle stretch to grow more ventrally where they can terminate on motor neurons. Semaphorin molecules have also been found to play important roles in the directional growth of axons from neocortex to various targets in the CNS; they play a role in the growth of other populations of axons as well.

In related studies, a role of the netrin molecules has been proposed for explaining the formation of the spinothalamic tract decussations. Netrin molecules were found to diffuse from the floor plate region by Tom Jessell and his co-workers at Columbia University. These molecules have tropic effects on axons of dorsal horn cells that form the spinothalamic tract. The axons are attracted to the floor plate region, where they cross to the other side of the spinal cord.

At this point, we have a problem: If the axons from the dorsal horn are attracted to the floor plate region by netrins, why do the axons cross and grow farther? (We know that they are destined to ascend toward the brain.) The answer is that the guidance mechanisms are not fixed. Once the axons have reached the midline region, they change in their response to the netrins; now they are repelled, and the growth cone advances away from the floor plate and into the lateral column of the cord.

This kind of flexibility in molecular effects on axons was reported in 1998 by a group of scientists led by M.-M. Poo. They described a study of axon growth in *Xenopus* spinal cord neurons in culture. The title of the paper states the major result: "Conversion of Neuronal Growth Cone Responses from Repulsion to Attraction by Cyclic Nucleotides." They found that they could change tropic effects of semaphorin III simply by addition of cyclic GMP or cyclic AMP to the culture medium.

Two Modes of Axon Growth

Much of this consideration of axon growth has been about the elongation mode of growth. In this mode, axons grow much faster and branch much less than during the subsequent arborization mode of growth. Figure 13.9A illustrates the two modes of axonal growth from a study of the growth of the optic tract in the Syrian hamster. In that study, my colleagues and I estimated the rate of growth in the elongation mode to be 80–100 micrometers per hour, whereas the average rate of advance of axons growing in the arborization mode was about a tenth of that rate. During elongation, the axons form fasciculi, with later-growing axons extending in close proximity to axons already present, although they do not appear to be growing along the surfaces of the earlier axons. When they start arborizing, they are supporting multiple growth cones simultaneously, with no fasciculation.

When axons arborize during development, they may sprout in an exuberant fashion, extending into some territories that will not be maintained into maturity. For example, the optic tract in the hamster grows transient collaterals in the thalamus that extend into the ventrobasal nucleus (the major somatosensory thalamic cell group) and into the medial geniculate body (the major auditory system thalamic nucleus). Within their normal terminal region, rudimentary branches form and later are lost, as illustrated for the optic tract in the midbrain tectum in figure 13.9A. Thalamocortical axons show a similar phenomenon. Later in development, axonal end arbors of thalamocortical axons in the visual neocortex grow larger and then regress as the animal matures (figure 13.9B). Exuberant transcortical axonal connections have also been described.

Formation of Maps in the Brain

The cellular and molecular mechanisms guiding the axons of the optic tract from the retina along particular paths to their various terminal locations in the brainstem are not

248   Chapter 13

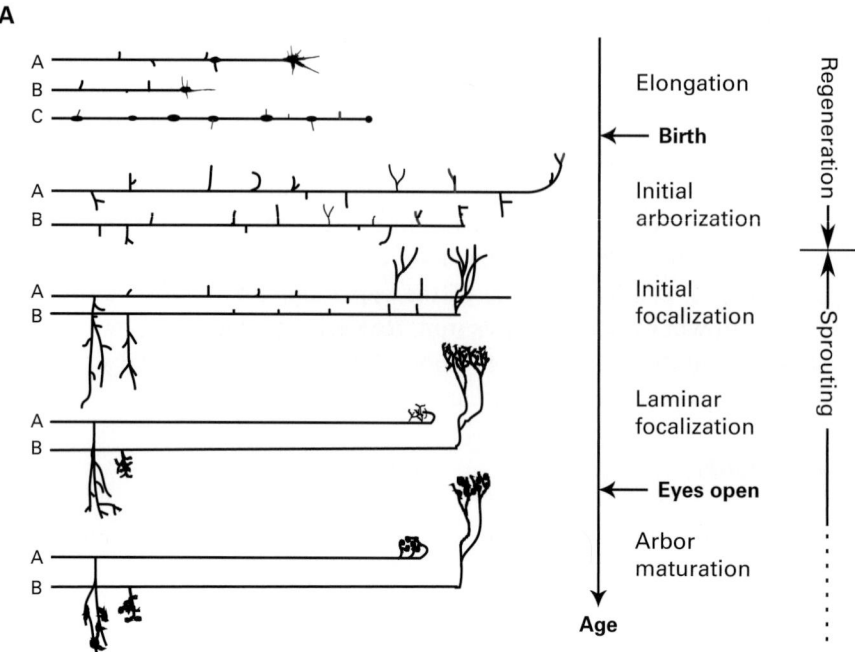

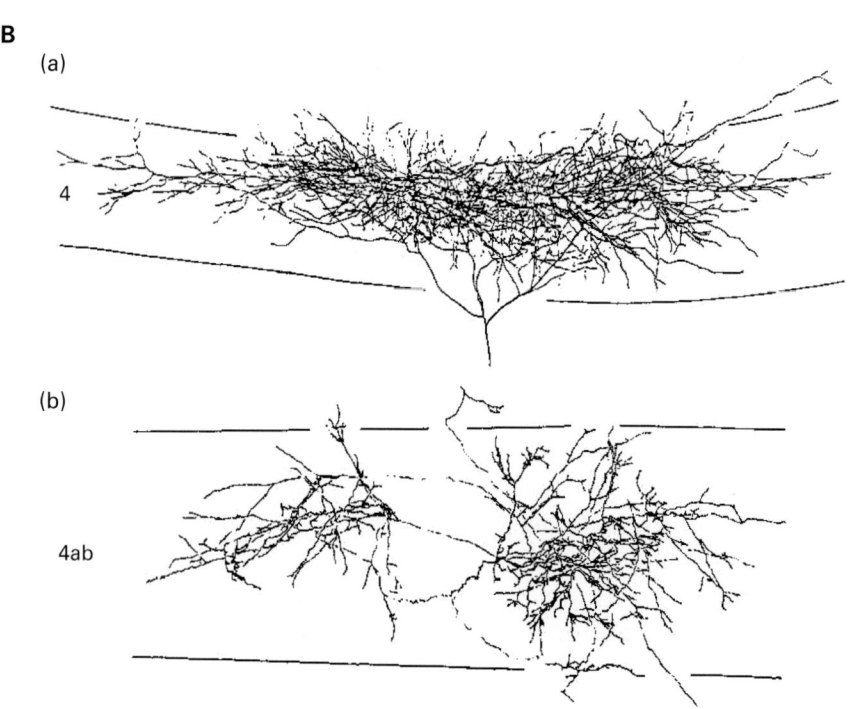

Figure 13.9
(A) Illustrative drawings of the appearance of axons of the hamster optic tract at five stages of development, illustrating two major modes of growth. At the youngest age (top) three elongating axons are shown, each with a growth cone at the tip and filopodia distributed along the length of the trunk. At the next stage, two axons that have reached their target are shown in the stage of initial arborization, with multiple small arbors, many of them transient. In the middle drawing showing the next stage, two axons show regression of most of the initial arbors and augmented growth of one arbor in each of two terminal areas. This is the initial stage of *focalization* of the end arbors, concentrated in one small region. Then, two axons are shown with growing end arbors that are becoming focalized in specific layers of the target tissue. Finally, in the bottom drawing, the arbors of two axons are continuing to mature, with small changes in ultrastructure and sometimes with pruning of some endings. Based on Jhaveri, Schneider, and Erzurumlu (1991). (B) Axons in the primary visual cortex of the cat that carry visual information from the thalamus. (Top) Tracing from Golgi-stained thalamocortical axon that is arborizing in layer 4 of a 17-day-old kitten. This is before axons carrying information from right and left eyes have separated into different columns. (Bottom) Tracing from a horseradish peroxidase (HRP)-filled thalamo-cortical axon in a normal adult cat. Axons carrying information from the two eyes have segregated into different columns; this axon has end arbors located primarily in two different columns. Presumably the space in between the arbors shown contains endings of axons from the other eye. Courtesy of Michael P. Stryker.

fully understood. Their pattern of terminal arbor formation in the midbrain tectum has been a topic of many studies directed at discovering how the terminal pattern maps the positions of the origins of these axons in the retina. This is the so-called retinotectal, or retinocollicular, map, formed in the most distant target of the optic tract. In animals like the frog or goldfish that can regenerate this connection after complete transection of the optic nerve, the topographically organized map re-forms even if the axons have been scrambled. Sperry and others have reported that the map re-creates the original pattern with respect to the retina even if the eyeball has been rotated in the socket, resulting in maladaptive orienting toward visually presented food objects. The map is clearly a result of innately determined mechanisms.

An important piece of the explanation of how this happens has come from discoveries in the developing chick by John Flanagan and co-workers at Harvard and by Friedrich Bonhoeffer and his team working in Germany. The nasotemporal retinal axis representation in the midbrain results from rostral termination of axons from temporal retina and caudal termination of axons from nasal retina. This pattern of distribution results from a gradient distribution of ephrin (Eph) receptors in the axons and a corresponding gradient of ephrin ligands in the developing midbrain tectum (figure 13.10). The mechanism, it has turned out, is not a selective affinity as previously proposed by Sperry and others, but rather a selective repulsion of axons from the more temporal retina by the cell membranes of the more caudal tectum.

Plasticity in Brain Maps

Molecular determination of map formation does not mean that a CNS map is incapable of plasticity. Plasticity in such maps after brain damage in developing animals indicates that there are other factors at work. For example, an ablation of the caudal portion of the

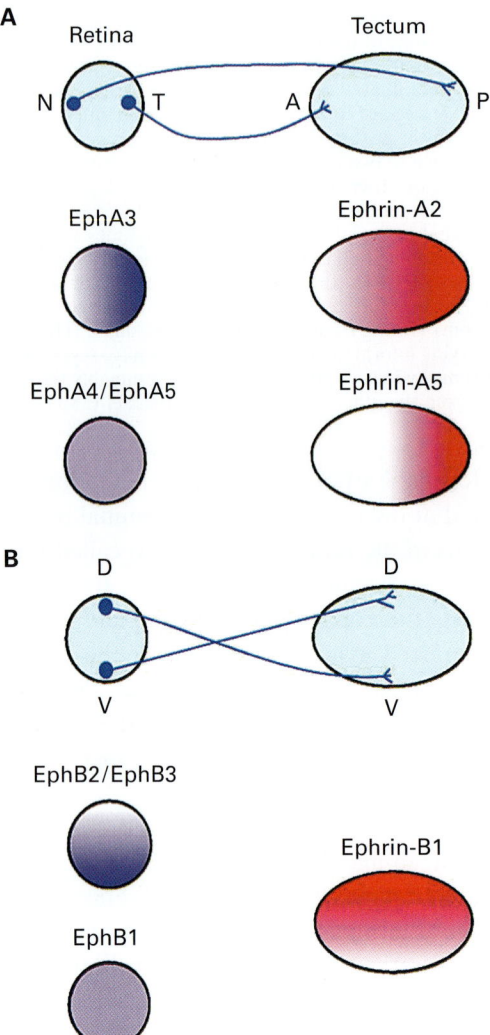

Figure 13.10
Schematic representation of the orderly arrangement of the chick retinotectal projection and the distribution of molecules that appear to guide the developing axons. This guidance depends on the distribution of ephrin (Eph) receptors in the retina and ephrin ligands in the midbrain tectum. Discovered initially by Bonhoeffer et al. and by Flanagan et al. with further discoveries by O'Leary et al. From O'Leary, Yates, and McLaughlin (1999). (A) The segregation of axons according to their origins along the nasotemporal axis of the retina, and the molecular gradients in retina and tectum that correspond to this axis. (B) The segregation of axons according to their origins along the superior-inferior axis of the retina, and the molecular gradients in retina and tectum that correspond to this axis.

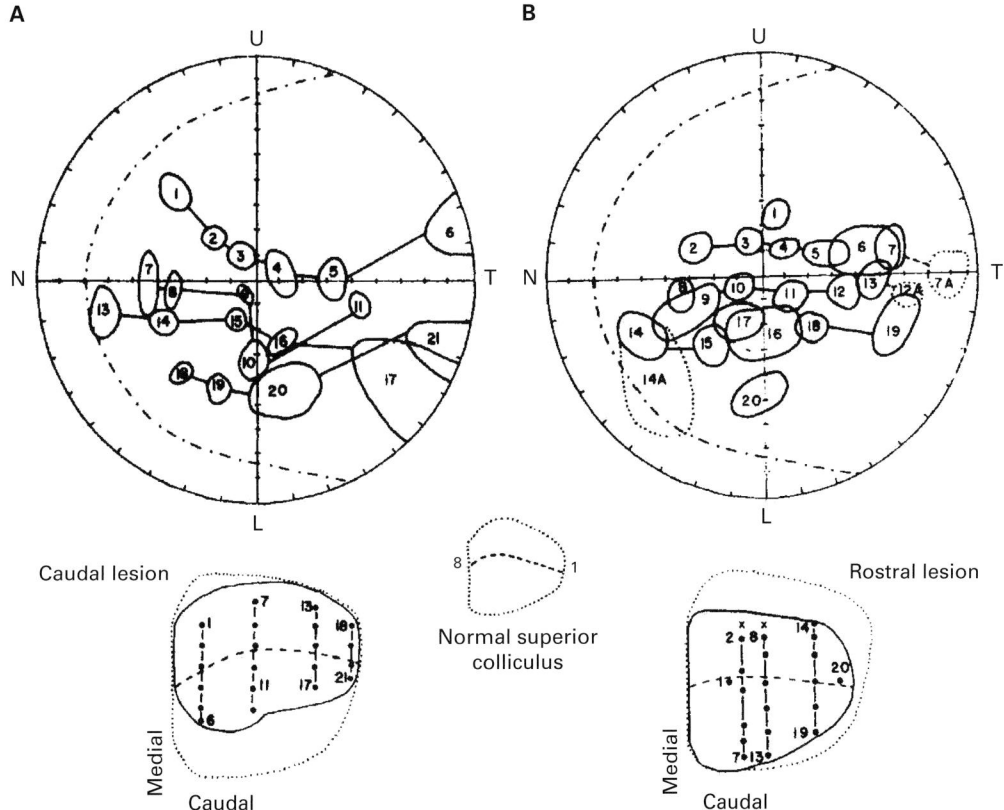

Figure 13.11
Illustrations of the finding that after ablation of either the rostral or the caudal part of the optic tectum (superior colliculus) in newborn hamsters, there is a compression of the entire retinotopic map in the remaining tectum. The visual field was mapped using electrical recordings from single neurons. Positions of numbered electrode penetrations in the superior colliculus are shown in the lower reconstructions for two different animals (A, B). Corresponding receptive fields are depicted in the visual field maps shown above. From Finlay, Schneps, and Schneider (1979).

superior colliculus in a newborn hamster is followed by the formation of a complete map in the remaining rostral colliculus. The map is compressed (figure 13.11). If, in contrast, part of the retina is ablated in the baby animal, the map does not form with an uninnervated area—a hole—in it, but rather, the map of the remaining retina expands into the denervated area. Such plasticity was discovered first in studies of regeneration of the optic nerve and tract of goldfish. Although the map shows the plasticity described in case of brain damage in hamsters during the first few days after birth, it no longer occurs when the damage is inflicted later in development.

This does not mean that axonal plasticity never occurs in older mammals. Some axonal systems retain much plasticity, whereas others become much more rigid in their structure after the developmental period. In recent years, changes in somatosensory neocortical maps have been found to result from functional activity caused by training situations (see chapter 34). The nature of any accompanying anatomical changes has not yet been made clear.

More Plasticity in the CNS: Collateral Sprouting

Other types of axon plasticity have been discovered to result from brain and spinal injury. *Collateral sprouting* was first discovered in peripheral nerves innervating skin and muscle and later in axons of the central nervous system of both mature and developing mammals. If axons of motor neurons innervating the extensor muscles of the hand in a human are partially damaged, the person may find it difficult to extend the fingers because of a weakness due to the loss of muscle fiber activation by axons that have been disconnected. However, before the axons have had time to regenerate, the affected muscles will regain much of their strength as the remaining, intact fibers innervating the same muscles sprout collateral branches that form synapses with the denervated muscle fibers. Sensory axons innervating the skin show a similar kind of collateral sprouting. In the late 1950s, a comparable kind of collateral sprouting of intact axons was found in the spinal cord of adult cats. A few years later, collateral sprouting of optic tract axons was found in the diencephalon of mature rats in regions that had lost innervation from the neocortex.

The sprouting of new axonal connections that replace terminals of axons that are degenerating after injury is not an invariable occurrence in the adult mammalian brain or spinal cord. Certain axon types show vigorous sprouting, but many long axons have terminal arbors that seem incapable of such plasticity or they are capable of only a meager amount. As for the short-axon interneurons, we have little information, but it is likely that some of them show much plasticity (see chapter 34).

Axons that show little or no sprouting in mature animals may show vigorous sprouting after damage inflicted very early in development. This sprouting can be great enough to cause developing axons to violate normal rules of regional specificity. For example, in the developing hamster or ferret, the optic tract can be induced to grow into the medial geniculate body of the thalamus—normally a part of the auditory system—or into the ventrobasal nucleus—normally the recipient of somatosensory system axons from the spinal cord (see chapters 11 and 12). Illustrations of results of a few of the experiments upon which these statements are based are shown in the drawings of figures 13.12 and 13.13.

Even in the absence of knowledge of the mechanisms at the molecular level, sprouting studies have uncovered some basic rules at the level of the cells. If we study these rules,

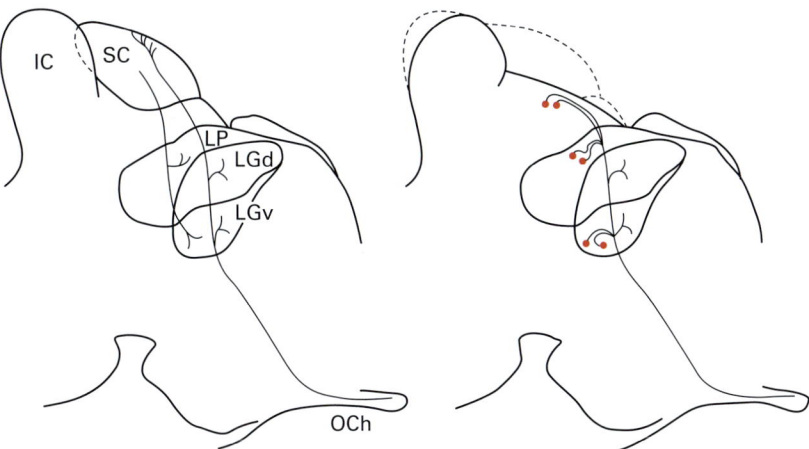

Figure 13.12
Summary of the effects of early ablation of the superior colliculus (SC) in the newborn hamster on the projections of the retina. Axon tracing was carried out using Nauta silver stains for degenerating axons. (Left) Projections of retina and of SC in normal hamster. (Right) After the early lesion of SC, there is sprouting of the optic tract axons. The most prominent sprouting, in lateral posterior nucleus of thalamus (LP) and ventral nucleus of lateral geniculate body (LGv) and in the remaining SC, is shown by the axonal endings in lines tipped in red. IC, inferior colliculus; LGd, dorsal nucleus of lateral geniculate body; OCh, optic chiasm. From Schneider (1973).

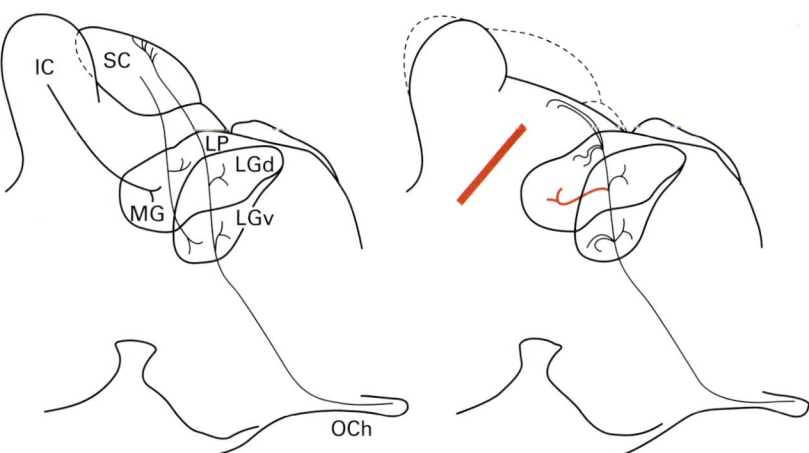

Figure 13.13
Summary of experiment with hamsters in which, in addition to early ablation of the superior colliculus (SC) in newborns, the axons carrying auditory information from the inferior colliculus (IC) to the medial geniculate body (MG) were surgically destroyed (thick red line). In the adult animal, in addition to the sprouting shown in figure 13.12, there was sprouting of optic tract axons into the medial geniculate body (red axon branch). The consequence of this abnormal pathway is a visual pathway to what is normally auditory neocortex. LGd, dorsal nucleus of lateral geniculate body; LGv, ventral nucleus of lateral geniculate body; LP, lateral posterior nucleus of thalamus; OCh, optic chiasm. From results of experiments by G. E. Schneider and R. Kalil.

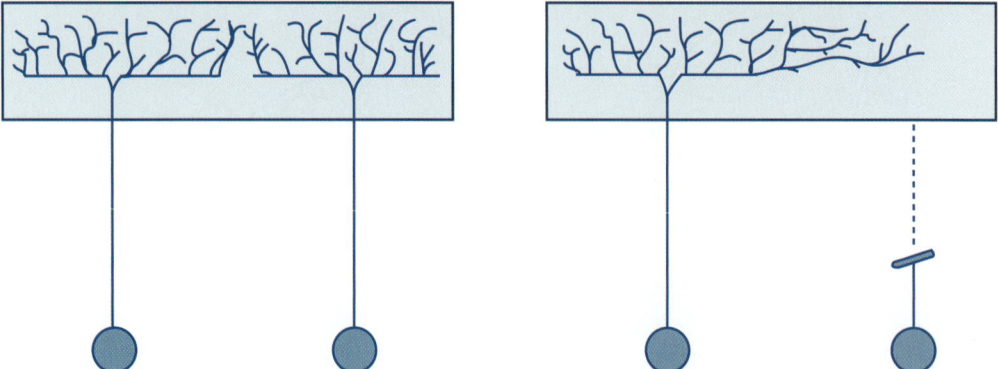

Figure 13.14
Illustration of a major factor that results in the sprouting of axonal endings to form abnormal connections after brain damage in young animals: Degeneration of some terminal arbors in a region can result in sprouting of remaining intact axons or spreading of their terminal arbors over a wider-than-normal territory. Based on Jhaveri, Schneider, and Erzurumlu (1991).

we can think about their implications for plasticity in the uninjured as well as in the injured brain. Two major reasons, or rules, for sprouting have been described.

First, multiple studies have found evidence that axons compete with other axons for terminal space, tending to spread into a termination region when other axons have been removed from it (figure 13.14), especially during the developmental period. But what are the axons competing for? "Terminal space" needs more definition. Evidence indicates that axons compete for the uptake of growth factors (chemicals such as NGF). To do this, they may have to compete for occupancy of specific synaptic sites on the membranes of the target neurons. The growth factors are taken up and transported back to the cell bodies where they promote cell survival and more axonal growth. In addition, growing axons may compete more directly during the contacts made by their growth cones: A growth cone that contacts another axon responds to the contact in a manner depending on the cell types involved. A common response is a retraction reaction, in which the growth cone collapses and pulls back a certain distance, then re-forms and the axon starts extending again, usually in an altered direction. Such retraction can be in response to specific molecules that cause this "contact inhibition of extension" (as it was described by Graham Dunn in 1973).

The second major factor can be described as a tendency of neurons to grow and maintain a certain quantity of axon terminals or number of synaptic connections. Its competitive growth vigor is greatest when it has formed few connections and becomes less and less as it forms more and more terminals, until it achieves the terminal quantity it is capable of. Evidence for this factor has been obtained in studies of sprouting responses after different types of injury sustained by the sprouting axons and by neighboring axons.

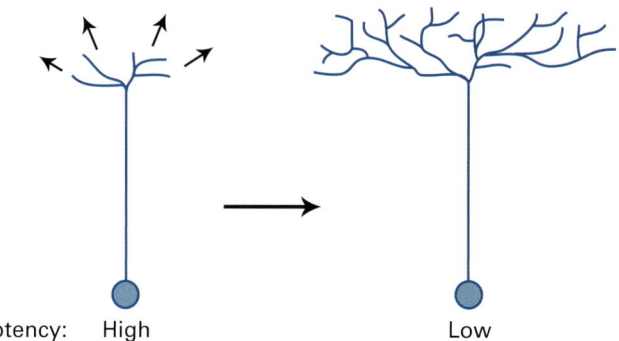

Figure 13.15
Illustration of postulated developmental changes in the intrinsic, competitive vigor of axon growth. As a cell's axon grows more and more terminal arbors and endings, its tendency to form these arbors becomes progressively weaker. It has an intrinsic tendency to form a certain quantity of endings and no more. Based on Jhaveri, Schneider, and Erzurumlu (1991).

Modulation of Competitive Axonal Growth Vigor

Changes in axonal growth potency, or growth vigor, during development are illustrated in figure 13.15. Figure 13.16 illustrates how this growth vigor can be modulated by "pruning" of the axonal arbors during development: Sprouting in one region can be induced by blockage of or damage to axons in another region. Such "pruning effects" have also been seen in some mature fiber systems (e.g., the widely branching norepinephrine-containing axons).

Growth vigor can be strongly influenced by chemical factors, especially by specific growth factors, as already mentioned. The photographs in figure 13.4 illustrate the effects of NGF on growth vigor of axons from a dorsal root ganglion in tissue culture. There are also molecular factors intrinsic to the cells that determine growth capacity, not only when the axon is arborizing but also when it is still in the elongation mode of growth.

Axonal plasticity can also occur without special chemical modulation and without a response to a lesion. It can be seen as a kind of learning, caused by electrical activity of neurons. More active axons (axons firing more action potentials) show more vigorous growth, so they out-compete less active axons of the same type. Thus, if one eye of a cat or monkey or human is closed or occluded in some other way during a critical period in development, the axonal arbors in the neocortex that are activated by that eye are smaller than those activated by the other, more active eye. Axons activated by the two eyes compete for terminal space in the cortex.

After Development: Sprouting Response to a Pruning Lesion

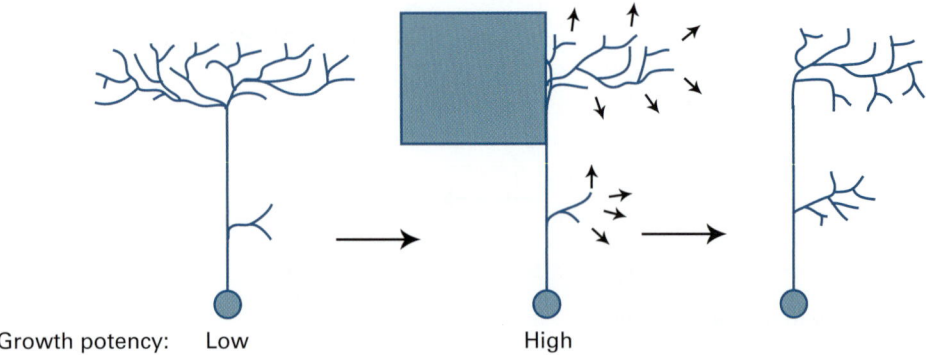

Growth potency: Low   High

Figure 13.16
Because of the postulated tendency of a neuron to form an end arbor or arbors of a certain size (or a certain number of terminals) during development, a lesion that "prunes" part of a terminal arbor will cause a compensatory increase in growth vigor in the remainder of the axon's arbor. It will tend to conserve its total terminal arbor size (or number of terminals). The phenomenon has been called a pruning effect because of its similarity to the consequences of pruning trees and plants. Adapted from Jhaveri, Schneider, and Erzurumlu (1991) and from studies of the developing optic and olfactory tracts by Schneider (1973) and Devor and Schneider (1975).

Rules of Sprouting Apply to Development, with Implications for Evolutionary Change

We have described two types of factors that could play roles in axonal development. They result in structural changes after brain damage during development and even later in life in the case of some systems. These factors can be summarized as follows: (1) Extrinsic factors that result in competition among axons for places to form terminal arbors and synapses. (2) Intrinsic factors that cause neurons to form a specific quantity of terminals and to conserve that quantity.

These two factors have implications for evolutionary changes in CNS pathways. Consider the following possible change: A mutation causes a reduction in the size of a terminal area without a reduction in the quantity of terminals that one of its afferent axon populations can form. This will result in a tendency of those axons to sprout collaterals elsewhere, especially in an area that has lost some afferents because of the mutation. Thus, a new projection appears in the animal with the mutation. Such animals would survive if the change is adaptive in some way, or at least if it is not maladaptive. See, for example, the type of sprouting illustrated in figures 13.12 and 13.13.

Plasticity in the Small Interneurons of the Adult Brain

Prominent neuroanatomists like Ramón y Cajal and Joseph Altman have proposed that the small interneurons of the neocortex, the granule cells, play a special role in learning by an individual. They may be the most plastic neurons in the mature brain (see also chapter 34). Their plasticity can be envisioned with the aid of the two cellular rules of axonal end-arbor formation we have been considering. Other rules of associative learning are no doubt involved as well, such as the kind of synaptic change proposed by Donald Hebb in 1949. His rule can be simplified in the phrase, "neurons that fire together wire together." Various studies of learning in recent years, although initially focused on molecular changes, have found local axonal sprouting as part of the engram (the physical changes underlying memory). Let's assume that this happens in a short-axon, locally connected granule cell of the cortex. When some terminals increase in number in one part of the axonal tree, and we assume that the cell conserves its total number of terminals, then the cell will withdraw some of its terminals in other parts of the tree. The conservation rule implies that there is a kind of homeostatic regulation of terminal quantity. It is clear that something like this must happen if there are synaptic increases underlying long-term retention, or else the number of synapses would keep increasing with more and more learning, *ad infinitum*.

Direct visualization of plasticity of dendritic arbors of neocortical interneurons has recently been accomplished using two-photon microscopy of cortical tissue in live mice. The studies made use of transgenic mice in which some neurons express a green fluorescent protein. Remarkably, neurons with dendrites undergoing changes from day to day were localized to a specific layer—a *dynamic zone* of superficial layers 2–3. The neurons express the inhibitory neurotransmitter GABA.

Earlier support for the plasticity of interneurons in the adult neocortex was reported from studies of the distribution of the growth-associated protein GAP-43. This protein is found along the axonal membranes of growing axons and has been implicated in filopodia formation. For example, it is found all along the optic tract during axon elongation, and it shifts to the end arbors during arborization. Its expression by the retinal ganglion cells is greatly reduced after these growth stages. However, there is evidence that even in adults it is expressed in short-axon interneurons in the thalamus and in the neocortex. In humans, it is expressed more in multimodal association areas than in other areas, and it is concentrated especially in the superficial layer 2 and also in the hippocampus.

Structural Regression during Development and Its Purposes

We mentioned earlier the observations of a growth of exuberant axonal connections and end arbors followed by a regression to the mature state. This amounts to a kind of self-pruning of the axonal processes. A more drastic form of structural regression is also common in development: entire neurons die. Many neurons depend on axonal contacts with a target for survival. In the target tissue, the axons take up trophic factors and transport them back to the cell bodies. Without sufficient trophic factor (growth factor), the cells undergo *apoptosis*, which means cell suicide or programmed cell death.

During the initial elongation stage of growth, before the axon reaches its target and begins to arborize, it is protected from apoptosis by intrinsic factors. Such a factor, found to be expressed strongly by retinal ganglion cells during the elongation period and at the onset of arborization, is Bcl-2. (This name was given after its discovery in a B cell lymphoma cell line.) At that time, synthesis of this *elongation protein* is downregulated. Concurrently with the downregulation, the neuron acquires a dependence on target contact for survival. At the same time, the majority of the neurons lose their ability to initiate elongation although they retain the potential for arbor formation.

One of the best-known examples of neuron-target dependence is that of spinal motor neurons. In experiments on the chick embryo, it has been found that extirpation of a limb bud during early development results in a great loss of motor neurons destined to innervate the leg muscles. Motor neuron loss is normal in development, but the loss is greater after the early limb removal. In contrast, if an extra limb is grafted onto the embryo at a very early stage, the motor neuron loss is reduced below the normal level. This is apparently because in the supernumerary limb, there are more muscles cells to innervate.

An important consequence of the naturally occurring neuronal cell death among spinal motor neurons is a population size matching. The genetic information does not have to determine the exact number of motor neurons, as excess cells will die. A second important purpose is error correction. For example, just when many retinal ganglion cells die, retinal axons that terminate in the wrong places in the midbrain map disappear. These cells may die because their axons cannot take up enough neurotrophin.

Axon Loss in the Damaged CNS: Is Regeneration Possible?

When a fish or an amphibian suffers brain damage, axons that have been cut off from their cells of origin by the damage undergo anterograde degeneration. However, in many cases the more proximal portions of the axons are able to regrow so that the original connections, after a period of regrowth, can become reconstituted. For many long axonal pathways, this kind of regeneration can occur in mammals only when the damage occurs very early in development, during the period of axonal elongation. Thereafter, the regenerative capacity becomes greatly reduced, and in most situations it is lost altogether. Much basic research on axon regeneration in mammals has been concerned with understanding this loss and discovering ways to recover the lost regenerative capacity in the CNS.

Many mature neurons simply lose their axonal growth vigor, and they may die after being cut off from their end arbors, from which they receive trophic factors. However, two kinds of studies have shown that many mammalian neurons do not completely lose their ability to regrow an axon after the developmental period. First, in tissue culture some neurons in a population can extend an axon into a favorable environment like that provided by embryonic brain tissue, although they may fail to show such growth into adult tissue. Second, many neurons in the brain can extend an axon into an implanted segment of peripheral nerve, although the axon will extend very poorly, if at all, into adult CNS tissue. The mature CNS tissue environment contains many inhibitory factors that slow or block axon growth.

It has become clear that obtaining sufficient axon regeneration in the mature mammalian brain in order to get functional recovery is a problem with multiple facets. One needs to preserve the damaged cells, as dying cells do not regrow axons. Procedures are needed to permit axon growth by circumventing the many inhibitory factors that block or stunt axon growth. Promotion of growth vigor may be needed after the period of early development is over. In addition, plasticity of regenerated connections can play an important, even a critical, role in functional recovery. This summarizes the "four P's of regeneration" reviewed by Ellis-Behnke in 2007; namely, Preserve, Permit, Promote, and Plasticity.

Recent research on CNS axon regeneration in mammals has yielded results that make it seem likely that regeneration sufficient for some functional recovery may be obtainable after some debilitating injuries of the brain and spinal cord. Injury sites have been bridged by peripheral nerve segments, allowing enough axon regrowth to bring functional return. Bridges have been formed by implants of several materials, including natural ones made of Schwann cells or olfactory bulb glial cells (ensheathing cells). Bridges of artificial materials have also been created successfully (e.g., a nanofiber meshwork formed by self-assembling peptides). Chemical methods have been used for inhibiting the formation of glial scars, which can form a barrier to growing axons, or for breaking up such scars. There is evidence that genetic transfections may be needed for some axon populations before they can regrow axons. Finally, intensive behavioral rehabilitation procedures may be essential for some systems to recover adequate function even after some axon regeneration has occurred.

Readings

Quotation from Ramón y Cajal: from p. 373 in Ramón y Cajal, S. (1989). *Recollections of my life*. Translated by E. H. Craigie with J. Cano. Cambridge, MA: The MIT Press. (Originally in Spanish, 1901–1917; first published in English in 1937 by the American Philosophical Society.)

Various pioneering studies of axon development are reviewed in the following book. Included are the first observations of living growth cones by Harrison and by Speidel, and the experiments by Bray who added carmine particles to cultures of rat sympathetic ganglion neurons growing axons. Also reviewed is the discovery of NGF by Rita Levi-Montalcini and Victor Hamburger, who found it in extracts of mouse salivary glands.

Purves, D., & Lichtman, J. W. (1985). *Principles of neural development*. Sunderland, MA: Sinauer Associates.

Microscopic studies of the developing grasshopper leg reveal non-neural cells in consistent locations along the axon trajectory:

> Bentley, D., & Caudy, M. (1983). Pioneer axons lose directed growth after selective killing of guidepost cells. *Nature, 304,* 62–65.

Studies of developing hamster optic tract:

> Jhaveri, S., Erzurumlu, R. S., & Schneider, G. E. (1996). The optic tract in embryonic hamsters: Fasciculation, defasciculation, and other rearrangements of retinal axons. *Visual Neuroscience, 13,* 359–374.

Discovery of the oligodendrocyte membrane protein Nogo and follow-up studies, reviewed by the head of the laboratory where this protein was discovered:

> Schwab, M. E. (2004). Nogo and axon regeneration. *Current Opinion in Neurobiology, 14,* 118–124.

Studies of axons, growing in tissue culture, that demonstrate chemoaffinity affects of NGF:

> Gundersen, R. W., & Barrett, J. N. (1979). Neuronal chemotaxis: Chick dorsal-root axons turn toward high concentrations of nerve growth factor. *Science, 206,* 1079–1080.

Description of two modes of axonal growth in a report of a study of the growth of the optic tract in the Syrian hamster:

> Jhaveri, S., Edwards, M. A., & Schneider, G. E. (1991). Initial stages of retinofugal axon development in the hamster: Evidence for two modes of growth. *Experimental Brain Research, 8,* 371–382.

The optic tract in the hamster grows transient collaterals in the thalamus that extend into the ventrobasal nucleus and into the medial geniculate body:

> Frost, D. O., So, K.-F., & Schneider, G. E. (1979). Postnatal development of retinal projections in Syrian hamsters: A study using autoradiographic and anterograde degeneration techniques. *Neuroscience, 4,* 1649–1677.

Axonal end arbors of thalamocortical axons in the visual neocortex grow larger and then regress as the animal matures:

> LeVay, S., & Stryker, M. P. (1979). The development of ocular dominance columns in the cat. *Society for Neuroscience Symposium, 4,* 83–98.
>
> Stryker, M. P. (1982). Role of visual afferent activity in the development of ocular dominance columns. *Neurosciences Research Program Bulletin, 20,* 540–549.

Similarly exuberant transcortical axonal connections have also been described:

> Innocenti, G. M., & Price, D. J. (2005). Exuberance in the development of cortical networks. *Nature Reviews, Neuroscience, 6,* 955–965.

"Contact inhibition of extension" in populations of growing axons described by Graham Dunn:

> Dunn, G. A. (1971). Mutual contact inhibition of extension of chick sensory nerve fibers *in vitro*. *Journal of Comparative Neurology, 143*, 491–507.

Bcl-2 and axonal elongation:

> Chen, D. F., Schneider, G. E., Martinou, J. C., & Tonegawa, S. (1997). Bcl-2 promotes regeneration of severed axons in mammalian CNS. *Nature, 385*, 434–439.

Spinal motor neuron survival depends on target innervations: reviewed in book cited above by Purves and Lichtman (1985).

Studies of axonal sprouting in response to early lesions in the visual and olfactory systems:

> Devor, M., & Schneider, G. E. (1975). Neuroanatomical plasticity: The principle of total axonal arborization. In Vital-Durand, F., & Jeannerod, M., eds., *Aspects of neural plasticity*, pp. 191–200. Paris: I.N.S.E.R.M.

> Schneider, G. E. (1973). Early lesions of superior colliculus: Factors affecting the formation of abnormal retinal projections. *Brain, Behavior & Evolution, 8*, 73–109.

The "four Ps" of CNS axon regeneration: reviewed by Rutledge Ellis-Behnke:

> Ellis-Behnke, R. (2007). Nano neurology and the four P's of central nervous system regeneration: Preserve, permit, promote, plasticity. *Medical Clinics of North America, 91*, 937–962.

Axonal regeneration through bridges formed by a nanofiber meshwork made of self-assembling peptides:

> Ellis-Behnke, R .G., Liang, Y. X., You, S. W., Tay, D. K. C., Zhang, S., So, K. F., & Schneider, G. E. (2006). Nano neuro knitting: Peptide nanofiber scaffold for brain repair and axon regeneration with functional return of vision. *PNAS, 103*, 5054–5059.

Evidence that the growth-associated protein GAP-43 is expressed in adult mammals in short-axon interneurons in the thalamus and in the neocortex.

> Moya, K. L., Benowitz, L. I., & Schneider, G. E. (1990). Abnormal retinal projections alter GAP-43 patterns in the diencephalon. *Brain Research, 527*, 259–265.

> Neve, R. L., Finch, E. A., Bird, E. D., & Benowitz, L. I. (1988). Growth-associated protein GAP-43 is expressed selectively in associative regions of the adult human brain. *PNAS, 85*, 3638–3642.

A dynamic zone of inhibitory interneuron dendrites in the superficial layers of neocortex:

> Chen, J. L., Flanders, G. H., Lee. W. C. A., Lin, W. C., & Nedivi, E. (2011). Inhibitory dendrite dynamics as a general feature of the adult cortical microcircuit. *Journal of Neuroscience, 31*, 12437–12443.

VI A BRIEF STUDY OF MOTOR SYSTEMS

14 Overview of Motor System Structure

We have looked at all levels of the central nervous system. Now we begin studies of specific functional systems, starting with the structures and pathways most directly involved in control of movements. We do not start with the motor cortex, which came relatively late in vertebrate evolution. In thinking about the evolution of motor control, we have to ask some basic questions: What were, and are, the major functional demands to be satisfied by motor systems of the central nervous system? These demands long preceded the vertebrates. What structures for motor control are present in all vertebrates? What were the mammalian motor system elaborations, and why did these evolve? In this chapter, we will try to come up with answers to these questions, and then we will begin a study of structural organization of motor systems by beginning with motor neurons, a study that will be expanded in the next chapter.

A Functional Starting Point for the Study: Three Major Types of Movement Critical for Survival

All organisms that can move about need certain general-purpose movements—movements that are used in many different action patterns.

First is *locomotion*. Animals need to be able to locomote toward or away from things or places: they make approach and avoidance movements. These are really two kinds of functions, but both involve locomotion in a direction defined by a location, therefore we can group them together. Escape from predators and other dangers is basic for survival no matter what current action the animal is engaged in. Foraging or exploring with seeking of and approach to a goal object or place is a basic ability serving various motivational states (drives).

The second general purpose movement ability, equal in importance to the first, is *orienting*: the turning of head and/or body toward or away from something. This ability is crucial for accomplishing the goals of approach or avoidance.

The third general purpose movement ability is *grasping* of an object, of obvious importance for consummatory actions. Grasping can be with the mouth or with the limbs, involving reaching and gripping something.

The motor neurons that control each of these movements are found in the ventral horn of the spinal cord and in the basal plate regions of the hindbrain, and also in the oculomotor nuclei of the midbrain and hindbrain. Higher control of all three types of movement is exerted by structures of the midbrain that have descending pathways to hindbrain and cord.

Midbrain Control of the Three Types of General-Purpose Movement

The first type of movement, locomotion, can be controlled by descending pathways from the caudal midbrain reticular formation. The neuronal structures in the midbrain, in a region designated as the midbrain locomotor area (MLA), have not been clearly defined by cytoarchitecture or fiber architecture. They have been defined by electrophysiological studies in which locomotor movements are elicited by electrical stimulation (figure 14.1). The same movements are seen in nature when an animal approaches or avoids something and when it is foraging for food or simply exploring the environment.

A major controller of the second type of movement, turning of head (and eyes) and/or body, is the tectospinal tract, originating in large neurons of the deep layers of the midbrain tectum, which is the superior colliculus of mammals. This pathway is illustrated in figure 11.2. It terminates in the medial hindbrain reticular formation and, more caudally, on cervical spinal cord neurons that control neck muscles.

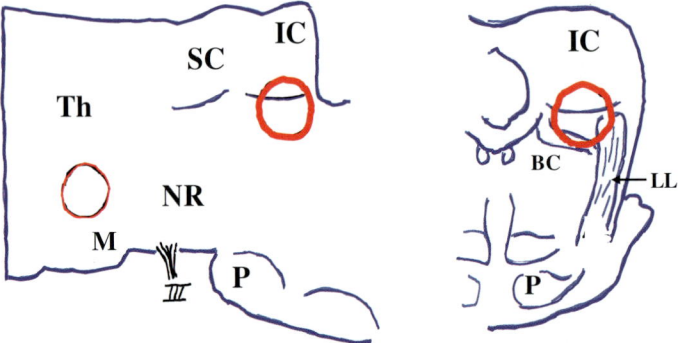

Figure 14.1
Position of the midbrain locomotor region (MLR) in the caudal midbrain reticular formation, as determined in one study by electrical stimulation that elicited locomotion in the domestic cat. (Left) On a parasagittal section, the MLR is shown beneath the inferior colliculus (IC). In addition, the position of the hypothalamic locomotor region, above the mammillary body (M), is shown (red circle). (Right) Frontal section through the IC shows the position of the MLR (red circle). Additional abbreviations: III, third cranial nerve; BC, brachium conjunctivum; LL, lateral lemniscus; NR, nucleus ruber; P, pontine gray; SC, superior colliculus; Th, thalamus. (Based on Orlovskii, 1970.) The MLR includes the pedunculo-pontine tegmental nucleus and the adjacent midbrain extrapyramidal region, and nearby areas. Projections from this region to the hindbrain reticular formation and the spinal cord have been found in axonal tracing studies.

The third type of movement, grasping, can have various purposes and can involve various movements. After orienting to a food object and approaching it, an animal frequently grasps it in its mouth before ingesting it. For oral grasping, there are connections from the midbrain tectal orienting apparatus to the motor nucleus of the trigeminal nerve for jaw opening and closing. However, jaw movements are more strongly controlled by somatosensory system inputs from the mouth area.

The midbrain also evolved a structure for control of limb movements for reaching and grasping. This structure, the red nucleus, and its descending output pathway, the rubrospinal tract, are illustrated in figure 11.2. It is called the red nucleus because in dissections of the unstained human brain, its vascularity gives it a reddish appearance. The term *rubrospinal* is based on the Latin name for "red nucleus": *nucleus ruber*.

The Midbrain Was the Connecting Link between the Primitive Forebrain and Motor Systems

These midbrain systems for motor control received various inputs from pathways connected to head receptors and from neurons with activity that varied according to motivational state. In addition, the primitive forebrain influenced movement primarily through connections with the midbrain structures. Early in its evolution, the forebrain of small aquatic animals, and later the forebrain of amphibians and early reptiles, most likely did not have axonal projections that extended caudal to the midbrain. This was pointed out in chapters 4 and 5 (see figure 5.4). It was the midbrain that contained neurons with projections directly to the premotor mechanisms of the hindbrain and the spinal cord.

The midbrain also controls the visceral nervous system and associated motivational states via its primitive core, the central gray area, strongly linked with the hypothalamus of the ventral 'tweenbrain. Motivational control also involves the midbrain's ventral tegmental area, a caudal extension of the lateral hypothalamus. These structures are depicted in figure 11.4. They receive inputs from more caudal structures, primarily those related to the viscera, as well as from forebrain structures that influence motivational states and autonomic nervous system control—most strongly from hypothalamus. The connections go in both directions. From the central gray and ventral tegmental area, there are outputs to autonomic nervous system structures.

The midbrain also evolved control and differentiation of special-purpose movement sequences for specific functions (specific fixed action patterns) that added to or influenced the repertoire of the hindbrain and spinal cord. Even after disconnection of the forebrain in a cat or a rat, electrical stimulation of the midbrain, particularly the central gray area, can elicit predatory attack, defensive aggression, or courtship movements.

Next we consider the various ways in which these midbrain-controlled movements can be elicited and how this shaped chordate evolution.

Head Receptors and Locomotor Approach and Avoidance

The elicitation of approach movements and escape or avoidance movements are basic consequences of sensory inputs. Head receptors have evolved with these functions playing crucial roles. This has shaped the chordate neural tube.

As already introduced in chapter 4, a major factor in the early expansion of the anterior end of the neural tube that came to be called the forebrain was the olfactory sense, which gave rise to the first cranial nerve. Smells that incite fear or entice approach required links to the neural apparatus of locomotor control in the caudal midbrain and/or in the hypothalamus. Such links were in the primitive corpus striatum, the portions we call ventral striatum in present-day mammals. They were also in the medial pallium (which evolved into the hippocampal formation). This part of the ancient pallium, judging from present-day animals, projected to the striatum (ventral portions) and to hypothalamus and related parts of the midbrain.

Vision was the other modality that strongly shaped the early evolution of forebrain. As we will study in chapters 20–22, there was an evolution of several different links between the retinal inputs and locomotor control systems, formed by synaptic connections in the subthalamus, the pretectal area, and the midbrain tectum.

Gustatory inputs, via pathways from hindbrain, played an important role in the animals' learning which sights and smells to approach and which to avoid. It paid to remember the places and things that yielded tasty food or that led to bad-tasting food or to dangerous encounters.

Even prior to the expansion of the forebrain, somatosensory inputs from the head, entering through the hindbrain, no doubt played a critical role in the initiation of approach and avoidance movements. We can surmise that the same was true for auditory inputs. These two modalities must have played especially important roles, along with olfaction, in the earliest mammals, living in the shadows of the forest floor to avoid larger and very visual reptilian predators.

Initiation of Foraging by Activity Intrinsic to the Brain

It is important to point out that locomotion is initiated not only by inputs that originate outside the organism. Foraging activity, for example, can be initiated and modulated by intrinsic activity in the hypothalamus. Hypothalamic cell groups control cyclic behaviors such as feeding, a preparation for which is the initiation of locomotion for foraging. We talk about internal "drives" that have an endogenous buildup of the level of neuronal activity. Such activity represents a motivational state and corresponds to an "action-specific potential" described by the ethologist Konrad Lorenz. When this activity rises to a certain threshold, it causes an initiation of the movements of appetitive behavior (e.g., searching for food).

Overview of Motor System Structure

The Motor System Hierarchy

In his book *Brain Architecture*, Larry Swanson has presented his conception of the motor system hierarchy in the CNS. At the bottom of the hierarchy are the motor neurons of the spinal cord and brainstem. The activity of those neurons directly causes behavior. Controlling the motor neurons are the central pattern generators, which are in the interneuronal networks of the cord and brainstem. Exerting higher control over the central pattern generators are the central pattern initiators, such as the circuits of the midbrain locomotor area. The central pattern initiators receive input from forebrain systems called central pattern controllers (e.g., in the hypothalamic neuronal systems of motivational control that can trigger or increase the likelihood of various fixed action patterns).

The hierarchic control of locomotor behavior is depicted in figure 14.2, modified from Swanson's illustrations to show connections of the forebrain. The olfactory system connection to the early corpus striatum was described earlier; this is illustrated in the figure, which also shows striatal connections (via the pallidal output structures) to both hypothalamic and midbrain locomotor areas. Non-olfactory inputs also reach the striatum, from paleothalamic projections directly and, in mammals, via neothalamic projections to neocortex, which projects to dorsal parts of the corpus striatum.

As you may well expect, forebrain structures are at the top of the motor system hierarchy. However, if we look further caudally, the first region to be designated a locomotor pattern controller has a caudal diencephalic location; the electrical stimulation studies in which this region has been defined have given it a hypothalamic location or a subthalamic location, or both subthalamic and hypothalamic. These are regions receiving inputs from the ventral striatal regions in mammals, regions that receive olfactory inputs. Some olfactory system projections bypass the striatum, and reach these regions directly in some animals.

The midbrain locomotor area, designated as a locomotor pattern initiator by Swanson, is a controller of the more caudal locomotor pattern generators. The MLA has been defined by electrophysiological stimulation studies in mammals as located at the level of the caudal tectum, below the inferior colliculus and caudal superior colliculus in the reticular formation: in nucleus cuneiformis and adjacent structures including the pedunculopontine nucleus. This region receives projections from the diencephalic locomotor region and from the corpus striatum via its output structure, the globus pallidus.

The postulated pathway for control of approach and avoidance locomotion by olfaction in early chordates is sketched in figure 14.3.

Locomotor Pattern Generation and Its Adjustments by Vestibular and Cerebellar Systems

Locomotion is a general-purpose fixed action pattern. It depends on the spinal cord's propriospinal network of connections (see figure 14.2). Its execution is strongly influenced by

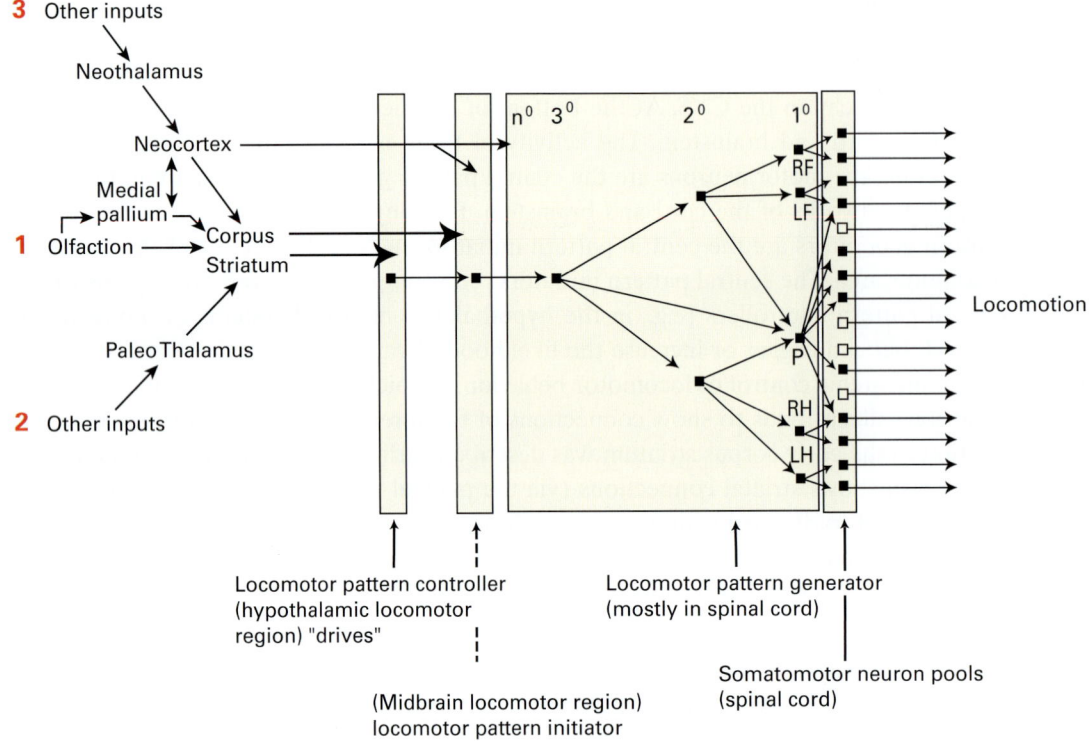

Figure 14.2
The hierarchical organization of neuronal systems that control locomotor behavior is depicted in this diagram. The right-hand portion of the information flow diagram, with the boxes, is based on Swanson (2003). At the left, connections from the endbrain, coming via corpus striatum and neocortex, have been added. LF, left forelimb motor neuron pool; LH, left hindlimb motor neuron pool; P, postural system motor neurons; RF, right forelimb motor neuron pool; RH, right hindlimb motor neuron pool.

activity in hindbrain structures, especially by the vestibular nuclei and by the cerebellum, part of which is closely connected to the vestibular nuclei. Its execution is modulated by reflex inputs coming into the spinal cord through the dorsal roots, especially from the feet.

The vestibular nerve is part of the eighth cranial nerve, and the secondary sensory neurons it projects to, the vestibular nuclei, are located in the alar plate region of the hindbrain. The large cells of the lateral vestibular nucleus (Deiter's nucleus) have direct projections to the spinal cord via the descending axons of the lateral vestibulospinal tract, which courses just below the lateral part of the ventral horn. Within the hindbrain and reaching rostrally into the midbrain is another fiber tract with many axons from the vestibular nuclei. It is called the medial longitudinal fasciculus (MLF). It interconnects the vestibular nuclei and the oculomotor nuclei (the nuclei of the third, fourth, and sixth cranial nerves)—for stabilization of eye position during head movements and

Overview of Motor System Structure

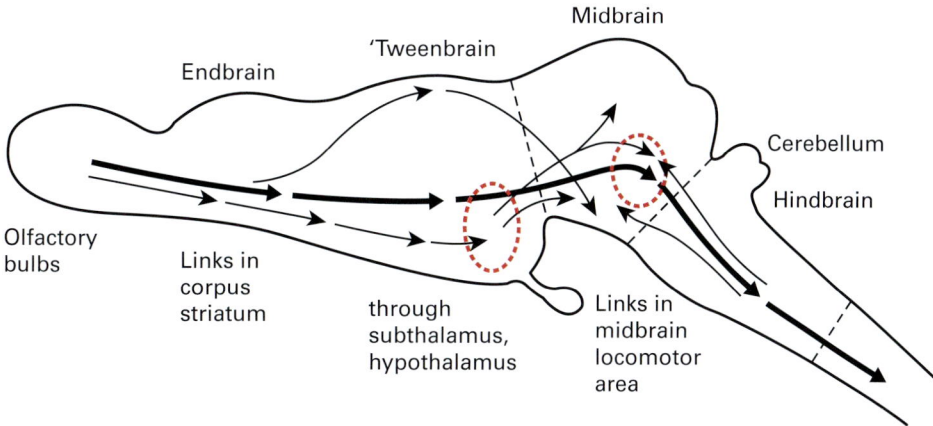

Figure 14.3
In this simplified side view of an early chordate brain after the evolution of olfactory bulbs, postulated pathways for control of approach and avoidance movements by olfaction are illustrated. Much of the hierarchical structure diagrammed in figure 14.2 can be seen.

coordination of the movements of the two eyes. The MLF has a caudal extension to the cervical spinal cord (often called the medial vestibulospinal tract), underlying coordination of head movements with the movements of the eyes. Coordinated stabilization of the body, eyes, and head during disturbances of equilibrium and during orienting movements, during locomotion or standing, is a function of the vestibulospinal axons.

The location in the hindbrain of the vestibular nuclei, within which four separate cell groups can be distinguished (medial, lateral, superior, and descending nuclei) is illustrated in figure 14.4.

The part of the cerebellum dominated by the vestibular system is probably the most ancient part of the cerebellum. It includes mainly the flocculus and the nodulus and uvula—the main parts of the vestibulocerebellum—which have direct input from the vestibular nerve (see figures 14.5 and 14.6). These parts of the cerebellar cortex project to the fastigial nucleus, the medial-most of the deep nuclei of the cerebellum, with direct projections to the spinal cord (fastigiospinal tract).

The direct vestibular input to the cerebellum and the vestibular nucleus projections to the spinal cord and to the oculomotor nuclei are summarized in figure 14.6. Note that primary sensory neurons project directly to the cerebellar cortex, and that the cerebellar cortex projects directly to the vestibular nuclei, bypassing the deep nuclei of the cerebellum.

The rapid route to the spinal cord and hindbrain from the cerebellum, from the fastigial nucleus, is one of the major controllers of the muscles of the body axis, influencing body posture and balance during locomotion and other activities.

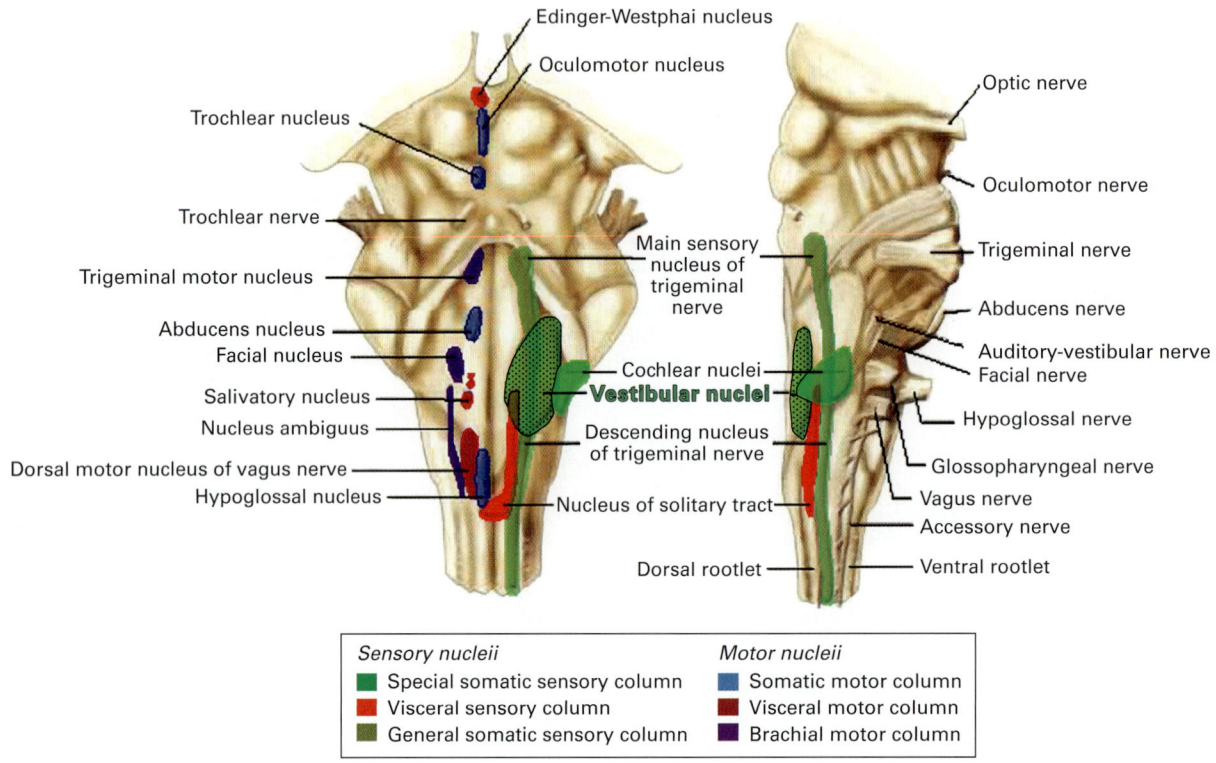

Figure 14.4
The human hindbrain and midbrain after removal of the cerebral hemispheres and the cerebellum are drawn in dorsal view. The secondary sensory cell groups and the motor neuron cell groups are depicted as if the rest of the brain were transparent, as in figure 10.11. Note the vestibular nuclei, which consist of the medial, lateral, superior, and descending nucleus. Based on Nauta and Feirtag (1986).

Orienting of Head and Body

Before an animal begins to approach something by initiating locomotion, it usually has to orient toward it. The sensory triggering of head and body turning long preceded the evolution of the refinements in motor control made possible by the neocortex. Orienting movements are of obvious importance for accomplishing the goals of approach and avoidance actions.

Orienting movements are elicited by somatosensory, visual, and auditory inputs, especially when a stimulus is novel and could indicate danger or a food source, but also when it is a key stimulus for releasing some other instinctive movement. Very early in evolution, we can assume that there were simple reflex-like pathways for escape movements and for

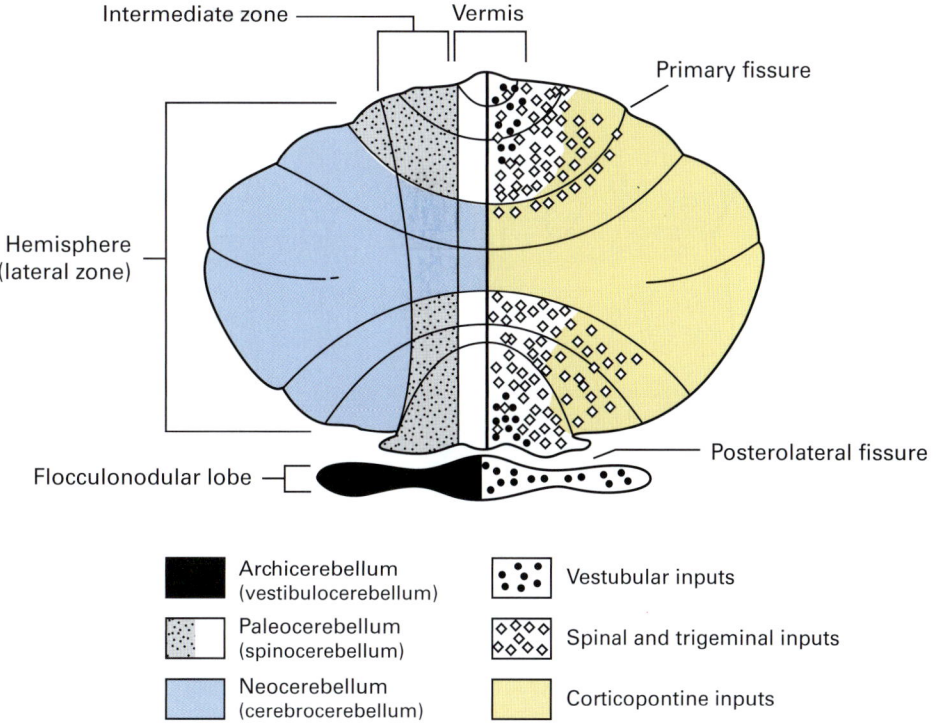

Figure 14.5
In this dorsal view of the primate cerebellum, the surface has been flattened so that all major lobes can be seen. On the right side, major inputs are depicted: from the vestibular nerve and nuclei, from the spinal cord, and from the pontine nuclei (which receives their major inputs from the neocortex). On the left side, the most primitive portions are shown in black and in white. Next to appear in phylogeny were the intermediate regions depicted in light gray with stipple; most recently evolved were the cerebellar hemispheres, depicted in blue. The hemispheres expanded as the neocortex expanded in mammals. Based on Goodlett (2008).

turning toward objects. In the early evolution of the visual pathways, control of escape and turning movements may have been via retinal inputs to the ventral thalamus (the lateral-most subthalamus, which is the location in mammals of the ventral nucleus of the lateral geniculate body), and from there to the brainstem and spinal cord. But projections to the midbrain tectum became dominant for these functions.

The optic tract will be studied in detail in chapters 20 and 21. Briefly, there was an evolution of a topographically precise projection from retina to the midbrain tectal correlation center, representing the visible world around the head. In this same structure, ventral to the visual map in mammals, there are auditory and somatosensory maps that correspond to the visual map. The outputs of the structure control precise turning toward objects. Even more important for the survival of small preyed-upon animals (and

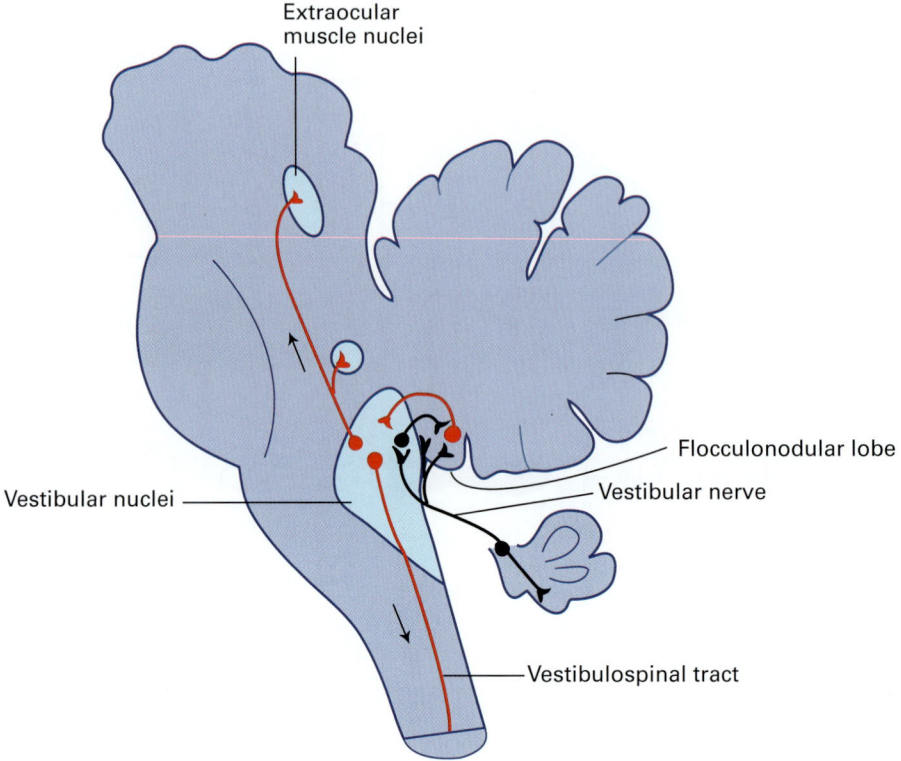

Figure 14.6
On this drawing of a sagittal section of the human cerebellum, connections of the vestibular nerve (part of the eighth cranial nerve) are depicted. Primary sensory neurons project directly to the cerebellar cortex as well as to the neurons of the vestibular nuclei. The vestibular part of the cerebellar cortex projects directly to the vestibular nuclei, bypassing the deep cerebellar nuclei (to which most of the cerebellar cortex projects). Also shown are projections of the vestibular nuclei to the spinal cord and to motor neurons controlling eye muscles. Based on Brodal (2004).

probably more primitive in evolutionary history), there are distinct outputs for triggering movements of fleeing (escaping) or freezing. The escape movements are the function of uncrossed descending pathways from the midbrain tectum, whereas the orienting movements are the function of the crossed tectospinal tract.

The pretectal area, which also receives a direct input from the retina, is relatively larger in non-mammalian animals. Even in mammals it plays important roles in protective responses: pupillary constriction, escape from rapidly approaching objects, and avoidance of barriers during locomotion. Its role in orienting movements and in antipredator escape or avoidance movements is not clear, as this structure has not been as well studied as the caudally adjacent, and generally much larger, optic tectum.

Grasping: The Third Major Type of Movement Controlled by the Midbrain

Orienting to an object and approaching it may be followed by grasping it (e.g., if it is food or nesting material). Long preceding the mammals and the neocortex, animals grasped things by mouth, and some of them also developed an ability to grasp with the limbs. Such movements were critical for consummatory action; when they involved a limb, they required reaching and control of distal muscles. Thus there were two kinds of grasping with midbrain control. They involved (1) connections from parts of the tectum (and pretectum and possibly from subthalamus) to neurons controlling jaw opening and closing via connections to the motor nucleus of the trigeminal nerve in the rostral hindbrain, and (2) connections from the red nucleus for control of limbs and hands.

Inputs to the red nucleus come from the ventral thalamus (subthalamus), from the cerebellum, and also from the hypothalamus and epithalamus. Evolution and expansion of the neocortex brought inputs from somatosensory and motor cortex, which also evolved output axons of greater length that bypassed the red nucleus and went directly to the spinal cord regions connected with the limbs. Forebrain control of the limbs became dominant, but you should remember that the more primitive midbrain structures did not disappear. Those forebrain controls are a subject of traditional neurology and will be discussed further in the next chapter.

Comparative Anatomy of the Red Nucleus and Its Projection to the Spinal Cord

Neuroanatomists have done interesting phylogenetic comparisons of the red nucleus (*nucleus ruber*). It is absent in the primitive, jawless vertebrates (hagfishes and sea lampreys), but it is present in frogs, with inputs from corpus striatum and cerebellum. The striatal connections tend to end mainly in the region of the ventral thalamus just rostral to the red nucleus, the so-called pre-rubral field, which projects to the red nucleus. The red nucleus is prominent in most reptiles and in birds and mammals.

Because of the function of the red nucleus in grasping with the limbs, you might expect it to be absent in animals without limbs. The rubrospinal tract is absent in pythons but not in many other snakes. It is present in teleosts. These comparative data indicate that nucleus ruber may also have functions other than limb control. Perhaps it adds dexterity to approach movements guided by intentions to acquire food.

The rubrospinal tract in mammals originates in the more caudal, large-celled (magnocellular) portion of the red nucleus. The more rostral small-celled (parvocellular) portion has projections directed rostrally to the thalamic cell groups that project to motor cortex. It is interesting to see how the relative sizes of these two portions of the red nucleus have changed with the evolution of greater dexterity in higher mammals. With greater dexterity, the parvocellular portion is larger. The change has a parallel in the cerebellum: With greater dexterity, the cerebellar hemispheres expand together with their output nucleus, the dentate nucleus. This is illustrated in figure 14.7.

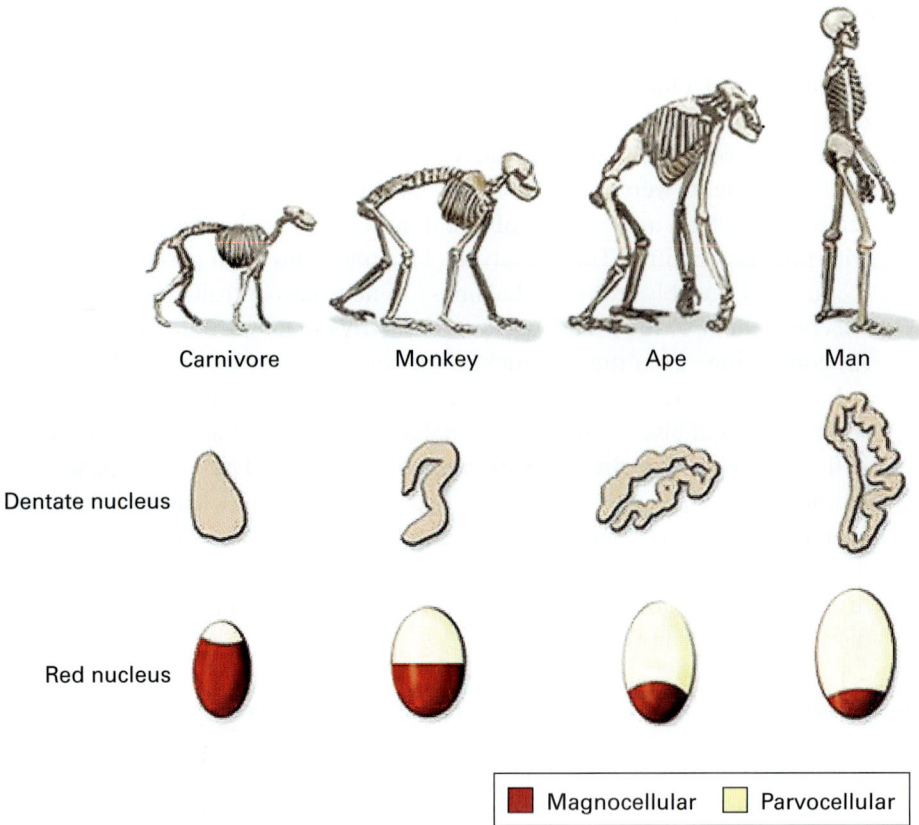

Figure 14.7
Comparison of the red nucleus of the midbrain with the lateral-most deep nucleus of the cerebellum, the dentate nucleus, in four mammalian groups. The large-celled part of the red nucleus—the source of the rubrospinal tract—is proportionately much larger in carnivores and smallest in apes and humans. Correlated with these differences, the dentate nucleus becomes proportionately larger from carnivores to monkeys to apes to humans. The small-celled part of the red nucleus, which receives input from the dentate nucleus, projects to the ventrolateral nucleus of the thalamus and to the inferior olive of the hindbrain. Based on Massion (1998).

Overview of Motor System Structure

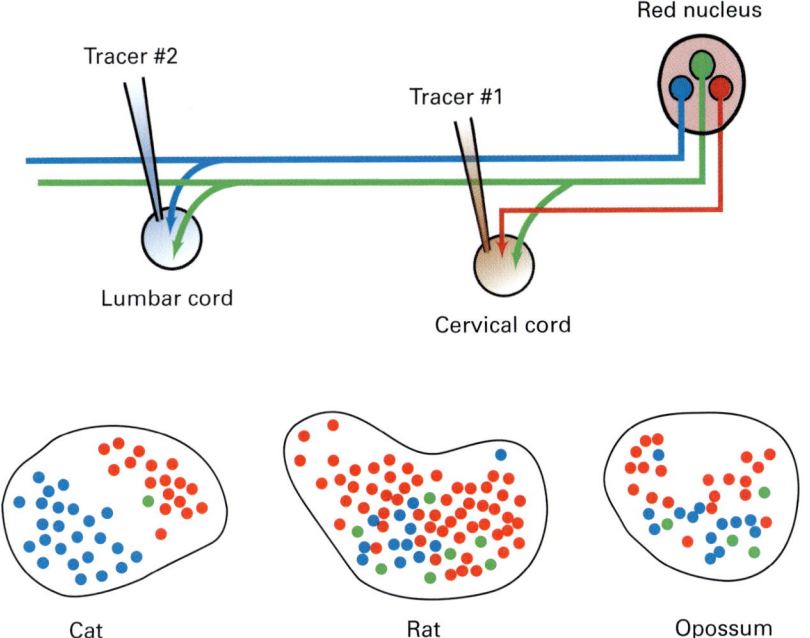

Figure 14.8
Results of studies of rubrospinal projections to the cervical and lumbar enlargements of the spinal cord in opossum, rat, and cat. Different retrograde tracers were injected into the cervical and lumbar cord. In the cat, the projections to the two enlargements come from opposite parts of the red nucleus, and few doubly labeled neurons were found (indicating neurons that have an axon that arborizes in both spinal enlargements). This pattern contrasts to what was found in the other two species, where the topographic separation of neurons projecting to the two levels of the cord is less complete. This organization is the least evident in the opossum. In addition, there are more doubly labeled red nucleus neurons in the rat and opossum. Figure based on Striedter (2005).

Comparative studies of the red nucleus' output pathway to the spinal cord illustrate another kind of change in evolution: a parcellation that has increased the topographic organization of the rubrospinal tract. Figure 14.8 shows results of a retrograde tracing study of rubrospinal projections in three species: the opossum, the rat, and the domestic cat.

A Structural Approach to Understanding Motor Control: Begin with the Motor Neurons

Thus far, this overview of motor system structure has been organized around the three major types of functional control we can distinguish in midbrain mechanisms. At this point, we will step much further back and look at the neurons that directly control the movements and other changes that result from CNS outputs.

In chapter 9, we discussed three types of axonal contact with effectors, as found in three major motor systems (see figure 9.12). You should review those three motor systems now: the somatic motor neurons of the spinal cord and brainstem with synaptic control, the autonomic nervous system's axonal control of preganglionic and then ganglionic motor neurons, and the neuroendocrine system's hormonal control. These three major systems of motor neurons are depicted in figure 14.9 as the output side of the central pattern generator networks.

Next, consider the locations in the central nervous system of all of the various groups of somatic motor neurons (we have discussed them previously in this book): the motor neurons of the ventral horn of the spinal cord, the groups of motor neurons in the motor nuclei of the hindbrain, and the motor nuclei in the midbrain. Their distribution is depicted in the mammalian CNS schematic of figure 14.10. Can you remember where these neurons are located in frontal sections?

The Spatial Arrangements of Somatic Motor Neurons in the Spinal Cord

Now we will go one step further in depicting the arrangements of motor neurons in an illustration of the spinal cord at one of the spinal enlargements where the neurons innervating the limbs are found. Inspect figure 14.11, which is from the work of the great Dutch neuroanatomist H. G. J. M. Kuypers (a compatriot and contemporary of W. J. H. Nauta) and his student D. G. Lawrence. It is a sketch of the spinal cord of a monkey and shows a proximal to distal topography in the arrangement of the motor neurons. Those innervating the most distal muscles are located laterally, and those innervating the muscles nearest the body axis are located medially. Also shown is the corresponding arrangement of the interneurons of the cord—the neurons most directly connecting with the motor neurons. Thus, there is a kind of radial arrangement of the spinal connections. Finally, the figure illustrates how the interneurons connected with the most medial motor neurons are closely connected with neurons on the opposite side of the cord.

In the next chapter, we will see how the descending connections of higher control, to the spinal cord from the brainstem and, in mammals, from the neocortex, retain this spatial organization.

Overview of Motor System Structure

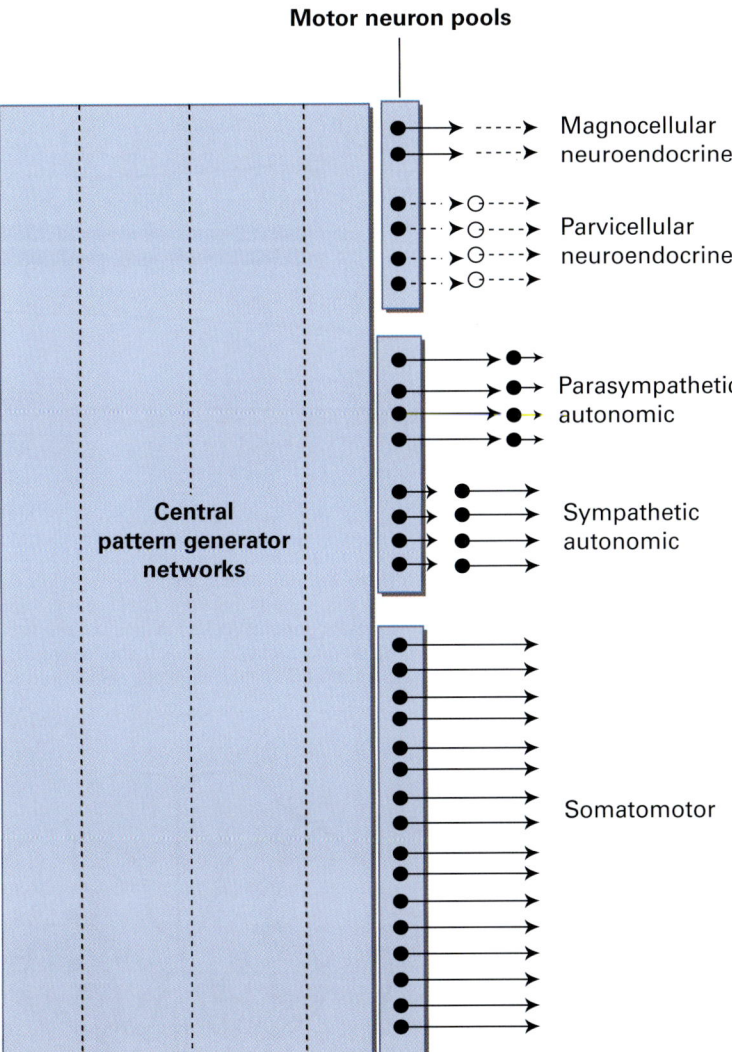

Figure 14.9
Diagram of the three motor systems, located on the output side of the central pattern generator networks of the CNS. (1) The somatomotor system consists of motor neurons with axons that synapse on striated muscle cells. (2) The autonomic nervous system outputs go through a preganglionic motor neuron in the CNS that projects to a ganglionic motor neuron in the PNS that has an axon contacting smooth muscles or glandular cells in apocrine fashion. In the innervation of the digestive tract, there is actually a large network of interconnected neurons that include the motor neurons—the enteric nervous system—not indicated in the figure. (3) The outputs of the endocrine system result from neuronal cells that secrete substances directly into blood vessels (in the posterior pituitary) or that result from apocrine contacts with gland cells that then secrete hormones into the bloodstream. Based on Swanson (2012).

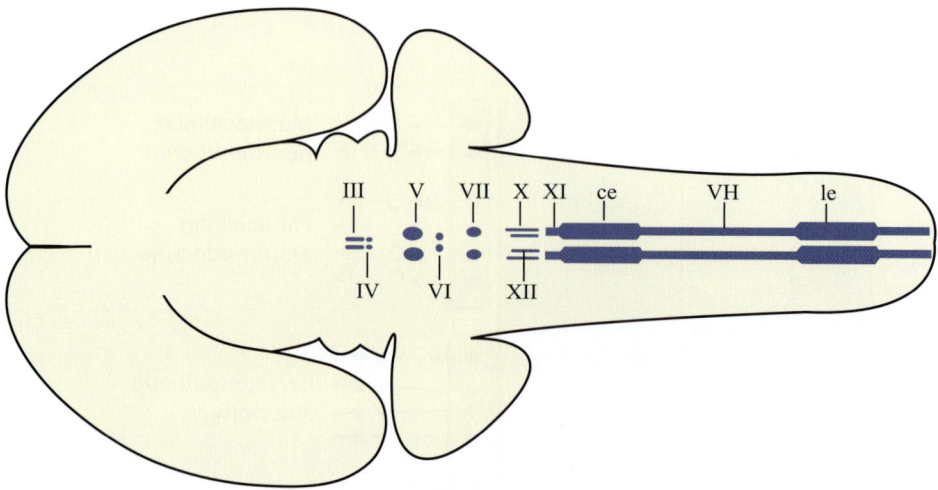

Figure 14.10
The distribution of somatic motor neurons is illustrated on a schematic horizontal section of a stretched out mammalian CNS. The locations of the cranial motor neurons are labeled according to the cranial nerves that contain their axons. The spinal motor neurons are located in the ventral horn (VH) of the spinal gray matter; there are many more of them in the cervical and lumbar enlargements (ce, le) than at other levels. III, oculomotor nucleus; IV, trochlear nucleus; V, motor nucleus of the trigeminal nerve; VI, abducens nucleus; VII, facial nucleus; X, nucleus ambiguus (with axons in the vagus nerve); XI, nucleus of the spinal accessory nerve; XII, hypoglossal nucleus. Based on Swanson (2012).

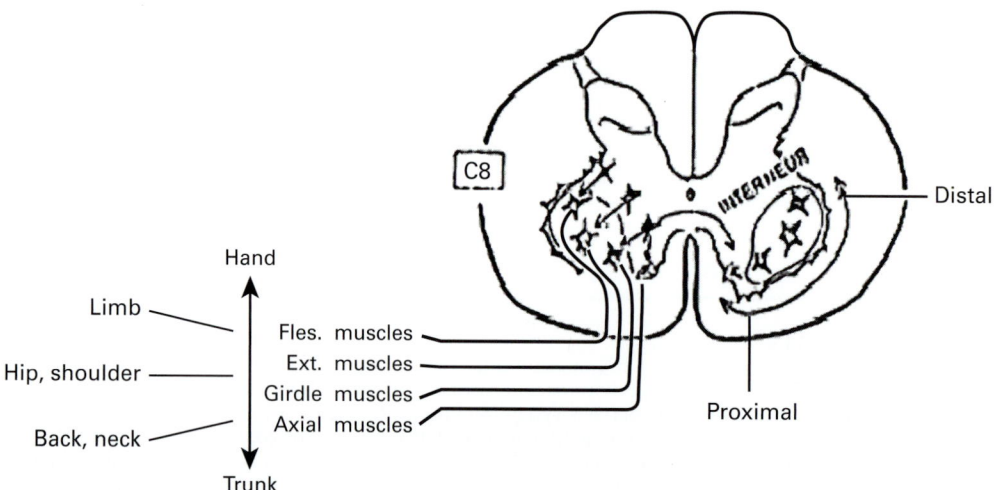

Figure 14.11
Section of a mammalian spinal cord at the level of the cervical enlargement illustrating the topographic distribution of somatic motor neurons and the interneurons that are their major sources of input. Distal muscles of the limb are innervated by the most lateral motor neurons; axial muscles are innervated by the most medial motor neurons; motor neurons in between innervate the muscles of the shoulder (the girdle muscles). Girdle muscles of the lumbar enlargement control movement of the hip. Axial muscles are the muscles of the neck and back. The most distal muscles of the limbs are in the fingers and toes. C8, level of the eighth cervical segment. From Lawrence and Kuypers (1968).

Readings

John Flynn and collaborators on predatory aggression in cats: importance of somatosensory input from whiskers and lips:

MacDonnell, M. F., & Flynn, J. P. (1966). Control of sensory fields by stimulation of hypothalamus. *Science, 152,* 1406–1408.

On the motor system hierarchy: See the book by Larry Swanson cited in chapters 1 and 10.

Even after disconnection of the forebrain in a cat or a rat, electrical stimulation of the midbrain, particularly the central gray area, can elicit predatory attack, defensive aggression, or courtship movements:

Bandler, R., & Shipley, M. T. (1994). Columnar organization in the midbrain periaqueductal gray: Modules for emotional expression? *Trends in Neurosciences, 17,* 379–389.

Konrad Lorenz on fixed action patterns:

Lorenz, K. Z. (1981). *The foundations of ethology.* New York: Springer.

Lorenz, K. Z. (1952). *King Solomon's ring: New light on animal ways.* New York: Thomas Y. Crowell.

The midbrain locomotor area (MLA) has been defined by electrophysiological stimulation studies in mammals as located just rostral to the hindbrain, below the inferior colliculus in the reticular formation that has been designated as part of nucleus cuneiformis, together with adjacent structures including the pedunculopontine nucleus: see the books by Butler and Hodos (1996, 2005) and a study by Larry Swanson and collaborators. (Many studies of this region can be found by searches with Google Scholar.) The books by Butler and Hodos contain much information about other motor system structures discussed in this chapter.

Butler, A. B., & Hodos, W. (1996). *Comparative vertebrate neuroanatomy.* New York: Wiley-Liss. This is the first edition of the book. The second edition was cited in chapter 3.

Swanson, L. W., Mogenson, G. J., Gerfen, C. R., & Robinson, P. (1984). Evidence for a projection from the lateral preoptic area and substantia innominata to the "mesencephalic locomotor region" in the rat. *Brain Research, 295,* 161–178.

Many interesting studies of locomotor control giving insights into physiological and anatomical mechanisms across a wide range of species have been carried out by Sten Grillner and co-workers, and by Edgar Garcia-Rill and co-workers. Examples:

El Manira, A., Pombal, M. A., & Grillner, S. (1997). Diencephalic projection to reticulospinal neurons involved in the initiation of locomotion in adult lampreys *Lampetra fluviatilis. Journal of Comparative Neurology, 389,* 603–616.

Garcia-Rill, E. (1986). The basal ganglia and the locomotor regions. *Brain Research Reviews, 11*, 47–63.

Ménard, A., & Grillner, S. (2008). Diencephalic locomotor region in the lamprey: Afferents and efferent control. *Journal of Neurophysiology, 100*, 1343–1353.

A recent study of locomotor pattern generators of spinal cord and hindbrain used light-activation of glutamatergic neurons:

Hägglund, M., Borgius, L., Dougherty, K. J., & Kiehn, O. (2010). Activation of groups of excitatory neurons in the mammalian spinal cord or hindbrain evokes locomotion. *Nature Neuroscience, 13*, 246–252.

The rubrospinal tract is absent in pythons but not in many other snakes: reviewed in the books by Butler and Hodos noted earlier.

There are many papers reporting experiments on the role of the midbrain tectum (superior colliculus) in the control of orienting movements, and also neuroanatomical and electrophysiological studies of this structure. Some papers also have reported studies of its role in antipredator behavior. Below are a select few of these papers:

Dean, P., Redgrave, P., & Westby, G. W. M. (1989). Event or emergency? Two response systems in the mammalian superior colliculus. *Trends in Neurosciences, 12*, 137–147.

Schneider, G. E. (1969). Two visual systems. *Science, 163*, 895–902.

Schiller, P. H. (2011). The superior colliculus and visual function. Comprehensive Physiology 2011, *Supplement 3: Handbook of physiology, The nervous system, sensory processes*: 457–505. First published in print 1984.

Papers on the motor system relevant to the end of this chapter and to the next chapter:

Lawrence, D. G., & Kuypers, H. G. J. M. (1965). Pyramidal and non-pyramidal pathways in monkeys: Anatomical and functional correlation. *Science, 148*, 973–975.

Lawrence, D. G., & Kuypers, H. G. J. M. (1968). The functional organization of the motor system in the monkey: II. The effects of lesions of the descending brain-stem pathways. *Brain, 91*, 15–36.

15 Descending Pathways and Evolution

In the previous chapter, the types, locations, and spatial arrangements of somatic motor neurons in monkeys were reviewed. Those motor neurons are the final common path whereby any CNS activity is able to influence the movements of an animal. It is a good place to begin a discussion of the pathways of motor control originating in more rostral brain structures.

In figure 15.1, we see illustrated the terminal distribution pattern of descending cortical and subcortical pathways in the lower cervical spinal cord of a rhesus monkey. Compare these patterns to the distribution of motor neurons connected with distal versus proximal muscles, and the corresponding topography of interneurons most closely connected with them. First of all, it is clear that the terminals of corticospinal axons are distributed to regions controlling both axial and distal muscles as well as the girdle muscles in between. In contrast, the lateral and the medial brainstem pathways have more limited distributions. Axons passing through, and also coming from, the lateral hindbrain reticular formation end primarily on interneurons connected with distal muscles of the limb—the muscles that move the hand and fingers. The axons from the medial hindbrain reticular formation have a contrasting distribution pattern. These axons, coming from the reticulospinal and from the tectospinal, vestibulospinal, and cerebellospinal pathways, distribute their terminals primarily, but not exclusively, to interneurons controlling axial muscles. In addition, some of these axons cross the midline to terminate in the ventromedial part of the spinal gray.

Axons Descending from Brain to Spinal Cord: Functional Groupings

To clarify the course of axons influencing axial and distal muscle control, the groups of axons projecting caudally from the brain to the spinal cord are sketched separately in figures 15.2 and 15.3. In figure 15.2, the course of the corticospinal axons in a monkey's brain is depicted in a side view of the CNS, which has the spinal cord enlarged for clarity's sake, and in a top view of the embryonic neural tube. In the two drawings of figure 15.2, the axons controlling muscles of the foot, hand, and face are separated from those

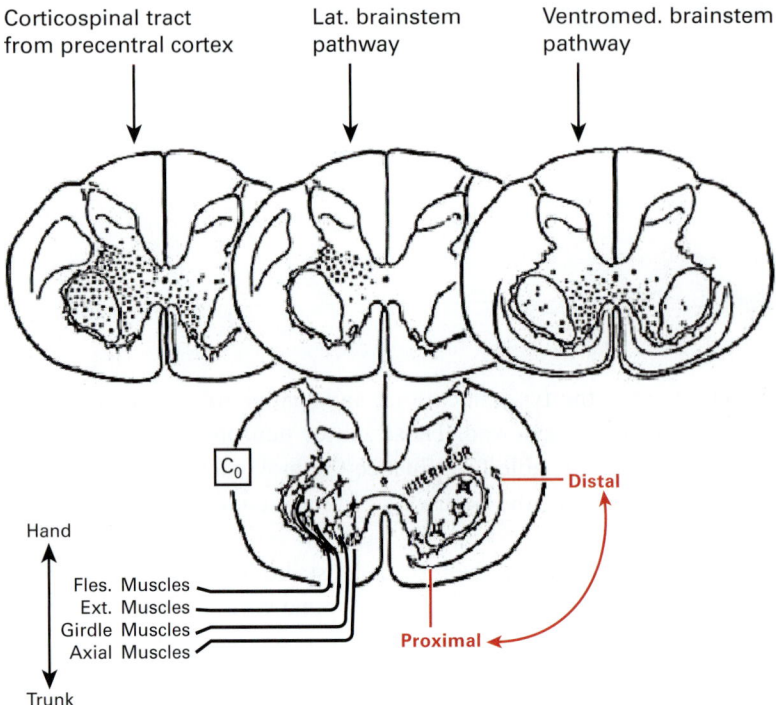

Figure 15.1
Frontal sections of spinal cord, lower cervical level, that show results of experimental studies of the rhesus monkey. (Top) Terminal distribution pattern of descending cortical and subcortical pathways in the monkey. (Bottom) The drawing from figure 14.11, showing the origins of the innervation of more proximal and more distal muscles in the limb, is repeated. (Top and bottom) Note the contrasting distributions of axons that course through the lateral hindbrain and the axons that course through the medial hindbrain, indicating that they have contrasting functions. From Lawrence and Kuypers (1968).

controlling muscles of the neck and trunk (the axial muscles). You can see that the descending pathway for movements of the distal musculature of the extremities terminate in the spinal enlargements. Those for facial movements end in the hindbrain. The motor cortical areas for trunk and neck control have endings throughout the entire length of the cord, distributing, as we have seen, in the ventral and medial parts of the spinal gray matter bilaterally.

In figure 15.3, the descending axonal systems of the medial and the lateral hindbrain are shown separately. There is nothing very complicated in this view of the anatomy except for some of the names! Many of the axons of the lateral system originate in the midbrain's red nucleus, and these are joined by axons from the lateral hindbrain reticular formation as they course through the lateral hindbrain below the nuclei of the trigeminal nerve. Within the spinal cord, the axons terminate mainly in the spinal enlargements on the dorsolateral interneurons that influence the distal muscles of hands or feet.

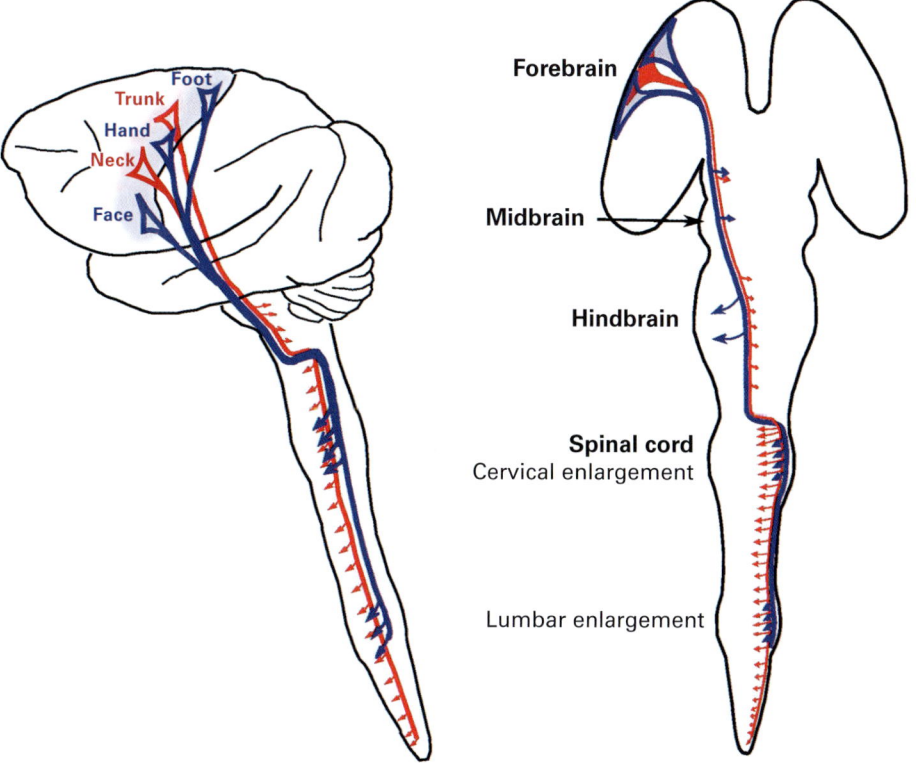

Figure 15.2
The corticospinal axons are illustrated according to studies of the rhesus monkey. In the motor cortex, the regions representing the neck and trunk are anterior to the regions representing the foot, hand, and face. (Left) Side view of CNS with exaggerated size of spinal cord. The axons that influence the axial and distal muscles are separated. (Right) The same axons are depicted on a sketch of the embryonic mammalian brain. Note the separation of lateral and medial terminations of the two groups of axons in both hindbrain and spinal cord.

On the right-hand side of figure 15.3, you see a sketch of the medial hindbrain system that runs in the ventral spinal columns and terminates bilaterally in the medial gray matter from the most rostral to the most caudal levels of the cord. Besides the large neurons of the medial hindbrain reticular formation, contributors to this system are the large output neurons of the midbrain tectum (superior colliculus of mammals) that cross and descend as far as the cervical spinal cord, many of them before they go that far as they end in the medial hindbrain reticular formation. Also contributing are neurons of the vestibular nuclei and neurons of the fastigial nucleus—the medial-most deep nucleus of the cerebellum.

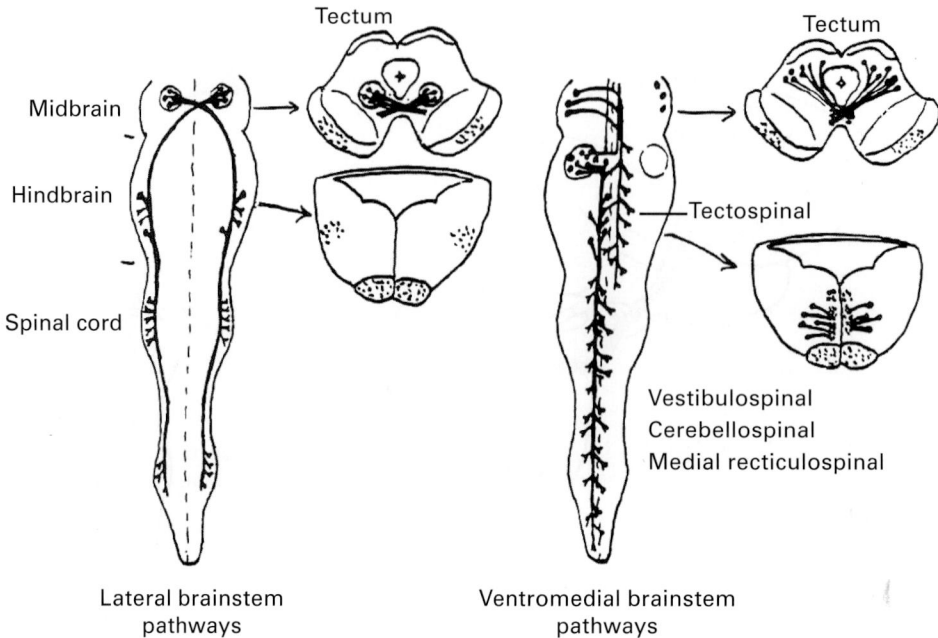

Figure 15.3
Axons that influence primarily axial and primarily distal muscles are depicted separately in the sketches shown on the left and the right. (Left) The lateral brainstem pathways are illustrated on a sketch of the embryonic midbrain, hindbrain, and spinal cord, and in frontal sections of the hindbrain and the midbrain. The upper section shows the origin of the rubrospinal tract (in the red nucleus of the midbrain), a major component of the lateral brainstem pathways. Other neurons that contribute to the pathway are found in the lateral hindbrain reticular formation. On the frontal sections, the corticospinal axons, involved in control of the same movements, are also shown (stippled). (Right) Similar illustrations of the origins and course of ventromedial brainstem pathways. Rostrally they originate in the midbrain tectum; the tectospinal tract descends only as far as the cervical spinal cord. Axons of hindbrain origin originate in the vestibular nuclei, in the medial deep nucleus of the cerebellum (the fastigial nucleus), and in the large neurons of the medial reticular formation. Again, the corticospinal axons are also indicated, as they can control the same movements.

Functions of the Descending Pathways: The Corticospinal Tract

Many of the relevant axonal tracing studies of these motor system pathways were carried out by Hans Kuypers and his collaborators. He, together with D. G. Lawrence, wanted also to verify the functional roles of these pathways, so they conducted a memorable laboratory experiment with rhesus monkeys. When they made hindbrain lesions that transected the medial or the lateral pathways alone, the monkeys recovered rapidly from the surgery and showed little remaining defects in motor control. They reasoned that this was because the corticospinal system was so important in this species (and no doubt also in humans) for both axial and distal muscle control. Therefore, they prepared a group of monkeys with complete bilateral transections of the pyramidal tracts, made at a caudal hindbrain level. These monkeys initially showed severe loss of motor control, affecting both axial and distal muscles, but as they recovered from the diaschisis effects (see chapter 7) of this drastic lesion, the residual motor defects that remained were

- loss of speed and strength, and
- loss of control of digits used one at a time.

Their knowledge of neuroanatomical pathways and connections led them to assume that after the pyramidotomy, the monkeys would be dependent on the lateral and the medial hindbrain axonal systems for motor control of distal and axial muscles (figure 15.4).

Functions of the Descending Pathways: The Medial Hindbrain Tracks

To test this hypothesis, Lawrence and Kuypers carried out lesion experiments on the monkeys that had recovered from the initial effects of the pyramidal tract lesion. The additional lesion in one group of monkeys—the lesion shown on the lower-right portion of figure 15.4—produced a destruction of the medial brainstem pathways. If this additional lesion was made on only one side, the effects were not long lasting, but when it was made bilaterally, the defects in motor control were dramatic. Control of the axial muscles in such a monkey was drastically reduced: Initially, the monkey was not even able to right itself if it fell over. Righting returned gradually, but only after 10–40 days. Even after recovery, falling did not elicit the usual corrective movements. When walking, only one of these monkeys could take many steps because of a severe disorientation. He veered from his course and bumped into obstacles.

The lack of motor control was not pervasive: These monkeys retained better control of distal muscles—the muscles of their hands and feet. If the monkey was strapped into a chair to help it overcome its poor axial control, it could grasp a chunk of apple or a peanut using the whole hand.

Functions of the Descending Pathways: The Lateral Brainstem Tracks

When the second lesion, added to a pyramidotomy, resulted in a destruction of the lateral brainstem pathways (the additional lesion shown on the lower-left portion of figure 15.4), the functional defects were very different. These defects depended largely on the completeness of the sectioning of the rubrospinal tract. This was verified by the neuroanatomists by inspecting Nissl-stained sections of the midbrain. Cutting of the rubrospinal axons on one side resulted in retrograde degeneration of the neurons of the magnocellular portion of the red nucleus on the opposite side. (Remember that the rubrospinal tract decussates before descending.) These lesions caused a loss of ability to grasp with the hand on the side of the hindbrain lesion: hand flexion seemed to be impossible unless the monkey was using his whole arm during whole-body movements. He retained good axial

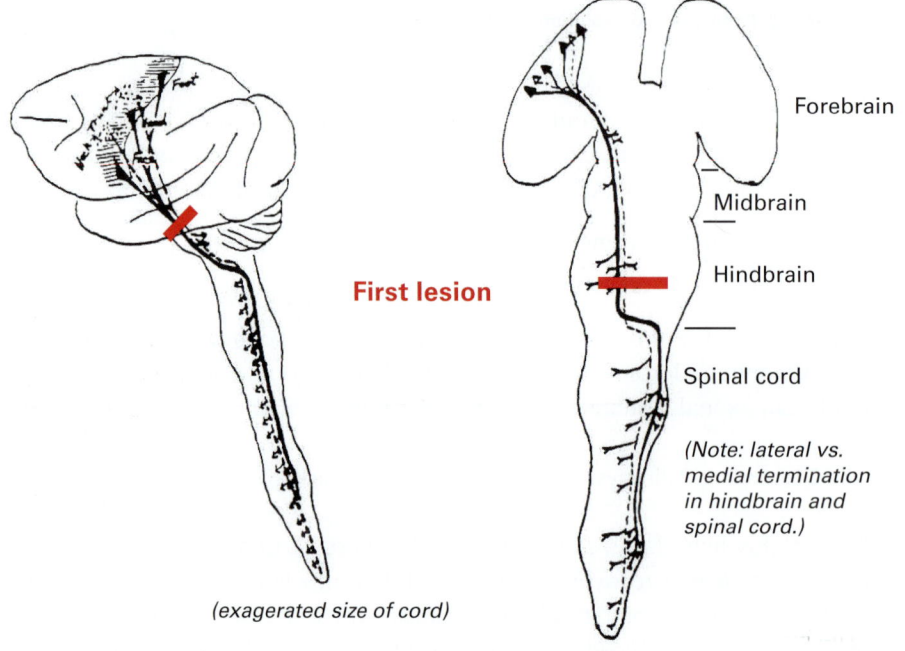

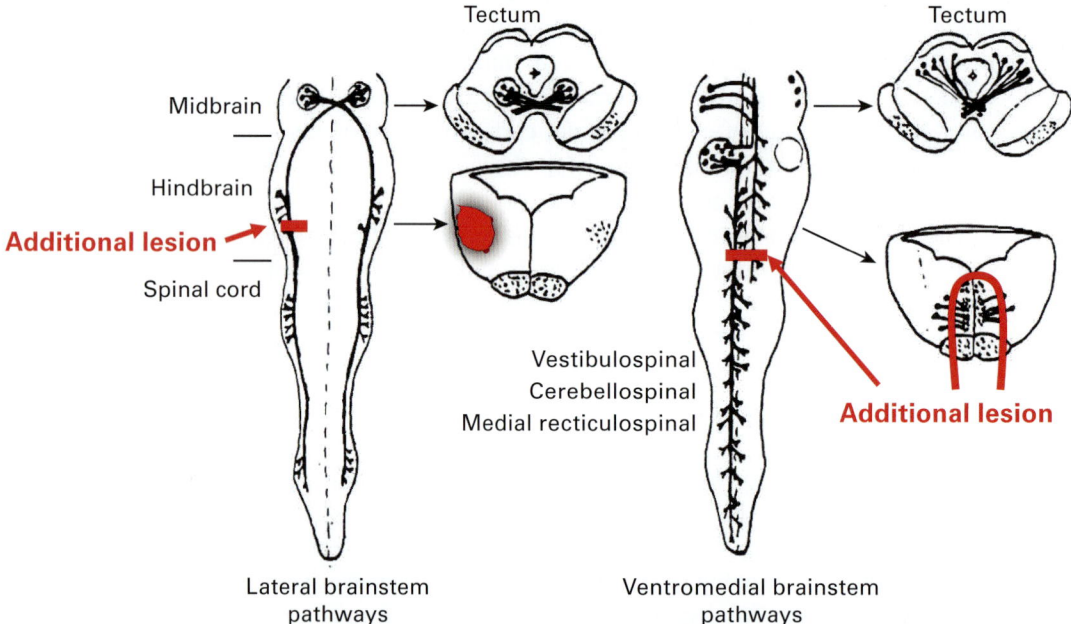

Figure 15.4
Selective lesions of the descending motor pathways. The lesions made by Lawrence and Kuypers in their study of the function of descending pathways in the monkey on motor control are depicted on these illustrations that are similar to the drawings of figure 15.2 (top) and figure 15.3 (bottom). The first lesion made in all of the experimental cases was designed to sever the pyramidal tract in the middle of the hindbrain as shown. The surgical knife entered from the ventral surface. The second lesion was different in different monkeys. In some cases, the axons coursing through the lateral hindbrain were severed on one side by a cut from a knife inserted from the side as shown. In other cases, the medial pathways were severed bilaterally by the insertion of a knife from the ventral surface, through the already degenerating pyramidal tract, as shown.

control: the hand was used well when he was running and climbing (e.g., climbing up the walls of his cage).

This is an important lesson on the nature of paralysis caused by CNS lesions: the paralysis may depend on the situation. In this case, the hand that appears to be paralyzed when the monkey tries to use it by itself to grasp a food object suddenly springs to life and moves well when he uses it in coordinated movements of the whole body as in running and climbing.

A Conclusion with Application to Humans

Thus, we can conclude that in the brainstem to spinal cord pathways, there is a separation of (a) control of axial muscles and whole-body movements and (b) the control of distal muscles and fractionated movements. However, as the neocortex evolved into a large structure, its motor outputs gained more and more control of both axial and distal movements.

The work with monkeys has clear application to humans: There is strong evidence that our species is very similar except that, with even larger neocortex compared to the spinal cord, the corticospinal diaschisis can be more severe after neocortical lesions or lesions of descending pathways. Recovery can result in considerable spasticity of the spinal reflexes. If a person suffers a lesion of descending pathways that is much more complete than a pyramidotomy because it is due to a spinal transection, he or she can become *tetraplegic*. (The term, from the Greek, means suffering from paralysis of all four limbs. The term often used is *quadriplegic*, based on both Latin and Greek roots.) In such an injured person, the motor cortex remains active, but it is cut off from the lower motor apparatus of the spinal cord.

The Brain Disconnected from the Motor Pattern Generators

Recall figure 14.2, which depicts the motor pattern generators (for locomotion) of the spinal cord and hindbrain and their control by more rostral brain structures. If you add to this figure the more direct descending connections from neocortex to those pattern generators, and also directly to motor neurons (see figure 12.7), you have a general picture of higher motor control. With that picture in mind, you can imagine the situation described in the previous paragraph: an injured person in whom the motor cortex and other controllers and initiators of movement have been cut off from the motor pattern generators. A spinal cord injury can do this, leaving the victim with the ability to think of a movement, even to execute it mentally, but without the ability to produce action except in dreams.

That such a person's motor cortex remains active has been demonstrated by functional magnetic resonance imaging studies (figure 15.5). Can the long pathways be reconnected in people or animals with such injuries? Such people and everyone around them want an

290   Chapter 15

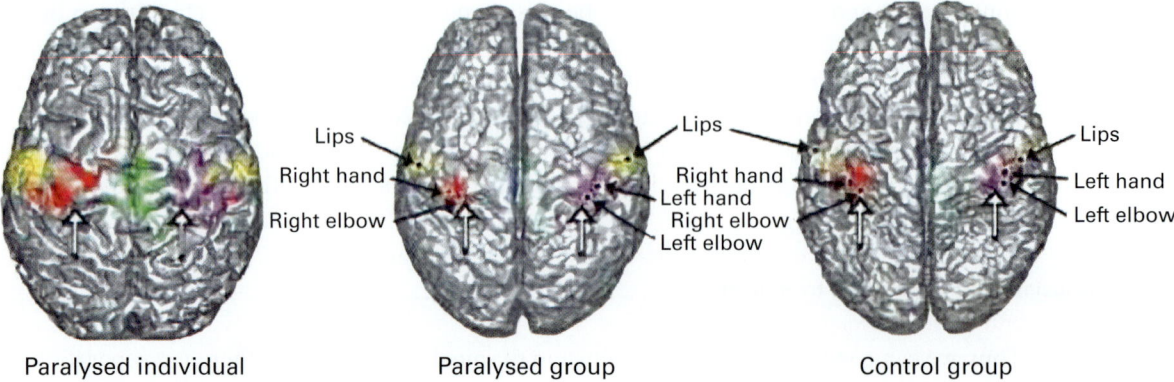

Figure 15.5
In a study published in 2001, functional magnetic resonance imaging was used to find evidence of activation in the motor cortex of tetraplegic human patients. Although these patients were unable to move their limbs because of damage to their spinal cords, voluntary efforts to move the lips or one of the hands or the arm at the elbow resulted in greater activity in corresponding parts of the motor cortex. Different colors show activation of different regions of the motor cortex of a patient who was mentally trying to move one or another of the body parts as indicated by the labels. From Shoham, Halgren, Maynard, and Normann (2001).

answer to this question. There has been much research on central nervous system axon regeneration; it was discussed briefly at the end of chapter 13. Some progress has been made, and experimental work with small animals has shown that some return of function is possible after regeneration has been produced or increased by various experimental treatments.

Importance of the Corticospinal Tract for Innate and Learned Movements That Require Special Dexterity

It is common to assume that innate patterns of movement depend on brainstem and spinal mechanisms, whereas learned movements depend on neocortex. The assumption has been reinforced by findings of plasticity in motor cortex, with formation of synaptic alterations during motor skill learning. This can lead one to assume that the corticospinal tract would not be of great importance in the execution of the motor component of fixed action patterns. To test this notion, together with one of my students I transected the corticospinal tract on one side at the hindbrain level in Syrian hamsters. We were particularly interested in the dexterous seed-shelling ability of these little animals. They can remove the shells from sunflower seeds faster than most humans—even if they have never encountered a seed previously in their lives. We found that after the unilateral pyramidotomy, the animals became very clumsy in the use of the paw opposite to the side of the small cut in the hindbrain; the deficit was enduring—still present a year after the surgery. Thus, it seems likely that dexterity, in unlearned movements as well as in learned, was a major reason for motor cortex and corticospinal tract evolution.

The Nature of the Spinal Motor Pattern Generators

Although the corticospinal tract has some direct connections with motor neurons in animals with large neocortex and very dexterous movements of the fingers (and toes and/or tail in some species), such direct connections are not the most numerous ones. Most connections are with the pool of interneurons, neurons that are highly interconnected, not only locally but with other levels of the cord. These interconnections form the propriospinal system, the organizer of innately determined motor pattern generation (see figure 9.10). When a limb withdraws automatically from a painful touch, the flexors contract and at the same time the extensors relax and the opposite limb tends to do the opposite by extending. These actions depend on segmental connections. Intersegmental connections control the various gaits of locomotion and the movements of grooming such as the scratch reflexes.

It is clear that descending pathways can initiate and modulate the execution of these innate spinally organized movements. But what about the many movement skills acquired during development by learning? There is evidence from experimental studies of both frogs and rats that the higher control is acting on distinct spinal modules that when activated result in limb movements toward specific locations around the body. It is possible that activation of various simple combinations of these spinal modules, not large in number, can produce a huge variety of movements.

Motor Cortex in Phylogeny

The motor cortex was originally defined by Fritsch and Hitzig (1870) as the neocortical region where the thresholds for eliciting movements by electrical stimulation were the lowest (i.e., where movements could be obtained by application of electric currents that were too weak to elicit them elsewhere in the cortex). Their experimental work was carried out with dogs; it confirmed observations made earlier by Hitzig on head-injured Prussian soldiers. Four years after the report by Fritsch and Hitzig, the neuroanatomist Betz (1874) described the location of very large pyramidal cells in the corresponding neocortical region in the human as well as in dogs, monkeys, and apes. After those early reports, a motor cortical region has been described for many other species of mammals. Somatosensory cortical regions have also been described, including in most mammals a primary somatosensory area located just behind the motor region. The two regions are often referred to as M1 and S1.

However, studies of the neocortex of the Virginia opossum have produced data indicating that M1 and S1 are mostly overlapping in that animal. The region appears to be a sensorimotor "amalgam" (figure 15.6). Similar studies of the rat have revealed motor and somatosensory regions that are mostly separate, each receiving projections from a separate part of the ventral thalamus except for overlap in the hindlimb region (figure 15.6).

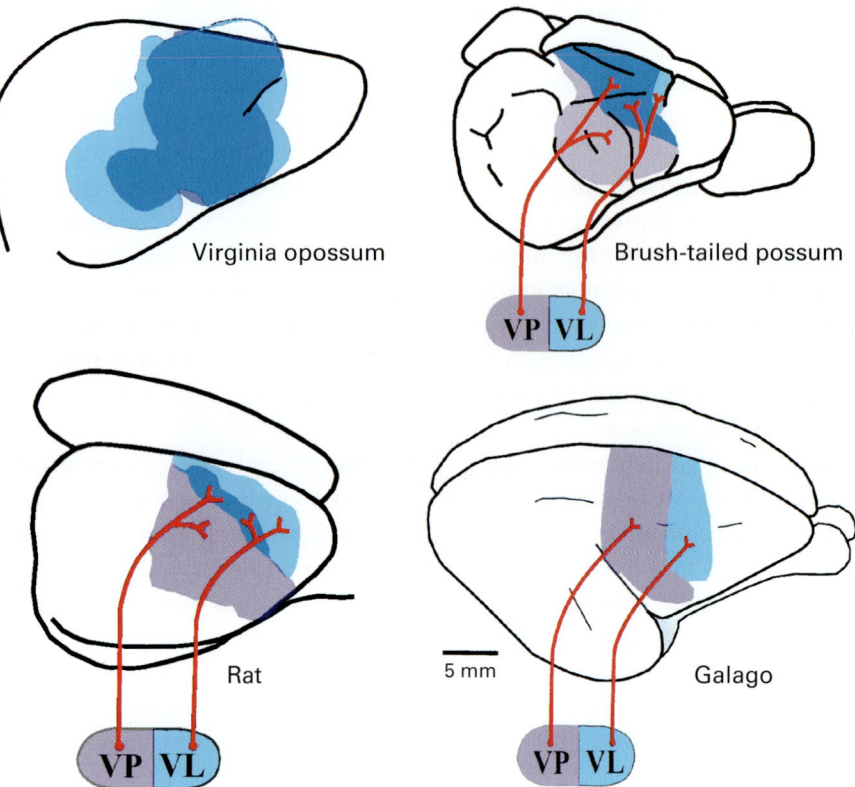

Figure 15.6
Overlap and non-overlap of primary somatosensory cortex and primary motor cortex in four species. The ancestral condition is proposed to be a lack of the separation of these two regions that is found in most mammals. Lende in 1963 reported an "amalgam" of somatosensory and somatomotor areas in the Virginia opossum using electrical stimulation and recording methods for defining motor and somatosensory regions, respectively. Anatomical studies of thalamic projections of the ventral posterior nucleus (VP) to somatosensory cortex and of the ventrolateral nucleus (VL) to the motor area have been carried out in various species. Such studies have found an incomplete overlap in the brush-tailed possum, greater separation in the rat where there is overlap in the hind limb and proximal forelimb representations, and an apparently complete separation of primary somatosensory and motor areas in primates like the galago. Based on the following: Virginia opossum: Lende (1963); brush-tailed possum: Haight and Neylon (1978, 1979); rat: Donoghue (1982); galago: Kaas (2004).

In the Virginia opossum, those two parts of the thalamus both project to the same cortical area, whereas in the brush-tailed possum the situation is intermediate between the rat and Virginia opossum. Similar studies in primates indicate a complete separation of M1 and S1 (figure 15.6). These studies support the idea that a parcellation of sensory and motor cortex has occurred in evolution.

It seems likely that the "motor cortex" is a specialization of a somatosensory neocortical area. It always receives somatosensory inputs, not always from thalamus but because of strong connections from the adjacent S1 area. However, it has been co-opted by other sensory systems as well, via connections through the association areas as we will see later when we turn our attention more specifically to neocortical anatomy.

Corticospinal Projections in Phylogeny

One of the behavioral functions described earlier in the monkeys with brain lesions was grasping with the hands. We know that grasping abilities have long been one of the functions of the midbrain, and that one of the contributions of the neocortex has been to add greater dexterity to this ability. Mammals in which more corticospinal axons terminate in the deeper spinal layers, in the region of the motor neurons, are generally the animals with greater dexterity. In the same animals, these axons can also be followed further caudally in the cord. There is a tendency for these animals to have greater neocortical volume as well, although there are exceptions.

The greater invasion of spinal cord with increasing neocortical size fits an approximate relationship that has been called "Deacon's rule," which can be summarized as "large equals well-connected." There are many other examples of this relationship, to be mentioned later. There is a fairly consistent tendency for animals with the greater neocortical area to have a larger number of neurons giving rise to corticospinal axons. This is illustrated in figure 15.7, which also shows the locations of corticospinal neurons in a number of different species. The areas of greatest density of these neurons are indicated in the figure.

It is believed that in most species, the corticospinal axons do not connect directly with motor neurons even when they terminate in the ventral horn. They connect with interneurons, many of which do connect with the motor neurons. Studies of monkeys have been carried out to uncover details of these connections, using retrograde transport of rabies virus injected into muscles of the fingers, elbow, or shoulder. The virus travels retrogradely to the motor neurons, then across synapses made by axons terminating on those neurons, then retrogradely to neurons that connect directly to the motor neurons, including any corticospinal neurons in the motor cortex that do that. The localization of the rabies virus in neocortical neurons indicated that the monkey's motor cortex includes two distinct regions: a more rostral region where the corticospinal neurons do not connect

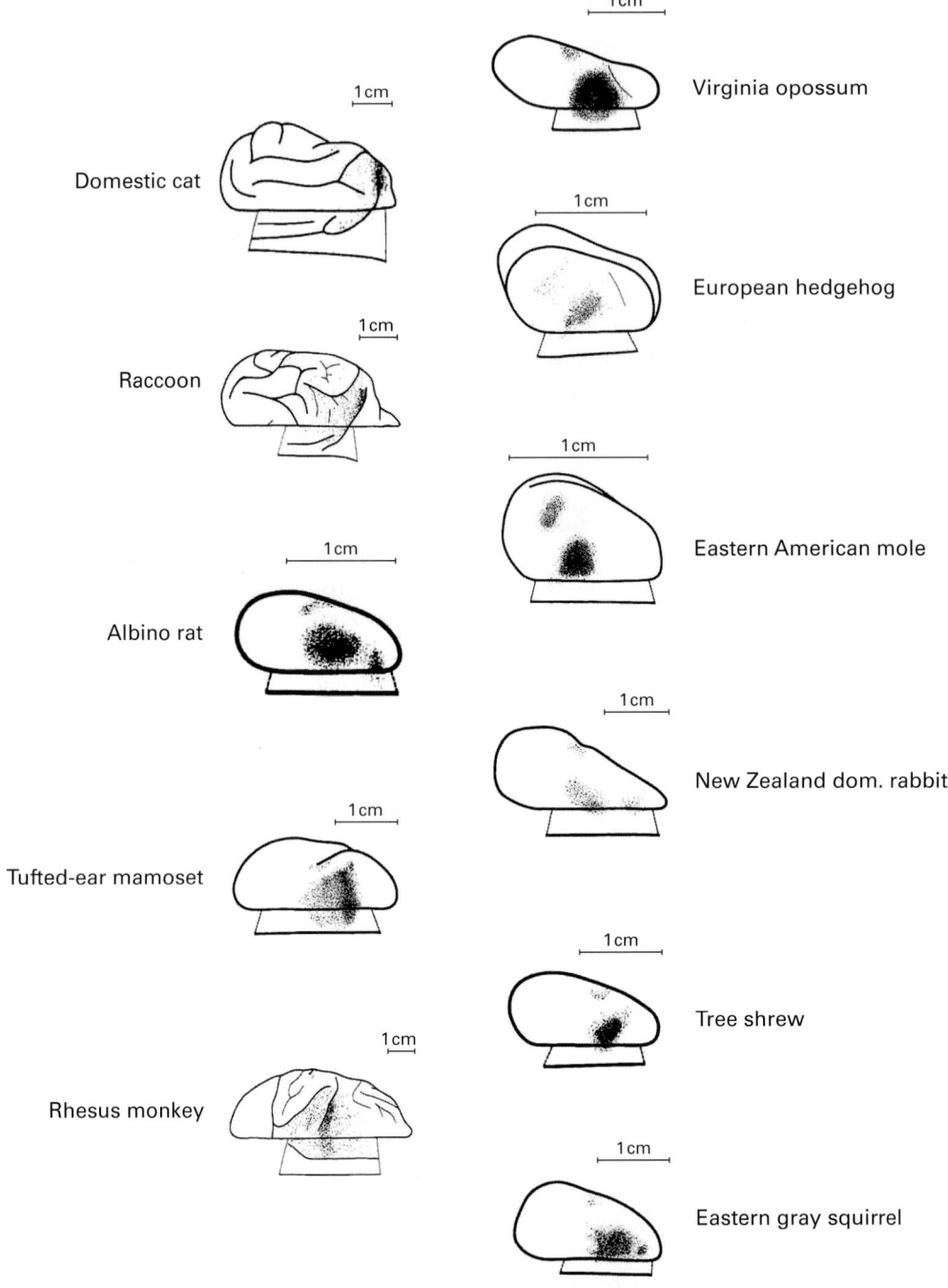

Figure 15.7
Dorsal view drawings of the left cerebral hemisphere of adult representatives of 11 species of mammal, with stippling indicating the location of layer-5 pyramidal cells that project to the spinal cord. Density of stippling corresponds to the density of neurons labeled with a retrograde tracer injected between the first and second cervical segments of the cord. The horizontal line near each drawing represents 10 millimeters; brains are thus drawn at different scales so that the drawings are of similar size. Adapted from Nudo and Masterton (1990).

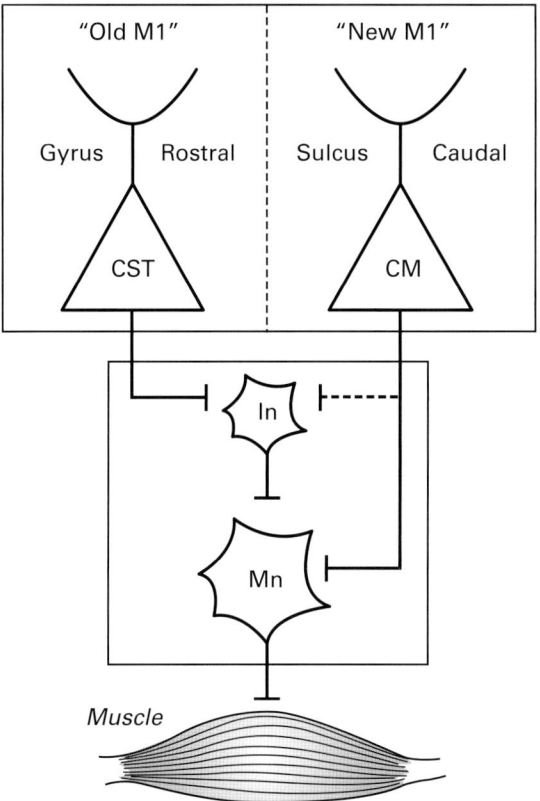

Figure 15.8
Diagram illustrating connections of "new" and "old" motor cortex regions in monkeys: New M1 is located caudally in the central sulcus and has pyramidal cells (CM cells, for corticomotoneuronal cells) with axons that make direct connections with motor neurons that innervate finger, elbow, or shoulder muscles. In contrast, old M1 is located rostrally on the precentral gyrus and lacks CM cells. However, old M1 has corticospinal tract (CST) neurons that influence motor neurons indirectly through connections with spinal interneurons. In, interneurons; Mn, motor neurons. From Rathelot and Strick (2009).

directly with spinal motor neurons, and a more caudal region where they do connect directly (figure 15.8).

Thus, the neocortex to motor neuron connections allow a direct control of muscle activity that bypasses the intermediate net of the spinal cord. Is this important for "sculpting novel patterns of motor output that are essential for highly skilled movements" as the scientists doing the study (J. Rathelot and P. L. Strick) concluded? In that case, a very dexterous instinctive movement pattern, like seed shelling by hamsters, would not require any neocortical connections direct to motor neurons. It would be interesting to confirm this.

The Highest Levels of Motor Control

As previously discussed (e.g., in chapters 4 and 12), at the lofty neocortical level of the central nervous system of mammals, there are sensory areas with projections to the striatum. These projections carry sophisticated information about the surrounding world that can result in learned habitual movement patterns. Furthermore, as just discussed, evolution has given rise to a motor cortex with direct projections to spinal cord, supporting fine control of extremities. This motor cortex has become a dominant structure in many mammals, by virtue of its size and the quantity of its connections.

Why has the motor cortex become so dominant in higher mammals? It is likely that it did not come to be so important just because of its strong control of habits via its connections to striatum or because of the dexterity it facilitates by its direct projections to motor neurons in the spinal cord. These functions are no doubt important elements in the story of its rise to prominence, but there is another major reason. The reason is that it is on the output side of a neocortical system for anticipation of and planning for the future—the prefrontal neocortex. Anticipating the immediate future, for example, are the frontal eye fields, located anterior to the motor cortex. Outputs of this cortex cause the eyes to move to positions of expected stimuli. When we do this, we often have the feeling of making voluntary eye movements. Other kinds of voluntary movements, made for achieving goals of motivational states, arise from other parts of prefrontal cortex, anticipating not only the immediate future but the more distant future as well.

Much of the lateral prefrontal cortex is closely connected to anterior cingulate cortex, and thence to the hippocampal region. It has evolved very extensive transcortical connections—not diffuse connections but very specific ones. We will discuss this type of cortex, the heteromodal association cortex, in part XII.

Readings

Shoham, S., Halgren, E. H., Maynard, E. M., & Normann, R. A. (2001). Motor-cortical activity in tetraplegics. *Nature, 413*, 793.

Kalil, K., & Schneider, G. E. (1975). Motor performance following unilateral pyramidal tract lesions in the hamster. *Brain Research, 100*, 170–174.

Rathelot, J.-A., & Strick, P. L. (2009). Subdivisions of primary motor cortex based on cortico-motoneuronal cells. *PNAS, 106*, 918–923.

For Lawrence and Kuypers (1968), see papers cited in the previous chapter.

There is evidence from experimental studies of both frogs and rats that the higher control works by descending pathways activating distinct spinal modules that when activated result in limb movements toward specific locations around the body:

Mussa-Ivaldi, F. A., & Bizzi, E. (2000). Motor learning through the combination of primitives. *Philosophical Transactions of the Royal Society of London, Series B, 355*, 1755–1769.

Bizzi, E., Cheung, V. C. K., d'Avella, A., Saltiel, P., & Tresch, M. (2008). Combining modules for movement. *Brain Research Reviews, 57,* 125–133.

Cheung, V. C. K., d'Avella, A., & Bizzi, E. (2009). Adjustments of motor pattern for load compensation via modulated activations of muscle synergies during natural behaviors. *Journal of Neurophysiology, 101*, 1235–1257.

"Deacon's rule," which can be summarized as "large equals well-connected," is discussed by Georg Striedter in the book cited in chapter 4. The idea was proposed by Terrence Deacon:

Deacon, T. W. (1990). Rethinking mammalian brain evolution. *American Zoologist, 30,* 629–705.

Functional imaging data on humans has indicated the important role of prefrontal cortex in anticipation of (or preparation for) movement. An example can be found in the following report:

D'Esposito, M., Ballard, D., Zarahn, E., & Aguirre, G. K. (2000). The role of prefrontal cortex in sensory memory and motor preparation: An event-related fMRI study. *NeuroImage, 11*, 400–408.

For a review of ideas of how several of the motor pathways and motor cortical areas emerged during the course of evolution, see the following chapter:

Nudo, R. J., & Frost, S. (2009). The evolution of motor cortex and motor systems. Chapter 33 in Kaas J. H., ed. *Evolutionary neuroscience.* New York: Academic Press, pp. 727–756.

Findings of plasticity in motor cortical maps and synapses during motor skill learning and after cortical damage:

Ramanathan, D., Conner, J. M., & Tuszynski, M. H. (2006). A form of motor cortical plasticity that correlates with recovery of function after brain injury. *PNAS, 103,* 11370–11375.

Sanes, J. N., & Donoghue, J. P. (2000). Plasticity and primary motor cortex. *Annual Review of Neuroscience, 23,* 393–415.

Xu, T., Yu, X., Perlik, A. J., Tobin, W. F., Zweig, J. A., Tennant, K., Jones, T., & Zuo, Y. (2009). Rapid formation and selective stabilization of synapses for enduring motor memories. *Nature, 462,* 915–919.

A speculation about the role of parietal cortical areas in manual dexterity in primates:

> Padberg, J., Franca, J. G., Cooke, D. F., Soares, J. G. M., Rosa, M. G. P., Fiorani, M., Jr., Gattass, R., & Krubitzer, L. (2007). Parallel evolution of cortical areas involved in skilled hand use. *Journal of Neuroscience, 27*, 10106–10115.

Studies of marsupial sensory and motor cortical areas with discussion of an integration with findings on other mammals:

> Karlen, S. J., & Krubitzer, L. (2007). The functional and anatomical organization of marsupial neocortex: Evidence for parallel evolution across mammals. *Progress in Neurobiology, 82*, 122–141.

16 The Temporal Patterns of Movements

In the previous chapter, the spatial organization of motor system pathways and connections was considered together with some information on relative quantities of neurons or axons in various animals and systems. This is the kind of information that neuroanatomists obtain. In addition to the peculiarities of structure they observe, neuroanatomists can also distinguish different types of cells and fibers by their chemical content. The most significant chemicals are the neurotransmitters and neuromodulators used by the neurons under study. Extending such studies to functional correlates adds many different demands. When thinking about motor system functions, an obvious, major topic is that of the generation of temporal patterns in animal movements. How can the production of this kind of patterning be explained? Trying to do so leads us to basic ideas of how the brain functions at the level of circuits.

Three Types of Mechanism

We have long known about three kinds of explanations of how a temporal pattern can be produced in the outputs of the central nervous system—when the pattern is not present in the stimuli coming from the outside. First, there are temporal patterns in the output of reflex pathways, a fact that has been used in stimulus–response (S-R) models of behavior. These models attempt to explain all or most behavior in terms of S-R connections. At the simplest level, reflex pathways can be conceptualized as "straight-through" processing (i.e., straight from the stimulus end through various stages to the response end of the pathways). Second, such models are extended to include various kinds of feedback circuitry, which can extend the duration of temporal patterns and add rhythmic changes to them. Third, many, perhaps most, neurons can generate spontaneous activity. Such endogenous activity can be generated rhythmically, and the period of such rhythms can vary widely.

Explaining Movement Dynamics in Terms of S-R Circuits

Consider first the timing of movements in a startle reflex. People as well as many animals are startled by a sudden loud noise: their current behavior is interrupted, at least briefly, by the movements of the startle reflex. First, the eyes blink, and in rapid succession there is a series of additional movements, always in the same sequence: contraction of other facial muscles, then neck flexion, then arm flexion, and in case of a very intense sound this is followed by leg flexion. The timing is attributable to conduction time in the reflex pathways underlying the movements. Axonal conduction rate depends on fiber size, and at the synapses there are synaptic delays and temporal summation times.

Many psychologists and physiologists in the nineteenth century and early twentieth century believed that reflex pathways are adequate to explain all behavior. The concept of the reflex had first been clearly formulated by René Descartes in the seventeenth century, and it became popular among some philosophers in the following centuries. As mentioned in chapter 2, Pavlov made the reflex model of behavior more complete by his demonstrations of plasticity in reflexes (conditional, or conditioned, reflexes). It was Ramón y Cajal who gave it anatomical reality, with his descriptions of Golgi-stained spinal cord sections from mammals (see chapters 2, 8, and 9).

But if all behavior is to be explained by reflex pathways, however complex, then the temporal patterns of movements, in complex serial order, have to be explained this way, too. To explain long sequences of such movements, a series of reflexes must be involved, and this led to the idea of the chaining of reflexes. One reflexive action pattern must result in the stimulus that triggers the next reflexive action.

Examples of reflex chaining are very hard to find, showing the inadequacy of the theory as presented. However, if the theory is expanded to include not only reflexes, as strictly defined, but also the *fixed action patterns* as defined by ethologists, the theory acquires some explanatory power. The term *fixed* does not mean that the behavior is rigidly invariable; it only means that a pattern of movement, together with an associated motivational system, is fixed in the genetic constitution of the species.

An example is not difficult for an animal behaviorist to imagine. Consider the foraging behavior of a Syrian hamster. The hamster continuously searches for pieces of food that it can put into its cheek pouches. A hamster will search for seeds that have been scattered about a space (e.g., in a field in Syria or in a room or in a cage if it is someone's pet animal). For each seed, the animal follows the same sequence of movements: find the seed using olfaction supplemented by vision and touch, take the seed into its mouth and push it back into a cheek pouch, then continue its searching. It continues with this pattern of behavior until the cheek pouches are full. The sensation of stretching of the skin of the pouches triggers a change in behavior: The hamster stops collecting seeds and moves toward its home base, and, once there, toward its hoard near its nest. Once in contact with

The Temporal Patterns of Movements

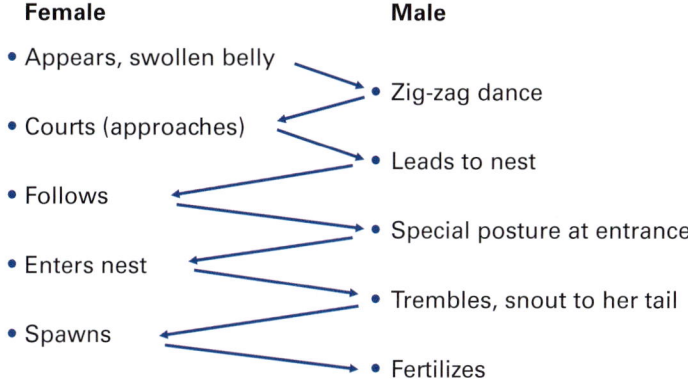

Figure 16.1
Stimulus-response chaining as observed in ethological studies by Niko Tinbergen of the courtship behavior of the three-spined stickleback fish: His description exemplifies the meaning of reflex chaining or the chaining of fixed action patterns. Each step results in the stimulus that is necessary for triggering the next step. In the example illustrated, the behavioral sequence begins when a female with a swollen belly appears before a male. The sight of this female causes the male to start his zigzag dance. Seeing this, the female, provided her courtship motivation is sufficiently high, approaches. The approach causes the male to lead her to his nest. She follows, and the sequence continues as shown, ending with the male's entering the nest after the female lays her eggs there, and he fertilizes the eggs. Based on Tinbergen (1951).

the hoard, it changes to a new pattern of movement: it pushes all the seeds out of the pouches. Each pattern of action in this sequence leads to the stimulus that initiates the next pattern.

Thus, the concept of chaining of S-R behaviors is made very clear. The behavior sequence may involve more than one animal of the species; for example, consider an example from studies of fish behavior by the ethologist Niko Tinbergen (1951). He described the three-spined stickleback's courtship pattern as a series of fixed action patterns. His description of the courtship behavior of a mating pair is summarized in figure 16.1. The sticklebacks' courtship behavior is a series of distinct steps, each step providing a necessary stimulus for the triggering of the next step. If a step is interrupted, the sequence cannot proceed.

Stickleback courtship and mating behavior is a true chaining of S-R events, the events being fixed motor patterns (largely innate); that is, the motor components of fixed action patterns (FAPs). Scientists who loved the reflex model have claimed that all behavior can be explained in a similar fashion. (We should note here that many of these scientists have looked at FAPs as complex reflexes. However, the ethologist Konrad Lorenz pointed out that FAPs include a motivational component that is not present in reflexes.)

Although some temporal patterns of behavior can be explained in terms of a chaining of S-R events, there are many data that are incompatible with a generalization of the model to all behavior.

Central Programs Rather than Reflex Chaining

In 1956, Doty and Bosma published an analysis of the swallowing reflex, or *reflex deglutition*. They tested the idea that the complex pattern of muscle contractions in swallowing is due to an S-R series—a chaining of reflexes. Swallowing involves about 20 muscles controlled by neurons from the midbrain to cervical spinal cord levels. They found that in this case, the pattern of contractions is centrally programmed. It is an inborn fixed action pattern that is executed rhythmically with a steady or varied stimulus. The peripheral sensory axons are located in the superior laryngeal nerve. The pattern is not changed by elimination of proprioceptive feedback during any stage of the movement once it is triggered. It is normally triggered by stimulation of the tongue in the back of the mouth, but there is also an endogenous input that builds up over time and lowers the threshold for eliciting the response.

Karl Lashley had come to the same conclusion much earlier in a 1917 paper entitled "The Problem of Serial Order in Behavior." He argued against the adequacy of the reflex chaining hypothesis. He used examples like the following: "The finger strokes of a musician may reach 16/sec in passages which call for a definite and changing order of successive finger movements." He noted that "The succession of movements is too quick even for visual reaction time." Therefore, there must be central generation of patterns of movement.

Many Fixed Action Patterns Are Centrally Generated

Similarly, many fixed action patterns in vertebrates are centrally generated, according to experimental evidence. They include inherited movement abilities just like reflex deglutition. Scientists have referred to such movements as "reflexes," but consider the evidence. John Fentriss at Dalhousie University in Nova Scotia has studied grooming by mice. He found that movement patterns continue even without the usual feedback stimulation. For example, when a mouse grooms an ear with a front paw after briefly licking the paw, the movements are providing stimulus feedback from tongue, paw, and ear. However, even if much of the limb has been amputated, the mouse executes the same movement pattern as part of a longer sequence, moving his body as if the limb were still there.

In Lorenz's and Tinbergen's studies of egg rolling in gulls, they found that the movements of rolling an egg toward the nest can occur in the complete absence of peripheral signals (i.e., even without the stimulus provided by an egg). This can happen in situations of very high motivation. Ethologists have described a number of other examples of movement patterns performed *in vacuo* (e.g., see the writings of Konrad Lorenz).

There is good evidence for central generation of locomotor movements in fish, involving rhythmic fin movements. Locomotor movements in mammals are similarly generated in the spinal cord, but they require activation from the periphery or from descending fiber systems (from the brain).

How are these action patterns generated, if not by reflex circuitry? We need one of the other types of explanation mentioned in the first page of this chapter.

Reverberating Circuits within the Brain and Spinal Cord

In the second type of explanation of temporal pattern generation, feedback circuitry of one sort or another complicates the straight-through processing of reflex pathways. Self re-exciting loops can lead to regular bursts of action potentials (a kind of central oscillator). This type of mechanism is illustrated in the oversimplified sketch of figure 16.2. With such a circuit, the nervous system can produce patterned output from unpatterned input. Alternatively, consider the behavior of neural circuitry in which there is negative

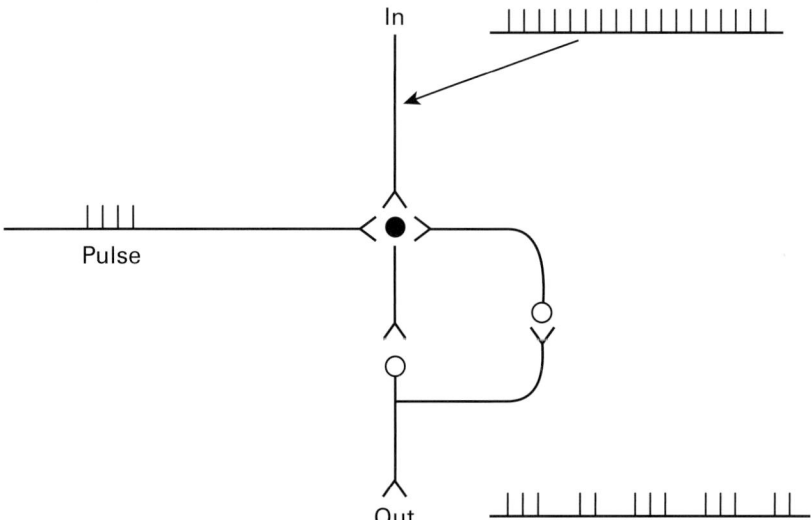

Figure 16.2
Diagram of a very simple, neural feedback circuit that can produce patterned output from unpatterned input. Assume that all synapses shown are excitatory. Assume also that the black neuron will initiate an action potential only when two different inputs converge simultaneously. This neuron receives continuous input in the form of a series of action potentials from the axon marked "in." When a single burst of action potentials occurs in the axon on the left (marked "pulse"), a small burst of action potentials will be generated by the black neuron. Its axon has a recurrent connection through two interneurons to itself. We assume that these interneurons, which act as a delay line, will fire action potentials when they receive a single short burst of excitatory postsynaptic potentials. A resulting burst of action potentials reaches the terminals of the interneuron that synapses on the black cell, and the process repeats. A series of short bursts of action potentials begins in the output neuron, as shown. One could reduce the circuit to include only the black neuron with a recurrent connection to itself, but the bursts of action potentials in the output axon would recur at a much faster rate.

feedback, as in homeostatic mechanisms. The response can oscillate around a certain level of input (the set point).

In theory, in order to get more complex patterns, more than one such circuit can be combined. An example supported by various experimental studies is the production of breathing rhythms.

Much of the best work in this area has been done on invertebrates, because of the smaller number of neurons involved and the ability of investigators to identify the same neurons in different members of the same species. The experimental studies have led to models of locomotor pattern generation in several species—models with much support from experimental studies.

Endogenous Activity of Single CNS Neurons

A third type of mechanism for temporally patterned activity generation is the endogenous activity of single neurons. Much of the earlier evidence in this area was obtained in studies of rhythmic potentials in single, identifiable neurons found in invertebrates like the sea slug, *Aplysia*. This was introduced at the end of chapter 1. It has been possible to record from such neurons after isolating them from other neurons so that any activity changes must be generated from within single cells. Even after isolation, cells have been found that generated action potentials in regular bursting patterns. These bursts were generated by periodic changes in membrane potential.

A neuron can generate more than one endogenous rhythm at the same time. The changes in membrane potential are generated by membrane proteins at so-called pacemaker loci in the membrane.

The existence of endogenously generated rhythmic depolarization of neuronal membranes that generate bursting patterns of spike potentials indicates the possibility of complex pattern generation by neurons or groups of neurons. This possibility is theoretical: Multiple oscillators added together could result in any pattern, as we know from the mathematics of Fourier analysis. However, the role of endogenous oscillators in the generation of movement patterns is not very well understood, especially in the case of mammalian movements.

The Endogenous Clock

One phenomenon where the evidence is abundant for a basis in endogenous activity is the biological clock, mentioned previously in chapters 1 and 7. This is the internal mechanism in vertebrates and other animals that causes them to enter a period of sleep about every 24 hours. The mechanism also causes animals to become active at about the same time every day (or night). The rhythm that is endogenously generated is not precisely 24 hours but is circadian (about a day in length); it is normally synchronized with—entrained by—the onset of daylight. The critical input for this entrainment comes from the retina via a projection to the hypothalamus (chapter 20). Animals depend on such "biological clocks" for their daily schedule of activities. These temporal schedules are important for animal survival.

How It All Works at the Circuit Level

Study of how organized sequences of movement are produced, and how they can change, has led to the development of the basic ideas we have been discussing in this chapter, about how the brain works at the circuit level. These ideas are important in theories of CNS functions when considered at the level of connections and circuits. We can summarize the ideas as follows.

First are the straight-through processing concepts, which come from the conceptualizations of reflexes, fixed action patterns, and stimulus–response models of behavior. When applied to input analysis, application of these concepts has led to the ideas of stimulus filtering by *cascade processing*, applied most successfully in studies of the visual system by Hubel and Wiesel and their successors. Theoretical models—computational models—of the functions of the visual pathways in the neocortex using cascade processing have been gaining in their explanatory power.

Next are the concepts of feedback control circuits, involving both positive and negative feedback pathways. We have mentioned reverberating circuits in which positive feedback occurs. With such circuitry in nervous system connections, a constant input can be converted to a rhythmic output pattern. Negative feedback can be used to construct homeostatic control circuits, which can go into oscillation and produce rhythmically changing output. In the complex interconnections of the brainstem reticular formation, or in the propriospinal interconnections, it is believed that many different innately determined action patterns are generated.

The next concept we reviewed was based on endogenous CNS activity, the "spontaneous" activity of the nervous system. We know that some neurons generate sustained excitatory states. We also know that some cells have rhythmically changing levels of membrane polarization, and that this can cause periodic bursts of action potentials. This is found in the periphery as well as in the CNS. Perhaps the best known is the beating of heart muscle cells. Within the brain, there are cells within the hypothalamus that change their activity in a circadian pattern—they constitute a biological clock.

The Circuits Are Not Always Fixed

We know that the mechanisms we have been discussing are not unchanging. Plasticity is not always possible, but many connections can show short-term or long-term changes. This may involve molecular changes at synapses, an active subject of much research by cell and molecular neuroscientists. Along with the synaptic changes, or independent of them, changes in the structure of end arbors may also occur, by axonal sprouting of various kinds, and also by dendritic changes.

Whatever plasticity may occur in endogenous rhythm generation is not so clear. It would be important to know more about possible changes in number and character of pacemaker loci in neuronal membranes: Are such changes possible? If so, what causes the changes? This is largely unknown.

It is important to address various combinations of all three mechanisms to explain behavior patterns in the real animal. For this, computational modeling takes on major importance.

How Adequate Are These Concepts?

When trying to explain how the brain works, how adequate are the concepts discussed above? There is little doubt that additional ideas are needed. For example, we have not yet tried to incorporate the actions of systems of diffusely projecting neurons in the brain. Some of these ideas will come up in the next chapter, concerning different states of the brain.

Readings

Doty, R. W., & Bosma, J. (1956). An electromyographic analysis of reflex deglutition. *Journal of Neurophysiology*, *19*, 44–60.

Lashley, K. S. (1951). The problem of serial order in behavior. In Jeffress, L. A., ed. *Cerebral mechanisms in behavior*. New York: Wiley, pp. 112–136. (Available on Google Books.)

Pavlov, I. P. (1928) *Lectures on conditioned reflexes: Twenty-five years of objective study of the higher nervous activity (behaviour) of animals*. W. H. Gantt, trans. New York: Liverwright Publishing.

The ethologist Konrad Lorenz pointed out that fixed action patterns include a motivational component that is not present in reflexes:

Lorenz, K. Z. (1981). *The foundations of ethology*. New York: Springer.

On the topic of fixed motor patterns n the absence of peripheral signals, see pp. 127–129, 236. Also see the review of the work of Erich von Holst on central generation of locomotor movements in fish involving rhythmic fin movements (pp. 136–144).

On stickleback courtship behavior as a series of fixed action patterns: See the book by Konrad Lorenz cited above. The work on stickleback fish was conducted by Niko Tinbergen and his students and collaborators. Two particular readable books, one by Tinbergen and one by Lorenz, are the following:

Tinbergen, N. (1958). *Curious naturalists.* New York: Basic Books.

Lorenz, K. Z. (1952). *King Solomon's ring: New light on animal ways.* New York: Thomas Y. Crowell.

The generation of locomotor patterns has been investigated in various species. Evidence for central pattern generation at spinal levels influenced by sensory input and by descending pathways from the brain has been obtained even in humans:

Duysens, J., & Van de Crommert, H. W. A. A. (1998). Neural control of locomotion; Part 1: The central pattern generator from cats to humans. *Gait and Posture, 7,* 131–141.

Barlow, S. M., & Estep, M. (2006). Central pattern generation and the motor infrastructure for suck, respiration, and speech. *Journal of Communication Disorders, 39,* 366–380.

Much literature on central pattern generation can be found by searching with Google Scholar for "endogenous rhythms" or "endogenous oscillators" and "circadian rhythms" (or oscillators).

VII BRAIN STATES

17 Widespread Changes in Brain State

In the previous chapter, we discussed temporal patterns and rhythms of activity including, briefly, the great changes in brain state that recur in a daily pattern. This leads us to a new chapter in the story of brain structure and its origins, a chapter that stands alone in a separate unit: What underlies such widespread changes in the state of so much of the CNS?

We know that ingesting certain chemical substances or being injected with them can result in widespread changes, sufficient, for example, to produce unconsciousness or anesthesia. Might this occur naturally, as by secretions into the cerebrospinal fluid (CSF)? This almost certainly does occur. Sleep in animals has been induced by transferring CSF from a sleeping animal to an awake companion. Pain can be reduced by natural opiates secreted by cells within the midbrain. This kind of state change has not been extensively studied.

Brain States Influenced by Widely Projecting Axon Systems

Neuroanatomical bases for brain state changes have been subjected to more research than have secretions into the CSF. The pathways involved are fundamentally different from the type of pathway we have been mainly discussing up to this point. They are the widely projecting axon systems (e.g., see figure 10.3). It shows a neuron of the hindbrain reticular formation that has an axon with very wide distribution in the brainstem, with terminal arbors in the thalamus, hypothalamus, midbrain tectum and tegmentum, hindbrain, and spinal cord. It is believed that such neurons of the hindbrain and midbrain constitute what electrophysiologists have described as the *reticular activating system*. Stimulation of this system can cause arousal from sleep and even arousal from coma. Because stimulation of the midbrain reticular formation can rouse the forebrain including the neocortex, it has been dubbed the "ascending reticular activating system," or ARAS.

In the enthusiastic early discussions of the ARAS in the 1950s and 1960s, many changes in overall brain state were attributed to this system. However, since that time a number of distinct neuron groups with very widespread projections, more diffuse than in the

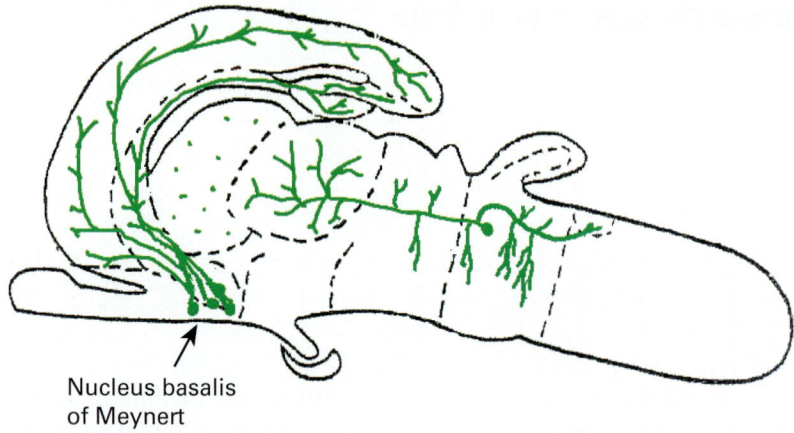

Figure 17.1
On this schematic sketch of the mammalian brain, the locations of certain groups of cholinergic neurons of the hindbrain and the basal forebrain are illustrated by drawings of only a few examples. Axons from these neurons are widely branching, and thereby they have broad influences on forebrain and brainstem activity. The branching of these axons is underestimated in the sketch.

illustration of figure 10.3, have been described. Included are three monoamine-containing groups originating in the brainstem and several other groups originating in the hypothalamus, each containing different peptides. Thus, changes in brain state seem to have more complex origins than recognized by the earlier investigators.

Cholinergic Systems

Some of the large neurons of the brainstem reticular formation use acetylcholine as their neurotransmitter. A group of these neurons, located in the caudal midbrain and rostral hindbrain reticular formation, projects widely to the thalamus (figure 17.1). Included in this projection are the midline and intralaminar cell groups known to have widespread projections to the neocortex. Electrical stimulation of these cell groups causes cortical activation, according to electroencephalographic recordings.

Widely projecting cholinergic axons to the neocortex are known to arise from neurons of the basal forebrain (figures 17.1), especially (but not only) from the nucleus basalis of Meynert. These neurons are activated during exposure to stimuli that are novel or particularly relevant to current motivations. The axons tend to end on cortical pyramidal cells, promoting excitation by other inputs. There is evidence that this system promotes the synaptic changes involved in learning. The cholinergic neurons of nucleus basalis show degenerative changes and losses in Alzheimer's disease, a phenomenon that helps explain the memory problems in Alzheimer's sufferers.

The Monoamine-Containing Systems

The monoamine-containing axon systems did not become known through the usual neuroanatomical methods. Three different monoamine-containing neuronal systems of the CNS were discovered by the Swedish neuroanatomists Dahlström and Fuxe in the early 1960s when they used a new method—developed by their mentors Falck, Carlsson, and Hillarp—that caused monoamines to fluoresce.

One of these systems uses the catecholamine **norepinephrine (NE)**, also known as noradrenaline, as its neurotransmitter. We have encountered this biogenic amine before, as it is the neurotransmitter of most of the postganglionic axons of the sympathetic nervous system. The NE-containing axons in the CNS come from relatively small groups of neurons in the hindbrain, some in the lateral tegmentum and others in a group of neurons in the rostral hindbrain central gray, below the cerebellum. The group of norepinephrine neurons below the cerebellum is marked in humans by a blue pigment. This is the *locus coeruleus*, meaning "the blue location." From these cells, numbering about 1,600 in the rat and 20,000 in humans, axons distribute themselves diffusely throughout the entire central nervous system—spinal cord, brainstem, cerebellum, and forebrain including the entire cortical mantle (figure 17.2). The norepinephrine is released, at least in many neocortical areas, in paracrine fashion according to electron microscopic evidence: Release of NE at a terminal enlargement can influence a number of cells in its vicinity.

By these widespread projections, the locus coeruleus plays a role in brain activation in response to novel stimuli, and during sleep it functions in the switching from one sleep state to another.

Dopamine is a molecule that is closely related to norepinephrine; both are catecholamines. The dopamine-containing neuron cell bodies are located in the ventral midbrain of mammals, in the ventral tegmental area and in the substantia nigra; there are others located more rostrally, in the hypothalamus. In fish and amphibians, dopamine neurons are located mostly in the caudal hypothalamus (the posterior tuberculum) and ventral thalamus. The dopamine (DA) axons are widely branching but they do not appear to distribute quite as widely as the NE axons. From the nigra they project heavily to the corpus striatum, the ventral portions of which receive dopamine axons from the ventral tegmental area (VTA). The VTA projects also to other limbic forebrain regions and to the neocortex, especially to prefrontal regions (figure 17.3).

The dopamine innervation is critical for the normal functioning of the striatum. The degeneration of dopamine neurons in Parkinson's disease causes devastating losses of function, especially affecting motor control. Activity of the dopamine neurons of the ventral tegmental area appears to correlate with the pleasurable experiences and positive reinforcement in the learning of new habits.

314   Chapter 17

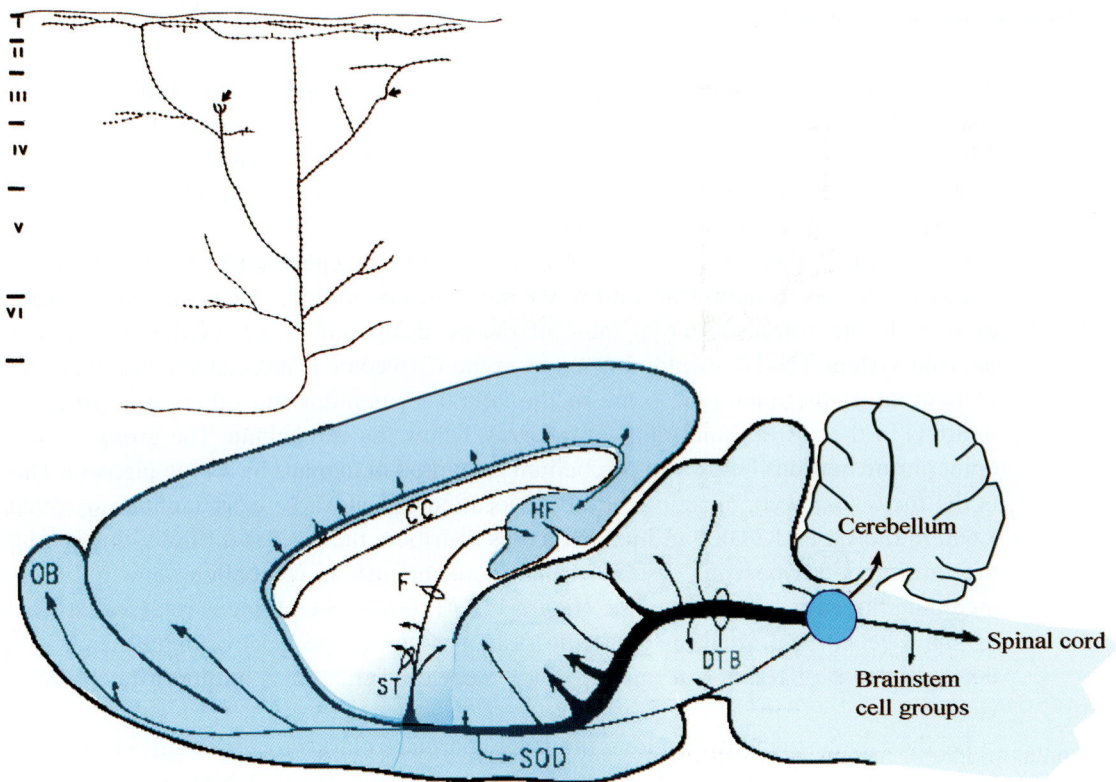

Figure 17.2
The noradrenergic system. Norepinephrine-containing axons originate mostly in the rostral hindbrain's *locus coeruleus* (shown as a blue disk), meaning "the blue location." Axons from this location distribute their terminals very widely throughout the CNS as indicated in the bottom diagram depicting a rat brain in a schematic sagittal section. Additional noradrenergic neurons are found in the lateral tegmentum, with axons reaching the hypothalamus and amygdala. The top diagram is a sketch of a single widely branching norepinephrine axon in the neocortex. It is studded with varicosities, most of which do not form specific synapses; they release norepinephrine into the intercellular space and thereby can influence many nearby cells. Top: from Levitt and Moore (1978); bottom: modified from Moore and Bloom (1979). CC, corpus callosum; DTB, dorsal tegmental bundle; F, fornix; HF, hippocampal formation; OB, olfactory bulb; ST, stria terminalis.

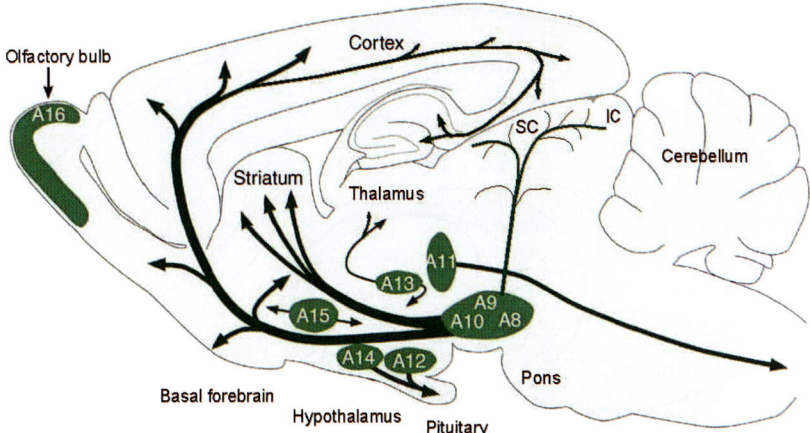

Figure 17.3
Dopaminergic systems. Diagrammatic representation of positions of dopamine-containing neuron groups and axon groups in the brain of a rat. The numbering of the cell groups is from Dahlström and Fuxe (1964). The long ascending dopamine axons arise mainly in the ventral midbrain, cell groups A8 (retrorubral area), A9 (substantia nigra), and A10 (the ventral tegmental area). In the endbrain, dopamine axon projections to the cingulate, retrosplenial, and hippocampal cortex have been added to the Björklund and Dunnett summary. The amygdala also receives dopaminergic innervation. Dopaminergic axon innervation of midbrain tectal areas have also been added; additional projections to some hindbrain structures are not shown. Modified from Björklund and Dunnett (2007). IC, inferior colliculus; SC, superior colliculus.

Serotonin, Another Monoamine Neurotransmitter Influencing Behavioral State

The third system of axons described by Dahlström and Fuxe uses 5-hydroxytryptamine, often called serotonin, as its neurotransmitter. The neuronal cell bodies are located in midline cell groups of the midbrain and hindbrain, and the axons project diffusely throughout the central nervous system (figure 17.4). Serotonin release in the CNS plays important roles in the regulation of sleep and in the modulation of arousal states of waking. Animals with lesions of these midline brainstem cell groups (the raphe nuclei) suffer insomnia to a degree proportional to the destruction of the serotonin neurons. The evidence that the serotonin system also plays more specific roles can be gathered from the widespread psychiatric use of serotonin reuptake inhibitors. Prolonging the action of serotonin at its points of synaptic release appears to increase self-confidence and to reduce anxiety.

Figure 17.5 illustrates some of the specificity that has been found in parts of the serotonin system. Thin, varicose axons from the dorsal raphe nucleus distribute to the striatum whereas a different kind of axon—thick, nonvaricose axons from the median raphe—reach the dentate gyrus of the hippocampal formation. Both types of axons distribute widely in the neocortex. The serotonin projections to the spinal cord come from the caudal hindbrain and form three distinct groups: The nucleus raphe magnus projects to the

316   Chapter 17

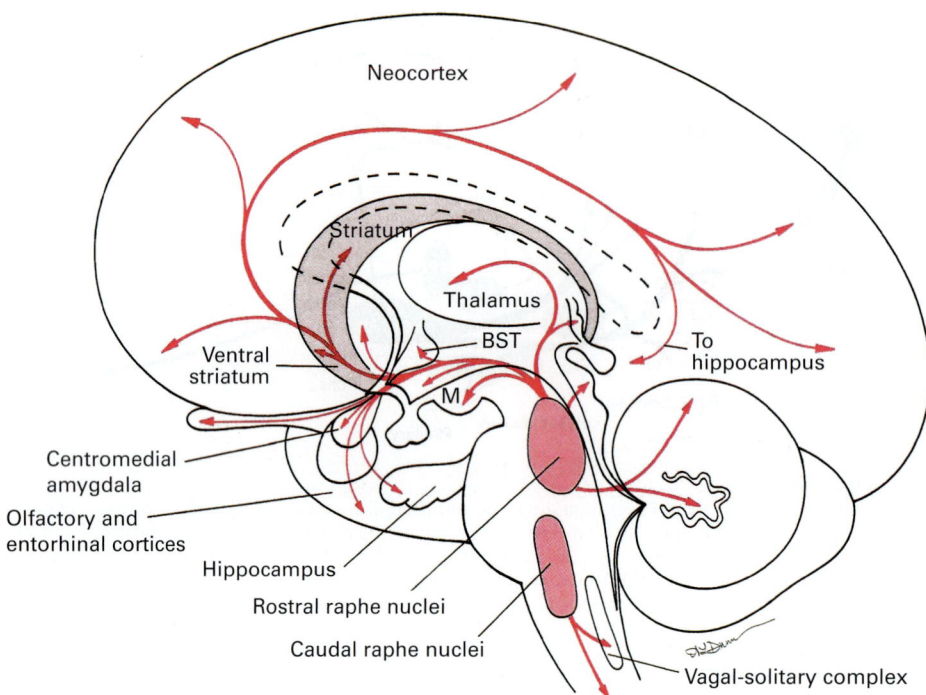

Figure 17.4
Schematic portrayal of the widespread distribution of serotonin-containing axons in the human brain. The serotonin axons that project to the spinal cord come not only from the caudal raphe nuclei of the hindbrain but also from more laterally placed cell groups and are more specific in their termination patterns than is the case for serotonin axon projections in the brain. The *rostral raphe nuclei* include the dorsal raphe and the median raphe nuclei. The *caudal raphe nuclei* include the nucleus raphe magnus, the nucleus raphe obscurus, and the ventrolateral medullary nuclei, each with a different pattern of projections to the spinal cord. BST, bed nucleus of the stria terminalis; M, mammillary body. From Heimer (1995).

superficial layers of the dorsal horn. This pathway modulates the input of pain stimuli. The nucleus raphe obscurus projects to the ventral horn (connecting with motor neurons), and a ventrolateral hindbrain group of serotonin cells projects to the lateral horn (preganglionic motor neurons of the sympathetic system).

Diencephalic Origins of Other Widely Projecting Axon Systems

In recent years, additional systems that influence behavioral state (e.g., mood, attentiveness, wariness, etc.) have been discovered to originate in diencephalic cell groups. In the mid-twentieth century, after the ascending reticular activating system became a popular concept, the widespread projections of the older parts of the thalamus—the midline and intralaminar nuclei—became part of the story. One cell group, the ventromedial thalamic nucleus, was found to have projections to layer 1 of nearly the entire cortical mantle.

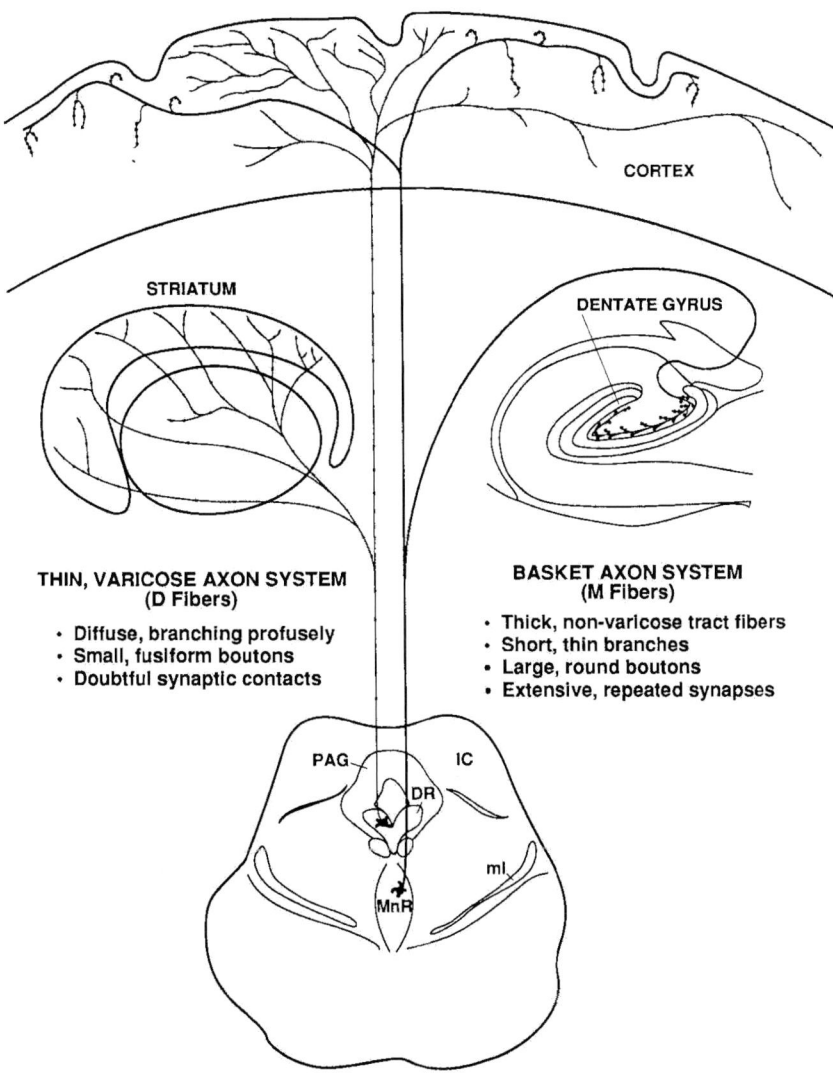

Figure 17.5
Serotonin axon distribution in the neocortex consists of two distinct axon types. One of these types also projects to the corpus striatum, and the other type projects to the dentate gyrus of the hippocampal formation. DR, dorsal raphé nucleus; IC, inferior colliculus; ml, medial lemniscus; MnR, median raphé nucleus; PAG, periacqueductal gray. From Törk (1990).

It has long been known that the hypothalamus can cause profound changes in behavioral state in relation to motivations and to emotional and arousal levels. In these effects, hypothalamic structures function as a forward extension of the midbrain central gray and ventral tegmental areas. Similarly the dorsally adjacent subthalamus is a forward extension of the midbrain reticular formation from the standpoint of both function and connections.

A group of cells in the lateral hypothalamus that has widespread projections has been discovered to use two peptides, hypocretin (orexin) and dynorphin, as neurotransmitters. This cell group must play a profound role in control of sleep states, judging from the finding that mutations in the gene for hypocretin/orexin or its receptor cause narcolepsy. Figure 17.6 shows one chart from a study of the overlap of hypocretin/orexin axons with norepinephrine axons.

Another hypothalamic cell group with very widespread axonal projections, one that appears to be involved in maintaining alertness, surrounds the mammillary nuclei. It is called the tuberomammillary nucleus. These cells are the only ones in the brain known to synthesize histamine. In addition to using histamine as a neurotransmitter, the axons of these cells also use GABA.

There is evidence for additional groups of widely projecting neurons in the lateral hypothalamus. One of them uses melanin concentrating hormone as one neurotransmitter. Another group uses corticotropin-releasing hormone.

How Many Different Brain States?

Consider for a few minutes the different states of the brain of a complex mammal like a human being. One moment, a person is feeling drowsy and she or he feels like doing nothing. Then the telephone rings, and someone announces that it is the person's boyfriend or girlfriend. This is an anxiously awaited call, and suddenly the person springs to life, drowsiness gone, motivational levels at high pitch. Another, equally profound change of state occurs if someone yells "Fire!" It is not difficult to imagine other striking examples of state change. Some of these changes can be taken as clear examples of "mind over body." One minute a teenager informs her mother that she is too tired to attend a martial arts class, and she settles down in front of the television with an extra helping of dessert. The next minute, after she has seen a much-admired young actress promoting physical fitness and diet, the teenager has forgotten about the food and is performing strenuous exercises on the living room floor. She no longer feels either hungry or tired. She was not lying about how she felt. Her brain state has changed.

How many such different states of the brain might be possible? We can think about that if we make a few simplifying assumptions, as follows. Assume that there are four systems of the brainstem that can alter the overall behavioral state, and that each of these systems—using acetylcholine, dopamine, serotonin, or norepinephrine as

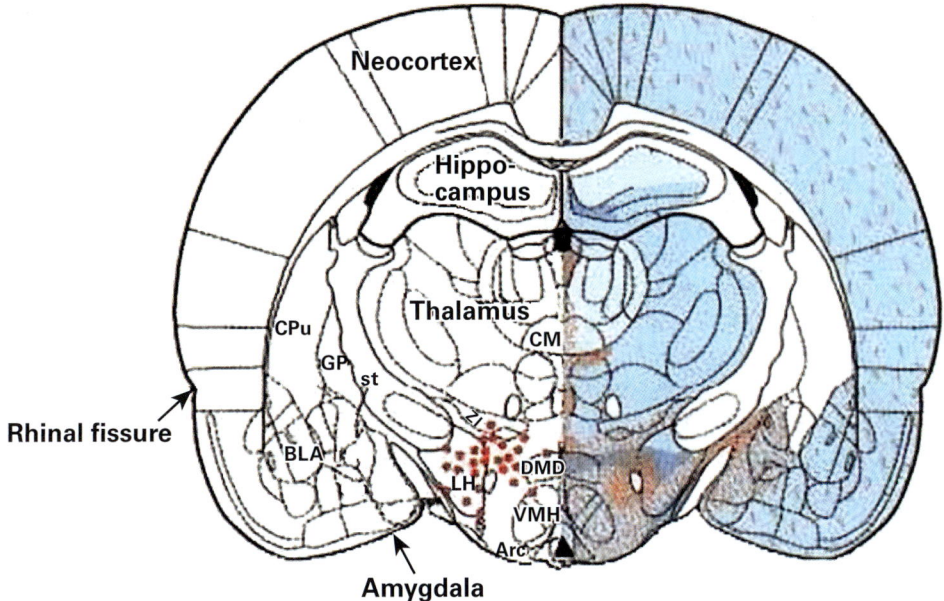

Figure 17.6
Chart of a single coronal section of the thalamus of a rat illustrating the widespread distribution of norepinephrine axons (blue color, indicated only on the right side) throughout the section except for striatal areas, and the more limited but overlapping distribution of orexin/hypocretin axons (red dots). On the left side, the larger red dots indicate positions of the cells, in the hypothalamus, containing orexin/hypocretin. Arc, arcuate nucleus; BLA, basolateral amygdaloid nucleus, anterior part; CM, central medial thalamic nucleus; CPu, caudate-putamen (striatum); DMD, dorsomedial nucleus of hypothalamus, dorsal part; GP, globus pallidus; LH, lateral hypothalamus; st, stria terminalis; VMH, ventromedial nucleus of hypothalamus; ZI, zona incerta of subthalamus. From Baldo, Daniel, Berridge, and Kelley (2003).

neurotransmitters—has three possible states (e.g., no activity, low level of activity, high level of activity) with noticeably different behavioral correlates. Assume further that there are four diencephalic state-changing systems, discussed earlier, each having three possible states. Assume that these eight systems can each operate independently and that they are not identical in their effects. (Just how independently they operate is unclear; we know that there are interconnections among these systems.) We simplify here by omitting the paleothalamic cell groups, including the neurons that project to the entire neocortex, as their activity may be dependent on inputs from the acetylcholine axons from the brainstem.

Now we can calculate the number of possible states:

$3^8 = 6,561$.

If the eight systems each have four possible states:

$4^8 = 65,536$.

In this small exercise, we have omitted a number of distinct cell groups of the rostral hindbrain, midbrain, and hypothalamus that also may be involved in controlling behavioral states. Larry Swanson, in the reading listed at the end of chapter 10, lists 15 different cell groups, including many of the ones mentioned above, involved in this kind of function.

This is a bare-bones introduction to brain mechanisms that can change the state of much or all of the brain of mammals. They were not all present in the earliest chordates. Investigations of amphioxus have shown serotonin and dopamine neurons but no norepinephrine system in the brain. Hagfish and lampreys have norepinephrine as well as the other two monoamines.

We should not conclude that these primitive animals have proportionately fewer diffuse axonal systems. The "primitive state" of many axons was that of being widely branched and having extensive distributions. In evolution, specific axon systems have become specialized for this kind of connection, or it may be more accurate to say that such axon systems have been retained while many more specialized neuronal systems have evolved.

Readings

The origins and distributions of widely projecting axonal systems are summarized in many neuroanatomy texts, including books listed in earlier chapters.

For an example of studies finding evidence for the role of brainstem neuronal systems using acetylcholine as the neurotransmitter in behavioral state control, see:

Steriade, M., Datta, S., Paré, D., Oakson, G., & Curró Dossi, R. (1990). Neuronal activities in brain-stem cholinergic nuclei related to tonic activation processes in thalamo-cortical systems. *Journal of Neuroscience, 10*, 2541–2559.

See also the book by Larry Swanson cited in chapter 10, in his chapter on "the behavioral state system." See especially his figure 9.5 in the second edition of his book.

Serotonin system functions, including effects of serotonin reuptake inhibitors like the drug Prozac, are discussed in the book by John Allman cited in chapter 3 (pp. 19–27). For physiological studies of serotonergic neurons in animals, see:

Jacobs, B. L., & Fornal, C. A. (1999). Activity of serotonergic neurons in behaving animals. *Neuropsychopharmacology, 21*, 9S–15S.

Extremely widespread projections to the neocortex from the ventromedial nucleus (one of the intralaminar and midline nuclei of the thalamus) was found by Miles Herkenham and Walle Nauta and later reviewed by Herkenham:

Herkenham, M. (1980). Laminar organization of thalamic projections to the rat neocortex. *Science, 207*, 532–535.

For reviews and theories of the function of the norepinephrine system of axons originating in the locus coeruleus of the hindbrain, see the following articles:

Berridge, C. W., & Waterhouse, B. D. (2003). The locus coeruleus—noradrenergic system: modulation of behavioral state and state-dependent cognitive processes. *Brain Research Reviews, 42,* 33–84.

Bouret, S., & Sara, S. J. (2005). Network reset: A simplified overarching theory of locus coeruleus noradrenaline function. *Trends in Neurosciences, 28,* 574–582.

Antidepressant drugs alter widely projecting neuronal systems. Problems this can cause have been discussed:

Moncrieff, J., & Cohen, D. (2006). Do antidepressants cure or create abnormal brain states? *PLOS Medicine, 3,* 961–965.

Evidence that neurons in somatosensory neocortex may process sensory inputs differently in different brain states as defined by different patterns of neuronal synchrony suggests that such states correspond to "different modes of cortical processing":

Poulet, J. F. A., & Petersen, C. C. H. (2008). Internal brain state regulates membrane potential synchrony in barrel cortex of behaving mice. *Nature, 454,* 881–885.

Some comparative studies in this area can also be found. For example, see the following review:

Yamamoto, K., & Vernier, P. (2011). The evolution of dopamine systems in chordates. *Frontiers in Neuroanatomy, 5* (Article 21), 1–21.

Modern experiments on arousal systems ascending from the brainstem:

Fuller, P., Sherman, D., Pedersen, N. P., Saper, C. B., & Lu, J. (2011). Reassessment of the structural basis of the ascending arousal system. *Journal of Comparative Neurology, 519,* 933–956.

VIII SENSORY SYSTEMS

In the previous chapters, we have been discussing sensory systems with regard to their major roles in the evolution of the central nervous system. Except for some aspects of somatosensory systems and the roles of distance senses (visual, olfactory, auditory) in brain evolution, details of sensory systems have not been the focus thus far in this study of brain structure and its origins. In part VIII, we change focus in order to consider the special sensory systems that have evolved at the head end of the chordates: gustatory, olfactory, visual, and auditory systems.

The somatosensory pathways have already been discussed, beginning with the study of the spinal cord with its somatosensory inputs entering mainly through the dorsal roots of the spinal nerves—see chapters 5 and 9, especially on the spinoreticular and spinothalamic tracts. In the overview of the forebrain in chapter 7, an important somatosensory pathway from spinal levels to forebrain was introduced—the dorsal column to medial lemniscus pathway, which expanded in a major way in the mammals. In chapter 10, we reviewed the somatosensory inputs from the head area into the hindbrain through the trigeminal nerve (fifth cranial nerve). The rostral projections of the trigeminal nuclei are major additions to the somatosensory pathways that pass rostrally from the spinal cord.

Now we consider the gustatory system before discussing the distance senses.

18 Taste

In this and the following chapter, we study the structure of the chemical senses as they are represented in the brain, with some notes on their evolution. We will see how new neurons are generated not only during the developmental period but also in the adult mammalian olfactory bulb, with the precursor cells migrating from the ventricular layer of the telencephalic hemispheres. Studies of the development of the olfactory tract, which carries olfactory information from the bulb to the olfactory cortex, have revealed some fascinating phenomena of anatomical plasticity that are relevant to an understanding of plasticity elsewhere in the CNS. We will also consider the major issue of the purposes of taste and olfaction as we outline some routes whereby these systems reach the motor system.

Both the taste and the olfactory systems show considerable phylogenetic variation, providing good examples of *mosaic evolution*. In chapter 4, we used examples of specializations in the taste organs of fish to illustrate this in speculating on early expansions of the hindbrain portion of the neural tube. The examples were of dramatic enlargements of secondary sensory neuron groups in the rostral part of the visceral sensory cell column (nucleus gustatorius, part of the hindbrain's nucleus of the solitary tract) as part of the specializations for feeding of certain fish (see figures 4.3, 4.4, 4.5, and 4.6). The olfactory system also provides examples of great variations in size, especially in the contrast between *macrosmatic* and *microsmatic* mammals.

In reviewing the olfactory system, we will include evidence that supports speculations by a number of comparative neuroanatomists on evolution of the CNS and the limbic system. It is believed that the functions of chemical senses have been important drivers of brain evolution. It is important to note here that in addition to the more familiar taste and olfactory systems, there are other chemosensory systems in vertebrates (see next section).

Pre-chordate Taste and Other Chemoreceptor Systems

Chemoreception is found to be nearly universal in the feeding mechanisms of animals according to studies of a great many species. It occurs in protists and in primitive multicellular animals like Porifera and Cnidaria. Thus, sponges have cells with chemoreceptor functions located around the pores in their body surface. Hydra have a peri-oral neural network with chemoreceptors that can be considered to be taste receptors.

Vertebrates have many nontaste chemoreceptors with innervation by the autonomic nervous system. There is detection of CO_2 levels in the blood by cells of the carotid body, activation of which causes a reflex increase in oxygen intake. Glucose levels in the blood are detected by glucose receptors in the hypothalamus. There is an ability to detect toxins in the blood also, by receptors in the hindbrain located in the area postrema, near the obex. These are exposed to circulating molecules that pass the blood–brain barrier in this particular region. Detection of toxins results in the triggering of a vomiting reflex. (For another example of this kind of penetration of the blood–brain barrier, see figure 26.4).

Vertebrates also have "solitary chemosensory cells" (connected to primary sensory neurons) that in some species play a role in predator avoidance or have taste functions in feeding. In addition, many free nerve endings have some chemoreceptive responses, constituting the "common chemical sense."

Olfaction or Taste?

Olfaction and taste appear to have evolved in aquatic vertebrates for detection of different classes of chemicals. Both have important roles in feeding and social behavior, not only in vertebrates but in many other animals as well, including the insects. The finding that taste buds are not found in hagfish but are found in all jawed vertebrates has been suggested to indicate that "the phylogenetic development of taste buds coincided with the elaboration of head structures at the craniate-vertebrate transition." (See the reading: Finger, 1997.)

It is well known that the complexity of taste perceptions cannot be attributed to the gustatory inputs alone. This is due mostly to the contributions of olfaction. For a long time, many descriptions of taste responsiveness in humans have included only four types: sweet, salty, sour, and bitter, with different localizations on the tongue and other parts of the oral cavities. However, there is strong evidence for another basic taste, umami (savoriness). Tastes experienced by animals could very well be different, considering the findings of a large number of different receptor molecules as well as a wide distribution of receptor locations.

Visceral and Taste Inputs to the Hindbrain

The secondary sensory column of neurons in the hindbrain known as the nucleus of the solitary tract is also known as the visceral sensory column. Its inputs come from axons of primary sensory neurons that are found in cranial nerves VII, IX, and X. The caudal portion of the nucleus of the solitary tract receives visceral sensory inputs from the body via the vagus nerve (cranial nerve X). The rostral portion, about a third of the nucleus in many mammals, receives taste inputs from the tongue via the seventh cranial nerve (facial nerve). Taste input also reaches the nucleus of the solitary tract from the throat via the ninth and tenth cranial nerves (glossopharyngeal and vagus nerves).

Variations in this pattern occur, especially in species with well-developed specializations in taste organs for feeding. Remember the brain of the freshwater buffalofish shown in chapter 4 (figure 4.4), with the relatively enormous vagal lobe. The enlargement is in the visceral sensory column of secondary sensory neurons receiving and processing taste input carried by the vagus nerve (tenth cranial nerve) from a specialized palatal organ for filtering water and trapping food particles. The catfish has taste specializations of a different sort (see figures 4.5 and 4.6); the taste receptor distribution over the body surface including the whisker-like barbels remind one of the chemoreceptors in insects like those that have been found on the feet of flies.

Innervation of the Tongue

We think of the tongue as the organ of taste. It is much more than that in humans and other mammals, as can be understood from its three distinct types of innervation by four cranial nerves. Very acute touch sensitivity of the tongue results from its dense innervation by the trigeminal nerve (cranial nerve V). This nerve does not innervate the specialized gustatory receptor cells of the taste buds. In mammals, the taste buds located from the tip of the tongue back to a row of papillae called the circumvallate papillae are innervated by the facial nerve (cranial nerve VII). Behind this distal/anterior region of the tongue are more taste buds, innervated by the glossopharyngeal nerve (cranial nerve IX). The exquisite movements of the tongue are controlled through the hypoglossal nerve (cranial nerve XII).

Distribution of Mammalian Taste Receptors

The distribution of taste receptors can change with age in some species, including humans. Young children may taste things differently than adults, at least in part because of taste receptors located in the throat innervated by the tenth cranial nerve. It has been reported that these disappear with maturity.

Because of the importance of taste in the control of food intake, it is not surprising that taste buds are distributed not only over the tongue surface but also throughout the oral cavity in many animals. In figure 18.1, we see the distribution of taste buds in the Syrian hamster's mouth and throat, shown in a drawing of a parasagittal section through the head.

328   Chapter 18

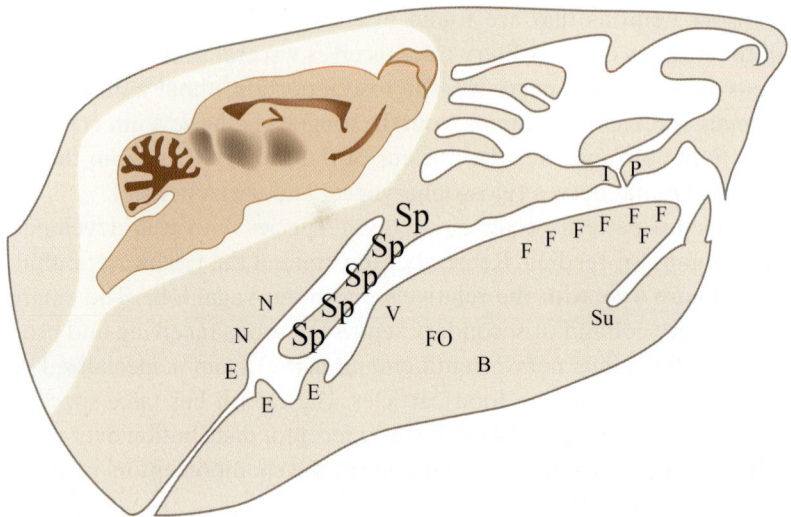

Figure 18.1
Hamster taste bud distribution illustrated on a drawing of a parasagittal section of the head. The position and relative size of the brain is shown with an appropriate placement of a parasagittal section of the brain. The locations of various taste papillae, many of which contain taste buds, and parts of the mouth are indicated by the abbreviations. B, buccal wall; E, epiglottis; F, fungiform papillae; FO, foliate papillae; I, incisive papillae; N, nasopharynx; Sp, soft palate; Su, sublingual organ; V, vallate papillae. Based on Zigmond, Bloom, Lands, Roberts, and Squire (1999).

From Tongue to Telencephalon

Figure 18.2 depicts axonal pathways for taste impulses in the rat, starting in the gustatory nucleus of the hindbrain, emphasizing ascending pathways. Figure 18.3 shows the same thing in a network diagram. A pathway from secondary sensory neurons (in the nucleus of the solitary tract) to the ventral tegmental area (VTA), the midbrain's reward center, has been added. This is a finding not often depicted in such diagrams. Most pathways reaching the VTA cells are less direct, but it is important to show this pathway here (although it is not a large pathway) because of the important role of taste in the rewards that are needed for habit formation. Thus, it can be considered to be of importance in evolution.

There are projections to the thalamus that go to *paleothalamic* cell groups (midline and intralaminar neurons), which project to corpus striatum and also in a fairly diffuse manner to neocortex. This is important in the context of evolution because of the primitive existence of the paleothalamic system and its projections to corpus striatum. This route to the striatum has become vastly overshadowed in mammals by the great expansion of the pallium in the form of neocortex, together with the dorsal thalamic cell groups that project to it. This cortex, then, evolved to become the dominant source of sensory input to the striatum.

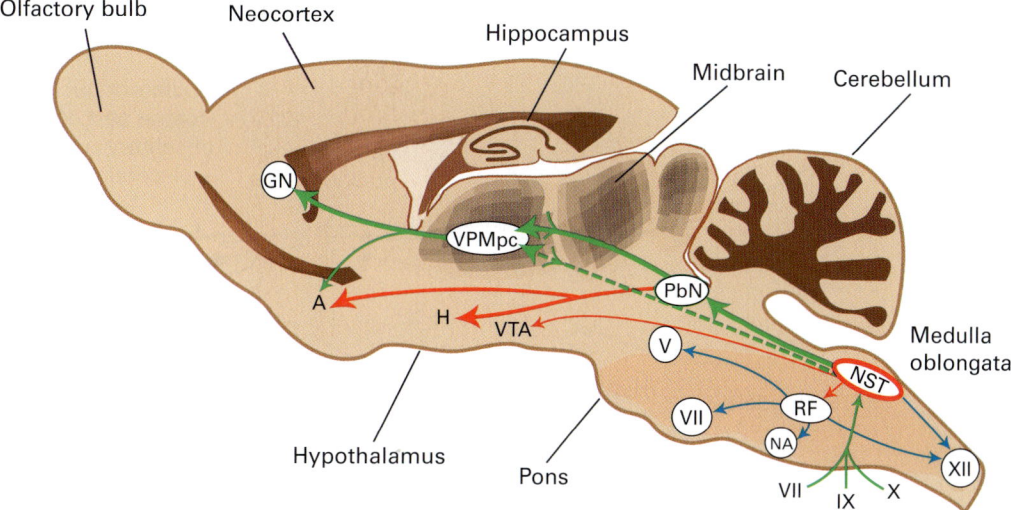

Figure 18.2
Ascending gustatory pathways are illustrated diagrammatically on this parasagittal section of a rodent brain. The pathway shown by the dashed line has been found in monkey and rat studies. A, amygdala (in temporal lobe) and bed nucleus of stria terminalis (in basal forebrain); GN, gustatory neocortex; H, hypothalamus; NA, nucleus ambiguus; NST, nucleus of the solitary tract; PbN, parabrachial nucleus of the rostral hindbrain—the pontine taste area; RF, reticular formation; VPMpc, ventral posterior nucleus, medial parvocellular part; VTA, ventral tegmental area. Roman numerals represent cranial nerves and nuclei. Based on Zigmond, Bloom, Lands, Roberts, and Squire (1999).

The major ascending route for taste impulses from the nucleus of the solitary tract of the hindbrain passes rostrally with many of the axons synapsing in the parabrachial nucleus of the rostral hindbrain in many mammals. The taste-receptive portion of this nucleus is often called the pontine taste area. Passing through the midbrain, the pathway reaches the dorsal thalamus (ventral posterior nucleus, medial parvocellular part). The parabrachial nucleus (one on each side) is located adjacent to the *brachium conjunctivum*, which consists of the axons that go from cerebellum to thalamus. There is a resemblance between the taste pathways and the auditory pathways, with a synaptic interruption between the secondary sensory neurons and the thalamus. Both of these pathways also have a more direct component, especially in primates, bypassing the parabrachial nucleus (see the dashed green line in figure 18.2). In monkeys, apes, and humans, the parabrachial nucleus may have little or no function in gustation.

Note also the pathways to limbic forebrain structures that ascend without a synapse in the dorsal thalamus. These go to the hypothalamus and to the amygdala. It is these pathways rather than the route through the thalamus that are most important in the rewarding effects of taste inputs during feeding in the rat.

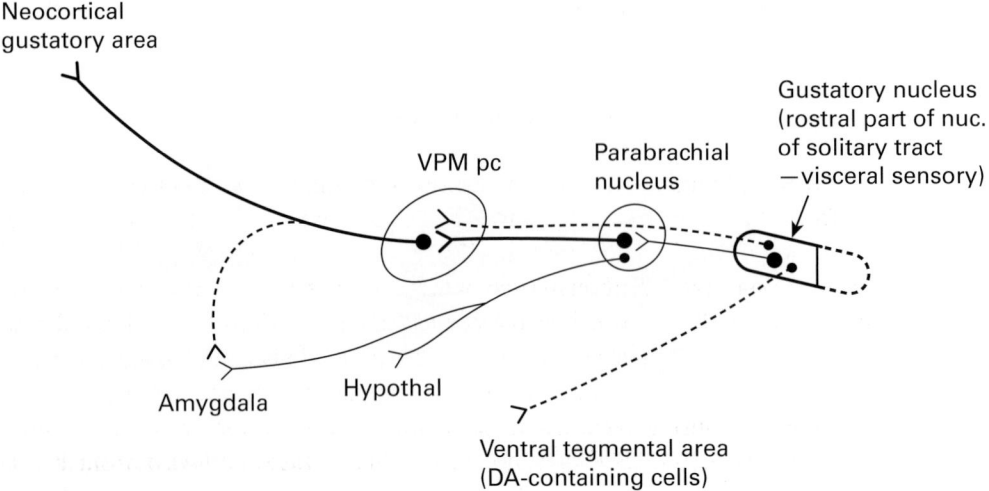

Figure 18.3
Diagram of mammalian taste pathways, beginning with the secondary sensory neurons of the gustatory nucleus of the hindbrain, which is the rostral part of the nucleus of the solitary tract—the visceral sensory cell column. VPMpc, ventral posterior nucleus of the thalamus, medial parvocellular part. Projections to the thalamus outside VPMpc go to paleothalamic cell groups, which project to the corpus striatum and also diffusely to the neocortex. (The latter pathways are omitted from the figure.) DA, dopamine; Hypothal, hypothalamus; nuc., nucleus.

Is taste a decussating pathway? Some textbooks are not clear about this issue. What would one expect? There is no obvious need for lateralization (e.g., for localization in orienting), unless taste receptors are distributed over the body surface so that taste inputs could be localized. In evolution, the taste system is likely to have preceded the development of a visual imaging system that could be used for localizing stimulus sources in space, so it is likely that the familiar decussations of ascending pathways had not yet evolved (see chapters 4, 10, and 20). Ascending pathways must have originally been bilateral with a larger ipsilateral component. The available data, based on the experimental tracing of axons from the nucleus of the solitary tract, which includes the gustatory nucleus, and from the parabrachial nucleus, indicate bilateral projections.

Purposes of Taste: Routes to Motor Control

The purposes of taste may seem obvious, centering on the selection of foods that are good for the body and the rejection of foods that are not good, but there is more to be considered. To begin with, there are taste-triggered reflex pathways within the hindbrain that reach the salivatory nuclei, which are preganglionic motor neurons of the parasympathetic nervous system (chapter 9). These nuclei are part of the visceral motor column (figures 10.9 and 10.11), with their axons leaving the brain through the facial nerve (VII). Salivation is important not only for swallowing but also in the initial stages of digestion. Other hindbrain pathways from the gustatory nucleus end in the hindbrain reticular formation and in nucleus ambiguus; these pathways may play a role in the preparation for or modulation of swallowing and in rejection of some food by innate responses.

We are all familiar with the pleasures of eating and with both selection and rejection of foods by taste, so it is not surprising that there are pathways to the limbic midbrain areas and to the hypothalamus itself—important for reward, punishment, and control of motivational states. Although such pathways may be mostly less direct (e.g., via the amygdala in the endbrain, which also receives taste projections from the parabrachial nucleus in the rat), there appear to be some direct projections from the gustatory nucleus to the ventral tegmental area, a major source of dopamine pathways to the forebrain known to be activated during pleasurable sensations. Pathways that reach the central gray area probably exist, but they may not be so direct. Taste pathways from the parabrachial nucleus reach not only the thalamus but also the hypothalamus. The latter pathway has been found to be particularly pronounced in various fish species.

These pathways that signal rewarding or punishing consequences of actions, especially of approach and intake of food, must have appeared very early in chordate evolution, as they were so critical for the formation of adaptive habits. These pathways, at such early times, must have reached the forerunners of the ventral striatal region, which from very early times also received olfactory inputs.

Taste is also important for functions other than feeding (e.g., in the shaping of feelings associated with objects and with other members of the same species). Underlying these affective responses are pathways from the parabrachial nuclei directly to the amygdala and to the hypothalamus. From the hypothalamus, approach and avoidance locomotor control can be effected via the hypothalamic locomotor area and thence to the midbrain locomotor area and midbrain extrapyramidal area.

With the evolution of mammals and the taste pathway to a gustatory neocortical area (figures 18.2 and 18.3), taste became important in object perception and hence in the cognitive functions of anticipation and planning. These functions have come to depend largely on transcortical pathways (see part XII). From gustatory cortex, there are also routes to the corpus striatum that could shape habit learning, passing from striatum to the orienting and locomotor mechanisms of the midbrain and to motor cortex via the thalamus (see chapters 29 and 30).

Readings

Dethier, V. G. (1962). *To know a fly*. San Francisco: Holden-Day, Inc. (Describes taste reception through the feet of flies, among many other observations. Many more recent studies of fly taste systems can be found by searches using Google Scholar.)

Finger, T. E. (1997). Evolution of taste and solitary chemoreceptor cell systems. *Brain, Behavior and Evolution, 50*, 234–243.

Glanville, E. V., Kaplan, A. R., & Fischer, R. (1964). Age, sex, and taste sensitivity. *Journal of Gerontology, 19*, 474–478.

Hajnal, A., & Norgren, R. (2005). Taste pathways that mediate accumbens dopamine release by sapid sucrose. *Physiology & Behavior, 84*, 363–369.

Small, D. M. (2010). Taste representation in the human insula. *Brain Structure & Function, 214*, 551–561.

19 Olfaction

The structures that very likely shaped the early evolution of the forebrain, if we judge from what can be seen in the most primitive living chordates, were the secretory region of the hypothalamus and the neurons responding to light and to olfactory inputs. It was the olfactory inputs that dominated the rostral portion, the endbrain. Here we will focus first on olfactory system pathways in primitive vertebrates and then in mammals. You will see that the nose brain (rhinencephalon) and associated structures have been pushed to the margins (the limbus, or fringe) of the cortex in higher animals. This fact led to the designation of these along with other structures closely connected to the hypothalamus as the limbic system (see part X). After an introduction to pathways of the olfactory system, we will illustrate briefly the nature of neural coding for odors at initial stages of the pathway. Also, we will see why development and plasticity of this system have been of special interest to neuroscientists.

Sections through the Forebrain of Vertebrates

Neuroanatomists commonly make use of frontal sections in their studies. This plane of section is also called transverse or coronal. In figure 19.1, we see simplified drawings of frontal sections through the forebrain of four species: an amphibian, a reptile, and two mammals—opossum and human. First, note that each section, except for that of a human, includes both 'tweenbrain and endbrain. The cerebral hemispheres of the endbrain, behind the olfactory bulbs, are a lateral outpouching, or ballooning out, of the embryonic forebrain vesicle, and in many species these endbrain structures protrude not only to the side and forward, but they are also folded back over the 'tweenbrain. In mammalian species where they have become very large, the hemispheres have come to lie back over the midbrain and part of the hindbrain as well.

Next, in each brain section, note the position of the piriform cortex. This is the cortical region, distinct from the neocortex, where axons originating in the olfactory bulbs terminate. *Piriform* means "pear shaped." When viewed from the ventral side in many animals, the olfactory cortex is pear shaped, as illustrated in figure 19.2. In mammals, a fissure or

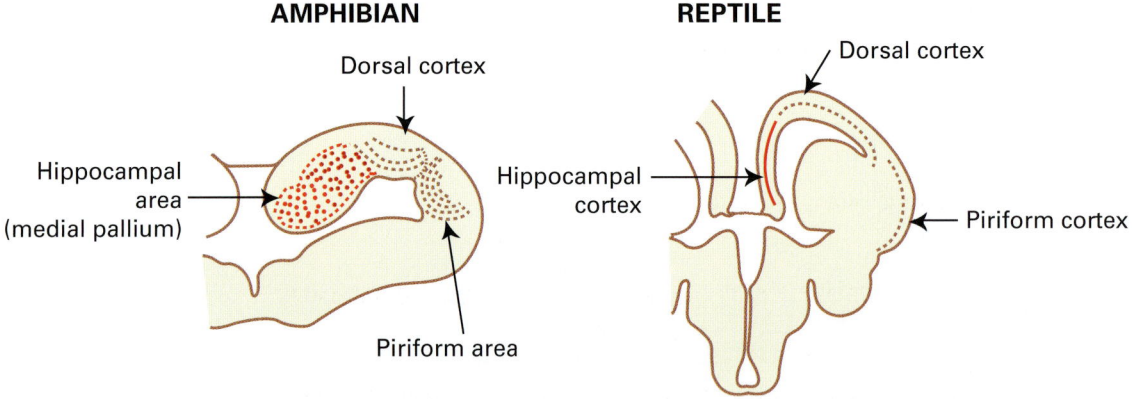

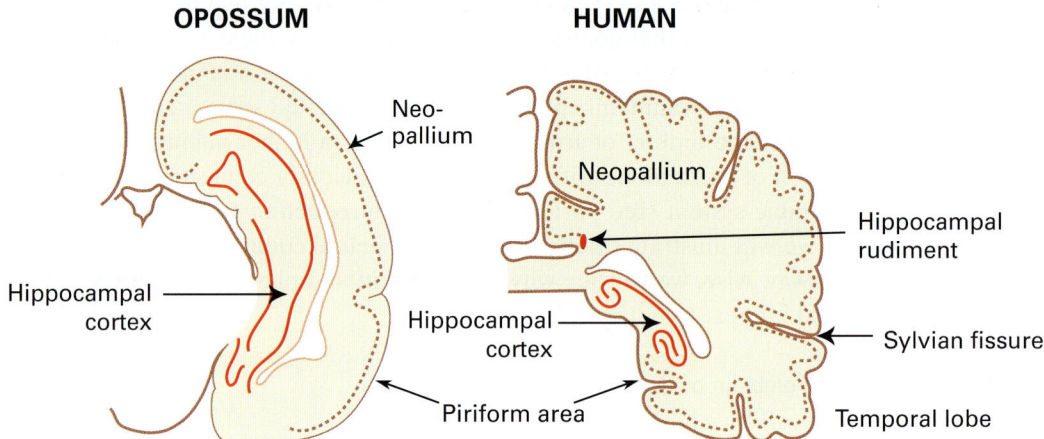

Figure 19.1
Simplified drawings of frontal sections selected from the brain of an amphibian, a reptile, and two mammals (opossum and human). The amphibian's medial pallium corresponds to the hippocampal formation of mammals. Lateral to a small dorsal pallium (dorsal area) is the piriform area, which receives olfactory bulb projections. In the reptile, the same three pallial regions are seen, but the medial and dorsal pallial areas look more cortical in their cellular structure. Again, the lateral pallium is the piriform cortex, recipient of direct projections of the olfactory bulb. In the mammals, the dorsal cortex has expanded to include neocortex, which has expanded the most in humans. The medial pallium is the hippocampal cortex, which has enlarged caudally, and in humans temporally, with only a small rudiment located in a medial position more rostrally. In relative terms, the piriform cortex is still in the same position at the ventrolateral margins of the hemisphere. Based on Herrick (1933).

Olfaction

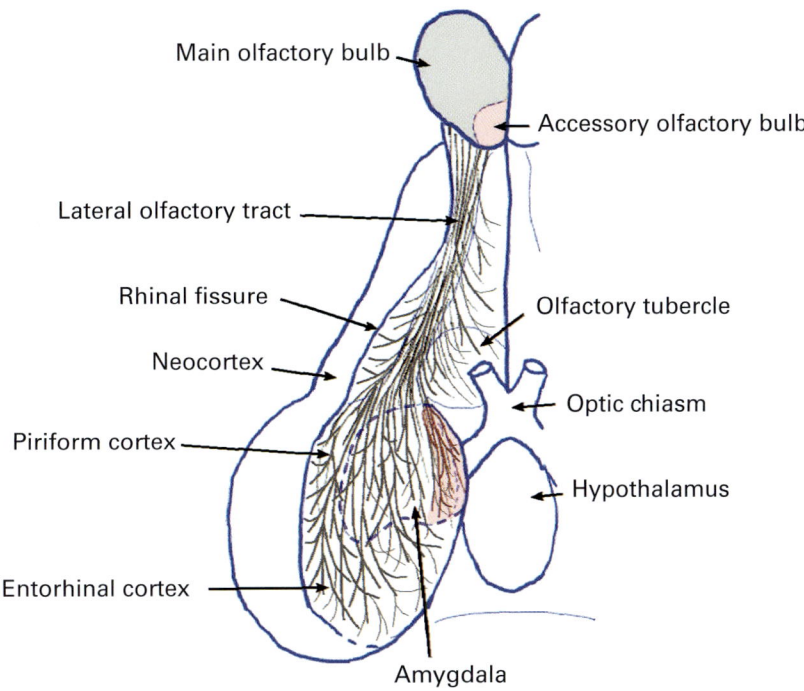

Figure 19.2
Ventral view of the brain of a small mammal showing projections of the olfactory bulb. The brain's shape is that of a rabbit. Projections based on experimental fiber tracing studies of hamster and rat as well as rabbit. The distinct projection of the accessory olfactory bulb, which receives input from the vomeronasal organ, is indicated by the pink color of the sites of origin and termination. The terminals are in the cortical and medial nuclei of the amygdala.

at least a shallow sulcus separates the piriform cortex from the adjacent pallium—the neopallium or neocortex. In reptiles and amphibians, cytoarchitectural features distinguish the piriform cortex from the precursor of neocortex known as the dorsal cortex or dorsal area (figure 19.1). In these non-mammalian species, the more medial region of the hemisphere, the medial pallium, is the equivalent of the mammalian hippocampal formation. At least some of the hippocampal region receives direct projections from the olfactory bulb in these species, whereas in mammals there are very few such projections.

Relating the medial pallium of reptiles and amphibians to the hippocampus of mammals in frontal sections can be a little confusing. Even relating the hippocampus in different mammals can be puzzling. One must think of the three-dimensional hemisphere and realize that the hippocampus is at the margin of the neocortex, where it is an infolded stretch of thinner, more primitive cortex. Because of the great expansions of the neocortex, this margin is pushed not just medially but to the caudal limits of the hemispheric surface, sometimes curving around to the ventral surface as well. We will return to this in

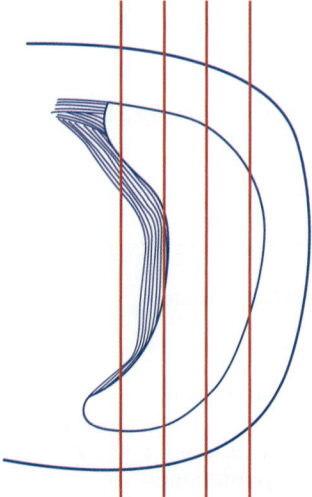

Figure 19.3
A lateral-view sketch of the hippocampus as it would appear in the caudal cerebral hemisphere of a rat, hamster, or mouse, if it were visible through the overlying neocortex. The vertical lines show the positions of typical coronal (frontal) sections. More rostral sections that include the hippocampus cut through it twice, both dorsally and ventrally. More caudally it is seen centrally in the section with cortex both above and below it. The latter kind of section is illustrated for the opossum brain in figure 19.1.

parts X and XII. At this point, just consider what happens to a curved structure when it is sectioned: Think of a standing banana that is sectioned as in figure 19.3, which, at a first approximation, has the shape of the hippocampus in an opossum or a rodent. This can help you understand what happens to the hippocampal banana shape in frontal sections. (As mentioned in chapter 7, the term *hippocampus* comes from the Latin for "seahorse," named for its appearance in dissections of the human brain.)

Olfactory Bulb Projections in Primitive Vertebrates

As mentioned in chapter 5, neuroscientists have had a special interest in two groups that appear to represent very primitive vertebrates. These are the sea lampreys and the hagfishes (jawless vertebrates). It was because of the evidence that these two groups resemble some of the earliest vertebrates that studies of their olfactory bulb projections were undertaken.

You may be wondering why we should care about where the olfactory bulbs project in such animals. The studies are important because of an idea about evolution that was being tested. It was a hypothesis that can be pictured as in the sketches shown in figure 19.4. The first sketch represents an olfaction-dominated endbrain in early chordate evolution. The next one depicts the postulated invasion of the endbrain by non-olfactory sensory

Olfaction

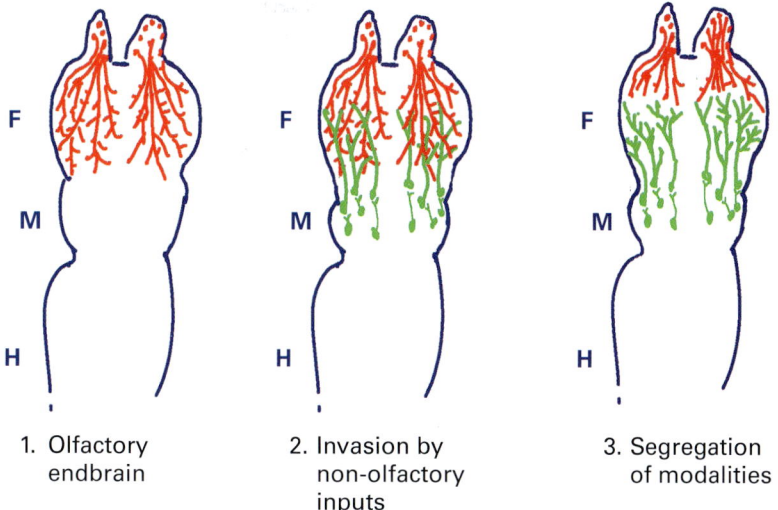

1. Olfactory endbrain
2. Invasion by non-olfactory inputs
3. Segregation of modalities

Figure 19.4
A sketch of the hypothesis concerning forebrain evolution presented in the text. (1) In very early chordates (now extinct), a forebrain began to enlarge as olfactory inputs developed. (2) With an invasion of the forebrain by non-olfactory projections (from the midbrain or after a synapse in the caudal forebrain), these projections overlapped with the olfactory projections. (3) As the non-olfactory inputs increased in density, they segregated from the olfactory projections. Similarly, axons of different modalities eventually segregated from each other. F, forebrain; H, hindbrain; M, midbrain.

systems. The axons are shown overlapping with the olfactory projections. The last sketch depicts a later segregation of modalities.

However, there was no strong evidence of any species with projections of the type shown in the left-hand sketches before axon-tracing studies were undertaken of olfactory bulb projections in the lamprey and hagfish (see box 4.1). Figure 19.5 illustrates results of a study of the sea lamprey endbrain in which projections of the olfactory bulb and of the thalamus were traced separately. The results indicate a situation like that in the middle sketch of figure 19.4. The same year this study was published (1993), an experimental study of olfactory bulb projections in the brain of the Pacific hagfish was published. Figure 19.6 shows a concluding illustration from that study: A cladogram of the two jawless vertebrates we have been discussing plus an amphibian is shown below charts of olfactory bulb projections to the forebrain in these three species. Thus, the idea of an olfaction-dominated endbrain invaded by non-olfactory sensory systems has found some support in these studies of living animals.

It is important to point out that these findings on overlap of thalamic and olfactory projections in the endbrain of lamprey and hagfish do not indicate anything about the origins of neocortex. It is not clear that either of these species has a dorsal cortex, which genetic studies have indicated to be a precursor of neocortex. Lamprey brains apparently

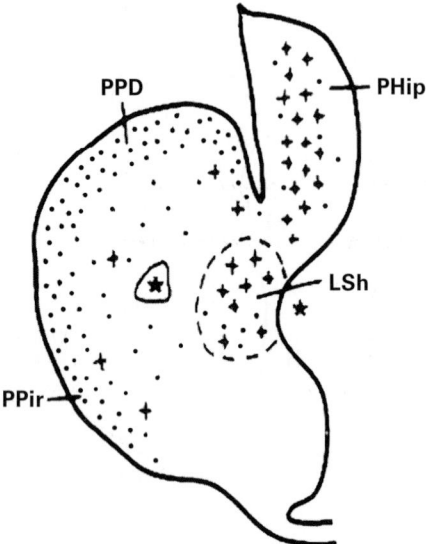

Figure 19.5
A drawing of a frontal hemisection of a sea lamprey endbrain, showing results of studies of olfactory bulb projections (small dots) and of projections from the dorsal thalamus carrying non-olfactory information to the endbrain (+). The non-olfactory inputs were found to overlap with the distribution of the axons of secondary sensory neurons of the olfactory bulbs. (Stars mark the forebrain ventricles.) LSh, lobus subhippocampalis; PHip, primordium hippocampi; PPD, primordium pallii dorsalis; PPir, primordium piriformi. Based on Polenova and Vesselkin (1993).

do have a genetic difference between dorsal and ventral pallial regions even though both regions receive direct olfactory inputs. In jawed vertebrates, the dorsal pallial region differentiates into medial and dorsal pallium (cortex), comparable to hippocampal and parahippocampal regions in mammals. The ventral pallial region differentiates into lateral and ventral pallium. The piriform cortex of mammals develops from the embryonic lateral pallium.

Variations in Relative Size of Olfactory Systems

Looking only at the whole brain of various mammals, it is clear that the relative size of the olfactory bulbs varies greatly. We can speak of macrosmatic and microsmatic extremes. For example, dogs, hamsters, and rats have olfactory bulbs that are relatively much larger than in humans, and the sensitivity of these three species to odors shows corresponding differences. Think of the use of dogs to track an animal or person or to detect hidden drugs using odor traces. The hamster or rat can quickly find a bit of food buried in cage bedding, although the food has an odor that is barely detectable by a human if it can be smelled at all when placed close to the nostrils.

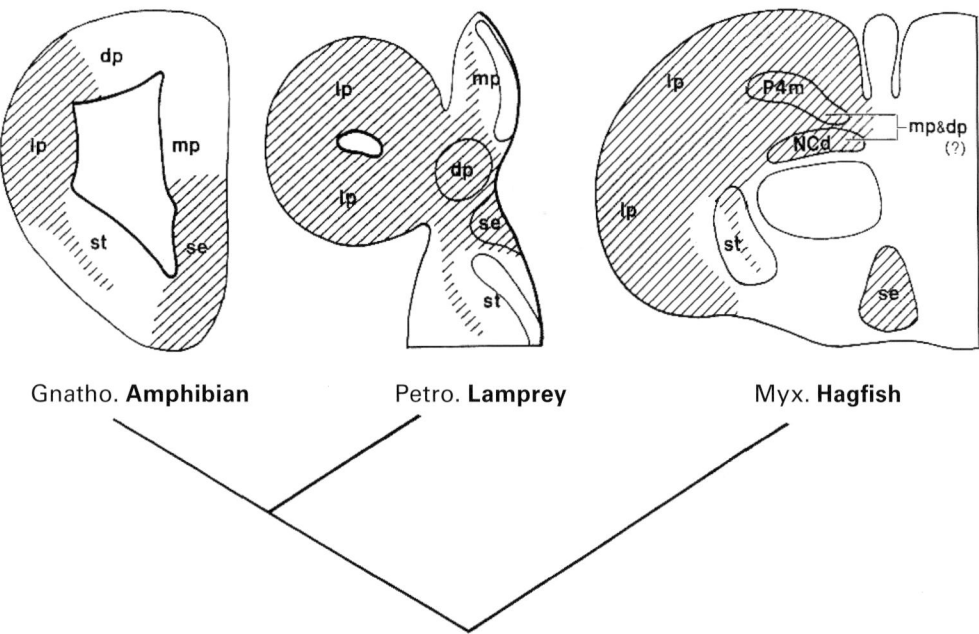

Figure 19.6
A cladogram illustrating the evolutionary relationships of jawless vertebrates (lamprey and hagfish) and amphibians is shown below sections through the endbrain of these animals. The sections depict the areas receiving olfactory bulb projections. In the hagfish, olfactory projections reach all parts of the endbrain — including structures that may correspond to medial and dorsal pallium. The lamprey is similar. In the amphibian brain, olfactory projections are much more limited, as they distribute mainly to the lateral pallium (lp) and to the basal forebrain and septal region (se), with a limited projection to the medial pallium (mp). dp, dorsal pallium; NCd, dorsal subnucleus of central prosencephalic nucleus; P4m, medial subdivision of pallial layer 4; st, striatum. From Wicht and Northcutt (1993).

Quantitative data on sizes of various brain structures in a large number of species was collected by Heinz Stephan and his co-workers beginning in the 1960s. These data show that the sizes of many brain structures differ from one species to another in a concerted way (i.e., when one structure is larger, another structure is also larger). This is evidence of concerted evolution. However, some structures, including those of the olfactory system, show evidence of mosaic evolution — where one structure may be disproportionately large or small. Recall the enormous size of the secondary sensory cell groups of hindbrain taste systems in some fish, mentioned and illustrated in chapter 4. Relative sizes of olfactory bulbs in mammals indicate mosaic evolution (figure 19.7).

Olfactory Bulb Projections in Mammals

The regions that receive the projections of the mitral cells, the secondary sensory cells of the olfactory bulbs, are found mostly at and near the surface of the brain. In viewing the

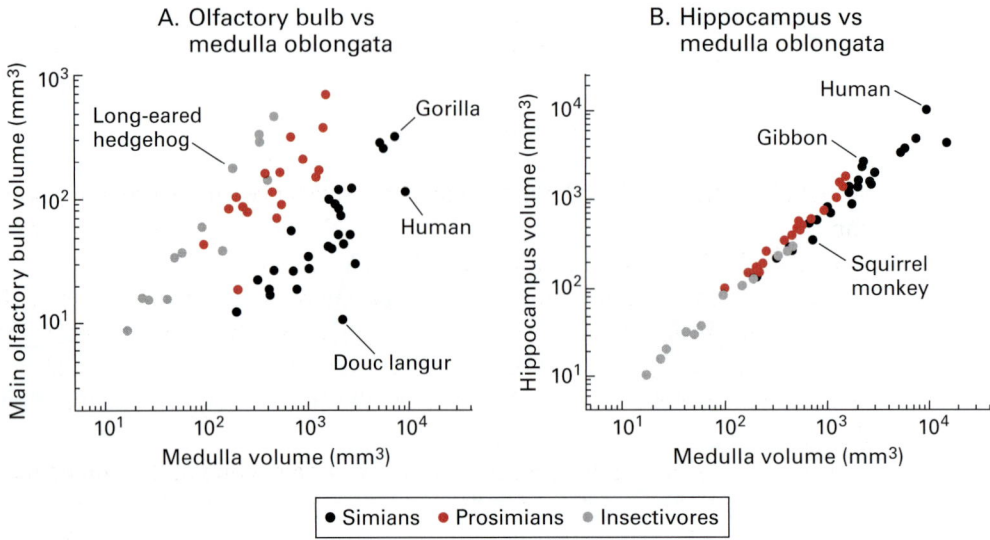

Figure 19.7
Evidence for mosaic evolution of mammalian olfactory structures has received some support from quantitative studies of the volumes of brain parts (Stephan et al., 1981). (A) The volumes of the olfactory bulb in various species are plotted against the volumes of the medulla oblongata. The relationship is not a strong one, as some species have relatively much larger olfactory bulbs. Of the three groups represented, the insectivores have relatively larger olfactory bulbs and the simians have relatively smaller bulbs. Prosimians are in between. (B) When the volume of the hippocampus is plotted against the medulla volume, there is a much stronger relationship, indicating concerted evolution of these two structures in the same three groups of species. From Striedter (2005).

whole brain, these structures are best seen by looking at the ventral surface. In figure 19.2, you can see a sketch of major olfactory bulb projections in a ventral view of a rabbit or rodent brain. The axons of the mitral cells are widely branched, with their terminal collaterals synapsing on branches of cortical cell dendrites near the pial surface. These terminals are distributed throughout much of the piriform cortex, located below (or medial to) the rhinal fissure, and not in the neocortex. This terminal field extends from rostral levels just behind the bulb to more caudal levels that include periamygdalar cortex and the entorhinal cortex. (In primates, only the most rostral parts of the entorhinal area receive olfactory bulb projections.)

Included in this pear-shaped expanse of olfactory cortex is a rostral region anterior to the hypothalamus, the olfactory tubercle. The olfactory tubercle is part of the ventral striatum, where the striatum extends to the brain surface. More caudally, in what becomes the temporal lobe in primates, the amygdala can be seen as an outpost of the ventral striatum. (Taken together, the various striatal components including the amygdala are sometimes called the basal ganglia or basal nuclei.)

The component of the amygdala that reaches the brain surface, known as the cortical nucleus of the amygdala, also receives a direct projection from the olfactory bulb, but the projection comes mostly from a specific subdivision of the bulb known as the accessory olfactory bulb (figure 19.2). The accessory bulb receives its input from primary sensory neurons that are not in the olfactory epithelium but in a structure called Jacobson's organ located in the base of the nasal cavity that in many species is connected by a channel to the roof of the mouth. It functions more like a taste receptor region, responding to substances brought into the mouth with the tongue. These substances are pheromones, known to play important roles in sexual behavior. Many other pheromones are detected by the olfactory system, beginning in the olfactory epithelium, in various species.

The major connections within the olfactory bulb are depicted in figure 19.8. Note the bush-like dendritic arbors of the mitral cells receiving the terminations of the olfactory epithelial neurons (the receptor cells). Note also the two kinds of small interneurons that are contacted by centrifugal axons coming from olfactory cortex and from the opposite bulb.

A simplified drawing of a parasagittal section of a portion of the human head is shown in figure 19.9. Note the position of the olfactory epithelium and the overlying skull bone known as the cribriform plate, through which the tiny primary sensory fibers—the olfactory filaments—pass into the olfactory bulb.

Human and Small Mammalian Brains

It should not surprise you that we have less reliable information about the olfactory bulb projections in humans than in the animals that have been used in experimental axon tracing studies. However, comparative cytoarchitectural studies have given us a good idea of the layout of equivalent structures for this system (as for many other systems). In figure 19.10, you can see a drawing of a ventral view of a human brain. For comparison, see the ventral view of a rabbit brain in figure 19.2. The stippling in figure 19.10 covers regions where the olfactory tract terminates, mostly the piriform cortex. Other olfactory areas, just behind the bulb and extending back over part of the olfactory tubercle, are also shown. Near the tip of the temporal lobe in the human brain we see a stippled region that corresponds to the periamygdalar cortex of the rabbit. The caudal-most limit of the olfactory projection reaches the entorhinal cortex, area 28. In humans, the entorhinal cortex is reached much less, if at all, by olfactory tract axons. Normally hidden from view in the human hemisphere is cortex in the fissure between the temporal lobe and the ventral-medial margins of parietal and frontal cortex; this area of olfactory projections is revealed in figure 19.10 by a retraction of the temporal lobe on one side (compare right and left sides).

Neuronal Organization as Depicted by Ramón y Cajal

To see these structures of the olfactory system in more detail, we can study some of the beautiful drawings of Ramón y Cajal. In figure 19.11, you see a drawing of a Golgi-stained

342 Chapter 19

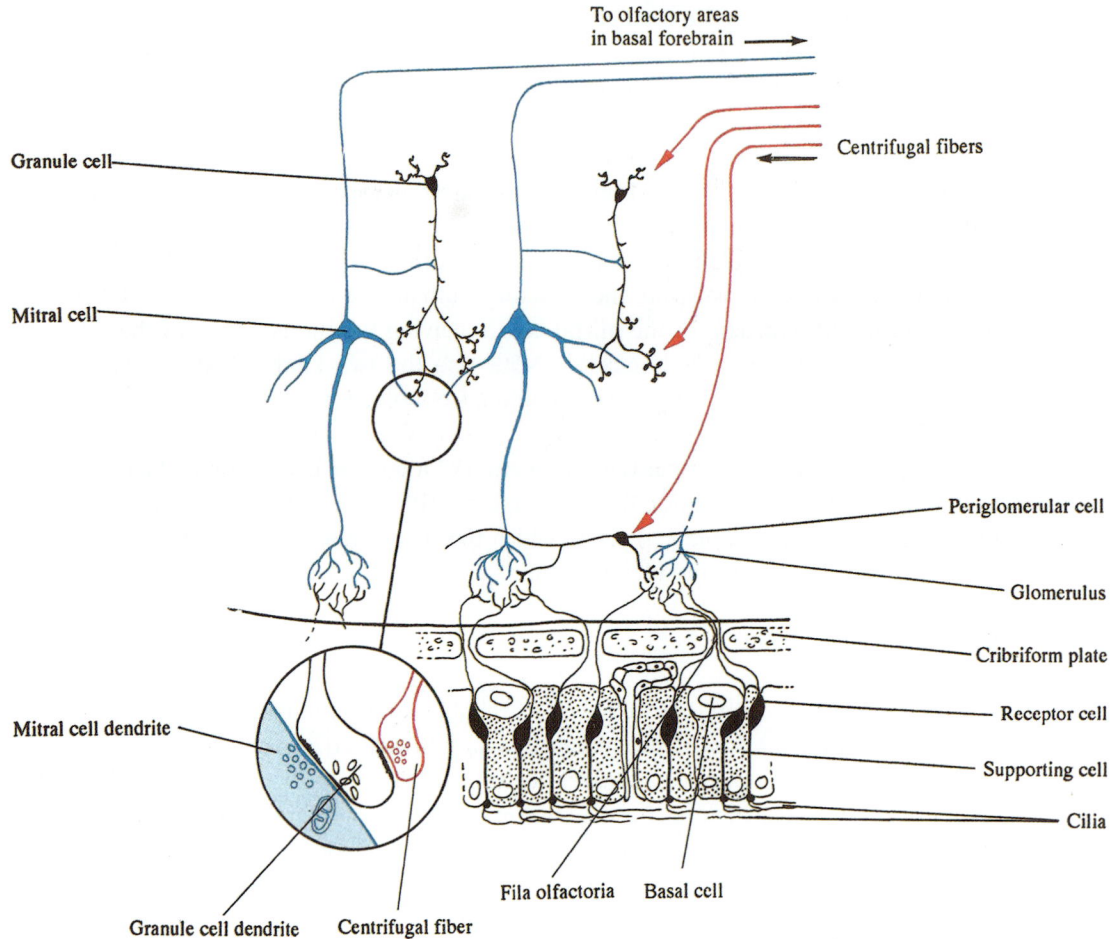

Figure 19.8
Schematic drawing that illustrates the organization of neurons in the mammalian olfactory bulb. From Heimer (1995).

Olfaction

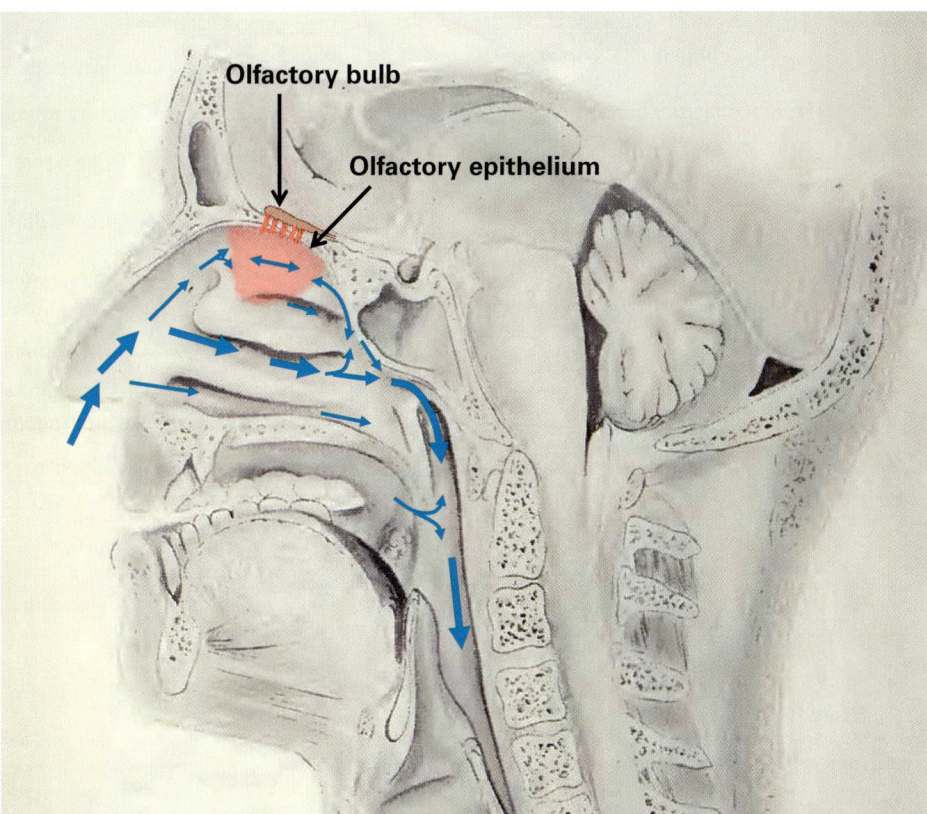

Figure 19.9
Human upper respiratory and olfactory apparatus showing air flow through the nasal cavity. The figure shows a drawing of part of a dissected human head that illustrates the positions of the olfactory epithelium (red color) and the position of the olfactory bulb underneath the frontal lobe. Primary olfactory neuron axons pass up through tiny openings in a part of the skull called the cribriform plate. Based on Parker (1906) and on other sources including studies of nasal air flow.

section of the olfactory bulb of a kitten a few days after birth. Cajal has drawn a number of axons of primary olfactory neurons, whose cell bodies are not shown. (They are outside the brain case in the olfactory epithelium of the nasal cavities.) Each axon has a compact end arbor within a *glomerulus*. Each glomerulus also contains similarly shaped dendritic arbors upon which the axon terminates with multiple synapses. These are dendrites of the large mitral cells or of smaller tufted cells, the neurons with axons projecting caudally through the olfactory tract (mostly the lateral olfactory tract). Two other cell types are also shown.

344 Chapter 19

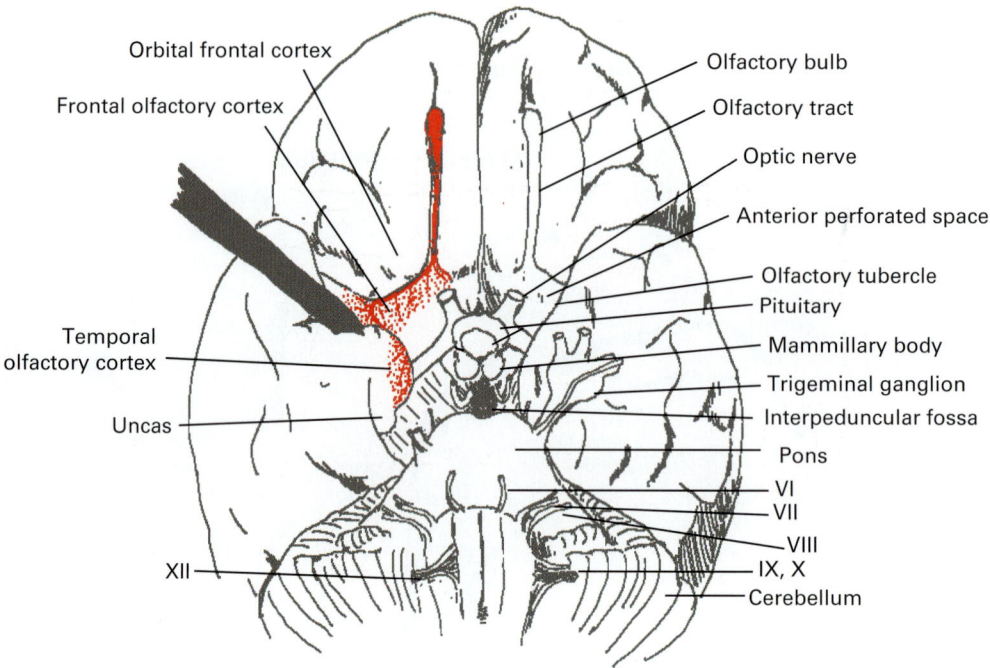

Figure 19.10
Human brain, ventral view with right temporal lobe retracted, showing olfactory bulb projections estimated from experimental studies of primates. Based on Heimer (1995).

The Axons of the Lateral Olfactory Tract

Figure 19.12 is a drawing of part of a section of the olfactory bulb and the olfactory peduncle just behind it in a 15-day-old mouse. At the left side you see the caudal end of the bulb, with mitral cells and olfactory glomeruli clearly drawn. You also can see the axons of mitral (and tufted) cells forming the lateral olfactory tract, which continues caudally (toward the right) over the surface of the cortex of the olfactory peduncle. Collateral branches of these axons are terminating in the molecular layer (layer 1, a superficial layer with very few neurons) of olfactory cortex, on dendrites of deeper lying neurons.

Figure 19.13 shows a drawing, at higher magnification, of a small part of a section through the piriform cortex, in this case from a Golgi stain of the brain of a 25-day-old rabbit. Here, the brain surface is at the top (the usual convention), where you can see the axons of the olfactory tract and their branches that are terminating among (and presumably forming synapses on) dendrites of cortical neurons. These neurons form three fairly distinct layers.

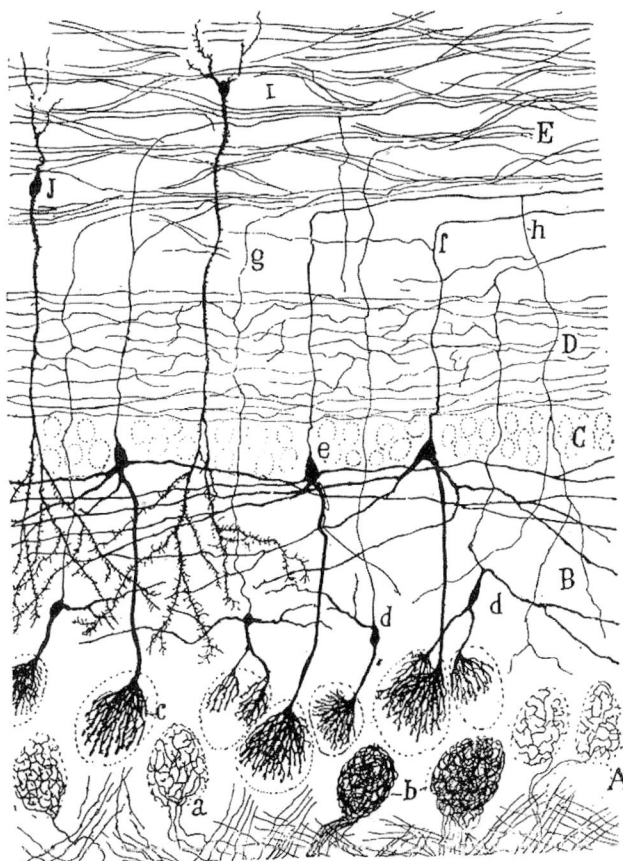

Figure 19.11
A drawing by Santiago Ramón y Cajal of a section of a Golgi-stained kitten olfactory bulb. The bottom of the picture is near the surface of the bulb, where one can see primary olfactory axons, from the olfactory epithelium, ending in egg-shaped bushes (a, b). They are terminating on dendritic arborizations (c) of mitral cell dendrites. Cajal had to draw axonal end-arbors and dendritic arbors of separate glomeruli as their intertwined processes could not have been discerned. Dendrites of tufted cells (d) also end in glomeruli. Note the mitral cell labeled "e": As is typical for this neuron type, its axon enters a deeper layer of fibers that turn caudally, entering the lateral olfactory tract. h, recurrent collateral of mitral cell axon; A, glomular layer; B, outer plexiform layer; C, mitral cell layer; D, inner plexiform layer; E, layer of granule cells and white matter; I, J, inner granule cells. From Ramón y Cajal (1995).

346   Chapter 19

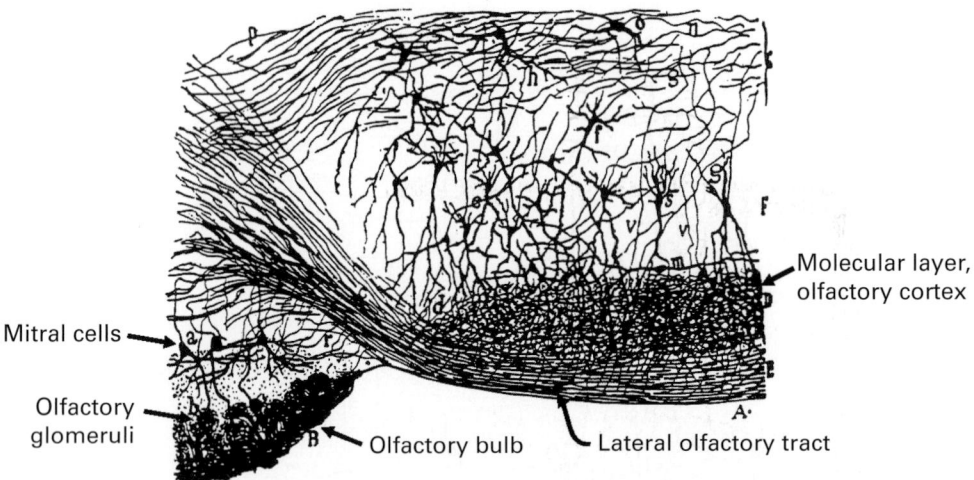

Figure 19.12
A drawing by Ramón y Cajal of a Golgi-stained parasagittal section of a 15-day-old mouse brain, seen at much lower magnification than that of figure 19.11. At the left is the caudal end of the olfactory bulb, with glomeruli and mitral cells included in the drawing. Mitral cell axons enter the lateral olfactory tract, with collaterals forming terminal arbors in the surface layer (molecular layer) of the olfactory cortex, which begins immediately caudal to the bulb. The axons are terminating on dendrites of pyramidal-shaped neurons with cell bodies lying deeper in the olfactory cortex. From Ramón y Cajal (1995).

The extensive branching of the axons of the mitral cells, also indicated in figure 19.2, means that each area of the olfactory bulb sends information very widely in the piriform cortex. The lack of topographic order is very unlike most of the projections of the somatosensory, auditory, and visual systems. Some of the synaptic recipients of the olfactory information, the pyramidal neurons of the piriform cortex, have axons that amplify this dispersal of information, as they branch widely throughout not only the piriform cortex but also in adjoining areas (figure 19.14).

Overview

In a low-power view (figure 19.15), Cajal uses a schematic drawing to show his conception of information flow (called "current flow" by Cajal) through the olfactory bulb. You see the primary sensory neurons—the bipolar cells of the olfactory epithelium—with axons ending in the bulbs in glomeruli, terminating on dendrites of mitral cells. You also see axons of secondary sensory neurons going to the bulb on the opposite side, ending on granule cells. Mitral cell axons can be seen forming the lateral olfactory tract with collaterals terminating in olfactory cortex.

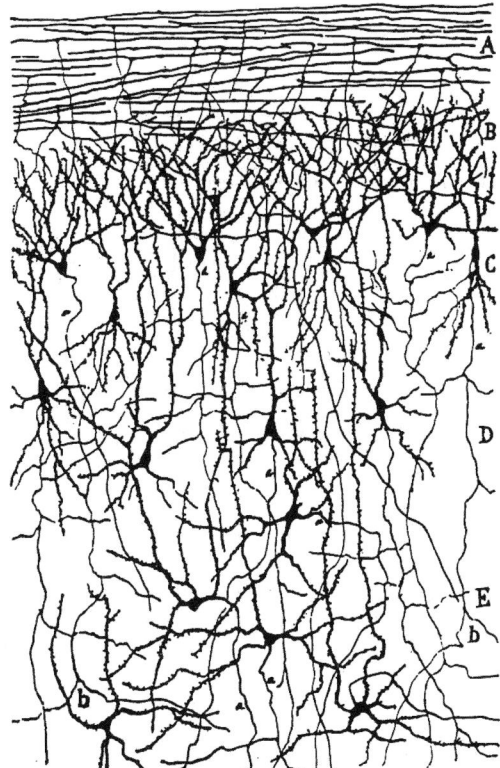

Figure 19.13
A drawing by Ramón y Cajal of a parasagittal section of the piriform cortex of a 25-day-old rabbit stained with the rapid Golgi method. The surface of the cortex is shown at the top, where one can see olfactory-tract axons that form the surface layer (A). In successively deeper layers, you can see (B) a layer of dendrites where axons from the bulb are terminating, then (C) a layer of cells of various shapes including pyramidal and fusiform neurons. Deeper still (D) is a layer of pyramidal cells, whereas even deeper (E) is a polymorphic layer (cells of multiple shapes). b, axon bifurcation. From Ramón y Cajal (1995).

Spatial Organization of the Primary Sensory Neurons

The topography of the inputs from olfactory epithelium to olfactory bulb has an organization that helps us understand something about the encoding of odor qualities. Figure 19.16 illustrates an important aspect of what has been found. First, it shows that longitudinal stripes or zones in the olfactory epithelium project to specific zones of the olfactory bulb. Second, it shows that different receptor types each project to one or a few olfactory glomeruli in the bulb. Different receptor molecules bind selectively to specific molecules from the air that are dissolved in the nasal mucosa, into which cilia from the dendritic processes of the primary sensory neurons protrude. Thus, different mitral cells of the

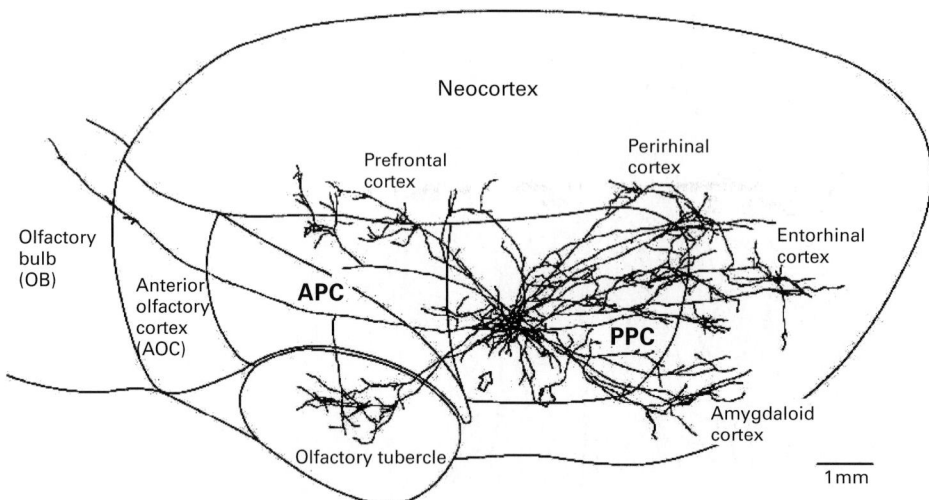

Figure 19.14
Reconstruction of the widely distributed axonal arbors of two pyramidal neurons of layer two of the posterior piriform cortex (PPC) of a rat. The arrow points to the position of the cell bodies. Axonal branches reach not only much of the piriform cortex but also various other regions: the olfactory tubercle, the olfactory bulb, the cortex overlying the amygdala, the entorhinal cortex, and the margins of the perirhinal and orbital prefrontal cortex. Scale bar, 1 millimeter. APC, anterior piriform cortex. From Haberly (2001).

olfactory bulb respond to stimulation of different receptors or combinations of receptors in the primary sensory neurons of the nasal epithelium.

This kind of extreme localization in the olfactory bulb applies best when odors are not very concentrated. With low concentrations of a single odor, several distinct regions over the surface of the olfactory bulb are activated. Stronger inputs due to higher concentrations of an odorant activate larger regions of the bulb. This spread of activation could reduce the contrast between distinct odorants, but the connections of small interneurons within the olfactory bulb counter this by increasing the contrast. This happens by lateral inhibition between neighboring secondary sensory neurons, through the granule cells, which form bridges between neighboring mitral cells and neighboring tufted cells. Periglomerular cells also form such bridges between neighboring mitral cell dendrites and receptor cell axon terminals.

It is interesting that the accessory olfactory bulb, focusing its projection on the amygdala, not only responds very specifically to socially relevant molecules—pheromones—but through the specificity of its projections drives instinctive behavior patterns related to reproduction. The main olfactory bulb, in contrast, although it receives highly specific odor information from many different receptor types and keeps this information separate in its various glomeruli, disperses this specific information widely in the surface layers or olfactory-recipient cortex at the base of the mammalian brain. The innate pattern of connections seems to be a random one. However, the olfactory stimuli acquire meaning very

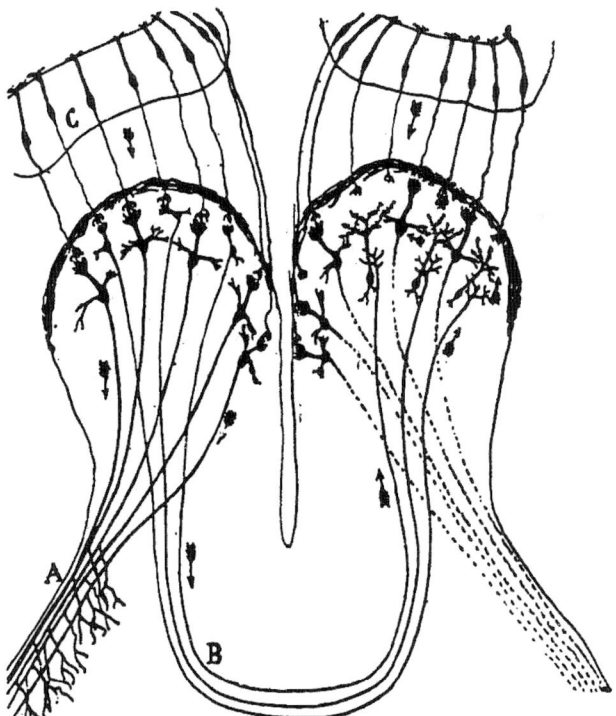

Figure 19.15
Ramón y Cajal's schematic picture depicting information flow through the olfactory bulb. A, Lateral olfactory tract axons with collaterals terminating in the olfactory cortex. B, Anterior commissure axons from tufted cells to the opposite bulb. C, Bipolar primary sensory neurons of the olfactory epithelium with axons passing through the cribriform plate (the skull beneath the front portion of the brain) to the glomeruli of the olfactory bulb. From Ramón y Cajal (1995).

differently: through individual experience. Positive and negative consequences of behavior associated with particular odors result in alterations in approach or avoidance tendencies—the objects and places associated with particular odors acquire valences. (See the reading listed at the end of this chapter by Choi et al., 2011.)

The great variety of odors that mammals can discriminate increased greatly very early in their evolution according to studies of the proliferation of different odor receptor molecules—numbering from almost 400 to about 1200, a much larger number than in non-mammalian vertebrates.

Beyond the Mitral Cells

We will return to the structures receiving olfactory bulb projections when we study the limbic system of the forebrain. At this point, we note that the olfactory cortex (piriform

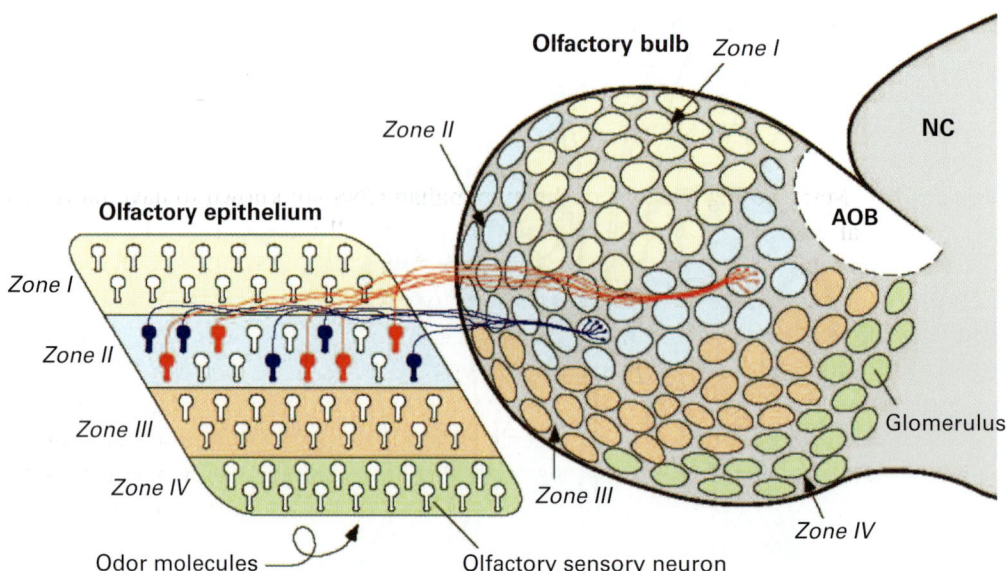

Figure 19.16
The organization of primary olfactory neurons in the olfactory epithelium and their axons is illustrated in this schematic drawing. Two receptor cell types, with different molecular specificity, are depicted within the second longitudinal zone of the olfactory epithelium, which projects to a corresponding zone of glomeruli in the olfactory bulb. Axons of one type of receptor cell converge on a single glomerulus, while axons of the other type converge on a different glomerulus. The figure does not indicate the full extent of the convergence and divergence in this system in mammals. In mice, several thousand receptor cells have axons that converge on single glomeruli. The glomeruli number about 1800. It is estimated that dendrites from 10 to 20 mitral cells and 50 to 70 tufted cells converge on each glomerulus. These numbers vary widely among different mammals. From Mori et al. (1999).

cortex) projects not only to limbic forebrain structures but also to the dorsal thalamus (medial part of the mediodorsal nucleus, which projects to the orbitofrontal areas of the frontal neocortex). Thus, input from the olfactory system, although passing directly from the olfactory bulb to cortical structures of the endbrain, reaches the newer portions of the endbrain pallium—the neocortex—less directly, via the thalamus.

Ongoing Plasticity in the Olfactory Bulb by Cell Turnover

Not only during the developmental period but also continuing in adulthood, new neurons produced in the cerebral hemispheres migrate into the olfactory bulbs where they replace older granule cells. This ongoing cell turnover was first discovered in the rat brain by Altman and Das at MIT in the 1960s, although other scientists were slow to accept its generality. The newly generated neurons move from the ventricular layer of the lateral ventricle in a small stream of migrating cells near the ventricle, surrounded by glial cells that appear to serve as guides. The role of the new neurons is discussed in the Intermission after chapter 31.

The function of this cell turnover is not understood. The other place where such vigorous turnover is known to take place is the dentate gyrus of the hippocampus. There, the turnover is greater in animals that are more active and apparently learning more.

Many young neurons of the mammalian CNS are known to have more growth potential than they have later in life. You may recall the discussion in chapter 13 of axonal growth after early brain lesions (see figure 13.16). Studies of the developing lateral olfactory tract have shown regeneration of severed axons after lesions that occur sufficiently early in life, equivalent to the prenatal period in humans. Even after regeneration fails, there are competitive interactions among developing axons, indicated by compensatory sprouting of axon collaterals into denervated regions by axons that have been pruned by a lesion, as depicted in figure 13.16. It has also been found that axons that have sprouted many new branches proximally will show a compensatory stunting of distant branches—a phenomenon observed in the olfactory tract projections in cases of early damage.

Such sprouting and stunting phenomena appear to be consequences of the same rule: a tendency to conserve the total size of the axonal end arbor or the number of synaptic endings. This rule may well apply also to axonal growth in newly generated neurons in adults. It may also be the case that the younger neurons are more plastic, more capable of changes necessary for learning, just as younger animals or persons show, in many respects, greater plasticity than older individuals (see also chapters 13 and 34).

Olfaction and Behavior

In a discussion of brain evolution in chapter 4 it was suggested that, after the earliest appearance of the brain vesicles, it was the olfactory inputs that led to a first expansion of the forebrain. The enlargement was mainly in the endbrain where the olfactory nerves connected and where the analysis of olfactory inputs occurred. Animals with better olfactory detection and discrimination survived more successfully because odorants alerted them by warning of dangers and by signaling nearby food sources and the presence and status of other members of the species. Olfaction was an excellent complement to touch sensations in that it was a distance receptor, responding to molecules carried through the air, especially downwind. Distances could be in time as well as in space, as odors could cling to objects and to the ground, fading slowly with time.

To enable the necessary escape and avoidance responses or approach and exploratory actions, the neurons activated by olfactory inputs had to reach motor systems. Originally, this required links to the midbrain and hindbrain via the ventral striatum/basal forebrain region. For influencing other motivational states (drives) involved in feeding and in sexual, parental, and other social behaviors, there were routes to the hypothalamus, directly and via the basal forebrain and amygdala, and thence through the hypothalamic locomotor area and midbrain locomotor area. With little doubt, epithalamic as well as hypothalamic links to these outputs systems existed.

In recent evolution, olfactory influences became able to reach systems of anticipation and planning in the neocortex—systems of object representation and images and of preparations for the future. There was an evolution of pathways from piriform cortex and basal forebrain structures, and from the amygdala, to the medial part of the mediodorsal nucleus of the thalamus, and thence to frontal neocortex. From there, other association cortical regions could be reached. (The amygdala, a structure that is important in social behavior, has also evolved direct projections to the prefrontal cortex.)

Olfaction, without doubt, played an important ancient role in the learning of approach and avoidance reactions. Habits could be formed and changed through alterable connections in the ventral striatum with neurons that were links to midbrain locomotor control. In addition, good and bad places could be remembered because of connections from olfactory cortex (via the entorhinal area or, more directly) to the hippocampus.

For us, smell plays an intricate part of our lives, of our sensory world, in the way we act. The process of recognition of a place is accomplished by our sense of smell.

—Essortment.com: "Our olfactory senses" (article by Shannon Demick, © 2002.)

Readings

Altman, J., & Das, G. D. (1965). Post-natal origin of microneurones in the rat brain. *Nature, 207,* 953–956. See also the Altman papers listed in the Intermission after chapter 31.

Buck, L., & Axel, R. (1991). A novel multigene family may encode odorant receptors: A molecular basis for odor recognition. *Cell, 65,* 175–187.

Choi, G. B., Stettler, D. D., Kallman, B. R., Bhaskar, S. T., Fleischmann, A., & Axel, R. (2011). Driving opposing behaviors with ensembles of piriform neurons. *Cell, 146,* 1004–1015.

Heimer, L. (1995). *The human brain and spinal cord: Functional neuroanatomy and dissection guide*, 2nd ed. New York: Springer-Verlag.

Medina, L., & Abellán, A. (2009). Development and evolution of the pallium. *Seminars in Cell & Developmental Biology, 20,* 698–711.

Mori, K., Takahashi, Y. K., Igarashi, K. M., & Yamaguchi, M. (2006). Maps of odorant molecular features in the mammalian olfactory bulb. *Physiological Reviews, 86,* 409–433.

Niimura, Y., & Nei, M. (2007). Extensive gains and losses of olfactory receptor genes in mammalian evolution. *PLoS ONE, 2*(8), e708. doi:10.1371/journal.pone.0000708

Polenova, O. A., & Vesselkin, N. P. (1993). Olfactory and nonolfactory projections in the river lamprey (*Lampetra fluviatilis*) telencephalon. *Journal für Hirnforschung, 34*, 261–279.

Stephan, H., Frahm, H., & Baron, G. (1981). New and revised data on volumes of brain structures in insectivores and primates. *Folia Primatologica, 35*, 1–29.

Striedter, G. (2005). Book listed in chapter 4 readings, p. 152, sections A, D on mosaic evolution (pp. 165–216).

Wicht, H., & Northcutt, R. G. (1993). Secondary olfactory projections and pallial topography in the pacific hagfish, *Eptatretus stouti*. *Journal of Comparative Neurology, 337*, 529–542.

Wicht, H., & Northcutt, R. G. (1998). Telencephalic connections in the Pacific hagfish (*Eptatretus stouti*), with special reference to the thalamopallial system. *Journal of Comparative Neurology, 395*, 245–260.

Circuits for lateral inhibition in olfactory bulb:

Shepherd, G. M. (2004). Olfactory bulb. In Shepherd, G. M., ed. *The synaptic organization of the brain* (5th edition). New York: Oxford University Press, pp 165–216.

Newly generated neurons in adult rat cerebral hemispheres (as discovered by Altman and Das) move from the ventricular layer of the lateral ventricle in a small stream of migrating cells near the ventricle, surrounded by glial cells that appear to serve as guides:

Alvarez-Buylla, A., & Garcia-Verdugo, J. M. (2002). Neurogenesis in adult subventricular zone. *Journal of Neuroscience, 22*, 629–634.

Pheromones: Numerous reviews can be found in both scientific and semi-popular literature.

20   Visual Systems: Origins and Functions

Very early in the evolution of the chordates, probably as early as the emergence of an olfactory system in the forebrain, an ability to detect light evolved. We will begin a discussion of visual systems with their origins in elementary light detection and explain why this was such an important adaptation and remains so in modern vertebrates. An image-forming ability evolved much later along with connections in the brain that made adaptive use of information from visual images. Even before true images could be detected, some location information about the environment around the animal's head could be used for orienting toward or away from particular places, e.g., places where a movement or a change in brightness caused by an approaching predator was detected. This information about things happening at a distance from the animal, however crude it must have been during the initial stages of evolution of vision, was clearly adaptive. It was an early warning system (see figures 10.13 and 10.14).

Images could also provide information about the identity of objects and places in the environment. This made the animal less dependent on olfaction for knowing where it is and what things were in that location. Thus, the eyes became useful for more than the most elementary orienting functions once there was an evolution of neuronal cells interconnected for analysis of the input. The information from the visual detectors was routed through connections in the 'tweenbrain and midbrain where it evolved linkages to output systems controlling highly adaptive functions: behavioral arrest or orienting in response to any novel object or movement, escape from potential predators, and approach or avoidance of specific objects or animals depending on their identity. To introduce the various connections involved, we will look at the axonal projections of the retina in vertebrates and take note of various specializations in this system.

Although we can construct some pieces of a plausible evolutionary history, we have to remember that we have only indirect knowledge of early chordate ancestors. We also have some evidence from fossil skulls and skeletons to suggest the nature of the brains of mammal-like reptiles representing transitions to the mammalian radiation. Most of all, we make inferences from comparisons of anatomical findings in studies of present-day vertebrates.

Origins of Vision, 1: Light Detection

Simple light sensitivity can be found even in some protozoa and primitive metazoa. In amphioxus, an invertebrate chordate that may resemble the earliest members of the phylum, one can find rudimentary light detectors in the forebrain—in tissues at the rostral end of the neural tube that appear to correspond to the diencephalon of vertebrates. Its frontal eye spot may correspond to the developing vertebrate eye, and a separate light-sensitive structure on the dorsal side of the rostral neural tube probably corresponds to the pineal eye found in some vertebrates (see chapters 3 and 7).

An important function that made light sensitivity adaptive was the ability to organize the activities of the day and night—to time them—for safety and efficiency. It was important to have this temporal organization without dependence on external light levels alone, but it was also important to synchronize the daily activity with the solar day–night cycle. The advantages of this function led to the evolution of light detection with a connection to a biological clock that controlled a daily rhythm of activity.

Such a clock is located, in vertebrates, at the base of the caudal forebrain, in the hypothalamus. Also important are effects of the level of light on the overall level of locomotor activity. In vertebrates, these functions are controlled mainly by the hypothalamus and epithalamus—subdivisions of the 'tweenbrain (diencephalon).

Near the ventral midline of the 'tweenbrain of all vertebrates, close to the hypothalamic region where axons from the retina join with the rest of the CNS, a small cell group can be found on each side of the midline (figure 20.1), where the neurons constitute a biological clock. These are the cells of the suprachiasmatic nucleus, so named because of their location just dorsal to axons from the retinae that cross from one side of the brain to the other. These neurons show spontaneous variations in their activity, variations that follow a circadian rhythm (i.e., with a cycle of changing activity that has a period of approximately 24 hours). The phase of this rhythm is normally entrained by the cycle of light and darkness determined by the sun. The entrainment results from activity in axons from the retina that connect directly to cells in the small nuclei. These endogenously driven cells with inputs from the retina have very widespread influences on the brain and autonomic nervous system by their projections to other hypothalamic areas and also to the basal forebrain—a rostral extension of the hypothalamic cell groups into the endbrain—and to midline portions of the thalamus, known to have very widespread projections to the endbrain. They also influence many tissues of the body outside the nervous system through endocrine secretions under the control of the pituitary organ (see chapter 25).

Sparse connections by a few axons from the retina can be found in other hypothalamic regions, but these have not been the subject of much scrutiny by investigators despite a potential significance for influencing the motivational states governed by this brain region.

The paraventricular hypothalamic (PVH) cells, which are dorsal to and strongly connected with the suprachiasmatic nucleus, have a major influence on the pineal gland via

Visual Systems: Origins and Functions

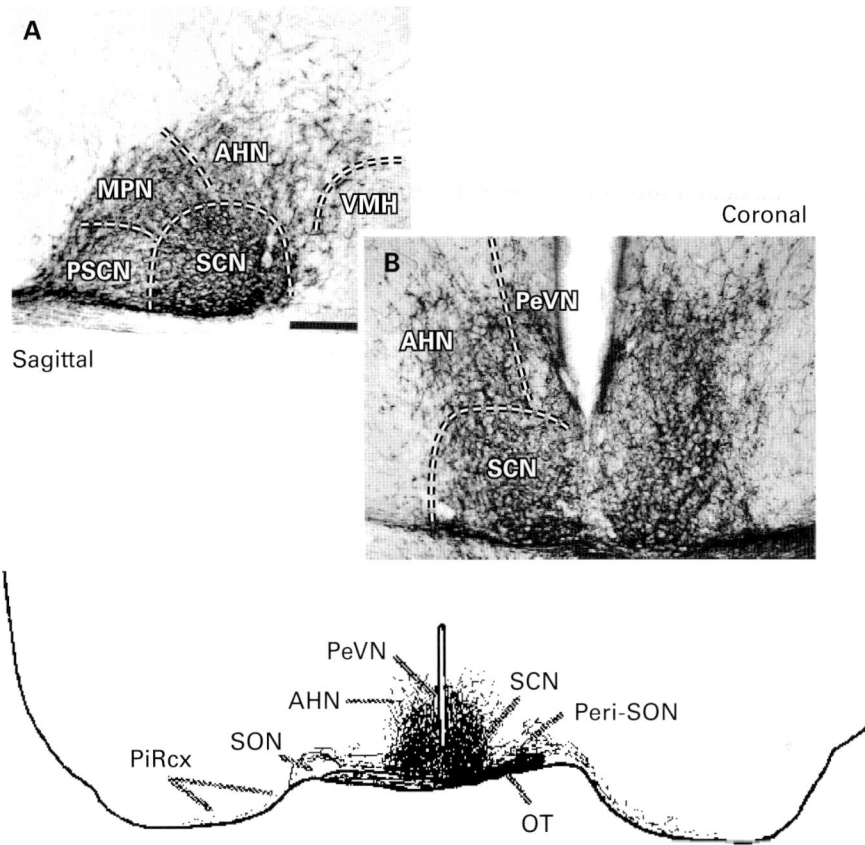

Figure 20.1
At the bottom is a chart of a coronal section through the middle of the suprachiasmatic nucleus (SCN) of a Syrian hamster. Axons from the left retina were labeled with cholera toxin fragment B, which fills entire axons of the retinal ganglion cells, including the terminals. Labeled axons are drawn in this low-power view, and terminals are drawn as black dots. The section is at the caudal end of the optic chiasm, where most of the axons in the optic tract (OT) have crossed the midline to the right side. The densest axon terminations are in the SCN bilaterally, but less dense terminations are found more dorsally in the periventricular nucleus (PeVN) and the anterior hypothalamic nucleus (AHN). Additional terminations are found more laterally in the peri-supraoptic nucleus (Peri-SON). A few axons were traced into the nearby piriform cortex (PIRcx). The photographs show more detail: (A) From a sagittally cut section about 200 micrometers lateral to the midline, rostral to the left, with densest terminations in the SCN, and additional terminations more rostrally in the preoptic suprachiasmatic nucleus (PSCN), rostro-dorsally in the medial preoptic nucleus (MPN), dorsally in the AHN; a few axons more caudally reach the ventromedial hypothalamic nucleus (VMH). (B) From a coronal section at the level of the chart. Scale bar (A, B), 200 μm. From Ling, Jhaveri, and Schneider (1998).

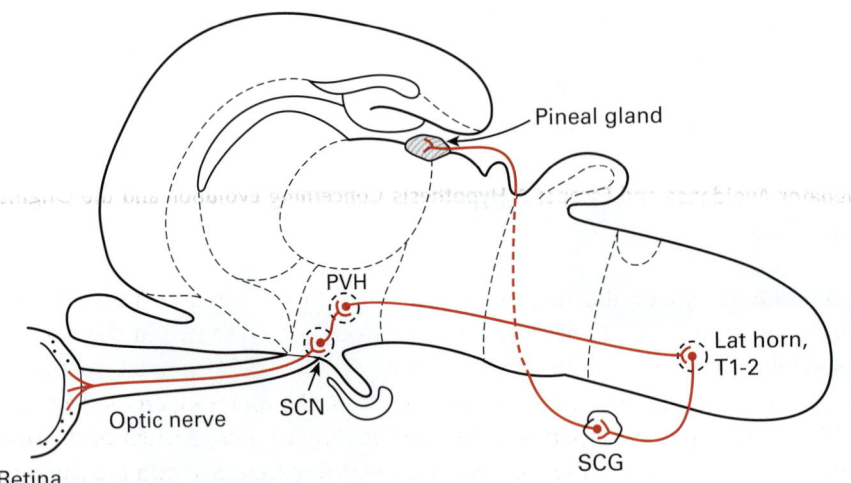

Figure 20.2
Schematic drawing of the pathway whereby light reaching the eyes controls the daily rhythm of melatonin production by the pineal gland. Axons from the retina terminate in the suprachiasmatic nucleus (SCN), which projects to the paraventricular nucleus of the hypothalamus (PVH) dorsal and caudal to it. The PVH projects to neurons of the lateral horn in the spinal cord at T1-T2 levels. The lateral horn neurons send axons to the sympathetic ganglia, including to the rostral-most of these paravertebral ganglia, the superior cervical ganglion (SCG). Axons of the SCG innervate structures in the head including the pineal gland as shown. In this roundabout manner, mammals, which do not have a pineal eye on top of the head, can still accomplish a major function of that third eye.

a lengthy indirect connection (figure 20.2): The PVH projects to the first two thoracic segments of the spinal cord. The axons connect directly with neurons of the lateral horn, the location of the preganglionic motor neurons of the sympathetic nervous system (see chapter 9). These neurons project to the cells of the superior cervical ganglion of the sympathetic chain of ganglia (one on each side of the cord). These ganglia send axons to the pineal gland and other glands and smooth muscles in the head region, using norepinephrine as a neurotransmitter. The connection to the visual inputs results in a regular daily rhythm of melatonin production and secretion, known to be important in influencing physiological differences between sleep and waking states.

Origins of Vision, 2: Image Formation

The most important function of object location information in retinal projections is, with little doubt, avoidance of or escape from predators. Importance for an individual in evolutionary terms is judged by increased likelihood of survival for passing genes onto the next generation. This function required detection of movement or brightness changes in one location near the animal, providing advanced warning and triggering sudden freezing or fleeing toward a safer place. Accomplishing this required an early evolution of connections carrying information about the sudden appearance of shadows on one side or the other that could trigger escape movements. Eventually the detection of locations became more precise with the evolution of lenses.

Predator Avoidance and Escape: A Hypothesis Concerning Evolution and the Origins of Crossed Projections

A hypothesis about the origins of visually triggered escape from predators in early chordates—the first role of visual images—was first presented in chapter 10 and will be reviewed now. The ideas that make up the hypothesis can account for the origins of the crossed representations of the external world in the midbrain and forebrain.

The starting point is a pathway that is probably the most ancient of the somatosensory pathways in vertebrates, the spinoreticular pathway together with the similarly organized trigeminoreticular system of axons carrying information from the face region. These axons ascend bilaterally but are predominantly ipsilateral (figure 10.12) even in mammals.

Escape movements in a simple aquatic animal begin with contractions of muscles on one side of the body causing a turning of the body. Contractions on the side opposite to the side of the stimulus result in turning away from that stimulus when the body is moving forward (figure 10.14). The neurons most directly connected to the necessary motor neurons are located in the brainstem on the same side as the motor neurons (figure 10.13). Rapid access to these neurons was important for survival. (Figure 10.13 shows a sketch of such neurons with both visual and somatosensory inputs.) The most direct relevant connection from the eye to this escape mechanism was crossed, so early bilaterally projecting axons shifted to a dominance of crossed projections (figure 10.13). This probably occurred in chordate ancestors of vertebrates or in the earliest vertebrates.

Later evolution of better imaging and of topographically organized projections from retina to the midbrain tectum retained the decussation. In this way, one side of the external world and the body came to be represented on the opposite side of the CNS. As the midbrain evolved projections to the forebrain, these projections were usually ipsilateral so that the representation of the right side of the space around the head on the left side of the midbrain was retained in the forebrain.

In modern vertebrates, the midbrain tectum (superior colliculus of mammals) is the dominant structure where visual inputs connect with outputs for escape reactions to a visual stimulus. However, this dominance is the result of a long evolutionary history. The optic tract reaches other structures before it reaches the midbrain roof (figure 20.3). After the suprachiasmatic nucleus, described earlier, it reaches the subthalamus where it terminates in the ventral nucleus of the lateral geniculate body. Then it reaches the thalamus, with terminations in the dorsal nucleus of the lateral geniculate body and to a much lesser extent in mammals in the adjacent lateral posterior nucleus. The axons continue by

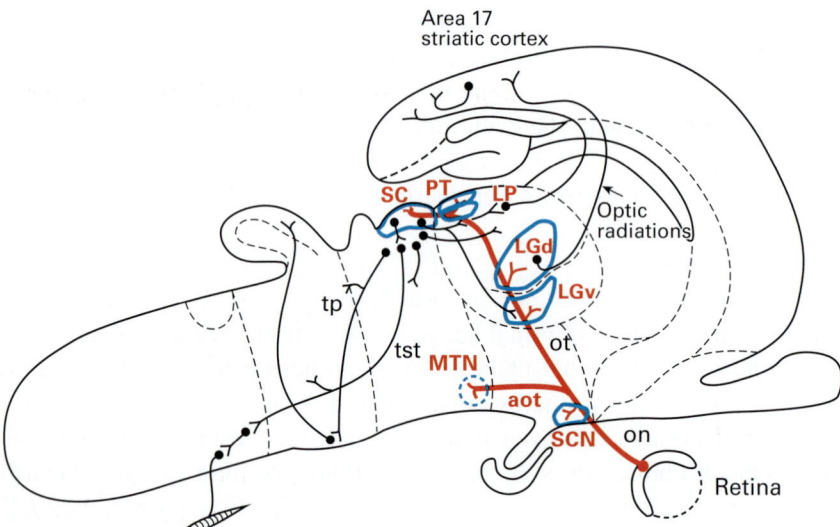

Figure 20.3
Schematic drawing of a mammalian brain depicting major subdivisions as defined in earlier chapters. The spinal cord is at the left and the endbrain at the right. The retinal projections are illustrated, and in addition some further connections of some of the retinorecipient cell groups are shown. The main optic tract has terminations in the hypothalamus (in SCN), in the subthalamus (in LGv), in the dorsal thalamus (usually just called thalamus) (in LGd), in the pretectal cell groups (PT) of the epithalamus, and in the midbrain tectum (the SC in mammals). Much smaller connections are also found in other parts of the hypothalamus and in the posterior part of the lateral thalamic nucleus (LP). The accessory optic tract axons leave the main tract at several points, terminating in several small cell groups, such as the medial terminal nucleus (MTN). Note two pathways to the neocortex, one via projections of the LGd, the other via projections of LP—which receives inputs from SC. An additional pathway to the neocortex passes from PT to a more anterior part of the lateral thalamus to extrastriate cortical areas. aot, accessory optic tract; LGd, dorsal nucleus of the lateral geniculate body; LGv, ventral nucleus of the lateral geniculate body; LP, lateral posterior nucleus of the thalamus; MTN, medial terminal nucleus of the aot; on, optic nerve; ot, optic tract; PT, pretectal area; SC, superior colliculus; SCN, suprachiasmatic nucleus; tpt, tecto-pontine tract; tst, tectospinal tract.

passing through the epithalamic region with terminations there, in the pretectal area, before they reach what in most animals is the major optic-tract terminal area in the roof of the midbrain (the optic tectum).

It will be noted below that the ventral nucleus of the lateral geniculate body and the pretectal area, both give rise to axons that terminate in the midbrain tectum and thereby must be involved in tectal functions. However, neurons of the lateral geniculate body (ventral portion) also project to the more medial part of the subthalamus—to the *zona incerta*—which has connections to the midbrain locomotor area; when stimulated electrically, rapid running movements can be elicited from the zona incerta, as in an escape response.

When escape reactions to visual inputs first evolved, the forebrain was probably very small, with little or no dorsal thalamus. Short axons of the optic tract reached the midbrain directly or after one synapse, and connections there activated neurons projecting ipsilaterally to escape mechanisms of the lower brainstem and spinal cord. However, it is also possible that early in evolution of the optic tract, escape reactions depended on a connection closer to the optic chiasm, like the connections in the subthalamus (see also the next chapter). The projections of the subthalamic region to the midbrain locomotor area, noted earlier, are consistent with this idea.

Orienting Toward or Around Visually Detected Objects and Other Responses

Next in importance for the use of information from visual images—a second role—was the eliciting of orienting movements toward sources of stimuli in the space around the organism and in guiding locomotion around stationary objects. The midbrain tectum became the dominant structure for orienting toward objects, using the crossed retinal projection that had evolved for control of rapid escape movements. But for the orienting toward objects, a different output pathway evolved—the crossed *tectospinal* pathway to the medial hindbrain reticular formation and to the cervical spinal cord (figure 20.4). This pathway brought the axon terminals close to the motor neurons most directly involved in turning toward a stimulus.

It is unclear how long it took in evolution for the midbrain tectum to become so dominant in controlling orienting functions. As suggested earlier, the retinal projections to the subthalamus, and perhaps also to the pretectal cell groups of the epithalamus, may well have played an early role, judging from some of their connections in extant vertebrates.

The retinal projection to the subthalamic portion of the 'tweenbrain goes to the ventral nucleus of the lateral geniculate body. In mammals, this structure projects to the midbrain tectum and to the pretectum, and it is also interconnected with the cerebellum. Whatever early role this structure played in turning movements, it clearly became overshadowed by the midbrain tectum (superior colliculus).

The retinal input to the pretectal cell groups in many vertebrates including mammals is known to be important for control of protective reflexes. One important response is the pupillary light reflex, with no requirement for orienting movements or even topography of the connections. Some role of this region in control of turning toward visual stimuli (orienting) may be indicated by its projections to the midbrain tectum. However, a different kind of movement that is also sometimes called orienting seems to depend specifically on the pretectal nuclei, according to evidence from studies of frogs and tree shrews, an insectivorous mammal living in the Malay peninsula. This kind of orienting involves the avoidance of stationary barriers partially blocking the animal's path by the animal side-stepping the barriers while it is locomoting.

A fundamental role of some pretectal cells, together with cells of the so-called accessory optic system, has been a topic of many studies in mammals, birds, reptiles, and amphibians: Some neurons there respond to movement of visual images simultaneously across the

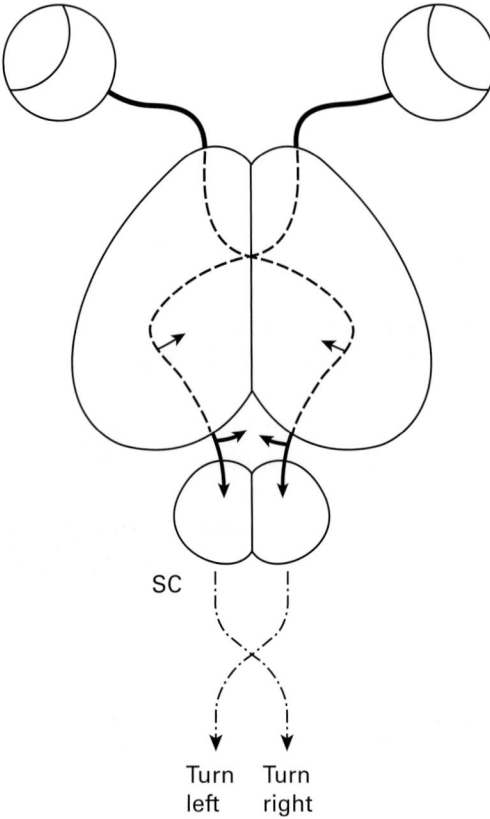

Figure 20.4
Schematic drawing showing the pathway on the two sides of a small mammal with lateral eyes that most directly elicits turning toward a visual stimulus source: from retina to contralateral midbrain tectum (superior colliculus; SC) to a crossed descending pathway to premotor neurons of the brainstem and cervical spinal cord.

entire retina—the type of movement that occurs during shifts in eye position (as during an orienting movement), or during locomotion, or during an unintended loss of position (e.g., during a loss of balance). The latter type of stimulus, which is not self-produced, triggers reflex compensatory movements of the eyes or the body as in optokinetic nystagmus or in reflex adjustments of posture. Thus, it has outputs with functions that overlap with those of the vestibular system. With regard to this kind of function, it is fascinating that hummingbirds, which maintain stationary positions for considerable times while hovering in the air, have a pretectal nucleus that is relatively much larger than in other birds.

These laboratory findings suggest that the pretectal area has diverse functions that have been only partially investigated.

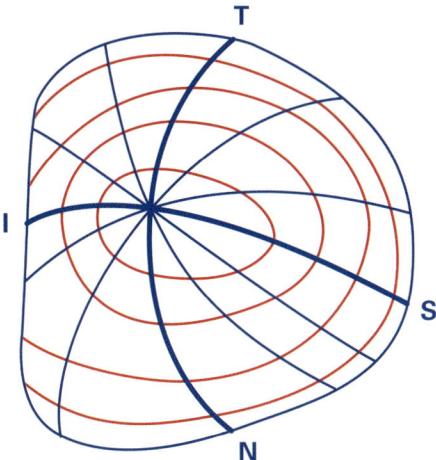

Figure 20.5
Map of the projection of the contralateral retina on the superior colliculus of the adult Syrian hamster as seen in a view of the dorsal surface of the midbrain. Experiments used both axonal tracing from discrete portions of the retina and electrophysiological mapping of receptive fields of single and small groups of neurons. The geometric center of the retina projects to the point where the meridians are shown to cross. Letters correspond to the midpoints of the attachments of the four rectus muscles: T, temporal pole of the retina defined by the lateral rectus; N, nasal pole of the retina defined by the medial rectus; I, inferior pole defined by the inferior rectus; S, superior pole defined by the superior rectus. The meridians defined by these points represent the major axes of the retina and visual field of the eye. The other meridians are drawn at 30-degree intervals. The closed curves are drawn in red at 20-degree increments from the representation of the center of the retina. Based on Finlay, Schneps, Wilson, and Schneider (1978) and on Frost and Schneider (1979) also using results from S. Jhaveri.

The Midbrain Tectum and Orienting Toward Novel Objects, Food, or Potential Mates or Rivals

Whatever were the roles of other structures receiving the optic tract projections, these roles have been added to and surpassed by the tectum of the midbrain (superior colliculus in mammals). The tectum may have enlarged more than the pretectal area and the ventral nucleus of the lateral geniculate body because of the importance for behavioral functions of the convergence of somatosensory, auditory, and visual inputs in the tectum. The retinal projection to the midbrain evolved topographic organization with greater precision than in the pretectum, enabling better novelty detection and more precise turning responses (figure 20.5). This triggering of turning responses has been termed the activation of the *visual grasp reflex* by researchers studying fish and other non-mammalian vertebrates.

The visual grasp reflex required a crossed descending pathway, the *tectospinal tract*, as already noted. Thus, the midbrain roof evolved to serve *two* major functions: escape from predators and orienting toward objects.

Identifying Animals, Objects and Textures

A third role of visual images was the role of identifying animals and objects and textures for making decisions about approach or avoidance, reaching, or other responses. This role has led to remarkable evolutionary changes in the central nervous system, particularly in the forebrain.

Identification of visual stimuli by analytic filtering cascades of neuronal connections begins in the retina, a portion of the 'tweenbrain that in early development forms, on each side, from an evagination of the rostral end of the neural tube. It extends out toward the surface of the head where it induces the formation of the lens. The filtering of information that begins in the functioning retinae continues in the optic tract's terminal areas. In the midbrain tectum and the pretectal region, the connections trigger innate patterns of movement in response to specific patterns of visual stimuli. Thus, these connections constitute some of the "innate releasing mechanisms" of the ethologists. For example, there are retinal ganglion cells in the frog that are activated by small dark moving objects—bugs—and their axons terminate in one layer of the optic tectum, where they can activate output cells and thereby elicit orienting movements that are followed by a flick of the tongue designed to catch a bug. Whether the orienting response is elicited depends also on the state of the animal that we can assume is correlated with other, modulating inputs with more diffuse connections to the tectum as well as other brain structures. The activity of these connections varies with behavioral states like hunger and with time of day and the level of perceived safety or danger from predators.

The ability to learn new visual stimulus identities followed, in evolution, the invasion of the endbrain by ascending pathways from brainstem structures, especially from the optic tectum. These pathways ascended, via the thalamic region, mainly to two regions of the endbrain, the corpus striatum and the neocortex (as first mentioned in chapter 4).

Consider two examples of the highly adaptive nature of this ability, both requiring fairly high visual acuity. A small rodent learns to use visual landmarks to find remembered sources of food or water and later to find its way back toward its home burrow. Many birds do the same over greater distances. Birds and other animals learn to recognize conspecifics by vision, even showing individual recognition. They also can recognize predators by visual cues.

Why was it that relatively so recently in evolutionary time these visual identification abilities evolved so much? It was made possible because the stage had been set by the evolution of precise topographic maps of an expanded retina. By various combinations of inputs from adjacent parts of the map, neurons gained selectivity to specific contour orientations and other configurations. This will become more evident in the next chapter. They did this in the midbrain, but it was only later as the visual pathways invaded the endbrain that this evolutionary trend really took off, as the ability to learn new associations and form visual memories of places, things, and other individuals became possible.

The Invasion of the Endbrain by Visual Pathways: Likely Evolutionary Steps in Pre-mammalian and Mammalian Ancestors

In a likely first step, axons from the midbrain tectum and from the pretectal area formed connections in the early thalamus. These connections came from what are now deeper layers of tectum and pretectum. They resembled many connections of the brainstem reticular formation, carrying multimodal information into what are now the midline and intralaminar thalamic cell groups. These thalamic cells evolved two kinds of ascending projections, in addition to some descending ones: Axons went rostrally to the corpus striatum and to the pallium (in mammals, the neocortex) (figure 20.6). Initially, the connections to the pallium were not to neocortex (which had not yet evolved); they were most likely fairly diffuse as well as multisensory.

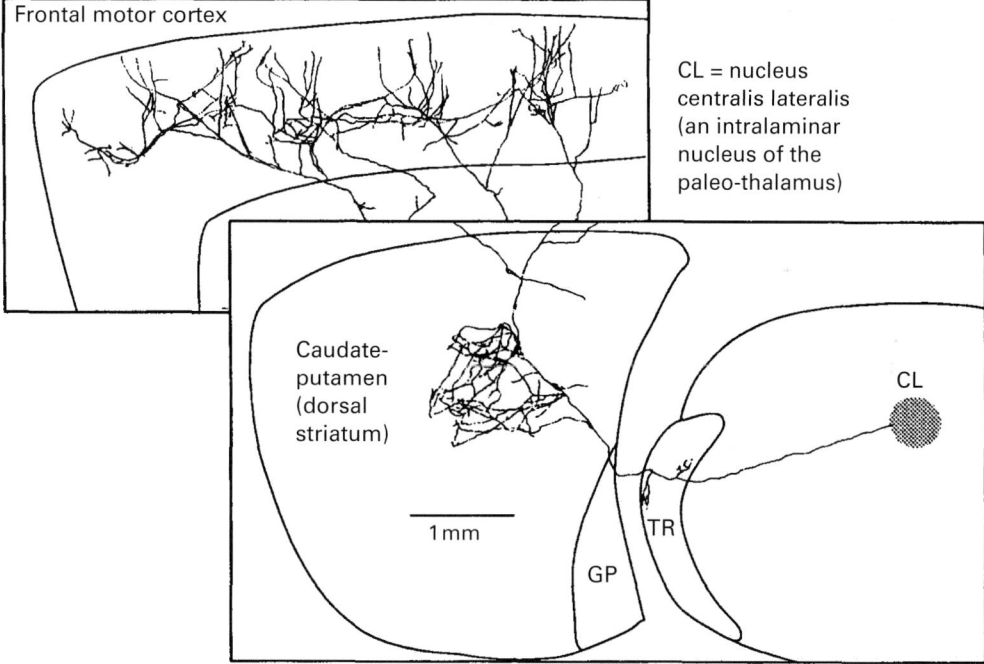

Figure 20.6
Drawing composed from three different sections of rat brain to illustrate the projections of a representative neuron of the intralaminar thalamic cell group, nucleus centralis lateralis (CL). These cells of the "old thalamus" receive sensory inputs (not shown) from the superior colliculus. Terminating collaterals of the axon are seen in the thalamic reticular nucleus (TR) and in the dorsal striatum (caudate-putamen). The axon continues into the neocortex where it forms a series of terminal arbors in the motor cortex. GP, globus pallidus. The axon was labeled with biocytin. From Deschênes, Bourassa, and Parent (1996).

Next, the specifically visual layers of tectum and pretectum formed projections to neuronal groups in what was evolving as the neothalamus. These can be found in current mammals in the lateral nuclei of the thalamus: in the lateral-dorsal nucleus and in the lateral posterior nucleus. These cells project to posterior regions of neocortex.

In another step, perhaps as ancient as the step noted above, a "short cut" from the optic tract into the lateral thalamus appeared or became stronger. This connection resulted in a more rapid route to the cortex that could evolve independently of midbrain and pretectal functions. This connection may have evolved in a region that already received visual input coming from the midbrain. In mammals, the termination area became the dorsal nucleus of the lateral geniculate body, which projects to the *primary visual cortex*, also known as the striate cortex or area 17. This projection appears to exist in all mammals that have been examined, and it is always topographically organized. In non-mammalian vertebrates, the equivalent cell group exists, but it is generally less developed.

It seems possible, then, that when a posterior cortical region first resembled what we call the primary visual cortex, its visual input came primarily from the midbrain tectum via a portion of the posterior part of the lateral thalamus that was the forerunner of the dorsal nucleus of the lateral geniculate body of modern mammals. Such a visual cortex was probably found prior to the evolution of mammals. In modern mammals (e.g., in the rat and the hamster), a tectal projection to the dorsal nucleus of the lateral geniculate body is found, overlapping with the direct retinal projection. Because of the importance in evolution of rapid conduction as well as advantages of independence from the midbrain, the direct retinal projection developed a greater importance.

More about the Third Role of Visual Images

The adaptive advantages of visual recognition abilities are not difficult to suggest. Imagining the specific steps by which this ability evolved is not as easy. In an initial step, we have suggested that visual information reaching the corpus striatum by an oligosynaptic route (a pathway with few synapses) had advantages for the learning of habits (see chapter 4). We just noted earlier that such a pathway can be found in modern mammals, passing from midbrain tectum to the paleothalamic cell groups: the thalamic midline and intralaminar nuclei. These cell groups also project, without precise topography, to the pallium—in mammals, the neocortex. In the rat, some of the very same axons passing rostrally from these parts of the thalamus have been found to send branches to both striatum and neocortex (figure 20.6).

In amphibians, the dorsal thalamus projects not only to the dorsal cortex—forerunner of neocortex—but also to the lateral amygdala and to the medial pallium, the forerunner of the hippocampal formation. Thus, visual information probably reached these structures long before the advent of mammals. The dorsal cortex also projects to the medial pallium. In mammals, visual information reaching the neocortex also follows visual routes to the amygdala and hippocampal regions. The visual inputs to these structures evolved in some mammals to become more dominant than the olfactory for object and place memories. The neocortex also provided another visual-system route to the striatum and also to output mechanisms of the pretectum, tectum, and subthalamus that, as a precise topographic organization evolved, enabled better acuity for orienting and for learned approach–avoidance choices.

Greater modality specificity led to an evolution of greater importance of the lateral thalamic cell groups that were dominated by inputs from the visual layers of the midbrain tectum. As suggested earlier, a portion of the lateral thalamus evolved a more precisely organized topographic representation of the retina, becoming the lateral geniculate body. This cell group became dominated more and more by direct inputs from the retina, resulting in faster conduction between eye and neocortex and an independence from midbrain tectal information processing.

Another factor leading to a retino-thalamocortical pathway that was independent of the midbrain tectum was found in animals with frontal eyes. These animals, mostly predators and animals dependent on manual reaching for food objects, were benefited by depth detection, which became more precise with greater acuity, enabling more accurate approach and grasping of objects. In mammals, the neocortical visual areas became important for depth vision as acuity improved.

The midbrain tectum/superior colliculus did evolve depth detection, as did the pretectal area. We know this from investigations that include recordings from single neurons in the cat. Very early in evolution, the tectum had been important for avoiding and escaping from predators as well as for orienting toward stimuli such as those coming from food objects. Precise depth detection is more important for the latter, but it is most important when reaching with the limbs is involved, and studies of neuronal connections indicate that the tectum had no easy access to reaching mechanisms (see chapter 11). This may have been an important factor that gave the advantage to the posterior cortex in the evolution of depth detection within a system of object and place recognition and the control of reaching.

Expansions and Specializations in the Visual System

Vision became of crucial importance for survival in a great many species, but different uses of vision evolved in different groups of animals. The species differences in adaptations created differences in the relative importance of different components of the visual system. These differences led in evolution to corresponding differences in size and cellular organization.

The midbrain tectum, or optic tectum, became an enlarged *optic lobe* in many teleost fishes (figure 4.8b) and in birds, with a superficial tectal cortex—a laminar pattern of cellular organization near the brain surface. Species differences are great. Just among mammals there is a great range of size and complexity of the homologous structure, the superior colliculus (see figure 11.3). The name means "little hill"—a term given this structure because of its appearance in dissections of the human brain. But such a name is truly a misnomer for this portion of the midbrain roof in some mammals where it has a great size and a laminar arrangement that make the term *optic lobe* a more relevant one. It is considerably enlarged (relative to other parts of the brain) in small visual animals like the squirrel and in small insectivorous animals like the Malayan tree shrew (figure 11.3).

All mammals have visual neocortex, including the region known as V1, or primary visual cortex (area 17). However, the dorsal thalamic visual system structures and their neocortical projection areas vary greatly in size and cellular arrangements. Primates have specialized in vision and visually guided behavior, resulting in great evolutionary expansions that included the optic tectum of the midbrain in some species, and especially in an expanded visual cortex and dorsal thalamic cell groups projecting to this cortex. In all modern mammals, the primary visual cortex receives input originating in the retina that passes through only one synapse in the dorsal thalamus, specifically in the dorsal nucleus of the lateral geniculate body. (The name refers to its appearance on the caudal thalamic surface as a knee-like bump, lateral to another bump caused by the auditory thalamic cell group.)

An anatomical method for judging the relative importance of midbrain vision and cortical vision involves making a comparison of the relative sizes of the optic-tract terminal areas in the midbrain tectum (the superficial gray layer of the superior colliculus) and the lateral geniculate body of the dorsal thalamus. In the course of teaching neuroanatomical techniques, I helped students make measures of the volumes of these terminal areas in the Syrian hamster, the Norway rat, and the rhesus monkey. The ratio of the volume of the superior colliculus terminal zone to the volume of the lateral geniculate body in the hamster was 3.0, in the rat it was 2.1, and in the monkey it was 0.1. The hamster is a slightly more visual animal than the rat, with great importance of subcortical visual mechanisms; the monkey is not only very visual, but also the relative importance of its neocortical visual system pathways for the learning of object and face identities and for various social behaviors is much greater. The midbrain in monkeys, with major inputs from neocortical areas as well as from the retina, retains its importance for avoiding predators and for orienting to novel stimuli.

The rhesus monkey is an old-world primate, and in primates the specializations in visual functions have resulted in great evolutionary expansions of visual cortical regions, with major transcortical pathways that begin in primary visual cortex. These pathways will be discussed in chapter 22.

The retina has evolved a great range of sensitivity and color detection in various species, and its relative size has enlarged in animals with a need for a large lens for light-gathering power or a need for high acuity with larger numbers of receptor cells and retinal neurons.

In other components of the subcortical visual system, there have been less noticeable expansions, probably because these structures played roles that made smaller demands on size.

All of the subcortical components can be defined, first of all, by the axonal projections from the retina. The pattern of these projections is very similar in all mammals, and in fact in other vertebrates as well. Next we illustrate and characterize the retinal projections.

Readings

Light detection in amphioxus: See Allman (2000), based on studies of T. C. Lacelli. Allman's book is listed in chapter 3. Also, see:

Stokes, M. D., & Holland, N. D. (1998). The lancelet. *American Scientist, 86*, 552–560.

Arctic animals, at least reindeer, no longer show daily rhythms of activity when it is dark throughout the day, but they do show regular daily activity patterns at other times:

van Oort, B. E. H., Tyler, N. J. C., Gerkema, M. P., Folkow, L., Blix, A. S., & Stokkan, K.-A. (2005). Circadian organization in reindeer. *Nature, 438*, 1095–1096.

The adaptive value of an endogenous circadian rhythm entrained by the day-night cycle:

Hut, R. A., & Beersma, D. G. M. (2011). Evolution of time-keeping mechanisms: Early emergence and adaptation to photoperiod. *Philosophical Transactions of the Royal Society, B, 366*, 2141–2154.

Bell-Pedersen, D., Cassone, V. M., Earnest, D. J., Golden, S. S., Hardin, P. E., Thomas, T. L., & Zoran, M. J. (2005). Circadian rhythms from multiple oscillators: Lessons from diverse organisms. *Nature Reviews Genetics, 6*, 544–556.

Klein, D. C., Moore, R. Y., & Reppert, S. M., eds. (1991). *Suprachiasmatic nucleus: The mind's clock*. New York: Oxford University Press.

Escape responses and the superior colliculus:

Dean, P., Redgrave, P., & Westby, G. W. M. (1989). Event or emergency? Two response systems in the mammalian superior colliculus. *Trends in Neuroscience, 12*, 137–147.

Merker, B. (1980). The sentinel hypothesis: A role for the mammalian superior colliculus. PhD thesis, Massachusetts Institute of Technology.

Not mentioned in the text is another route carrying visual information to the endbrain of the rat—and probably in other mammals. It passes from the superior colliculus, superficial as well as deeper layers, to the so-called suprageniculate nucleus of the thalamus (a caudal extension of the lateral-posterior nucleus located above the medial geniculate body), to the lateral nucleus of the amygdala.

Much information comparing structures and connections in a wide range of species can be found in the book by Butler and Hodos cited at the end of chapter 3.

Two interesting discussions of visual system evolution from insects to vertebrates:

> Horridge, G. A. (1987). The evolution of visual processing and the construction of seeing systems. *Proceedings of the Royal Society of London, Series B, 230,* 279–292.
>
> Sanes, J., & Zipursky, S. L. (2010). Design principles of insect and Vertebrate visual systems. *Neuron, 66,* 15–36.

Several interesting additional papers on brain evolution:

> Barton, R. A., & Harvey, P. H. (2000). Mosaic evolution of brain structure in mammals. *Nature, 405,* 1055–1058.
>
> Katz, P. S., & Harris-Warrick, R. M. (1999). The evolution of neuronal circuits underlying species-specific behavior. *Current Opinion in Neurobiology, 9,* 628–633.
>
> Redies, C., & Puelles, L. (2001). Modularity in vertebrate brain development and evolution. *BioEssays, 23,* 1100–1111.

Interesting findings concerning pretectal area functions can be found in the publications of David Ingle on his research with frogs:

> Ingle, D. (1977). Detection of stationary objects by frogs (*Rana pipiens*) after ablation of optic tectum. *Journal of Comparative and Physiological Psychology, 91,* 1359–1364.
>
> Ingle, D. J. (1980). Some effects of pretectum lesions on the frogs' detection of stationary objects. *Behavioral Brain Research, 1,* 139–163.

The hypertrophied pretectal nucleus of hummingbirds:

> Iwaniuk, A. N., & Wylie, D. R. W. (2007). Neural specialization for hovering in hummingbirds: Hypertrophy of the pretectal nucleus lentiformis mesencephali. *Journal of Comparative Neurology, 500,* 211–221.

21 Visual Systems: The Retinal Projections

We have noted that the projections of the retina follow a similar pattern not only in all mammals but also in all vertebrates, although the relative sizes of optic-tract termination areas and their cellular architectures show great variability. The basic layout of the retinal projections was illustrated in the schematic of the mammalian brain shown in figure 20.3. This projection pattern will now be examined in more detail, with a look at some species differences. Included are some illustrations of the laminar architecture of the midbrain tectum and an overview of the topographic organization of the entire optic tract in relation to its termination patterns.

Two Views of the Optic Tract and Its Terminations

Figure 21.1 illustrates frontal sections of the 'tweenbrain as it would appear in an embryonic human brain or in an adult hamster or rat brain. On the left side of panel A, the optic tract is seen coursing over the surface from the hypothalamus ventrally to the pretectal region of the epithalamus dorsally. It is because the optic tract envelopes the caudal thalamus that this part of the dorsal thalamus is called the *optic thalamus* in some of the neuroanatomical literature.

Note the structures that the optic tract traverses in these brain sections. As the axons extend dorsally after many of them have crossed the ventral midline at the optic chiasm, they cover part of the hypothalamus and reach the ventral thalamus, or subthalamus, where some of them terminate in the ventral nucleus of the lateral geniculate body (LGBv, sometimes abbreviated as LGv). (Sometimes a neuroscientist will use the term *nucleus* instead of *body*, so the LGBv becomes the ventral part of the lateral geniculate nucleus, or LGNv.) As the optic-tract axons continue dorsally, they reach and terminate heavily in the dorsal part of the lateral geniculate, the LGBd, or LGd. Many optic tract axons continue beyond the geniculate nuclei, passing over (or through) the lateral posterior nucleus (LP) where there is some much sparser termination. Finally, they reach and pass over or through the pretectal area with some dense terminations. It is in this region that the axons are turning in a more caudal direction. Just behind (caudal to) the pretectal

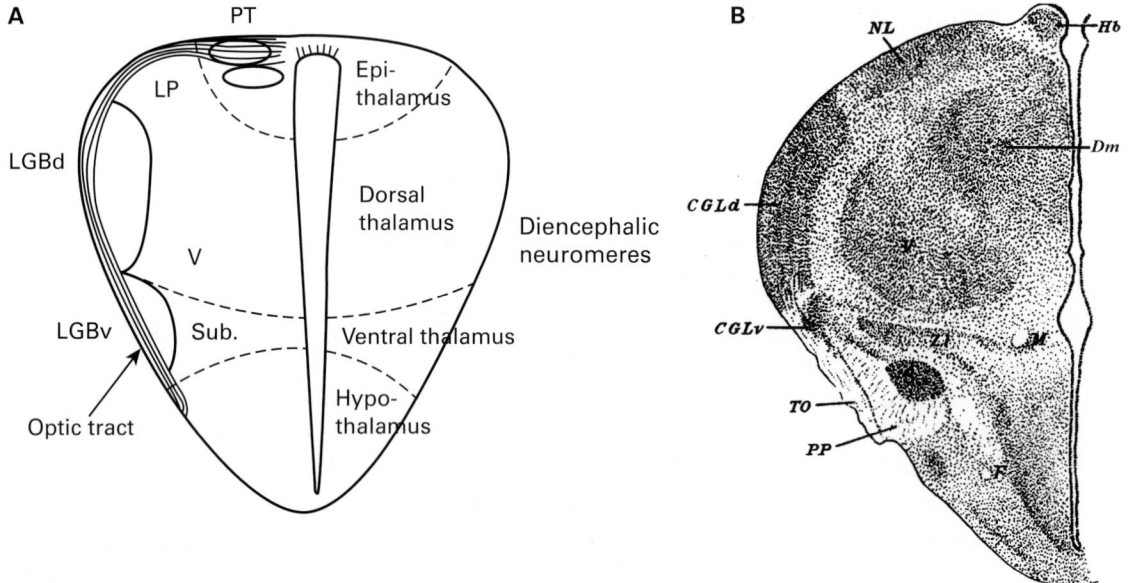

Figure 21.1
(A) Drawing (simplified) based on frontal sections of the embryonic human caudal diencephalon showing the optic tract (similar to other developing mammalian brains). The level is caudal to the optic chiasm and rostral to the superior colliculus. This plane of section passes through four segments of the forebrain neuraxis—the diencephalic neuromeres, named on the right side—as illustrated in figure 21.2. These segments had been described by structural studies preceding genetic evidence for segmentation. On the left, the positions of adult structures are indicated. LGBd, dorsal nucleus of the lateral geniculate body; LGBv, ventral nucleus of the lateral geniculate body, at the lateral edge of the ventral thalamus—called the subthalamus in the adult (Sub). LP, lateral posterior nucleus; PT, pretectal nuclei; V, ventral nucleus of the thalamus. Note that the "ventral thalamus" of the embryo is not the same as the ventral nucleus of the dorsal thalamus. (B) Drawing by W. E. Le Gros Clark of a cell-stained coronal section through the thalamus of a 92-millimeter human embryo (12 weeks postconception), showing the position and shape of the lateral geniculate nuclei. Labels are abbreviations of the Latin names. CGLd, the dorsal nucleus of the lateral geniculate body; CGLv, the ventral nucleus of the lateral geniculate body; Dm, the mediodorsal nucleus of the thalamus; F, fornix bundle; Hb, habenular nucleus; M, mammillothalamic tract; NL, lateral nucleus (lateral posterior portion); PP, cerebral peduncle; TO, optic tract. Panel (B) from Le Gros Clark (1932).

area, the axons enter their major midbrain terminus, the superficial layers of the superior colliculus (the optic tectum of non-mammalian vertebrates).

On the right side of figure 21.1A, to the right of the third ventricle, four subdivisions are demarcated. These are four embryonic segments or neuromeres. (The root word *mere* means "part," from the Greek *meros*.) These neuromeres are segments of the developing neural tube, arranged as shown in figure 21.2. In that figure, note the position of the neuromeres of the hindbrain, commonly called the rhombomeres (discussed in chapter 10) and two additional subdivisions corresponding to the midbrain and the isthmus. Note also that because of the bending of the developing neural tube—the forebrain and midbrain flexures—a dorsal to ventral cut just rostral to the midbrain will show the epithalamus at

Visual Systems: The Retinal Projections

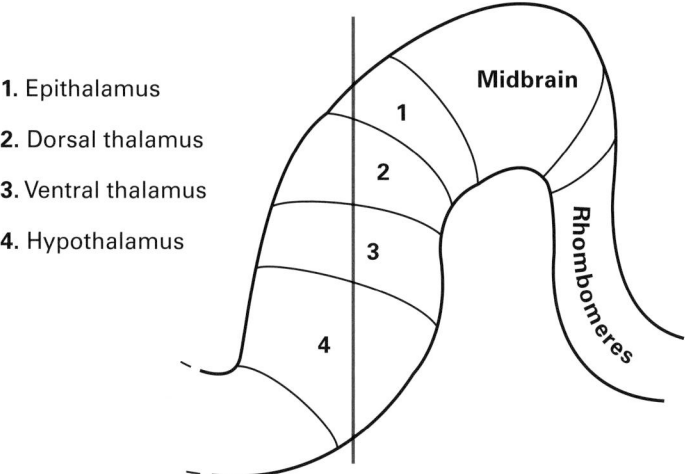

1. Epithalamus
2. Dorsal thalamus
3. Ventral thalamus
4. Hypothalamus

Figure 21.2
Sketch of a side view of the cranial portion of the neural tube caudal to the developing cerebral hemispheres. The numbered segments are the diencephalic neuromeres: 1, epithalamus; 2, dorsal thalamus; 3, ventral thalamus; 4, hypothalamus. The vertical line indicates the plane of the section sketched in figure 21.1. (Note: In some recent interpretations of genetic expression evidence, there are two or three diencephalic neuromeres that become the hypothalamus.) The segment just caudal to the midbrain segment is the isthmus.

the top and the hypothalamus at the bottom, the positions of these regions in the frontal sections in figure 21.1.[1]

Next we look at the optic tract and its terminal areas in a different way (figure 21.3). To understand this figure, imagine slicing through the optic tract at right angles to the brain surface, beginning at the optic chiasm at the base of the diencephalon, and continuing this cut all along the tract, always parallel to the direction of the fibers, all the way through the superior colliculus. Thus, we get a section that has been stretched so the optic tract is straightened. The sketch illustrates the nature of terminal arbors of different types that branch from the main tract in each of the terminal areas.

The complete stretched section illustrates the optic tract terminations in a rat, mouse, or hamster.[2] Some of the retinofugal axons (from both eyes) terminate in the hypothalamus, mostly, but not only, in the suprachiasmatic nucleus discussed in the previous chapter. Beyond that region, the axons shown in the upper drawing are axons that have crossed at the chiasm from the retina of the opposite eye. Note that in each of the lateral geniculate nuclei, the terminal arbors avoid a medial region. This is because in this part of the geniculate nuclei there are two distinct layers, one related to each eye. The medial region receives optic tract axons originating in the eye on the ipsilateral side—axons of the uncrossed portion of the optic tract.

In some other mammalian species there are more than two layers. For example, in carnivores like the cat and ferret there are at least four layers as shown in the lower part

374   Chapter 21

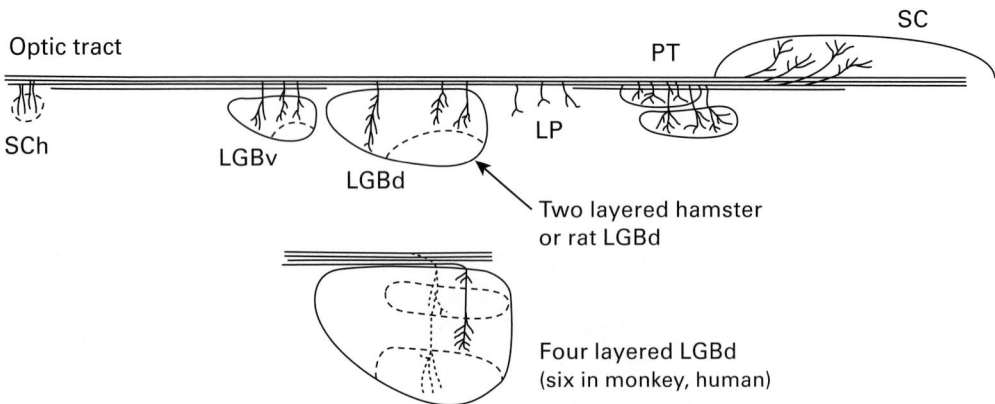

Figure 21.3
Schematic drawing of the mammalian optic tract: Consider the tract as straightened out from the optic chiasm to the superior colliculus and sectioned at right angles to the surface. A group of axons from the contralateral eye is depicted to show the terminal arbors formed by the axons and their collaterals. Terminations are formed, in order from chiasm to midbrain, in the suprachiasmatic nuclei (SCh); the ventral nucleus of the lateral geniculate body (LGBv); the dorsal nucleus of the lateral geniculate body (LGBd); the lateral posterior nucleus (LP) where there is only weak termination; nuclei of the pretectal area (PT)—the drawing shows the more superficial nucleus of the optic tract and the deeper-lying posterior pretectal nucleus; and finally the superior colliculus (SC). Only in the superior colliculus do most axons turn toward the surface rather than away from the surface before terminating. In the LGBd of the rat or hamster or mouse, there is an area of the nucleus where axons only from the ipsilateral eye terminate. The lower drawing shows how in some animals the axons from the contralateral eye terminate in two separate layers, whereas axons from the ipsilateral eye end in two other layers. In the rhesus monkey, in apes and in humans there are six layers: three layers for terminals from the axons of each eye.

of figure 21.3. Note that the axons from the contralateral eye terminate in two distinct layers of the geniculate, and axons from the ipsilateral eye end in the other two layers. In monkeys and in humans, there are six distinct layers of this type; the contralateral retina sends axons to three (layers 1, 4, and 6, numbered from the optic tract surface) and the ipsilateral retina to the other three. In each of these species, the terminating end arbors within a single small bundle in the lateral geniculate body, containing axons of both contralateral and ipsilateral origin, come from corresponding spots in the two retinae: These spots "look at" the same part of the visual field when the animal is focused on one particular place. The channels of information flow from the two eyes remain separate through the dorsal lateral geniculate to their initial terminations in the visual cortex.

Distortions in Large Primates

Now, look at the photograph shown in figure 21.4A. It shows the laminated structure of the dorsal nucleus of the lateral geniculate body on the left side of the thalamus of a macaque monkey brain in a frontal section stained for neuronal cell bodies. The six layers are very clear. But why are they curved in this way? The optic tract at the surface is at the bottom, next to the first two layers that have larger cells than the other four layers. (Thus, we talk about the two magnocellular layers and the four parvocellular layers.) The shape of the entire cell group is very different from what it was in the embryo, when it looked like the section illustrated in figure 21.1B. As the embryo grows and develops, the growth of the lateral posterior portions of the thalamus is much greater than in the rodents we have mentioned. This growth continues and results in a pushing of the lateral geniculate nucleus from the medial and the dorsal sides. There is also an expansion of the thalamus on the rostral side, so the geniculate becomes somewhat rotated. It is not so difficult to imagine how these changes result in the "scrunched" appearance seen in figure 21.4A. The enlargement of the posterior part of the lateral thalamic region results in the very large and complex thalamic nucleus that has been named the *pulvinar*, meaning "a pillow," because of the large bulge it makes at the caudo-dorsal thalamus in the human brain.

Figure 21.4B shows the LGBd on the right side of a human brain. The photograph was taken at a somewhat lower magnification, so more of the very large pulvinar nucleus is seen above and to the left (to the medial side) of the geniculate body.

Some of the variety in the appearance of this structure in primates and the tree shrew (once considered to be the most primitive primate, but now considered to be an insectivore) is illustrated in figure 21.4C, in drawings by the great British anatomist W. E. Le Gros Clark published in 1932.

How the Optic Tract Looks in the Brain of an Adult Animal

Now you can understand more about the appearance of a real brain and the retinal projections it includes. Figure 21.5 shows a reconstructed outline view of the right side of the upper brainstem (midbrain and 'tweenbrain) of an adult hamster. The borders of several structures that are near the surface are shown. The illustration depicts the course of a small bundle of axons of the optic tract as they course from the optic chiasm up the side of the 'tweenbrain, over the thalamus and into the midbrain. The terminations of these axons in the lateral geniculate cell groups and in the superior colliculus are depicted schematically.

Also in the illustration is another strong projection in the subcortical visual system: from the visual layers of the superior colliculus (SC) to the lateral posterior nucleus and to the outer layer of the ventral nucleus of the lateral geniculate body. The pathway to LP is very much stronger than the weak retinal projection to this structure. The LP projects to the posterior neocortex, including a strong projection to the area adjacent to the primary visual cortex. Later, we will have further discussions of these pathways carrying visual information to the neocortex.

Neuroanatomical data that have provided strong evidence for the projections just discussed have been gathered from experiments with axon tracing methods. Figure 21.6 shows the reconstruction of the upper brainstem of the adult hamster again, this time with a reconstruction of the entire optic tract and the areas where it has strong

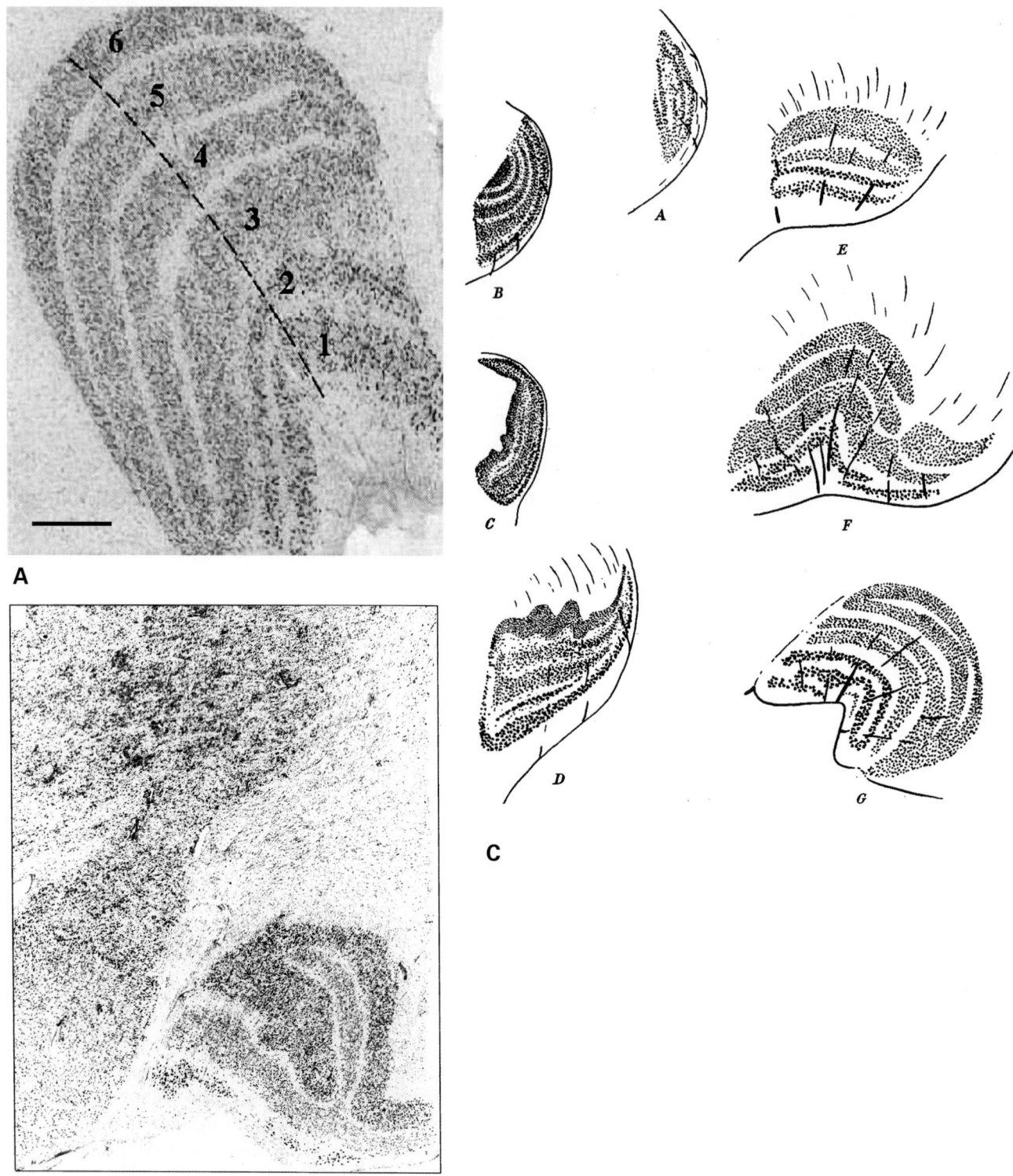

Visual Systems: The Retinal Projections

Figure 21.4
(A) Frontal section through the dorsal nucleus of the lateral geniculate body of the rhesus macaque monkey. The geniculate layers are usually numbered beginning at the optic tract at the surface (at bottom right). Axons from the eye on the contralateral side terminate in layers 1, 4, and 6. Axons from the ipsilateral eye terminate in layers 2, 3, and 5. Axons that terminate along a line that is approximately at right angles to the layers are from the same or corresponding parts of the retinae (note the dashed line in the figure). Scale bar (1 mm) has been inserted. (B) Human lateral geniculate body on the right side, showing lamination, and parts of the very large pulvinar nucleus to the left and above the geniculate. At this level, four small-celled layers are clearly discernible, plus one of the two large-celled (magnocellular) layers. A pathology of the ipsilateral eye has resulted in a reduced staining of ipsilateral layers 3 and 5. The optic tract is just below the layers at the edge of the thalamus. Fibers are not very visible in this high-contrast method of staining Nissl substance of cell bodies using a silver-staining method developed by Bjorn Merker. (c) Lateral geniculate body, dorsal nucleus, on the right side of the thalamus, drawn from cell-stained coronal sections of seven species: A, tree shrew; B, mouse lemur; C, Coquerell's dwarf lemur; D, *Lemur catta*; E, orangutan; F, human; G, *Cercopithecus* (old world monkey, often called a guenon). The arrangement of cell layers and the direction of entering blood vessels are illustrated. Panel (A) from Hubel (1988); panel (B) courtesy of Bjorn Merker; panel (C) from Le Gros Clark (1932).

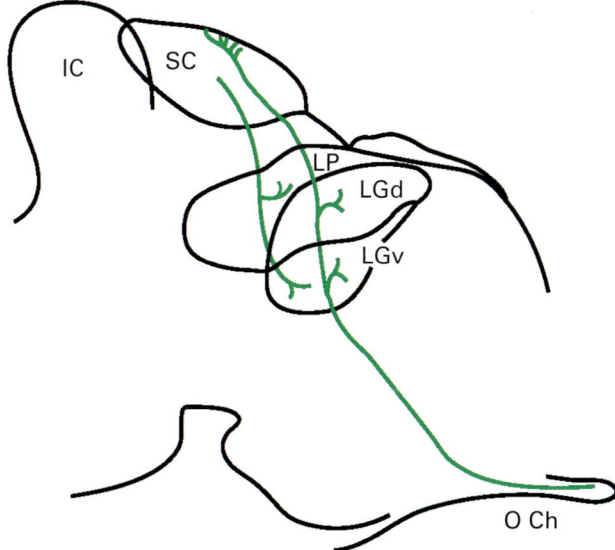

Figure 21.5
On this reconstructed view of the right side of the midbrain and 'tweenbrain of a hamster brain, two small bundles of axons are depicted in a simplified manner, one coming from the retina and one from the superficial layers of the superior colliculus. Axon terminals are schematically indicated. Retinal projections to the suprachiasmatic nucleus, the pretectal nuclei, and the nuclei of the accessory optic tract are omitted. IC, inferior colliculus; LGd, dorsal nucleus of lateral geniculate body; LGv, ventral nucleus of lateral geniculate body; LP, lateral posterior nucleus; O Ch, optic chiasm; SC, superior colliculus.

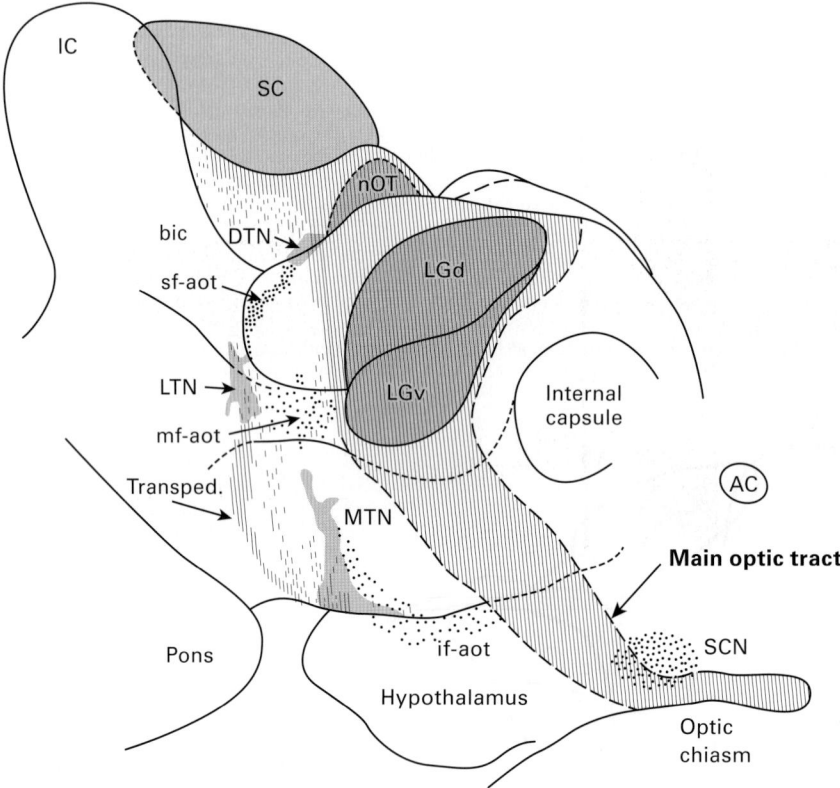

Figure 21.6
The optic tract of an adult Syrian hamster, reconstructed in this view of the right side of the midbrain and diencephalon from serial, frontal sections. Retinal projections were made visible in the sections with a Nauta silver-staining method for degenerating axons and their terminals. AC, anterior commissure at the midline; bic, brachium of the inferior colliculus; DTN, dorsal nucleus of the accessory optic tract; if-aot, inferior fasciculus of the accessory optic tract; LTN, lateral terminal nucleus of the accessory optic tract; mf-aot, middle fasciculus of the accessory optic tract; MTN, medial terminal nucleus of the accessory optic tract; nOT, nucleus of the optic tract; transped., transpeduncular component of the accessory optic tract; other abbreviations as in previous figures.

terminations. These findings have been verified with several different tract-tracing techniques. Note that the main optic tract has several smaller offshoots that terminate in three small cell groups located near the surface of the brain. The smaller offshoots are called the superior, middle, and inferior fasciculi of the accessory optic tract. These axons provide retinal input to the nuclei of the accessory optic tract (the dorsal, lateral, and medial terminal nuclei). This system of axons and these small nuclei are found in many vertebrates, but there is variability in the relative prominence, and even the existence, of one or more of them in some species.

The photograph in figure 21.7 shows a similar view of the right side of the brain of an adult hamster that had been perfused with a saline solution (after a lethal overdose

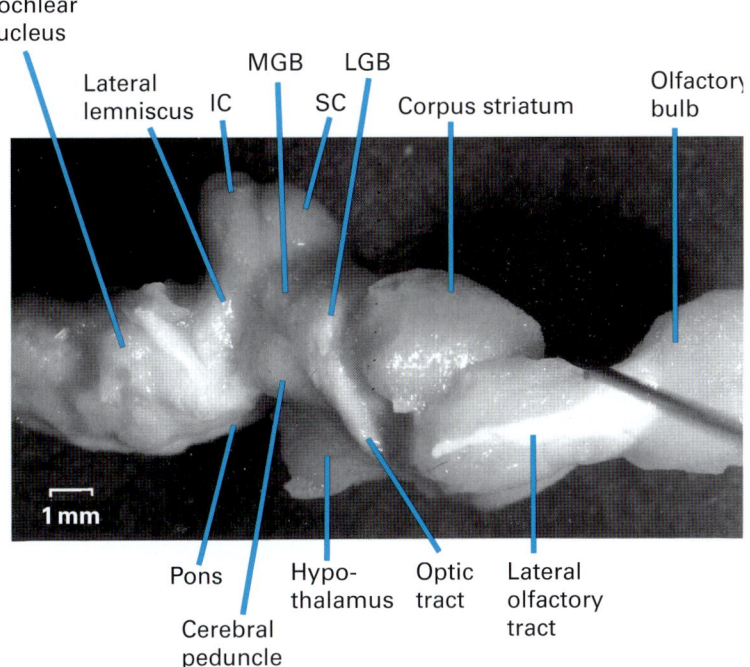

Figure 21.7
Photograph of a hamster brain with the hemispheres and cerebellum removed, seen from the right side. Olfactory bulbs are to the right, hindbrain is to the left. Major tracts are visible because of the whiteness of myelinated axons: the lateral olfactory tract behind the olfactory bulb; the optic tract sloping upward and caudally in the middle; the middle cerebellar peduncle caudally, cut off in the removal of the cerebellum. At the top, the inferior and superior colliculi are prominent. At the bottom, the hypothalamus is prominent below the optic tract. IC, inferior colliculus; LGB, dorsal nucleus of lateral geniculate body; MGB, medial geniculate body; SC, superior colliculus.

of anesthetic) and fixed with a formaldehyde solution. The cerebral hemispheres and cerebellum have been removed, exposing the corpus striatum and the surface of the diencephalon, the midbrain, and the hindbrain. The dissection spared the olfactory bulbs and some of the olfactory cortex (the piriform cortex) with the lateral olfactory tract showing up whiter than the tissue around it because of its myelinated fibers. Behind the striatum, another white band of fibers is very distinct: This is the optic tract as it ascends along the surface of the 'tweenbrain and covers the lateral geniculate body. Caudal to the optic tract ventrally is the hypothalamus with the stump of the pituitary (which was torn away during the removal of the brain from the skull). Further dorsally behind the optic tract, you can see the cerebral peduncle—the fibers carrying the output of the neocortex—covering the ventral third of the midbrain. At the top of the midbrain, you can see clearly the surfaces of both the superior (anterior) and the inferior (posterior) colliculi. Further details are noted on the figure. (The auditory pathway can be seen extending dorsally from the ventral part of the rostral hindbrain up to the inferior colliculus. In the photograph, this band of axons, called the lateral lemniscus, is as prominent as the optic tract. The auditory nerve stump is also visible as it joins with the brainstem at the cochlear nucleus.)

Looking at the Exposed Brain from Above

Next, we look at some views of the brain from above, looking down on the dorsal surface. A comparison of the brains of an adult hamster and a newborn is shown in the photographs of figure 21.8, both shown to the same scale. The only optic-tract terminal area seen in this view of the adult brain is part of the superior colliculus. The pretectal and thalamic surfaces are covered by the cerebral hemispheres. However, these hemispheres are much less developed in the newborn, so that the entire surface of the SC and part of the pretectal area are visible.

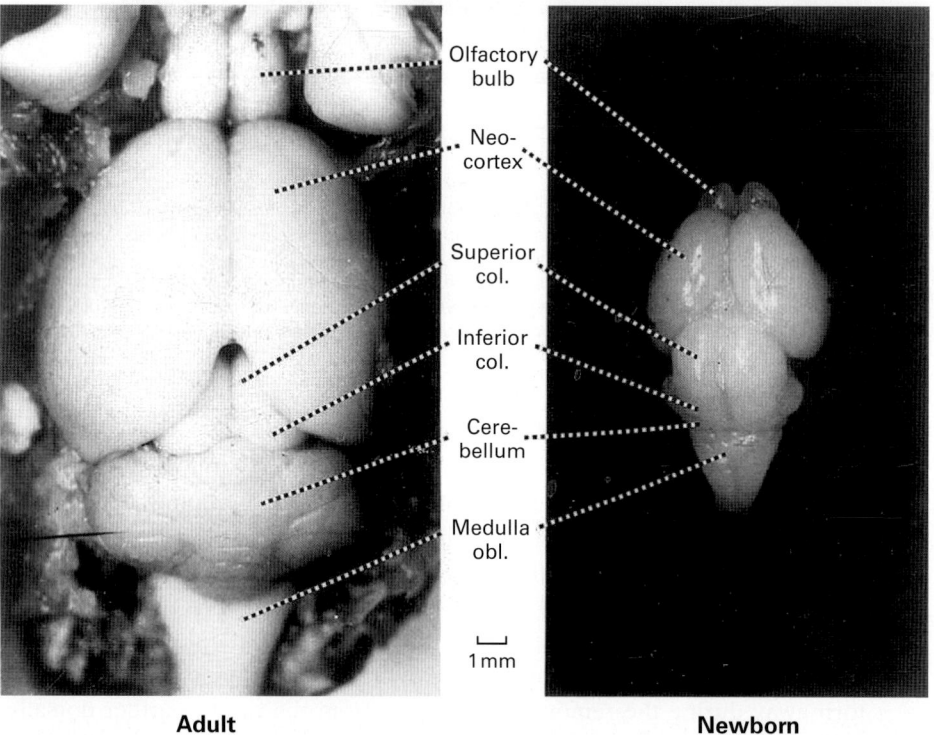

Figure 21.8
Photographs showing dorsal views of the adult hamster brain (left panel) and newborn hamster brain (right panel). col, colliculus; obl, oblongata.

Just behind the SC in both brains is the inferior colliculus, which looks whiter in the adult because of myelinated axons. Behind the colliculi in both brains is the cerebellum, but this structure in the newborn is so early in its development that it is only a small collar behind the midbrain.

If we remove the two cerebral hemispheres from the adult hamster brain, we can expose the subcortical structures beneath them as shown in figure 21.9. In this brain, small segments of the cortex were left in place rostrally just behind the olfactory bulbs. The cerebellum was removed. You can see the white, myelinated optic tract coursing over the thalamus on both sides. It crosses the lateral geniculate body and then the lateral posterior nucleus, with the border region marked by a slight change in lighting. Reaching the pretectal area, the brain surface becomes darker because more of the axons dive to a deeper location as they approach the midbrain. At the rostral border of the superior colliculus there is a slight shadow; the caudal border with the inferior colliculus is very obvious.

The brain shown in figure 21.10 has been dissected similarly, but on the right-hand side of this brain, much of the corpus striatum was also removed, exposing the anterior commissure. The borders of the lateral posterior nucleus with the geniculate and with the pretectal area are easier to see because of a change in lighting.

The Embryonic Optic Tract

The upper brainstem of a human or a great ape appears to be very complex compared with the hamster brain in the previous figures. However, keep in mind that early in embryonic development, it was very similar. Look again at figures 21.6 and 21.7: Notice how the optic tract bends around the internal capsule fibers in a curve that takes the rostral edge in the anterior direction before it turns caudally once again in its pathway toward the midbrain. The caudal edge of the tract also does not follow a perfectly straight course. Now, look again at the section of the lateral geniculate body of the monkey pictured in figure 21.4: The lateral border of the geniculate in the rat or hamster, similar to what is found in the human or monkey embryonic thalamus, corresponds to what has become the ventral border, folded over on itself, in the adult monkey or human. The distortion is the consequence of the growth of other structures, and this is much greater in the large primates, including humans, than in the rodents.

Now take a look at the hamster optic tract at an embryonic stage, reconstructed in figure 21.11. The optic nerve and tract are seen to grow very directly to the dorsal surface of the midbrain. After most of these fibers cross the ventral midline, they grow mostly in the caudal direction, along the side of the neural tube. The direction of the axons appears to be mainly dorsal only because of a strong bend in the tube — the mesencephalic flexure. The human brain passes through a very similar stage. However, in the brains of humans and other large primates, the course of the optic tract is altered more by the greater growth of various components of the forebrain.

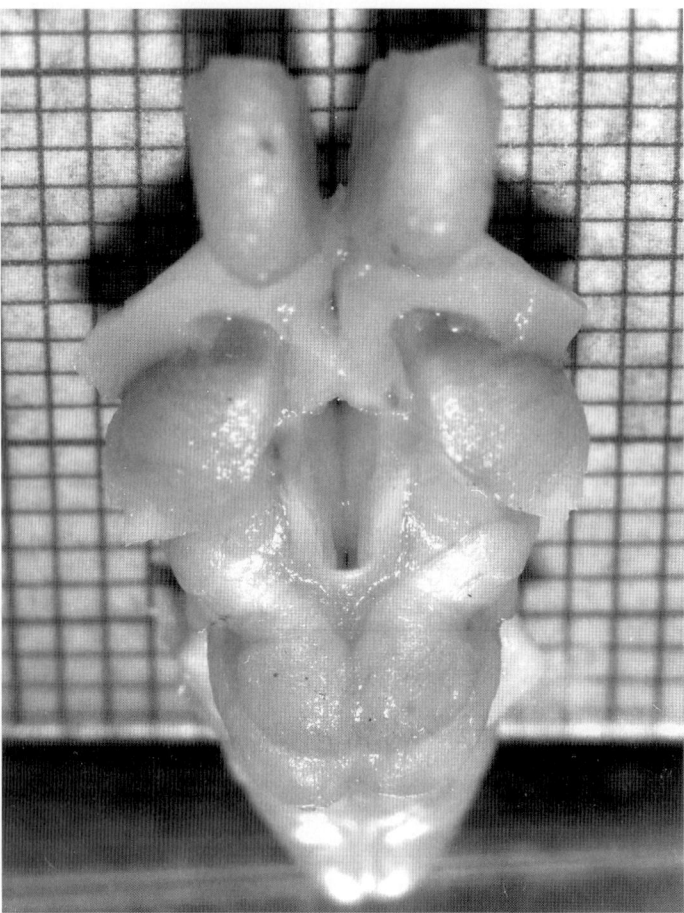

Figure 21.9
Photograph showing a dorsal view of the adult hamster brain with the cerebral hemispheres and cerebellum removed. The photograph was taken with the brain arranged so that the surface of the midbrain tectum was horizontal. The olfactory bulbs are at the top, the hindbrain at the bottom. The optic tracts are the white ribbons coursing from the lateral sides toward the pretectal area and midbrain tectum. The brain is resting on a millimeter grid.

Figure 21.10
Photograph of adult hamster brain similar to that of figure 21.9, but with the additional removal of the right corpus striatum and the exposure of the anterior commissure. The commissure is coursing toward the right olfactory bulb. The white diagonal band above the thalamus is the stria terminalis, coursing around the internal capsule from the amygdala to the basal forebrain (see chapter 29).

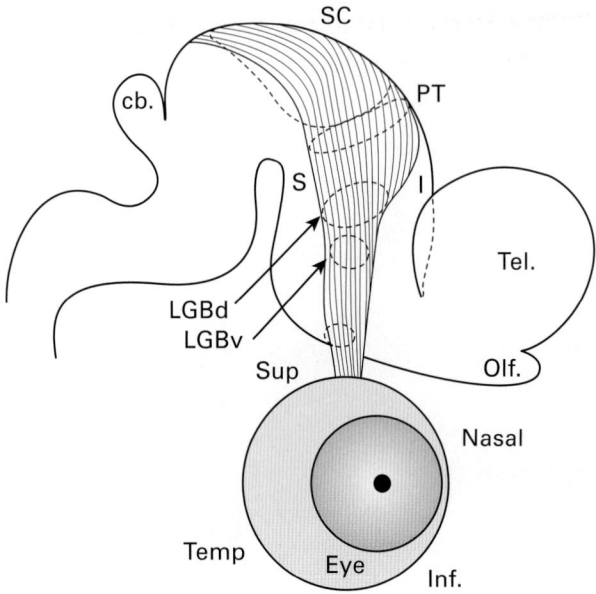

Figure 21.11
A reconstruction of the right side of the brain of an embryonic hamster showing a side view of the optic tract. At this early stage, about 3 days before birth, the caudal edge of the tract follows a straight course from chiasm to superior colliculus (SC). Cb, cerebellum; I, axons from the inferior retina; Olf., olfactory bulb; PT, pretectal area (the nucleus of the optic tract is outlined); S, axons from the superior retina; Tel., telencephalon. Retinal coordinates (Sup., superior; Inf., inferior; Temp., temporal; Nasal, nasal) are defined by the attachments of the four rectus muscles. Reconstruction by K. Hsiao and G. E. Schneider.

Midbrain Tectum: Species Differences

We have just seen an indication of the marked species differences in the lateral geniculate body region. There are some impressive species differences in the optic tectum as well. (When we use this term for mammals, remember that it is the same as the superior colliculus, or SC.) Look again at chapter 11, where we discussed the midbrain and find figure 11.3. In that figure, we see three frontal sections through the midbrain. At the top is depicted a typical midbrain section from the brain of a hamster, rat, or mouse. The middle section is a sketch of a human midbrain (see also figure 2.4). The most obvious difference is a huge increase in the size of the fiber bundles at the base—the cerebral peduncles, which include the axons coming from the greatly enlarged neocortex on their way toward the lower brainstem and the spinal cord.

The bottom of the figure, a sketch of a midbrain section from a small mammal like a squirrel or a tree shrew, is also very different from the small rodent section at the top of the figure. In this case, it is not the cerebral peduncle that is enlarged; it is the superior colliculus, which has enlarged so much that it can be called an optic lobe. The mammals with such an enlarged midbrain tectum have evolved much greater specialization for vision. Their visual neocortex has also evolved, but not to the great degree that it has in the large primates including humans. In these primates (old world monkeys, apes, and humans), it is not "midbrain vision" that has increased, but cortical vision (see the next chapter).

Lamination of the Midbrain Tectum

In neuroanatomy, the word *cortex*, meaning "bark" in Latin, refers to layered structures near the brain surface. Although the term usually refers to the cortex of the endbrain (both neocortical and paleocortical regions), it can be applied to some brainstem structures as well, especially to the midbrain surface structure usually called the optic tectum in many species of vertebrate. Figure 21.12 shows a section through the midbrain of a ray-finned fish, a bowfin. A dense distribution of neuronal cell bodies is near the ventricle, but additional cell bodies are seen in more superficial layers. The dendrites of the deep-lying cells extend toward the surface. On those dendrites, the axons from the retina terminate more superficially, and axons carrying somatosensory information terminate in a deeper position (figure 21.13). Thus, the lamination is not only of cell body locations but also of the terminals of populations of axons.

The next figure shows a midbrain section of a teleost fish, a sunfish, with a more elaborate laminar distribution of neurons (figure 21.14). Findings from the use of the Golgi method for study of the optic tectum of another teleost species reveal that dendritic arbors, as well as cell bodies and axonal end arbors, show a laminar pattern (figure 21.15). Studies of the optic tectum of amphibians and of reptiles have also found complex laminar patterns of neurons with relatively cell-free layers of axons and dendrites. Figure 21.16A illustrates a coronal section through the midbrain of a reptile, the iguana, stained for cell bodies. It shows a well-defined laminar pattern. Figure 21.16B depicts the layers of the frog's optic tectum. Neurons shown with dendrites and axons are simplified drawings from Golgi-stained material. At the far left, the layers are numbered beginning at the ventricle, based on Nissl and fiber stains. Some of the afferent axons coming from the retina are colored in red to highlight different types.

The mammalian superior colliculus likewise shows a well-defined laminar pattern. In many mammals, this pattern is most obvious in a fiber stain (figure 21.17).

Topographic Organization of the Retinal Projection to the Midbrain Surface

A very important aspect of any sensory pathway in the CNS is its topographic organization. For the visual system, the topography is expressed as an orderly termination of axons from different retinal locations in different parts of a terminal region, resulting in a map-like central representation of the receptor surface. The organization of these maps can be

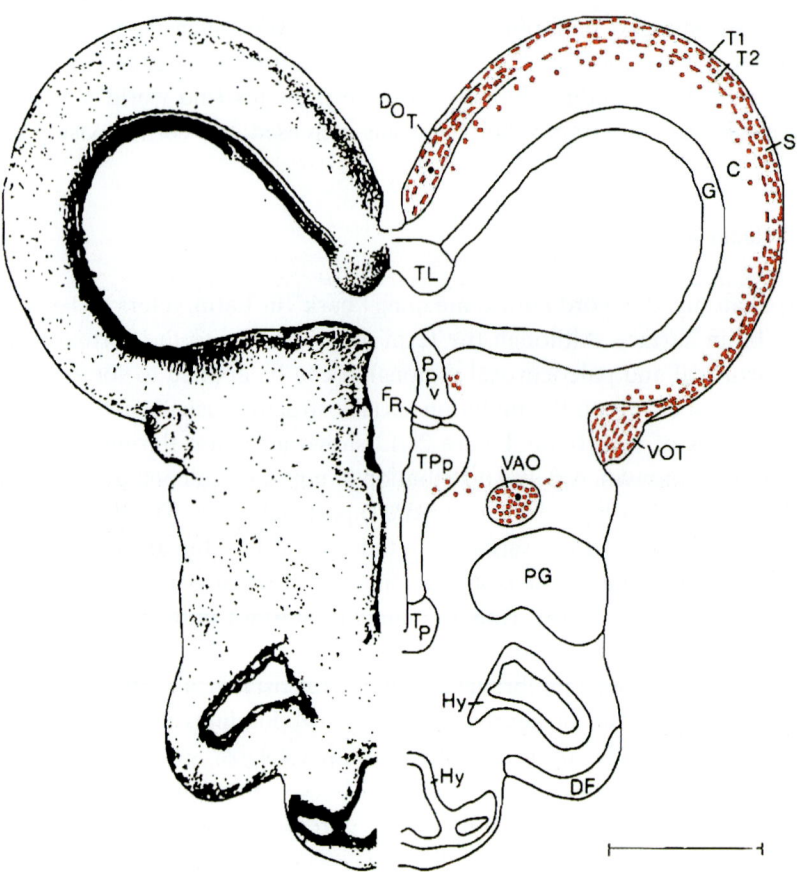

Figure 21.12
Photograph (left side) and drawing (right side) of frontal section through the midbrain of a bowfin fish. On the right, axons from the left eye and axon terminals labeled with radioactive proline or with HRP are charted in red. C, central zone of optic tectum; DOT, dorsal optic tract; DF, nucleus diffusus of inferior lobe; FR, fasciculus retroflexus; G, periventricular gray zone of optic tectum; Hy, hypothalamus; PG, preglomerular nuclear complex; PPv, nucleus pretectalis periventricularis pars ventralis; TL, torus longitudinalis; Tp, nucleus tuberis posterior; TPp, periventricular nucleus of the posterior tuberculum; S, superficial zone of optic tectum; T1, T2, fasciculi and terminal fields of optic tectum; VAO, ventral accessory optic nucleus; VOT, ventral optic tract. From Butler and Northcutt (1992).

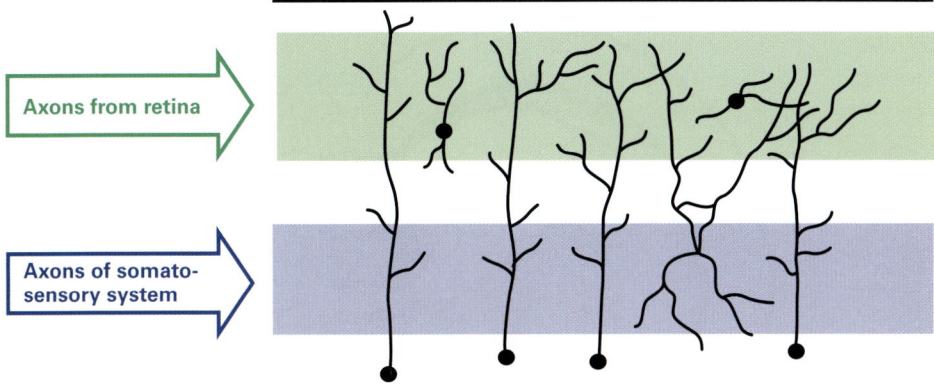

Figure 21.13
In vertebrates like the bowfin, with a midbrain tectum like that shown in figure 21.12, dendrites of periventricular neurons extend toward the pial surface as shown in this figure. Visual and somatosensory inputs enter the tectum by axons that terminate in distinct layers as illustrated here. Axons carrying somatosensory inputs terminate in a deeper layer, closer to the cell bodies of the neurons. The axons from the retina terminate in a more superficial layer, populated mostly by dendrites of the deeper lying cells. Based on Butler and Hodos (2005).

worked out by anatomical tracing methods, but a more common method is the use of electrophysiological recording, moving the electrode systematically from one position to another and finding the visual field location of the optimal visual stimulus. Results of such a study of the hamster optic tectum are shown in figure 20.5.

A review of many such studies of topographic organization of the axons from the retina and their termination patterns yields a consistent picture for all vertebrates.

Look back at figure 21.11: Axons of the optic nerve, just after they can be followed across the midline, spread themselves into a ribbon that can be followed toward the ventral nucleus of the lateral geniculate body; the axons from the superior retina move to the caudal edge of the ribbon, and axons from the inferior retina move to the rostral edge. This is the organization they maintain over the entire distance to the SC.

What about the nasotemporal axis of the retina? Axons from these two sides of the retina are intermixed, at least partially, along the diencephalic optic tract, but their terminals become sorted as they form in each terminal area from the ventral part of the lateral geniculate to the colliculus. They do this sorting with the aid of molecular interactions, and later there is a refinement influenced by patterned retinal activity. How they become organized is illustrated in the next figure.

Figure 21.18 is a flattened and straightened-out view (diagrammatic) of the entire optic tract surface and also the optic nerve and the retina. The topographic organization of the terminal arbors as they penetrate the cell groups from the surface is indicated by the letters: N for nasal retinal origin, T for temporal retinal origin. The superior–inferior (S-I) axis of retinal origin is indicated at the right in the figure. The organization is precise in

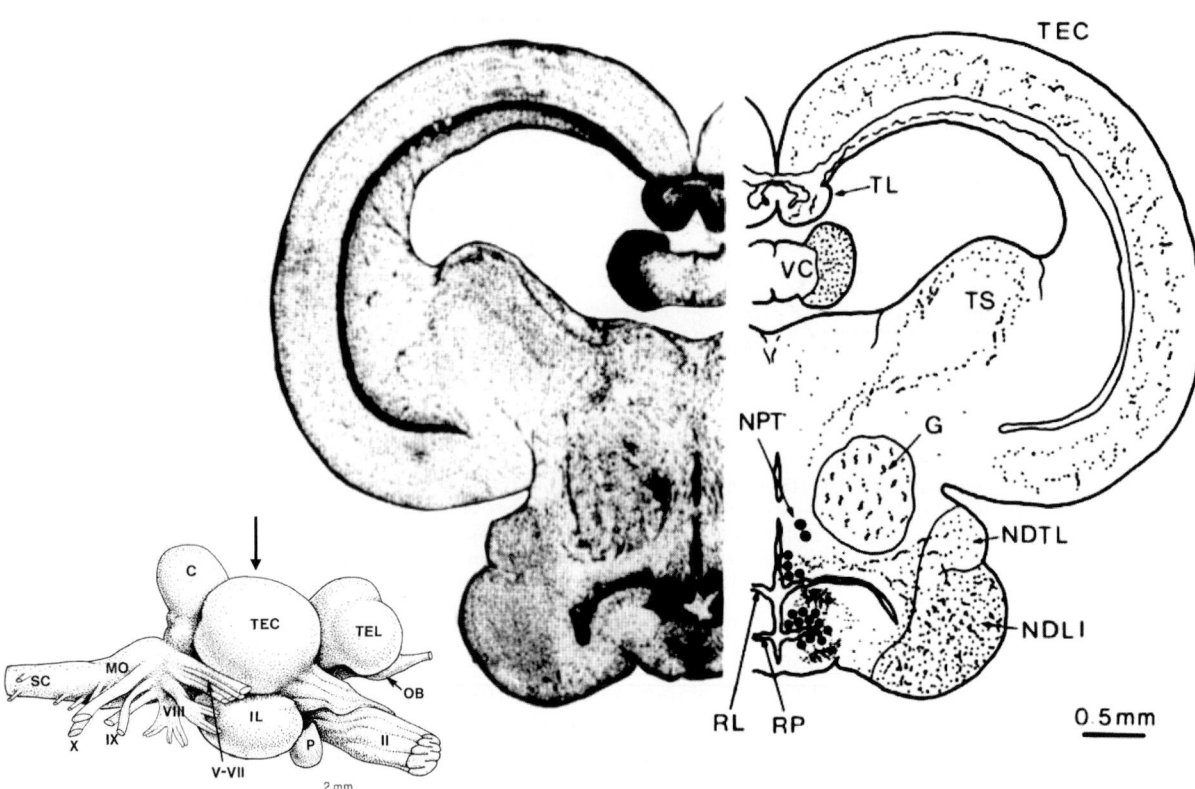

Figure 21.14
Frontal section through the optic tectum of a teleost fish. The left hemisection is Nissl-stained for cell bodies. The right hemisection is a mirror-image drawing with a few structures and tectal layers labeled. G, nucleus glomerulosis; NDLI, nucleus diffuses of the inferior lobe; NDTL, nucleus diffuses of the torus lateralis; NPT, nucleus posterior tuberis; RL, lateral recess of third ventricle; RP, posterior recess of third ventricle; TEC, optic tectum; TL, torus longitudinalis; TS, torus semicircularis; VC, valvula of cerebellum. From Parent, Dube, Braford, and Northcutt (1978).

the two geniculate cell groups and in the SC and less precise in the pretectal cell groups. Students should try to relate the central maps depicted in this figure to the topography data illustrated in figure 20.5.

Note also in figure 21.18 the branching origins of the accessory optic tract components. The accessory optic tract, mentioned earlier, can be seen in the more realistic, reconstructed surface view of the 'tweenbrain and midbrain shown in figure 21.6. In the hamster, as in the rat and mouse, there are three terminal nuclei of the accessory optic tract (dorsal, lateral, and a larger medial terminal nucleus). There are good indications that one of the pretectal cell groups that receive axons from the retina, the superficially

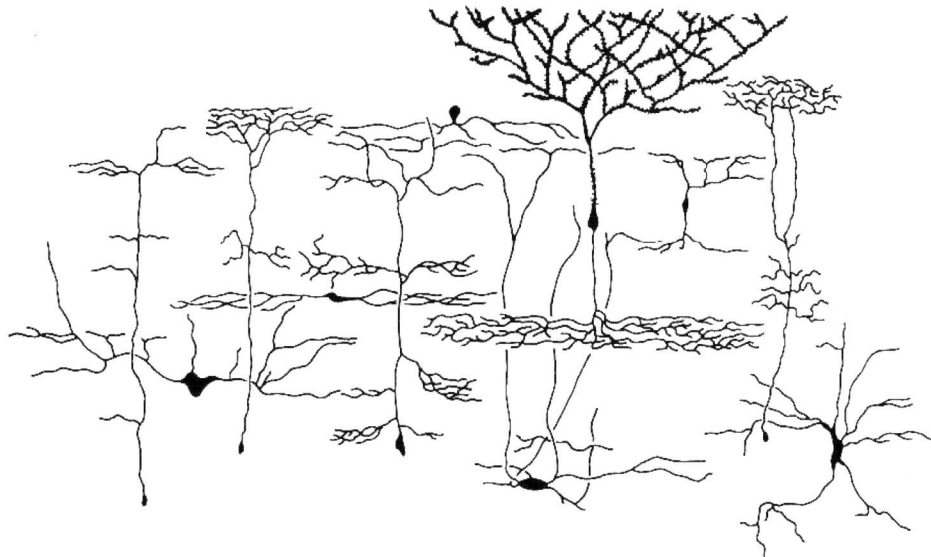

Figure 21.15
Neurons in the optic tectum of the teleost fish *Eugerres plumieri* drawn from Golgi-stained sections. Distinct neuronal types are found in different layers, and dendrites and axons are seen to arborize in laminar fashion. Adapted from Vanegas, Laufer, and Amat (1974) by Butler and Hodos (2005).

located nucleus of the optic tract, should be classified together with the terminal nuclei of the accessory optic tract, anatomically and functionally. These neurons respond to synchronous movements of objects and textures covering large parts of the visual field—the kind of stimuli produced on the retina when an animal or person is falling or rotating. Thus, they provide visual input that has the same meaning as stimulation of the vestibular system.

Finally, look back at figure 21.3, a stretched-out and straightened (diagrammatic) section through the entire optic tract cut at right angles to the surface as described earlier. You can imagine that in order to get this picture, you have cut at right angles to the paper in the diagrammatic flattened view of figure 21.18. Note that the terminating axons branch from the main tract at approximately right angles to the brain surface and penetrate a terminal area in orderly lines or columns. In each geniculate nucleus, the terminals nearest the optic chiasm are from the nasal retina. The same is true in the pretectal nuclei, but in the SC this pattern reverses so the axons from temporal retina terminate nearest the rostral border.

In the colliculus of mammals that have been studied in detail, the developing axons grow over or near the surface, but early in subsequent development neuronal cell bodies migrate dorsally toward the surface and form the superficial gray layer. In many

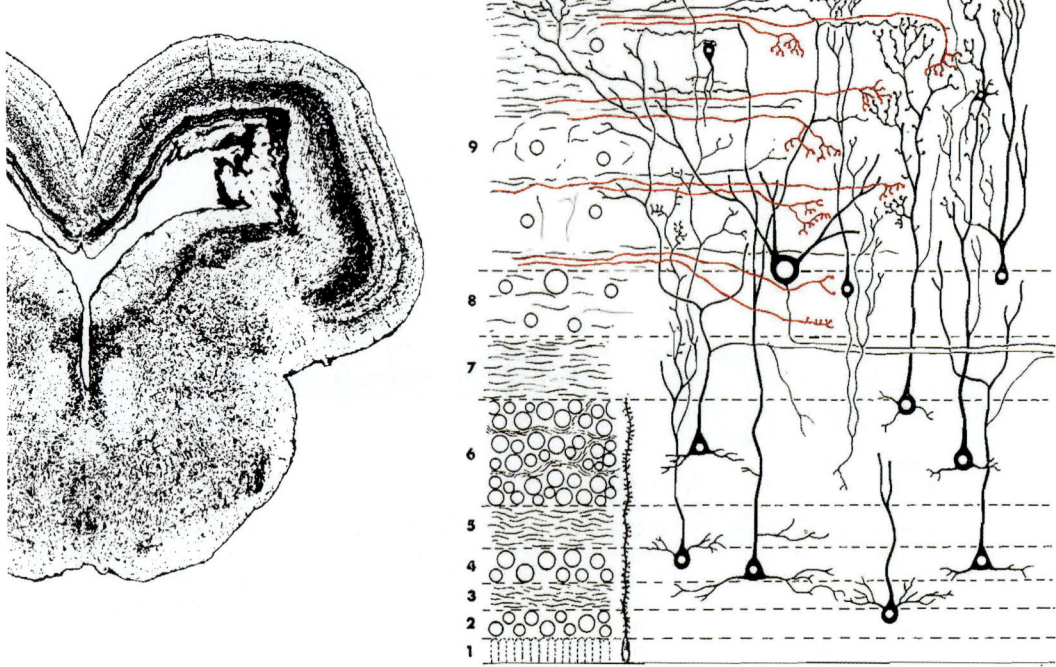

Figure 21.16
(A) Photograph of coronal section through the midbrain of a reptile, *Iguana iguana*, stained for cell bodies with a Nissl method. (Much of the left side is not shown.) The layered structure seen dorsally is the optic tectum. (B) Drawing of a small part of a section made perpendicular to the surface of the frog's optic tectum. At the left is a drawing of the cell bodies and fiber layers. The remainder of the drawing is based on Golgi impregnated tissue showing major cell types with their dendrites and axons. The superficial layers, labeled layer 9 in the figure, contain optic fibers from the retina. These have been traced in red color. Other fibers in these layers are not in color, as some fibers in the superficial layer come from the pretectal area. Panel (A) from Foster and Hall (1975); panel (B) from Szekely, Setalo, and Lazar (1973) (color has been added).

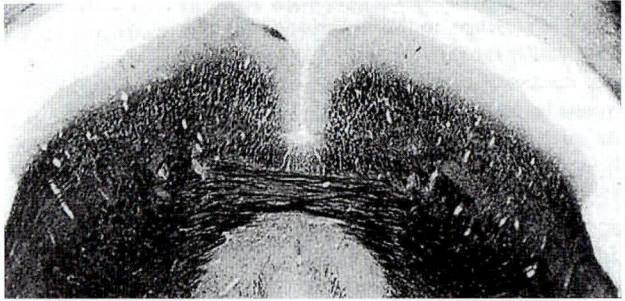

Figure 21.17
Mammalian superior colliculus in frontal section: Stain for myelinated axons of the superior colliculus of the dorsal midbrain of a hamster, showing the lamination of the structure. The superficial gray layer, where most of the axons from the retina terminate, is the most superficial layer with little myelin staining. Below it is the layer of fibers running at right angles to the section; it includes axons from the opposite eye and from visual cortex. Below this layer is the intermediate gray layer, which includes many axons as well as cells. The next layer includes bundles of myelinated axons running parallel to the more superficially located optic-tract axons, as well as many large cells that give rise to the tectospinal tract. The deepest layers cannot really be separated: the deep gray layer and the deep white layer—the axons of the commissure of the superior colliculus joined by output axons that will enter the tectospinal tract. From Schneider (1973).

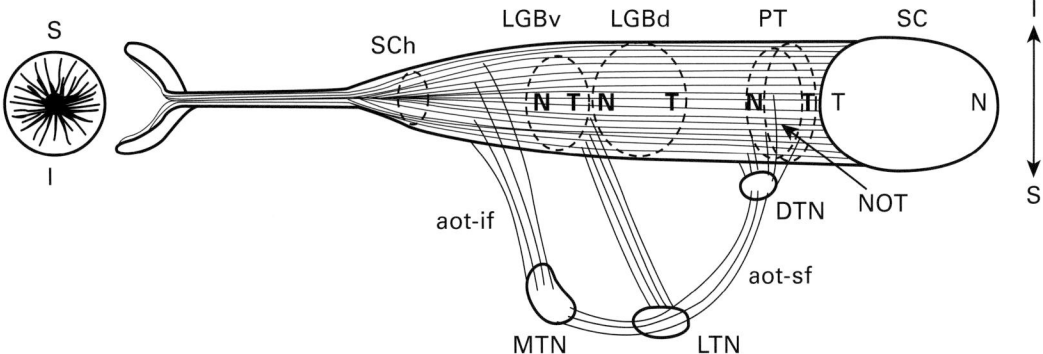

Figure 21.18
Flattened and straightened view of the mammalian optic tract showing topographic organization (cf. the companion figure 21.3, also figure 21.6). Axons of the retinal ganglion cells form a broad ribbon after they pass the optic chiasm, as shown in figure 21.6. The axons from the inferior retina are located rostrally (at the top in this figure), and axons from the superior retina are located more caudally as the optic tract courses dorsally over and near the surface of the diencephalon. Past the geniculate body, the tract turns caudally before coursing over and through the pretectal area and entering the superior colliculus. After this change in direction, the axons from inferior retina are located medially and those from superior retina are located laterally. Axons leave the tract, many as collaterals of axons continuing in the tract, to form terminal arbors in each terminal area. The terminations are arranged in order of their retinal origin, from nasal retina (N) to temporal retina (T) in each geniculate nucleus and in the pretectal area. In the superior colliculus, the order is reversed as the axons terminate from temporal retina rostrally to nasal retina caudally, as illustrated in the figure. DTN, LTN, MTN, dorsal, lateral, and medial terminal nuclei of the accessory optic tract; aot-if, inferior fasciculi of the accessory optic tract; aot-sf, superior fasciculi of the accessory optic tract. The middle fasciculus of the accessory optic tract is also shown, as is the transpeduncular tract to the MTN. NOT, nucleus of the optic tract (in pretectal area); other abbreviations same as in figures 21.3 and 21.11.

non-mammalian species, this superficial layer of cells does not form above the optic tract fibers. As indicated in the figure, axons from one retinal locus, terminating in the same column in the terminal regions, are not all identical. Not only are they from two different eyes, which segregate into layers as noted earlier, but also they come from different neuronal types. The axons of these different cell types may also segregate into different layers. The different axonal types may carry different kinds of information from the retinae, just as in the frog one class of axons is activated by the visual stimuli coming from a bug-like object whereas other classes bring different kinds of information.

Notes

1. In some recent interpretations, there are five or six diencephalic neuromeres.

2. Additional retinal projections have been reported—projections that may be remnants of ancient pathways that have been supplanted in evolution by less direct connections that allow more processing of information. Thus, in some species there appear to be sparse projections from the retina directly to the anterodorsal and the laterodorsal thalamic nuclei, two structures that project to posterior limbic cortical areas—areas that are closely connected to parahippocampal cortex.

Readings

Altman, J., & Bayer, S. A. (1979). Development of the diencephalon in the rat. VI. Re-evaluation of the embryonic development of the thalamus on the basis of thymidine-radiographic datings. *Journal of Comparative Neurology, 188,* 501–524.

Butler, A. B., & Hodos, W. (2005). *Comparative vertebrate neuroanatomy: Evolution and Adaptation*, 2nd ed. New York: Wiley-Liss.

Herrick, C. J. (1917). The internal structure of the midbrain and thalamus of Necturus. *Journal of Comparative Neurology, 28,* 215–348.

Senn, D. G. (1972). The ontogenesis of the optic tectum of a frog (*Rana temporaria* L.). *Acta anatomica, 82,* 267–283.

Szekely, G., Setalo, G., & Lazar, G. (1973). Fine structure of the frog's optic tectum: Optic fibre termination layers. *Journal für Hirnforschung, 14,* 189–225.

Wiesel, T. N., & Hubel, D. H. (1966). Spatial and chromatic interactions in the lateral geniculate body of the rhesus monkey. *Journal of Neurophysiology, 29,* 1115–1156.

The diencephalic neuromeres from structural data:

Coggeshall, R. E. (1964). A study of diencephalic development in the albino rat. *Journal of Comparative Neurology, 122,* 241–269.

A publication that includes reconstructions of the hamster optic tract with topography based on neuroanatomical work by Sonal Jhaveri as well as electrophysiological work by Barbara Finlay (students of G. S.):

Frost, D. O., & Schneider, G. E. (1979). Plasticity of retinofugal projections after partial lesions of the retina in newborn Syrian hamsters. *Journal of Comparative Neurology, 185,* 517–568.

22 The Visual Endbrain Structures

In chapter 4, we summarized the story of the earliest evolution of what became the crown of the mammalian endbrain—the neocortex. Non-olfactory inputs initially reached the endbrain primarily via connections in the midbrain, which evolved connections to the thalamic region of the 'tweenbrain. This was probably true for the visual system as well as for the somatosensory and auditory systems, although the retina evolved as a diencephalic structure and connected directly to the ventral part of this region. The importance of the functions of the optic tectum for escape from predators and for orienting led to its evolutionary elaboration in size and connections, and some of its outputs reached the thalamic region. More direct retinal projections to the dorsal thalamus were no doubt present very early in evolution also, but these remained small until an increasing adaptive importance of the dorsal pallium (cortex) led to its elaboration and expansion in the mammals. A portion of the dorsal cortex of the early reptiles was transformed into the neocortex of mammals.

In this chapter, we look first at the multiple routes from the retina to the endbrain, with a focus mostly on mammals. Some of these routes are little studied but are nevertheless important to note in order to understand the brain's evolution. Studies of neocortical visual areas have revealed evidence for older and newer areas in evolution. We see amazing evolutionary increases in the number of distinct visual areas in the cortex. An expansion of the total area occupied by visual regions of neocortex in some species, including our own, has been a major contributor to the formation of the temporal lobe. We will also note some major transcortical pathways that begin at the primary visual cortex of the occipital lobe, and we will discuss the nature of interconnections of cortical areas.

Multiple Routes from Retina to the Endbrain

Early electrophysiological studies using recordings with fairly large electrodes were the first to give evidence for multiple parallel pathways taking visual information from the retina to the endbrain. Evidence suggested at least six distinct pathways. There are neuroanatomical data that support such a claim. Neuroscientists have focused most of their attention on only the most prominent pathways in mammals. I will go through them from the ones likely to have been earlier to those that became prominent later in evolution. The fifth and sixth pathways in the list to follow are generally considered to be the most significant ones.

It should not be surprising that there are so many routes for visual information to take in reaching the endbrain. It became adaptive for visual information to reach the endbrain not only to modulate the effects of olfactory pathways on behavior, but also to take advantage of the plasticity of connections that had evolved there. Many early projections were in all likelihood formed by widely branching axons so that various pathways appeared. Some of them became enlarged much more than others simply because of differences in their efficiency in controlling adaptive behavior.

1. The first structure encountered by the optic tract after it traverses the surface of the hypothalamus is the ventral thalamus, more commonly called the subthalamus. The most superficial structure of this segment of the diencephalon is the ventral nucleus of the lateral geniculate body, and some of the optic tract axons terminate there. By way of projections to the more medial parts of the subthalamus (the so-called zona incerta), it also projects to paleothalamic structures, which have projections to both striatum and neocortex. Axons also ascend directly from the subthalamus to the striatum and pallidum; there may be some neocortical connections as well. These last connections are not prominent in modern mammals and their role in evolution is not clear, but it is possible they had an early importance.

2. and 3. Both the pretectal area and the optic tectum have projections from their deeper layers to the paleothalamus—the older portions of the thalamus that constitute the midline and intralaminar cell groups (see figure 11.4). These parts of the thalamus project to both corpus striatum and neocortex. The projections are less topographically precise than are the projections of the newer portions of the dorsal thalamus.

4. The superficial parts of the pretectal area, dominated by visual inputs including retinal inputs, project to anterior parts of the lateral nucleus of the thalamus (often called LD—the lateral dorsal nucleus). This part of the thalamus projects to posterior neocortex next to the primary visual area. As the primary visual cortex is called striate cortex, the areas receiving major visual inputs from the lateral nucleus are called juxtastriate (or extrastriate) cortical areas. In the rat and hamster, it is a region medial and also rostral to the primary visual cortex. (For results in the mouse, look ahead to figure 22.10.) The anterior part of the lateral nucleus also projects to parahippocampal cortical areas in the rat.

5. The superficial, visual layers of the superior colliculus also have a projection to the thalamus, connecting densely to the lateral posterior nucleus (LP), which projects to posterior neocortex as does the remainder of the lateral nucleus of the thalamus. A projection of the optic tectum of non-mammalian species likewise projects to a

thalamic nucleus, and this nucleus projects to the entopallium in birds and, in a similar location, to the dorsal ventricular ridge in reptiles (see following section).

6. The ascending projections from the SC also go to the lateral geniculate body (LGB), both dorsal and ventral parts. But the dorsal nucleus of LGB is dominated by a larger input directly from the retina via the optic tract, as already described. This thalamic cell group projects to the striate cortex of the occipital lobe—considered to be the primary visual cortical area. The retino-geniculo-striate pathway is seen as the dominant visual pathway to the endbrain of large primates and many other mammals. An equivalent pathway exists in birds and reptiles as well, but the names are different (see discussion of figure 22.2 later).

In the early evolution of the thalamus and the precursors of neocortex, it is likely that the pretectal area and the entire tectum projected widely to the rostrally adjacent posterior portions of the thalamus (figure 22.1). The retina probably had sparse and fairly diffuse projections to the same region. We can suggest a likely progression: As the axon terminals increased in density in animals in the line of evolution leading to mammal-like reptiles and the earliest mammals, they became more segregated, replicating the pattern of their origins. Auditory tectum dominated the most caudal posterior thalamus, optic tectum the adjacent lateral posterior thalamus, and (visual) pretectum the more anterior lateral thalamus. Superficial unimodal tectum terminated more dorsally than the deeper multimodal tectum. As this parcellation of the lateral thalamus evolved, retinal projection density increased nearest the thalamic border where it was first reached by the optic tract, and this became the lateral geniculate body. Tectal input remained there but the retinal input became dominant.

The Visual System's Two Major Routes to the Endbrain in Phylogeny

Figure 22.2 diagrams what are usually called the two major pathways whereby information from the retina reaches the endbrain in mammals, reptiles, and birds. These are the pathways numbered 5 and 6 in the list of the previous section, where the focus was on mammals. In the diagrams, the names of the apparently homologous structures in reptiles and birds are given. In a later chapter on the neocortex, these structures will be discussed further. A frontal section through the endbrain of a turtle is shown in figure 22.3.

Figure 22.4 presents a photograph of the primary visual neocortex, area 17 of Brodmann, in a rhesus monkey. The photograph is of a section that was stained for cell bodies with a Nissl method. You can see the distinctiveness of this visual area and can get an indication of why it is called the striate area as you can see a prominent white band, due to a fiber layer, located above a very dense layer of small cells (i.e., closer to the surface). The cortex adjacent to area 17 is also visual cortex, but the areas there get their most direct thalamic input from parts of the pulvinar and the lateral posterior nucleus. These areas also receive transcortical input from the primary visual cortex.

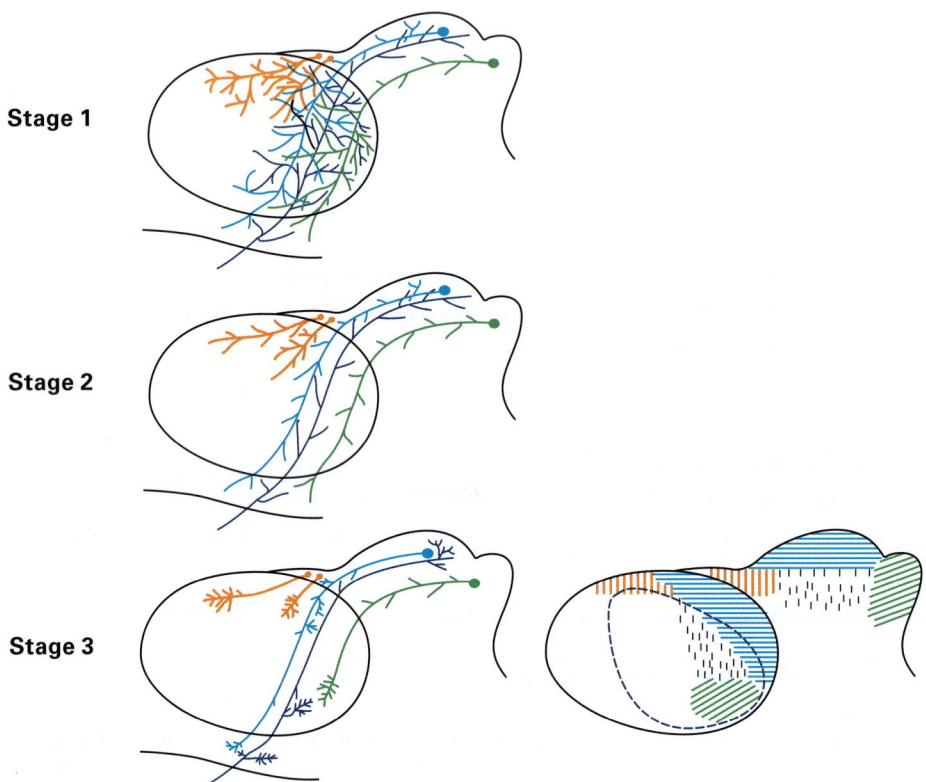

Figure 22.1
Schematic illustration of hypothetical evolution of connections from midbrain tectal regions to the thalamus. The thalamic region, with subthalamus below it, is shown at the left in each drawing, and the dorsal midbrain at the right. An axon from the retina, coming in from below, is also illustrated. Early in the evolution of the pathways (stage 1), the axons arborize widely, overlapping extensively. At later stages, the axons reduce their overlap as they arborize less widely, maintaining an order determined by their trajectories of entry (stage 2). As the density of their end arbors increases in evolution (and during development), they compete for exclusive occupancy of the terminal space and separate from each other almost completely. This view of thalamic evolution can be called parcellation by competition. In the sketch at the right of the drawing for stage 3, the projections to the thalamus from four major parts of the pretectum and tectum of the hamster are illustrated by the distinct markings: The regions of axon termination form a spatial pattern that corresponds to the regions of the cell bodies of origin. The dashed line shows the borders of the dorsal nucleus of the lateral geniculate body.

The Visual Endbrain Structures

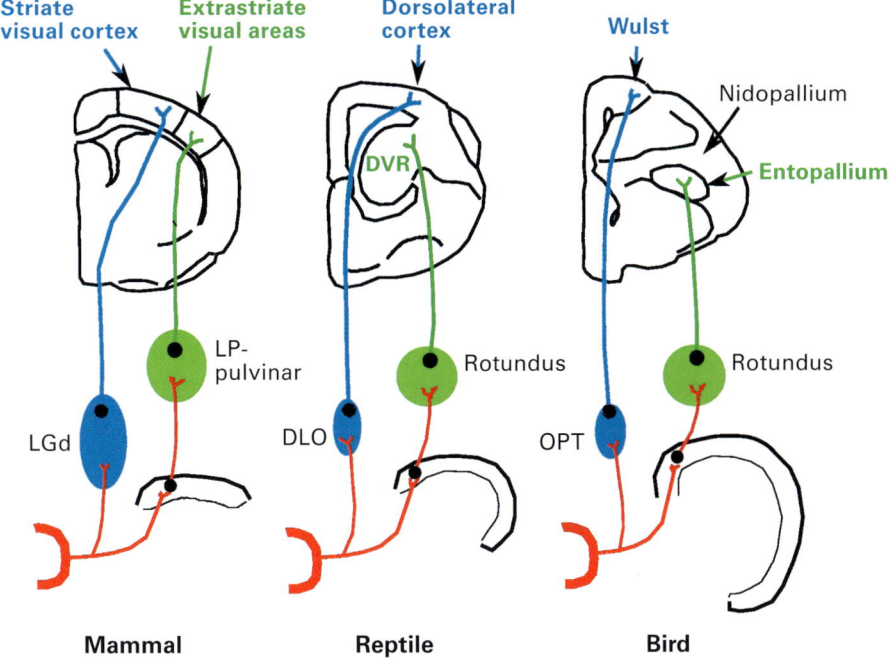

Figure 22.2
Two major routes from retina to pallial structures of the endbrain are depicted for a mammal, a reptile, and a bird. In each group of animals, the retina projects directly to one thalamic cell group, illustrated by the blue oval: the lateral geniculate in mammals (LGd), the dorsal lateral optic nucleus in reptiles (DLO), and the nucleus opticus principalis thalami in birds (OPT). In each group there is another nucleus, shown in green, that receives visual input from the optic tectum of the midbrain: the lateral posterior nucleus (LP) in mammals (and part of the pulvinar in some species) and nucleus rotundus in reptiles and birds. The retinorecipient nucleus projects to the striate cortex in mammals, a lateral part of the dorsal cortex in reptiles, and the visual Wulst (of the hyperpallium) in birds. In some mammals including the hamster, the retina projects to both of the thalamic cell groups, but much more densely to the one in blue. In the hamster, the tectal projection also goes to both thalamic cell groups, but much more densely to the one in green. DVR, dorsal ventricular ridge. Based on the work of Harvey J. Karten, Ann B Butler, and others.

The Route through the Lateral Geniculate Body

The most direct visual pathway to the neocortex is the retino-geniculo-striate route, with only one synapse between the retinal ganglion cells and cells in the visual cortex. This route thus has the advantage of speed over routes that go first from retina to the midbrain or to the pretectal area, because of synaptic delay times. In the large primates, with well-developed visual systems, the geniculate is the largest optic-tract termination area, and the geniculostriate axons are large in number. These axons form a large ribbon of fibers called the optic radiations. Axons from the dorsal nucleus of the lateral geniculate body exit the nucleus medially and rostrally and then pass laterally out of the thalamus, through

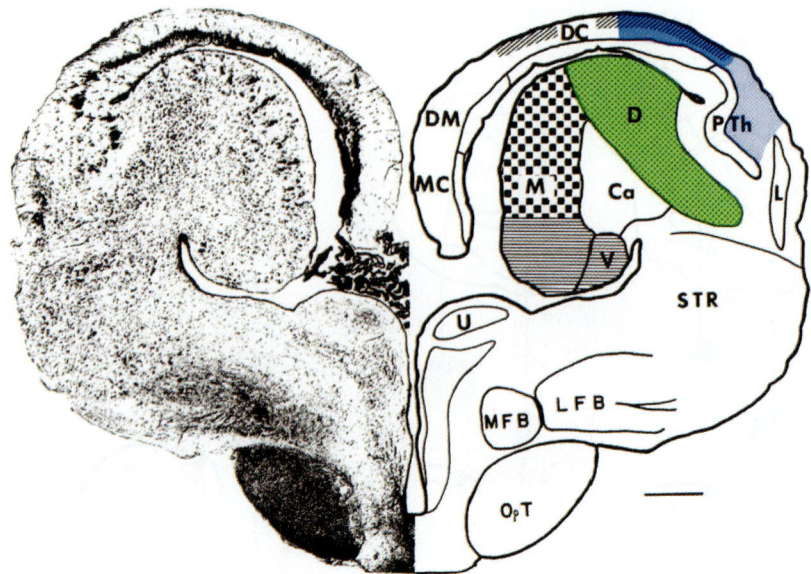

Figure 22.3
Transverse section through the telencephalon of a turtle. Left hemisection stained for cell bodies with a Nissl stain; right side is a tracing. Colors added using the scheme in the previous figure. Green color: the visual part of the anterior dorsal ventricular ridge. Blue parts: the lateral part of the dorsal cortex, and the pallial thickening lateral to it. D, dorsal area (of dorsal ventricular ridge); Ca, central area; ; DC, dorsal cortex; DM, dorsomedial cortex; L, lateral cortex; LFB, lateral forebrain bundle; M, medial area; MC, medial cortex; MFB, medial forebrain bundle; OpT, optic tract; PTh, pallial thickening; U, umbonate nucleus; V, ventral area. From Balaban and Ulinski (1981); colors added using findings of Heller and Ulinsky (1987).

the reticular nucleus where they have some collateral terminations. From there the axons can be followed up through the corpus striatum where they are part of the caudal portion of the internal capsule. Reaching the white matter of the neocortex, they turn caudally and proceed among the most superficially located fibers of the white matter and continue into the occipital area where they reach area 17 from its rostrolateral side and begin terminating, mostly by turning toward the surface and then arborizing in layer 4.

You may sometimes hear the term *secondary optic radiations*. In the hemispheres, this refers to axons mixed in with the optic radiation fibers but going in the opposite direction. Many output axons from visual cortex project to other cortical areas and to the corpus striatum, whereas the secondary optic radiations continue toward the brainstem. Some of these descending fibers terminate in the lateral geniculate body, others go much farther, reaching the pretectal area and the superior colliculus. Some of the projections from the visual cortex do not follow the route toward the colliculus; they remain in the cerebral peduncle and thereby reach and terminate in the pontine gray matter, their gateway to the cerebellum.

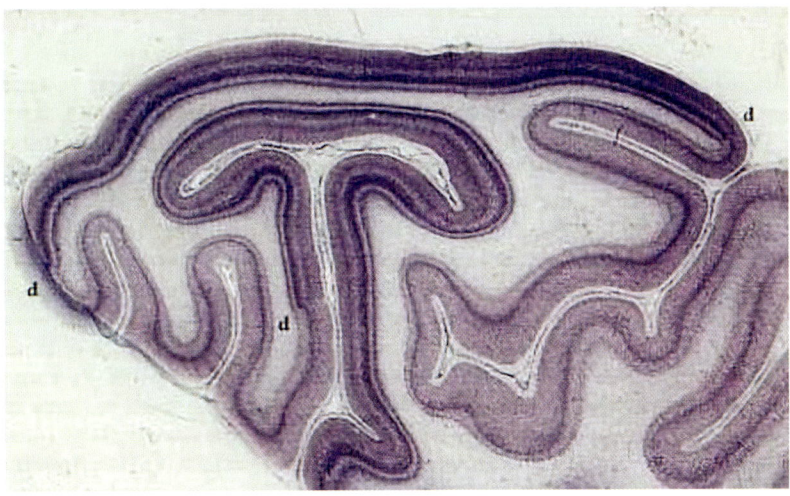

Figure 22.4
The occipital lobe of a rhesus monkey in a Nissl-stained section. The primary visual cortex—the striate cortex—is identifiable even at low power by a light stripe superficial to a darkly stained layer of cells. This cortical area can be followed deep into the calcarine fissure (part of which is seen in the mushroom shape at the left). An abrupt change in cytoarchitecture (d) marks the transition from the striate area (area 17) to area 18. From Hubel (1988).

Early Myelination of the Optic Radiations

The geniculostriate axons of the optic radiations become myelinated earlier in development than other fibers around them. This enables us to visualize the radiations histologically in sections of the infant human brain stained for myelin. The neuroanatomist Paul Flechzig used this method in a major study of myelination of the human brain published in 1920. You can see this in figure 22.5, which shows a horizontal section of the brain of a 7-week-old human. The section shows the early myelinating optic radiations passing out of the thalamus and curving caudally in the direction of the occipital cortex. Two other thalamocortical pathways that acquire myelin early are also seen in the figure: the auditory radiations reaching the dorsal portion of the temporal lobe and the more rostral somatosensory radiations.

The Brain and Neocortex in Human Development and in Phylogeny

Figure 22.6 pictures the human brain during fetal life. At 10 weeks postconception, the entire brain is smaller than a walnut; the neocortex is smooth and does not yet cover the midbrain. Four weeks later, early in the second trimester of pregnancy, the brain has more than doubled in length, and the neocortex has expanded so much that it covers the

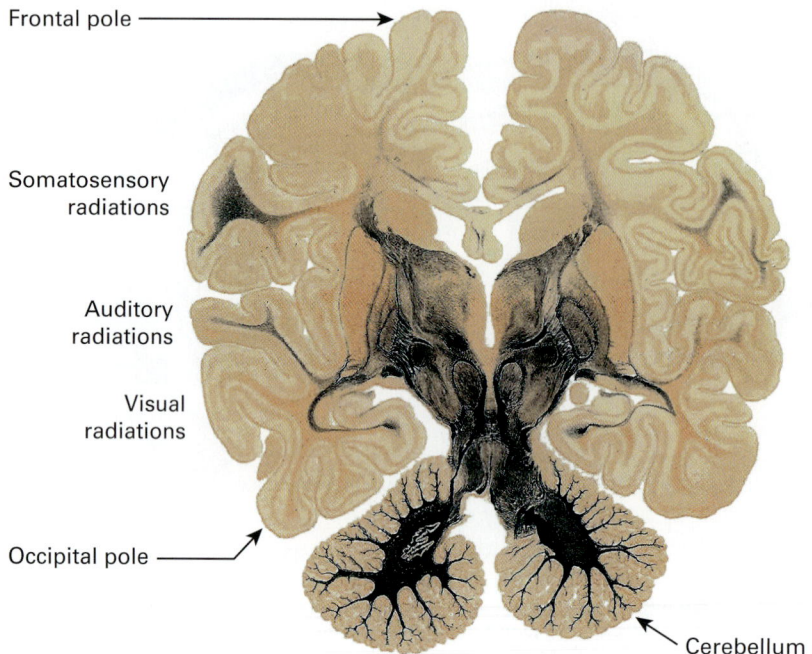

Figure 22.5
Myelin-stained horizontal section of a postnatal 7-week human brain. At this stage, the major myelin in the cerebral cortex is in primary sensory pathways, including the optic radiations (geniculostriate pathway) closest to the occipital pole, the auditory radiations, and, more rostrally, the somatosensory radiations. The brainstem and cerebellum contain many more myelinated axons. From Flechsig (1920).

midbrain and is beginning to develop the folds we call gyri and sulci. At 16 weeks, some of the primary visual cortex is already becoming hidden from a surface view as the calcarine fissure begins to infold in the caudomedial hemisphere. (The 14- and 16-week stages are not illustrated in the figure.) This process continues with further growth of the hemispheres. The subcortical regions are undergoing much less growth as the endbrain continues to expand throughout fetal life.

One consequence of the huge increase in neocortical area is called temporalization: What was caudolateral cortex in the first trimester expands so much that it comes to occupy the temporal lobe. A temporal lobe can be seen in a side view of the brain of any large primate, and to a more limited extent in the brain of a cat or dog (figure 22.7A). Thus, a resemblance of the human brain to the brain of a small rodent like the hamster, rat, or mouse can be seen most easily during the first third of the gestation period. See figure 22.7B for illustrations of progressive temporalization with neocortical expansion.

An effect of the expansion of the temporal lobe is that some of the optic radiation axons—the geniculostriate axons—are pulled downward and forward so that they come to lie in the white matter of the temporal lobe. This is illustrated in figure 22.8. Because

401 The Visual Endbrain Structures

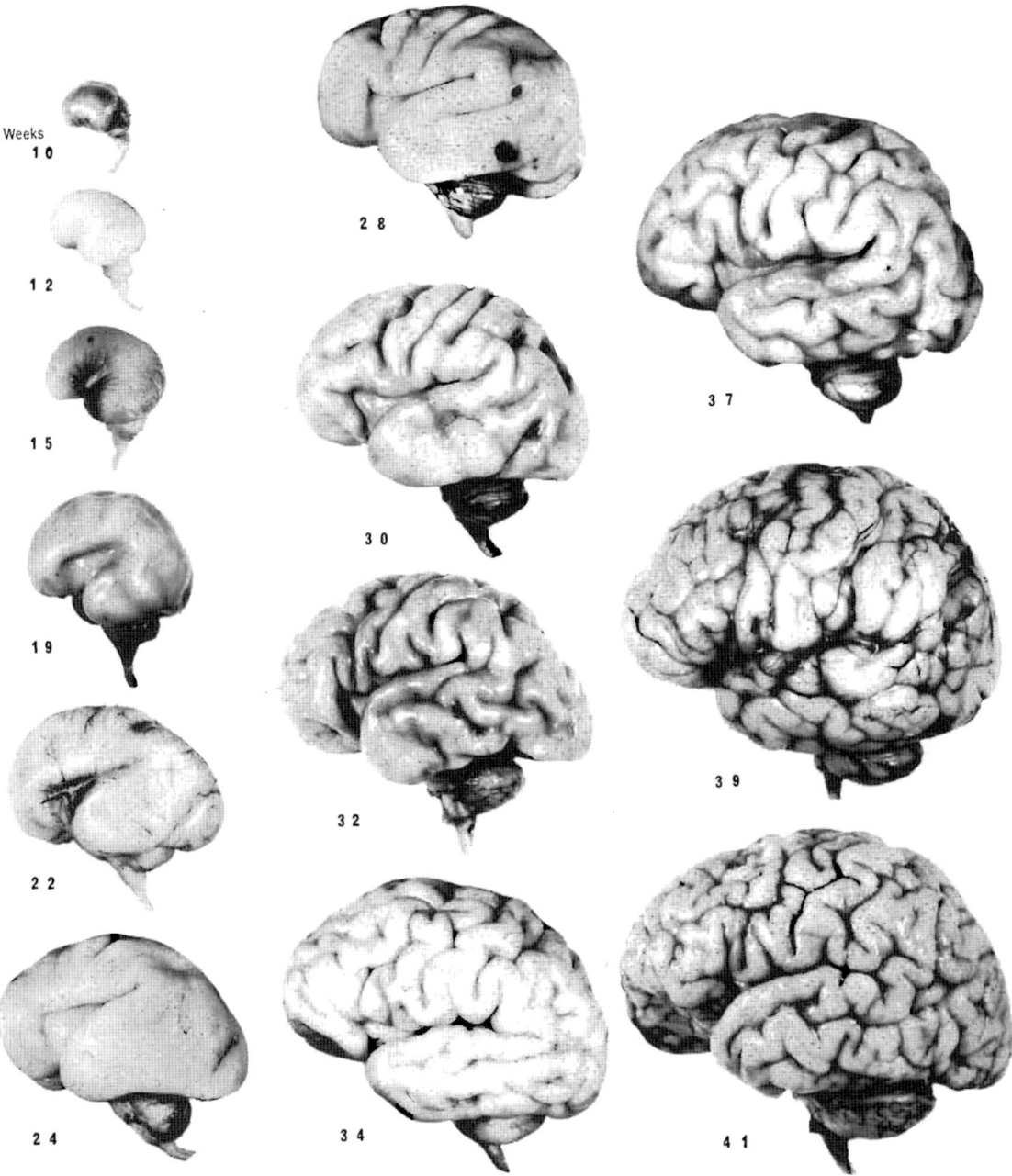

Figure 22.6
Photographs of developing human brain. (Top) Lateral views at successive postconception ages (weeks); the oldest brain is that of a newborn. (Bottom) The medial surface of the bisected human brain at five gestational ages (given in weeks postconception). From Larroche (1966).

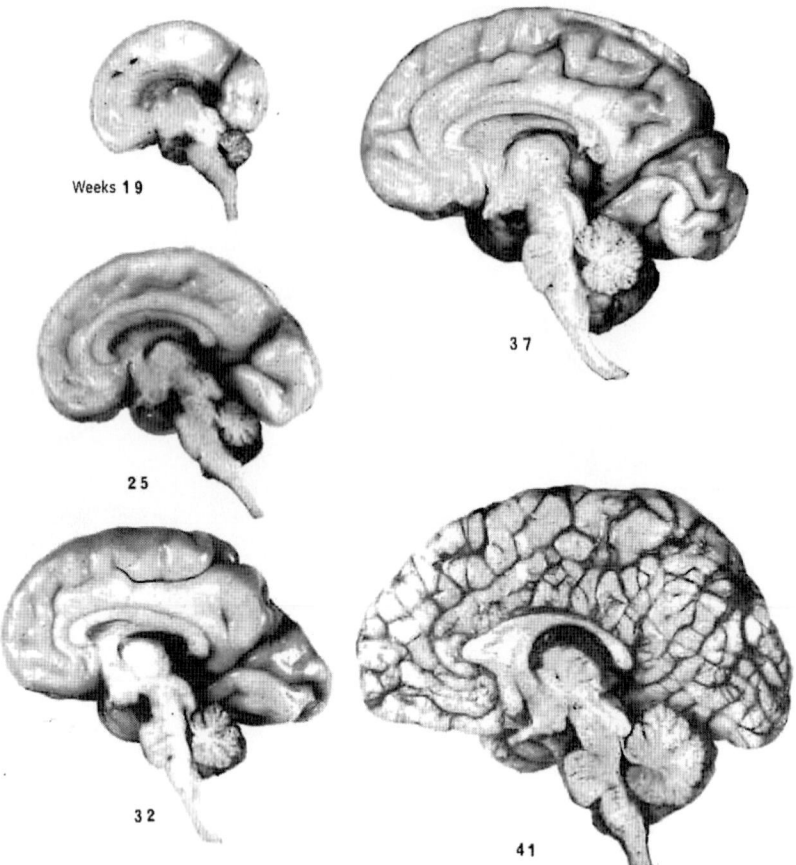

Figure 22.6
(*Continued*)

Figure 22.7
(A) Drawings of the left hemisphere of the brains of four mammalian species, not drawn to the same scale. The only sulcus visible in the brain of the rat is the rhinal sulcus, whereas each of the larger species has many sulci because of much greater surface areas of neocortex. Note the temporalization of the hemisphere in the animals with larger neocortex. (B) An illustration of progressive temporalization of the neocortex in evolution from an ancestral smooth-brained mammalian hemisphere. Views are from the left side with the olfactory bulbs at the far left in the pictures. An outline of the main body of the hippocampus within the hemisphere is shown with a depiction of the progressive shift into the temporal lobe as posterior association areas expand. Panel (A) based on Nauta and Feirtag (1986).

A

Rat
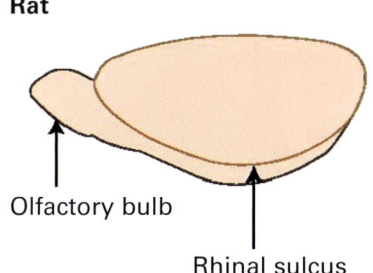
Olfactory bulb
Rhinal sulcus

Monkey
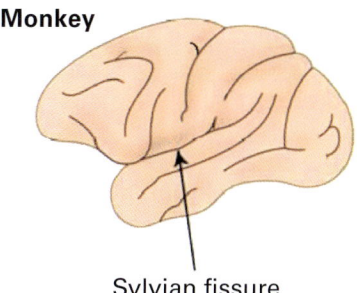
Sylvian fissure

Cat
Rhinal sulcus
Pseudo-sylvian fissure

Human
Frontal lobe
Parietal lobe
Sylvian fissure
Temporal lobe
Occipital lobe

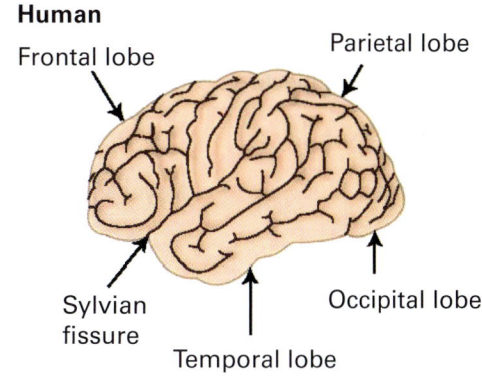

B

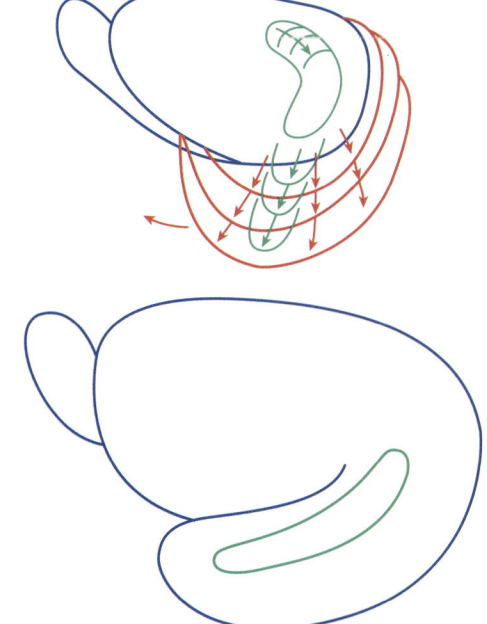

Ancestral, smooth-brained mammalian hemisphere

Expansion of posterior association neocortex

Main body of hippocampus at the medial margins of the hemisphere: pictured as if hemisphere were transparent

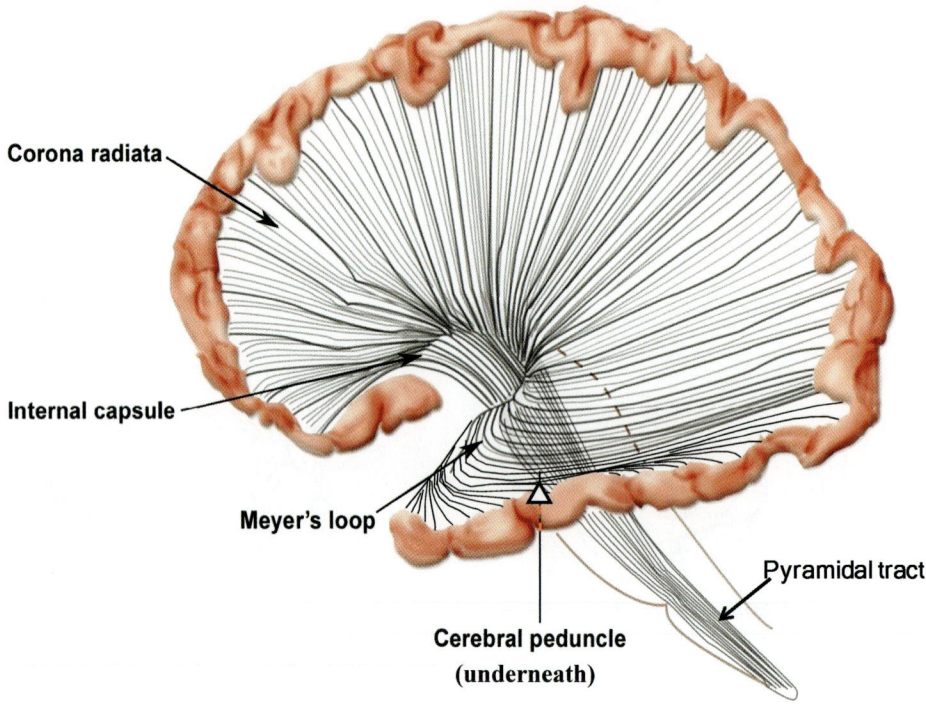

Figure 22.8
Drawing of a left hemisphere of the human brain together with the brainstem, dissected to reveal the course of axons that descend to the brainstem and spinal cord. These axons pass through the corona radiata into the internal capsule (within the corpus striatum), then into the cerebral peduncle, and through the pontine gray to form the pyramidal tract of the caudal hindbrain. With the growth of the hemisphere, axons of the optic radiations (to and from the visual cortex) are pulled into the temporal lobe. This part of the radiations is called Meyer's loop; it carries information that originated in the upper visual field. Based on Nauta and Feirtag (1986).

of this, damage to the temporal lobe in an adult human can cause cortical blindness in part of the upper visual field. The same is true for other large primates.

Evolutionary Multiplications of Cortical Representations of the Retina

In the initial portion of this chapter, we outlined multiple routes from retina to neocortex in mammals via visual system connections with thalamic cell groups. The cortical receiving areas each contain one or more representations of the retina. These retinotopic maps in the posterior neocortex are also a consequence of transcortical projections, for example, from the striate cortex—the primary visual cortex. The number of these maps varies from species to species. Physiological mapping studies of the posterior neocortex indicate that in mammalian evolution, particularly in the primates, there has been a multiplication of

The Visual Endbrain Structures

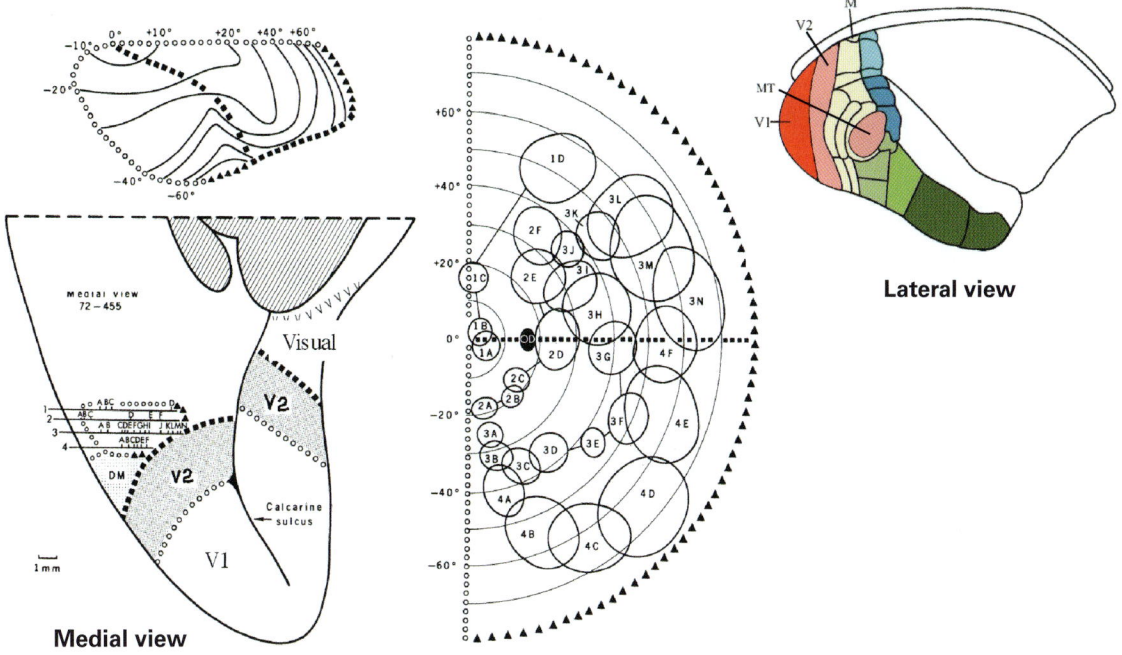

Figure 22.9
Multiple cortical representations of the retina are illustrated for the owl monkey. The picture of the owl monkey hemisphere shown in figure 6.4 (seen here at the right) depicts many distinct retinotopic maps in that species. At the left is an illustration of results of mapping of visual area M of the owl monkey by John Allman in microelectrode recording experiments. From Allman and Kaas (1976).

these cortical representations of the retina. Many electrophysiological studies have focused on distinguishing the functions of these various representations.

Figure 22.9 illustrates how a visual area in the neocortex is mapped with electrophysiological methods and also shows a pictorial summary of the many representations of the visual field that have been discovered in the neocortex of the owl monkey (see also figure 6.4).

Anatomical methods have also been used to distinguish different representations of the retina in the neocortex, by tracing projections from primary visual cortex to other cortical areas. Thus, small injections of different anterogradely transported tracers can be made in different parts of the striate cortex. The method has been used to show that the striate area projects in topographic fashion to nine different nearby areas in the mouse (see figure 22.10). It also projects in less organized fashion to six additional areas.

Looking at many different mammals in which visual areas of the cortex have been mapped, one finds that the number of neocortical representations of the retina is correlated with the total area of the neocortex. In the macaque monkey, 32 visual areas have been distinguished. The relationship leads to the prediction that the human brain should have even more such areas, as the total area of neocortex is much larger.

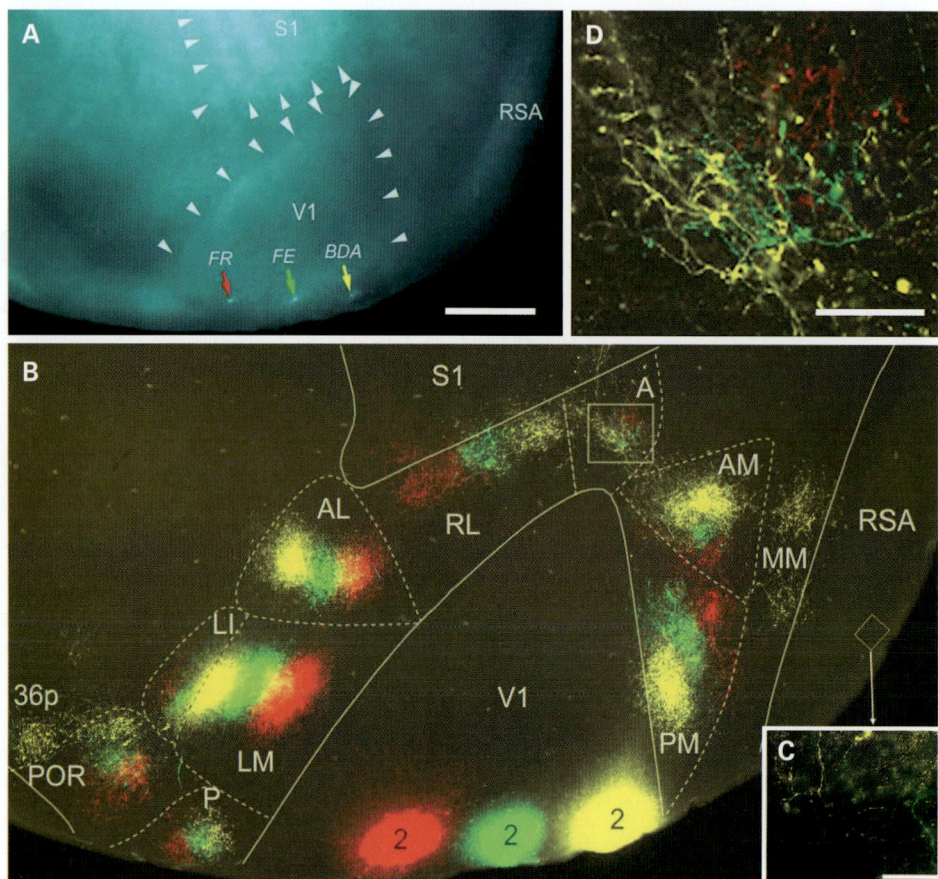

Figure 22.10
Anatomical evidence for multiple topographic representations of the retina in mouse visual cortex. (A) Low-magnification photograph of flat-mounted section with an outline of the striate cortex (V1), showing the positions of three injections of fluorescent dyes (red, green, yellow). (B) Photograph of flattened section showing the injections (2) and multiple spots where the labels were transported by anterograde axonal transport to nearby cortical areas (outlined with abbreviations of their names). (C) Enlarged photograph of labeled axons in the agranular retrosplenial area in the caudomedial hemisphere. (D) Labeled axons in the anterior area labeled *A*, corresponding to area 18b of older studies based on cell and fiber architecture. A, anterior area; AL, anterolateral area; AM, anteromedial area; BDA, biotinylated dextran amine; FR, fluororuby; FE, fluoroemerald; Ll, laterolateral area; LM, lateromedial area; MM, mediomedial area; P, posterior area; PM, posteromedial area; POR, postrhinal area; RSA, retrosplenial agranular cortex; S1, somatosensory area 1; V1, visual area 1; 36P, posterior area 36. From Wang and Burkhalter (2007).

Comparing Species: Evidence for Older and Newer Visual Cortical Areas

In the evolution of mammals, two neocortical representations of the retina seem to have appeared very early. In mammals, occipital visual areas 1 and 2 (V1 and V2) have been mapped even in marsupials and in the hedgehog, an animal with a relatively very small neocortex. Studies have indicated that these two areas probably existed in the common ancestor of all mammals (figure 22.11). A third, more rostral, visual area known as MT in primates may have homologs in most placental animals that have been examined.

The location of V1 and V2 in owl monkey, rat, and hedgehog neocortex can be seen in figure 22.12. This figure also shows other neocortical areas likely to be very old: the first and second somatosensory areas and a ventral parietal area, and the primary motor and premotor areas. In the figure, note also the remaining regions of the cortex—the location of areas that have expanded more recently. The hedgehog does not have much. These additional areas are somewhat larger in the rat. But look at the much greater extent of these territories in the owl monkey. This territory is still larger in the apes and larger still in humans. It is occupied not only by additional sensory areas, the unimodal association areas, but also by heteromodal association areas that communicate by transcortical connections with more distant areas and receive inputs from more than one sensory modality (see part XII).

And Where Do We Go from Here?

Once sensory information has reached the neocortical mantle, we have not reached any end of the line, as naïve views of brain functioning may portray it. We have to ask a key question as we did for the midbrain tectum: What are the outputs of the visual cortex and what are the functions that provided adaptive advantages leading to its presence and enlargement in evolution?

A more direct pathway to neocortex from the retina with precise topographic organization facilitated high-acuity discrimination of distant visually detected patterns and objects. The higher acuity and more elaborate sensory filtering made possible by expansions in the neocortex provided adaptive advantages because it enabled more precise orienting of head, eyes, and hands, and because it enabled more accurate recognition of objects and scenes. To support these functions, there was an evolution of specific output pathways that reached the relevant motor systems. But the output pathway that may have been the earliest in evolution, we suggest, added something in addition to sensory and motor refinements; namely, plasticity of connections to the motor system.

This axonal output pathway from the visual cortex went to the corpus striatum. The striatum had evolved in the endbrain, according to the evolutionary schema presented in chapter 4, as a link between the olfactory system and the output systems of the midbrain—a link that was plastic. The strength of connections there could change so that habits could be formed. Visual inputs to the striatum took advantage of this plasticity. Very early in

408   Chapter 22

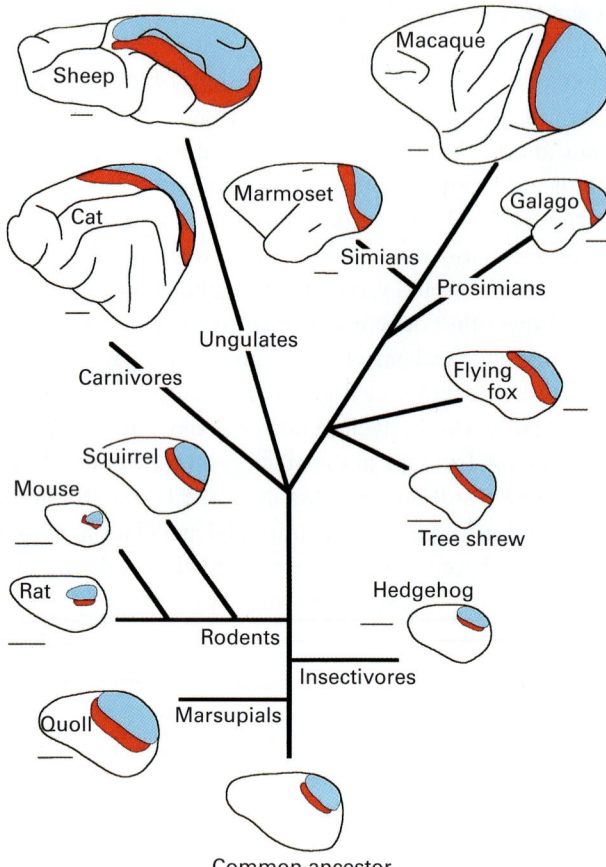

Figure 22.11
A phylogenetic tree of visual neocortical areas. Investigators have found a primary visual cortex (V1, labeled in blue) and also a V2 (labeled in red) in all mammals. Not shown is a third visual area called area MT in primates, which may have a homolog in other placental mammals but not in marsupials. From Rosa and Krubitzer (1999).

The Visual Endbrain Structures

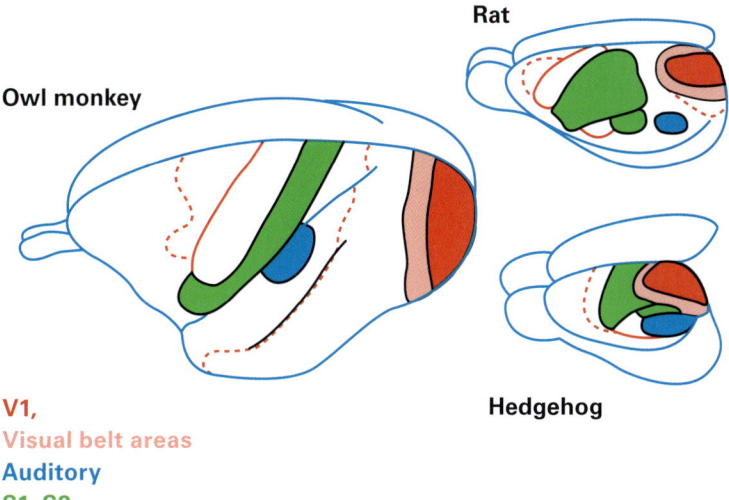

V1,
Visual belt areas
Auditory
S1, S2

Figure 22.12
Views of the left hemisphere showing sensory and motor areas of three species. The hedgehog is believed to resemble some of the earliest mammals; most of the neocortex in hedgehogs is composed of primary and secondary visual areas (V1, V2) shown in red and pink, primary auditory cortex (A1) shown in blue, primary somatosensory cortex (S1) shown in green, secondary somatosensory areas (S2, PV) shown in lighter green, and primary motor cortex (M1) outlined in red, according to Catania et al. (2000). These same areas and also a premotor area (anterior, dashed red line) are found in other placental mammals, as shown here for rat and owl monkey. Based on: Sereno and Allman (1991); Stepniewska, Preuss, and Kaas (1993); Catania, Collins, and Kaas (2000). Additional and hence "newer" areas have appeared as neocortex has expanded in phylogeny: additional visual, auditory, and somatosensory areas, and larger multimodal areas (see chapters 32–34).

evolution they came into the striatum from the older parts of the thalamus and from the subthalamus, and also from the dorsal cortex. The route from the cortex became the dominant one as the neocortex evolved from the dorsal cortex and expanded.

The outputs of the striatum to the midbrain provided an important link to mechanisms controlling orienting movements and to mechanisms controlling approach and avoidance movements, as discussed in chapter 7. This link enabled the learning of visuomotor habits.

Another visuomotor pathway from the visual cortex simply enhanced the orienting movements controlled by the midbrain tectum. Projections from the original visual cortical areas V1 and V2 went directly to the superior colliculus, enhancing the reflex-like turning movements of head and eyes and enabling the following of a moving stimulus source without losing it in the confusion of retinal background changes caused by the receptor movement (as the head moved).

Transcortical Pathways from Visual Cortex

Thus far, the outputs we have mentioned are involved in reflex-like functions and sensorimotor habits. Beyond these kinds of behaviors are movements initiated by internally generated plans, plans influenced by various motivational states. The adaptive advantages of being able to plan the next move led to the expansion of prefrontal neocortex, and an early part of this expansion was in the *frontal eye fields*. This part of the cortex receives input from a visual area rostrally adjacent to the striate cortex (as well as other areas); it probably did so very early in evolution. (The juxta-striate area that projects to this frontal cortex in the hamster receives inputs not only from striate cortex but also from the lateral thalamic nucleus [LD], which receives projections from the pretectal area. See chapter 33.)

The frontal eye fields also evolved many other transcortical inputs. The area is important for working memory of object locations (and object movements) around the head. Such transient memory is used for what we call voluntary eye movements. In addition to the pathways to frontal eye fields, primates have similar transcortical pathways to the *ventral premotor area*, involved in voluntary reaching and grasping movements. The transient retention (working memory) of "where" objects are, and where they are headed, with respect to the eyes, head, and body is critical for rapid decisions, internally initiated, about both eye-head turning and reaching movements.

The transcortical pathways follow a dorsal trajectory from visual cortex to juxta-striate areas to posterior parietal cortex. From posterior parietal areas, or more directly from areas in direct receipt of visual cortex projections, longer transcortical pathways to the frontal areas arise. These are illustrated for the monkey in figure 22.13.

A second major type of transcortical pathway has a different nature; it does not follow a dorsal route from juxta-striate areas, but passes ventrally into the temporal neocortical areas before it links with pathways into the prefrontal cortex. Much attention has been focused on this pathway in monkeys. In the large primates, this more ventral pathway can be followed from the striate area through prestriate areas into the temporal lobe expansion where it reaches the inferotemporal cortex (figure 22.14). Much evidence indicates the importance of this pathway in object vision—the "what" rather than the "where" of a visually detected object. The functions of the pathway include discrimination of objects and of conspecific individuals. The latter function was of such importance that it led to the evolution of a distinct, specialized visual area—a "face area"—in the temporal lobe of humans, adjacent to the parahippocampal gyrus.

This more ventral pathway can be followed beyond the inferotemporal region of neocortex to the amygdala (to be considered further in chapter 29), a limbic system structure, part of which can be thought of as a caudal extension of the striatum. The connections to the amygdala, like those to the more anterior striatum, are involved in a kind of habit formation—in this case the learning of approach or avoidance tendencies accompanied by positive or negative affect (feelings). Axonal projections from the amygdala go not only to the hypothalamus and more rostral basal forebrain, but also to ventral portions of the prefrontal cortex. Anterior inferotemporal cortex also gives rise to projections into ventral prefrontal areas. By means of these pathways, especially those from the amygdala,

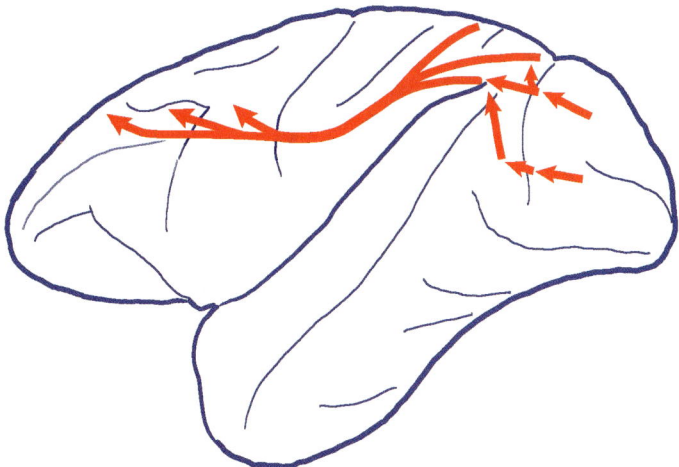

Figure 22.13
Visual pathway 1: "Where is it? Where is it heading?" (object localization). Transcortical pathways in monkeys carry information from visual cortex (area 17, the physiologists' V1) to prestriate areas, thence to posterior parietal areas. Information on object location goes from there to various cortical areas, including (as shown) premotor and prefrontal cortex, including the frontal eye fields. Based on Schmahmann and Pandya (2006).

the emotional meaning of visually identified objects—the feelings associated with them—reach an area important for planning and consequent internally initiated approach or avoidance and other socially relevant movements.

Note the contrast between the outputs of the cortex at the ends of these dorsal and ventral transcortical pathways (figures 22.15 and 22.16). These connections represent the contrasting functions of the two pathways.

Three Visual Pathways and Their Functions

The transcortical pathways discussed above represent a relatively small portion of the many interconnections in the hemispheres. They are singled out because of their relative importance among functions that have impressed investigators of brain-behavior relationships. The contrast in functions associated with the "where and whither" and the "what" of a visually sensed object is a fundamental one that had been discovered earlier in studies of subcortical vision.

The "what" (identity) of a particular object or individual, or a feature of an object such as its surface pattern or texture, and so forth is conceptually straightforward, at least initially.

The "where" idea quickly raises more questions: It can mean a direction from the organism, specified as an angular direction with respect to head or eyes (i.e., it is an "egocentric" direction), together with a distance from the organism. This is the "where" that the

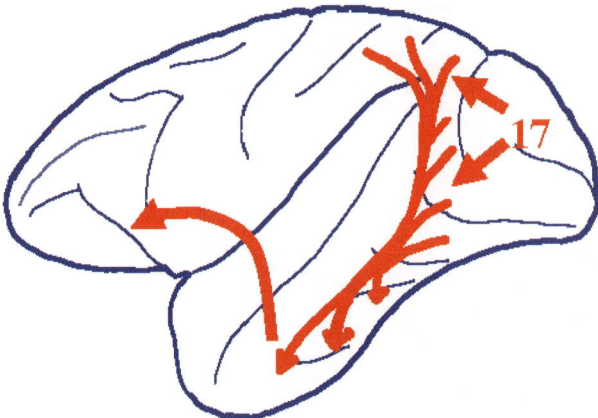

Figure 22.14
Visual pathway 2: "What is it?" (object identification). Transcortical pathways carry information from visual cortex to the cortex of the inferior temporal gyrus by way of prestriate and posterior parietal areas. Functions include object identification and discrimination of conspecific individuals (e.g., by identification and discrimination of faces). A pathway from the inferotemporal cortex leads to ventral prefrontal areas. Based on Schmahmann and Pandya (2006).

midbrain tectum is involved with. It can also mean the position of an object within a local environment or of the organism within that environment. It may even refer to a position in time or in both time and space.

The dorsal pathway leading from the visual cortex to the frontal eye field appears to carry information about egocentric directions with respect to the eyes. This directional information is critical for planning and control of eye and head movements and for predicting changes in position of a locomoting organism within an environment. However, representation of place of the organism within an environment does not appear to be a function of either the frontal eye field or of the midbrain tectum, but it is a function of neurons in the hippocampal formation, to be considered in chapter 28. (We will not try here to deal with positions in time from a brain-structure perspective.) Parietal lobe association cortex functions in a related way, as it has been found to be important for route finding, including the ability humans use in following a map.

These considerations raise many questions, of obvious importance to an organism, about the integration of these different aspects of spatial information. Next we consider the third visual pathway.

The Third Major Transcortical Pathway

Limiting our discussion to the "what" and the "where" of a visually sensed object omits important aspects of adaptive vision. Recent work by cognitive scientists has been

The Visual Endbrain Structures

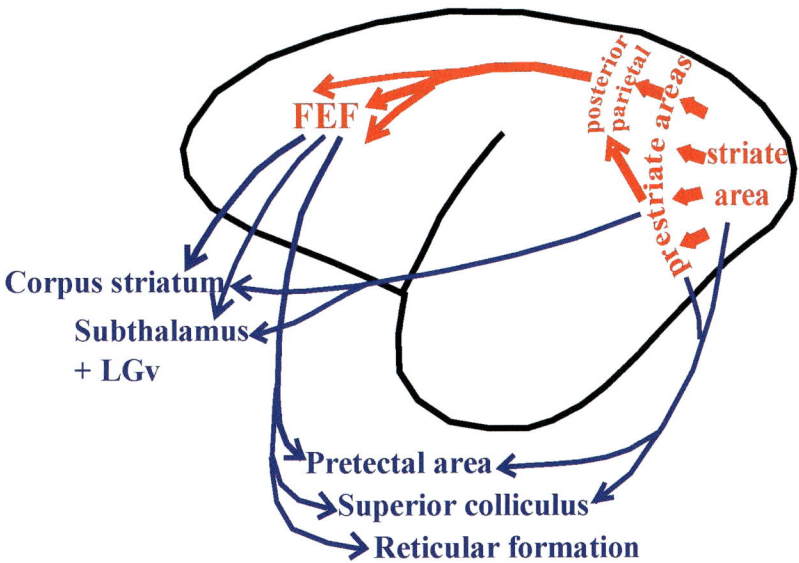

Figure 22.15
Some neocortical outputs controlling spatial orientation toward visually perceived object locations in primates. Functions of area 8 include anticipatory and voluntary orientation movements of the eyes. Cortical outputs are indicators of function. FEF, frontal eye field in prefrontal cortex (area 8); LGv, ventral nucleus of lateral geniculate body.

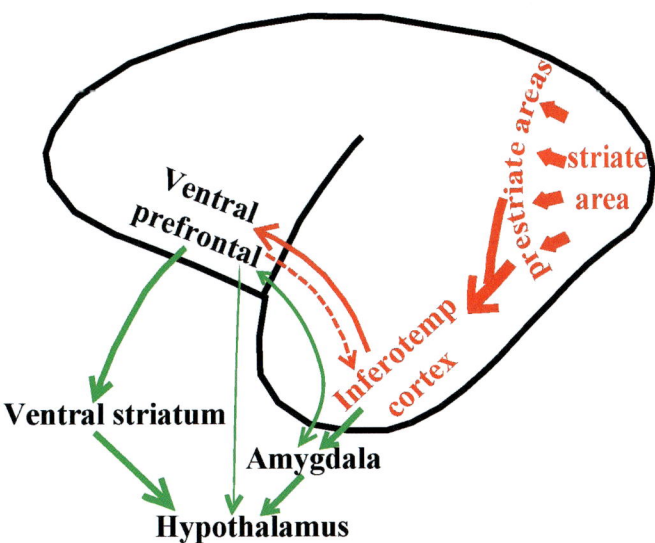

Figure 22.16
Outputs from the inferotemporal cortex and from its projection area in ventral prefrontal cortex. Cortical outputs are indicators of function. The inferotemporal cortex and the ventral/orbital prefrontal areas have outputs indicating affective associations, mood and motivational changes, and approach–avoidance tendencies.

generating interesting findings on *scene perception* as a kind of rapid acquisition of broad aspects of the visual world. The "gist" of a scene contains low-resolution information that can be important for rapid organization of possible priorities and plans. This kind of information gives the organism a rapid answer to the question, "Where am **I**?" (as contrasted with "Where is **it**?" and "What is **it**?"). The parietal association cortex of large-brained mammals appears to be important for learning and retaining more specific information about locations in the environments traversed by the animal—places where it has been and where it could be again. Visual information is of obvious importance for this function in many animals. However, the more ancient pathways important for this function were olfactory, especially connections to the medial pallium (see chapter 7). The medial pallium evolved into the hippocampal formation, which retains its ancient function in all mammals and is a critical partner in the learning and retaining of "Where am I" information (see chapter 28).

The adaptive importance of this kind of information in perception makes it likely that it appeared early in the evolution of the visual system, as a subcortical function and/or a pathway to the medial pallium. The pretectal area has characteristics and connections that suggest it as a likely participant in this function, as described below.

Within the neocortex, there are specific pathways involved in "Where am I" information. They are distinct from the two visual pathways already described (see figures 22.13 and 22.14). They can be followed from neocortex surrounding the striate cortex to posterior parietal areas and to retrosplenial cortex more medially, and from there, or more directly, to posterior parahippocampal cortex (postsubiculum in rodents; parahippocampal gyrus in large primates), which projects into the hippocampal formation (figure 22.17).

In rats, and possibly also in primates, there is a more direct pathway to the hippocampal formation that may carry this kind of information about where the organism is located or how its motion is changing its location. The retina projects directly to the pretectal cell groups, which project to the rostral part of the lateral thalamic nucleus, often called the lateral dorsal nucleus. This nucleus projects to visual cortical areas rostral and medial to area 17—the primary visual area: Included are area 18b and the adjacent retrosplenial dysgranular (or agranular) area. These areas project not only to more rostral motor areas, but the retrosplenial area projects to the postsubiculum, which connects to hippocampus. Even more direct is a route from the lateral dorsal nucleus directly to the postsubiculum. (See chapter 28 for more details about the hippocampus in mammals.)

Comments on Transcortical Interconnections

Within the neocortical mantle itself, the nature of transcortical connections is worthy of particular attention, especially when we consider evidence that the volume of such connections has increased greatly in more intelligent mammals with larger brains. However, the volume of transcortical interconnections has not increased to such a degree that "everything is connected to everything else" in the cortex—very far from it. The increase

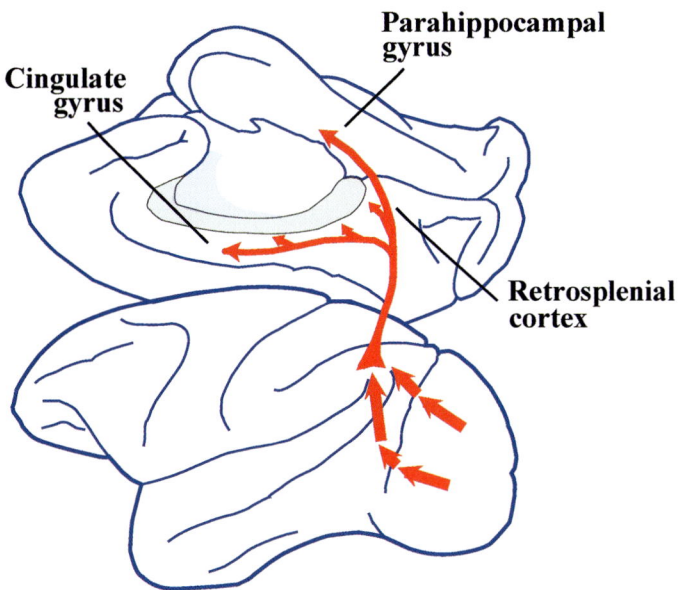

Figure 22.17
Visual pathway 3, "Where am I?" Lateral view of left hemisphere of a monkey is drawn below an upside-down medial view of the same brain. The red arrows depict transcortical pathways from the primary visual cortex to prestriate areas, which project to multisensory cortex in the posterior parietal area. From there, pathways can be traced to the parahippocampal gyrus (hippocampal gyrus) and to retrosplenial and cingulate cortex. Not shown are posterior parietal projections to temporal lobe association areas and to prefrontal association areas. The retina also has more direct projections to parahippocampal areas via the pretectal area and lateral-dorsal nucleus of the thalamus. Figure based on Nauta (1971).

is only slightly greater than what would be expected if neurons in the cortex were interconnected only locally. The longer transcortical connections are selective and specific. They appear to have been formed, or at least retained, not randomly but rather according to functional advantages determined by natural selection.

Readings

Kanwisher, N., McDermott, J., & Chun, M. M. (1997). The fusiform face area: A module in human extrastriate cortex specialized for face perception. *Journal of Neuroscience, 17,* 4302–4311.

Kolster, H., Peeters, R., & Orban, G. A. (2010). The retinotopic organization of the human middle temporal area MT/V5 and its cortical neighbors. *Journal of Neuroscience, 30,* 9801–9820.

Mishkin, M., Ungerleider, L. G., & Macko, K. A. (1983). Object vision and spatial vision: Two cortical pathways. *Trends in Neurosciences, 6,* 414–417. (Distinction between two transcortical pathway important in object vision.)

Montero, V. M. (1993). Retinotopy of cortical connections between the striate cortex and extrastriate visual areas in the rat. *Experimental Brain Research, 94,* 1–15.

Schneider, G. E. (1967). Contrasting visuomotor functions of tectum and cortex in the golden hamster. *Psychologische Forschung, 31,* 52–62.

Schneider, G. E. (1969). Two visual systems. *Science, 163,* 895–902. (Dissociation of two visuomotor pathways that can function somewhat independently. This paper extends the work described in the 1967 paper listed above.)

Review of connections in non-mammalian vertebrates: see the Butler and Hodos (2005) book listed in chapter 3.

Projection of thalamic nucleus LP carrying visual information from the superior colliculus of the midbrain to the amygdala:

Linke, R., De Lima, A. D., Schwegler, H., & Pape, H.-C. (1999). Direct synaptic connections of axons from superior colliculus with identified thalamo-amygdaloid projection neurons in the rat: Possible substrates of a subcortical visual pathway to the amygdala. *Journal of Comparative Neurology, 403,* 158–170.

Connections from visual pathways to the hippocampal formation in the rat:

Burwell, R. D., Witter, M. P., & Amaral, D. G. (1995). Perirhinal and postrhinal cortices of the rat: A review of the neuroanatomical literature and comparison with findings from the monkey brain. *Hippocampus, 5,* 390–408.

Van Groen, T., & Wyss, J. M. (1990). The postsubicular cortex in the rat: Characterization of the fourth region of the subicular cortex and its connections. *Brain Research, 529,* 165–177.

Van Groen, T., & Wyss, J. M. (1992). Connections of the retrosplenial dysgranular cortex in the rat. *Journal of Comparative Neurology, 315,* 200–216.

Wyss, J. M., & Van Groen, T. (1992). Connections between the retrosplenial cortex and the hippocampal formation in the rat: a review. *Hippocampus, 2,* 1–12.

Transcortical connections in monkeys: see publications of Deepak Pandya and collaborators.

In theoretical terms, the pattern of transcortical interconnections fits the idea of "small-world" networks. References: Watts, D. J., & Strogatz, S. H. (1998). Collective dynamics of "small-world" networks. *Nature, 393,* 440–442; reviewed by Striedter, G. F. (2005) with regard to brain connections.

23 Auditory Systems

When we think of our auditory abilities, many of us think of their importance for communicating with language or we think of the love we have for music. Both of these functions depend on the genetically determined development of the connections of the auditory system and on a large amount of learning. In this chapter, we look at what is behind these functions, as we review the basic components with a little about likely evolutionary origins.

We begin a discussion of the auditory systems of the CNS with the embryonic origins of the receptors and primary sensory neurons, followed by a summary of the pathways carrying the sensory information from the periphery to the endbrain. Next, we take note of functions of auditory systems of some animals for escape from predators or other dangers—the kind of functions that illustrate the importance of this distance sense for survival of preyed-upon animals. For predators, the evolution of audition was also important, but for other functions: the need to identify and to localize prey. These two functions led to the evolution of distinct ascending auditory pathways from the hindbrain. We will illustrate these pathways after some notes on transduction and initial coding and a summary of major channels of conduction of auditory information. When pathways for identification of auditory stimuli reached the neocortex, there was an evolution of specialized circuits for identification of temporally patterned sound stimuli. These circuits have made possible the evolution of speech in humans. Temporally patterned sound stimuli are also analyzed by hindbrain circuitry for sound localization using binaural cues and in some species for echolocation.

Embryonic Placodes Give Rise to Auditory and Vestibular Nerves

In the head region, some peripheral nerves (cranial nerves) arise from placodes, which are thickenings of the surface ectodermal cell layer. The so-called dorsolateral placodes are present in all vertebrate groups during development and give rise to a group of sensory cranial nerves including the auditory and vestibular nerves—which together form the eighth cranial nerve of humans and other mammals. Other cranial nerves originating

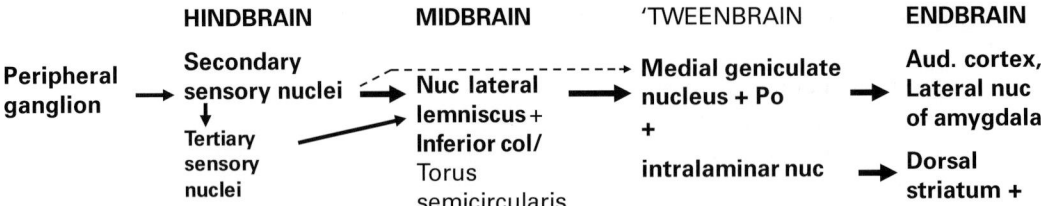

Figure 23.1
Diagram of ascending pathways in auditory systems of vertebrates, simplified and with an emphasis on mammalian structures. There are clear similarities to pathways taken by taste information in reaching the endbrain (see figures 18.2 and 18.3). Aud., auditory; Col, colliculus; nuc, nucleus or nuclei; Po, posterior nuclear group in the lateral thalamus.

from these placodes are the lateral line nerves. Lateral line receptors are innervated by up to six separate cranial nerves. Mechanosensory lateral line nerves were present in early vertebrates but are absent in terrestrial vertebrates. Fewer species, all of them aquatic, possess electrosensory lateral line receptors that give rise to lateral line nerves.

The Auditory Pathway (in Brief)

Figure 23.1 presents a simplified summary of ascending axonal pathways for the auditory and lateral line systems. Our focus will be on the auditory pathways. Receptors activate the primary sensory neurons of the eighth nerve. The axons of these cells enter the rostral hindbrain where they make synaptic contacts with secondary sensory neurons. In mammals, these neurons are in the cochlear nuclei. From cells of these nuclei, axons reach the dorsal midbrain. In at least some mammals, there are also axons that proceed directly to the thalamus. Back in the hindbrain, there are additional connections of the cochlear nuclei, to be discussed later, to cells that send processed information rostralward in the pathway. A major output of the auditory midbrain is to cells in the thalamus, many of which project their axons to the endbrain. What is apparently the most recently evolved auditory region of the thalamus projects to auditory neocortex. Older portions of the thalamus receiving auditory information project to the dorsal part of the corpus striatum as well as to cortex.

Some Special Functions of the Auditory System: Antipredator Behaviors

Detection and avoidance or escape from predators has been a basic function of auditory circuits that must have guided much of their evolution. Detection of sounds produced by predators, with a triggering of escape behavior, has evolved in many species. A simple example is the triggering of a diving response by moths in response to cries of bats hunting for insects. An example in vertebrates is the rapid escape movements in fish and tadpoles from sources of vibrations. The escape movements result from activation of the large Mauthner cells of the hindbrain in these animals. The response is one that makes a large difference in whether the animal survives to pass on its genes. It is dependent mainly on hindbrain and spinal mechanisms.

In adult amphibians, hearing can also trigger escape behavior, probably via the same hindbrain mechanisms used for visually elicited escape behavior, mechanisms that earlier in evolution may have been triggered by somatosensory input from the face. Rapid escape is so critical for survival that it has been a strong priority in evolution.

Mammalian examples abound as well. Many small mammals, like Syrian hamsters, when they are on a foraging trip respond to novel sounds by freezing or running for protective cover. The running response, triggered by sound, is usually guided by visual cues such as the shadow of an opening of a crevice or tunnel. The fixed action patterns appear to be organized primarily by brainstem mechanisms. They depend on midbrain and hindbrain with outputs to spinal circuits.

The output of the tectum that reaches motor neurons the most rapidly in rabbits is a pathway to the facial motor nucleus that turns the ears toward the source of a sound. With this movement, the animal is better able to detect faint sounds that may signal a need to escape.

Escape responses have also become highly specialized in some species. Consider the example of the kangaroo rat. Detection of a rattlesnake causes initial immobility as the rat awaits the attack. Then the auditory system is able to detect the sound of the rapidly onrushing mouth of the snake, which triggers a sudden leap out of harm's way as the strong rear legs propel the animal rapidly backward in a somersaulting motion to a place beyond the reach of the striking viper. This instinctive response, so beautifully engineered by natural selection, is usually successful.

Aversiveness of Noise and the Role of the Limbic System

A small animal is more likely to be safe if it avoids areas of loud noise. Laboratory experiments have revealed that a rat's aversion to loud noise does not depend on the forebrain but rather on a central core of the midbrain. This has been investigated by experimental work with brain-damaged rats.

Experimenters have found that auditory intensity thresholds are very difficult to affect permanently by CNS lesions, including ablation of the inferior colliculus or of the entire auditory cortex. There is no loss of the aversion to loud noise after lesions that destroy the entire inferior colliculus (the auditory tectum); adding an ablation of the superior colliculus, which also receives auditory input, still causes no loss of the aversion. Such brain-damaged rats can still discriminate different intensities of sound in a test where the animal is rewarded with food for pressing a bar when it hears a sound. However, if the brain lesion includes the ventral part of the midbrain central gray area (CGA) and

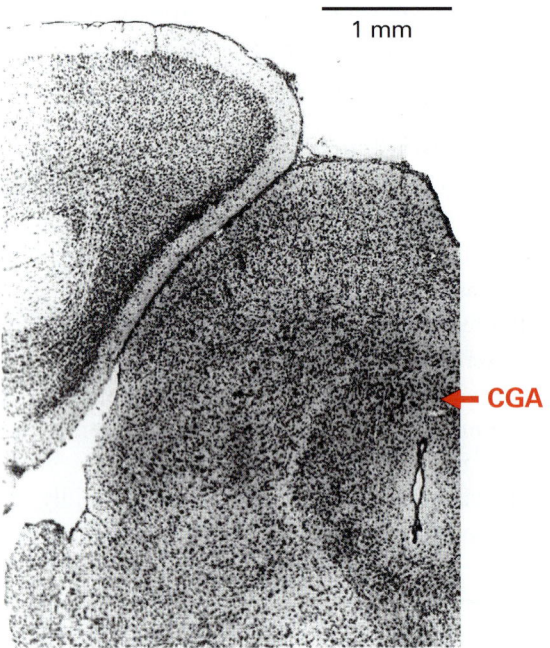

Figure 23.2
Portion of a frontal section through the hamster midbrain stained for cell bodies with a Nissl stain. The dorsal two-thirds of the left half is illustrated together with overlying occipital neocortex. The central gray area (CGA) surrounding the ventricle, called the aqueduct of Sylvius at this level, is labeled. A lesion of the ventral part of the CGA has been observed to cause a loss of aversion to loud noises although the ability to detect sounds is not affected.

the adjacent reticular formation, then there is a loss of noise aversion even though the animal does not suffer a loss in its ability to detect sounds. For the location of the CGA, see chapter 11, figures 11.2 and 11.4, and the photograph in figure 23.2.

Humans also find certain sounds to be more fear-inducing and others to be happier or soothing and relaxing. Sounds can influence our moods not only because of innate preferences but also because of our experiences. The possible neural bases for these influences in humans can be guessed at with some confidence from experiments using animals.

Learned Fear: Importance of the Forebrain

Early in evolution, non-olfactory information entered the CNS and was processed at hindbrain and midbrain levels and, in the case of light detection, the 'tweenbrain. The olfactory endbrain, however, acquired a great advantage with its connections in the corpus striatum: the advantage was plasticity, the ability to form and change habits. A likely reason for the invasion of the endbrain by non-olfactory senses was the adaptive advantage of this ability: A route to the striatum was very useful as it provided a pathway, modifiable by learning, to the substrates of instinctive behaviors and the motivational systems associated with those behaviors.

In the case of vision and somatic sensation, early routes to the striatum from the intralaminar and midline thalamic nuclei became mostly overshadowed by routes to the cortex. But in the case of the auditory system, another pathway evolved and has persisted: a route from the auditory thalamus to the caudal outpost of the striatum known as the amygdala (figure 23.3). (A similar, apparently smaller pathway has been reported for the visual system.) This caudal portion of the basal ganglia has been shown to be very important in the learning of fear responses to auditory stimuli. With this neural pathway to the endbrain, a small animal learns to respond with appropriate fear to sounds that could indicate danger, as from a predator.

Predation Had Other Requirements: Identifying and Localizing Prey Animals

For a small preyed-upon animal, it is most urgent that danger be detected, even before the source of that danger is actually known. Thus, a warning of possible attack can be provided by a novel sound or sight or any sudden change or by something that has been previously experienced as dangerous. It is less important that the source of the danger be identified or that the exact location be known, and much more important that the little creature find a safe haven or concealment very rapidly. But in the evolution of predators there were other requirements, important for success in finding food. Beyond the initial detection, the predator needed to identify prey animals so it could judge the quality of the meal and the likelihood of a successful attack, and it had to be able to localize its prey for an efficient attack. These two requirements, for increasing the abilities to identify and to localize beyond the simplest untutored hindbrain reflexes, were served by evolution of distinct ascending pathways from the hindbrain.

How were these two requirements for efficient predation met? There had been an evolution of neural equipment for detection of a range of sound frequencies and for discrimination of differences in frequency. This led to brain circuits for detection and discrimination of sound patterns—especially temporal patterns. Distinct cues for spatial localization were exploited by evolution of neural apparatus for using those cues.

Audition came to serve other abilities, too, besides the abilities to avoid capture by predators or to be a successful predator: various abilities to communicate with other members of the species. The adaptive advantages of communication abilities led to further expansion of forebrain auditory circuits together with the apparatus for emitting sounds.

With these things in mind, we are ready for a closer look at the apparatus of the eighth nerve and the cochlear nuclei and their connections.

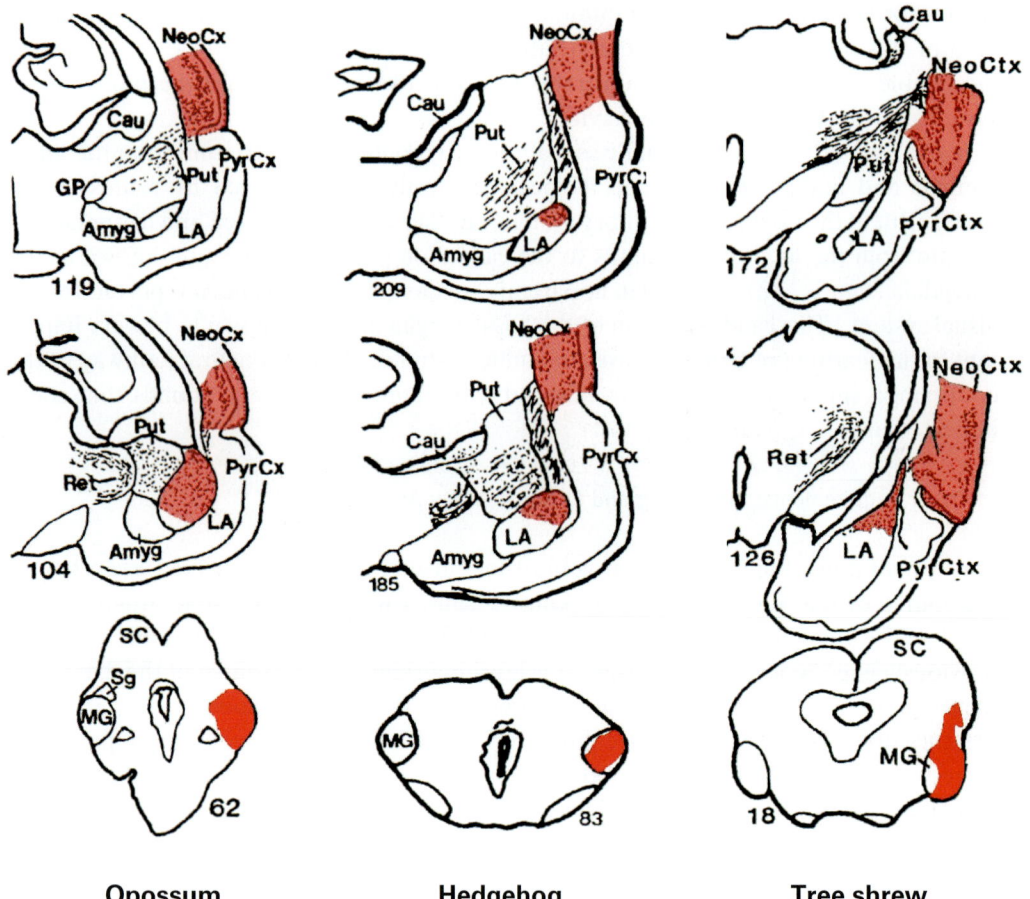

Figure 23.3
The projections of the main structure of the auditory thalamus—the medial geniculate body (MG)—to the endbrain in three mammals. Axonal tracers were injected into the MG in opossum, hedgehog, and tree shrew. As shown by the color representing the label, the MG has two main projections: the lateral nucleus of the amygdala and the neocortex of part of the temporal lobe. In the opossum, the projections to amygdala and cortex are about equal in quantity. In the hedgehog, the projection to the amygdala is relatively smaller, about half as large. In the tree shrew, the projection to the lateral amygdala is much less, with less than 5% of the MG cells projecting there. Amyg, amygdala; Cau, caudate nucleus; Ctx, cortex; Cx, cortex; GP, globus pallidus; LA, lateral nucleus of amygdala; MG, medial geniculate body; NeoCx, neocortex; Put, putamen; PyrCx, piriform cortex; Ret, reticular nucleus; SC, superior colliculus; Sg, suprageniculate nucleus. Extracted from Frost and Masterton (1992) with coloration added.

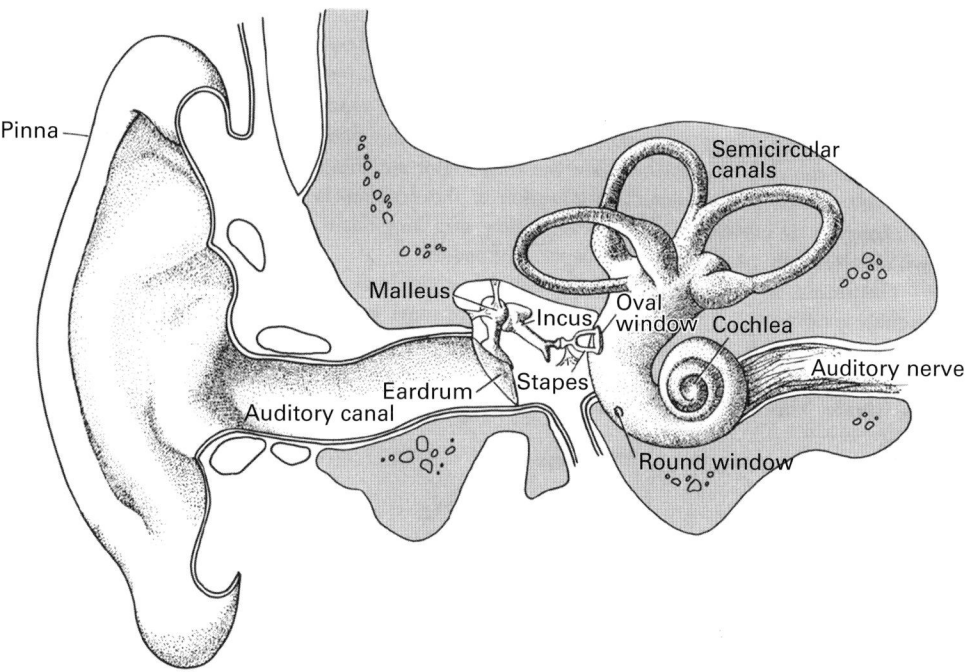

Figure 23.4
The human ear: external, middle, and inner. From Lindsay and Norman (1977).

Transformations of Sound Vibrations in Middle Ear and Cochlea

In the cochlea of the inner ear, there is a transduction of the mechanical energy of sound vibrations to nerve impulses in the eighth nerve. Studies of the entire nerve and its terminations reveal an encoding of sound frequencies in topographically organized brain maps. There is also the encoding of sound intensity.

But first we should take a look at the middle ear, where there is a preparation for the transduction process by an impedance matching, as variations in sound pressure level in the external auditory meatus (the tunnel from the external ear to the tympanic membrane—the eardrum) are transferred very efficiently to vibrations in the fluid of the cochlea (figure 23.4). This transformation is accomplished in mammals by a tiny trio of middle-ear bones. The vibrations in the cochlear fluid result in vibrations in the basilar membrane of the inner ear. Tiny movements in this membrane produce shearing forces in the inner hair cells, so named because of hair-like protrusions into the overlying tectorial membrane. These forces cause depolarization of the hair cell membranes that trigger action potentials in the axons of the spiral ganglion cells of the eighth nerve.

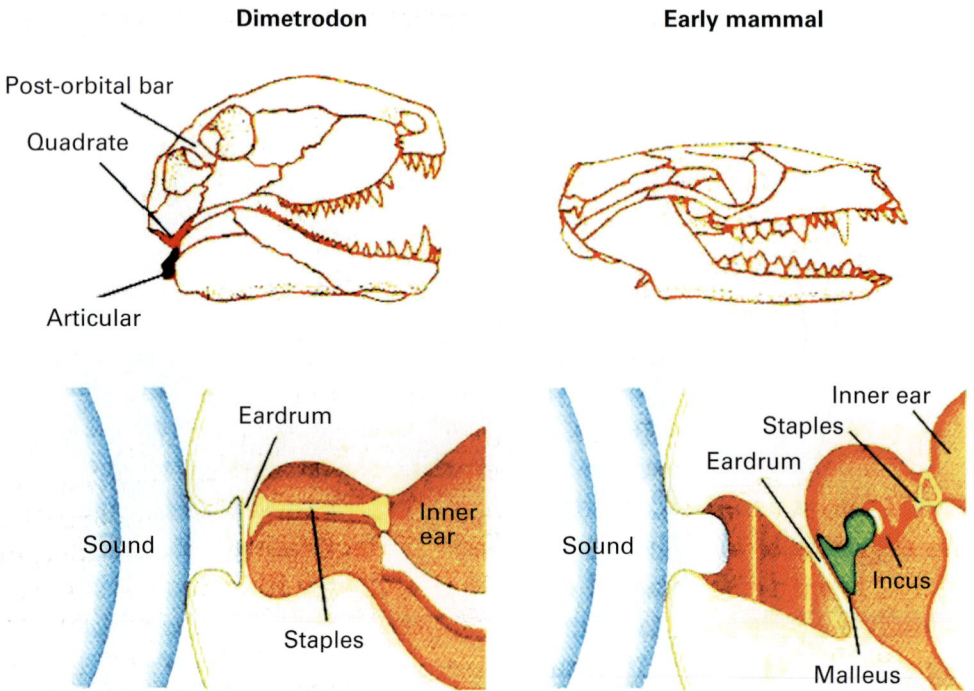

Figure 23.5
Drawings that illustrate evidence that jaw bones evolved into the ossicles of the middle ear in mammals. Also shown in the drawings of Dimetrodon (a predatory mammal-like reptile) and an early mammal, there was also a loss of the postorbital bar in the early mammals, indicating a reduction in the eyes. From Allman (2000).

Evolution of Jaw Bones of Ancestral Reptiles into the Ossicles of the Middle Ear in Mammals

Paleontological findings on the skulls of ancestral vertebrates show that in certain mammal-like reptiles that preceded mammals, and in the earliest mammals during the time when dinosaurs still roamed the earth, there was a remarkable evolutionary change in the hearing apparatus. The change gave mammals improvements in their hearing ability that made it easier for them to avoid the many reptilian predators that dominated large regions of the earth when mammals first evolved. Bones that had been part of the jaw evolved into two ossicles of the middle ear, the malleus and incus (figure 23.5), which joined to the stapes—the middle-ear bone already present in the ancestor reptiles. The changes not only allowed a very marked extension in the range of frequencies that could be heard by mammals (figure 23.6), but also it made hearing possible during chewing. The hinged construction of the trio of tiny bones accomplished the impedance matching mentioned earlier. This contributed to the ability of mammals to hear at frequencies far above the range of their non-mammalian predecessors, a difference that remains today between mammals on the one hand and birds, reptiles, and amphibians on the other.

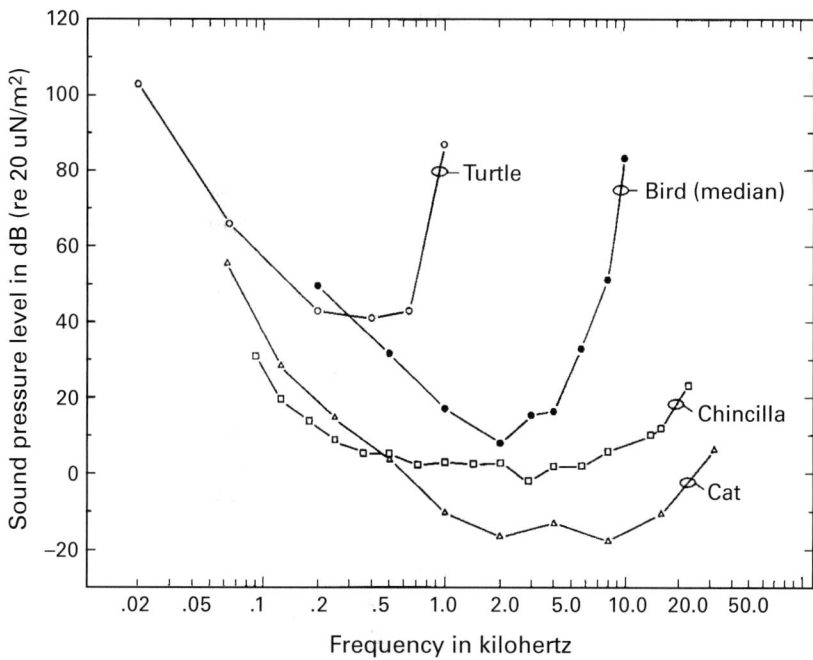

Figure 23.6
Plots of the outer envelope of frequency ranges of three groups of vertebrates, showing that mammals can hear much higher frequencies than reptiles (as well as amphibians) and birds. Note the logarithmic scale on the abscissa. From Dooling (1980).

Also contributing to the ability to detect higher frequencies was the appearance of outer hair cells among the cells located between the basilar and tectorial membranes in the cochlea. These additional hair cells function by their changes in shape that alter the mechanical properties of the membrane and affect the responses of the inner hair cells, the cells that are the transducers activating the eighth nerve axons (figure 23.7).

Mammalian Ear Structures and Dynamics

The ossicles of the middle ear connect to the tympanic membrane on the outer side and to the oval window of the cochlea on the inner side (figure 23.8). The cochlea is an elongated tube rolled into a coil, with the basilar (also the tympanic) membrane running down the center; side chambers form the vestibular apparatus. The figure shows a simplified schematic of the ear structures, with the cochlea unrolled. Figure 23.9 illustrates how the position of maximum movement of the basilar membrane changes with sound frequency. Note that this position moves further and further from the oval window (and stapes) as

426   Chapter 23

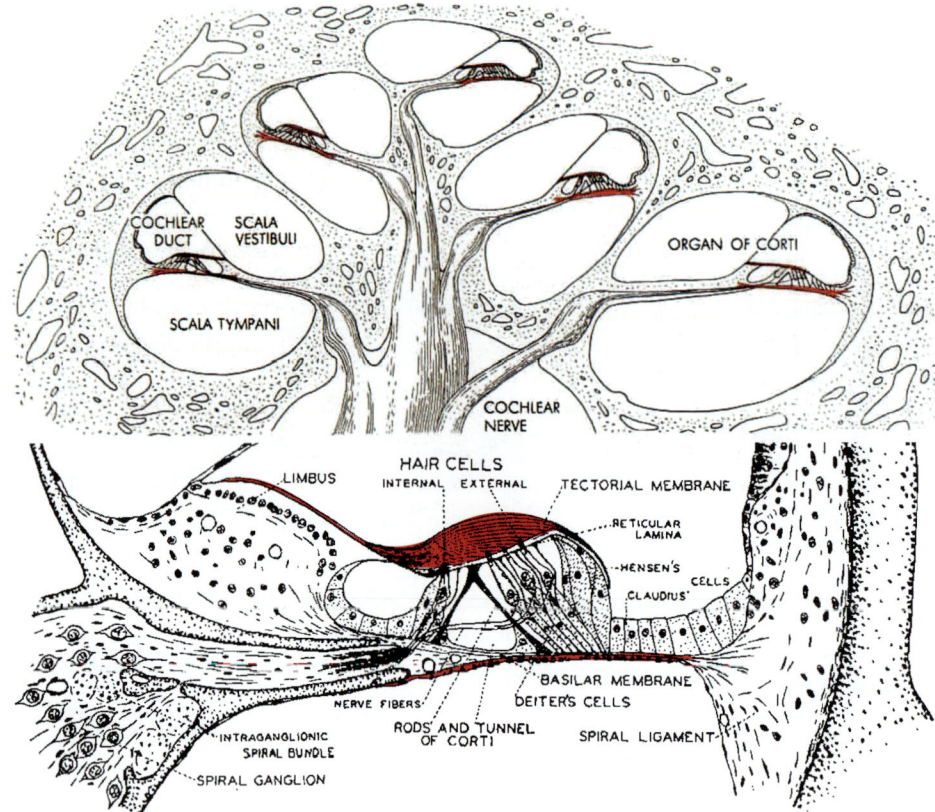

Figure 23.7
(Top) A section through the human cochlea, enclosed in the mastoid bone, shows the coils of the cochlea cut through at five places. At each section, the compartments of the cochlear duct are seen, with the organ of Corti in the middle. (Bottom) Enlargement of the organ of Corti from a guinea pig, showing the outer and inner hair cells. The auditory nerve, composed of the peripheral axons of spiral ganglion cells, can be seen with connections to the hair cells. Top panel from von Békésy (1957); bottom panel from Davis (1961).

frequencies become lower. The result is that when the axons activated by these displacements terminate in a topographically organized map in the brain, they form a two-dimensional frequency map (see later).

Initial Coding of Information

Different axons in the auditory nerve are stimulated maximally by different frequencies of sound, as just noted. Recordings of cell responses in the cochlear nuclei of cats has demonstrated that the eighth nerve axons terminate in a precise order there. As one advances a tiny electrode through the ventral cochlear nucleus, the higher frequencies are

Auditory Systems

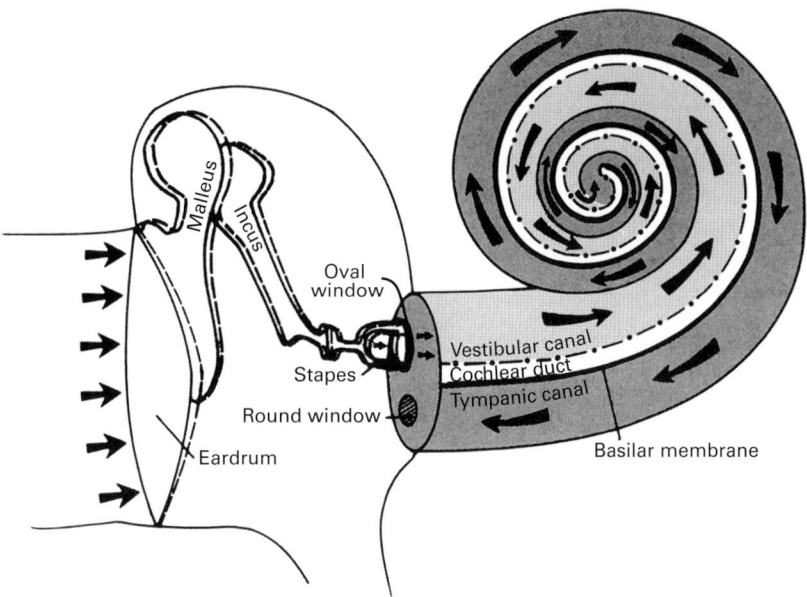

Figure 23.8
Simplified sketch of a section through mammalian ear structures, showing the rolled-up shape of the cochlea. The basilar membrane runs down the center. From Lindsay and Norman (1977).

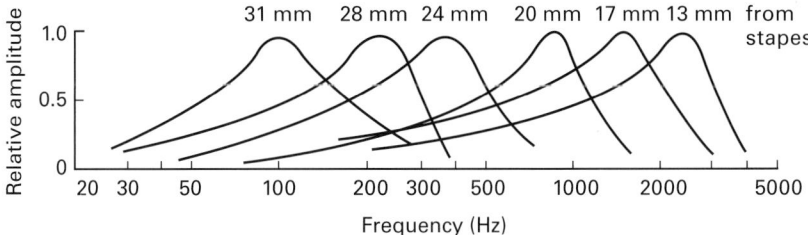

Figure 23.9
Illustration of basilar membrane dynamics. The relative amplitude of basilar membrane vibrational movements caused by different tone frequencies is greatest at different locations along the length of the cochlea. From von Békésy (1949), with frequency designation changed from cycles per second to hertz (Hz).

found to activate the most dorsal cells, and then lower and lower frequencies are the most effective as one advances the electrode more and more ventrally. The two-dimensional map of sound frequencies corresponds to a representation of the cochlea from positions near the base (near the oval and round windows) to the apex (at the far end of the tube).

The encoding of different intensities of sound is not simply a result of different firing frequencies in the nerve. Different fibers representing the same best frequency not only have different thresholds, but they have different best intensities as well, so that different fibers encode different intensities. Thus, the CNS has evolved a place code for frequencies and a place code for intensities. Differential responses to different temporal patterns, however, depend on further processing.

The Flow of Auditory Information in the CNS

The further processing of the auditory information reaching the cochlear nuclei of the rostral hindbrain, just below the lateral wing of the cerebellum, depends on the auditory pathways, which have been traced using experimental neuroanatomical methods. A simplified summary of these pathways is shown in figure 23.10.

The diagram shows a single primary sensory neuron, a bipolar cell of the spiral ganglion with its receptive end in the cochlea and its axon going to the two cochlear nuclei (dorsal and ventral). The secondary sensory neuron axons can be followed into distinct channels, functionally different: a reflex channel, a cerebellar channel, and some ascending lemniscal channels. The lemniscal channels include two main routes to the inferior colliculus of the midbrain (or its equivalent in non-mammals), which has a major projection to the medial geniculate body (MGB) of the thalamus. In some mammals, there is a smaller route from the cochlear nuclei directly to MGB, discovered in chimpanzees and found later also in rats, ferrets, and other species. The MGB sends axons to the auditory areas of the neocortex.

Note that the diagram also depicts a route, usually considered secondary but probably very ancient, that does not go through the inferior colliculus and terminates in the posterior nucleus of the thalamus, an area where there is a convergence of inputs of different modalities. The pathway has been called the lateral tegmental pathway, as it passes through the lateral midbrain tegmentum (below the tectum). The thalamus's posterior nucleus also receives inputs from both superior and inferior colliculi.

Terminations of the Auditory Nerve

The eighth nerve axons have a branch terminating in the dorsal cochlear nucleus and a branch terminating in the ventral cochlear nucleus. The axonal branches entering the ventral cochlear nucleus have multiple branches, terminating at different anteroposterior levels, forming columns or tiers of cells that respond to the same frequency.

Auditory Systems

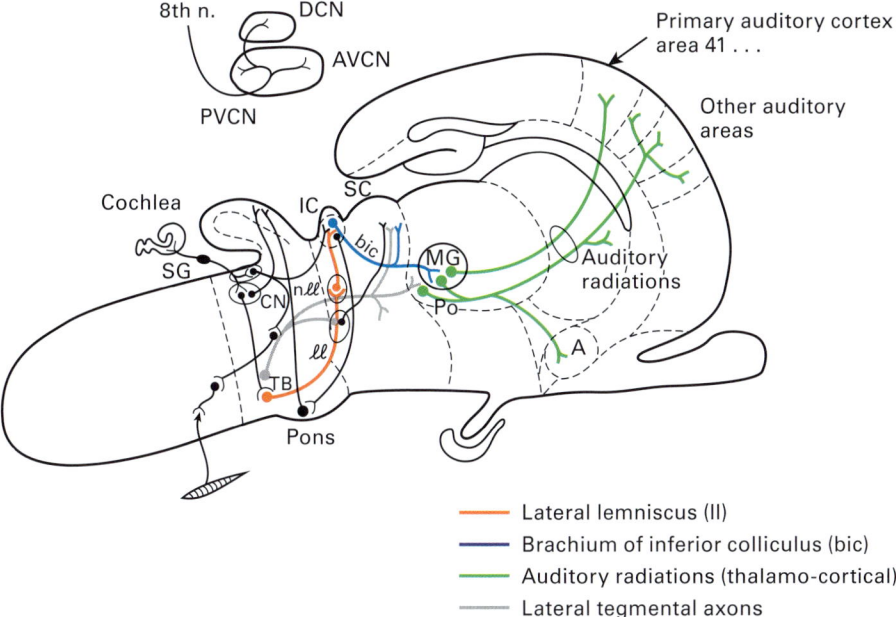

Figure 23.10
Auditory pathways are sketched for the mammalian brain. Major pathways from the spiral ganglion of the eighth nerve to the auditory areas of neocortex are illustrated. At the top is an enlarged sketch of the cochlear nuclei. In addition to major ascending pathways in most mammals, some additional brainstem pathways are included in order to show reflex and cerebellar channels of conduction. Not included are direct projections from the dorsal cochlear nucleus to the medial geniculate body in some species. Also omitted are other projections that include those that carry auditory information to intralaminar thalamic cell groups of the thalamus and hence to corpus striatum. A, amygdala; AVCN, anteroventral cochlear nucleus; bic, brachium of inferior colliculus; CN, cochlear nuclei; DCN, dorsal cochlear nucleus; IC, inferior colliculus; ll, lateral lemniscus; MG, medial geniculate nucleus; nll, nuclei of lateral lemniscus; Po, posterior nuclear group; SC, superior colliculus: SG, spiral ganglion; TB, trapezoid body.

Electrophysiological studies show that the secondary sensory cells have different response patterns, including one type of cell that fires in the very same way that the primary sensory axon fires. How can it be that every action potential arriving at such a neuron causes a single action potential in the postsynaptic cell?

To answer this question, we have to know some anatomical details: we have to know about the *endbulb of Held*. A postsynaptic neuron generally will not fire an action potential because of the action of a single synapse. It requires much convergence, spatial and temporal. This can be a result of nearly simultaneous inputs from multiple axons ending on the cell or from multiple synaptic contacts made by one axon (or both). The auditory system evolved a way to preserve precise timing of the eighth nerve inputs in some of the output axons of the cochlear nucleus—so the timing could be used for localization of sound sources (see later). The requirement was the generation of many simultaneous excitatory postsynaptic potentials resulting from arrival of a single action potential of the primary sensory axon. This was enabled by an enlargement of the terminal—the bouton—so it formed a cup shape abutting the membrane of the cochlear nucleus cell, with formation of a large number of excitatory synapses.

The axons of the cells contacted by the endbulbs of Held project to the *trapezoid body* of the ventral hindbrain, more specifically to the cells of the *superior olive*, so named because in humans it corresponds to an olive-sized bump on the hindbrain surface. From the superior olive, axons project, via the lateral lemniscus, to the caudal midbrain—ending in the inferior colliculus and in the nuclei of the lateral lemniscus. Additional cell groups of the trapezoid body have similar projections.

Thalamic Projections Carry Auditory Information to the Limbic System and to Neocortex

The auditory pathways go from the medial geniculate body via the internal capsule to auditory areas of neocortex. This cortex is found in the dorsal part of the temporal lobe of primates. Auditory information that reaches the posterior nucleus of the thalamus, adjacent to MGB, goes to several additional auditory areas in the neocortex.

In addition, the medial part of MGB and nearby thalamic cells have been discovered to project directly to the limbic endbrain—to the amygdala in the temporal region (see figure 23.3). It is likely that this projection evolved very early. Also, the amygdala as well as more rostral portions of the basal ganglia known as the dorsal striatum receive multimodal inputs from the midbrain by way of thalamic cells near the MGB, including caudal portions of the intralaminar nuclei.

Thus, the projection from MGB to amygdala appears to be part of a system of projections that carry multiple modalities of sensory inputs from paleothalamic cell groups—including the midline and the intralaminar nuclei and the posterior nuclear group—to the basal ganglia (striatum and amygdala). Some have branches reaching neocortex as well. The advantage of learning about sounds that warn of danger no doubt gave a special impetus to the evolution of the amygdala connection as it has been found to be important in the learning of fear responses to sounds that warn of pain.

Review: Multiple Routes Carrying Auditory Information to the Forebrain

Table 23.1 summarizes major projections whereby auditory information reaches the thalamus. These projections are indicated in figure 23.10. In the figure, you can also see depicted the projections of the medial geniculate, via the internal capsule, to auditory areas of the neocortex, located in the temporal region. Also indicated are projections carrying auditory information from the adjacent posterior nucleus of the thalamus to several auditory neocortical areas. Also indicated are the projections to the amygdala and to the striatum.

Table 23.1
Major projections through which auditory information reaches the thalamus

- Dorsal cochlear nucleus (DCN) to midbrain to medial geniculate body (MGB)
  Midbrain: inferior colliculus (IC) and nuclei of the lateral lemniscus (nLL)
- Ventral cochlear nucleus (VCN) to *trapezoid body* to midbrain to MGB
  Trapezoid body includes the *superior olive*
  Midbrain: not only IC but also middle layers of the superior colliculus, which project to posterior portions of the lateral nucleus of the thalamus
- Cochlear nuclei (CN) to reticular formation and/or nLL to posterior group of nuclei of thalamus (Po) and medial part of MGB
  These are components of the *lateral tegmental system* of Morest
- DCN directly to MGB
  Found in some mammals but possibly not all

Distinct Pathways for Two Major Functions: Orienting and Identification

Functional anatomists have traced two distinct pathways going rostrally from the hindbrain for two functions: One is for sound localization and orienting toward different parts of the space around the head. The other is for auditory pattern detection and discrimination. The first has connections with the superior colliculus for the control of orienting movements. These connections can also trigger escape movements. The second pathway, for pattern identification—which generally means temporal patterns—goes more directly to the thalamus and hence to the endbrain.

The ascending auditory pathways are depicted in a simplified schema in figure 23.11. The diagram separates pathways from the ventral cochlear nucleus from those from the dorsal cochlear nucleus. It is some of the projections beginning in the ventral cochlear nucleus that are involved in spatial localization, whereas those originating in the dorsal cochlear nucleus are more involved in conveying information about auditory patterns that identify the sound source—warning cries of fellow creatures, the voice of a mate, sounds of a baby crying, words. The pathways from the ventral cochlear nucleus are doubtless involved in auditory functions other than spatial localization and orienting movements, but the ascending pathway from the more dorsal secondary sensory neurons do not appear to be have location specificity.

Location Specificity Arises in Cells of the Hindbrain's Trapezoid Body

Many of the axons coming from the secondary sensory neurons of the ventral cochlear nucleus grow ventrally in mammals and make synaptic contact with neurons in several cell groups. Some of the terminations are on the same side and others on the opposite side of the brainstem. This system of axons is called the trapezoid body. The trajectories of axons and the cell groups are not the same in other groups of animals. For example, in birds there are homologous dorsally located cells receiving inputs from the cochlear

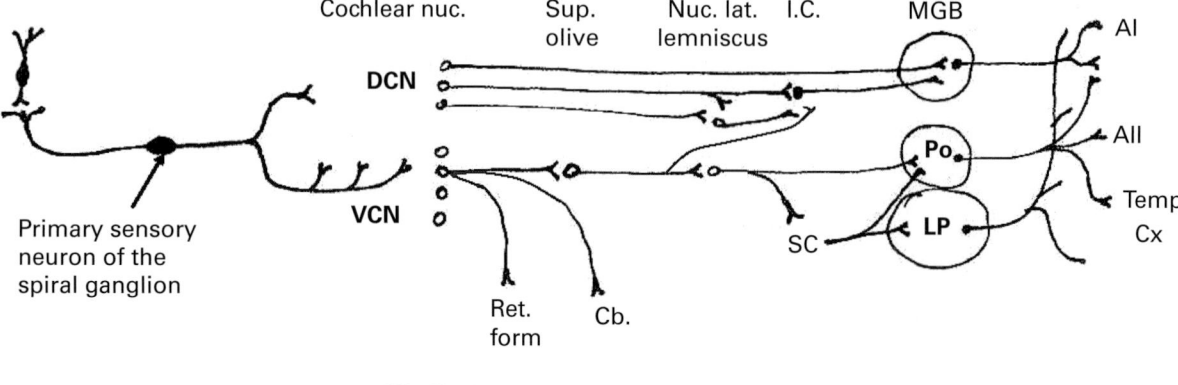

Figure 23.11
Ascending auditory pathways to the neocortex are summarized diagrammatically, adding details to figure 23.1. Pathways leading from the dorsal cochlear nucleus (DCN) and the ventral cochlear nucleus (VCN) are separated in this schema. Axons from the DCN do not appear to be involved in spatial localization of sound sources and include axons that ascend directly to the medial geniculate body (MGB) of the thalamus. Axons from the VCN constitute several short pathways to hindbrain structures including structures of the trapezoid body—which includes the superior olive, known to contain neurons that respond specifically to sounds at particular spatial positions. This spatial information reaches the superior colliculus (SC) as well as thalamic structures. The central nucleus of MGB projects to the primary auditory neocortex (AI); other parts of the nucleus project to adjacent areas. Cb, cerebellum; IC, inferior colliculus; Temp.Cx., temporal neocortical areas.

nucleus cells. In both groups of animals, these cells arise in the alar plate but differ in their postmitotic migration patterns.

The functional anatomy of this auditory subsystem can be summarized as follows. The neurons in one cell group located along the trajectory of the trapezoid body axons constitute the medial superior olive. The neurons here receive inputs that originate in both of the ears, and they are sensitive to precise differences in time of arrival at the two ears such that their firing patterns indicate azimuthal position in the space around the head. There is also a lateral superior olive where the cells are sensitive to the relative amplitudes of a sound reaching the two ears; this difference is also used to derive the azimuthal position of a sound source. Location in the vertical axis is also decoded in the system, using several different cues.

The location information reaches the midbrain's superior colliculus. Lesions there (in studies of cats and hamsters) can cause a considerable loss of orienting toward sound sources, especially when the orienting is toward an overhead sound. Location information also reaches auditory areas of the neocortex via the caudal thalamus.

How Direction Can Be Derived from Precise Time-of-Arrival at the Two Ears

A combination of detailed neuroanatomical studies and careful electrophysiological studies of the chicken brain has revealed how the azimuthal position of a sound source can be

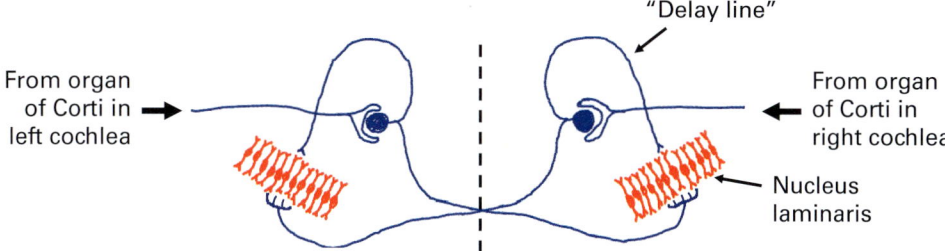

Figure 23.12
Illustration of neurons in nucleus magnocellularis of the chicken, part of the cochlear nucleus, and their axonal connections. Only one of these neurons is sketched on each side of the hindbrain. The midline is indicated by the vertical dashed line. Axons of the auditory nerve terminate on nucleus magnocellularis cells by forming a cup-like enlargement called an *endbulb of Held* after its discoverer. The many excitatory synapses of an endbulb trigger one action potential in the postsynaptic neuron for each action potential in the afferent axon. The axon of the magnocellularis neuron divides into two main branches, each branch terminating on dendrites of nucleus laminaris on opposite sides of the brain as shown. Thus there is a convergence of inputs that originate in right and left ears on the nucleus laminaris neurons; these neurons are activated only by nearly simultaneous inputs to their dorsal and ventral dendrites. The axons of the magnocellularis neurons function as delay lines that vary in length in a systematic manner. The result is an activation of different nucleus laminaris neurons when the location of a sound source changes around the head in the horizontal plane. The axons (not shown) of the nucleus laminaris neurons carry the location information to the midbrain.

derived. First of all, there is preservation of timing information through the cochlear nucleus because of the giant terminals described earlier in this chapter. These terminals are found on the neurons of a secondary sensory cell group known as *nucleus magnocellularis* because of its large cells (figure 23.12). As shown in the figure, the axons of these large cells branch, with one branch going to one side of the hindbrain and the other branch going to the other side. With this arrangement, the neurons of another cell group called *nucleus laminaris* in chickens receive a convergence of inputs from right and left sides. The inputs from the two sides are found on different dendritic trees of the same neurons.

The anatomy reveals how coincidence detection can result in an azimuthal map of sound direction in *nucleus laminaris*. The axons from the magnocellular nucleus to the same side take an arching trajectory that appears to constitute a delay line, making the axonal conduction time from the two sides nearly the same. It is not difficult to see how systematic changes in the amount of delay in different axons can result in a coincidence of action potential arrivals from the two sides, the cell location of this coincidence varying with location in the laminar nucleus according to position of the sound source in the space around the animal's head. (The variable delays have been reported to occur in the axons from the contralateral side.)

In mammals, the medial superior olive is similar to *nucleus laminaris* in its connections and function.

A Second Hindbrain Mechanism for Sound Localization

It is very likely that another hindbrain mechanism occurred earlier in evolution: a much simpler one, described briefly above. A sound from a location to the right of an animal's head is slightly louder in the right ear than in the left ear. Thus, relative amplitude at the two ears can be used as a simple cue to location. This cue appears to be processed by another cell group in the trapezoid body: the lateral superior olive.

How Did This Spatial Localization Apparatus of the Hindbrain Evolve?

We can imagine a plausible sequence of steps in the evolution of the spatial localization mechanisms described above. When hearing began to evolve, a primitive cochlea or its predecessor probably had diffuse connections to the nearby hindbrain with connections to secondary sensory neurons that varied in relative strength. These secondary sensory neurons must have projected widely, with axons reaching both sides of the nearby brainstem. Some cells received convergence of inputs originating on two different sides of the head. Certain cells must have fired more when a sound came from one side rather than the other side—either because of amplitude differences or from timing differences. The timing differences were quite variable due to variations in fiber size (and conduction rate) and variations in axon length. The resulting crude localization effect was exploited in evolution and became much more precise. Use of timing differences, of course, required the evolution of very strong synaptic connections with secondary sensory neurons.

Judging a Sound's Direction in the Vertical Axis

Computing the location of a sound on the vertical axis is a problem different from that of determining the azimuthal location. Head movements or external ear movements in many animals are used to aid vertical localization—sounds are louder when the ears are facing the stimulus. The mobile ears of some animals, like the rabbit, exploit the sound collection ability of a large external ear. In addition, there are cues that do not require head movement. Different frequencies are attenuated differently depending on elevation, because of external ear asymmetries. In the owl this effect is especially increased by this animal's external ear mechanics.

Location Information Reaches the Midbrain's Superior Colliculus

Both the lateral and the medial nuclei of the superior olive project to the midbrain tectum (see figure 6.3), carrying to that structure the information on spatial location. Electrophysiological recordings from the superior colliculus have been used to find a map of auditory space there. This map is in register with visual and somatosensory maps. The superficial layers receive a topographic projection from the retinae. The map of auditory space is found in the intermediate layers, and the organization of this map matches the overlying map of visual space around the head. There is also a somatosensory map located in the deepest collicular layers; in rodents, this map is dependent largely on input from the facial vibrissae.

The auditory map has been found to be plastic. Experiments in which prisms in front of the eyes displace the visual inputs have found that the auditory map changes over time to match the change in the visual field representation.

Experiments with hamsters and cats have shown the importance of the midbrain tectum for auditory orienting. For example, ablation of the superior colliculus has been found to cause large defects in orienting toward sounds.

Location Information Also Reaches the Endbrain

The existence of the midbrain tectal map of auditory space around the head suggests that this spatial information also reaches the neocortex, as the middle, auditory, layers of the superior colliculus project to the lateral posterior portion of the dorsal thalamus. The projection terminates more deeply in the thalamus than do the inputs from the visual layers. Spatial information from the auditory system may also reach the posterior thalamus by routes bypassing the superior colliculus.

Like the visual system, the auditory system has evolved multiple neocortical areas— although the number of such areas that have been discovered thus far in higher primates does not appear to be as large. An area in the superior temporal gyrus of the macaque located caudal to the *primary* auditory area (A1) has been found to be activated by spatial location information. The area receives input from the posterior part of the lateral thalamus and from A1. It projects axons to the posterior parietal region and to lateral prefrontal cortex.

The Second Function of Ascending Auditory Pathways: Pattern Identification

Return to the sketch of ascending auditory pathways shown in figure 23.11. We have been discussing a major function of the pathways sketched in the lower portion of the diagram, originating in the neurons of the ventral cochlear nucleus. The other major function of the ascending pathways is the detection of auditory patterns and discrimination of differences in such patterns. The aspects of auditory patterns being distinguished are primarily temporal. Studies of auditory areas of the mammalian neocortex have found much evidence for auditory pattern selectivity in the responses of the cortical neurons. As is true for the visual system, auditory patterns are important determinants of approach–avoidance decisions, both innate and learned. After learning experiences occur, such

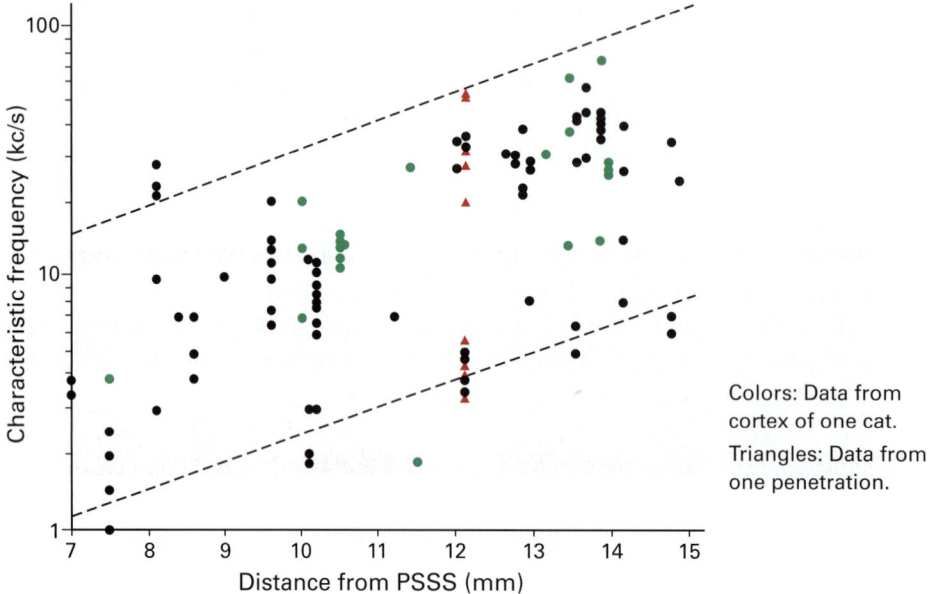

Figure 23.13
Characteristic frequencies of neuronal units in area A1 of cat: Points on the graph at one position along the abscissa show units located at one neocortical location. Many of the cells at a single location were found at different depths of single-electrode penetrations. PSSS, posterior suprasylvian sulcus. From Evans (1968).

patterns become associated with positive or negative motivational tendencies of various kinds—good or bad feelings—about the sources of the sounds.

Auditory cortical areas have been mapped according to sound frequency—tonotopic maps that resemble in this respect the representations of sound frequencies found in the cochlear nuclei. However, recording at any single location in A1, the first or primary auditory cortical area, can result in recordings from neurons that have a variety of characteristic frequencies—the pitches of sounds that stimulate them at the lowest thresholds. But the overall envelope of all the best frequencies indicates progressive changes across the cortical area (figure 23.13).

The cat has been used for many electrophysiological studies of the auditory system. Cats have well-developed auditory systems, with a large portion of their neocortex devoted largely to the processing of auditory inputs. In addition to the primary auditory cortex (A1), investigators have found four other cortical areas with tonotopic organization. In addition, there are three auditory areas that lack this tonotopic organization (see illustration in figure 23.14). Auditory responsiveness is also found in three posterior areas that respond to multimodal inputs and in two ventral areas that are closely tied to the limbic system.

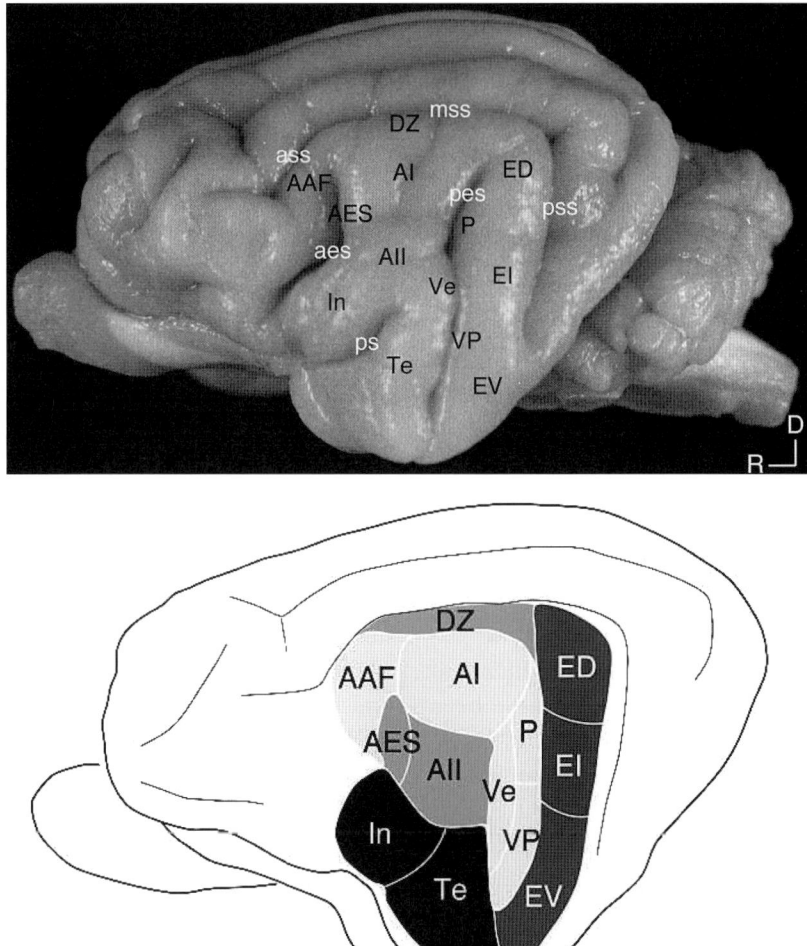

Figure 23.14
Left hemisphere of an adult domestic cat with identification of 13 auditory cortical areas and the major gyri and sulci. Both the photograph and the drawing show the auditory areas identified as follows. Tonotopic areas: AAF (anterior auditory field), AI (primary auditory cortex), P (posterior auditory sulcal cortex), Ve (ventral auditory area), VP (ventral posterior sulcal cortex). Non-tonotopic areas: DZ (dorsal auditory zone), AES (anterior ectosylvian field), AII (secondary auditory cortex). Multisensory areas: ED (posterior ectosylvian gyrus, dorsal part), EI (posterior ectosylvian gyrus, intermediate part), EV (posterior ectosylvian gyrus, ventral part). Limbic areas: In (insular cortex), Te (temporal cortex). The photograph at the top also identifies the sulci as follows: ass (anterior suprasylvian), aes (anterior ectosylvian), mss (middle suprasylvian), pes (posterior ectosylvian), ps (pseudosylvian), pss (posterior suprasylvian). From Lee and Winer (2008).

Despite the representations of sound frequencies, the neural unit response properties indicate that frequency discrimination is not the major function of the auditory neocortex. Studies of the cat and the monkey have found many neurons that are specific for particular temporal patterns. Effects of ablating these neocortical regions in the cat support a role in temporal pattern identification rather than in mere frequency discrimination.

Temporal Pattern Selectivity: Examples

Early in the neuroscience history of single-unit recording studies, the British investigators Whitfield and Evans found numerous examples of simple auditory pattern selectivity in the cat auditory cortex. Figure 23.15 shows data from their recordings of an auditory cortex neuron sensitive to frequency modulation. Maximally sensitive to frequencies of about 13.6 kilohertz, the neuron responded to a range of nearby frequencies as well. More interesting was its preference for upward frequency sweeps—tones rising in pitch. Other auditory cortical neurons responded to other temporal patterns. These scientists presented a simple model of how cells in an array with tonotopic organization could generate responses specific to upward frequency sweeps by having asymmetric inhibitory connections formed by short-axon interneurons (figure 23.16).

These findings on unit response properties fit well with results of studies that have found deafness for temporal patterns after auditory cortex lesions. Lesions of primary or of secondary auditory areas have been found to preserve frequency and loudness discrimination while disturbing or even abolishing discrimination of temporal patterns. Similar findings have been obtained using habituation to novel sounds.

Neuron sensitivity to simple frequency modulation or to the onset or offset of tones is easier to model, but what about responses to natural sounds, which can be far more complex? A nice example is the specific response of certain neurons in the auditory system of bullfrogs to the sound made by the splash of another bullfrog when it jumps into the water.

There are good examples of specificity for natural sounds in primate auditory system neurons. For example, in the auditory cortex of squirrel monkeys, some neurons respond preferentially to specific vocalizations of other squirrel monkeys. Different cells respond best to different cries in the repertoire of vocalizations of these small primates. This specificity is apparently inborn.

A more recent but closely related discovery has been reported for macaques, based on functional magnetic resonance imaging. One region in temporal lobe neocortex becomes more active when the monkey is hearing the voice of another monkey, and the activation is greater when the other monkey is an unexpected individual. The region is not activated in this way by other natural sounds and noises.

This discovery parallels the activation of part of the human temporal lobe cortex during listening to human voices. It is not very far-fetched to postulate that the human brain has neurons that are selective to specific phonemes, the elements of speech that when put

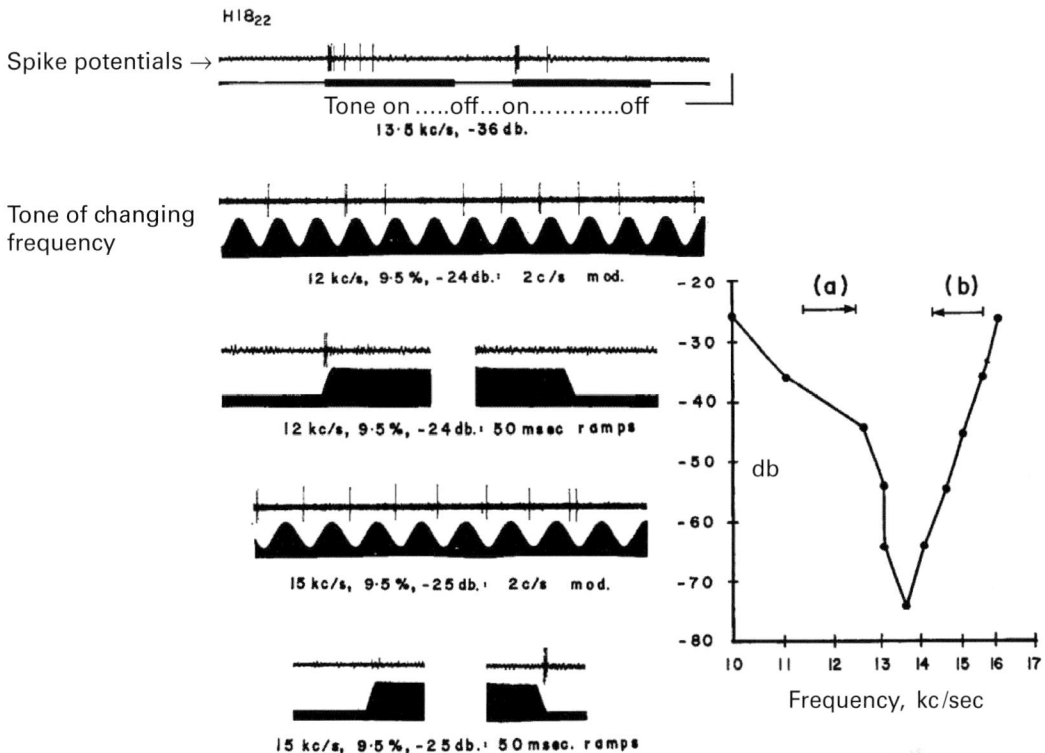

Figure 23.15
Extracellular recordings from a single unit (neuron) in the auditory cortex of a cat. The cell is sensitive to frequency modulation. (Left) Top lines: Responses to two presentations of a steady tone of 12 kilohertz. Second lines: Responses (upper) to a frequency that is oscillating continuously in the range shown by the arrow marked "a" in the right-hand figure. Frequency is depicted on the ordinate of the lower graph, time on the abscissa. The unit responds only during upward frequency sweeps. Third lines: Similar responses to a sudden ramp upward in frequency, but not to a sudden ramp downward. Fourth and fifth lines: Similar depictions of responses to downward frequency sweeps in the range shown by the arrow marked "b" in the right-hand graph. (Right) Minimum intensities of steady tones that elicit unit responses. The sound intensities are on the ordinate, shown in decibels relative to 100 db sound pressure level; intensities were increasing upward during presentations. The frequencies in kilohertz are shown by the points in the graph. The best frequency of this unit was found to be 13.6 kilohertz. From Whitfield and Evans (1965).

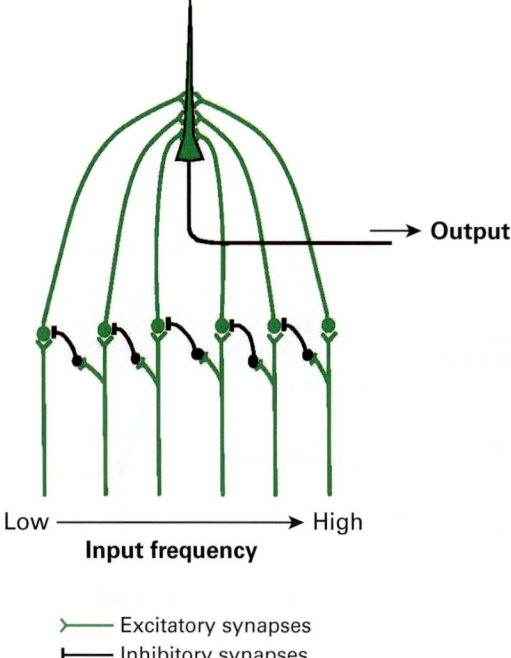

Figure 23.16
A simple model to show how tonotopically organized projections with high-frequency *sideband inhibition* could predict some of the observed patterns of responses of cortical units. The drawing is based on a figure of Evans and Whitfield. They were explaining units that respond specifically to upward frequency sweeps and also showed delayed inhibition to steady tones. The output neuron responds to upward frequency sweeps over a wide range of frequencies, even if the frequency changes were not great. Based on Whitfield (1969).

together in certain sequences make up meaningful words. These ideas are supported by the finding of *word deafness* in humans with lesions of auditory neocortex—findings that have a correspondence with lesion effects in cats.

Imaging studies of human auditory regions of neocortex have also confirmed the right–left differences expected from lesion effects. An example is shown in figure 23.17. As expected from the dominance of the left hemisphere that is usually found in studies of speech processing, human auditory cortical areas in the left hemisphere are activated more than these areas in the right hemisphere by temporal changes in tone stimulation. When spectral changes are presented, the auditory areas in the right hemisphere are activated more than in the left hemisphere.

Functionally Distinct Auditory Pathways in the Neocortex

Functional brain imaging studies have revealed regions in the monkey temporal lobe that respond specifically to monkey voices (as noted above) and other regions of temporal

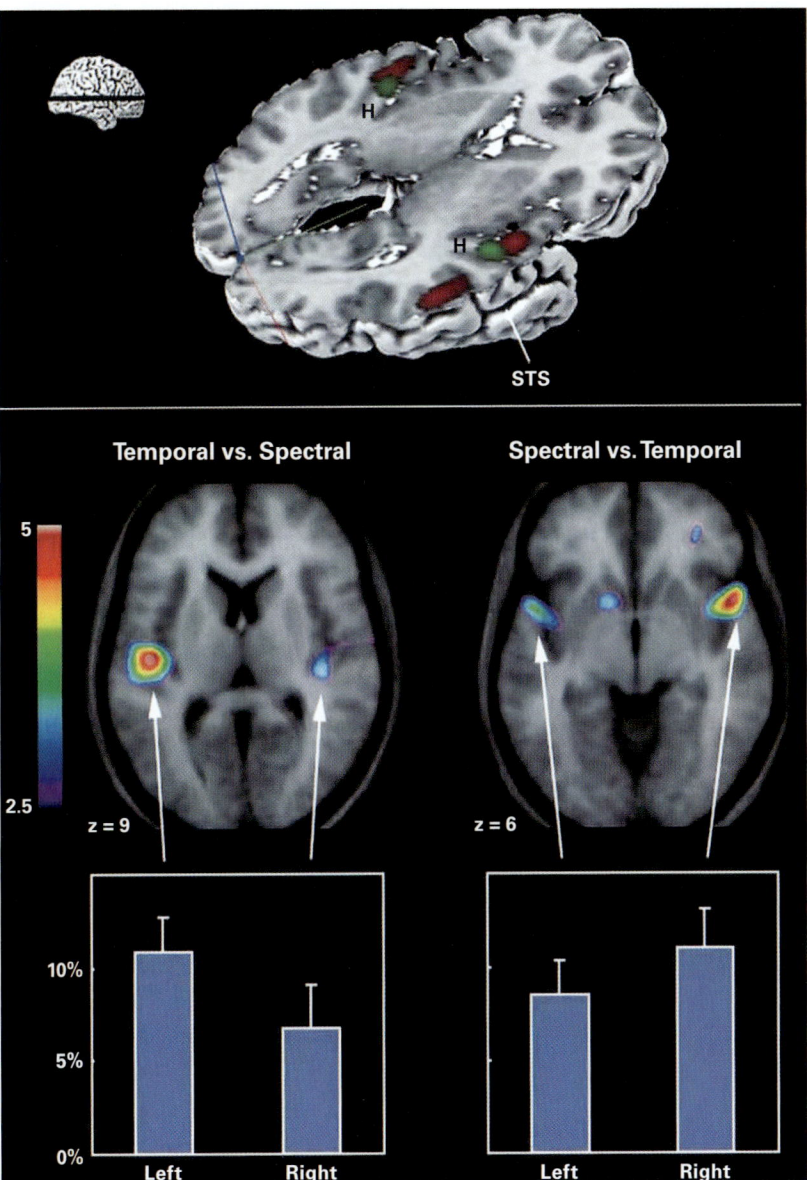

Figure 23.17
Human auditory cortex responds in both hemispheres to temporal and spectral changes in tone stimulation. However, left hemisphere responds more to temporal variations, and right hemisphere responds more to spectral changes. Results from positron emission tomography (PET) measures of cerebral blood flow. At the top, the right hemisphere is nearer. Red color indicates significant changes during stimulation with spectral complexities. Green color indicates significant changes during stimulation with temporal complexities. At the bottom, right–left differences in changes during the two types of stimulation indicate that temporal complexity increases the responses in the left side more than the right, whereas the right–left differences are reversed when spectral complexity characterizes the stimulation. From Zatorre and Belin (2001).

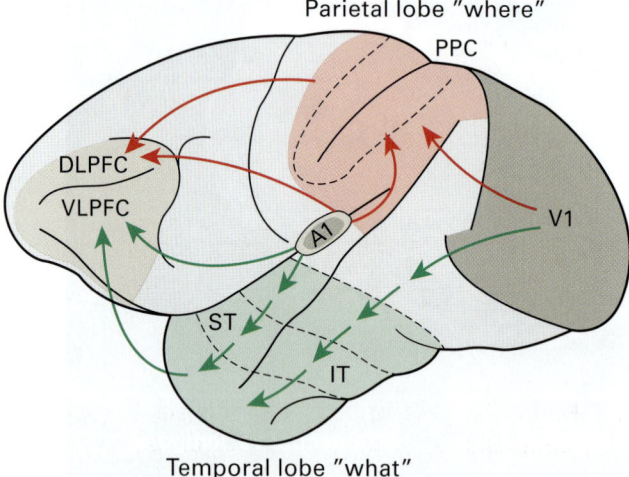

Figure 23.18
In primates, the major processing streams followed by outputs of auditory association areas (that receive input from the primary auditory cortex, A1) are similar to those in the visual system (see chapter 22). A third important stream that reaches the hippocampal formation is not shown. DLPFC, dorsolateral prefrontal cortex; IT, inferior temporal gyrus; PPC, posterior parietal cortex; ST, superior temporal gyrus; VLPFC, ventrolateral prefrontal cortex; V1, primary visual cortex. From Rauschecker and Scott (2009).

and parietal lobe cortex that respond to spatial location information. These studies indicate that there are distinct "what" and "where" transcortical pathways in the auditory system, thus resembling the two transcortical visual-system pathways discovered earlier (chapter 22). Paralleling the visual system, these pathways can be traced through posterior parietal areas and from temporal lobe areas to prefrontal regions (figure 23.18). With these connections, a mammal—particularly a mammal with highly developed frontal lobe cortex—is able to anticipate sounds and other actions of sound sources and to plan responses to those sounds and their sources.

Auditory System Specializations

The transcortical pathway for processing location information has been elaborated in bats with specializations based on their echolocation ability. We can speculate that the same must be true for the less-studied cetaceans. Both groups of mammals show a remarkable expansion of the major auditory midbrain structure, the inferior colliculus (figure 6.3). In electrophysiological studies of the neocortex of bats, investigators have found regions with neurons that respond specifically to the range and velocity of prey. The tuning of these neurons depends on the processing of the bat's cries and their echoes, in the pathways from inferior colliculus to medial geniculate nucleus to auditory cortex.

Auditory Systems

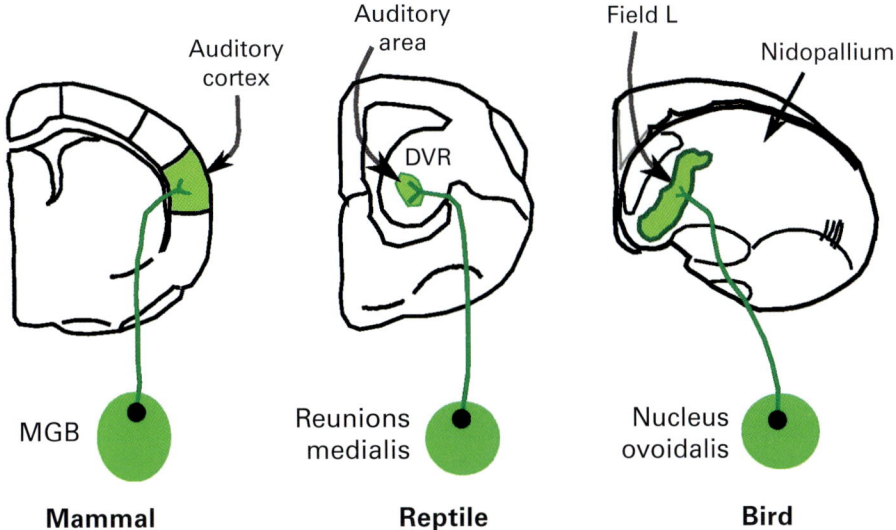

Figure 23.19
Pathways from the major auditory nucleus of the thalamus to the endbrain in mammal, reptile, and bird. MGB, medial geniculate body (nucleus). (The projection to the lateral nucleus of the amygdala in mammals is not shown.) DVR, dorsal ventricular ridge. Based on: bird, Karten (1968); reptile, Balaban and Ulinski (1981).

Another fascinating specialization of the auditory system together with the vocal control system is found in songbirds. In this system, the emphasis has been very different. Birdsong control systems remind us of the even more complex systems for control of, and understanding of, human speech. Birds certainly possess systems for spatial localization of sounds as well, although these have been less studied above the midbrain. Figure 23.19 presents a simplified sketch of ascending auditory pathways in the mammal, reptile, and bird. Figure 23.20 shows more detail of the areas in the pigeon forebrain that correspond to the auditory cortex of mammals. They are called *field L* or *area L pallii* of the nidopallium. The structure develops from the embryonic dorsal ventricular ridge. Area L is composed of three layers, designated L1, L2, and L3, which are like the binaural and monaural columns of mammalian auditory cortex. Area L receives auditory input from the thalamic nucleus ovoidalis, equivalent to the mammalian medial geniculate body.

Axons of neurons in area L of songbirds go to the *higher vocal center* (HVC) of the mesopallium. This structure is more developed in males and is important for the control of singing (see chapter 27). It has been discovered to be larger in species with a more complex song repertoire. The most direct pathway to the motor system from HVC goes to another endbrain structure, the nucleus robustus of the arcopallium (the avian amygdala region), also more developed in males. From the robust nucleus, axons descend directly or indirectly to a part of the hindbrain's hypoglossal nucleus, containing motor

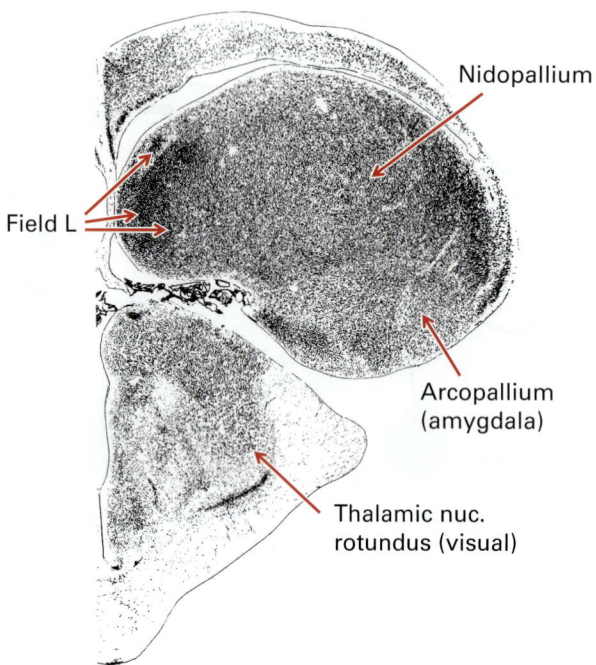

Figure 23.20
Photograph of a frontal hemisection through the endbrain of a pigeon. Field L is where the auditory projections from the thalamus terminate. They are in the nidopallium—formerly called the neostriatum. The position of a region similar to the amygdala in mammals is also identified in the photograph—the arcopallium, formerly called the archistriatum. From Karten and Hodos (1967).

neurons that innervate the syrinx. (The syrinx is the vocal control structure in the throat of the bird.) The indirect pathway goes to the midbrain region equivalent to the mammalian inferior colliculus; from there, axons go to the hypoglossal nucleus.

There is also an indirect vocal control pathway from HVC in songbirds that is important for the learning of song patterns (chapter 27). The recipient structure is located more anteriorly in the endbrain, a sexually dimorphic cell group in the bird's corpus striatum. The striatal structure projects to the thalamus, which projects to a nidopallial cell group with outputs to the arcopallial vocal output structure. The nidopallial cell group's functional importance reminds one of the importance of Broca's area for producing speech in humans.

In songbirds, as in human speech, there is a strong hemispheric dominance in the control of singing, with the greater importance of the left side extending down to the hindbrain.

The understanding and production of language is, of course, the most important specialization of the auditory system in the temporal lobe and related structures in the parietal and frontal lobes in our own species. In birds as well as in humans, learning is critical in the early development of vocalizations and responses to them. Structures known to be critical in speech and in the understanding of language have become known from studies of people with various kinds of aphasia. In recent years, functional imaging has cast new light on the older findings and has been adding to our knowledge of the brain regions most involved.

Readings

Relevant to the kangaroo rat's escape from rattlesnake strike:

Webster, D. B. (1962). A function of the enlarged middle-ear cavities of the kangaroo rat, Dipodomys. *Physiological Zoology, 35,* 248–255.

On the transformations of the hearing apparatus in early mammals:

Allman, J. (2000) (book listed in chapter 3).

Auditory system of the cat:

Lee, C. C., & Winer, J. A. (2008). Connections of cat auditory cortex: I. Thalamocortical system. *Journal of Comparative Neurology, 507,* 1879–1900.

Winer, J. A. (2011). A profile of auditory forebrain connections and circuits. In Winer, J. A., & Schreiner, C. E., eds. *The auditory cortex,* pp. 41–74. New York: Springer.

Evolution of auditory cortex:

Frost, S. B., & Masterton, R. B. (1992). Origin of auditory cortex. In Webster, D. B., Fay, R. R., & Popper, A. N., eds. *The evolutionary biology of hearing,* pp. 655–671. New York: Springer-Verlag.

Kaas, J. H. (2011). The evolution of auditory cortex: The core areas. In Winer, J. A., & Schreiner, C. E., eds. *The auditory cortex,* pp. 407–427. New York: Springer.

Chick nucleus magnocellularis and nucleus laminaris:

Jhaveri, S., & Morest, D. K. (1982). Neuronal architecture in nucleus magnocellularis of the chicken auditory system with observations on nucleus laminaris: A light and electron microscope study. *Neuroscience, 7,* 809–836.

Projections from auditory thalamus to the amygdala: See Frost and Masterton (1992) above.

"What" and "where" pathways in neocortex:

Rauschecker, J. P. (2011). An expanded role for the dorsal auditory pathway in sensorimotor control and integration. *Hearing Research, 271,* 16–25.

Rauschecker, J. P. (2013). Processing streams in auditory cortex. In Cohen, Y. E., Popper, A. N., & Fay, R. R., eds. *Neural correlates of auditory cognition,* pp. 7–43. Handbook of Auditory Research 45. New York: Springer.

Rauschecker, J. P., & Scott, S. K. (2009). Maps and streams in the auditory cortex: Nonhuman primates illuminate human speech processing. *Nature Neuroscience, 12,* 718–724.

Reviews of auditory system anatomy and function:

Butler, A. B., & Hodos, W. (2005). *Comparative vertebrate neuroanatomy: Evolution and adaptation*, 2nd ed. New York: Wiley-Interscience.

Evans, E. F. (1992). Auditory processing of complex sounds: An overview. *Philosophical Transactions of the Royal Society of London, Series B, 336,* 295–306.

Petkov, C. I., Kayser, C., Steudel, T., Whittingstall, K., Augath, M., & Logothetis, N. K. (2008). A voice region in the monkey brain. *Nature Neuroscience, 11,* 367–374.

Petkov, C. I., & Jarvis, E. D. (2012). Birds, primates, and spoken language origins: Behavioral phenotypes and neurobiological substrates. *Frontiers in Evolutionary Neuroscience, 4,* 1–24.

Trainor, L. (2008). The neural roots of music. *Nature, 453,* 598–599.

IX THE FOREBRAIN AND ITS ADAPTIVE PRIZES: A SNAPSHOT

24 Forebrain Origins: From Primitive Appendage to Modern Dominance

The various components of the forebrain have evolved for specific purposes as we have sketched in chapters 4 and 5, culminating in the specialty of the mammals—the neocortex, as we discussed in chapter 7. We added to this picture of the mammalian forebrain in chapter 12, and we began to fill it in some more when we discussed forebrain components of the motor system (chapters 14 and 15) and of the visual and auditory systems (chapters 20–23).

Life without a Forebrain

Discussions of the human brain are very often centered on the endbrain, especially on the neocortex because it has become so dominant for the most unique functions of our species. With such special treatment, we have to remind ourselves that animals, even humans, can do many things without much of a forebrain. Let's review this briefly and then review the nature of the primitive forebrain with a specific question in mind: Why did the forebrain appear in evolution in the first place, ages before the advent of today's mammals?

Without a forebrain, the cranial nerves that attach to this forward-most division of the CNS are absent. The one most important for humans is the optic nerve. The retina develops as an outgrowth of the caudal portion of the forebrain, the 'tweenbrain (diencephalon) as noted at the beginning of chapter 22. The rostral-most cranial nerves, the olfactory and the vomeronasal, have supported functions just as crucial for survival, more crucial for many species. Also attaching to the forebrain are the terminal nerve, associated with the olfactory, and the epiphysial nerve (related to the pineal organ).

These forebrain nerves were and are important not only for control of various brainstem motor functions but also because of their inputs into a major integrative center for higher control of endocrine systems and the autonomic nervous system and related motivational states—the hypothalamus. The hypothalamus, located in the caudoventral forebrain, exerts this influence both by its control of the attached pituitary gland and by its neural connections, not only to forebrain structures but also directly to the midbrain, hindbrain, and spinal cord.

Rather than asking what animals can do without a forebrain, we can ask a better question: What can animals do without a forebrain if the hypothalamic-pituitary axis is preserved and the functions of the forebrain's cranial nerves can be discounted? This was the major question in chapter 7, where we surveyed effects of forebrain removal in pigeons, rats, and cats—decerebration with sparing of an island of hypothalamic tissue and the attached pituitary.

Decerebrate animals show a great reduction in self-initiated behavior because they do not have normal motivations. They fail to initiate actions without prodding. They also show poor sensory acuity for many functions and a lack of fine motor control, although this was not evident, especially in the pigeons, when it came to the execution of instinctive actions (see next paragraph). The forebrainless animals lost their learned habits and generally failed to anticipate changes in the flux of events happening around them.

However, once the animals in these experiments had recovered from the major diaschisis effects of the decerebrations, they were found to retain much consummatory behavior: In response to oral stimulation they showed eating responses such as opening of the mouth, chewing, and swallowing. Pain-elicited aggressive responses were seen, as were sexual postures and reflexes (lordosis in females, and in males penile erection in response to genital stimulation, and also thrusting and ejaculation). Righting responses and walking could occur. In the pigeons, if the optic nerves were spared together with the optic tract leading to the midbrain tectum, throwing the birds into the air resulted in flying with some normal navigational responses. All of this was reviewed in chapter 7.

These studies give us some orientation to the lives of ancient animals with very primitive forebrains (see figures 4.8b, 24.1, 24.3).

In Search of Ideas about Origins: Primitive Forebrains

Before modern cladistics, a popular view of brain evolution was formulated by Ludwig Edinger (1908). Edinger was a comparative neuroanatomist who relied on methods for visualizing cytoarchitecture. He noted that the spinal cord, hindbrain, and midbrain had a fairly similar structure across a large variety of vertebrates, whereas the forebrain showed much greater variation. Relying on sizes of structures and forebrain differentiation, he arranged selected vertebrates into an estimated phylogenetic scale and discussed this scale in terms of evolution. He illustrated the appearance in cartilaginous fish of what he termed a "neencephalon," and then he pictured its expansion in tetrapods. A small dorsal cortex in a lizard is depicted as an expansion of a tiny dorsal pallium in a fish; this cortex was superseded by the neocortex of small mammals like the rabbit, and the neocortex expands into a very large and dominant structure in primates, especially in the great apes and the hominids (represented now by our own species). A modern version of Edinger's picture of cerebral hemisphere evolution has been arranged on a cladogram by the comparative neuroanatomist R. Glenn Northcutt—see figure 24.1.

Forebrain Origins

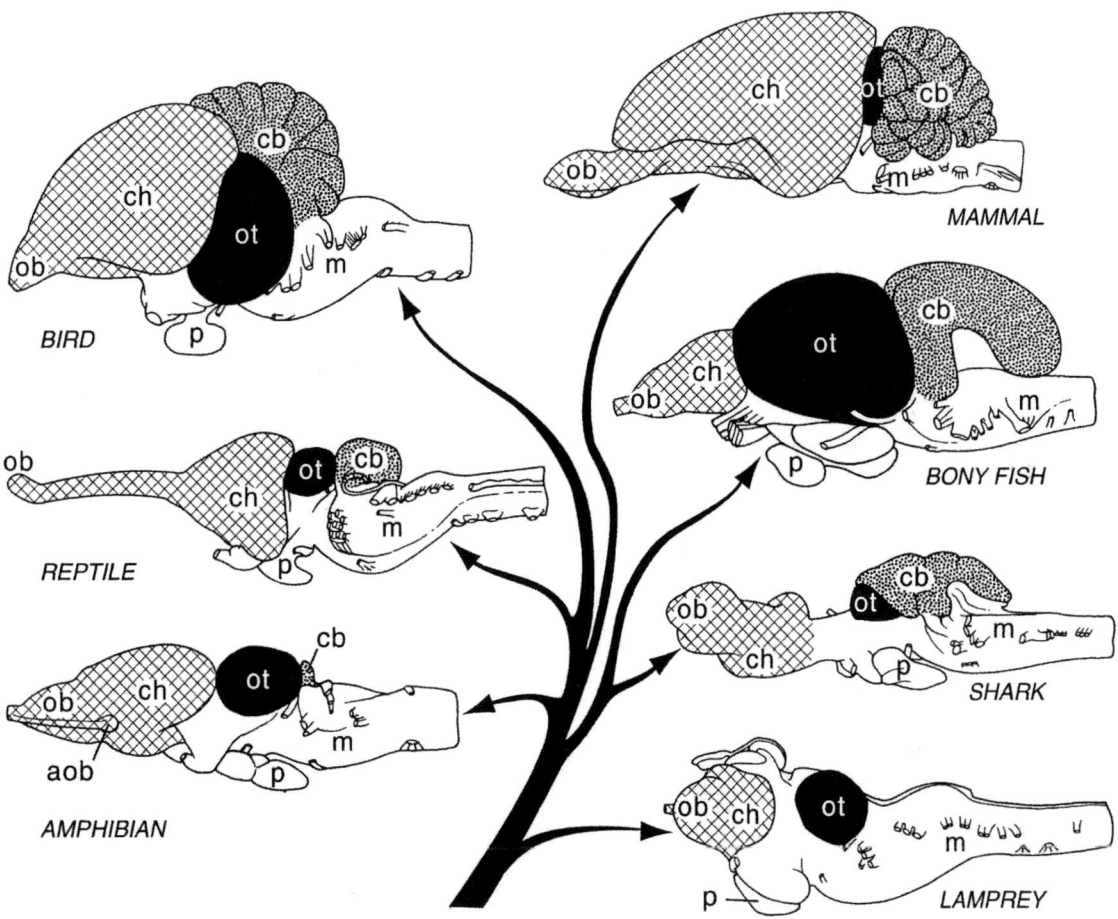

Figure 24.1
The brains of living species of widely different groups, placed on a cladogram showing evolutionary lines. Note the changes in relative size of the cerebral hemispheres (ch)—the endbrain structures—with respect to the size of the hindbrain. The optic tectum of the midbrain (ot) and the cerebellum (cb) also vary widely in relative size. aob, accessory olfactory bulb; m, medulla oblongata; ob, olfactory bulb; p, pituitary gland. From Northcutt (2002).

As promoted by such modern anatomists, the place to start our search is suggested by the somewhat more extensive cladogram of chordates, emphasizing vertebrates, illustrated in figure 5.11. We see three groups of animals that contain species that have survived without much change from a time before the appearance of jawed vertebrates. These groups are represented by hagfishes and sea lampreys and by the non-vertebrate chordates. The latter have been discussed in chapter 3, with illustrations of the cephalochordate amphioxus; the brains of hagfish and lamprey have also been discussed. The bodies and brains of these three animals are depicted in figure 24.2.

Even in these primitive chordates, one can find evidence of two major direct sensory inputs to the forebrain plus evidence of one major controller of outputs, suggesting that these were what shaped its earliest evolution. The sensory inputs are optic and chemosensory (via olfactory or terminal nerve). Light detection enables a daily cycle of actions and endocrine secretions that varies with time of day. Chemical substances in the water can influence avoidance and approach actions critical for safety and for finding food or mates (just as odors influenced the animals that were evolving into mammals).

Brain Expansions in the Vertebrates

Figure 24.3 shows the brains of three fish in order to illustrate expansions of particular parts of the brain in non-mammalian vertebrates. In the mormyrid electric fish, the cerebellum has become huge (as illustrated previously in figure 6.1). In the trout, you can see an enlarged midbrain tectum. In the bichir, the diencephalon and the telencephalon show some enlargement, but far less than what has occurred in mammals (figure 24.1).

In thinking about size increases in the brain, it is the size relative to body weight that is the most interesting, as body size is such a major determinant of brain size. The illustrations cited in the above paragraph focus on enlargements of particular parts of the brain relative to other parts rather than on the size of the entire brain relative to body size. Figure 4.11, in contrast, illustrates the range of sizes of the entire brain relative to body weight in major groups of animals. Thus, for animals of a particular body weight, the average mammal has a larger brain than that of an average bird, the average bird has a larger brain than that of an average reptile, and the average reptile has a brain larger than that of a jawless vertebrate.

It seems evident that in the course of evolution, adaptive changes and additions to the repertoire of behavioral abilities and other functions are related to increases in brain size. However, there have been decreases in relative brain size as well. Probable decreases as well as increases in relative brain size in evolution are noted on the cladogram of the vertebrates shown in figure 24.4, based on an illustration by the comparative neuroanatomist Georg Striedter. Evidence indicates that both salamanders and eels have shown relative decreases, whereas increases, some of them very large ones, have occurred early in

Amphioxus (branchiostoma)

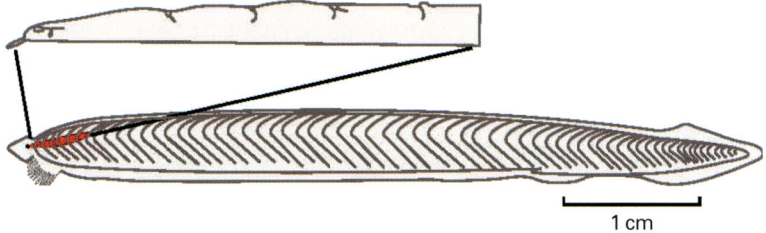

Hagfish (myxine)

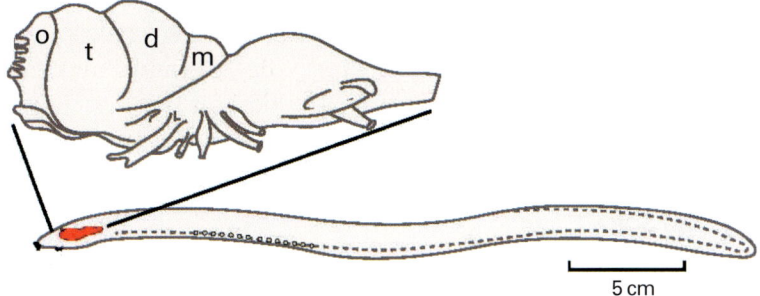

Lamprey (petromyzon)

Figure 24.2
Drawings of the bodies and brains of three small animals (one cephalochordate [amphioxus] and two jawless vertebrates [hagfish and sea lamprey]) with special places on the cladogram of chordates shown in figure 5.11. The brain is depicted above each body, enlarged as indicated. d, diencephalon; m, midbrain; o, olfactory bulb; p, pineal organ; t, telencephalon; t-mp, medial pallium. Based on: Striedter (2005); Nieuwenhuys, Donkelaar, and Nicholson (1998); with additional information from Romer and Parsons (1977).

Mormyrid electric fish (gnathonemus)

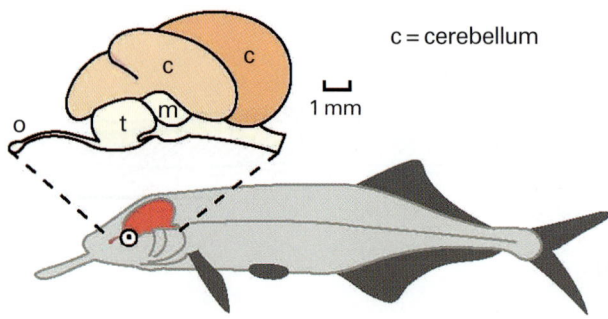

Trout (salmo)

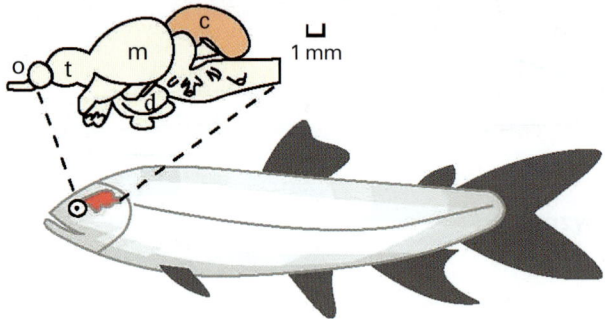

Bichir (polypterus)

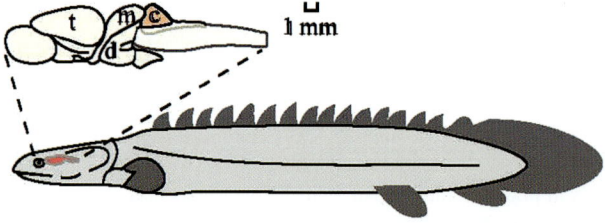

Figure 24.3
Views of the bodies and brains of three ray-finned fish. They illustrate expansions of brain regions compared with the brains shown in figure 24.2, especially in the cerebellum and midbrain, and also in the forebrain. c, cerebellum; d, diencephalon; m, midbrain; o, olfactory bulb; t, telencephalon. Based on: Striedter (2005); Nieuwenhuys, Donkelaar, and Nicholson (1998); with additional information from Papez (1929).

Forebrain Origins

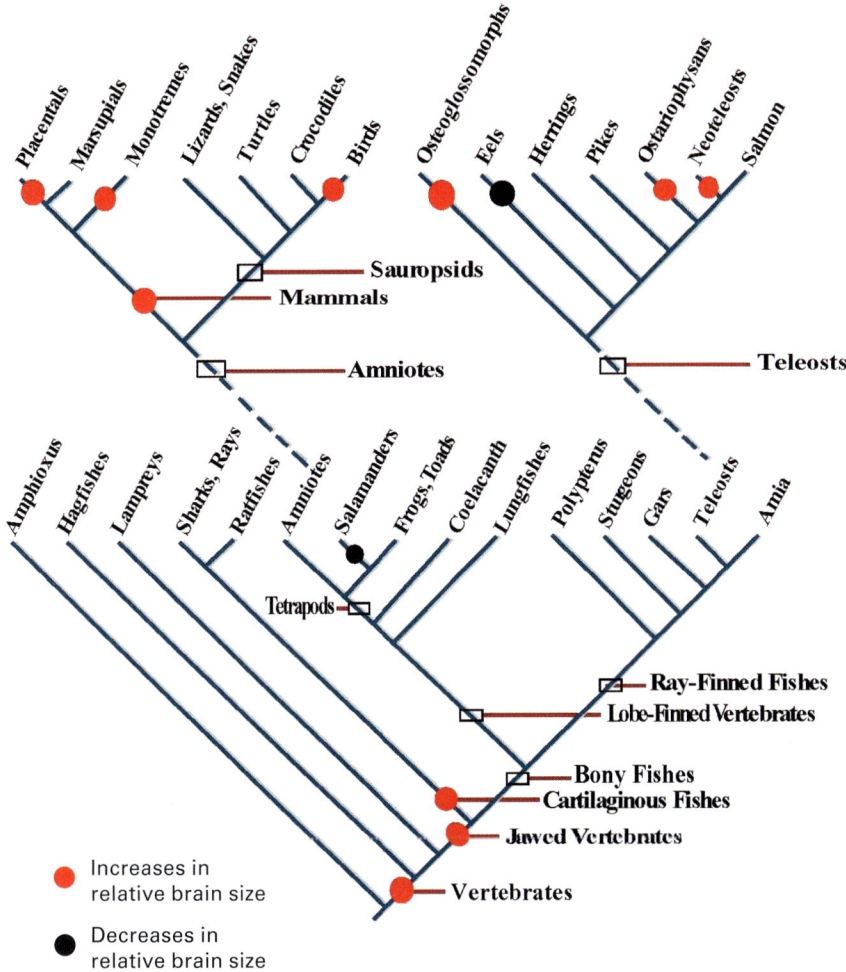

Figure 24.4
Cladogram of the vertebrates showing probably evolutionary increases (red dots) and decreases (black dots) in relative brain size. Increases have occurred more often than decreases. They have probably evolved independently in different lineages. Based on Striedter (2005).

the jawed vertebrates, and then further not only in mammals but also in birds, in several teleost groups, and in cartilaginous fish (including the sharks and rays).

More about the Early Forebrain

The earliest forebrain was only a small appendage with no cerebral hemispheres: it was a precursor of the diencephalon. The earliest inputs were probably optic, from pigmented light-detecting cells, and from detectors of chemical substances in the blood. A very early output was from neurosecretory cells, precursors of cells of the hypothalamus–pituitary region. With the evolution of olfactory inputs, the tiny forebrain showed considerable expansion in the evolution of the jawless vertebrates, represented by the hagfish and the sea lamprey.

Comparative anatomical studies suggest that olfaction and its links to motor systems caused the earliest endbrain appearance and expansion. Olfaction brought adaptive advantages over other chemical senses by enabling responses to sources at a distance in space and in time. At this point, it would be useful to review the studies of olfactory bulb projections in hagfish and lamprey that were summarized in chapter 19. In these species, the olfactory bulbs have been found to project to most or all of the endbrain and also to more caudal parts of the forebrain.

How did this olfactory input influence behavior in the early chordates? The endbrain just behind the olfactory bulbs must have been dominated by olfactory inputs—the ventral, lateral, dorsal, and medial portions of the caudally adjacent neural tube. The ventral part—just rostral to the hypothalamus—evolved into the corpus striatum and amygdala of mammals. The dorsomedial part was the medial pallium, which evolved into the hippocampus of mammals. The lateral part has remained olfactory cortex in its later evolution, whereas the dorsal part evolved into the dorsal pallium, part of which became transformed when it evolved into the neocortex of the mammals. These distinct areas differentiated as they did because of the adaptive functions they served.

The Scene Is Set for the Early Expansion of the Striatum

With the appearance of the olfactory epithelium that contained receptor cells, and the olfactory bulbs with their secondary sensory neurons of the rostral-most CNS, the scene was set for the evolution of the caudally adjacent endbrain. Its surface layer received axons coming from the bulbs. A portion of the region became the olfactory tubercle, at the ventral surface of the pre-hypothalamic endbrain. The superficial layer of the tubercle was and is directly connected to the olfactory bulbs. Deeper layers of this structure connect with the hypothalamus and related cell groups. This linking region between the olfactory bulbs and the hypothalamus evolved into the ventral striatum of mammals. With the evolution of the cerebral hemispheres, a caudolateral extension of this striatal region became major parts of the amygdala.

The primitive striatal outputs to the hypothalamus influenced endocrine and autonomic nervous system control. Links to the midbrain also evolved for the control of approach and avoidance locomotion and for influencing the orienting movements dominated by the midbrain tectum.

The synaptic links in the primitive striatum between the olfactory input system and the outputs through the hypothalamic region and through the midbrain were—it is very likely—plastic, responding to reward so that responses to objects in the environment could be learned (conditioned) and thus become habitual. It is also very likely that the adaptive advantages of this plasticity were why there was an evolutionary invasion of non-olfactory projections to the striatal region. These projections, coming from the primitive thalamus and subthalamus, must have overlapped with olfactory inputs at the outset, but then they became segregated, and the dorsal striatum differentiated from the ventral striatum.

In this way, the non-olfactory sensory systems of vision, audition, and somatic sensations probably took advantage of the endbrain apparatus for plastic connections—for habit learning. The projections of the auditory thalamus to the amygdala in mammals, most striking in primitive mammals, also fit with this idea. The adaptive advantages of these connections resulted in an expansion of the corpus striatum. Later in the mammals, inputs to the striatum from the neocortex made use of the plasticity. The improved object differentiation represented in the inputs from neocortex provided the adaptive advantages that ensured their persistence and expansion in evolution.

Another Kind of Plasticity in the Primitive Endbrain

Olfaction provided something to the primitive endbrain other than information that enabled object identification. Places in the environment, not just the isolated objects in those places, could be remembered as good or bad places by remembering the odors associated with the places where events good or bad for the individual occurred. One part of the ancient dorsal endbrain began to specialize in the forming of such memories. This part was the medial pallium, located dorsomedial to the striatal portion of the neural tube. The medial pallium has strong, direct olfactory inputs in hagfish and sea lampreys, animals that are believed to resemble very primitive vertebrates (see figures 19.5, 19.6). The medial pallium in early vertebrates evolved into the hippocampal formation. In mammals (e.g., in the Syrian hamster), only the most rostral hippocampus, the rostral end of the *hippocampal rudiment*, receives direct olfactory projections. However, in hamsters and in other mammals, olfactory projections also reach at least part of the

458   Chapter 24

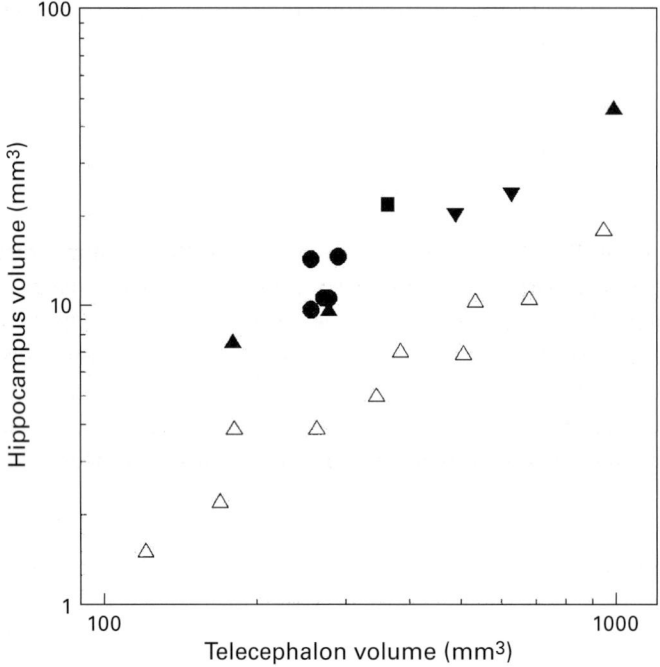

Figure 24.5
Relative size of hippocampus in various species of passerine birds that store food in remembered places (filled symbols) and birds that do not store food (unfilled symbols). The hippocampus volume is plotted against the volume of the entire telencephalon on log-log scales. Storing and nonstoring species can be in the same families. The hippocampus is consistently larger, relative to the entire telencephalon, in species that make food caches for later retrieval. From Sherry and Duff (1996).

entorhinal cortex, which is a parahippocampal region that provides major inputs to the hippocampus.

This speculation about evolution is supported by a major function of the hippocampus of modern, advanced vertebrates: the learning and recall of locations in the environment and the locations of objects as landmarks within local regions. Evidence for this in mammals comes from the effects of hippocampal ablations and from unit recording studies. Other evidence comes from species variations in hippocampal size: there are strong correlations between the size of the hippocampus and hoarding abilities in both birds and mammals. Figure 24.5 illustrates this correlation from studies of songbirds. Species that store food in caches, at specific places that they must remember, have a relatively larger hippocampus. (The correlations by themselves indicate nothing about causes, but together with the lesion effects and the recording studies, they are certainly supportive.)

The role of the medial pallium in remembering places was not a simple one. Animals roam from one region to another, each region having its own spatial arrangements to be remembered. This issue will be discussed further in a later chapter.

Parallel Evolution of Pallium and Subpallium

The innervation of the primitive ventral portion of the endbrain—the striatum—by non-olfactory inputs was paralleled by a similar innervation of the pallium—resulting in a differentiation of the dorsal cortex of primitive vertebrates. The non-olfactory inputs must have resulted in a parcellation of this cortex (as suggested by the hypothesis pictured in figure 19.4). As this occurred, the only pallial regions that had evolved were probably most similar in connections to the limbic and paralimbic regions that include the hippocampal and cingulate regions on the medial side and the olfactory and parolfactory regions on the lateral side. It was prior to the appearance of a neocortex.

The non-olfactory modalities brought clear adaptive advantages to the animals. In the case of the corpus striatum, they enabled the non-olfactory elicitation of habits involving locomotion, orienting, and grasping movements. The non-olfactory inputs came initially from paleothalamus (midline and intralaminar nuclei of mammals) and subthalamus; much later they came mainly from neocortex. The striatal outputs went to the 'tweenbrain and to the midbrain.

In the case of the dorsally located cortex, the evolution of a thicker cortex eventually occurred with the addition of new cell layers in the regions dominated by non-olfactory modalities; this thicker cortical region became the neocortex of mammals, which has expanded in its surface area and volume more and more as new species appeared. It arose in between the medial pallial and adjacent regions on one side and the olfactory and adjacent regions laterally—the two cortical regions that became *limbic* (meaning "at the fringe"). (There are traces of evolutionary influences of both of these adjoining cortical regions in the neocortex.) The result, initially, was an increasing sophistication of sensory analysis and object discrimination. Outputs from this cortex went to the striatum and to the medial pallium, with the functions discussed above. More direct neocortical control of behavior was also achieved by connections to the 'tweenbrain and midbrain and later by evolution of longer connections to the hindbrain and spinal cord.

The expansion of the pallium in early mammals can be illustrated by a look at the brains of small mammalian insectivores that are alive today. In the brain drawings shown in figure 24.6, note the relatively small expanse of neocortex in the European hedgehog and in the West African hedgehog. Other species with a relatively very small neocortex are the tenrecs and opossums. As species with more neocortex evolved, the neocortex became folded into multiple gyri and sulci.

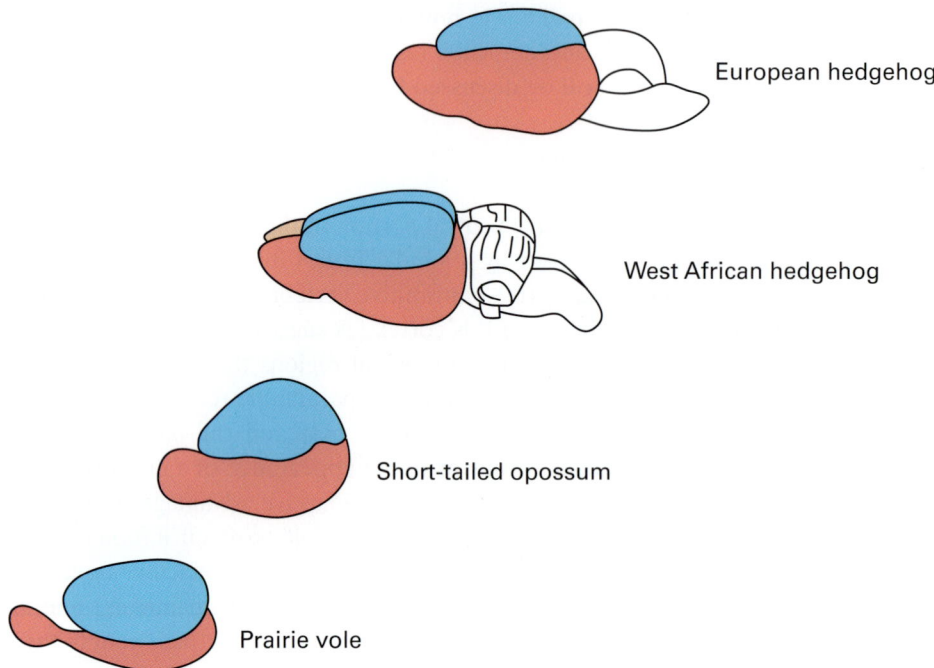

Figure 24.6
A relatively small neopallium of the endbrain in two small mammalian insectivores (hedgehogs), a small marsupial (short-tailed opossum), and a cricetid rodent (the prairie vole) is depicted in blue, and the olfactory bulb and non-neocortical pallial areas are in red. Based on figures in various reports.

Visualizing the Early Striatum and Pallium

The speculative ideas presented in this chapter about the early evolution of the endbrain, including the basal ganglia (corpus striatum and striatal amygdala) and the pallial regions, can be sketched in simple pictures of the changing brain (figure 24.7). The pictures and the sequence of changes they depict can help you remember the major endbrain structures and how they are arranged. This arrangement becomes more difficult to visualize in the large-brained vertebrates, especially in humans and other large mammals.

The first sketch depicts the beginnings of the striatum as a link between olfactory bulbs and output structures, in particular the hypothalamus and epithalamus. This link became the ventral striatum of vertebrates. The connections of this link were modifiable—capable of experience-induced change and thus supporting the learning of olfactory-guided habits of approach and withdrawal.

Such learning required feedback from the sensory consequences of responses. It can be suggested that this feedback activated pathways from the hypothalamus and

Stage 1. Postulated beginnings in primitive chordates

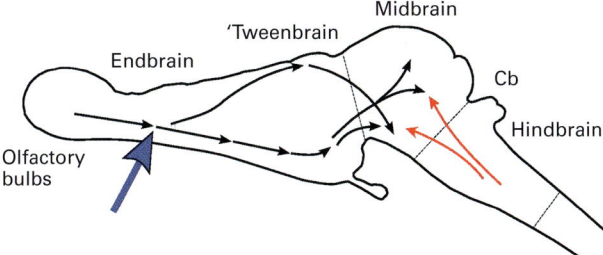

Stage 2. Other inputs reached the striatum

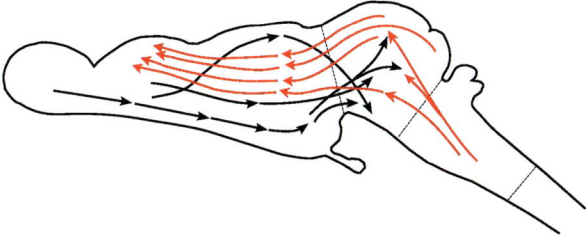

Stage 3. Early expansion of striatal and adjacent pallial areas

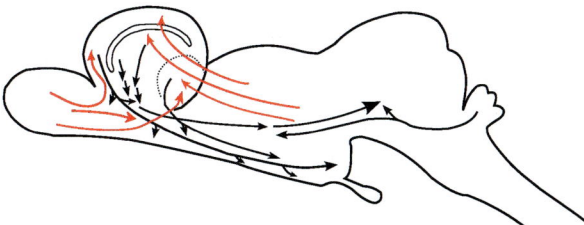

Stage 4. Pre-mammalian, and then mammalian expansions

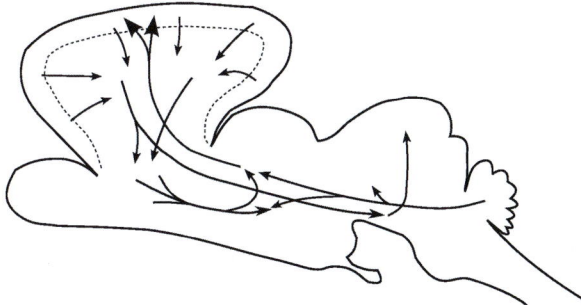

Figure 24.7
Evolution of corpus striatum and other parts of endbrain: suggested major stages. Stage 1: Beginnings: a link between olfactory inputs and motor control: The link becomes "ventral striatum." It was a *modifiable* link (arrow), capable of experience-induced change. Reward pathways not shown. Stage 2: Non-olfactory inputs invade the striatal integrating mechanisms (via paleothalamic structures). Stage 3: Early expansions of endbrain: striatal and pallial. Stage 4: Pre-mammalian and then mammalian expansions of cortex and striatum. For the striatum, the earlier outputs and inputs remain as connections with neocortex expand.

midbrain, including dopamine pathways, that responded to indicators of success in reaching goals. In lampreys and cartilaginous fish, and also in ray-finned fish, lungfish, and amphibians, there are known to be dopamine-containing axons ascending from neurons of the posterior tuberculum and other cells in the hypothalamus. In the amniotes and in sharks, skates, and rays, such axons come from the midbrain's ventral tegmental area and the substantia nigra. Activation of the ventral tegmental area is known to be associated with positive, rewarding feelings. It is interesting to note that even in some mammals, pathways from the taste system of the hindbrain have been traced to the ventral tegmental area.

The second sketch in figure 24.7 illustrates the invasion of the striatal integrating mechanisms by non-olfactory pathways coming from paleothalamic structures. Initially, these inputs must have overlapped with the olfactory inputs, but as they increased there was a parcellation of the striatum as the dorsal portion became non-olfactory. Thus, the other sensory systems would have been able to take advantage of the plastic links to output systems. The outputs were evolving also, as the projections to the ventral thalamus (subthalamus) increased and longer axons extended into the midbrain. In bypassing the hypothalamus, they had more direct influences on locomotion (toward or away from objects) via the midbrain locomotor area—a region that also received projections from the caudal hypothalamus. Other axons to the midbrain influenced orienting movements. This influence is found in mammals in a strong pathway to the substantia nigra and thence to the superior colliculus. Axons from the striatum also came to influence grasping movements via pathways that probably evolved later.

The adaptive advantages of the changes sketched in figure 24.7, stages 1 and 2, no doubt resulted in an early expansion of the ventral portion of the endbrain, the striatal portion (figure 24.7, stage 3). Meanwhile, the pallial portions were also expanding as non-olfactory portions were segregating from the olfactory cortical regions, or simply expanding independently of the olfactory regions. Pallial output neurons had axons that went to the basal ganglia, with longer axons extending into the 'tweenbrain and the midbrain.

The last sketch in the figure simply depicts the pre-mammalian and mammalian expansions of cortex and striatum (figure 24.7, stage 4). For the corpus striatum, the earlier outputs and inputs remained as the connections with an enlarged dorsal pallium—evolving into the neocortex—expanded. The figure illustrates how a striatal output to the thalamus expanded in mammals. This region of the thalamus also received input from the cerebellum, which was also enlarging as the neocortex enlarged in evolution. These changes resulted in major pathways to a cortical area (a somatosensory area) that was evolving into the motor cortex. As the neocortex enlarged, somatosensory regions including one that became the motor cortex, as we have discussed in chapter 15, evolved long connections to the brainstem and spinal cord, bypassing the midbrain. In this way, the midbrain control of locomotion, orienting, and grasping has been diminished in the evolution of the large-brained mammals, particularly in the large primates.

Next Steps into the Forebrain

Next, we will examine the hypothalamus and the structures most closely connected with it, the limbic system structures. Then we will return to the corpus striatum, emphasizing but not limiting ourselves to the mammals. Finally, we will turn our attention to the crown of the central nervous system, the neocortex of mammals and the related structures found in other highly evolved species, together with the thalamic cell groups that provide the major inputs to these brain structures that underlie what we think of as the highest functions.

Readings

Butler, A. B., Reiner, A., & Karten, H. J. (2011). Evolution of the amniote pallium and the origins of mammalian neocortex. *Annals of the New York Academy of Sciences, 1225*, 14–27.

Jarvis, E., et al. (The Avian Brain Nomenclature Consortium.) (2005). Avian brains and a new understanding of vertebrate brain evolution. *Nature Reviews: Neuroscience, 6*, 151–159.

Jerison, H. J. (1985). Animal intelligence as encephalization. *Philosophical Transactions of the Royal Society of London, Series B, 308*, 21–35.

Karten, H. J. (1969). The organization of the avian telencephalon and some speculations on the phylogeny of the amniote telencephalon. *Annals of the New York Academy of Sciences, 167*, 164–179.

Medina, L., & Abellan, A. (2009). Development and evolution of the pallium. *Seminars in Cell & Developmental Biology, 20*, 698–711.

Moreno, N., Gonzalez, A., & Retaux, S. (2009). Development and evolution of the subpallium. *Seminars in Cell & Developmental Biology, 20*, 735–743.

Northcutt, R. G. (2001). Changing views of brain evolution. *Brain Research Bulletin, 55*, 663–674.

Smeets, W. J. A. J., Marin, O., & Gonzalez, A. (2000). Evolution of the basal ganglia: New perspectives through a comparative approach. *Journal of Anatomy, 196*, 501–517.

Striedter, G. (2005). *Principles of brain evolution*. Sunderland, MA: Sinauer Associates.

Striedter, G. F. (2006). Précis of principles of brain evolution. *Behavioral and Brain Sciences, 29*, 1–12.

Friedrich Sanides has used cytoarchitecture as a basis for speculation about neocortical evolution:

> Sanides, F. (1972). Representation in the cerebral cortex and its areal lamination pattern. In: Bourne, G. H., ed. *The structure and function of nervous tissue*, vol. 5, pp. 329–453. New York: Academic Press.
>
> Sanides, F. (1969). Comparative architectonics of the neocortex of mammals and their evolutionary interpretation. *Annals of the New York Academy of Sciences, 167,* 404–423. (Special issue: *Comparative and Evolutionary Aspects of the Vertebrate Central Nervous System.*)

See also the note on the work of Sven Ebbesson at the end of chapter 4, and his paper cited here.

X THE HYPOTHALAMUS AND LIMBIC SYSTEM

25   Regulating the Internal Milieu and the Basic Instincts

When the chordates were beginning to evolve, the predecessors of the hypothalamus very probably dominated the anterior end of the neural tube—the primitive forebrain vesicle. Secretory cells are found there in the simplest living chordates, cells that in the ancient cousins of these little creatures were the forerunners of portions of the pituitary and its hypothalamic controllers. In the non-vertebrate chordate amphioxus, pigmented cells that respond to light are also found in this region, which has been identified by gene expression as a primitive diencephalon (chapter 3). The responses to light can help us see the logic of controlling circulating hormones from this location in the CNS. Diurnal rhythms in hormone levels and general physiological and locomotor activity became influenced by the cycle of light and dark and by an endogenous clock with inputs from light detectors for synchronization (entrainment) by the day–night cycle (chapter 20). Later, the retinae evolved as outgrowths of this forerunner of the 'tweenbrain.

In contrast, the differentiation and enlargement of a more anterior part of the forebrain—endbrain, or telencephalon—was largely a result of the evolution of another sensory input, the olfactory. The olfactory inputs strongly influenced the early hypothalamus. The connections to the hypothalamic region came largely by way of the primitive striatum and by way of the more dorsally located pallium. Like the early striatum ventrally, the walls of the more dorsal neural tube at this level were dominated at their evolutionary outset by olfactory inputs (chapter 19).

Nature of the Hypothalamus and Affiliated Structures: The Limbic System

Working at the time of Ramón y Cajal, at the beginning of the twentieth century, the famous British physiologist Charles Scott Sherrington called the hypothalamus the "head ganglion of the autonomic nervous system." It is certainly that, and more than that, despite its small size. The hypothalamus and closely connected structures are controllers of the motivational states that lead to eating, drinking, reproductive and agonistic behaviors (defensive and aggressive), as well as other activities. The activation of hypothalamic cell groups underlies motivational states, which we also call drives, and the affects (feelings)

468   Chapter 25

accompanying them. These cells appear to function as *central pattern controllers*—near the top of a motor system hierarchy—by their connections to the midbrain and to a lesser extent to the hindbrain and spinal cord (chapter 14). The affects are often accompanied by emotional expressions that communicate the motivational state to other members of the species.

Strong influences on these hypothalamic neurons originate at still higher levels, in the endbrain. The structures of the endbrain with strong connections to the hypothalamus have been grouped together and called the *limbic* system (originally by Paul MacLean), because in the endbrain they are located at the fringe or edge (the limbus) of the cerebral hemispheres. The term came from a phrase used by Pierre Paul Broca: "le grand lobe limbique." Why it can be called a lobe of the endbrain is apparent in a medial view of the human cerebral hemisphere shown in figure 25.1, where colors are used to distinguish major functional modules of the entire CNS. Note the portions in red that are labeled "motivation." These are the regions that are most directly connected to the hypothalamus.

Approaching the Limbic System from Below: Two Kinds of Arousal from the Midbrain

In 1949, two neurophysiologists made a discovery that captured the imagination of many neuroscientists of that time. Neuroscientists then were not grouped together into a single

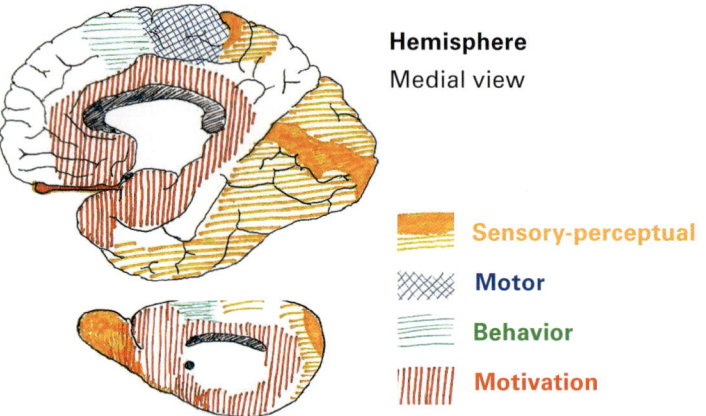

Figure 25.1
Drawings of a medial view of the right cerebral hemisphere of a human (top) and of a rat, mouse, or hamster (bottom). Major functional divisions have been separately marked as indicated in the key. The blank area in the center of each hemisphere represents the surface of a cut that separated the hemisphere from the diencephalon. The cortical regions nearest the medial margins of the hemisphere, labeled the "Motivation" module, are the regions most closely connected with the hypothalamus. These regions are also known from effects of damage and of electrical stimulation to play important roles in influencing motivational states, feelings accompanying those states, and emotional expressions. Together, the structures of this module have been called the limbic lobe (first by Paul Broca in 1878). See figure 33.16 for additional definitions. The drawing of the human brain is based on a figure of Mesulam (2000), with added functional interpretations.

field. They were studying the nervous system from within various disciplines such as medicine, anatomy, physiology, neurology, and psychology. Giuseppe Moruzzi and Horace Winchell Magoun, physicians and scientists, discovered that electrical stimulation of the reticular formation of relaxed or drowsy cats caused a remarkable change in state: a pervasive arousal response. In recordings of electrical activity in the cortex, the response appeared to be identical to the arousal produced by sensory stimulation. The brain system they had stimulated became known as the ascending reticular activating system (ARAS). It was part of a brainstem core of neurons extending through the hindbrain and the midbrain, many of them cholinergic, with very widely branched axons (see figures 10.3 and 17.1).

The physiological and behavioral arousal evoked by stimulating the ARAS, which activates the sympathetic division of the autonomic nervous system, is the kind of arousal we experience when we encounter a very novel or unexpected stimulus. It is part of the orienting response to novelty.

But there is a type of arousal that differs from the first type in several ways. Although like the first kind of arousal it also involves activation of the sympathetic nervous system, it is accompanied by increases in motivation and by strong positive or negative affects—pleasure or displeasure. The midbrain areas where electrical stimulation causes this second kind of arousal were named the *limbic midbrain areas* by the neuroanatomist W. J. H. Nauta in the 1950s, because of studies by him and others that demonstrated connections of these regions with the hypothalamus and with endbrain regions that also projected axons to the hypothalamus—the limbic system structures. Therefore, the second kind of arousal from the core of the midbrain is a **limbic system arousal**, in contrast to the non-limbic arousal caused by stimulation of the ascending reticular activating system.

For an illustration of the midbrain areas of the two arousal systems, see figure 11.4. In that illustration, the structures underlying nonlimbic arousal (the ARAS) are termed *somatic regions*, and the structures underlying limbic arousal are simply called *limbic regions*—Nauta's limbic midbrain areas that correspond to the central gray area around the ventricle and the ventral tegmental area.

The two arousal systems in the midbrain differ not only in their locations and in the effects of electrical stimulation. As you might expect (because structure always underlies function), they differ also in types of anatomical inputs and outputs. Summarizing this information can help you understand the forebrain, as these two regions of the midbrain are in structural continuity with the rostrally adjacent diencephalic structures (figure 11.4)—with many interconnections—and this organization can be used to define major components of the endbrain.

Arousal is usually accompanied by changes in electrical patterns throughout the brain, as can be measured by electrodes that are too large to detect activity of individual neurons. Behavior of the animal or person also changes: He or she becomes more active and attentive (or, if the arousal is extreme, manic).

In the case of **nonlimbic arousal**, forced turning often occurs in response to midbrain stimulation, and the effects show habituation with repeated stimulation, as long as the location of the electrode does not change. These effects contrast with effects of limbic system arousal, which are very resistant to habituation.

The anatomical inputs from the brainstem to the nonlimbic reticular formation are of mixed sensory nature, coming from vestibular, somatosensory, auditory, and visual systems; they also come from cerebellum and from other parts of the reticular formation. Many descending axons also terminate in the midbrain reticular formation, coming from neocortex, corpus striatum, subthalamus, and more ancient portions of the dorsal thalamus (paleothalamic structures—the midline and intralaminar nuclei).

In contrast to this system, stimulation of the **limbic midbrain areas**—central gray area (CGA) and ventral tegmental area (VTA)—causes an arousal that is resistant to habituation (as mentioned earlier) and has strong motivational consequences. The animal shows signs of strong dislike, even fear, in response to stimulation of the more dorsal component, the central gray, whereas it seems to love stimulation of the ventral tegmental area. If allowed to turn the stimulation on or off by its own actions, a rat or cat will quickly learn to turn off the more dorsal stimulation and turn on, repeatedly, the stimulation of the VTA, even if the stimulation is not strong. Stimulation of nonlimbic areas does not have such strong motivational consequences.

Ascending anatomical inputs to the limbic midbrain areas are dominated by projections from visceral sensory structures. Axons from more rostral structures come from the limbic forebrain: hypothalamus, septal area and basal forebrain, amygdala and hippocampal formation. We will encounter these endbrain structures again in later chapters.

Thus, we can say that the first arousal system is somatic in a broad sense. It repeatedly responds to sensory novelty, keeps us wide awake and attentive as we explore the world, plan our next actions, or work on new ideas. The second arousal system is "gut level," a system dominated by instinctive attractions and aversions and by learned likes and dislikes. All the various affective tags we attach to objects and to living beings we have encountered are constantly activating this core of the midbrain one way or another.

These arousal systems function not only as transient activators, responsive to external conditions. They are also subject to regulation by the forebrain, primarily by hypothalamic systems that follow a regular daily rhythm of change as we will note later when we mention structures controlling the sleep–wake cycle.

Autonomic and Endocrine Functions of the Hypothalamus

If we follow the pathways from the limbic midbrain areas rostrally, we reach the hypothalamus, in many ways the core structure of the limbic system (see figure 11.4 and figure 2.2).

Although the hypothalamus is the "head ganglion" of the autonomic nervous system, we should also remember its sympathetic division can also be activated at many different levels of the CNS. Within the hypothalamus, activation of neurons of the posterior nucleus wakes an animal and results in sympathetic arousal with increased secretion of epinephrine (adrenaline) from the adrenal glands. At sites within the more rostral end of the hypothalamus, neuronal stimulation can have opposite effects, making the parasympathetic system dominant. Electrical activation in this region can even cause an animal to prepare for sleep and then lie down and go to sleep.

Closely allied with the autonomic nervous system (ANS) is the endocrine system, which centers on the pituitary. We should review how the hypothalamus controls the pituitary—the structure that presides over endocrine activity (figure 25.2). The posterior pituitary is part of the CNS itself, a little appendage protruding from the base of the 'tweenbrain. It is called the neurohypophysis. It contains a special bed of capillaries. Axons of two groups of hypothalamic neurons terminate on these capillaries rather than on neurons. The hypothalamic neurons with axons that terminate this way are in the supraoptic and the

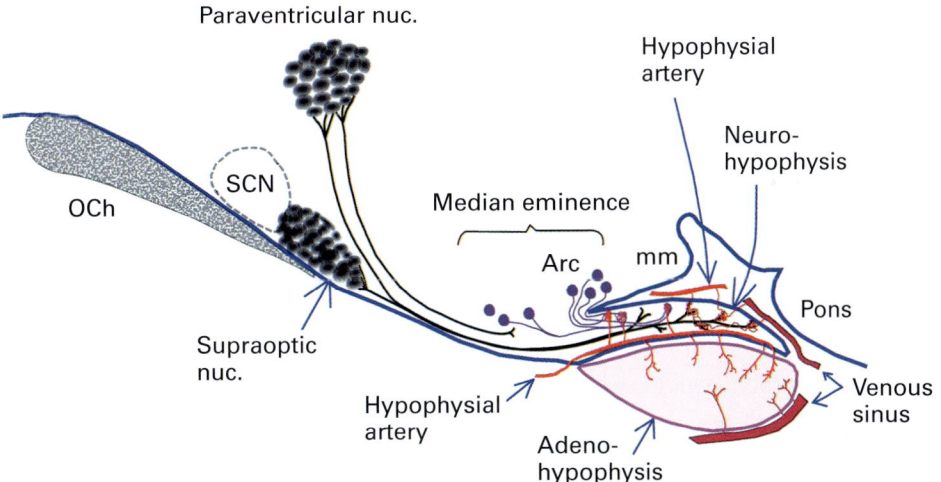

Figure 25.2
Hypothalamic control of the pituitary depicted in an anatomical drawing. The two parts of the pituitary of a small rodent are illustrated together with the adjacent hypothalamus. Neurons of two hypothalamic cell groups—the paraventricular and the supraoptic nucleus—have axons that terminate in the posterior lobe of the pituitary, an appendage of the 'tweenbrain. These axons form terminals on capillaries, secreting hormones into the general circulation (the neurohormones oxytocin and vasopressin). The anterior lobe of the pituitary, the adenohypophysis, is actually a gland, with secretions that are controlled by *releasing factors* that arrive via the bloodstream from the nearby hypothalamus as shown. The factors are released into capillaries of the *median eminence* by axons in the arcuate nucleus of the hypothalamus and other nearby neurons. The pituitary hormones secreted by the gland enter capillaries that drain into a venous sinus and thereby reach the general circulation. Arc, arcuate nucleus; mm, mammillary nuclei; OCh, optic chiasm; nuc, nucleus; SCN, Suprachiasmatic nucleus. Based on Brownstein, Russell, and Gainer (1980).

paraventricular nuclei. The axon terminals release one of two substances: oxytocin, which stimulates uterine contractions and milk secretion, and vasopressin, which causes vascular constriction and also water retention. The last function gives this substance its other name, the antidiuretic hormone (ADH). These hormones enter the bloodstream and exert their actions on the end organs when they reach them via the general circulation.

If an individual suffers damage to the posterior lobe of the pituitary, a lack of ADH secretion causes diabetes insipidus. Lacking normal water retention, a person with this problem shows polydipsia and polyuria (i.e., copious drinking of water and very frequent urination).

The anterior part, or lobe, of the pituitary contains secretory cells, and thus it is considered to be a gland rather than part of the CNS. It is not innervated by hypothalamic neurons but is controlled by the hypothalamus in a different way (figure 25.2). Neurons in the median eminence region including some in the arcuate nucleus produce hormones known as releasing hormones; these hormones move from axonal contacts with capillaries, as is true also for the neurohypophysis, but these capillaries flow into a portal vein that flows into a second capillary bed in the anterior pituitary, or adenohypophysis. There the releasing hormones cause the release of specific hormones into the general circulation: These include growth hormone, sexual hormones, adrenal corticotrophic hormone, and thyroid stimulating hormone.

Homeostatic Regulation of the Internal Milieu

Hypothalamic neurons play critical roles in maintaining the relative constancy of the internal milieu (e.g., temperature, blood levels of glucose, and levels of several hormones). We can describe these roles as the homeostatic functions of the hypothalamus. For example, the endothermic animals—the mammals and birds—generally maintain body temperature within narrow limits. Specific neurons in the anterior hypothalamus change their activity with the temperature of the blood in a precise way. If the temperature drops a small amount below a set point—usually about 37°C in humans—neuronal activity is triggered that causes piloerection (to obtain increased insulation by erection of body hair), and increased muscular activity including shivering. It also causes the motivation to seek shelter, and in our species to put on more clothes (accompanied by the feelings of cold). If the temperature rises a small amount above the set point (a point that varies slightly with time of day), activities that have evolved to cool the body are triggered (sweating or panting, seeking of shady places, removal of clothes, etc.).

Temperature regulation is important in ectothermic animals, too, animals that include the reptiles and amphibians. However, lacking the internal processes that have evolved in birds and mammals for raising and lowering blood temperature, the ectotherms must rely on behavior (e.g., lying in the sun and becoming active mostly at certain times of the day). Body temperatures can vary greatly in these animals, depending on external temperatures.

Because so many biochemical processes in the body are temperature dependent, it is easy to understand the critical nature of temperature regulation in animals. It is also not surprising that animals depend on more than a single level of the nervous system for temperature control. Even if the hypothalamus has been completely disconnected from the more caudal brainstem in animal experiments, some regulation of body temperature remains. In such animals, however, body temperatures fluctuate more than in the intact animal. The animal's thermoregulatory responses triggered by changes in environmental temperatures have been found to be organized at spinal, hindbrain, and midbrain levels.

Regulation of Cyclic and Episodic Behaviors

The most obvious cyclic behavior is one we have mentioned above, the cycle of sleeping and waking. Neurons in the region of the rostral-most hypothalamus, when they become active under the influence of the *biological clock* in the nearby suprachiasmatic nucleus, trigger the onset of various fixed action patterns that serve as preparations for going to sleep. The motor patterns subject to this hypothalamic control are organized by more caudal brainstem circuits. Electrical stimulation of the critical anterior hypothalamic region in a cat can cause a reduction of locomotor speed and may elicit the onset of sleep. Lesions there can result in insomnia in rats. By contrast, neurons in the posterior hypothalamus appear to play the opposite role. Lesions there can result in somnolent animals, and stimulation causes waking and arousal.

Various cyclic behavior patterns are necessary to support the homeostatic processes discussed above: feeding, drinking, foraging, hunting, and predatory attack are major examples. All these behaviors fit into a daily cycle of activity regulated by the endogenous clock, entrained by the light–dark cycle. They show species-typical patterns. The motivational states that drive these patterns of behavior are influenced not only by the temporal cycle but, of course, also by the internal needs of the animal, which depend on the degree and the length of deprivation (e.g., of food, water, opportunity to hunt).

Other behavior patterns subject to hypothalamic control are not generally cyclic. We call them episodic behaviors. They are also instinctive, but this does not mean that they are without individual, learned components. Examples are the agonistic behaviors: defending, fleeing, and some kinds of fighting. Fleeing from predators, as discussed in earlier chapters, was so important for survival that it was supported by very rapidly conducting pathways with inputs from the body surface and from the eyes and ears. But the olfactory sense brought something different to the scene: an animal could learn to associate an odor with a dangerous place or a safe place or with a specific predator or prey animal, and thus the odor could be used to heighten sensitivity of escape responses or of feeding or attack responses. These settings of mood were the province of the hypothalamus as a high-level modulator of responses organized by the brainstem and spinal cord.

Because of the importance of olfactory signals in the early evolution of sexual and parental behavior, the forebrain came to play critical roles in control of reproductive behavior. Central to this control was the hypothalamus, especially with its regulation of sexual hormones controlled by pituitary secretions. These hormones, in turn, affect various neurons in the brain that control various behavior patterns—the patterns of courtship and mating, hatching or birthing, and parental behaviors.

The Center of Motivational State Control

Thus, we can see from the above discussion that the hypothalamic region is a "center" for basic drives—the motivational states that have evolved to promote the survival of the genes of the animal. They are sometimes referred to as the biological drives, the inherited motivations. They are unlearned, although the details of their expression are also influenced by learning during development.

But what actually is a drive? You will get various answers from psychologists and psychiatrists, but here we take a neuroanatomical view. Recall the discussion in chapter 14, an overview of motor system structure. Following Larry Swanson, we characterize the hypothalamus as neuronal groups situated at the top of a motor-system hierarchy, acting as central pattern controllers via connections with more caudal structures (see figure 14.2). Increased activity in these neurons increases the likelihood of specific kinds of action patterns and also causes specific affects (internal feelings) and their expression in emotional displays, thus communicating the inner state to other members of the species.

We can see two distinct kinds of behavior when we study the actions promoted by any drive state: appetitive behavior and consummatory behavior. Three kinds of studies have made the distinction very clear. The first two have involved brain lesions in animals.

Distinguishing between Appetitive and Consummatory Behaviors Involved in a Drive

First, recall the effects of forebrain removals in cats and rats, the *decerebration* experiments reviewed in chapter 7. The decerebrate animals showed no seeking of food or a mate and generally lacked spontaneous behavior. They lacked appetitive behavior—the goal-seeking behavior of a normally motivated animal. In contrast, these same animals retained much consummatory behavior: they showed eating responses in response to oral stimulation (mouth opening, chewing, swallowing); they showed pain-elicited aggressive responses, and sexual postures and reflexes in response to genital stimulation.

Next, imagine what would happen if the hypothalamus were partially but not completely disconnected from the lower brainstem networks that control fixed motor patterns (e.g., the movements of predation and eating). If higher control of the actions instigated by drive states were partially lost, we may see evidence of abnormalities in the relative dominance of various motor components.

Walter Randall reported such experiments in 1964, describing effects of large lesions of the caudal midbrain in cats. He observed that fragments of cat consummatory behavior were abnormally organized and sometimes just omitted. The animals did not lose their regulation of food intake, but their prey-catching and eating became very abnormal. For example, cats showed abnormally constant approach-to-prey behavior and treated each mouthful of food as a prey object, carrying it with head held high as if dragging a kill, and then administered what looked like killing bites. An empty bowl of water could elicit fishing behavior. However, they showed no positioning for ambush, no pouncing, no grooming after eating, and none of the normal searching, digging, and scratching associated with cat elimination behavior.

Thus, Randall found evidence of disruption of a patterning system of the forebrain and rostral midbrain. Swanson may suggest that a behavioral pattern controller was partially disconnected from a midbrain pattern initiator (see chapter 14). The first view stresses an organizing aspect of motivational systems. The second is a complementary view coming from the hierarchical view of motor control as sketched in figure 14.2. We should add that to the extent that the sequence of instinctive actions has been linked together by learning, it is likely that the corpus striatum is involved as well.

Computational Neuroethology

Another type of work that clarifies the distinction between appetitive and consummatory actions as components of motivational states has used computer modeling of behavior. Randall Beer in 1990 reported simulations of the cockroach's neural control of feeding. To get realistic behavioral patterns, the simulations used distinct appetitive and consummatory components.

At the Massachusetts Institute of Technology, Bruce Blumberg and his students (at the MIT Media Laboratory from 1994 to 2005) have used *synthetic screen characters* controlled by programs inspired by ethology and also by the anatomy of the brain's limbic system. Active characters are seen on a computer monitor. The characters are not simply animated cartoons. Underlying their actions are specific drives that build up over time and activate appetitive behaviors (locomotion, searching). Specific stimuli, separately generated in the simulation, trigger consummatory behaviors. In addition, characters show habit learning as specific stimuli become associated with "good" or "bad" effects, and this changes future responses to them. The characters come to respond to different objects and other characters as if those objects and characters had acquired *affective tags* based on previous encounters.

In general, the characters in the Blumberg group's simulations are very animal-like, and in some projects even human-like, despite a lack of "higher" cognition. Details of the results of a simulation sequence are not predictable. Moreover, the personalities of the characters can be consistently different—according to the settings of certain parameters—although the structure of the underlying circuitry is the same.

Hunger, Feeding, and Brain Circuits

Recalling the decerebration experiments with animals, we know that if the midbrain is the highest level of the CNS, there is no hunger or satiety, no active seeking of food (no appetitive behavior). Eating reflexes and consummatory patterns are seen, but there is little regulation of eating. Forebrain mechanisms appear to be critical for motivational control.

In human infants born with a failure of the neural tube to close at the rostral end, the forebrain fails to develop. As expected from the animal work, such anencephalic infants show no regulation of hunger. Both normal and anencephalic infants eat reflexly, showing rooting, sucking, and swallowing reflexes, but only the normal babies show satiety by responding to stomach distention.

Later in development, the hypothalamic feeding control mechanisms develop responses to other signals of satiety. (Note: Subthalamic involvement in hunger control has also been implicated.) Some signals come by way of inputs through the vagus nerve (cranial nerve X). These signals indicate not only stomach distention but also the sensations of food passing through the throat, which can be used to compute meal size. Some hypothalamic neurons respond to factors in the blood including blood glucose and certain hormones secreted by the digestive tract. In the short term, signals influence length and quantity of a meal. Longer-term signals can influence total food intake in a day.

Electrically Elicited Drive States

Fascinating studies were conducted, mostly in the 1960s and 1970s, of remarkably normal-looking drive states elicited by electrical stimulation of the hypothalamus in rats; studies were also conducted in cats and a few other species. For example, it was found that hunger can be elicited in satiated rats by stimulation through small wire electrodes implanted with their uninsulated tips at points in the lateral hypothalamus. During stimulation with fairly weak currents, the rats show many of the characteristics of normal hunger.

The electrodes were not microelectrodes, and it is clear that the stimulation was activating many axons of the medial forebrain bundle, a diverse population, and a group of cell bodies in the lateral hypothalamic area as well. Thus, it is surprising that generally only one drive state seemed to be activated. However, small changes in position of the electrode could cause other effects to become dominant. For example, if investigators made a slight change in a stimulation site where hunger was being elicited, they might be able to elicit a drive for gnawing or they might see an antipredator response.

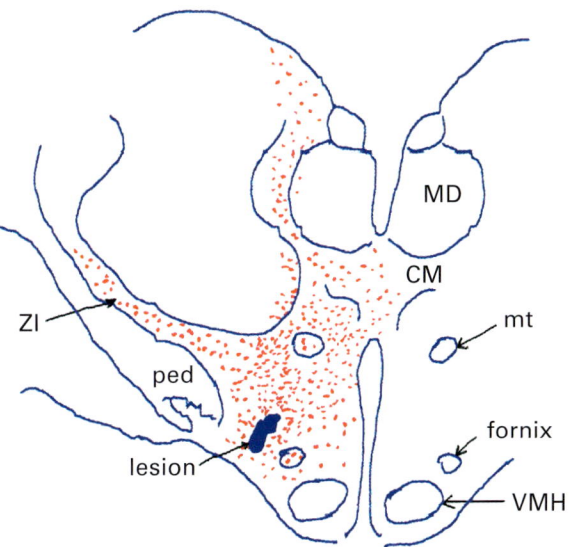

Figure 25.3
Drawing of a frontal section through the diencephalon of a cat. During its life, neuroscientists had implanted an electrode into the hypothalamus and applied electrical stimulation that elicited a dramatic mood change: the cat immediately showed strong motivation to stalk and attack prey (like a mouse or rat). Before withdrawing the electrode, a small lesion was made at the stimulation site, using stronger current to inflict damage. The cat lived long enough after the lesion for the axons from the destroyed cells to degenerate enough to be stained with a modified Nauta method. In this drawing, the location of degenerating axons and terminals is shown by the red dots. The results provided evidence that this region of the hypothalamus sends axons not only to other parts of the hypothalamus but also to the zona incerta (cells between thalamus and hypothalamus, in the subthalamus) and to the midline and intralaminar nuclei of the thalamus. (Additional terminal degeneration was found rostral and caudal to this level.) CM, nucleus centralis medius (the region probably includes the rhomboid and reunions nuclei); MD, mediodorsal nucleus of thalamus (lateral to which is the central lateral nucleus of the intralaminar nuclei); mt, mammillothalamic tract; ped, cerebral peduncle (continuous with the internal capsule); VMH, ventromedial nucleus of hypothalamus; ZI, zona incerta of the subthalamus. Based on Chi and Flynn (1971).

At Yale University, John Flynn and his students conducted many such studies using cats. The focus was on electrically elicited aggression. Hypothalamic sites were found where the electrical current caused what appeared to be defensive aggression, with hissing and an arched back and striking out with the claws of the front feet. At other sites, usually slightly more laterally placed, predatory aggression could be evoked (figure 25.3). The cat became more alert and watchful, and soon it became obvious that it was looking for mice and rats. The cat commenced stalking the prey animal if one was detected, then it pounced in the typical cat fashion and administered a killing bite. This biting attack sequence was not simply a result of hunger, as a cat would bypass a dish of tuna fish, its favorite food, in order to get at a small rodent prey animal.

The neuroscientists conducting these experiments found that the mood of predatory attack (inferred from behavioral signs) could also be elicited from midbrain stimulation, generally within the central gray area. How critical, then, is the hypothalamus for such a motivational state? Experiments with circumsections of the hypothalamus—severing its connections with lower structures—showed that midbrain limbic structures (in the central gray area) are sufficient. However, without the hypothalamus connections, there is reduced initiation of the behaviors and presumably less modulatory regulation by other forebrain activities.

Cats that had never developed a preference for hunting, lacking the presence of a mother they could imitate, showed the electrically stimulated biting attack just as did cats that had hunted. Thus, the innate nature of the predator motivation and fixed action patterns was clear.

(Note: New work on hypothalamic control of motivational states is being done with much greater specificity using "optogenetics"—methods of inserting genes for modified channelrhodopsin, enabling stimulation of specific cells with exposure to certain wavelengths of light.)

Connections with Other Systems

Flynn and his co-workers did not stop with these findings. Because they were able to control the predator drive with specific hypothalamic stimulation, they could ask further questions about the motivational state. How does the change in mood controlled by the hypothalamus affect the activity of neurons in the brainstem? How does it affect activity of cells in sensory systems of the thalamus and the neocortex? Some answers were clear: As the stimulation alters the mood to the predatory state, the threshold for activation of the jaw-opening reflex goes down, and the area of the mouth where a touch elicits mouth opening increases. Thus, the probability increases that movements that are part of the biting attack sequence will occur. In the dorsal thalamus, the activity of neurons of sensory pathways changes, as does the activity of various neurons in sensory neocortex.

The axonal pathways from the hypothalamus involved in this modulation of thalamus and cortex are at least partially known. They include the widespread projections of several small groups of neurons in the lateral hypothalamus, recently discovered, that were mentioned in chapter 17. In addition, there are interconnections of the lateral hypothalamus with midline and intralaminar thalamic nuclei (part of the paleothalamus) as illustrated in figure 25.3: the pattern of dots in the figure shows the sites of axons, and axon terminals, coming from a hypothalamic stimulation site where predatory attack motivation had been elicited. Cells of some midline nuclei have axons that connect widely in the dorsal thalamus. The cell groups of these parts of the thalamus also have interconnections to the midbrain reticular formation. The reticular formation also receives inputs from hypothalamus via the limbic midbrain areas. As we have seen in chapter 17, reticular formation cells can have axons that distribute their terminations widely both rostrally in the 'tweenbrain and caudally in the midbrain and hindbrain.

It is not certain which of these various connections are most involved in the modulatory activity associated with any specific motivational state. However, the connections do imply that changes in hypothalamic activity, associated with arousal of a strong motivation, can cause widespread changes in the state of the forebrain and alter the activity of sensory pathways ascending through the thalamus to the neocortex.

Drive and Reward Involve Distinct Axon Populations in the Medial Forebrain Bundle

When a specific drive like hunger or predatory attack can be elicited by electrical stimulation of the hypothalamus, usually in the lateral portion where the medial forebrain bundle axons are found, investigators have found that the stimulation is usually also rewarding to the animal. The animal will learn to turn the stimulation on if given access to a lever that does this. Evidence that drive and reward, elicited in this way from the same hypothalamic site, may involve different axon populations was provided by clever investigations where trains of electrical pulse pairs were used. The time between the two pulses in each pair was varied. If the time gap between pulses was less than the refractory period for stimulating action potentials, then a pair of pulses would act like a single pulse. If the time gap was longer than the refractory period of an axon, then a pulse pair acted like two pulses, not one. It is known that axons of different sizes have different refractory periods. Therefore, progressive increases in the interpulse interval made it possible for experimenters to find out the critical pulse pair timing when the behavioral effect of the stimulation changed from an effect of one rate of stimulation to an effect of a rate twice as high. Results of such investigations have indicated that drives, like hunger, elicited by hypothalamic stimulation, and positive reward resulting from stimulation of the same sites, are attributable to axons with different refractory periods.

Readings

Brodal, P. (2004). *The central nervous system: Structure and function*, 3rd ed. New York: Oxford University Press.

Chi, C. C., & Flynn, J. P. (1971). Neural pathways associated with hypothalamically elicited attack behavior in cats. *Science, 171,* 703–706.

Deutsch, J. A. (1964). Behavioral measurement of the neural refractory period and its application to intracranial self-stimulation. *Journal of Comparative and Physiological Psychology, 58,* 1–9.

Flynn, J. P. (1969). Neural aspects of attack behavior in cats. *Annals of the New York Academy of Sciences, 159,* 1008–1012.

Gallistel, C. R., Rolls, E., & Greene, D. (1969). Neural function inferred from behavioral and electrophysiological estimates of refractory period. *Science, 166,* 1028–1030.

Moruzzi, G., & Magoun, H. W. (1949). Brain stem reticular formation and activation of the EEG. *Electroencephalography and Clinical Neurophysiology, 1,* 455–473.

Swanson, L. (2011). *Brain architecture: Understanding the basic plan,* 2nd ed. New York: Oxford University Press.

Modern experiments on arousal systems ascending from the brainstem: a reading from chapter 17:

Fuller, P., Sherman, D., Pedersen, N. P., Saper, C. B., & Lu, J. (2011). Reassessment of the structural basis of the ascending arousal system. *Journal of Comparative Neurology, 519,* 933–956.

The term *limbic system* was first used by Paul MacLean:

McLean, P. D. (1952). Some psychiatric implications of physiological studies on frontotemporal portion of limbic system (visceral brain). *Electroencephalography and Clinical Neurophysiology, 4,* 407–418.

Origin of the term *ascending reticular activating system*:

Moruzzi, G., & Magoun, H. W. (1949). Brain stem reticular formation and activation of the EEG. *Electroencephalography and Clinical Neurophysiology 1,* 455–473.

Walle Nauta described evidence for *limbic midbrain areas*, which, like the limbic areas of the endbrain, were closely connected to the hypothalamus:

Nauta, W. J. H. (1958). Hippocampal projections and related neural pathways to the midbrain in the cat. *Brain, 81,* 319–340.

Aryeh Routtenberg brought together convincing evidence for two distinct arousal systems of the midbrain:

Routtenberg, A. (1968). The two-arousal hypothesis: Reticular formation and limbic system. *Psychological Reviews, 75,* 51–80.

Walter Cannon, in his 1932 book *The Wisdom of the Body* (W. W. Norton), described extensive evidence for regulatory mechanisms that result in homeostasis of the internal environment. See the following paper:

Cooper, S. J. (2008). From Claude Bernard to Walter Cannon. Emergence of the concept of homeostasis. *Appetite, 51,* 419–427.

Evelyn Satinoff conducted extensive experimental studies of the mechanisms of body temperature regulation in mammals:

Satinoff, E. (1978). Neural organization and evolution of thermal regulation in mammals. *Science, 201,* 16–22.

Experiments on cats with large midbrain lesions that caused partial disconnection of the forebrain from brainstem and spinal mechanisms:

Randall, W. (1964). The behavior of cats (*Felis catus L.*) with lesions in the caudal midbrain region. *Behaviour, 23,* 107–139.

At the MIT Media Laboratory, Bruce Blumberg and his students used *synthetic screen characters* that were controlled by programs inspired by ethology and also by the anatomy of the brain's limbic system:

> Yoon, S., Schneider, G. E., & Blumberg, B. M. (2000). Motivation driven learning for interactive synthetic characters. In Carles, S., Gini, M., & Rosenschien, J. S., eds. *Agents '00, Proceedings of the fourth international conference on autonomous agents*, pp. 365–372. New York: ACM.

> Yoon, S., Burke, R. C, Blumberg, B. M., & Schneider, G. E. (2000). Interactive training for synthetic characters. *AAAI/IAAI, 2000*, 249–254. See also Randall Beer's simulations of the cockroach's neural control of feeding:

> Beer, R. D., Chiel, H. J., & Sterling, L. S. (1991). An artificial insect. *American Scientist, 79*, 444–452.

It has been known since the middle of the twentieth century that the lateral hypothalamus is involved in the body's energy regulation. The old methods of using electrical stimulation and lesions made this evident, and now newer techniques using biochemistry and molecular biology promise to lead to more detailed understanding of feeding, weight regulation, and related brain and hormonal states at the level of specific cell types and connections:

> Berthoud, H.-R., & Münzberg, H. (2011). The lateral hypothalamus as integrator of metabolic and environmental needs: From electrical self-stimulation to opto-genetics. *Physiology & Behavior, 104*, 29–39.

Cells of midline thalamic nuclei, which receive projections from lateral hypothalamus, have axons that connect widely to dorsal thalamic neurons (as well as to more caudal reticular formation and to endbrain structures):

> Scheibel, M. E., & Scheibel, A. B. (1972). Input-output relations of the thalamic non-specific system. *Brain, Behavior and Evolution, 6*, 332–358.

26 Core Pathways of the Limbic System, with Memory for Meaningful Places

In the previous chapter, we approached the hypothalamus from below, by way of the midbrain and the arousal systems that are centered there. In this chapter, we will add some important details to our picture of the hypothalamus. Then we will focus on forebrain pathways in which this structure plays a very important role. These pathways are of special significance in the regulation of social life and emotions. They have also been found to play crucial roles in the formation of episodic memories—so important for our sense of location and orientation in the world, in both space and time.

Cell Groups in the Hypothalamus

Le Gros Clark in 1936 showed how the neuroanatomical study of cytoarchitecture reveals multiple distinct cell groups in the medial part of the hypothalamus. In this part, there are fewer long axons passing through and denser collections of neuronal cell bodies than in the more lateral part. Look back at figure 2.2 for an illustration of how cytoarchitecture reveals various fairly distinct groupings of neuronal cell bodies, including the ventromedial hypothalamic nucleus. The photograph is from the brain of a rat in a frontal section. Figure 26.1 shows a drawing of a thick parasagittal section based on Le Gros Clark's illustrations of the human medial hypothalamus. More details have been added in various subsequent studies using neurochemical as well as anatomical data from several species.

The shape of the human hypothalamus in parasagittal section may be a little confusing because of the funnel shape and with the close proximity of the optic chiasm to the stalk of the pituitary (the infundibulum). Perhaps figure 26.2 will be a little easier to understand. It shows a sketch of a similar view of the medial hypothalamus in a small rodent like the mouse or hamster.

The cyctoarchitecture of the lateral hypothalamus (LH) shows a very different picture. This region has the medial forebrain bundle running through it, mixed with neurons of lesser density than in the more medial part. However, neurochemical studies have revealed special groups of neurons in the LH. In chapter 17, we mentioned three groups of neurons

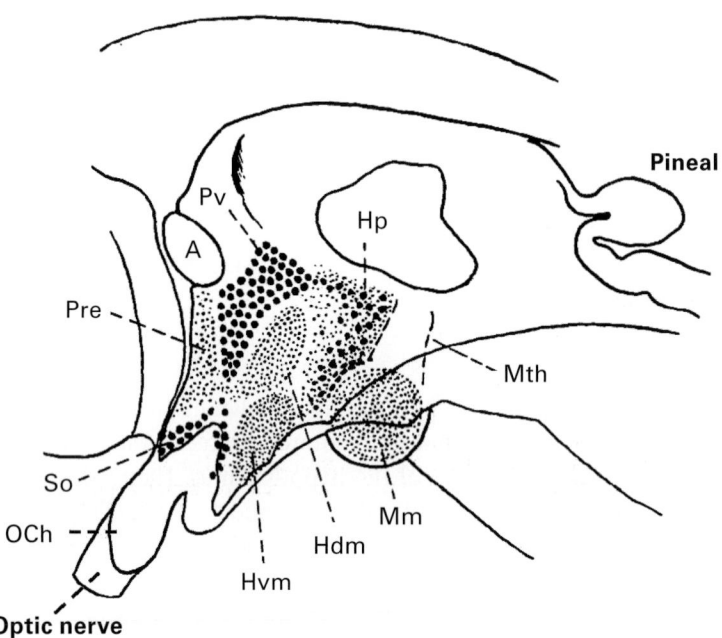

Figure 26.1
A slightly modified version of an illustration of Le Gros Clark (1936) showing the positions of hypothalamic nuclei in a human brain. The nuclei shown are medially located. Their outlines were projected onto the wall of the third ventricle. Rostral is to the left in this version. "This diagram has been made by taking a tracing from a photograph of the brain which was afterwards serially sectioned in a sagittal plane. The extent of the nuclei was reconstructed from the serial sections." A, anterior commissure; Hdm, dorsomedial nucleus of hypothalamus; Hp, posterior hypothalamic nucleus; Hvm, ventromedial nucleus of hypothalamus; Mm, mammillary body (group of nuclei); Mth, mammillothalamic tract; OCh, optic chiasm; Pre, preoptic nucleus (area); Pv, paraventricular nucleus; So, supraoptic nucleus. Caudal to the optic chiasm, the funnel shape of the infundibulum is evident but the pituitary attachment is not shown. From Le Gros Clark (1936).

there that have widely projecting axons and that thereby can influence the overall state of the brain. One of them contains melanin concentrating hormone. Another contains corticotropin-releasing hormone, and a third such group uses hypocretin (orexin) and dynorphin as neurotransmitters. The third group is implicated in the control of sleep.

Feedback from Visceral Afferents and from Blood Chemistry

Hypothalamic functions (e.g., regulation of temperature and of hydration) require feedback from the body. Information arrives from the periphery via visceral sensory afferents, mostly through the vagus nerve (the tenth cranial nerve). Sensory axons from the viscera travel through the vagus and reach the nucleus of the solitary tract in the hindbrain (see chapters 10 and 18). From there, axonal connections not only form reflex connections in

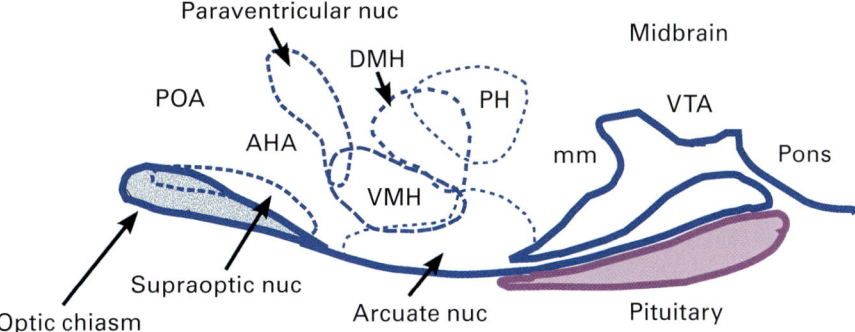

Figure 26.2
Schematic side view of the hypothalamic region of a small rodent (rat, mouse, or hamster). Rostral is to the left. Positions of neuronal cell bodies of some structures mentioned in the text are indicated. Note the difference in shape of the structure in comparison to the human hypothalamus sketched in the previous figure. AHA, anterior hypothalamic area; DMH, dorsomedial hypothalamic nucleus; mm, mammillary nuclei (body); PH, posterior hypothalamic nucleus; POA, preoptic area; VMH, ventromedial hypothalamic nucleus; VTA, ventral tegmental area.

the hindbrain but, mostly via polysynaptic pathways, reach the hypothalamus. However, there are direct pathways to the hypothalamus from the dorsal horn of the spinal cord as well. Such axons have been revealed by both electrophysiological and neuroanatomical methods. These pathways appear to carry pain information, and they may bring other information to the hypothalamus as well.

But neural pathways are not the only means of feedback to the hypothalamus. We know that neurons there, as well as in a few other locations in the brain, can respond directly to factors in the bloodstream. We have mentioned detection of blood temperature and blood glucose levels; in addition, several proteins found in the bloodstream can cause changes in the brain, such as the circulating hormone angiotensin II and some sexual hormones. Unlike smaller molecules like glucose and oxygen, angiotensin does not pass through the blood–brain barrier, so how can levels of this hormone control hypothalamic neurons?

The answer to the question is simply that the blood–brain barrier is lacking or much weaker at certain sites in the walls of the third and fourth ventricles (figure 26.3). Without that barrier, created by the pial–glial membrane, some larger molecules can penetrate the nearby brain tissue. An example of this penetration is illustrated in figure 26.4, which illustrates an experiment demonstrating that the blood–brain barrier is weak in the median eminence region of the hypothalamus, close to the stalk of the posterior pituitary.

All this is background if we want to begin to understand the neuronal circuits underlying feeling and emotion. In the past chapter, we summarized various functions for which the hypothalamus is particularly important. Next, we will follow pathways from that

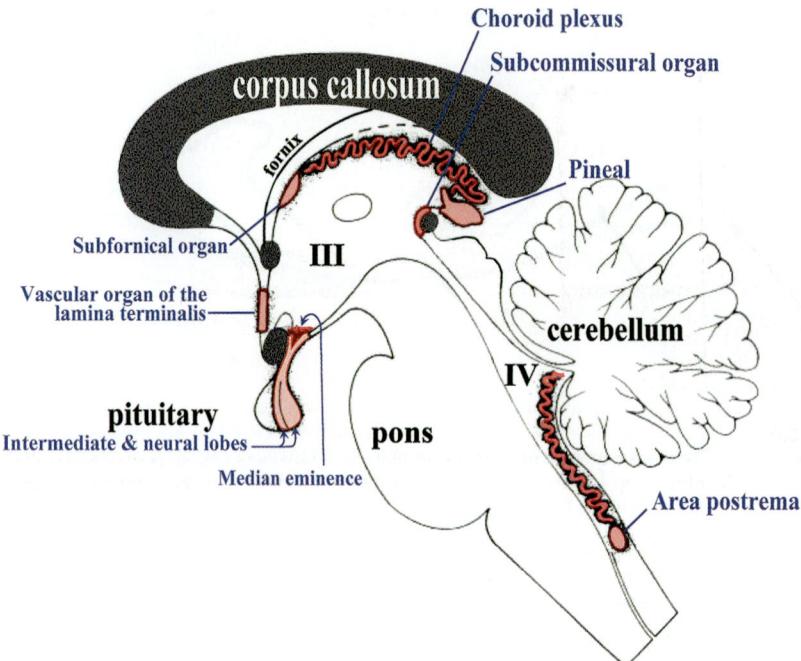

Figure 26.3
Sites in the walls and roof of the third and fourth ventricles where a blood–brain barrier is lacking or weak are shown (pink) and named in this drawing of a midsagittal section of the human brain. The subfornical organ is located at the rostral end of the third ventricle near the openings into the lateral ventricles. There is evidence that it is involved in control of thirst and fluid balance. The vascular organ of the lamina terminalis has similar functions and may also be involved in neuroendocrine functions. The area postrema, as mentioned in chapter 18, is at the caudal end of the fourth ventricle. Cells there detect blood-borne toxins and can trigger vomiting. The choroid plexuses of the lateral ventricles and in the roof of the third and the fourth ventricles (shown in red) also have leaky capillaries. This is also true of the median eminence of the hypothalamus and of the pineal gland. III, third ventricle; IV, fourth ventricle. Based on Johnson and Gross (1993).

central structure to the endbrain and from there back to the hypothalamus. It is in activities of circuits of the endbrain that the truly mammalian affects (feelings), as well as cognitions, are shaped.

Limbic System Interconnections within the Forebrain: The Circuit of Papez

From the caudal end of the hypothalamus, cells of the mammillary bodies have a major projection to the dorsal thalamus, the mammillothalamic tract. This tract terminates in the anterior nuclei of the thalamus. Neurons of those nuclei project to the cingulate gyrus of the cerebral hemisphere. This *paralimbic* cortex, located above the corpus callosum in mammals, projects to the structural complex we call the hippocampal formation (the

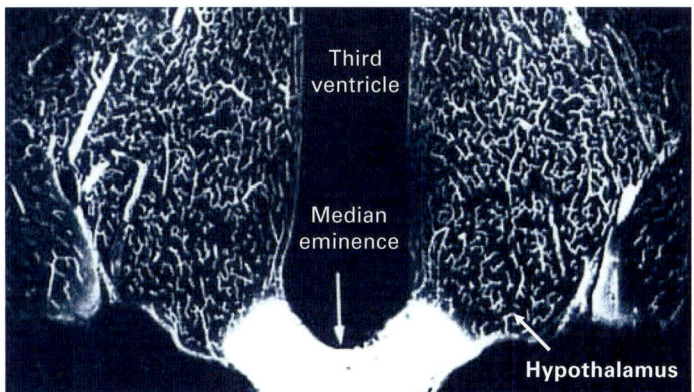

Figure 26.4
Illustration of results of an experiment, using a monkey, in which horseradish peroxidase (HRP) was injected intravenously. The photograph is of part of a frontal section showing the median eminence of the hypothalamus, treated with histochemical procedures to reveal the presence of the HRP. In the median eminence region, the HRP is found both inside and outside the brain capillaries, so it has penetrated the brain tissue, indicating a lack of a blood–brain barrier. In nearby tissue of the hypothalamus, the HRP is found only within the blood vessels, indicating an intact blood–brain barrier. From Broadwell, Balin, Charleton, and Salcman (1987).

medial pallium of non-mammalian animals). The hippocampal formation has a strong projection back to the hypothalamus (as well as to other structures), especially to nuclei of the mammillary bodies. The complete loop we have just described is known as *the circuit of Papez*. It was first described by the neuroanatomist James Papez (pronounced Pay-ps) in 1937.

Papez's publication was entitled "A Proposed Mechanism of Emotion." When he published this paper, it was really just a theory because many of the connections he proposed were not certain, as the experimental techniques for demonstrating them were lacking at that time. Papez began his argument by outlining the forebrain structures that were believed to be dominated by olfaction. In fact, these structures together were called the *rhinencephalon*, or nose brain. Then he pointed out that this olfactory dominance was not the case for humans. Reviewing human clinical cases of brain lesions and of seizure symptoms, it appeared that disturbances of the relevant medial structures of the hemispheres caused changes not so much in olfactory senses but rather in feelings and emotional expressions. It was also noteworthy that these structures have a threshold for seizures that is lower than that for other regions, and that an entire "circuit" tended to be involved together in seizures that began in one of them. Therefore, he reasoned, they must be strongly interconnected.

The ring of connections proposed by Papez can be illustrated in clearer fashion using sketches of the rodent cerebral hemisphere and brainstem (figure 26.5). The originally proposed circuit is pictured in figure 26.6. Experimental neuroanatomical studies

488   Chapter 26

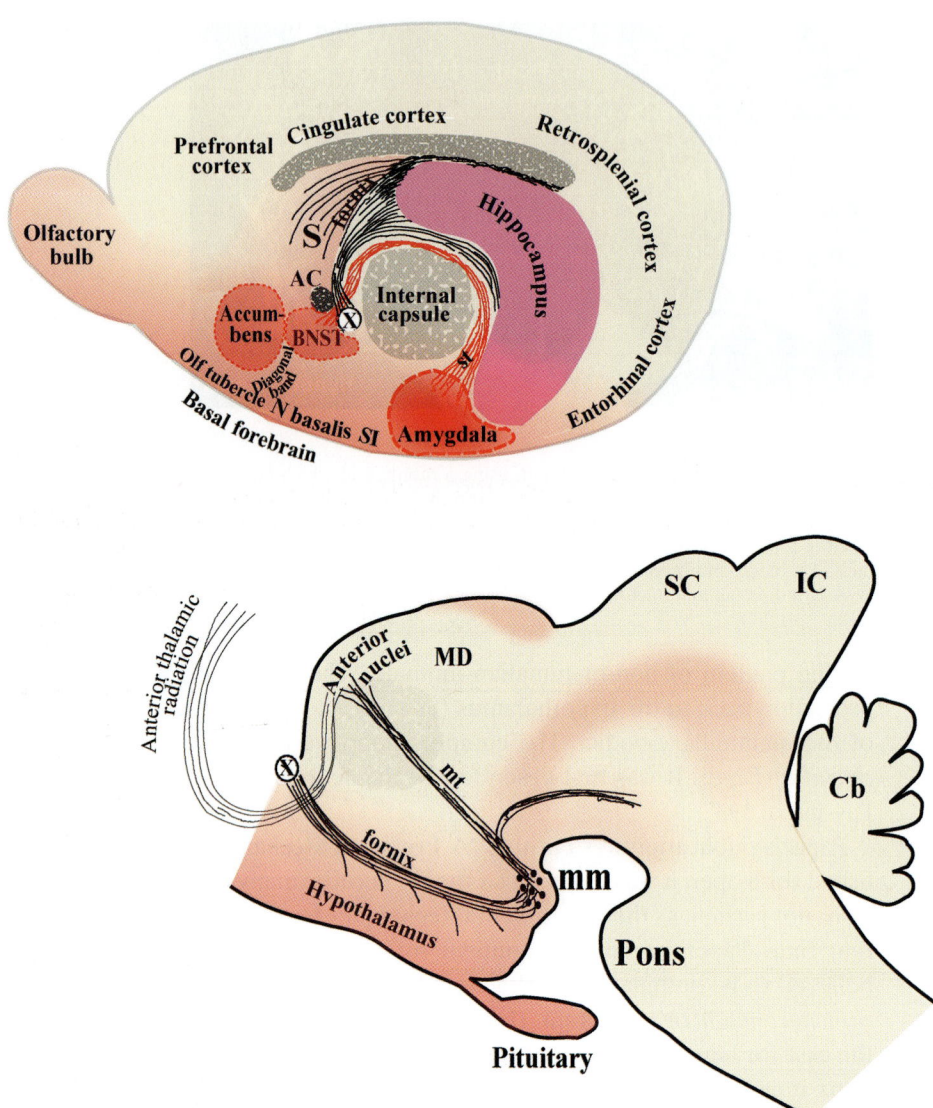

Figure 26.5
A medial view of the right cerebral hemisphere of a small rodent is shown in the upper drawing. Various structures of the limbic endbrain are labeled. This hemisphere has been separated from the underlying brainstem by severing the internal capsule and the fibers of the fornix (circled "X" in the drawing), and also cutting through the basal forebrain and septum so the 'tweenbrain is separated from the endbrain. (The hippocampus is depicted as if it could be seen through brain structures partially covering it.) In the lower drawing, a sketch of the brainstem, somewhat schematic, is shown with the point where the fornix was severed in the upper drawing again marked with a circled "X." The severed internal capsule is omitted from the lower drawing in order to depict the course of the mammillothalamic tract. AC, anterior commissure; BNST, bed nucleus of the stria terminalis; Cb, cerebellum; MD, mediodorsal nucleus of thalamus; mm, mammillary body; mt, mammillothalamic tract; N, nucleus; Olf, olfactory; S, septal area; SC, superior colliculus; SI, substantia innominata; st, stria terminalis.

Core Pathways of the Limbic System

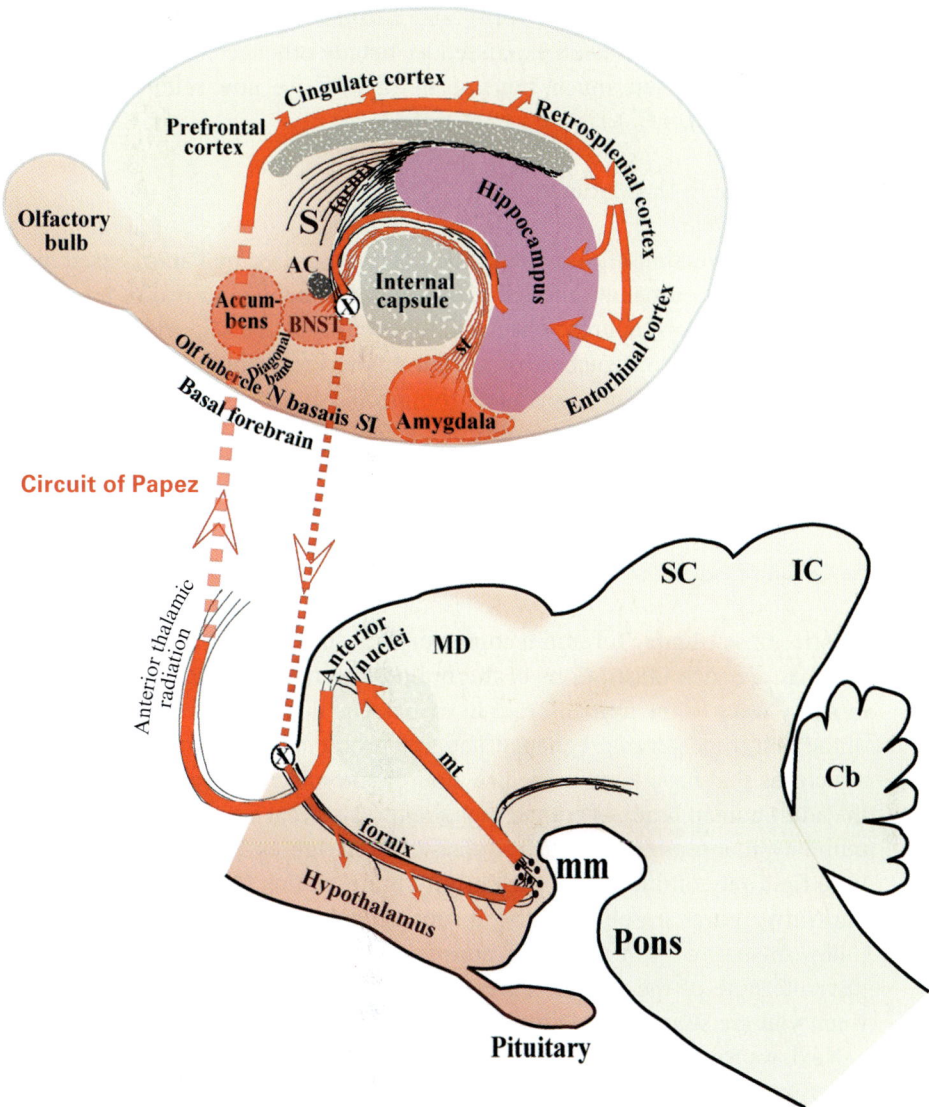

Figure 26.6
The drawings of figure 26.5 are repeated here with the circuit of Papez depicted with red arrows. BNST, bed nucleus of the stria terminalis; Cb, cerebellum; MD, mediodorsal nucleus of thalamus; mm, mammillary body; mt, mammillothalamic tract; S, septal area; SC, superior colliculus; SI, substantia innominata; st, stria terminalis.

have allowed investigators to verify and add details to the Papez's circuit. The original group of structures has been expanded to include others closely connected to them. Altogether they constitute much, but not all, of what we now refer to as the limbic system (see chapter 25).

Why the Revival of Interest in This Circuit?

Despite the oversimplifications in the picture described by Papez, based more on clinical cases than on neuroanatomical studies, there has been a revival of interest in the circuit of Papez in modern neuroscience. A major reason is the importance of the hippocampal formation in the construction of spatial memories. In humans, this role extends to the laying down of long-term memories for specific events, but not of *procedural* memories, the habits that depend on connections in the corpus striatum. This functional role has added a special importance to the study of hippocampal inputs, outputs, and internal circuitry. This will be discussed in more detail in chapter 28.

Visualizing the Circuit of Papez

Illustrations of limbic forebrain connections in the human brain are not as easy to understand as are such illustrations of the brain of a rat, mouse, or hamster, which are similar in many ways to the human brain in embryonic development, especially toward the end of the first trimester of pregnancy. The first problem is simply the large number of distinct structures that have been named by neuroanatomists. The second problem in visualizing the adult human brain, of course, is the relatively great size of the neocortex and its major outputs and inputs. In the illustrations shown in figures 26.5 and 26.6, you should concentrate first only on the major parts of the brain, noting how the cerebral hemisphere, for illustrative purposes only, has been separated from the brainstem. Then, in figure 26.6, follow the pathways in red, which constitute the circuit of Papez described earlier. Finally, just take note of the various other named structures. This will be useful as a reference when you are studying the limbic forebrain.

Next, we want to bring this circuit more up to date, drawing on various neuroanatomical experiments. Study figure 26.7a in which the structures included by Papez are named in red. Note that many of the connections are not just one-directional as described by Papez; they are bidirectional. Next, it should be obvious that the connection from the cingulate cortex to the hippocampus is complex. Most of the projections are not direct ones. They go from the cingulate to other paralimbic areas including the entorhinal area—cortex that extends caudally from olfactory cortex. (Most of it in monkeys does not receive direct olfactory input.) The entorhinal cortex projects axons to two parts of the hippocampal formation. These two parts are the dentate gyrus and the subiculum. These will be pictured in chapter 28. It is worth noting also that the posterior cingulate cortical area known

Core Pathways of the Limbic System

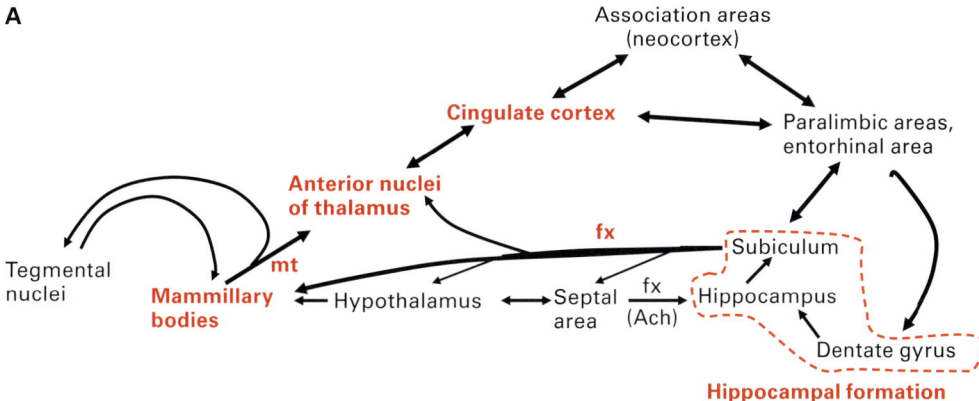

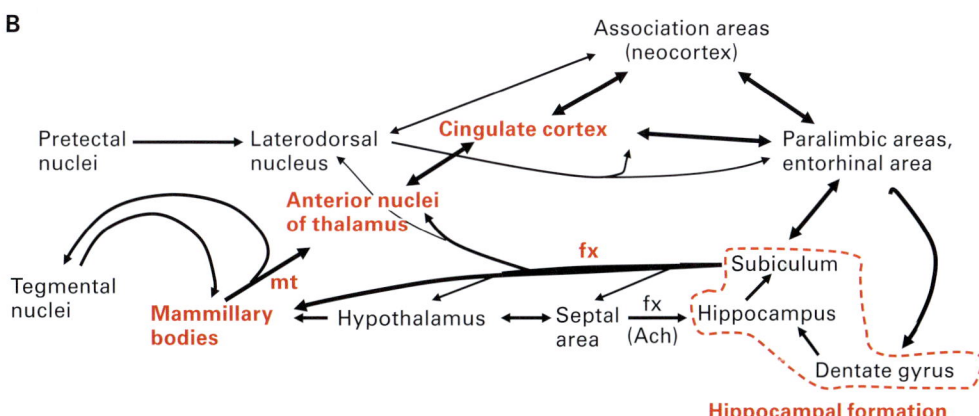

Figure 26.7
(A) Information-flow diagram of the circuit of Papez brought more up to date than in figure 26.6. The structures originally included by Papez are shown in red font. Neuroanatomical studies have supported the connections as shown by the arrows (as well as others not shown), including connections not postulated by Papez. Structures included in the definition of the hippocampal formation are circled in the red dashed line. Note also two major inputs into the circuit of Papez that come from the association areas of the neocortex and from the tegmental nuclei, as discussed in the text. Both connections are reciprocated. (B) Further additions to the circuit of Papez are essential for a more complete view of its functions. Geometric or landmark orientation information in each local environment provides the animal with a sense of direction—an allocentric coordinate system. This information comes into the hippocampus as a head/body orientation signal, which is sent to the mammillary bodies. This "sense of direction" signal is constantly being updated by inputs that signal changes in head direction. These inputs come from the hindbrain and from the visual system as shown. We return to this in chapter 28. Ach, acetylcholine; fx, fornix bundle; mt, mammillothalamic tract.

as the retrosplenial cortex (behind the splenium, which is the caudal end of the corpus callosum) also projects axons (not illustrated) to these same two parts of the hippocampal formation.

A major output of the hippocampal formation to the hypothalamus is by way of the fornix, or fornix bundle. This group of axons goes not only to the mammillary bodies but also to other parts of the hypothalamus and to the septal area, located rostral to the 'tweenbrain. The fornix also contains some axons going in the reverse direction, from the septal area to the hippocampus. (Some axons pass through the septal area and continue ventrally and caudally, terminating throughout the medial hypothalamus.)

Now we can ask the following question, a crucial one if we are to understand something about the functions of this circle of limbic system connections. What are the major routes into and out of this circuit, and what information do they carry? In the diagram, we can see two such routes. One is located caudally, and is a connection between the mammillary bodies and more caudal structures. The other is a complex of connections that have expanded considerably in the evolution of higher primates. These connections are between the association areas of the neocortex and the paralimbic cortical areas that are the gateways to the hippocampus itself.

What Is the Functional Significance of the Return Pathway?

The major pathway within the circuit of Papez from the hypothalamus back to the hippocampus begins with the mammillothalamic tract. As you can see in the figure, the thalamic targets are the anterior nuclei, which project axons to the cingulate cortex, the cortex found medially in the hemisphere just above the corpus callosum. The big question here is about the information carried by this pathway. Note that the mammillary bodies receive inputs not only from the hippocampus via the fornix and from the more anterior hypothalamus, but also from the tegmental nuclei of the caudal midbrain.

An important neurophysiological discovery about the information from the tegmental nuclei has shown that when we are examining the circuit of Papez, we are not just dealing with the cells and circuits of feeling and emotion. The discovery was that one kind of information carried by the pathway directed rostralward from the caudal hypothalamus is about the direction faced by the head. Changes in this direction are signaled by vestibular inputs and by proprioceptive information from the neck originating in the hindbrain and relayed from tegmental nuclei.

It is not difficult to imagine how information signaling changes in direction could be useful in the pathway leading to the hippocampus. Neurons that encode head-direction information are found all along the pathway to the hippocampus and in the hippocampus itself. We know that the hippocampus is crucial in remembering places in the local environment. The place of the animal or person in an internal map of the environment has to be continuously anticipated and updated. Changes in position can be anticipated from information about locomotion and head direction (see chapter 28). Shifts in head direction are combined with information representing the animal's sense of direction in its local environment, information that for many animals comes from the visual sense. The result is the direction the animal is facing with respect to its sense of direction—gathered, in most animals, from visual landmarks.

Additional pathways have been added to the circuit of Papez in figure 26.7b. Note that the fornix fibers from the hippocampus project not only to the mammillary nuclei but also directly to cell groups of the anterior-dorsal thalamus, named the anterior nuclei and the laterodorsal nucleus (LD). Head-direction cells are recorded in all of these cell groups as well as in the mammillary nuclei. It is postulated that the descending inputs from the hippocampus convey sense of the direction in the environment faced by the animal, while the information from the hindbrain, or from the visual system in case of the LD, alters this by signaling changes in direction.

Place Memories in the Neuronal Pathways of Feeling and Emotion

Our understanding of the functions of these pathways is still in flux, under investigation by scientists in many laboratories—and in need of ideas to be tested in new experimental work. At present, we can only speculate about the role of the hypothalamus in the updating of the internal map, beyond serving as a conduit for head-direction information from the hindbrain and midbrain.

We can speculate about what the hippocampus is sending to the various parts of the hypothalamus (other than the mammillary nuclei). It may alter the level of various drives according to information about location of the individual in the current or anticipated frame of reference. Memory for places is associated with the relative values of those places from past experiences. Furthermore, each change in direction anticipates the places that can be reached by moving in that direction, and the values of those places influences the decision of whether to proceed in that direction.

Studies have shown that in the ventral part of the hippocampus (the part nearest the amygdala, the anterior part in large primates) the places represented are larger in size—more like regions than small locations—and the neurons there are much more interconnected with limbic system structures underlying affect and motivational states than in the dorsal part (the posterodorsal part in large primates).

Review of Structures in the Limbic System

At this point, you have been introduced to the limbic midbrain areas and the hypothalamus and to portions of the limbic endbrain—the olfactory system and the circuit of Papez structures. You can begin to familiarize yourself with the locations of these structures, and to some additional parts of the limbic system, in frontal sections of the mammalian brain.

For this, we will not use the mature human brain because of the distortions caused by the very great enlargement of the neocortex. Instead, we will use simple sketches based on the brains of small rodents, which are similar to human brains at early stages of embryonic development (figure 26.8). The figure includes a top view of a brain as it would appear earlier in embryogenesis for orientation to the levels of the frontal sections. See if you can relate these illustrations to the sketch of a medial view of the right hemisphere shown in figure 26.6.

We find a useful and considerably more schematic view of limbic system structures and their relationship to sensory and motor areas of the mammalian endbrain in a review paper by the neuroanatomist and neurologist Marek-Marcel Mesulam. His figure (figure 26.9) indicates only transcortical pathways of the neocortex and how they reach the adjacent *paralimbic* cortical areas, which connect with limbic system structures; these connect to hypothalamus. Note that the major connections from the neocortex to the limbic system are from the so-called heteromodal association areas of neocortex. These are the areas shown in white in figures 25.1 and 33.16. The transcortical connections schematized in Mesulam's diagram have increased in quantity and importance in evolution of mammals with larger brains.

A Variety of Ways the Hypothalamus Sends Its Influences to the Neocortex

The diagram of figure 26.9 indicates that the connections between hypothalamus and other limbic system structures are two-way connections, as are the connections between limbic endbrain structures and neocortex. Such connections are only one of the ways that this core structure of the limbic system can influence mental states by influencing the neocortex. We have already mentioned other connections that do this; for example, the widespread projections of certain lateral hypothalamic neurons that contain histamine or hypocretin/orexin (chapter 17).

The hypothalamus has a more specific projection to the medial part of the mediodorsal nucleus of the thalamus (MD), a structure that also receives input from olfactory cortex. Note the location of MD in the third frontal section shown in figure 26.8. This nucleus projects to the orbital prefrontal neocortex, part of the anterior association cortex in mammals.

Hypothalamic neurons also have a gating influence on information processed by other thalamic structures via connections to midline and intralaminar thalamic structures, which give rise to some intrathalamic axons of a fairly diffuse sort. These connections have not been highly studied but may have played important roles in thalamic evolution.

The thalamic cells that receive the mammillothalamic tract, the neurons of the anterior nuclei, project directly to the cingulate cortex, positioned in between the neocortex and limbic system structures as we have seen. As shown in the updated circuit of Papez diagram of figure 26.7, the cingulate not only receives synaptic connections of axons from association cortex but also sends axons back to the association areas. Thus, here is one

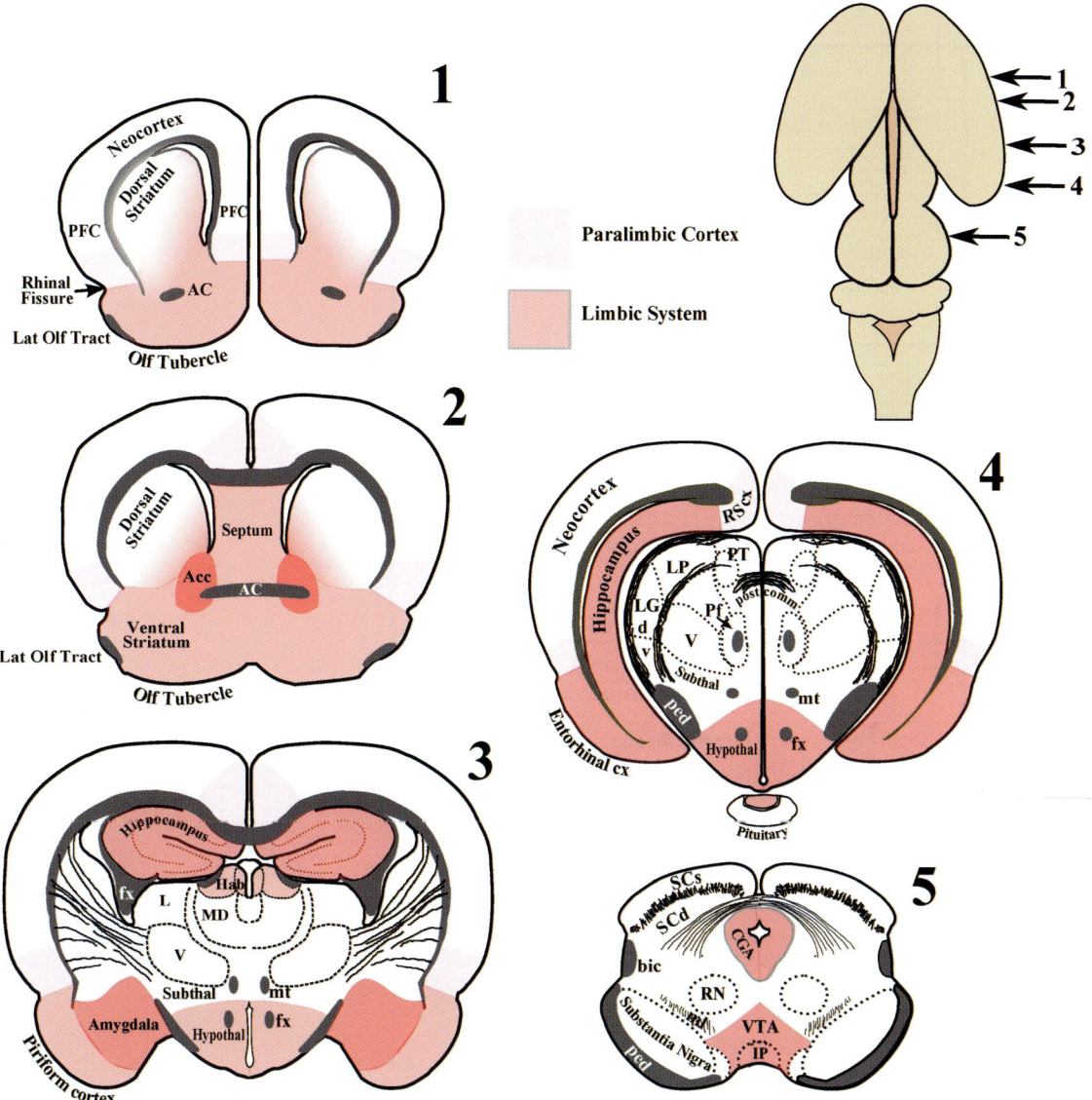

Figure 26.8
Structures of the limbic system of the endbrain, 'tweenbrain, and midbrain are illustrated in drawings of five frontal sections through the neuraxis at the levels indicated in the top-view drawing. Note that if a hemisphere is removed by freeing it from the brainstem and by separating it from the other hemisphere by a midline section, the limbic structures could be seen in a view from the medial or ventral side. Acc, nucleus accumbens; AC, anterior commissure; bic, brachium of inferior colliculus; CGA, central gray area; cx, cortex; fx, fornix; Hab, habenula; Hypothal, hypothalamus; IP, interpeduncular nucleus; L, lateral nucleus of thalamus; LGd,v, dorsal and ventral nuclei of the lateral geniculate body; Lat Olf, lateral olfactory; LP, lateral posterior nucleus of thalamus; MD, mediodorsal nucleus of thalamus; ml, medial lemniscus; mt, mammillothalamic tract; Olf, olfactory; ped, cerebral peduncle; Pf, parafascicular nucleus; PFC, prefrontal cortex; PT, pretectal area; RN, red nucleus; RS cx, retrosplenial cortex; SCs, superficial layers of superior colliculus; SCd, deep layers of superior colliculus; Subthal, subthalamus; V, ventral nucleus of thalamus; VTA, ventral tegmental area.

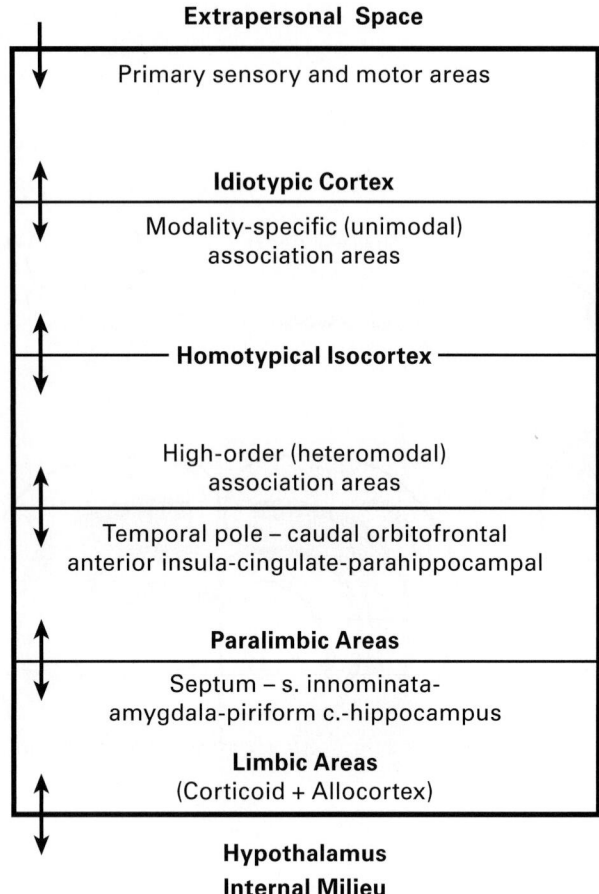

Figure 26.9
Transcortical pathways from the specialized neocortical areas (sensory and motor) through association cortical areas to the limbic system. Such transcortical connections increased in quantity and importance in larger mammalian brains. c, cortex; s, substantia. Based on Mesulam (2000).

Mental State and the Hypothalamus

Profound influences of the hypothalamus on mental state are exerted by its controlling influences on motivational states such as hunger, aggression, or the urge to sleep. Effects of disturbances of the hypothalamus during neurosurgical procedures when the patient is conscious are also very telling. For example, irritation of tissue there has caused uncontrolled crying. A person starting to experience this symptom may not identify with the crying by feeling sad—he may feel he is an observer—or the person may identify with it and experience a shift in mood to one of sadness.

Such phenomena can lead to questions about what underlies mental state or questions about what is voluntary and what is not. The very name *autonomic* implies involuntary, and we think of the hypothalamus as the major controller of the autonomic nervous system. However, these terms can be misleading in light of great individual differences in degrees of self-control of autonomic activities like heart rate, vigilance, and even body temperature. Certain yogis have demonstrated considerable control of such processes and have been able to slow their metabolic rate to extreme degrees and thereby have been able to survive long periods of greatly reduced oxygen intake.

Influence of the Hypothalamus on the Brainstem and Spinal Cord

We have been considering the influences of the hypothalamus on higher structures and have also discussed its connections with the midbrain limbic areas. It also has many influences on hindbrain and spinal cord, mostly the autonomic nervous system elements at those levels. Direct connections from hypothalamus to spinal cord have been discovered (e.g., a pathway controlling urination). There are also direct connections from the dorsal horn to hypothalamus (carrying somatosensory information including responses to pain). However, as discussed by Nauta and Feirtag, there is probably a relatively greater importance of polysynaptic, shorter axon connections in this system. Autonomic nervous system functions are so critical for survival that controls of these functions have evolved at every level of the CNS, with each higher level adding some increase in precision or efficiency.

Support for this claim comes from the effects of surgical disconnections of the hypothalamus from lower structures. If a complete disconnection is suffered by an animal all at once, in a single-stage lesion, the disability is so devastating that it is very difficult to keep the animal alive. However, if the lesion is made in small steps, the result is surprisingly different. This was accomplished by Rudolf Thauer in his experiments with rabbits. He implanted a wire loop that went around the upper end of the midbrain and could be pulled upward bit by bit, allowing recovery from the additional lesion after every small pull. The rabbit seemed to recover very well each time, and Thauer was able to complete the transection of the critical pathways descending from the hypothalamus with the rabbit remaining quite healthy. The midbrain structures were able to function well as the highest level of control of autonomic functions.

Thus, Nauta's claim about the relatively greater importance of short-connection pathways was supported by the multistage lesion findings.

Review

This is a good time to check your memory for several major points concerning the limbic system. You should be able to contrast the pathways for hypothalamic control of the two divisions of the neurohypophysis (pituitary) (see figure 25.2). Review the origins of the two major pathways from the limbic and nonlimbic portions of the mammalian endbrain: the medial and lateral forebrain bundles (see figures 12.4 and 12.5). Check again the limbic midbrain areas as first defined by Nauta and their relation to the diencephalon (see figure 11.4).

Functional Specificity in the Limbic Midbrain

The ability of the midbrain to control autonomic and motivational functions has indicated that the limbic midbrain areas must contain considerable specificity, but studies directed at finding evidence for this have been more limited than those on the hypothalamus and limbic endbrain. There are exceptions, however. For example, experimental studies of rats have led to reports of functional specificity in the midbrain central gray area. We know that the ventral tegmental area is very important in positive reward states and that much of the central gray is involved in negative reward and in pain states. The central gray area has been found to have a lateral region that when stimulated can elicit different behavioral states associated with sympathetic nervous system arousal: defensive behavior patterns and increased tension with rapid heart beat. Stimulation there can also result in analgesia via a descending connection that inhibits pain inputs in the spinal cord. The connection goes to nucleus raphé magnus in the hindbrain (containing serotonin). A ventrolateral column through the central gray is associated with parasympathetic functions: quieting down, hyporeactivity and reduced tension, slower heart rate, and also release of endorphins, which are like opiates in causing analgesia. Investigators have in addition defined a column associated with sexual behavior.

We stand in the wake of this chattering and grow airy.

How can anyone say what happens, even if each of us

Dips a pen a hundred million times into ink?

—from *The Steambath*, by Rumi (1207–1273), translated by Coleman Barks

Readings

Brodal (2004): listed at the end of chapter 25.

Giesler, G. J., Jr., Katter, J. T., & Dado, R. J. (1994). Direct spinal pathways to the limbic system for nociceptive information. *Trends in Neuroscience, 17,* 244–250.

Johnson, A. K., & Gross, P. M. (1993). Sensory circumventricular organs and brain homeostatic pathways. *FASEB Journal, 7,* 678–686.

Mesulam, M.-M., ed. (2000). *Principles of behavioral neurology*. Philadelphia: F. A. Davis. (See Mesulam's chapter (chapter 1), "Behavioral neuroanatomy: Large-scale networks, association cortex, frontal syndromes, the limbic system, and hemispheric specializations," pp. 1–120.)

Swanson, L. (2011). *Brain architecture: Understanding the basic plan*. New York: Oxford University Press.

The experiments of Rudolf Thauer are described by W. J. H. Nauta in his book with Feirtag (1986), on pp. 114–115. Thauer's report with a co-author is written in German: Thauer, R., & Peters, G. (1938). Wärmeregulation nach operative Ausschaltung des "Wärmezentrums." *Pflügers Archiv für die Gesamte Physiologie 239,* 483–514.

The roles of various hypothalamic nuclei in the integration of the various daily rhythms timed by the biological clock cells in the suprachiasmatic nucleus have become better understood since the discoveries summarized in earlier chapters:

Saper, C. B., Lu, J., Chou, T. C., & Gooley, J. (2005). The hypothalamic integrator for circadian rhythms. *Trends in Neurosciences, 28,* 152–157.

Readings on the hippocampus and pathways connected with it can be found at the end of chapter 28.

Axons with connections resembling those of the medial forebrain bundle course through the stria medullaris over the dorsal surface of the diencephalon. Some terminate in thalamic nuclei, and many end in the habenular nuclei of the epithalamus. An interesting paper on this system of axons is the following:

Sutherland, R. L. (1982). The dorsal diencephalic conduction system: A review of the anatomy and functions of the habenular complex. *Neuroscience & Biobehavioral Reviews, 6,* 1–13.

27 Hormones and the Shaping of Brain Structures

There are various behavioral differences—some obvious, some more subtle—between male and female animals. Underlying these differences are differences in the brain. There are structural differences resulting, at least in part, from effects of hormones during development. Sexual hormones play a different kind of role as well, modulating the functioning of neurons that have receptors for those hormones. In many studies of sex differences in the brain, investigators have, not surprisingly, focused on the hypothalamus and on the roles of hormones on brain development. In this chapter, we will take note of sexual differentiation in the hypothalamus, particularly in humans, and possible functional correlates of this differentiation. We will also introduce you to the fascinating story of bird song and the brain, a story that includes seasonal changes in both singing and brain structures.

Experiments with brain development in animals have found specific sex differences that are a result of the actions of circulating hormones like testosterone and estrogen. Effects of these hormones have also been observed in tissue culture. The culture experiments indicate that it is not only hypothalamic tissues that are influenced by sexual steroids: Other tissues are also influenced, including neocortex. From the viewpoint of evolution, this should not be surprising, as the entire organism can be considered to be sexual in the sense that it has evolved to pass genetic material to the next generation.

Hormones other than the gonadal hormones can influence brain development. It is well known that thyroid hormone abnormalities result in abnormal brain development. There are substantial changes in thyroid function during a pregnancy; chorionic gonadotropin stimulates not only the gonads but also the thyroid gland, which results in increased demand for iodine.

It is relevant to note here that certain pollutants can cause brain abnormalities because of their structural similarities to thyroid hormones: Polychorinated biphenyls (PCBs) and dioxin can bind to thyroid hormone receptors and disrupt normal hormone actions.

Table 27.1
Sex differences in human neurological and psychiatric diseases

Disease	% Female:Male
Anorexia nervosa	93:7
Bulimia	75:25
Schizophrenia following Dutch hunger winter	72:28
Anxiety disorder	67:33
Depression	63:37
Multiple sclerosis	58:42
Severe mental retardation	38:62
Autism	29:71
Stuttering	29:71
Schizophrenia	27:73
Dyslexia	23:77
Sleep apnea	18:82
Tourette syndrome	10:90

Statistics from seven published studies on sex differences in humans with the diseases listed at the left. Diseases are arranged in order of ratios of female to male. The dashed line separates the diseases in which female percentages are higher than male percentages (above the line) from diseases in which the percentages of males are higher than those of females (below the line.) Adapted from Swaab and Hofman (1995).

Sex Differences in the Human CNS: Evidence from Pathologies

Dramatic sex differences have been shown in studies of human neurological and psychiatric diseases (table 27.1). Why might this be? Because behavior depends on the brain and spinal cord, we have to conclude that there must be differences in the central nervous system of men and women, whatever the origins of those differences. The differences must be other than simply in size. (Male brains, in humans and a number of other species, are significantly larger in proportion to body weight.) Sex differences in the central nervous system could result from direct effects of genetic differences, from developmental effects of hormones, and also from modulatory influences of learning.

What Determines an Individual's Sexual Orientation?

It may seem very likely that an individual's sexual orientation would be determined by the sex hormones mentioned above—the same hormones that cause so many changes in body and brain at puberty. However, if this were true, then why do some individuals develop a sexual preference for the same rather than for the opposite sex? There are other possibilities to be considered—genetic and environmental influences. Many studies have been carried out, with attempts to distinguish the contributions of genes, hormones, and environment (including parental influences and the family environment).

Although there have been claims for social influences, these have generally not been supported, as most studies have resulted in strong support only for genetic and hormonal influences.

When one thinks about such issues, there are two questions that soon arise: When in development are gonadal hormone differences between males and females at a peak? When do anatomical differences in the hypothalamus appear?

Hormone Peaks and Brain Differentiation

Studies of gonadal hormone levels have revealed three distinct periods in human development when sex differences are greatest. The first period is during the first half of gestation—the period when the genitalia become differentiated. The second period is around the time of birth, and the third—as expected—is at puberty.

Brain differences have been more difficult to specify. The first type of evidence for brain differences has come from behavioral studies. Clear statistical differences in behavior of males and females, of various mammalian species including human, have been documented. Experiments have shown these differences to be related to differences in exposure to the male hormone testosterone very early in development, in the prenatal period. Some of the behavioral differences have been seen early in postnatal development.

After the discovery of specific morphological differences in certain cell groups in the hypothalamus of men and women, studies were carried out with quantification of the differences in brains of various ages. Differences in one particular group of cells—the sexually dimorphic nucleus of the preoptic area (of the hypothalamus)—emerged after age 4 years (figure 27.1). The difference was generally maintained throughout life.

It is not always an easy task to find convincing evidence of a structural difference between different groups of human brains using tissues obtained at postmortem. The problem is the variable degrees of deterioration of the tissues, as tissue fixation is carried out at various times after death, and never immediately. Deterioration can cause volume changes in a cell group, as can the nature of the fixative used. The only real solution is to avoid simple volume measures for the comparisons and to use counts of neurons instead. This is not an easy task, but it can be done by reliable techniques for estimation.

After investigations had established sex differences in hypothalamic cell groups, another question was raised about these differences. Are the same cell groups different in homosexual individuals when compared with heterosexuals? Studies indicated that the answer was negative for the sexually dimorphic nucleus of the preoptic area. (Note in figure 27.1 that the falloff in cell number found in female brains was not found in the brains of male homosexuals.) However, a different answer was found for certain other structures.

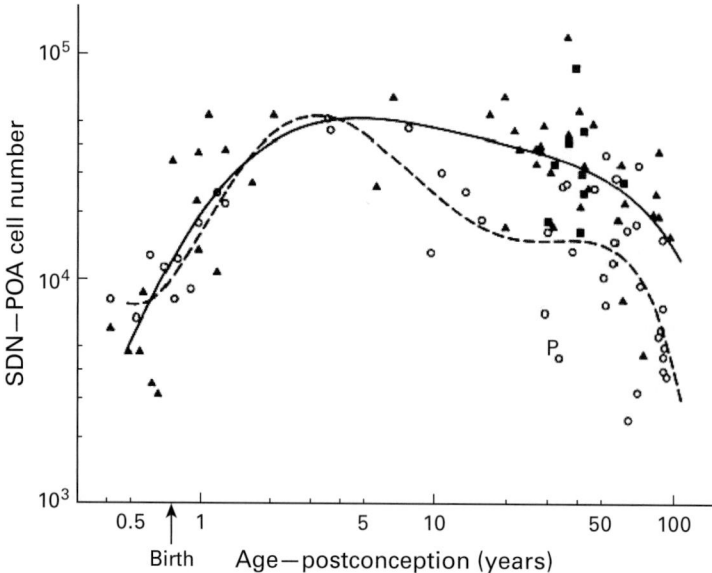

Figure 27.1
Growth of the sexually dimorphic nucleus of the preoptic area of human hypothalamus in males (filled squares and triangles) and females (unfilled circles). Data from homosexual males are shown as squares. Neuron numbers are plotted against age, using log-log scales. SDN-POA, sexually dimorphic nucleus of the preoptic area. Adapted from Swaab and Hofman (1995).

More Sex Differences in the CNS

Another small hypothalamic cell group, not far from the sexually dimorphic nucleus of the preoptic area discussed above, is different in male and female rats. This result led to new examinations of human brain tissues, and it was found that the nucleus was larger in gay men than in non-gay men. The cell group is the vasopressin-containing subnucleus of the suprachiasmatic nucleus (SCN) (figure 27.2). The section shown in the figure is between levels 2 and 3 in figure 26.8a. (You may remember that the suprachiasmatic nucleus is the site of neurons with an endogenous activity cycle of about 24 hours—a biological clock.)

Differences in relative size of specific CNS areas of males and females, or of other groups, have not been easy to establish. Obtaining the brain tissues from sufficiently large numbers of representative brains is much more difficult in the case of humans than for other species. However, advances in our ability to obtain images of the living brain have brought a new potential to this type of study. Imaging studies have confirmed that the corpus callosum, the large mass of axons that interconnect the two hemispheres of the endbrain, is relatively larger in females. In some respects at least, the brains of women

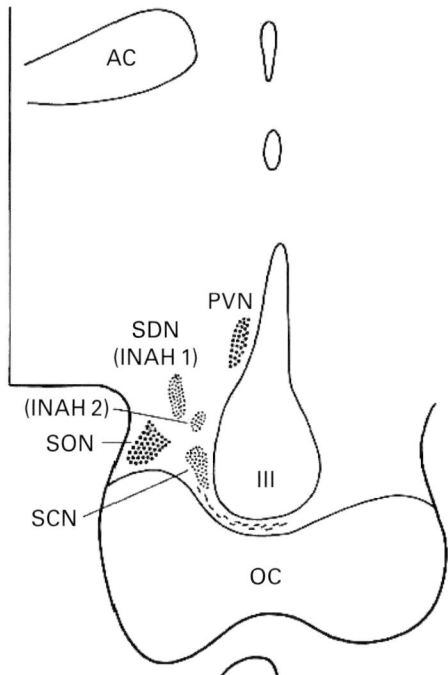

Figure 27.2
Drawing of frontal section through the human hypothalamus showing the locations of cell groups where studies have shown anatomical differences correlated with sex and, in some cases, sexual preferences. For some of these cell groups the data are not strong, but the findings are particularly consistent for two of them: The SDN (sexually dimorphic nucleus of the preoptic area) and one part of the SCN (the vasopressin-containing subnucleus of the suprachiasmatic nucleus). The SDN of gay men has not been found to be different from that of heterosexual men, but the SCN subnucleus has been found to be larger in homosexual men. III, third ventricle; AC, anterior commissure; INAH, interstitial nucleus of the anterior hypothalamus (numbered 1 and 2); OC, optic chiasm; PVN, paraventricular nucleus; SON, supraoptic nucleus. Adapted from Swaab and Hofman (1995).

may function in a more integrated, less specialized way. Other differences have been found as well, although the meaning of these differences is not clear.

Another difference discovered in rats is a particularly dramatic one. It is in the sacral spinal cord, in a group of motor neurons located medially in the ventral horn. This cell group stands out clearly in the male sacral spinal cord in histological sections stained for cell bodies, because of its large and darkly stained neurons, but it is more difficult to find in females. Neuroanatomical and neurophysiological experiments demonstrated that these cells in males are responsible for penile erection. (They are also involved in control of sphincter muscles.) The cell group is called Onuf's nucleus after its discoverer, but it has also been named the nucleus of the bulbocavernosis, after one of the names of the muscle that it innervates, located in males at the base of the penis. In females the muscle is also present and plays similar roles. As with other functions, the larger size of the structure being innervated in males has resulted in a larger size of the corresponding CNS structure.

One other reported finding of a male–female difference in mammalian brains is worth special mention because it has been found to be different in male-to-female transsexual humans. The cell group is one that will be discussed further in chapter 29: the bed nucleus of the stria terminalis, a part of the basal forebrain that receives a strong projection from the amygdala and which projects to the anterior hypothalamic region known to be important in sexual behavior. The central part of this nucleus is much larger in males than in females. It is also larger in male homosexuals but not in male-to-female transsexuals.

Next we go to a related but even more dramatic story. It is about songbirds.

Sexual Dimorphism Underlying Singing in the Canary and Other Songbirds

Discoveries at Rockefeller University in the 1980s have led to fundamental changes in our picture not only of sex differences in brains but also of brain plasticity. The first discovery came out of straightforward examinations of Nissl-stained sections of the canary brain. Fernando Nottebohm and his co-workers were studying the canary and its singing abilities, and they wanted to map the forebrain as a first step in relating the behavioral abilities of this little songbird to its brain structures. They saw specific groups of cells that stood out very clearly but their identity was not immediately clear, and furthermore, they could not be found in every canary brain! The variation from brain to brain was particularly intriguing, and the scientists soon found that the differences depended on the sex of the birds. Subsequently, they made similar findings in the zebra finch.

It is the male canaries and zebra finches that do most of the singing, and the cell groups that stood out so clearly were found only in males. One of them is known as the *higher vocal center* (HVC) of the mesopallium. The HVC receives auditory input from field L of the nidopallium, mentioned in chapter 23 (see figure 23.20). This endbrain structure projects to another sexually dimorphic cell group, the nucleus robustus of the arcopallium (figure 27.3), which projects directly both to the midbrain auditory structures and to a part of the hypoglossal nucleus of the hindbrain—the part that controls the bird's syrinx in the throat. The bird's syrinx controls the sounds of its song like the vocal cords control the sounds of speech in humans.

The pathway underlying birdsong described above is known now as the *direct vocal control pathway* (figure 27.4). Additional vocal control structures, also more pronounced in males, are located more anteriorly in the endbrain of songbirds, functioning as part of the *indirect vocal control pathway* (figure 27.5). This region functions as a kind of bird "Broca's area."

Comparison of the songbird's brain structures to the structures underlying human speech goes further. Just as in humans, there is a marked lateral asymmetry in these

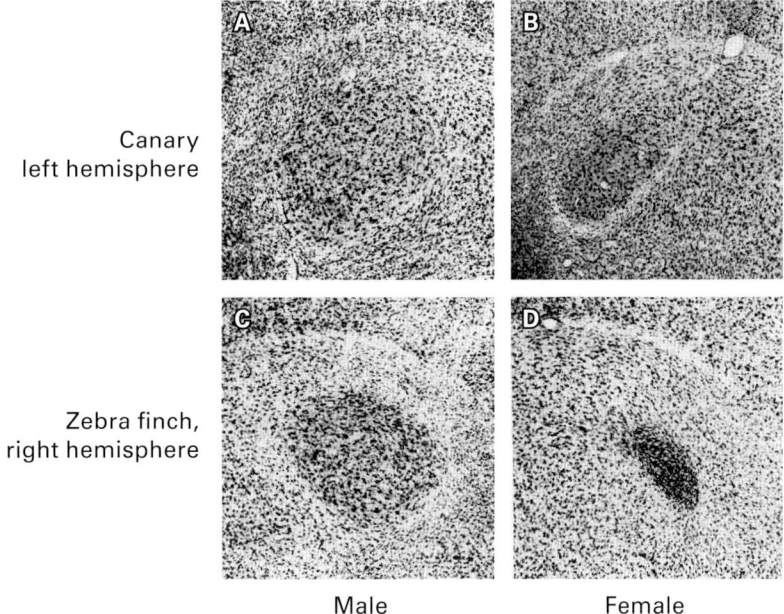

Figure 27.3
Sexual dimorphism in the canary and zebra finch brains is illustrated by photographs of sections through the largest part of the nucleus robustus of the arcopallium (formerly called the archistriatum) of the endbrain. The sections are stained for cell bodies with a Nissl method. Panels (A) and (B) are from the brain of a male and female canary, respectively. Panels (C) and (D) are from a male and female zebra finch, respectively. From Nottebohm and Arnold (1976).

structures in songbirds resembling the commonly found left-hemisphere dominance in brain control of speech in humans.

"A Brain for All Seasons"

A second discovery of the Nottebohm group was even more surprising than the first. The extent of the male–female differences changed from season to season! The volumes of the higher vocal center and of the nucleus robustus gradually increase during the period of *plastic song*, when the birds are adding new syllables to their song repertoire, and remain at maximum volume during the breeding season, the period of *stable song*. Then, these volumes decline rapidly. The pattern repeats over the next seasonal cycle.

Measures of the blood levels of the male hormone testosterone revealed a strong relationship between hormone levels and changes in song. Testosterone levels were lowest during periods of plastic song and highest at the beginning of the breeding season when song became stable. Few new syllables are added to the songs when testosterone levels

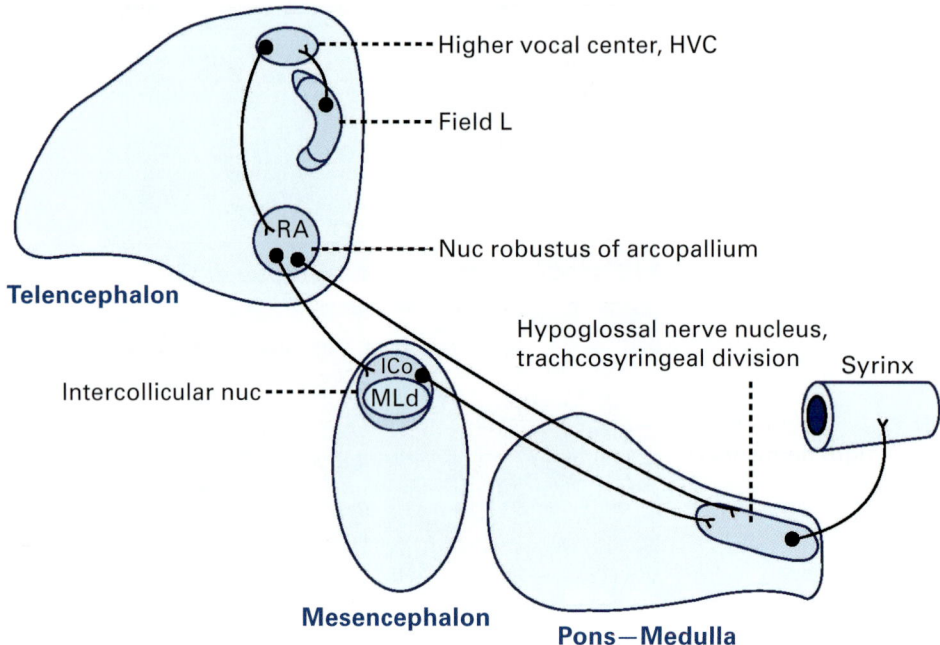

Figure 27.4
Diagrammatic illustration of the direct vocal control pathway in a songbird. Field L of the nidopallium receives auditory input from the bird's thalamus and projects to the higher vocal center (HVC) of the mesopallium. The HVC projects to nucleus robustus of the arcopallium. The robust nucleus projects both directly and via the midbrain to a portion of the hypoglossal nucleus of the hindbrain—the tracheosyringeal division. This nucleus contains motor neurons that innervate the muscles of the syrinx (the bird's vocal apparatus in the throat). ICo, nucleus intercollicularis; MLd, nucleus mesencephalis lateralis pars dorsalis; RA, nucleus robustus of the arcopallium. Based on Butler and Hodos (1996).

are increasing or high. Later research by Sarah Bottjer modified this picture when she provided evidence that it was not only testosterone levels but also the bird's learning experiences that caused the changes.

Next came the biggest surprise. The investigators injected tritiated thymidine, a DNA precursor given a radioactive tag, at various stages in order to label cells undergoing mitosis. Results soon revealed what was causing the seasonal changes. Neurogenesis was occurring on a large scale in the adult canary brains, and the newborn neurons were migrating to final locations throughout the endbrain (figure 27.6).

How Many More Sex Differences in the Brain Will Be Found?

Tissue culture studies have indicated that sex differences in the brain are probably much more widespread than has been found in studies of cell-stained postmortem material. An

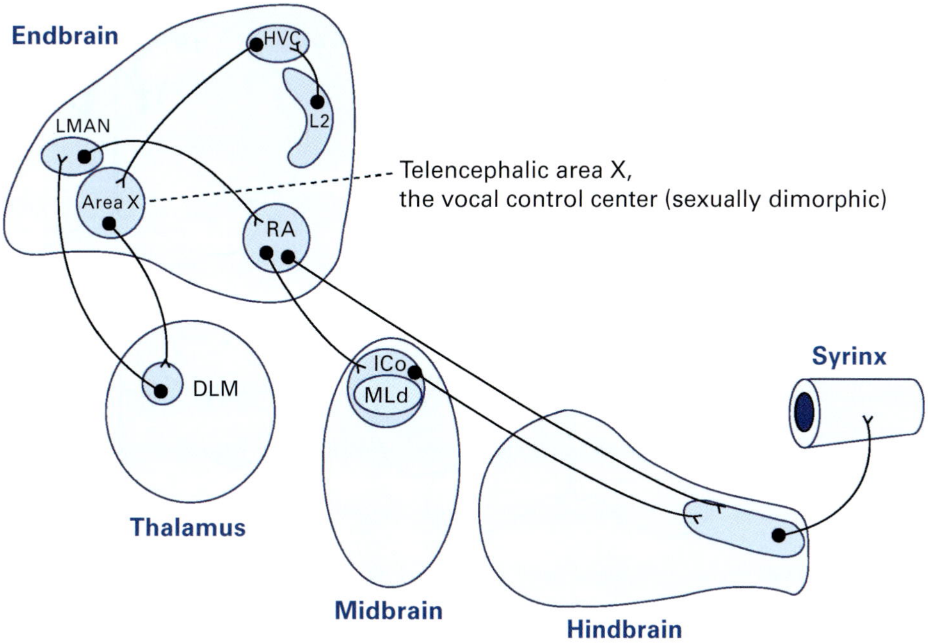

Hormones and the Shaping of Brain Structures

Figure 27.5
Diagrammatic illustration of the indirect vocal control pathway in a songbird. The higher vocal center projects to "area X" of the striatum. Area X is a sexually dimorphic vocal control center that is important in the learning of new song syllables. It projects to a cell group in the thalamus that projects back to the endbrain, to the lateral magnocellular nucleus of the anterior nidopallium (LMAN). LMAN, in turn, projects to nucleus robustus of the arcopallium, which projects to the midbrain and hindbrain structures that control the motor neurons of singing. DLM, medial part of thalamic nucleus dorsolateralis anterior; ICo, nucleus intercollicularis; MLd, nucleus mesencephalis lateralis pars dorsalis; RA, nucleus robustus of the arcopallium. Based on Butler and Hodos (1996).

example comes from in vitro studies by Toran-Allerand. She has shown that sex steroids affect the differentiation of midbrain dopamine-containing neurons. She also found an interaction between estrogens and neurotrophins, indicating that the steroids may cause alterations in growth factor levels.

Since the 1970s, neuroscientists have believed that the gonadal hormones testosterone and estrogen are the causes of sexual differentiation of the brain and of the circuits that organized sexual differences in behavior. The fetal brain has been viewed as essentially female, and then later in development more testosterone results in a masculination of the male brain. More recently, evidence from gene expression studies indicates that brain differences between the sexes may have more complex explanations.

In 2003, Vilain and colleagues at the University of California at Los Angeles compared the expression of more than 12,000 genes in embryonic mice, male and female, before the differentiation of genital organs. To their surprise, these researchers found 54 genes

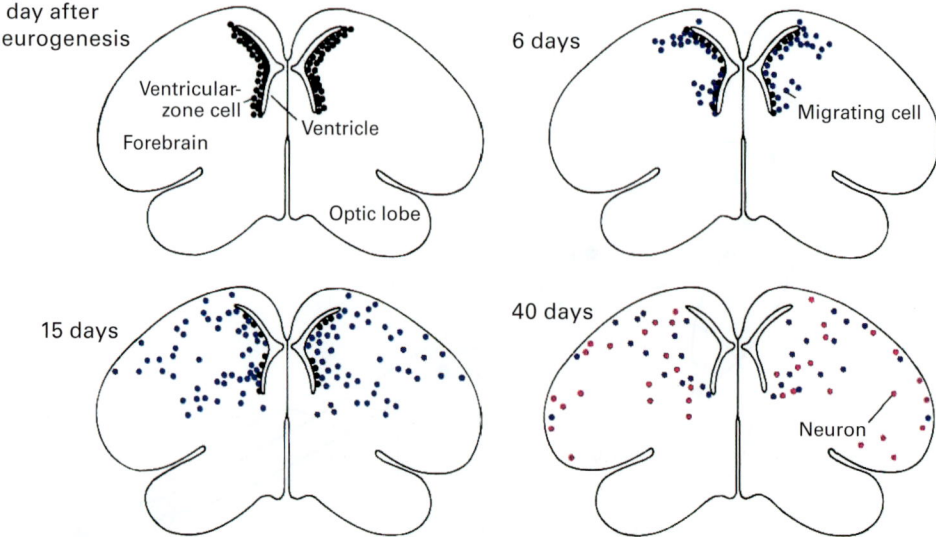

Figure 27.6
Sections of the canary brain taken four times after injection of the birds with tritium-labeled thymidine in order to label cells undergoing mitosis. The label remains in the DNA and is passed on to all daughter cells. At 1 day after being labeled during neurogenesis, the cells are all in the ventricular zone (upper left). After 6 days, some of the labeled cells are migrating away from the ventricular zone. After 15 days, many of the labeled cells have migrated widely over the endbrain. After 40 days, some of the labeled cells have begun to differentiate into neurons (shown in red). From Nottebohm (1989).

that were expressed at different levels in males and females. In males, 18 were expressed more, while 36 were expressed at higher levels in the females.

The gene expression result was a surprise for many neuroscientists because the differences were found so early in fetal development. Was it really before there were effects of hormonal differences? How would these results change if testosterone levels were manipulated? Many questions remain unanswered. We do not even know the proteins of many of the genes involved or the functions of these proteins. It is obvious that an understanding of sexually related brain differences will require much more research. Its importance is great. Remember that the CNS can be considered to be the major means the genes have for reproducing and surviving into the future.

Back to the Anatomy of the Limbic System

Before proceeding to the next chapter, it will be useful for you to return to the previous chapter to review the structures of the limbic system (figure 26.8), the Papez circuit (figures 26.6 and 26.7) including its major inputs and outputs, and also the schematic of transcortical pathways leading to the hypothalamus and the internal milieu, shown in figure 26.9.

They try to say what you are, spiritual or sexual?

They wonder about Solomon and all his wives.

In the body of the world, they say, there is a soul

and you are that.

But we have ways within each other

that will never be said by anyone.

—by Rumi (1207–1273), translated by Coleman Barks

Readings

Arnold, A. P. (1992). Hormonally-induced alterations in synaptic organization in the adult nervous system. *Experimental Gerontology, 27,* 99–110.

Chen, Z., Ye, R., & Goldman, S. A. (2013). Testosterone modulation of angiogenesis and neurogenesis in the adult songbird brain. *Neuroscience, 239,* 139–148.

Nottebohm, F., & Arnold, A. P. (1976). Sexual dimorphism in vocal control areas of the songbird brain. *Science, 194,* 211–213.

Sengelaub, D. R., & Forger, N. G. (2008). The spinal nucleus of the bulbocavernosus: Firsts in androgen-dependent neural sex differences. *Hormones and Behavior, 53,* 596–612.

Zhou, J.-N., Hofman, M. A., Gooren, L. J., & Swaab, D. F. (1995). A sex difference in the human brain and its relation to transsexuality. *Nature, 378,* 68–70.

There are websites with interesting information about thyroid hormones and brain development; for example, the Colorado State University website: http://arbl.cvmbs.colostate.edu/hbooks/pathphys/endocrine/thyroid/thyroid_preg.html.

On the topic of thyroid hormones and brain development, see the following reviews:

Darras, V. M., Van Herck, S. L. J., Geysens, S., & Reyns, G. E. (2009). Involvement of thyroid hormones in chicken embryonic brain development. *General and Comparative Endocrinology, 163,* 58–62.

Horn, S., & Heuer, H. (2010). Thyroid hormone action during brain development: More questions than answers. *Molecular and Cellular Endocrinology, 315,* 19–26.

Porterfield, S. P. (1994). Vulnerability of the developing brain to thyroid abnormalities: Environmental insults to the thyroid system. *Environmental Health Perspectives, 102* (Suppl 2), 125–130.

Eric Montie has authored and co-authored a series of papers relevant to the topic of environmental pollutants and development. His work stems from his PhD thesis (written at the Woods Hole Oceanographic Institute) on marine mammals, their brain development, and effects of environmental pollutants in the ocean, especially polychlorinated biphenyls. The work requires knowledge of the normal brains, as described in the following papers, showing how much work can be done on intact animals:

> Montie, E. W., Schneider, G. E., Ketten, D. R., Marino, L., Touhey, K. E., & Hahn, M. E. (2007). Neuroanatomy of the subadult and fetal brain of the Atlantic white-sided dolphin (*Lagenorhynchus acutus*) from in situ magnetic resonance images. *Anatomical Record, 290,* 1459–1479.

> Montie, E. W., Schneider G. E., Ketten, D. R., Marino, L., Touhey, K. E., & Hahn, M. E. (2008). Volumetric neuroimaging of the Atlantic white-sided dolphin (*Lagenorhynchus acutus*) brain from in situ magnetic resonance images. *Anatomical Record, 291,* 263–282.

> Montie, E. W., Pussini, N., Schneider, G. E., Battey, T. W. K., Dennison, S., Barakos, J., & Gulland, F. (2009). Neuroanatomy and volumes of brain structures of a live California sea lion (*Zalophus californianus*) from magnetic resonance images. *Anatomical Record, 292,* 1523–1547.

Many studies have been carried out on the determination of sexual preferences. Most studies have resulted in strong support only for genetic and hormonal influences.

> Swaab, D. F., & Hofman, M. A. (1995). Sexual differentiation of the human hypothalamus in relation to gender and sexual orientation. *Trends in Neuroscience, 18,* 264–270.

> Swaab, D. F. (2007). Sexual differentiation of the brain and behavior. *Best Practice & Research Clinical Endocrinology & Metabolism, 21,* 431–444.

Magnetic resonance imaging studies of human brains have found sex differences: see reports by Judith Rapaport, of the NIMH, and her co-workers.

See papers by Eric Vilain and his colleagues for studies of gene expression and the determination of sexuality and sexual orientation.

28 The Medial Pallium Becomes the Hippocampus

When we use the term *limbic telencephalon*, we mean the endbrain structures strongly connected to the hypothalamus. The limbic telencephalon includes the medial pallium. *Pallium*, you may recall, is a word that means a cloak or mantle. It includes all the nonstriatal surface structures of the endbrain—in mammals, the neocortex and the other types of cortex (allocortex). The medial pallium is an allocortical portion of the pallium; it is part of the limbic telencephalon. It includes the hippocampus. In this chapter, the hippocampal formation is our focus.

In any discussion about the hippocampal formation, there are related functional topics that must be included: hippocampal *place cells*, synaptic changes in the hippocampus, spatial learning, and information storage. Also, concerning memory processes, we will mention a special role of connections using the neurotransmitter acetylcholine.

Evidence of an Internal Map: Place Cells

The importance of the hippocampus for remembering places has been indicated by behavioral studies of brain lesion effects. Rats with destruction of the hippocampus, for example, cannot even remember how to find the goal—the location of food—in a simple maze, although they had previously found it repeatedly. With an intact hippocampus, rats can learn such spatial tasks very readily. This kind of finding led to the discovery of the *place cells* of the hippocampus.

A place cell is a neuron that increases its activity only when the animal is in one particular part of a familiar space. This is well illustrated in figure 28.1. Each small square in the figure shows a map of a small room where a rat is walking around with multiple electrodes that are recording action potentials of neurons in its hippocampal formation. One square depicts results for one neuron. When the firing of a neuron is high, the rat's location is shown in red; blue areas represent low activity of the same neuron when the rat was in those places. Note that some of the cells are very specific in their activity—they are active only when the rat is in one small region. Others are less specific, and some are not responsive in this situation.

514   Chapter 28

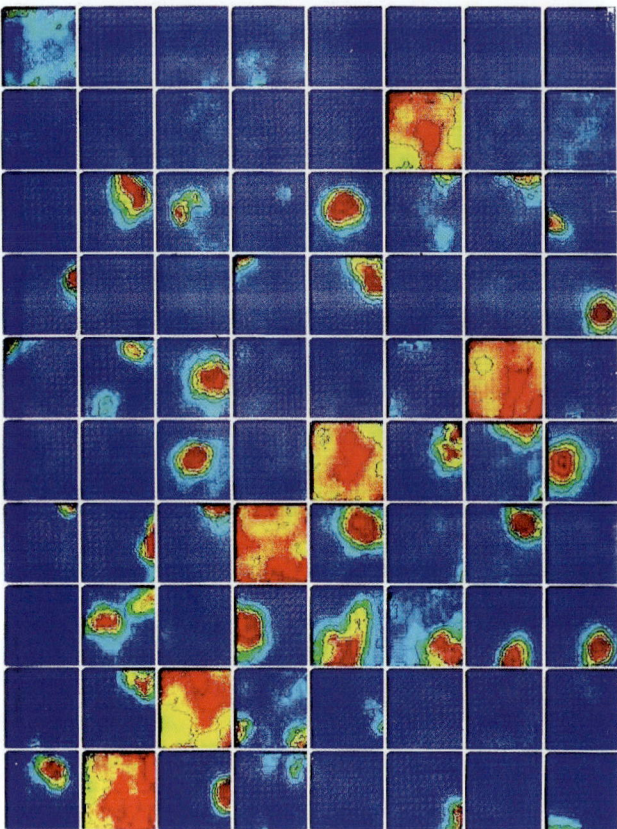

Figure 28.1
Display of neuronal activity in hippocampal *place cells*. Responses of 80 simultaneously monitored hippocampal cells were recorded in a rat during exploration of a rectangular environment. Each square depicts that environment and the activity of a single cell as the rat moves through it: Red means high activity and blue means low activity at that location. Lighter reds and yellow indicate decreased activity; lighter blue means slightly higher activity than darker blue. Note that some cells respond only when the animal is in a particular region of space, others respond over the entire environment, and many are nearly silent throughout. From Wilson and McNaughton (1993).

Whenever the rat is placed in a different environment, the pattern is disrupted. However, after a period of familiarization, the cellular activity changes, and place cells appear again. Many of the cells in the hippocampus are not place cells, but it is clear from the electrophysiological results that spatial representations are an important aspect of hippocampal function—in agreement with effects of lesions of this tissue in rats.

Head-Direction Cells and the Circuit of Papez

Other cells in the hippocampus play a different role in spatial representations: Their activity encodes the allocentric direction of the head (figure 28.2). This direction is sensed by the visual system as an angle from a direction based on local landmarks. Changes in head direction are detected in the brainstem, and the changes alter the head-direction information that is fed back to the hippocampal formation.

Head-direction cells are found throughout the circuit of Papez: in the hippocampus and in the mammillary bodies of the caudal hypothalamus, in the anterior nuclei of the thalamus, in the adjacent lateral-dorsal nucleus, and in the cingulate and retrosplenial cortex and areas these cortices connect with. Look back at the schematic drawing of the circuit of Papez in Figure 26.7 and note the return pathway ascending from the mammillary bodies. An important origin of information about changes in head direction is in the hindbrain. In the figure, one can see an input to the circuit of Papez coming from the tegmental nuclei of the caudal midbrain. Presumably, the information comes from the vestibular and related proprioceptive systems in the hindbrain. Another signal of shifts in head direction comes from the pretectal area of the visual system; this information follows a parallel route to the endbrain, as indicated in the figure.

It appears that the hippocampal representations of the location of the animal are within an internal map of the current environment. As the rat is locomoting in the real world, its place in the internal map—the map of the current "frame of reference"—is constantly being updated, as discussed briefly in chapter 26. The role of head-direction cells can be understood most easily as anticipative—they signal a prediction of location of the body, or possible location, in the near future. To accomplish this, information on head direction and locomotion is crucial. The knowledge of head direction, and thus a potential direction of future locomotion, can be used to anticipate changes in location.

The importance of this ability to anticipate is not difficult to imagine. A decision to move in the momentary direction of the head, as the head is moving during locomotion or in an animal about to initiate locomotion, has consequences. What places will be encountered in that direction? What has happened there in the past, and were the consequences good, bad, or neutral? Thus, movements of the individual can be guided by past experiences—guided through the environmental maze.

But how could head direction affect the motivation of the animal to move in one direction and not in others? The head-direction signal may activate hippocampal place cells

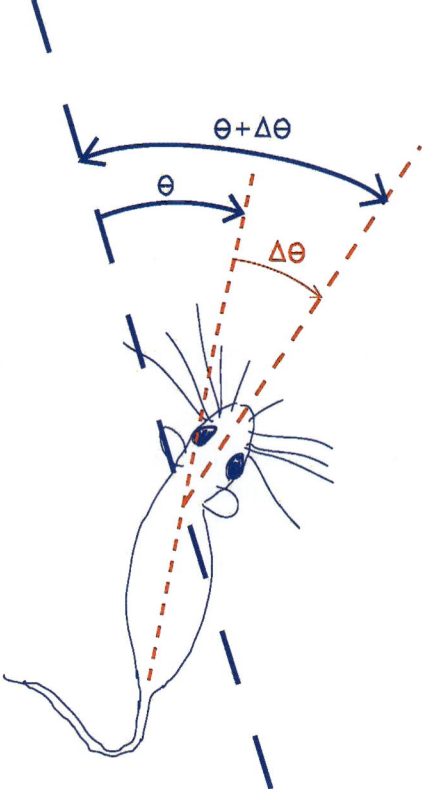

Figure 28.2
The activity of head-direction (HD) cells encodes the specific angle of the head from one's sense of direction. Thus, HD cells are very different from place cells. In this figure, the rat usually knows the direction of its body axis with respect to the local scene, especially from visual landmarks detected by its visual system. This direction is represented by the angle θ in the figure. Any turn of the head, represented by $\Delta\theta$, corresponds to a potential change in direction. Changes in $\Delta\theta$ are detected by the vestibular system and by neck proprioceptors, and also by the visual system. The brain is always computing $\theta + \Delta\theta$, represented in blue at the top of the figure. HD cells fire at specific angles ($\theta + \Delta\theta$) that correspond to allocentric directions faced by the animal.

representing the regions ahead together with remembered values of being in those places. The place cells in the rat represent much smaller areas in the dorsal hippocampus than in the ventral parts. In the ventral part, closer to the amygdala, the place cells in the rat have been found to represent areas more than ten times as large. If we assume that the rat can anticipate regions ahead, at least the larger regions, in the direction of its locomotion, the place cell activity representing those areas may become more active, and this activity could arouse prior learned associations with those regions. The cells representing the larger areas are in the ventral hippocampus, and this temporal end of the structure is interconnected with the amygdala, olfactory cortex, and orbital and ventral prefrontal cortex; it also projects to anterior parts of the hypothalamus rather than to the mammillary bodies. These connections, from the parts of the hippocampus representing larger portions of the animal's territory, could change motivational states in an anticipatory manner. (Details of the mechanisms have not been analyzed. Those details could vary among different species.)

Remember that many axons of the fornix bundle, from hippocampus to the 'tween-brain, appear to be conveying an allocentric sense of direction to cells of the mammillary bodies and to anterior and laterodorsal thalamic nuclei. These axons come predominantly from the dorsal parts of the hippocampus (caudodorsal in humans and other large primates) where the places represented by place cells are much smaller than in the ventral hippocampus. Very early in the evolution of the hippocampus, there was probably only one frame of reference for the internal map, but later animals evolved the ability to remember various localities. There is evidence that these memories in mammals are stored in cells and circuits of association areas of neocortex, in particular the parietal neocortex. This cortex projects via parahippocampal cortical areas to the hippocampus. (See figure 22.17.)

In humans (and perhaps in other advanced species), memories within various conceptual spaces may use the same kind of circuitry. Place cells in these species may sometimes be locations in conceptual environments.

Thoughts about the Origins of the Medial Pallium

This brings us back to an earlier discussion of the evolutionary origins of the endbrain and the medial pallium. The evolution of an olfactory sense at the forward end of very early chordates led to an early expansion of the forebrain as endbrain structures evolved—the structures underlying olfaction. There were two major links, it is postulated, between olfactory sensory structures and the motor systems of the midbrain. One was through what became the ventral endbrain, which became corpus striatum and basal forebrain. The other link was through the dorsally located pallium, which at early stages was comparable to what became the medial pallium—evolving into the hippocampal formation and nearby cortical areas.

In chapter 24, it was suggested that, in the evolutionary history of the chordate CNS, the plasticity of these two kinds of links had strong adaptive advantages that led to expansions of both striatum and pallium.

The first link What became the basal forebrain and ventral striatum had outputs to the hypothalamus and midbrain and also to the epithalamus and subthalamus. These outputs affected locomotion and orienting movements, as well as motivation and the endocrine system. Plasticity of the links would have allowed habits to form according to rewarding outcomes of actions (positive or negative).

The second link What became the medial pallium, and eventually the hippocampal formation, had outputs to the ventral striatum and hypothalamus (and epithalamus). These links were also plastic, but the memories formed were different: They were associations between place in the environment and the good or bad consequences of being in that place or of approaching or avoiding that place.

The Expansion of the Pallium, with a Focus on the Medial Pallium

Some pictures from comparative neuroanatomy of the telencephalon illustrate the pallial expansion. First, look at figure 28.3, which shows two frontal sections of a shark endbrain. Note the simple hemispheres with a thickened medial pallium. The next illustration, shown in figure 28.4, shows sections of a lungfish endbrain with a pattern of thickening

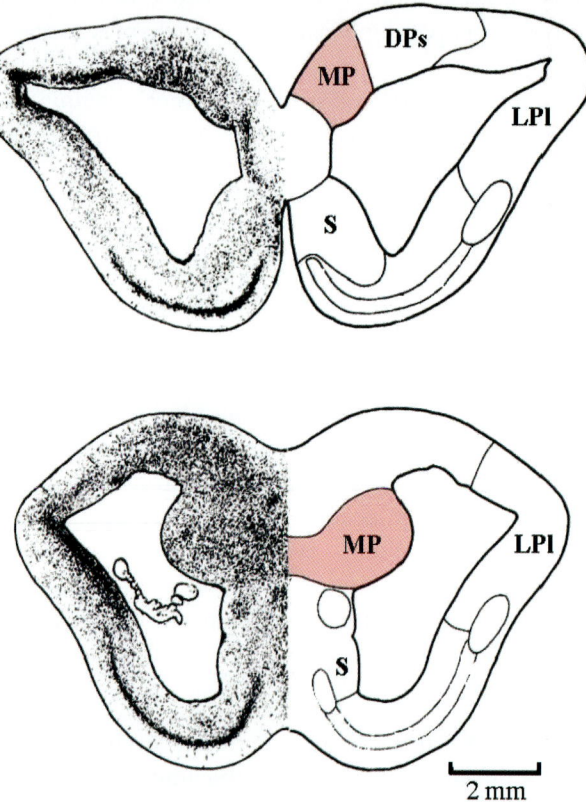

Figure 28.3
Frontal (transverse) sections of a shark telencephalon (the spiny dogfish, *Squalus acanthias*). The left sides are stained for cell bodies; the right sides are mirror-image drawings. The medial pallium, shown in red coloration, is thicker than other pallial regions. DPs, dorsal pallium, pars superficialis; LPl, lateral pallium, pars lateralis; MP, medial pallium; S, septal nucleus. Adapted from Northcutt, Reiner, and Karten (1988).

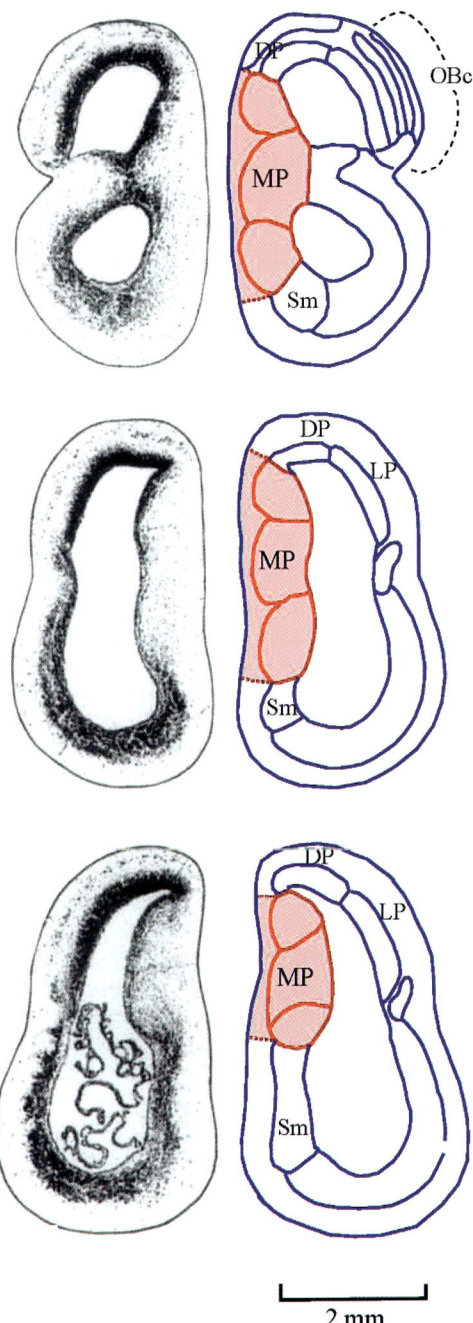

Figure 28.4
The telencephalon of an African lungfish (*Protopterus annectens*) in frontal sections, prepared as in figure 28.3; the most rostral section is at the top. The medial pallium (MP) is shown in red coloration. DP, dorsal pallium; LP, lateral pallium; Obc, caudal end of olfactory bulb; Sm, medial subpallium. Adapted from Reiner and Northcutt (1987).

that is different in some of the details. There is less thickening but more differentiation within the medial pallium and greater thickening ventrally, a region that includes a primitive striatum. The bullfrog endbrain, illustrated in figure 28.5, shows a related pattern of thickening and differentiation, but different in details.

A similar picture of a mammalian endbrain, illustrated by a section from a marsupial in figure 28.6, shows some remarkable differences. Both the medial pallium and the corpus striatum (caudate nucleus and putamen) have expanded, and the dorsal pallium (at least part of it) has evolved into a neocortex. The medial pallium has become infolded with the differentiation of a hippocampus.

Cells and Circuits of Ammon's Horn (*Cornu Ammonis*)

The infolded cortex of the medial pallium in mammals is the so-called archicortex. It looks like very simple cortex in its cellular architecture. There is only one layer of cell bodies; their dendrites form fan-shaped bushes that extend up to the surface. This infolded lip of cortex is found all along the posterior margins of the hemisphere in embryonic mammals and many adult mammals. With the growth of a temporal lobe (see chapter 22), this infolded lip extends along the inner margin of temporal cortex. The entire structure is the hippocampus. It was named Ammon's horn by an early anatomist—in Latin, *cornu Ammonis*—because of its overall shape in the human brain. (The name apparently comes from the horn of the ancient Egyptian god Amon or Amun.) If one adds to this one-cell-layered structure the adjacent non-neocortical areas that are closely connected to it, we call the whole complex the hippocampal formation.

Note in figure 28.6 how the infolded hippocampus is continuous with the cingulate cortex, which is continuous with neocortex. The six-layered neocortex does not change abruptly to hippocampus. Moving from neocortex toward hippocampus, one finds a series of changes as the number of layers becomes reduced from six to only one. This is seen most clearly in horizontally cut sections through the hemisphere of a rat, as shown in figure 28.7. In that picture, the major subregions of the hippocampus and adjacent structures have been labeled. The abbreviation CA stands for cornu Ammonis. (There are four CA regions. The largest are CA1 and CA3. CA2 and CA4 are less studied.)

The location of the hippocampus in a hamster or rat brain is depicted in the drawings of figure 28.8. A cross section made at a right angle to the long axis of the hippocampus is shown in a drawing. Such a section looks very similar when it is made at very different locations along the long axis. A basic circuit that passes through the hippocampus is also repeated at every location (see figure 28.10 later in text).

Further Notes on Hippocampal Anatomy

Figure 28.9 shows a simplified drawing of the right hemisphere of a human brain with the position of the hippocampus and some related structures indicated. Note again the

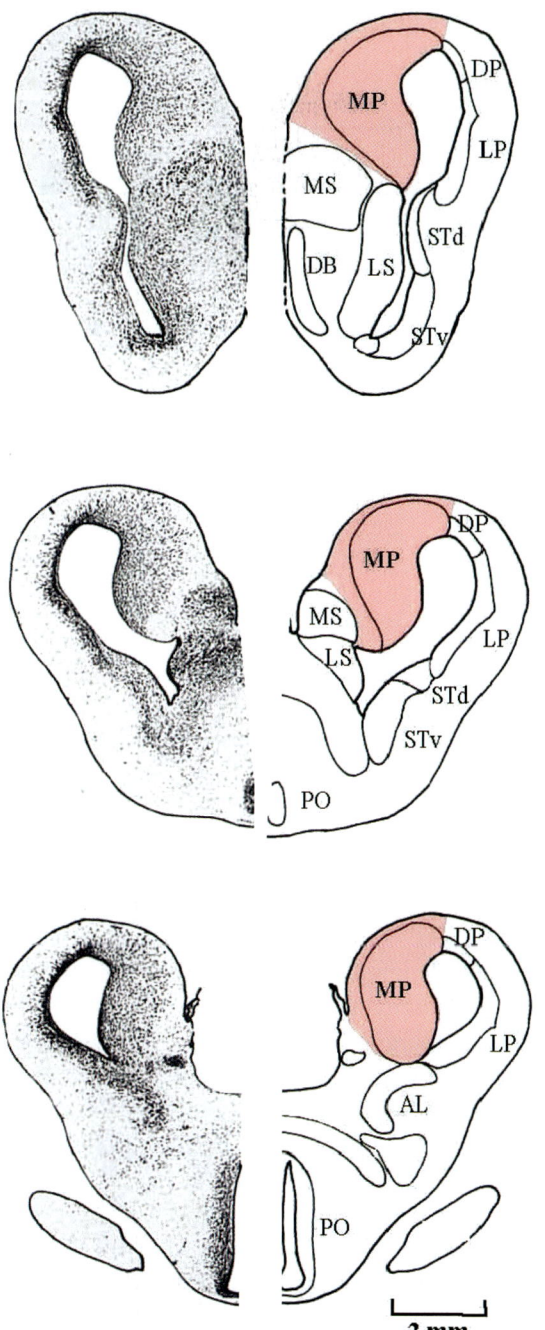

Figure 28.5
The telencephalon of a bullfrog (*Rana catesbeiana*) in frontal sections, prepared as in figures 28.3 and 28.4. The most rostral section is at the top. The medial pallium (MP) is shown in red coloration. AL, lateral amygdala; DB, diagonal band; DP, dorsal pallium; LP, lateral pallium; LS, lateral septum; MS, medial septum; PO, preoptic area; STv, ventral striatum; STd, dorsal striatum .Adapted from Wilczynski and Northcutt (1983).

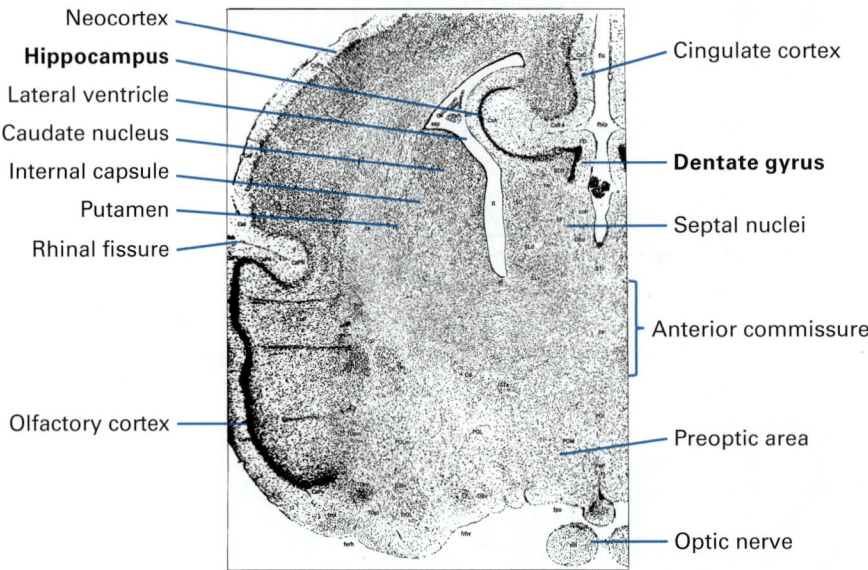

Figure 28.6
Photograph of a Nissl-stained frontal hemisection through the telencephalon of a marsupial, a Virginia opossum. Note the infolded medial pallium—called the hippocampus in mammals; it merges with the septal nuclei anteriorly (at the level illustrated). In marsupials, there is an absence of a corpus callosum, but fibers still connect the two hemispheres; they pass through a greatly enlarged anterior commissure. The white area to the left of, and below, the putamen contains myelinated fibers of an enlarged extreme capsule, made up of fibers to and from the anterior commissure. From Oswaldo-Cruz and Rocha-Miranda (1968).

temporalization, as explained in chapter 22. The apparent position of the hippocampus was greatly affected by the growth of the temporal lobe, although topologically it was not changed. What changed the most in mammals was the relative size of the hippocampus at various positions along the medial margins of the hemisphere. It became enlarged in the portions of the hemisphere nearest the visual cortical regions. More anteriorly, it is a tiny rudiment just above the corpus callosum—the hippocampal rudiment. This rudimentary structure can be followed anteriorly in this position until, rostrally, it joins with cells of the septal area.

Note the limbic cortical regions adjacent to the hippocampus and its rudiment. Anteriorly you see the cingulate cortex. Posterior to the callosum this region is called the retrosplenial cortex—behind the splenium, or posterior limit, of the callosum. Continuing into the temporal lobe in the human brain, we reach the hippocampal gyrus and then the entorhinal cortex. Between each of these and Ammon's horn are the parasubiculum, presubiculum, and subiculum. The subiculum can properly be included in the hippocampus itself, and the other regions are often included in the hippocampal formation.

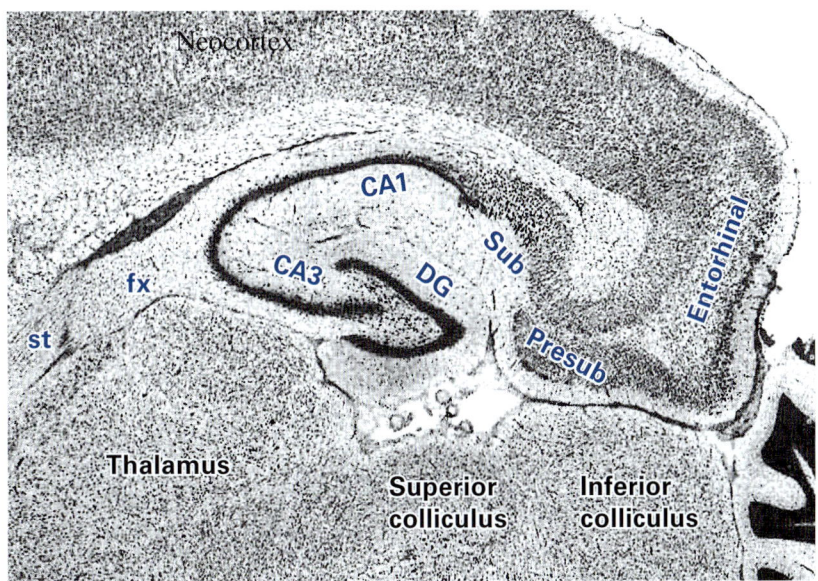

Figure 28.7
Nissl-stained horizontal section through the hippocampal formation on the right-hand side of a rat brain, with parts identified in blue letters. CA1 and CA3, two major subdivisions of the hippocampus or cornu Ammonis; DG, dentate gyrus; Entorhinal, entorhinal area; fx, fornix axons (to and from the hippocampus); Presub, presubiculum; Sub, subiculum; st, stria terminalis axons (from the amygdala). Note also a part of the cerebellum at the far right (the caudal side). From Paxinos and Watson (1982).

Major Connections between the Mammalian Hippocampus and Neocortex

At this point, you should look again at figure 26.7, the diagrammatic sketch of the circuit of Papez. It shows the hippocampal formation with its major inputs and outputs, not only with subcortical structures but also with structures within the hemisphere. A major bidirectional pathway can be followed between association areas of neocortex, to be discussed more in part XII, and the hippocampus. These pathways include synapses in the entorhinal cortex and other paralimbic areas (like the postsubiculum mentioned in chapter 22, equivalent to part of the parahippocampal gyrus of humans) and/or cingulate and retrosplenial areas. Many pathways reach the cells of Ammon's horn via the dentate gyrus.

The dentate gyrus was given its name because of its appearance in human brain dissections. It is found at the extreme margin of the cortical expanse that ends in hippocampus: It is like the hem of a skirt, where the nearest part of the skirt is the CA4 region of the hippocampus, and parts of the skirt more distant from the edge are cortical areas further from the hippocampus. It can be seen in figures 28.7 and 28.10A.

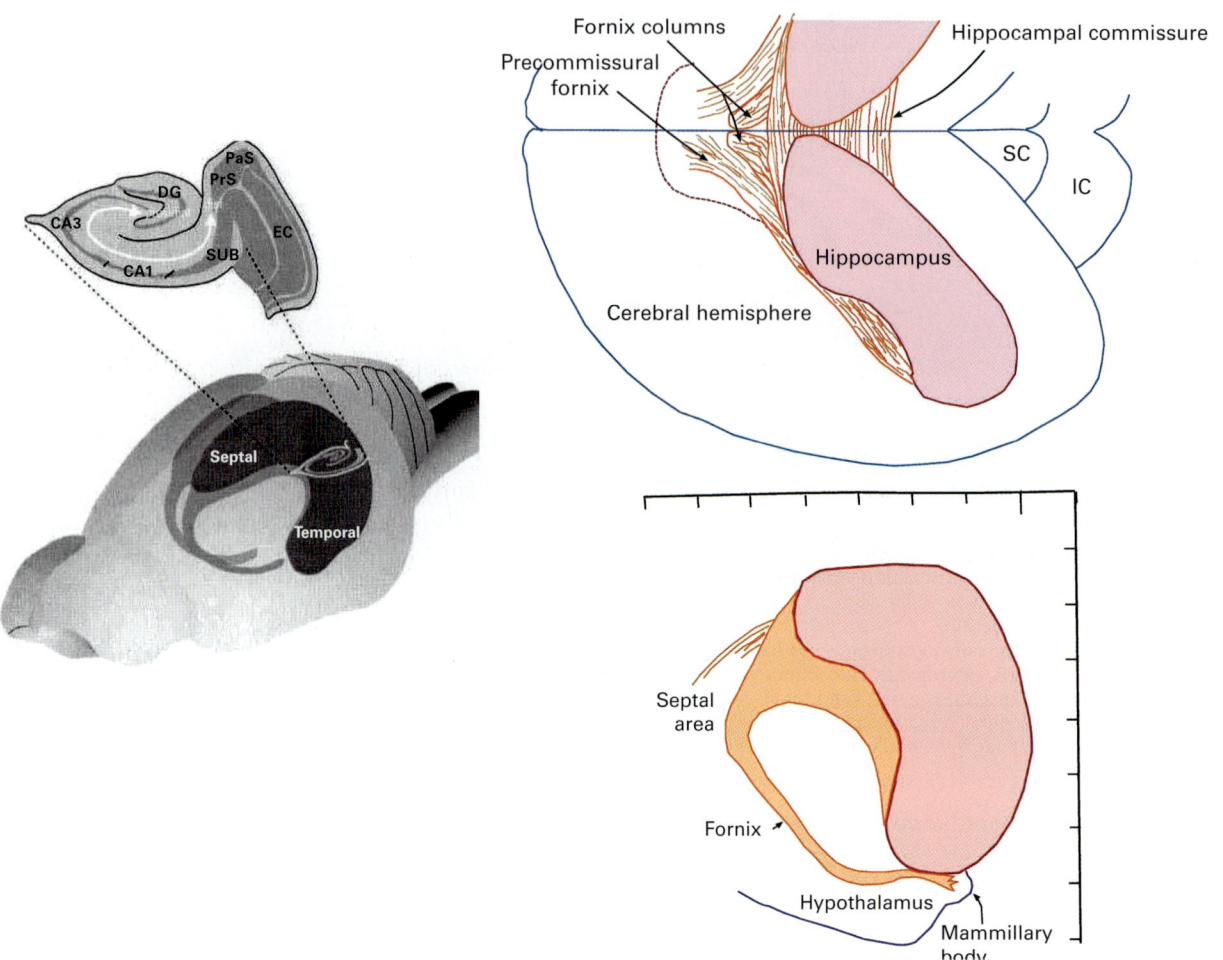

Figure 28.8
Right-hand side: Reconstructions of Syrian hamster endbrain structures, emphasizing the hippocampus and its major output to subcortical areas via the fornix. Left-hand side: Sketches of the rat brain showing the subcortical positions of the hippocampus and the fornix on the left side, and also a horizontal section of the hippocampal formation at a mid-hippocampal level. Note that the two ends of the hippocampus are designated as being nearer to the septum or nearer to the temporal lobe surface. The drawings at the right are anatomically accurate reconstructions showing the position of the hippocampus and fornix in top and side views. The millimeter scale shown applies to both views. The longer tick mark on the horizontal scale is the position of the lambda point on the skull. The brain is positioned so the cortical surface is horizontal. On the side view, the scale is positioned at the level of the neocortical surface overlying the hippocampus. SC, superior colliculus; IC, inferior colliculus. Abbreviations for the hippocampus section: CA1 and CA3, major subdivisions of the hippocampal pyramidal-cell layer; DG, dentate gyrus; EC, entorhinal cortex; PaS, parasubiculum; PrS, presubiculum; SUB, subiculum (can be considered part of the hippocampus itself). Rat sketches from Witter, Wouterlood, Naber, and Van Haeften (2000); hamster reconstructions by G. Schneider.

The Medial Pallium Becomes the Hippocampus

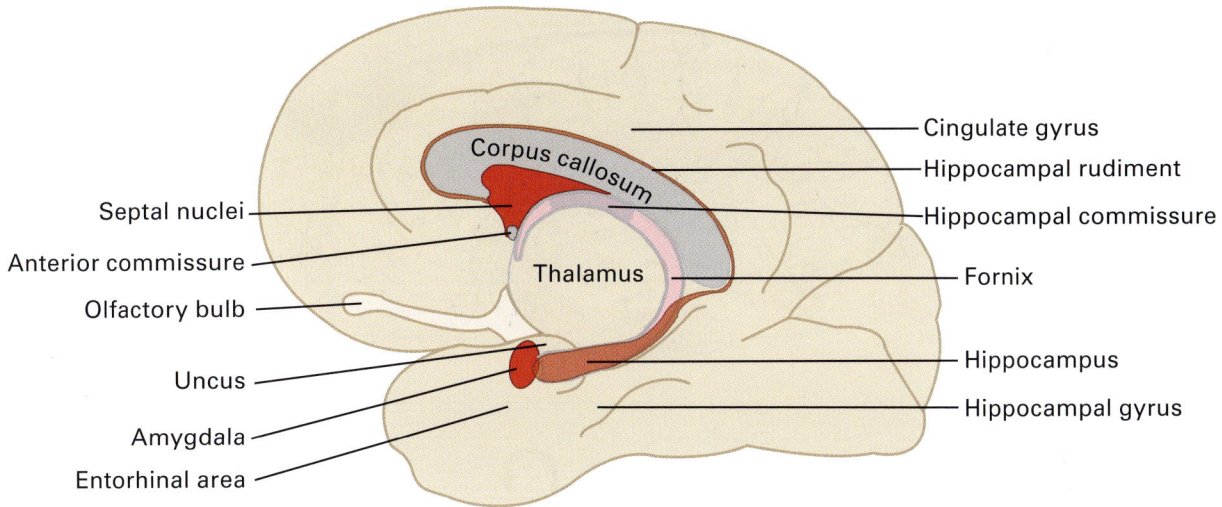

Figure 28.9
Drawing of a medial view of the right hemisphere of a human brain. The hemisphere has been separated from the upper brainstem (diencephalon). The hippocampus and amygdala in the temporal lobe and the septal nuclei rostrally are pictured in red and brownish red. The thin hippocampal rudiment can be followed continuously from the hippocampus caudally to the septal area rostrally. The temporal lobe position of the main body of the hippocampus can be better understood by referring to figure 22.7. Based on Brodal (2004).

Now look at the Cajal drawing shown in figure 28.10A. You can see what should be becoming a familiar cross section through the hippocampus and immediately adjacent areas. The drawing includes axons of the so-called perforant path (see axon 1 in the Cajal drawing, highlighted in red), passing from the entorhinal cortex to the dentate gyrus. It is called "perforant" because it penetrates what was, earlier in evolution, the surface of the dentate. This pathway carries information from association neocortex because the entorhinal area receives many axons from areas that interconnect with neocortex (cingulate and retrosplenial areas), as indicated in figure 26.7.

The pathways from the neocortex provide the major inputs from sensory systems into the hippocampus in mammals. This was not true in ancestral tetrapods. In amphibians, projections from the thalamus go directly to the medial pallium, and these are the major sensory inputs. In reptiles there are still sensory inputs from the thalamus to the medial pallium but also inputs from the dorsal cortex (which is similar to mammalian parahippocampal areas in many of its connections). In mammals, little sensory information reaches the hippocampal formation directly from the thalamus. An exception, according to discoveries in rats and in hamsters, is visual information from the pretectal area, which projects to the anterior part of the lateral thalamus, which in turn projects not only to neocortex medial and rostral to the striate area but also directly to parahippocampal areas—the postsubiculum and retrosplenial areas. This route may be carrying information on locomotion and changes in head direction.

526 Chapter 28

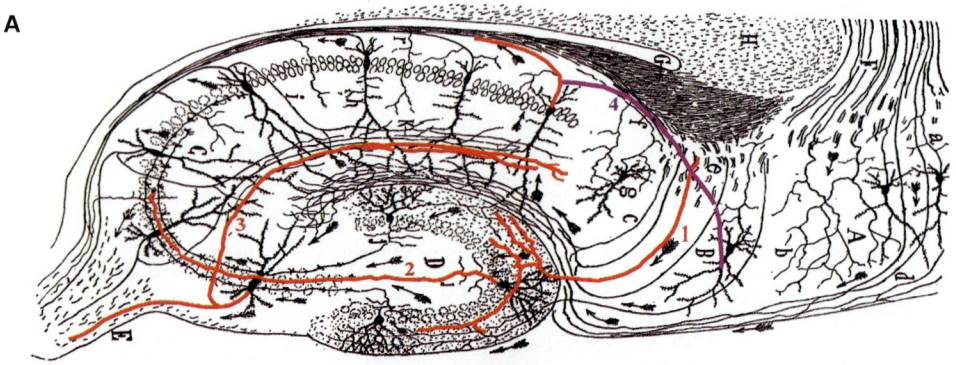

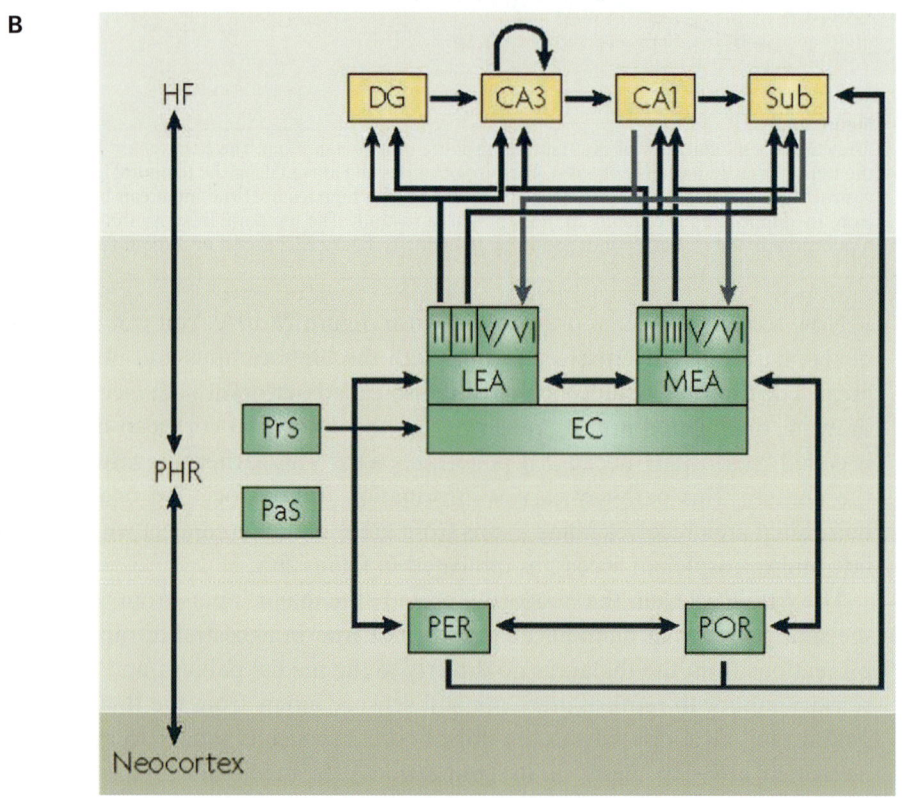

Figure 28.10
(A) Section of the hippocampus drawn by S. Ramón y Cajal, with the additions of red color tracings of single axons of a major local circuit that is present at every position along the longitudinal axis. Axons of this circuit are numbered in order (1–4). The purple axon shows a pathway from CA1 to subiculum that has been described in modern tracing studies. The circuit: from entorhinal cortex to dentate gyrus to CA3 (via mossy fibers) to CA1 (via Schaffer collaterals of CA3 cell axons) to subiculum. Axons of the perforant path into the dentate gyrus also come from the cingulum bundle of the cingulate gyrus. Cajal's labels: A, retrosplenial area; B, subiculum; C, Ammon's horn (threshold); D, dentate gyrus; d, perforant fibers from cingulum bundle; E, fimbria of fornix; F, cingulum bundle; G, angular bundle (fibers from entorhinal area) or hippocampal commissure; H, corpus callosum; K, recurrent collaterals (Schaffer's); collateral of fibers in alveus (hippocampal white matter). (B) Summary of experimental findings on projections to the hippocampus in mammals. The major picture is summarized at the far left: Connections between neocortex and parahippocampal region (PHR), and between that region and the hippocampal formation (HF). Many of the neocortical inputs carrying spatial location information go to the postrhinal cortex of small animals (POR), whereas inputs carrying object identity information go to perirhinal areas (PER). These areas project to the medial entorhinal area (MEA) and lateral entorhinal area (LEA) as shown. Inputs to entorhinal cortex also come from other regions, including the presubiculum (PrS). The entorhinal area projects to the hippocampal formation: dentate gyrus (DG), the CA3 and CA1 regions of hippocampus, and the subiculum (Sub). The subiculum gives rise to major outputs of the hippocampal formation; other outputs come from CA3 and CA1. PaS, parasubiculum. Panel (A) from Ramón y Cajal (1995); panel (B) from van Strien, Cappaert, and Witter (2009).

Local Circuits within the Hippocampus

Figure 28.10A illustrates a major pathway through the hippocampus. Figure 28.10B shows more of the pathway in diagrammatic fashion. In the following description of this pathway, one should keep in mind that many axonal connections are omitted, including other pathways from paralimbic areas, intrahippocampal pathways (association connections), and connections from the hippocampal formation of the opposite hemisphere.

The same connections are found in every section made across the hippocampus (figure 28.10A). In the diagram of these connections shown in part (B) of the figure, you can see pathways from the entorhinal cortex that follow the perforant path into the dentate gyrus. They form synapses on the dendrites of the granule cells of the dentate. The axons of these neurons are called mossy fibers. The mossy fibers connect with the CA3 pyramidal neurons (figure 28.10A). These CA3 cells have axons that join the output axons in the hippocampal white matter (the alveus). In addition, they have axon collaterals known as the Schaffer collaterals. The collaterals connect to nearby CA1 pyramidal neurons, and those pyramidal neurons project to the adjacent subiculum. Finally, the subiculum neurons give rise to the major hippocampal output pathway to the hypothalamus.

Memory and the Hippocampus

Neurophysiologists have recorded from each of the neuron types depicted in figure 28.10A, in living slices cut as shown in the figure. Their experiments have demonstrated long-term potentiation at each of the connections illustrated. They report robust alterations in synaptic strength as a consequence of repeated stimulation. Although these findings do not really explain the complexity of the learning processes for which the hippocampus is critically important, they do give us a model of a kind of neuronal plasticity that probably plays an important role.

The role of the hippocampal formation in memory functions has been a topic of frequent study by neurologists and neuroscientists. It has been found to play a critical role that is quite different from the role played by the corpus striatum. The plasticity of striatal connections is most important for the kind of habit formation we know from studies of reinforcement learning during various conditioning procedures in animals. In humans, such *procedural learning*, including such things as learning to ride a bicycle and to make coordinated movements in sports, or the formation of other sensorimotor habits, is still possible after extensive hippocampal lesions.

Hippocampal lesions, in contrast, disrupt nonprocedural types of learning. In studies of animals, learning of spatial tasks is commonly found to be disrupted..In humans, the effects on memory formation have been really dramatic. A person with lesions affecting this structure or its major connection pathways can lose the ability to form new long-term memories of episodes in his or her life. It appears that such episodic memory is closely linked to the ability to remember places and the things and actions closely associated with those places. We can be certain that episodic memory did not suddenly appear in humans but evolved with the evolution of the medial pallium and its connections.

Recordings of hippocampal neurons have found not only place specificity but also specificity for particular objects encountered at those places within the remembered environment. (This is more characteristic of ventral hippocampus in rodents than for dorsal portions—i.e., for anterior hippocampus in humans more than for posterodorsal hippocampus.) The objects, of course, give a place part of its special character and identity—its landmarks.

Subcortical Projections to and from the Hippocampus

When you were looking at the connections of the circuit of Papez—see figure 26.7—you may remember that the connections were generally two-way connections. A connection from hippocampus to the septal area is a strong one, and there is a strong return connection as well, from neurons of the septum to the hippocampus. The return connection uses acetylcholine as its neurotransmitter. The axons follow the fornix bundle, the major route from hippocampus to subcortical structures including the hypothalamus. The septohippocampal connections are important modulators of hippocampal neuronal activity.

The fornix, part of the medial forebrain bundle (chapter 12), is shown in the drawings of figures 26.5 and 26.6. Figure 28.11 depicts the fornix in a human brain in a simplified sketch with the inclusion of septohippocampal fibers. Early Italian anatomists visualized this bundle in the human brain, seeing it as an arch that reminded them of the large

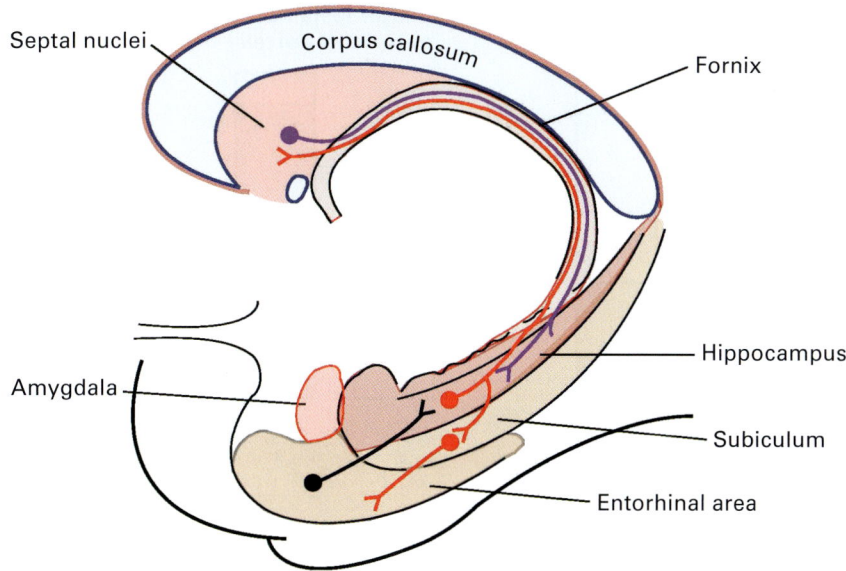

Figure 28.11
Drawing of hippocampus and related structures in the human brain, shown in a medial view of the right hemisphere as in figure 28.9. Connections to and from the septal nuclei, just anterior to the thalamus, are illustrated schematically. The septohippocampal projection uses acetylcholine as its neurotransmitter. Also shown in this schema are connections of hippocampus including the subiculum with the entorhinal area. Based on Brodal (2004) and Nauta and Feirtag (1986).

archways at the entrances to Rome. The arches were called the arches of fornication because prostitutes were often found there soliciting business. Hence the name *fornix*. The arched fornix bundles on the two sides of the brain appear to be connected by a broad band of fibers posterodorsally. These are the axons that interconnect the right and left hippocampi (see figure 28.8, "hippocampal commissure"). The appearance of this formation reminded early anatomists of a harp, so they called it the dorsal psalterium or the harp of David.

Anatomical Plasticity in the Hippocampus

It was noted above that hippocampal connections are altered during long-term potentiation. Another kind of plasticity has also been reported: major changes in axonal end arbors after elimination of specific pathways by brain damage. The changes are a kind of collateral sprouting (see chapter 13). The sprouting in the hippocampal formation appears to be greater than any sprouting that has been observed in sensory and motor system pathways of the mature CNS, where such plasticity is much more likely to occur after lesions in very immature brains. The pattern of sprouting after an extensive lesion of the

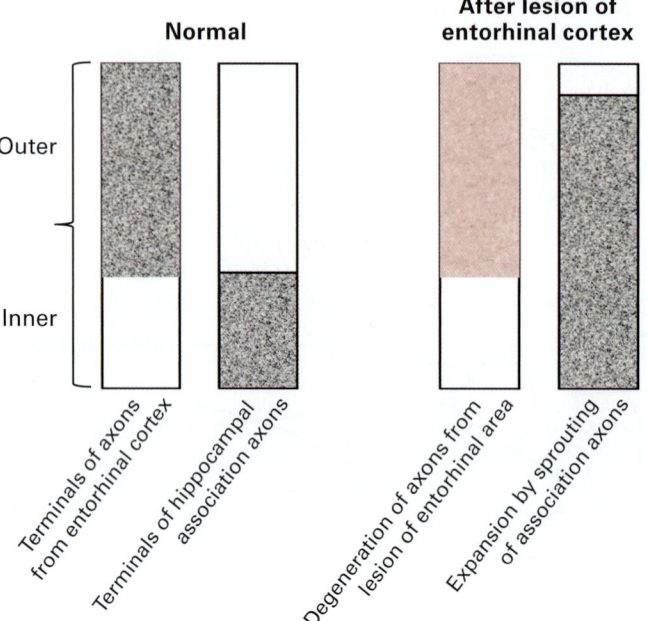

Figure 28.12
Projections to the dentate gyrus show considerable plasticity in response to entorhinal cortex ablation. This figure depicts the distribution of axonal endings on dendrites present in the molecular layer. Each box represents a small column that passes from the surface of the cortex down to the base of the cell dendrites. The positions of the granule cell bodies and their dendrites in this cortex are illustrated in figure 28.13. In separate boxes at the left are shown projections from the ipsilateral entorhinal cortex and projections from association fibers. Illustrated at the right are changes in the association fiber projections a short time after removal of the entorhinal cortex. Based on experiments by G. Lynch and C. Cotman. Based on Lund (1978).

entorhinal cortex on one side in an adult rat is illustrated in figures 28.12 and 28.13. The patterns of terminals of two of the axon groups under investigation, in the dentate gyrus, are illustrated for normal rats on the left side of figure 28.12. Each box represents a small slice of the cortex of the dentate gyrus, a slice that encompasses the layer of dendrites of the dentate gyrus cells. The position of the cell bodies is just below the box. The fine dots represent terminals of the axons named at the bottom of the figure. The box at the left end shows the dense terminations of axons from the ipsilateral entorhinal cortex, on the distal two-thirds of the dendrites. These are the terminals eliminated by the entorhinal lesion. The adjacent box shows terminals of intrahippocampal association fibers. On the right side of the diagram, you can see that these terminals of association axons increase greatly after the lesion: they show extensive sprouting of new terminals, spreading throughout most of the thickness of the dendritic layer. Sprouting of other connections to the dentate gyrus occurs as well. There is an increase in density of connections from

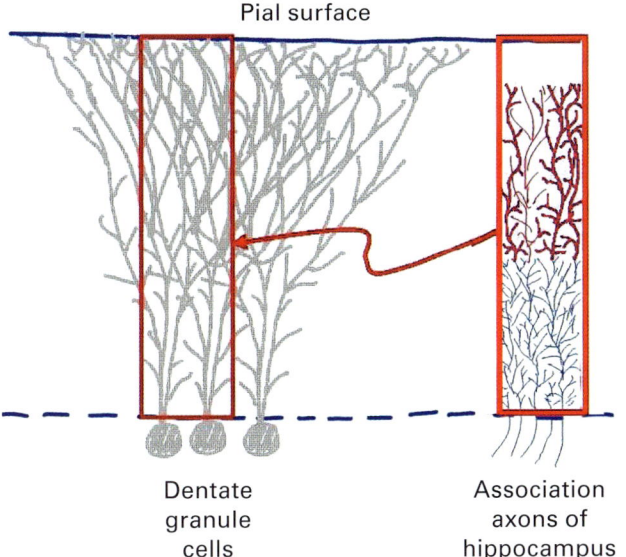

Figure 28.13
Sketches of granule cells of the dentate gyrus of the hippocampal formation are shown at the left. At the right is a rectangle that corresponds to the boxes in figure 28.12. At the same scale, axons of association axons are sketched in the rectangle. The axons normally terminate on the proximal portions of the granule cell dendrites. Drawn in red are axon collaterals that have sprouted after a lesion of the entorhinal cortex.

the entorhinal cortex on the opposite side of the brain. There is also an increase in the connections from the septum.

Figure 28.13 may help with an understanding of figure 28.12. It shows, at the same scale, a picture of a few of the granule cells and their dendrites. It also shows, in a box, a sketch of terminating axons coming from other parts of the hippocampus, of both sides, and an example of how these axons may sprout after adjacent axons are caused to degenerate after an entorhinal cortex lesion.

These results show the hippocampal formation's great capacity for plasticity in the adult mammal. The plasticity displayed by the large amount of collateral sprouting of the major afferents of the dentate gyrus may correspond to a capacity of the neurons involved to change in learning situations. However, the nature of anatomical changes correlated with learning is not as well understood.

Brain States and Memory Consolidation during Sleep

One of the surprises reported by neuroscientists in recent years has come from evidence that the consolidation of the memories for which hippocampus is important is increased during sleep. How might this occur? We know that overall states of the brain are altered

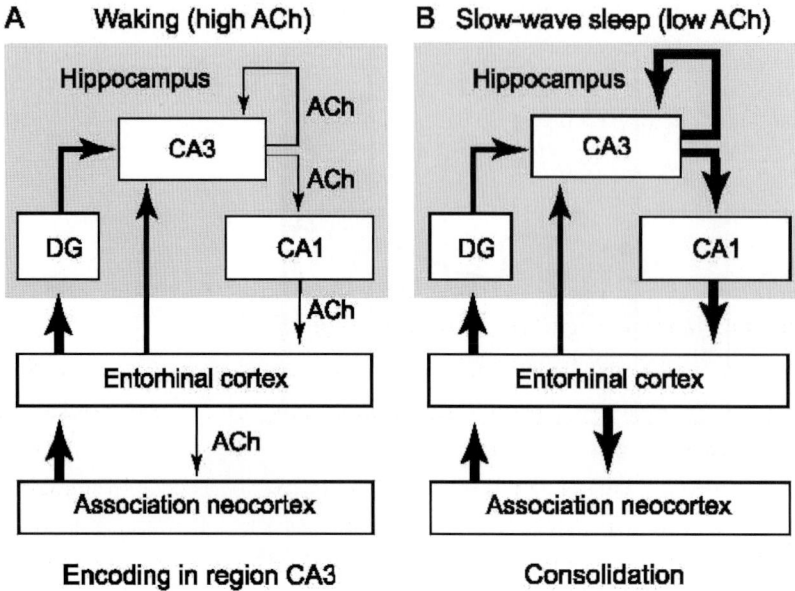

Figure 28.14
Hippocampus during sleep and waking: Consolidation of long-term episodic memory during slow-wave sleep may be promoted by reduced levels of acetylcholine (ACh) in the hippocampal formation and neocortex. (A) During waking, connections less sensitive to modulation by ACh are the most active, carrying information from association neocortex into the hippocampal formation. (B) During slow-wave sleep, when ACh levels are low, more information flows (thicker arrows) from hippocampus to neocortex, promoting consolidation. DG, dentate gyrus. From Hasselmo (1999).

by widespread neuromodulatory connections from the brainstem (see chapter 17). Brain states change greatly when a person or animal goes to sleep. There is good evidence that acetylcholine (ACh) levels in the neocortex drop considerably from waking to slow-wave sleep. (Norepinephrine and serotonin levels also drop, but not as much as the ACh levels.) Rapid-eye-movement sleep (correlated with what we usually call dream sleep) is very different, as the ACh levels rise and the norepinephrine and serotonin levels drop to very low levels. How might these changes affect the information flow between hippocampus and neocortex?

A hypothesis concerning how the acetylcholine changes could influence memory consolidation is illustrated in figure 28.14. It is proposed that reduced ACh levels could result in increased hippocampus-to-neocortex activity and increased intracortical activity, and these changes could increase memory consolidation. The flow of information from neocortex to hippocampus through the entorhinal cortex and dentate gyrus is not expected to be affected by the ACh change. Other changes are also illustrated in the figure: During waking, neocortex is more dominant over hippocampal activity as hippocampus to cortex activity is reduced.

The development of such theories is very important in enabling us to reach an understanding of the dramatic and surprising importance of this region of the brain in major cognitive functions like memory for places and events. We can hope that experimental work designed to test the theories will lead to their refinement and to a more complete understanding.

Readings

Many modern studies of the hippocampal system followed the experimental work of O'Keefe and Nadel on *cognitive maps* in the brain:

O'Keefe, J., & Nadel, L. (1974). Maps in the brain. *New Scientist, 62,* 749–751.

O'Keefe, J., & Nadel, L. (1978). *The hippocampus as a cognitive map.* Oxford: Clarendon Press.

Place cell existence has been strongly supported by electrophysiological evidence, mostly involving rats. Related cell types have also been described, for example, in the papers that follow. They include details of entorhinal cortex anatomy, and the recordings of *grid cells* there, which may yield clues to spatial functions of the hippocampal formation:

Canto, C. B., Wouterlood, F. G., & Witter, M. P. (2008). What does the anatomical organization of the entorhinal cortex tell us? *Neural Plasticity*, article ID 381243, 18 pages.

Fyhn, M., Molden, S., Witter, M. P., Moser, E. I., & Moser, M.-B. (2004). Spatial representation in the entorhinal cortex. *Science, 305,* 1258–1264.

Moser, E. I., Kropff, E., & Moser, M.-B. (2008). Place cells, grid cells, and the brain's spatial representation system. *Annual Reviews Neuroscience, 31,* 69–89.

Examples of experimental studies of head-direction (HD) cells:

Bassett, J. P., & Taube, J. S. (2001). Neural correlates for angular head velocity in the rat dorsal tegmental nucleus. *Journal of Neuroscience, 21,* 5740–5751.

Clark, B. J., Harris, M. J., & Taube, J. S. (2012). Control of anterodorsal thalamic head direction cells by environmental boundaries: Comparison with conflicting distal landmarks. *Hippocampus, 22,* 172–187.

Yoder, R. M., & Taube, J. S. (2011). Projections to the anterodorsal thalamus and lateral mammillary nuclei arise from different cell populations within the postsubiculum: Implications for the control of head direction cells. *Hippocampus, 21,* 1062–1073.

Yoder, R. M., Clark, B. J., & Taube, J. S. (2011). Origins of landmark encoding in the brain. *Trends in Neurosciences, 34,* 561–571.

Two electrophysiological studies by John Lisman and collaborators are of special relevance to ideas in the chapter. One presents evidence that hippocampal activity anticipates future locations of a locomoting rat. The other discusses the role of dopamine pathways to the hippocampus in the forming of new memories.

Lisman, J., & Redish, A. D. (2009). Prediction, sequences and the hippocampus. *Philosophical Transactions of the Royal Society B: Biological Sciences*, *364* (1521), 1193–1201.

Lisman, J. E., & Grace, A. A. (2005) The hippocampal-VTA loop: Controlling the entry of information into long-term memory. *Neuron, 46*, 703–713.

Comparative neuroanatomy:

Butler, A. B., & Hodos, W.: book listed at the end of chapter 3.

Striedter, G.: book listed at the end of chapter 4.

Studies of hippocampal anatomy:

Amaral, D. G., Scharfman, H. E., & Lavenex, P. (2007). The dentate gyrus: Fundamental neuroanatomical organization (dentate gyrus for dummies). *Progress in Brain Research, 163,* 3–22, 788–790.

Brown, M. W., & Aggleton, J. P. (2001). Recognition memory: What are the roles of the perirhinal cortex and hippocampus? *Nature Reviews: Neuroscience, 2,* 51–61.

Burwell, R. D. (2000). The parahippocampal region: Corticocortical connectivity. *Annals of the New York Academy of Sciences, 911,* 25–42.

Burwell, R. D., Witter, M. P., & Amaral, D. G. (1995). Perirhinal and postrhinal cortices of the rat: A review of the neuroanatomical literature and comparison with findings from the monkey brain. *Hippocampus, 5,* 390–408.

Lavenex, P., & Amaral, D. G. (2000). Hippocampal-neocortical interaction: A hierarchy of associativity. *Hippocampus, 10* (4), 420–430 (Special Issue: The Nature of Hippocampal-Cortical Interaction: Theoretical and Experimental Perspectives).

Van Strien, N. M., Cappaert, N. L. M., & Witter, M. P. (2009). The anatomy of memory: An interactive overview of the parahippocampal-hippocampal network. *Nature Reviews Neuroscience, 10,* 272–282.

Contrasts between the dorsal and ventral parts of the hippocampus in rodents and the equivalent parts in primates (posterodorsal and anterior parts):

Kjelstrup, K. B., Solstad, T., Brun, V. H., Hafting, T., Leutgeb, S., Witter, M. P., Moser, E. I., & Moser, M.-B. (2008). Finite scale of spatial representation in the hippocampus. *Science, 321,* 140–143.

Popenk, J., Evensmoen, H. R., Moscovitch, M., & Nadel, L. (2013). Long-axis specialization of the human hippocampus. *Trends in Cognitive Sciences, 17,* 230–240.

Hippocampus and memory formation: Case H. M.

    Bohbot, V. D., & Corkin, S. (2007). Posterior parahippocampal place learning in H.M. *Hippocampus*, *17* (9), 863–872 (Special Issue: Hippocampal Interactions within the Medial Temporal Lobe).

    Eichenbaum, H. (2012). What H. M. taught us. *Journal of Cognitive Neuroscience, 25,* 14–21.

    Milner, B., Corkin, S., & Teuber, H.-L. (1968). Further analysis of the hippocampal amnesic syndrome: 14-year follow-up study of H. M. *Neuropsychologia, 6,* 215–234.

Work of Gary Lynch and Carl Cotman on changes in connections within the adult hippocampus after destruction of specific afferent pathways has been summarized in a book by Ray Lund:

    Lund, R. D. (1978). *Development and plasticity of the brain: An introduction.* New York: Oxford University Press.

Memory consolidation during sleep:

    Gais, S., & Born, J. (2004). Declarative memory consolidation: Mechanisms acting during human sleep. *Learning and Memory, 11,* 679–685.

    Hasselmo, M. E. (1999). Neuromodulation: Acetylcholine and memory consolidation. *Trends in Cognitive Sciences, 3,* 351–359.

    Rasch, B. H., Born, J., & Gais, S. (2006). Combined blockade of cholinergic receptors shifts the brain from stimulus encoding to memory consolidation. *Journal of Cognitive Neuroscience, 18,* 793–802.

Pathway from pretectal area to parahippocampal areas via the lateral thalamic nucleus: See also figure 22.10 showing visual inputs to areas medial to V1 in mouse; also see figure 32.4 showing for hamster the projections from the pretectal area to the anterior part of the lateral nucleus.

    Van Groen, T., & Wyss, J. M. (1990). The postsubicular cortex in the rat: Characterization of the fourth region of the subicular cortex and its connections. *Brain Research, 529,* 165–177.

    Van Groen, T., & Wyss, J. M. (1992). Connections of the retrosplenial dysgranular cortex in the rat. *Journal of Comparative Neurology, 315,* 200–216.

    Wyss, J. M., & Van Groen, T. (1992). Connections between the retrosplenial cortex and the hippocampal formation in the rat: A review. *Hippocampus, 2,* 1–12.

    Van Groen, T., & Wyss, J. M. (1990). Projections from the laterodorsal nucleus of the thalamus to the limbic and visual cortices in the rat. *Journal of Comparative Neurology, 324,* 427–448.

540   Chapter 29

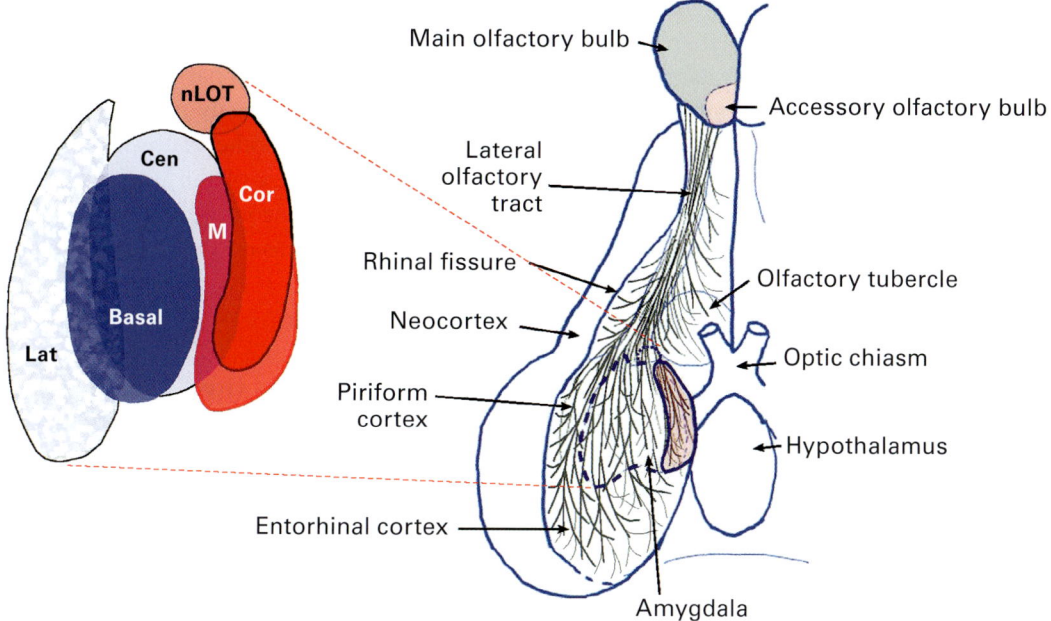

Figure 29.2
At the right is a ventral view of the rabbit forebrain (from figure 19.2) showing the position of the amygdala and its subcomponents and the projections to parts of it from the olfactory bulb (based on studies of rodents and rabbits). At the left is an enlarged view of the amygdala with outlines of its separate cell groups: Basal, basal nucleus; Cen, central nucleus; Cor, cortical nucleus; Lat, lateral nucleus; M, medial nucleus; nLOT, nucleus of the lateral olfactory tract. Based on Lammers (1972).

amygdala, receives many direct projections from the olfactory bulb. Some of the axons coming from the olfactory bulb also reach the medial nucleus. Notice in the figure that the cortical and medial nuclei receive an exclusive direct line from a distinct part of the bulb: the accessory olfactory bulb. The accessory olfactory bulb gets its input from the vomeronasal organ in the roof of the mouth. The input through this organ is crucial for mating behavior in many species.

We will have more to say about the other structures and connections later. We should note here that there are also many short-axon connections within the amygdala that are without doubt crucial for the functioning of this complex (box 29.1).

An indication of the cytoarchitecture of the amygdala and related structures of the rat brain can be seen in figure 29.3. The picture is of a Nissl-stained coronal section of the right-hand side of the brain of a rat, cut through the middle of the diencephalon. Although the section is stained for cell bodies and not axons, you can make out the positions of many axons coursing between the thalamus and the neocortex, passing through the corpus striatum. These are the axons of the internal capsule. Follow the corpus striatum ventrally and you reach the cell groups of the amygdala. Medially and ventrally you can see the

In the figure, I have sketched in the connections from the neocortex that most directly connect with the hypothalamus. The only portion of the neocortex that has any direct projections to the hypothalamus appears to be the caudal and ventral part of the prefrontal cortex—the orbitofrontal cortex. These connections are apparently not strong ones. However, the neocortex can influence the hypothalamus through other pathways, involving more than one synapse. For example, the association cortex of the temporal lobe (inferotemporal portion) projects axons directly to the underlying amygdala, which projects axons directly to the hypothalamus, as we shall see. We focus on the amygdala next.

What Is the Amygdala?

The word itself means "almond" in Latin (coming to Latin from the Greek language). The amygdala is an almond-shaped structure—actually a collection of structures—located at the tip of the temporal lobe in humans and other large-brained mammals, beneath the white matter of the neocortex. Part of it reaches the surface of the brain, but this part of the surface is hidden unless we view the temporal lobe from the medial side. We can think of the amygdala, as did Ramón y Cajal, as part of the corpus striatum—a modified part, to be sure: It is a caudoventral part that abuts the tail of the *dorsal* striatum in the temporal lobe. We can think of it as a differentiated caudal portion of the ventral (limbic) striatum. Together, the striatal structures and the amygdala are often called the basal ganglia.

Recent studies by neuroanatomists have found evidence that only the central and medial components of the amygdala are really striatal in nature, whereas the other portions have differentiated from the pallium. Even if this is true, it is also the case that studies of function and of interconnections within the amygdala seem to justify calling it a single structure with multiple subdivisions and functions. Studies using gene-expression data indicate that an anterior portion called the nucleus of the lateral olfactory tract is formed by migration of neuroblasts from the caudoventral pallial ventricular layer that gives rise also to neocortex. However, that nucleus is not neocortical in its structure or connections. (Neocortex itself contains cells that migrate from the ventricular layers of both dorsal pallium and striatum. The amygdala is also found to contain neurons of such diverse origins.)

As in the striatum, connections in the amygdala are plastic. Connections there underlie associations between perceived objects and sounds on the one hand and affects and autonomic changes on the other. The learning of these associations gives objects in the world valences, or *affective tags*. The connections formed in the amygdala are important structural underpinnings of the meaning of the object or sound to the individual.

The amygdala of a rabbit, viewed from the underside, is depicted in figure 29.2. It is a small collection of cell groups differentiated from each other by cytoarchitecture and by major connections. It is located underneath the caudal-most part of the piriform cortex (olfactory cortex). In fact, one portion of the amygdala, the cortical nucleus of the

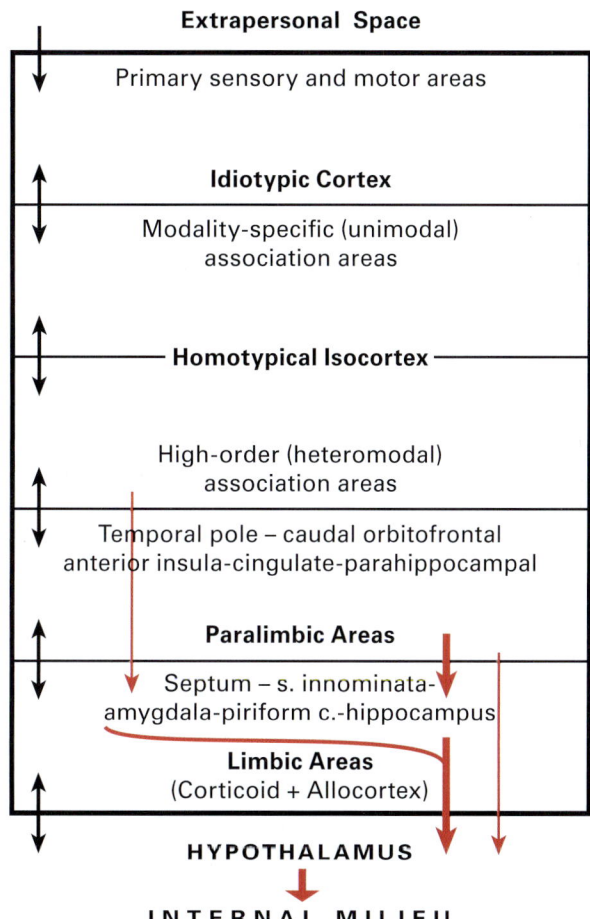

Figure 29.1
In this modification of figure 26.9, showing the basic groups of interconnected neocortical structures, red arrows have been added to illustrate connections of the endbrain to the hypothalamus. The largest source of such connections is the limbic areas, including the septal area and the amygdala. Inputs to these structures come mainly from the paralimbic areas as shown. Note that the only neocortical projections that go directly to the hypothalamus arise in the orbitofrontal cortex. Caudal orbitofrontal cortex is often classified as a paralimbic area. Note also the connection from neocortical association cortex to the amygdala. Such connections carry visual object information from the inferotemporal cortex. Adapted from Mesulam (2000).

29 The Limbic Striatum and Its Outpost in the Temporal Lobe

In the discussion of ideas about evolution of the forebrain in chapter 24, we called attention to the differentiation of the ventral striatum. With expansion of the hemispheres, a caudal portion of this region differentiated into portions of the amygdala. You may also want to check again the forebrain picture described in chapter 12 and also the mention of basal forebrain in chapter 28. (The basal forebrain—extending rostrally from the hypothalamus in the direction of the olfactory bulbs—includes the structures that have become recognized as ventral [or ventromedial] striatum by neuroanatomists.)

A Brain System of Forebrain and Midbrain Underlying Motivation, Emotion, and Autonomic Regulation

In our description of cells and connections of the hypothalamus in chapter 26, we cited the well-known 1937 publication of Cornell neurologist and neuroanatomist James Papez. He described evidence that various structures that had been included in the term *rhinencephalon*, or nose brain, are not actually dominated by the olfactory system in humans. Effects of brain damage and of seizures involving these structures led Papez to propose a circuit of interconnected structures concerned with feelings and emotional expressions. His ideas led to new thinking among neuroscientists and resulted in the resurrection of a term that had been used by Broca in his description of "the great limbic lobe": Paul MacLean at the U.S. National Institutes of Health laboratories in 1952 gave the name *limbic system* to the structures of Papez's circuit. This term was broadened by Nauta and others to include additional structures of the forebrain closely connected with the hypothalamus. Nauta's research on connections of limbic forebrain structures led him later to designate limbic midbrain structures as well—the central gray area and ventral tegmental area.

An intriguing question in this regard concerns the influence of the neocortex on the hypothalamus. Evidence indicates that the vast neocortex of higher primates, so important for perceptual, motor, and cognitive functions, has mostly indirect influences on this center of control of motivational states and autonomic functions. What are the most direct connections for such influences? The findings of Nauta and others on this question can be illustrated briefly using the diagram of figure 29.1 (based on figure 26.9).

The Limbic Striatum and Its Outpost in the Temporal Lobe

> **Box 29.1**
>
> A summary of some major connections of the amygdala in mammals (from Brodal, 3rd ed.; Pitkänen, Savander & LeDoux, 1997):
>
> **Corticomedial nuclei** Inputs from olfactory bulbs, hypothalamus, lateral amygdala; outputs to hypothalamus, influencing motivation, emotion, and autonomic nervous system.
>
> **Basolateral nuclei** Inputs from thalamus, neocortex, hippocampus; outputs to prefrontal cortex, ventral striatum.
>
> **Central nucleus** Intra-amygdalar inputs; outputs via stria terminalis, a major route to the basal forebrain and hypothalamus.
>
> (Some amygdala projections that reach hypothalamus also continue into the midbrain limbic areas.)

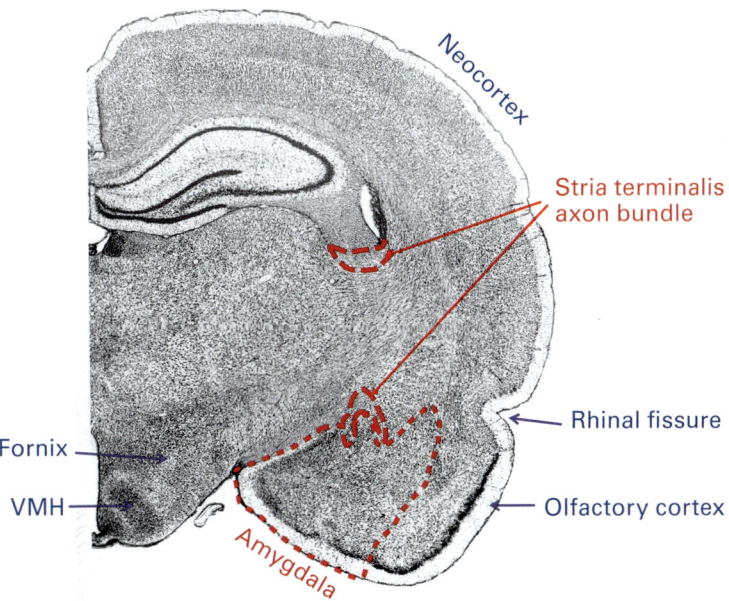

Figure 29.3
Nissl-stained coronal section of the rat 'tweenbrain and endbrain showing the amygdala and one of the major targets of its projections, the ventromedial nucleus of the hypothalamus (VMH). One can also see details of the cytoarchitecture of part of the neocortex (at the top) and the dorsal hippocampus as well as structures of the diencephalon dorsal to the hypothalamus. Adapted from Swanson (1992).

cortical nucleus, with a surface layer that appears to be neuron-free. In that most superficial layer are the terminations from the olfactory bulb, which are also found in the similar superficial layer more laterally, in the piriform cortex.

The Stria Terminalis: A Major Output of the Amygdala

The locations of the stria terminalis in this photograph are also evident to a neuroanatomist. This axon tract has been outlined in the figure. They are also labeled in figure 29.4 using drawings shown earlier, in chapter 26. They are mostly axons of neurons in the central nucleus of the amygdala.

This compact bundle of axons, the stria terminalis, is a major output pathway from the amygdala. It goes to the hypothalamus by a route through the basal forebrain, following a trajectory similar to that of the fornix fibers coming from the hippocampal formation and going to the hypothalamus.

By this connection, the amygdala strongly influences motivational states.

The stria terminalis axons (figure 29.5) begin their trajectory by going dorsally from the central nucleus, passing around the caudal-most axons of the internal capsule, coursing over the capsule as shown (separated by the lateral ventricle from axons coming from the hippocampus), and then diving down in front of the thalamus and passing through the septal area and the basal forebrain before turning caudally and coursing into the hypothalamus. Some axons terminate in the basal forebrain on neurons located among this group of axons. Because of this localization, these neurons are called the *bed nucleus of the stria terminalis*. The longer axons extend further, turning caudally from the basal forebrain and terminating in various parts of the hypothalamus. They form endings most densely in a cell-sparse shell surrounding the ventromedial nucleus (VMH) as you can see clearly in figure 29.3. That shell around the cell bodies of the VMH is made up largely of dendrites of the VMH neurons. In this way, the amygdala projects directly to ventromedial nucleus neurons.

Some of the stria terminalis axons continue their caudal course, reaching into the midbrain where they terminate in the limbic midbrain areas discussed previously (chapters 11, 25, and 26). Remember that these areas of the midbrain are, like the hypothalamus, strongly involved in control of motivation and emotion. We will discuss output connections of the amygdala further, but first we return to its inputs from sensory systems.

Sensory Pathways to the Amygdala

As is also true for the larger anterior portion of the corpus striatum (to which we will return shortly), where the amygdala reaches the brain surface it receives direct projections from the olfactory bulb, as described above (figure 29.2) and in chapter 19. This most ancient connection reaches the cortical and the medial nucleus of the amygdala. This

543 The Limbic Striatum and Its Outpost in the Temporal Lobe

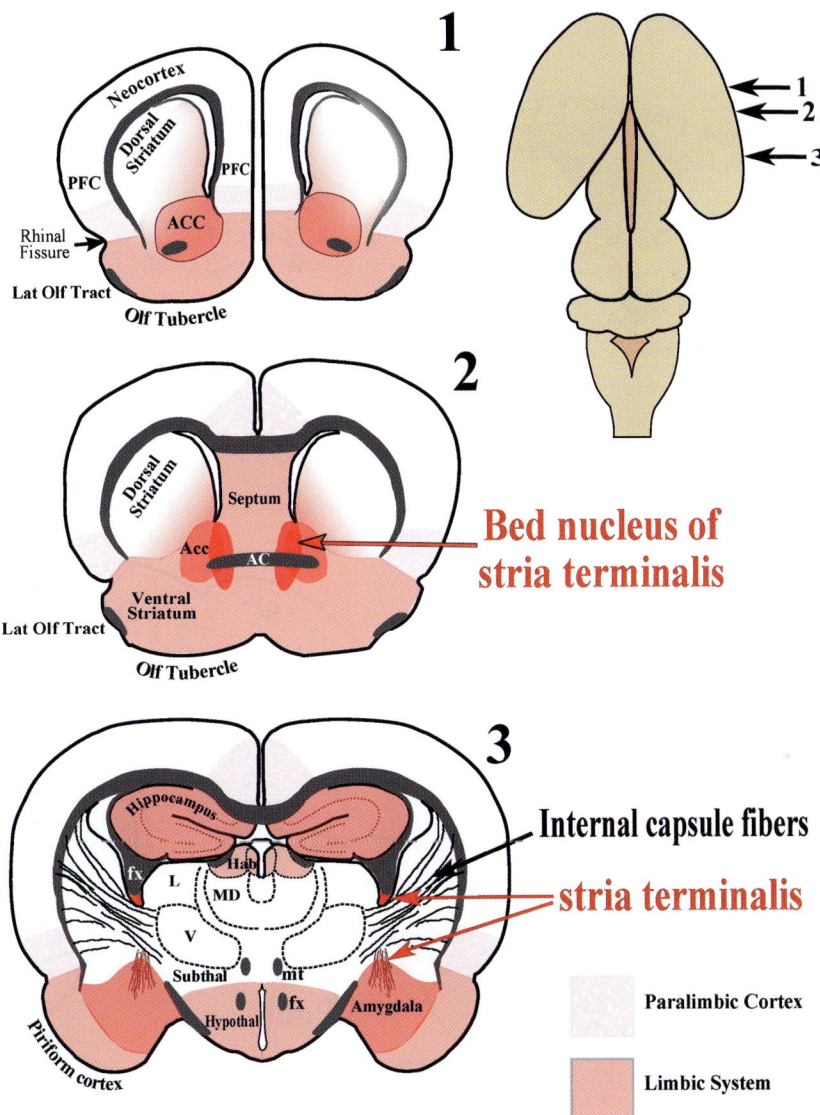

Figure 29.4
Structures of the limbic system are shown, with additions to part of figure 26.8 to show the amygdala with its projection through the stria terminalis to the bed nucleus of the stria terminalis in the basal forebrain. One can also see, at level 3, the nearby fornix fibers from the hippocampus—which follow a similar trajectory around the internal capsule. Acc, nucleus accumbens; AC, anterior commissure; Hab, habenula; Hypothal, hypothalamus; L, lateral nucleus of thalamus; Lat Olf, lateral olfactory; MD, mediodorsal nucleus of thalamus; PFC, prefrontal cortex; Subthal, subthalamus; V, ventral nucleus of thalamus.

544   Chapter 29

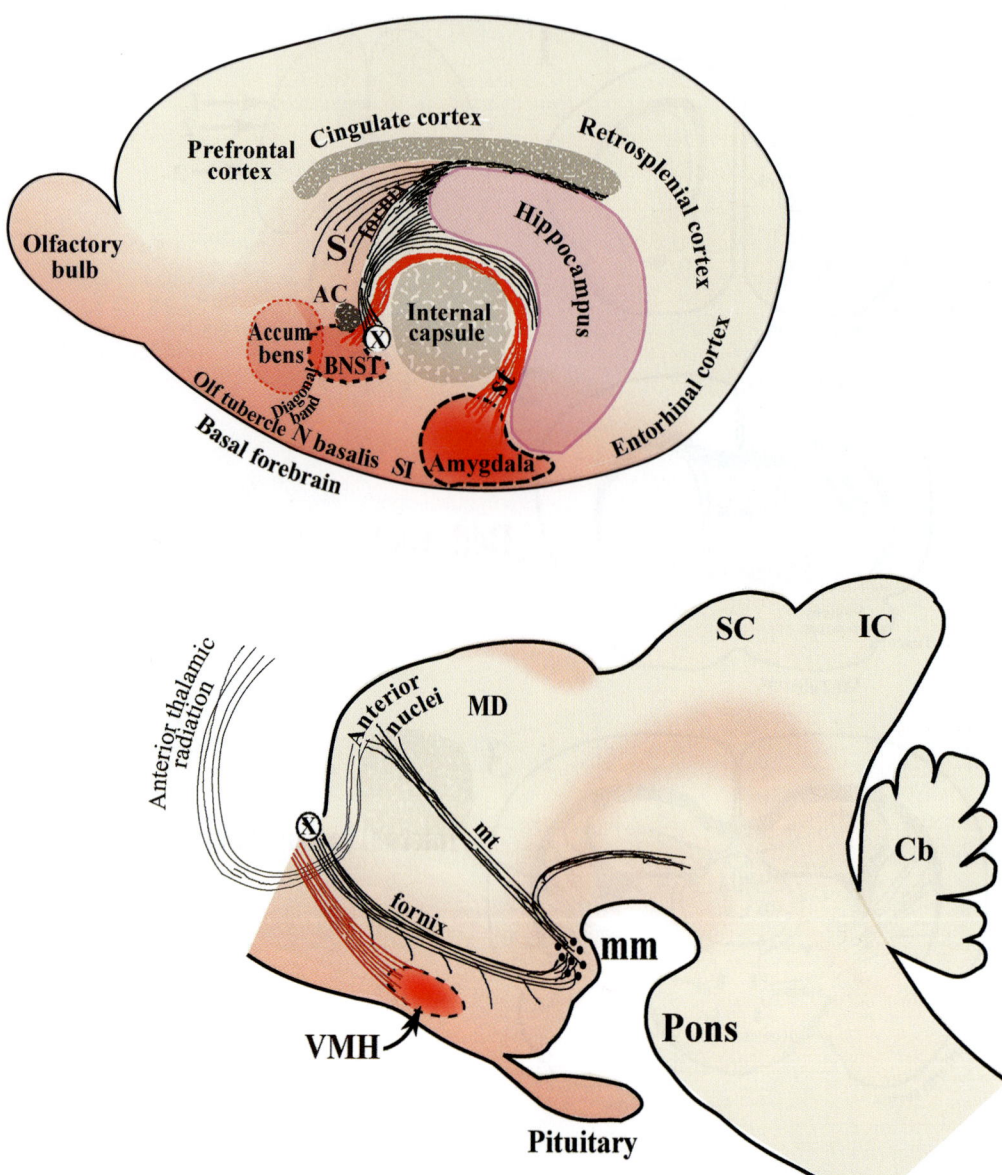

Figure 29.5
In the medial view of the hamster right hemisphere and the underlying brainstem from figure 26.5, the amygdala and its output through the stria terminalis are illustrated, with two major targets of this pathway: the bed nucleus of the stria terminalis (BNST) and the ventromedial nucleus of the hypothalamus (VMH). AC, anterior commissure; Cb, cerebellum; IC, inferior colliculus; MD, mediodorsal nucleus of thalamus; mm, mammillary body; mt, mammillothalamic tract; N basalis, nucleus basalis of Meynart; Olf Tubercle, olfactory tubercle; S, septal area; SC, superior colliculus; SI, substantia innominata; st, stria terminalis.

portion of the amygdala also receives axons coming directly from the taste nucleus in the hindbrain called the parabrachial nucleus; it likely receives inputs also from the cortical taste area (located in the anterior insula in humans). Pain inputs also reach the amygdala by a route from the hindbrain (and with little doubt from paleothalamic cell groups as well); these terminate in the central nucleus. Comparative studies indicate that the taste and olfactory inputs are ancient inputs to the amygdala.

Other inputs to the amygdala have been discussed previously: The lateral nucleus receives auditory inputs directly from the thalamus—a projection that appears to be relatively larger in mammals that we call "more primitive." (See figure 23.3 for an illustration of this pathway from the medial geniculate body of the thalamus found in studies of the opossum, hedgehog, and tree shrew.) Information about visual object identities reaches the amygdala via temporal lobe association neocortex, as discovered in the monkey by Nauta. Electrical recording gives evidence of somatosensory inputs, not only those representing pain, as well.

Amygdala Connections with Systems of Higher Cognitive Functions

Sensory inputs to the amygdala include those carried by neocortical projections from association cortical areas, indicating cognitive relevance, and outputs pass to the hypothalamus and to the limbic core of the midbrain. Thus, the inputs and outputs of the almond-shaped structure partially buried within the tip of the temporal lobe in primates fit a picture of circuits that underlie visceral and emotional responses to objects, individuals, or places. In fact, there is good evidence that the learning of such responses can depend on the amygdala (see later).

Neuroanatomists have also traced axonal output pathways from the amygdala to structures known to be involved in higher cognitive functions. One of these pathways goes to prefrontal cortical areas in mammals. These areas have been found to be crucially important for voluntary control and planning of movements. Other pathways from the amygdala have been traced to temporal portions of the hippocampal formation and to the entorhinal cortex, structures of major importance in the formation of spatial memories and more generally in remembering past events.

The position of the amygdala in the human brain is topologically similar to its position in other mammals. As in other primates, it is found at the tip of the temporal lobe on the medial side (figure 29.6). The frontal section reveals the proximity of the human amygdala to the corpus striatum. Such sections reveal its proximity to the basal forebrain and hypothalamus as well.

Behavioral Change Caused by Artificial Activation of the Amygdala

The projections of the amygdala to the hypothalamus and limbic midbrain areas indicate an involvement in visceral and motivational reactions. A direct, and somewhat crude, way

546   Chapter 29

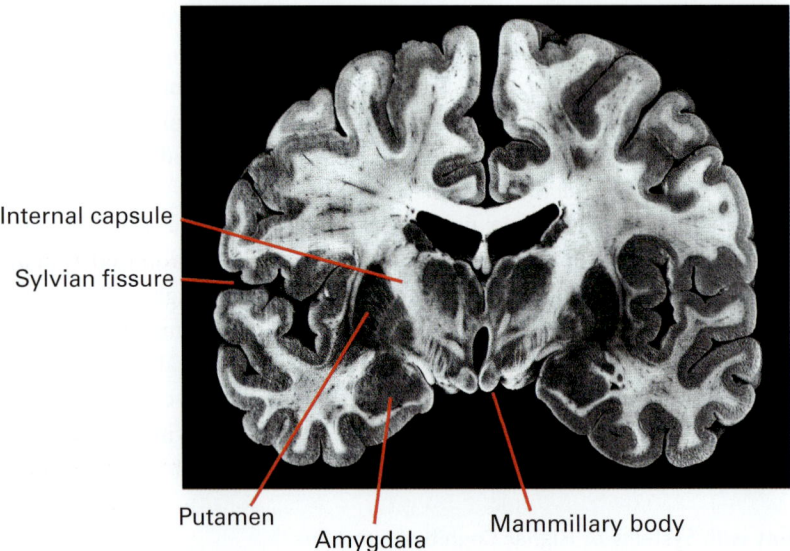

Figure 29.6
In this section through a human forebrain, cut at the level of the anterior temporal lobe, gray matter is darker, white matter is light. The position and shape of the amygdala can be seen. Note also the structures of the corpus striatum (caudate nucleus and putamen, separated by the internal capsule). The claustrum can be seen as an island of gray matter in the white matter beneath the cortex of the insula (neocortex hidden in the depths of the Sylvian fissure). From Gluhbegovic and Williams (1980).

to test this is to apply electrical stimulation through electrodes placed into the structure. When this has been done in primates and also humans during neurosurgical procedures, the expectations have been confirmed.

When the electrodes are placed in the basolateral nuclei of the amygdala, one sees a more aroused and attentive animal or sometimes a fearful or enraged animal. Humans receiving such stimulation may feel great anxiety or other strong emotions. Or they may have the illusory feeling that what they are perceiving is something they have perceived before; for example, a place or a face may seem strangely familiar—a *déjà vu* experience. This suggests an importance of the amygdala in memory for past experiences.

Activation of the cortical and medial nuclei has resulted in oral effects like lip smacking, salivation, and chewing movements. It has also caused elimination. An inhibition of voluntary movements has also been reported; this could result from stimulating amygdala projections to prefrontal neocortex or it could be from less direct effects on motivational states.

As we shall see further below, the amygdala is the site of learned links between perceptions of world events or objects and internal feeling or motivational states. These links underlie what you could call emotional habits, or the affective tags attached to remembered faces and objects.

Behavioral Change after Amygdala Lesions

Destruction of the amygdala or its connections are even more revealing about the function of this part of the basal ganglia. It is well known that rhesus monkeys with amygdala lesions act "tame": they lose their normal defensiveness and aggressiveness in interactions with people as well as with other monkeys. They become socially isolated with a loss of sensitivity to social dominance relationships. It is obviously very maladaptive when such a monkey fails to defer to a dominant male. The amygdalectomized monkey shows no fear in response to visual inputs.

Humans with amygdala lesions have shown loss of autonomic reactions accompanying fear and anxiety. Altered dietary preferences and hypersexuality have also been reported as effects of these lesions in monkeys, and also in some humans with amygdala damage.

In 1962, an investigator described how he had produced rhesus monkeys that showed amygdala lesion symptoms when looking with one eye but not with the other eye. He produced what is known as a disconnection syndrome. Downer's animals had the lesions depicted in red color in figure 29.7: unilateral amygdalectomy plus *split-brain* surgery that had cut through the crossing optic tract fibers at the chiasm and had also severed the corpus callosum. The callosum lesion was necessary to produce the strange syndrome, because information from the affected half of the visual field would otherwise have been able to reach the temporal lobe on the opposite side.

How Sounds That Warn of Danger Come to Cause Fear and Avoidance

In previous chapters, the early evolution of pathways enabling organisms to escape from and avoid dangerous predators has been pointed out. Thus, major outputs of the midbrain control escape and withdrawal behavior or defensive attack, and at the same time the autonomic nervous system can shift into a *fight or flight* mode of sympathetic arousal. Olfactory inputs, you may also recall, provided signals that could anticipate danger, and this led to forebrain systems that greatly enhanced the ability of animals to survive. Sensory pathways to striatal structures could trigger either withdrawal or approach responses, and the connections in the striatal structures could change—the animal could learn to withdraw or to approach places or objects. Initially, the sensory signals were olfactory, but with the evolutionary invasion of the endbrain by other senses, better and better anticipation of danger or reward became possible.

Early in mammalian evolution and probably long before mammals appeared, auditory information reached the amygdala structures. The information followed pathways from the evolving 'tweenbrain. Thalamic structures receiving axons from the auditory midbrain expanded as a neothalamus differentiated from the older, more diffusely connected paleothalamus. In mammals that appear to be the most similar to the earliest mammals, the medial geniculate body (nucleus) receives auditory input and projects heavily to the

548   Chapter 29

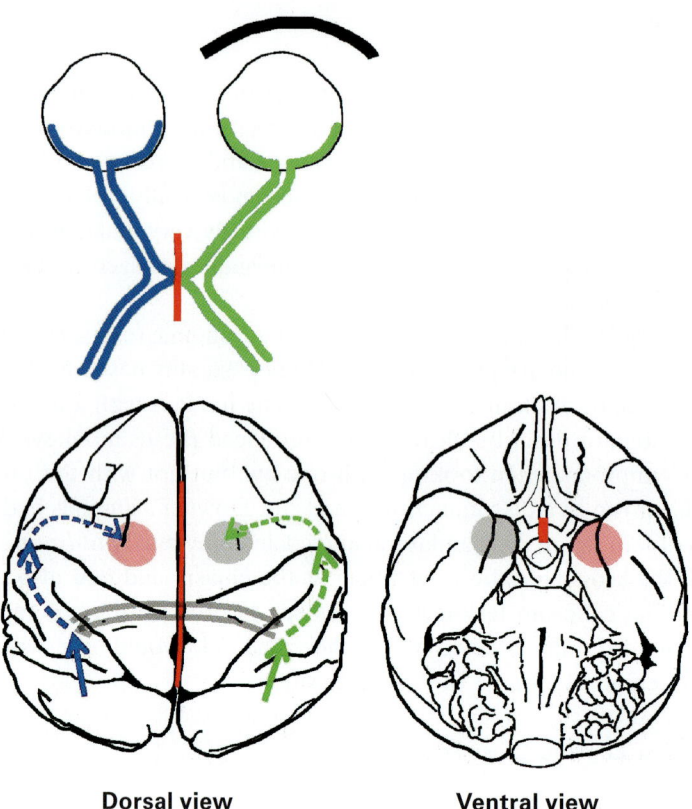

Dorsal view   **Ventral view**

Figure 29.7
Illustration of the surgery carried out in Downer's experiments with behavioral effects of unilateral amygdalectomy in split-brain monkeys. In addition to a unilateral ablation of the amygdala, the optic chiasm was cut along the midline, and the posterior part of the corpus callosum was sectioned. In monkeys with these lesions, symptoms of amygdalectomy were seen when the animal was looking with one eye and not when it was looking with the other eye. Based on descriptions in Downer (1961).

amygdala as well as to neocortex. This was mentioned in chapter 23. Experiments using rats have demonstrated the importance of this pathway to the limbic endbrain for remembering sounds that signal danger. The experiments have shown a kind of *fear conditioning* that depends on the amygdala. A fear conditioning experiment is illustrated in figure 29.8.

The More Rostrally Located Ventral Striatum, in the "Basal Forebrain"

Structures located in the ventral endbrain just rostral to the hypothalamus are of very primitive origins, as discussed in chapter 24. These structures are closely related to the amygdala, with similarities in functions and close neuroanatomical connections. They are

The Limbic Striatum and Its Outpost in the Temporal Lobe

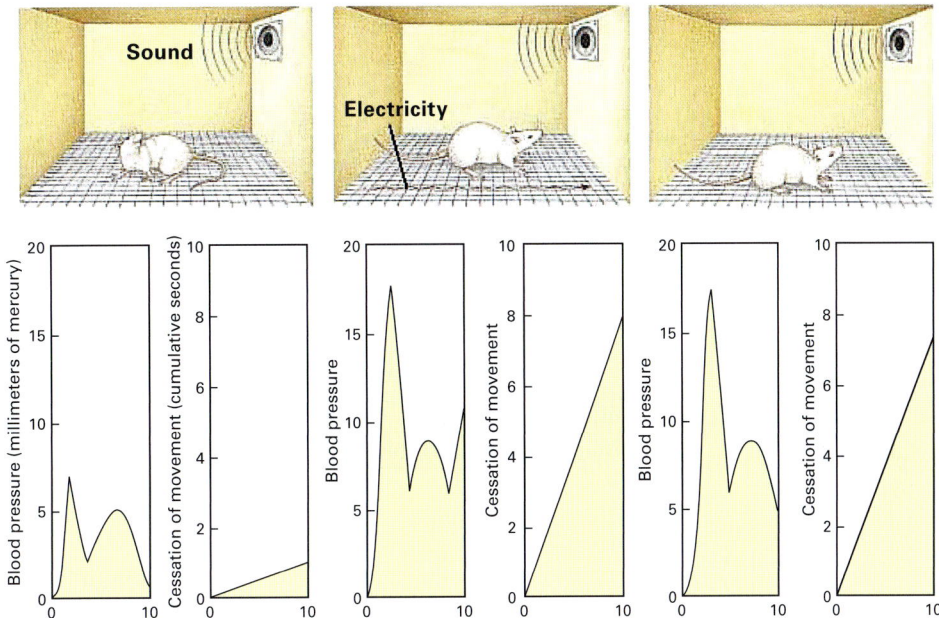

Figure 29.8
Illustration of a behavioral experiment that has been used to provide evidence that fear conditioning to a sound is dependent on the amygdala in rats. The pictures show the training procedure. The rat is placed in a small room with a grid floor. The rat shows little blood pressure change or freezing in response to a neutral sound (left-most illustrations). When the same sound is paired with a shock to the feet through the grid, the rat's blood pressure shows a sharp increase, and the animal freezes for about 8 seconds (middle illustrations). Subsequently, the sound alone results in a blood-pressure rise and a prolonged freezing response (right-most illustrations). The memory that results in this conditioned fear response can be abolished by a lesion of the amygdala. From LeDoux (1994).

often referred to as the basal forebrain. Since the work of the neuroanatomist Lennart Heimer while he was working with Nauta at MIT, and then later when he was at the University of Virginia, we consider these structures to be the ventral striatum (introduced in chapter 12), not only because of the nature of their connections but also because they are located ventral to the dorsal striatal structures—the caudate nucleus and the putamen. (In the next chapter, we will summarize the proposed major steps in evolution of the striatum with greater focus on the more recently evolved dorsal striatum.)

To visualize the ventral striatal brain region, we return to the medial view of the rat's cerebral hemisphere used in picturing the circuit of Papez in chapter 26. In figure 29.9, the basal forebrain structures are highlighted. The olfactory tubercle receives, at its surface layer, direct projections from the olfactory bulb. Medial to the tubercle is the basal nucleus of Meynart, with acetylcholine-containing cells that have axons projecting very widely to the neocortex. (These large cells show degenerative changes in people with Alzheimer's disease.)

550   Chapter 29

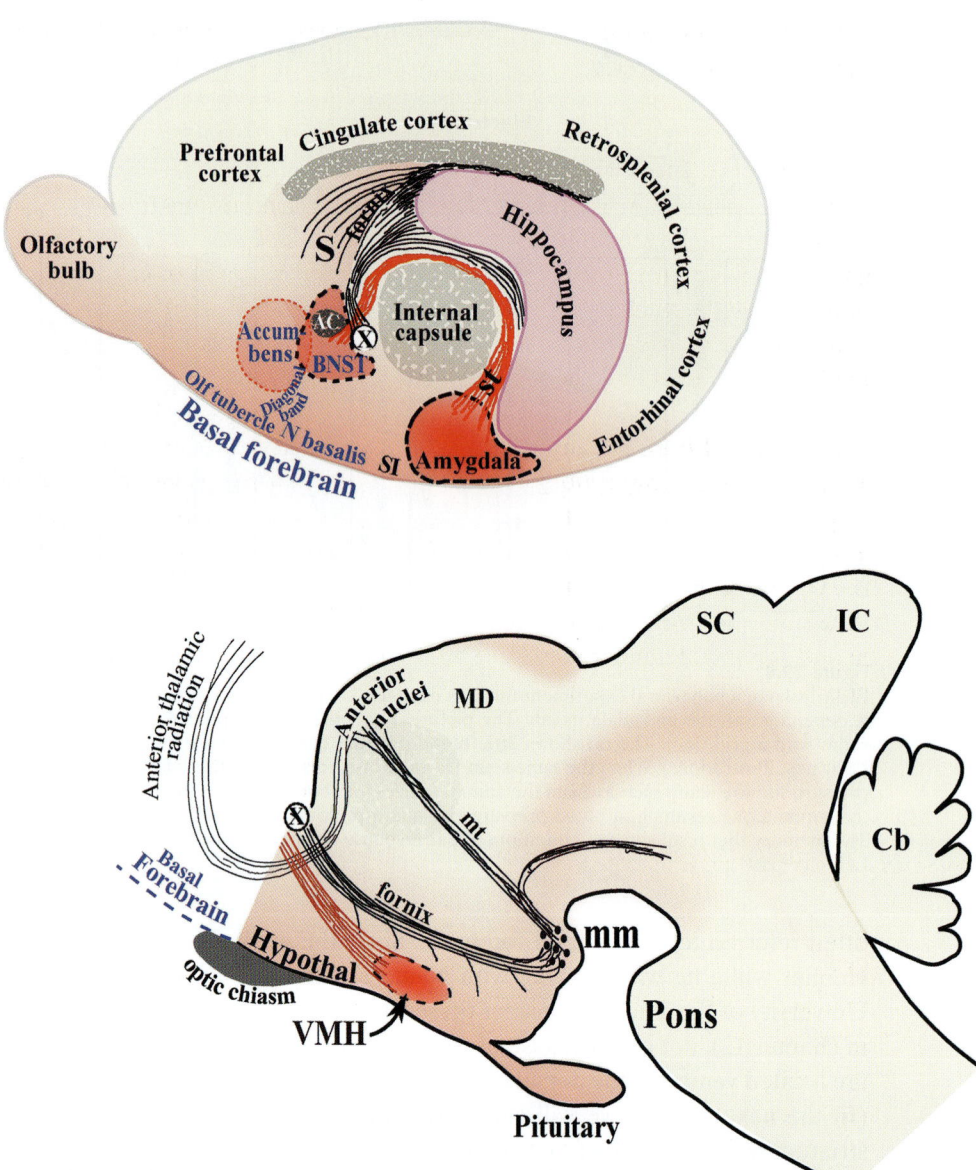

Figure 29.9
In this copy of figure 26.5, the structures of the basal forebrain region are emphasized by the blue font. They are subpallial structures located at the base of the forebrain rostral to the diencephalon. The septal area (S) is sometimes included, as is the substantia innominata (SI) of the "extended amygdala." Abbreviations as in figure 29.5.

More medially, and extending dorsally along the medial edge of the hemisphere, are the cells of the diagonal band of Broca. Located farther from the ventral brain surface are major cell groups of the ventral striatum: the nucleus accumbens, "leaning" against the septal area and abutting the bed nucleus of the stria terminalis, discussed earlier in the chapter. These cell groups are indicated in the frontal-section sketches of figure 29.10. The accumbens, like the bed nucleus, receives input from the amygdala: An important contributor to the projection to accumbens is the amygdalo-hippocampal area of the amygdala, adjacent to the cortical nucleus and the recipient of projections from several other amygdalar nuclei—nuclei that also project to the central nucleus.

Also in the medial view of the hemisphere, you can see the location of tissue between the basal forebrain and the amygdala and lateral to the hypothalamus. This tissue has come to be referred to as the substantia innominata (SI). The region got this name because it had been left unnamed by an early anatomist who was studying the region; studies of axonal connections have shown it to be an important link in the output pathways from the limbic striatum. Together with part of the anterior hypothalamus known as the magnocellular preoptic nucleus, the SI receives projections from major portions of the ventral striatum—the nucleus accumbens and the olfactory tubercle and nearby cells. Axons descending from the SI and the magnocellular preoptic nucleus project to the hypothalamus and midbrain and thereby influence endocrine and motivational states and approach or avoidance locomotion.

These functions typify the basal forebrain, affecting the viscera through the endocrine system and the autonomic nervous systems and affecting basic motivations and more directly the approach to and withdrawal from objects and places.

Disordered Functions of the Limbic System

The nature of this array of functions has made the basal forebrain and amygdala the foci in studies of human mental and emotional disorders. These disorders include the abnormal motivational states of drug addiction and obsessive-compulsive behavior and complex psychotic states as in some types of schizophrenia, involving both cognitive and emotional-motivational disorders. This kind of psychosis has various causes, none of them free from controversy, but it seems clear that genetic anomalies contribute to many cases and that early brain damage contributes to some cases. The latter type of cause is of special interest here because of the possible neuroanatomical abnormalities.

Neuropathological studies have shown that an enlargement of the cerebral ventricles is common in many difficult-to-treat schizophrenic patients (figure 29.11). This anomaly indicates early brain injury. A reduction in amygdala volume is a frequent accompaniment. One study is of special interest because it was based on a kind of "natural experiment." A neurosurgeon was treating cases of temporal lobe epilepsy by removal of the abnormal brain tissue in the temporal lobe, mostly in the amygdala on one side, where

552 Chapter 29

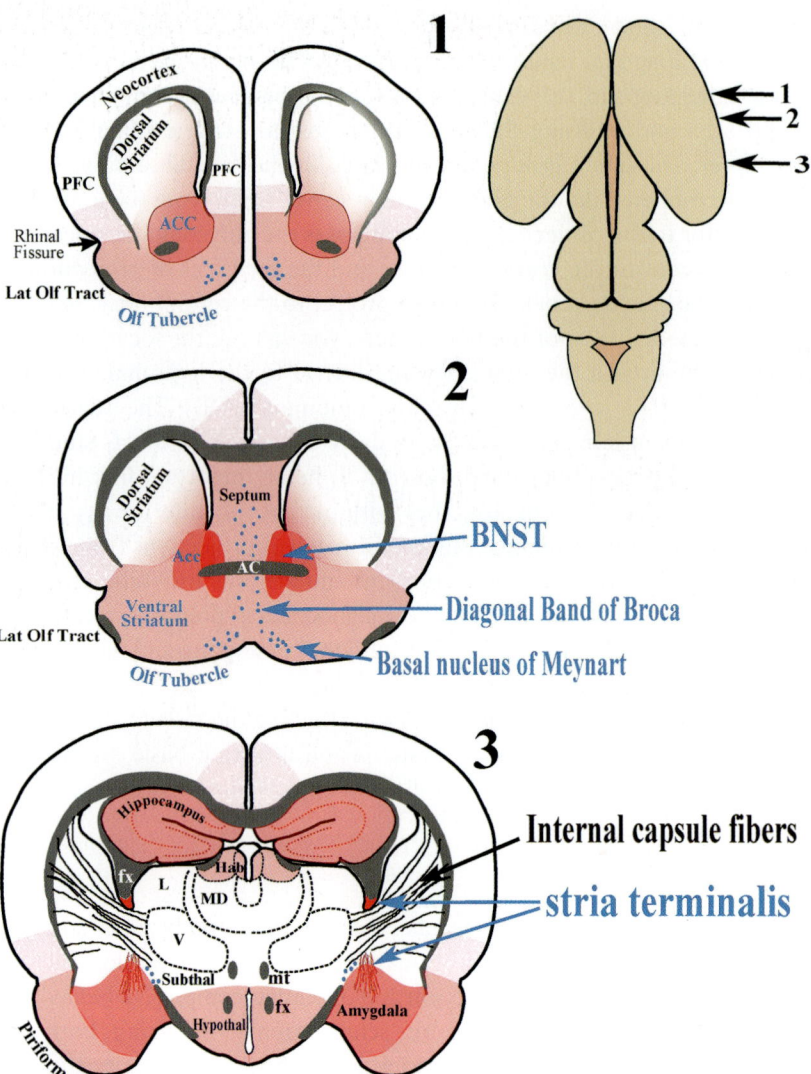

Figure 29.10
The drawings of frontal sections shown in figure 29.4 are shown again here; in addition to the amygdala and bed nucleus of the stria terminalis, additional structures of the basal forebrain are listed. Acetylcholine-containing neurons are indicated by blue dots, extending from the medial septal nucleus through the nuclei of the diagonal band of Broca and the basal nucleus of Meynart and extending into the substantia innominata. Abbreviations as in figure 29.4.

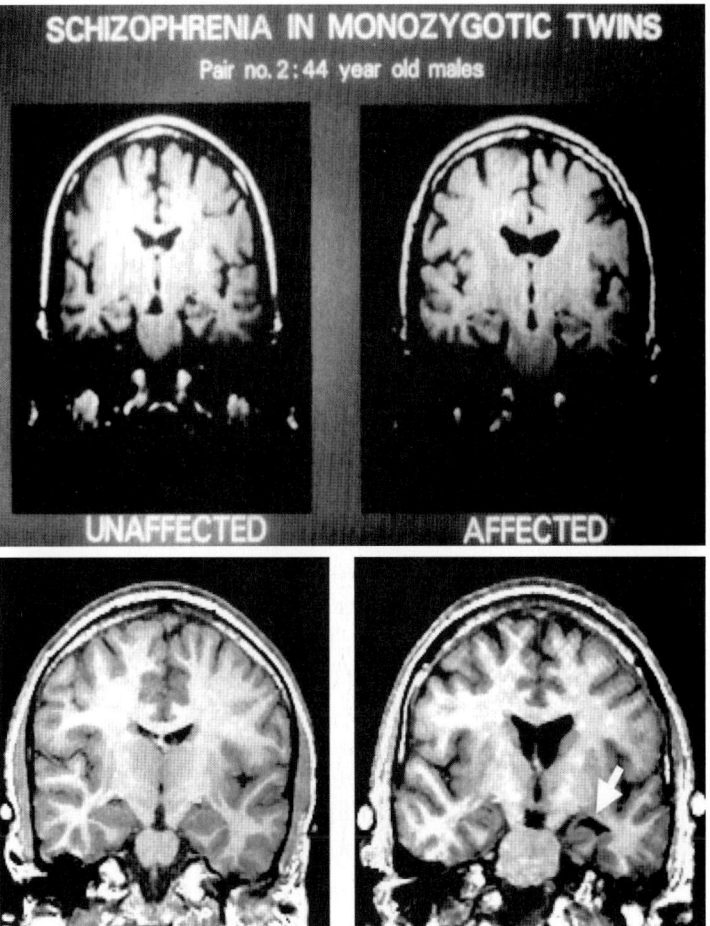

Figure 29.11
MRI scans showing coronal sections, comparing patients with and without schizophrenia. (Top) Scans of a pair of monozygotic twins, one with schizophrenia (right), and one without this pathology (left). The enlarged lateral ventricles of the affected twin are obvious. There is also some enlargement of the third ventricle. (Bottom) Similar coronal sections showing a typical result from a normal patient (left) and one from a schizophrenic patient (right). In this case, ventricular enlargement was greater in the left hemisphere — the right-hand side of the scan. The white arrow points to the temporal horn above the amygdala and hippocampus. From (top) Hyde and Weinberger (1990); (bottom) Shenton et al. (1992).

the seizures were originating. Histological examination of the tissue provided results that divided the patients into two groups: a group where the abnormality was a hamartoma and a group where the abnormality was a result of fever-induced seizures early in life. A hamartoma is a tumor-like malformation of prenatal origin, so the damage in these cases was considerably earlier in life. Thus, there were two types of amygdala lesions that resulted in the seizures, one of them an abnormality of much earlier origin.

The question was whether the behavioral history of these two groups, before the neurosurgical treatment, was different. Examination of medical records revealed a difference very relevant to the current discussion. The patients with the earlier amygdala lesions showed a significantly higher incidence of admissions for treatment of schizophrenia-like symptoms. To explain this, look at figure 29.12A. Illustrated are projections of the amygdala to prefrontal neocortex and to the basal forebrain, and also catecholamine axons that are part of the widely projecting axon systems originating in the midbrain and hindbrain (especially, dopamine-containing axons from the ventral tegmental area of the midbrain, or norepinephrine-containing axons from the locus coeruleus). The catecholamine axons just mentioned project to all three structures as well as to other structures (see figures 17.3 and 17.4). A lesion of the amygdala results in degeneration of projections to prefrontal cortex and to basal forebrain, and it also results in a kind of pruning of catecholamine axons. Predicting from studies of small animals (see chapters 13 and 19), an expected consequence is a sprouting of catecholamine axons as shown in figure 29.12B. An effect of the increased innervation of the prefrontal cortex and of the basal forebrain by dopamine and/or norepinephrine axons is an alteration of their functions. One more important piece of information: Some studies have shown abnormally high levels of these two catecholamines in both prefrontal and basal forebrain regions in some schizophrenic patients. Thus, we have a prenatal lesion hypothesis of the etiology of some types of schizophrenia.

It is important to note that many schizophrenic patients do not appear to fit this picture but do show abnormalities in density and distribution of catecholamine receptors.

Drug treatments for psychoses like schizophrenia have greatly reduced the population of these patients in mental hospitals since the last half of the twentieth century. Many of the effective antipsychotic drugs bind to neuronal receptors for the monoamines. (The monoamines include the catecholamines norepinephrine and dopamine and the indolamine serotonin. The prenatal lesion hypothesis described above could probably be extended to serotonin axons from brainstem midline cell groups.)

Other Behavioral Difficulties in People Are Tied to Amygdala Functions

As you may anticipate by now when thinking about the functions of the amygdala that we have reviewed, this structure is very involved in behaviors that can cause social and personal problems in humans. Various emotional problems that lead people to medical

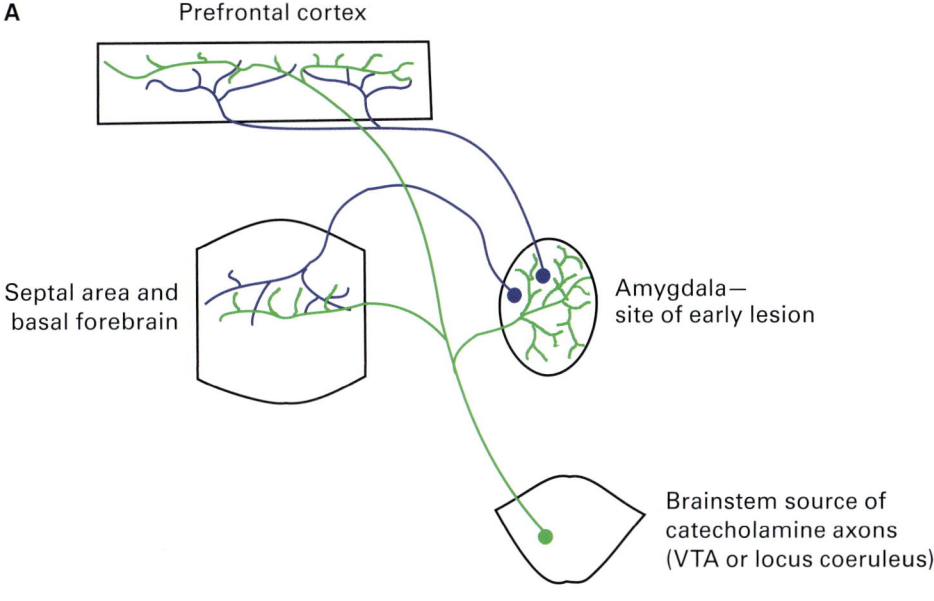

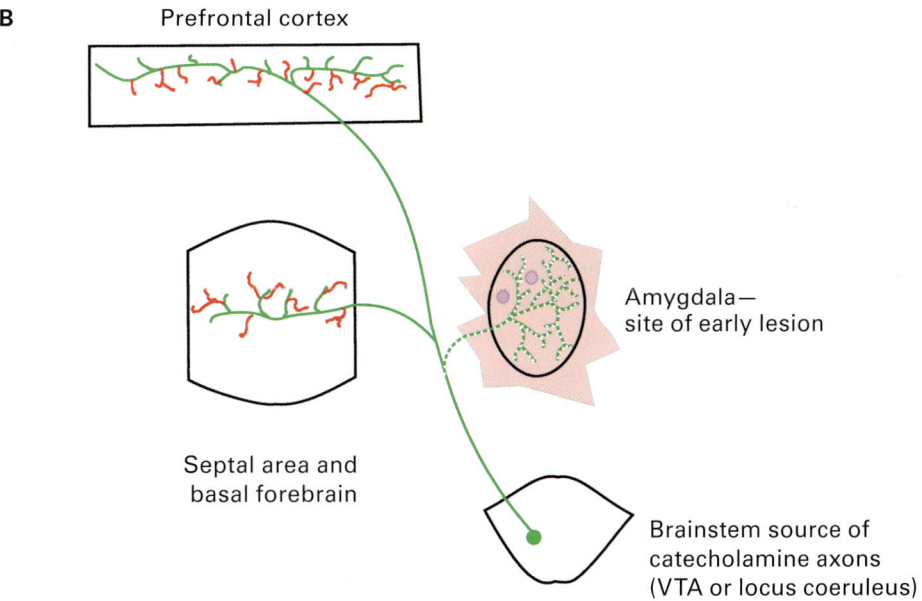

Figure 29.12
The diagrams illustrate the hypothesis that prenatal or perinatal brain damage is one cause of schizophrenia. Findings from effects of early brain injury in the visual and olfactory systems are applied to known pathways of the forebrain to illustrate postulated sprouting of axons in prefrontal cortex and basal forebrain after early damage to the temporal lobe, particularly the amygdala. (A) Converging axonal projections from the amygdala and from the sources of dopaminergic or noradrenergic projections are shown for a normal human brain. (B) Early damage to the temporal lobe with lesions of the amygdala remove projections to both prefrontal cortex and basal forebrain structures. Such lesions also eliminate projections of widespread axonal systems that normally reach the amygdala. A predicted consequence is the sprouting of dopaminergic (DA) and/or noradrenergic (NE) axons in both prefrontal cortex and the basal forebrain. A consequence of such sprouting would be abnormal activity in the areas of abnormal innervation by DA and/or NE axons. Based on Schneider (1979).

clinics can result in the prescription of drugs that alter the function of the amygdala and other limbic system structures. Examples are learned fears, which can cause a condition known as posttraumatic stress disorder. Extreme aggression—especially in males—can be strongly influenced by the connections of the amygdala. Many studies with animals like rats have demonstrated a strong role of androgens in aggressive behavior. The androgens exert effects on behavior by direct influences on limbic system neurons; besides the hypothalamus, a strong focus of these influences is the amygdala.

Central Roles of Basal Forebrain and Amygdala for Learned Affects

The world of objects and places is seldom a neutral world from a subjective perspective. It is a world of feelings because of the valences acquired by those objects and places. These valences correspond to various degrees of good or bad impressions as well as indifferent ones. They are the feelings, attitudes, and moods a person or any animal acquires from his experience with the places and objects in his or her environment.

Previously, you learned about the hippocampus and place memories (chapters 26 and 28), but that story is very incomplete without the circuits outlined in this chapter, the circuits of the basal forebrain and amygdala. The connections in these structures are plastic: They are connections of pathways from the olfactory system early in evolution and then, later in evolution, they are connections of cortical pathways of object representation. Changes in those connections occur according to what happens to the animal or person in the presence of those stimuli. The changes constitute the affective tags acquired by objects and living beings and places, and they underlie what amounts to emotional habits, the feelings and moods elicited by places and by the things located in those places. These feelings steer an individual through his or her environment (or through the inner memory of that environment)—both the social and the nonsocial environments.

For each place, the animal or person organizes his or her navigation according to good and bad, safe and dangerous, okay and not okay, or neutral, as the previously formed connections have it. Where the individual is located at the moment is remembered from the past as he or she updates his or her position in the inner model—the function of the hippocampus and its connections (including connections to the ventral striatum). The objects within his or her environment, if stable, are recalled in that model. If the objects are moving or have moved, then the model is subject to more active updating, so the individual's path is optimized according to a best judgment, a judgment that comes out of his or her already formed affective associations.

To the extent that the environment or the objects in it are novel, an animal explores to build up its inner world and establish the guide maps for the future. Each encounter affects her or him, but the effects are strongest when the consequences of an encounter evoke the strongest pleasures or the most negative reactions. Thus, in every living vertebrate those connections in the ventral striatum including the amygdala change and shape the affective meaning of the environment.

Readings

Brodal, P. (2004). *The central nervous system: Structure and function*, 3rd ed. Oxford University Press.

Downer, J. D. C. (1961). Changes in visual gnostic function and emotional behavior following unilateral temporal lobe damage in the "split-brain" monkey. *Nature, 191*, 50–51.

Ehrlich, I., Humeau, Y., Grenier, F., Ciocchi, S., Herry, C., & Lüthi, A. (2009). Amygdala inhibitory circuits and the control of fear memory. *Neuron, 62*, 757–771.

LeDoux, J. E. (2000). Emotion circuits in the brain. *Annual Review of Neuroscience, 23*, 155–184.

Mesulam, M.-M. (2000). Patterns in behavioral neuroanatomy: Association areas, the limbic system, and hemispheric specialization. In Mesulam, M.-M., ed. *Principles of behavioral and cognitive neurology*, 2nd ed. (pp 1–120). New York: Oxford University Press.

On connections of the amygdala:

Pitkänen, A., Savander, V., & LeDoux, J. E. (1997). Organization of intra-amygdaloid circuitries in the rat: An emerging framework for understanding functions of the amygdala. *Trends in Neurosciences, 20*, 517–523.

Pitkänen, A., Pikkarainen, M., Nurminen, M., & Ylinin, A. (2000). Reciprocal connections between the amygdala and the hippocampal formation, perirhinal cortex, and postrhinal cortex in rat: A review. *Annals of the New York Academy of Sciences, 911*, 369–391. (Issue entitled The Parahippocampal Region: Implications for Neurological and Psychiatric Diseases.)

Electrical stimulation of human amygdala:

Stevens, J. R., Mark, V. H., Erwin, F., Pacheco, P., & Suematsu, K. (1969). Deep temporal stimulation in man. Long latency, long lasting psychological changes. *Archives of Neurology, 21*, 157–169.

Schaltenbrand, G., Spuler, H., Wahren, W., & Wilhelmi, A. (1973). Vegetative and emotional reactions during electrical stimulation of deep structures of the brain during stereotaxic procedures. *Zeitschrift für Neurologie, 205*, 91–113.

Formulation of the concept of a ventral striatum was initiated by Heimer and Wilson at MIT. For more recent reviews and studies, see papers by Henk J. Groenewegen.

Heimer, L., & Wilson, R. D. (1975). The subcortical projections of the allocortex: similarities in the neural associations of the hippocampus, the piriform cortex, and the neocortex. In Santini, M., ed. *Perspectives in neurobiology* (pp. 177–193). Golgi Centennial Symposium. New York: Raven Press.

An interpretation of early temporal lobe lesion effects in humans is included in the following review by Schneider of studies of the effects of brain damage suffered early in life. The study of human patients summarized in the chapter is reported in the paper by Taylor:

Schneider, G. E. (1979). Is it really better to have your brain lesion early? A revision of the "Kennard Principle." *Neuropsychologia, 17,* 557–583.

Taylor, D. C. (1975). Factors influencing the occurrence of schizophrenia-like psychosis in patients with temporal lobe epilepsy. *Psychological Medicine, 5,* 249–254.

XI CORPUS STRIATUM

30 The Major Subpallial Structure of the Endbrain

We know from previous chapters that the corpus striatum has expanded considerably in the evolution of the chordates with the invasion of non-olfactory inputs, first from the thalamus and then by way of the neocortex. Primitive cortical structures, present long before the evolution of mammals and dominated by olfactory inputs, projected to the ventral striatal structures discussed in the previous chapter. But as the neocortex appeared and expanded in mammals, its projections to the striatum went mostly to the dorsal striatum, which, with its non-olfactory inputs, was segregating from the ventral portion.

In chapter 24, I outlined a proposed evolutionary progression in which the appearance and expansion of the corpus striatum played a central role. Check that story again by inspecting the diagrams of figure 24.7. Now I want to go through the steps in the proposed evolutionary progression again and point out some connections found in modern vertebrates—especially in mammals including primates—that fit with this picture of striatal evolution.

Beginnings: A Link between Olfactory Inputs and Motor Control

It is likely that the ventral striatum originated as a link between olfactory inputs and motor control. In mammals, the structures in front of the optic chiasm known as the olfactory tubercle and the adjacent piriform cortex receive, in their surface layer, direct projections from the olfactory bulb. Deeper portions of these structures have longer projections, and some of the connections made are links to outputs, some of which long predate the evolution of mammals. These outputs go to the hypothalamus and subthalamus, including the region that became the hypothalamic and subthalamic locomotor areas; projections also go to the epithalamus.

Outputs from these targets, as well as more direct connections, pass to the midbrain for two major types of motor control (chapters 11 and 14): locomotion toward or away from something, and orienting of head and eyes. The midbrain controller of locomotion, in receipt of striatal, subthalamic and hypothalamic outputs, is called the midbrain locomotor area. (Many of the striatal outputs are not direct, as noted later.) The midbrain controller of orienting movements is the midbrain tectum (known in mammals as the superior colliculus). Striatal influences on orienting in many vertebrates depend on connections to the substantia nigra in the ventral midbrain, which has a major output to the dorsally located tectum.

The early links to the motivational and motor systems were (most probably) modifiable links—capable of experience-induced change. Habits of responding to specific odors could form. The advantages of this modifiability were doubtless of great importance and are most likely why, in early evolution, the non-olfactory modalities began to invade the striatal area and expand. This must have begun very early in chordate evolution, before the advent of the jawed vertebrates.

The modification of connections occurred not because of mere repetition but because of feedback from sensory systems monitoring the consequences of the behavior. There evolved an efficient means of signaling good consequences—an activation of what we call reward systems (sometimes called pleasure systems). For example, good tastes as well as good odors, and other sensory effects of approach or consummatory activity, activated dopamine pathways originating in caudal hypothalamus or ventral midbrain that ascended to the striatum, briefly altering the state of that part of the forebrain.

Figure 30.1A illustrates some of the main connections of the mammalian substantia nigra. This illustration shows some of the connections just mentioned. Note the projection to the optic tectum (superior colliculus) and the dopamine pathway to the corpus striatum (caudate nucleus and putamen). A dopamine pathway to the ventral striatum probably evolved earlier. In mammals, much of it comes from the ventral tegmental area, located just medial to the substantia nigra.

Figure 30.1B shows the same connections with two additional inputs to the nigra worth special mention because they arise in structures of the limbic system and could be involved in the signaling of the good or bad consequences of a movement. One comes from the central nucleus of the amygdala. The other originates in the lateral hypothalamus. In addition, because nigrothalamic axons have been shown to connect strongly to the ventromedial thalamic nucleus (VM)—a nucleus of the midline group—I have depicted the projections of that nucleus, which are fairly widespread; they go also to the dorsal striatum. This projection may have an arousal function.

Non-olfactory Inputs Invade the Striatum

Looking again at figure 24.7, we see depicted the second major stage of striatal evolution. Non-olfactory inputs began to reach the striatum from the 'tweenbrain, and over evolutionary time these inputs became segregated in the more dorsal parts, the parts that became the caudate nucleus and the putamen. It is likely that this happened because of the adaptive advantages of the habit formation that these connections made possible. Then, these advantages led to an expansion of the subpallial structures.

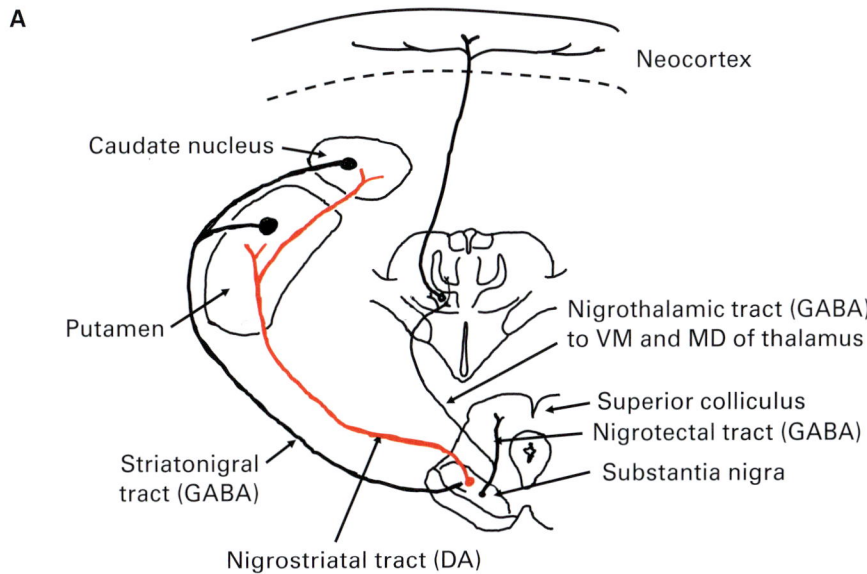

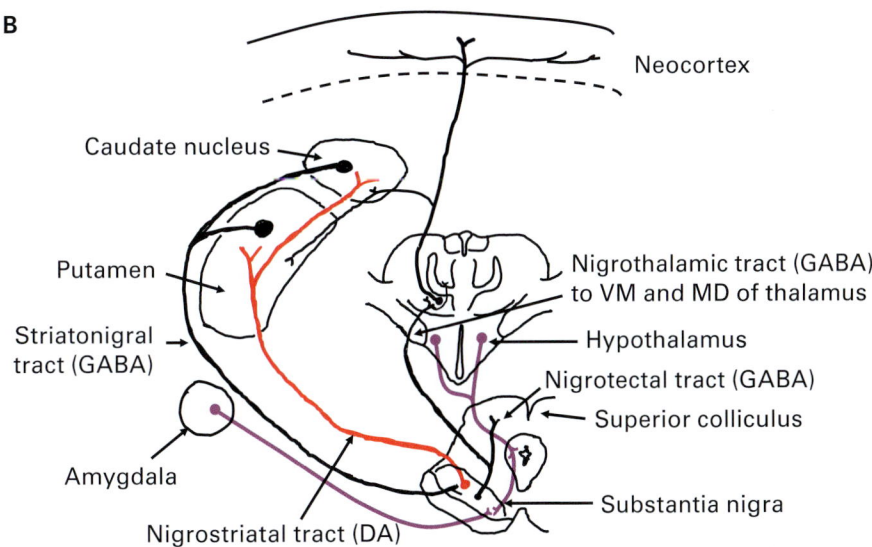

Figure 30.1
(A) Diagrammatic illustration of some of the main connections of the mammalian substantia nigra. Omitted, among other connections: nigral projection to the pedunculopontine nucleus of the mesencephalic locomotor region. The ventromedial nucleus of the thalamus (VM) projects widely to central and anterior cortical areas and may function in arousal situations. The nigrothalamic pathway also projects to the laterodorsal thalamic nucleus and to the caudal intralaminar cell group, the parafascicular nucleus. (B) The same connections as shown in (A) are repeated together with some additional connections discovered in experimental studies of the rat brain. The ventromedial thalamic nucleus projects also to the striatum; the nigra receives projections from hypothalamus and from the central nucleus of the amygdala.

Figures 30.2 and 30.3 illustrate these diencephalic projections, which have persisted in modern mammals. The first is a sketch that shows a convergence of dopamine axons from the substantia nigra and axons carrying sensory information from neurons of the so-called intralaminar thalamus. The next (figure 30.3) is a drawing of a Golgi-stained neuron from this part of the thalamus in a rodent. The ascending axon arborizes widely, reaching not only the dorsal striatum but also the neocortex. Similar neurons have been found in other intralaminar thalamic cell groups and also in midline thalamic groups. Some neurons projecting to both neocortex and striatum have also been found in other cell groups of the thalamus.

The evolutionary story of non-olfactory inputs invading the endbrain can be told not only for the corpus striatum but also for the pallium. Thalamic projections to nonstriatal structures of the endbrain have been found in sea lampreys as well as in tetrapods. These inputs to the endbrain became focused on dorsal ventricular ridge structures, including the amygdala, in amphibians and reptiles, and on the dorsal pallium (or dorsal cortex). In mammals, part of the dorsal cortex evolved into the neocortex. In reptiles and mammals, the endbrain structures became the sources of greater input to the striatum than the inputs from the dorsal thalamus. The expansion of the endbrain accompanying these changes was greater for neocortex than for the striatum.

Visualizing the Striatum and Other Major Structures in the Cerebral Hemisphere

We can visualize the positions of these endbrain structures in a mammal like the rat, mouse, or hamster in a single picture. The picture would be similar for an embryonic human brain 2–3 months postconception. Imagine the right hemisphere after the entire brainstem including thalamus has been removed (figure 30.4). Because the view is from the medial side, our view of neocortical areas is very limited, but, in the figure, visible portions of motor cortex, somatosensory cortex, and visual cortex are labeled. Auditory cortex cannot be seen in a medial view, but a label points to its position that is hidden from view. The infolded dorsomedial and caudal edge of the cortex is the hippocampus, labeled in the figure, which omits the fibers passing between the hippocampus caudally and the septal area rostrally. The dorsal and ventral portions of the corpus striatum are separately labeled, and the "caudal outpost" of the ventral striatum, the amygdala, is shown. (Much of the amygdala is of striatal origin, but some parts are pallial derivatives.)

Three coronal sections are also illustrated in the figure, with the sections made at the levels indicated by vertical lines above the lower drawing. It will be useful for your learning about brain structure to study these sections and the medial view drawing. You can find sections made at similar levels in brain atlases of rodents or on the Web.

Major Outputs of the Mammalian Endbrain: The Position of the Striatum

In the teaching of human anatomy, it has been common to divide the motor system into pyramidal and extrapyramidal systems (see chapters 12, 14, and 15). The term *pyramidal*

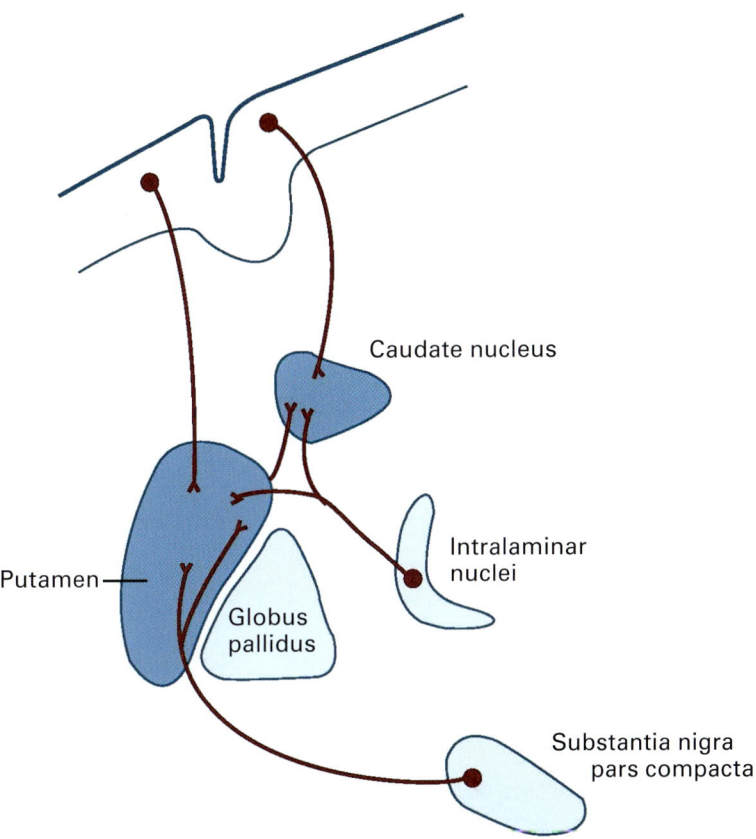

Figure 30.2
Illustration of major inputs to the dorsal striatum (caudate nucleus and putamen) in mammals. These inputs come from the neocortex (all regions), from the intralaminar nuclei of the thalamus (as exemplified by the centromedian nucleus in large primates), and from the substantia nigra. The latter projection comes from the dopamine-containing neurons of the pars compacta. The intralaminar thalamic nuclei receive sensory projections from multimodal cells of the midbrain tectum, also projections from the hindbrain (including vestibular and cerebellar), and somatosensory projections from the spinal cord. Based on Brodal (2004).

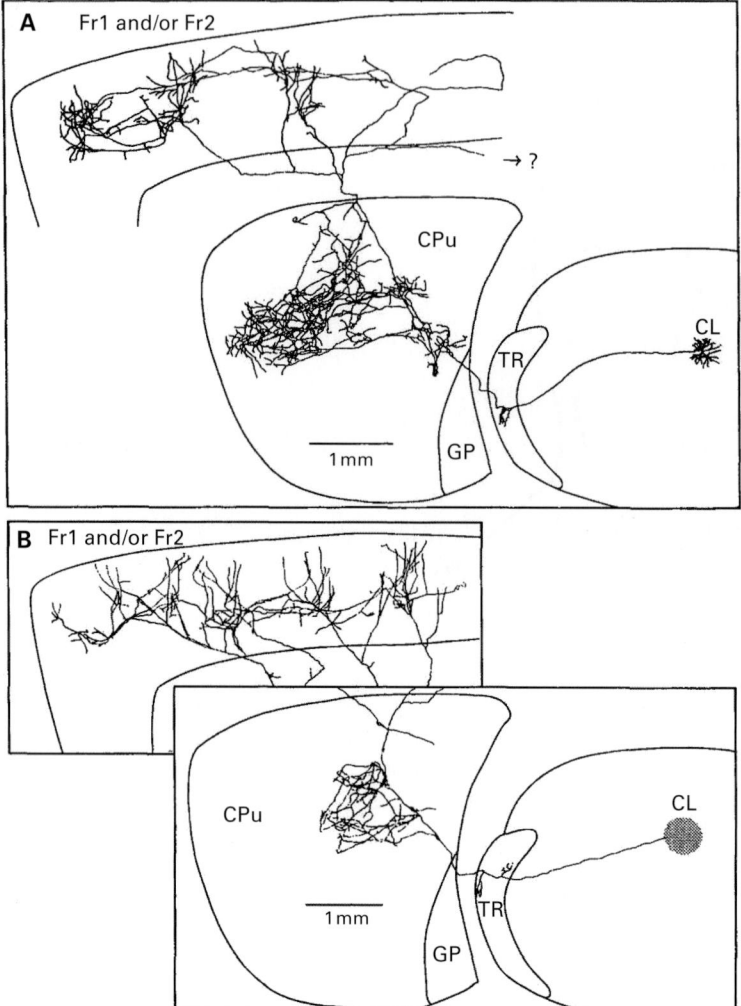

Figure 30.3
Examples of axons from the diencephalon of a mammal that form terminal arbors in the dorsal striatum and then continue into the neocortex. The cell bodies of the two axons illustrated (panels A and B) are in nucleus centralis lateralis (CL), an intralaminar cell group. The axons have collateral projections to the thalamic reticular nucleus (TR), then form extensive end arbors in the caudate-putamen (CPu) before continuing into frontal regions of neocortex. (Panel B is separated into two parts as part of the axon's trajectory to the neocortex is not included.) Other axons of intralaminar neurons may form additional arbors in other cell groups, including the subthalamus. Other studies have indicated that only a minority of intralaminar cells project to both striatum and neocortex. GP, globus pallidus; Fr, frontal neocortex. From Deschenes, Bourassa, and Parent (1996).

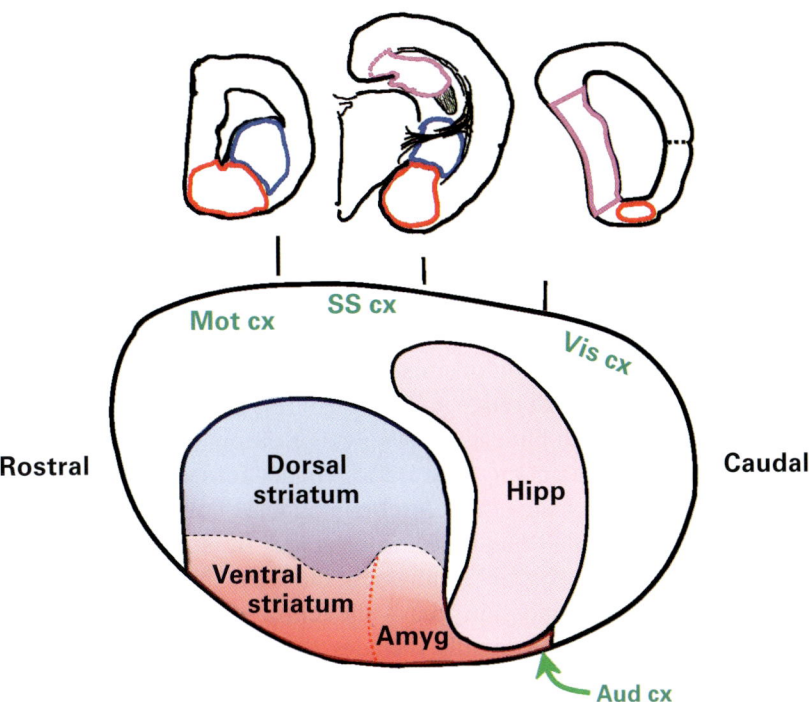

Figure 30.4
Sketches of the right hemisphere of a small rodent, illustrating the positions of the dorsal striatum (indicated in blue), the ventral striatum including the amygdala (in red), and the hippocampus (medial pallium, in violet). At the top are three frontal sections with the three structures outlined. At the bottom is a view of the right hemisphere from the medial side, rostral to the left, with outlines of the positions of the same three structures. (In the case of the amygdala, for simplicity, parts derived from the embryonic pallium are included together with parts derived from the embryonic striatum.) Amyg, amygdala; Aud, auditory; Cx, cortex; Hipp, hippocampus; Mot, motor; SS, somatosensory; Vis, visual.

refers to the pyramidal tract, the name for the corticospinal fiber bundle at hindbrain levels below the pons. These axons play a dominant role in the human movements we often refer to as voluntary. The extrapyramidal system is a grab-bag term that includes everything else that affects movement. Its largest structures in large mammals are the corpus striatum and the cerebellum. This division of the motor system becomes very confusing when we hear a neuroanatomist state that "the major output of the extrapyramidal system is the pyramidal system." The statement reflects the fact that major outputs of the corpus striatum and the cerebellum in mammals go to the thalamus, and the thalamus projects to the neocortex. The parts of the neocortex involved are motor cortical areas, the areas that project most heavily to the spinal cord through the pyramidal tract. See the summary diagram of figure 12.7. In that figure, the two major groups of axons descending from the endbrain are depicted in a simplified manner using different colors.

The figure is based on the neuroanatomy of modern mammals, especially the large primates. It omits smaller pathways left over from earlier stages of chordate evolution. Thus, it stresses neocortical control. It also omits the portions of the striatum that we believe evolved first; namely, the ventral striatum. In figure 30.5, we have added the ventral striatum to another version of this kind of summary network diagram, an addition that makes it more complex but more informative. (All structures depicted are found on both left and right sides of the brain.) Take special note of the output pathways from the neocortex and the limbic cortex: control of more caudal structures by neocortex is much more direct. Control of limbic system functions depends heavily on links in the hypothalamus and midbrain.

For another type of illustration of some of the major outputs of both cortical and the striatal components of the endbrain, see figure 30.6. This figure illustrates the origins of the medial and lateral forebrain bundles, the two major output pathways of the endbrain. In the previous, more diagrammatic figure 30.5, connections of these two pathways are depicted on the left side for the lateral forebrain bundle and the right side for the medial forebrain bundle. The medial bundle of axons comes from the limbic structures of the endbrain and courses through and terminates in the lateral hypothalamus. Some axons go further, reaching the limbic midbrain areas—the central gray area and the ventral tegmental area. The lateral forebrain bundle includes the axons from neocortex that course through the striatum on their way to the brainstem and spinal cord (chapter 12). It also includes the outputs of the dorsal striatum and dorsal pallidum. The "pallidum" here refers to the globus pallidus, the location of cells that are major recipients of outputs of the dorsal striatum (caudate nucleus and putamen); these cells project caudally, serving as the major output cells of the striatum. (The ventral striatum has a pallidal component also; we have included this component together with the ventral striatum in figures 30.5 and 30.6.)

Simplified Pictures of Dorsal Striatal Connections

Figure 30.7 presents another picture of striatal connections, even more simplified. The purpose of this redundancy is to reinforce in your mind some major connections of the dorsal striatum in mammals. It omits the more ancient projections to the striatum, from the thalamus, as well as the related but different connections of the ventral striatal components. In studying this figure, remember the large and numerous outputs of the neocortex to all levels of the CNS, although these are not shown in the figure. You can see that the dorsal striatum (the caudate nucleus and the putamen) has two major outputs—to the pallidum and to the substantia nigra of the midbrain. Each of these structures projects to the thalamus and thereby influences the neocortex, including the motor cortex. Each of them also projects to the midbrain, thereby influencing both locomotion and orienting movements. The nigra provides major feedback to the striatum by way of dopamine projections.

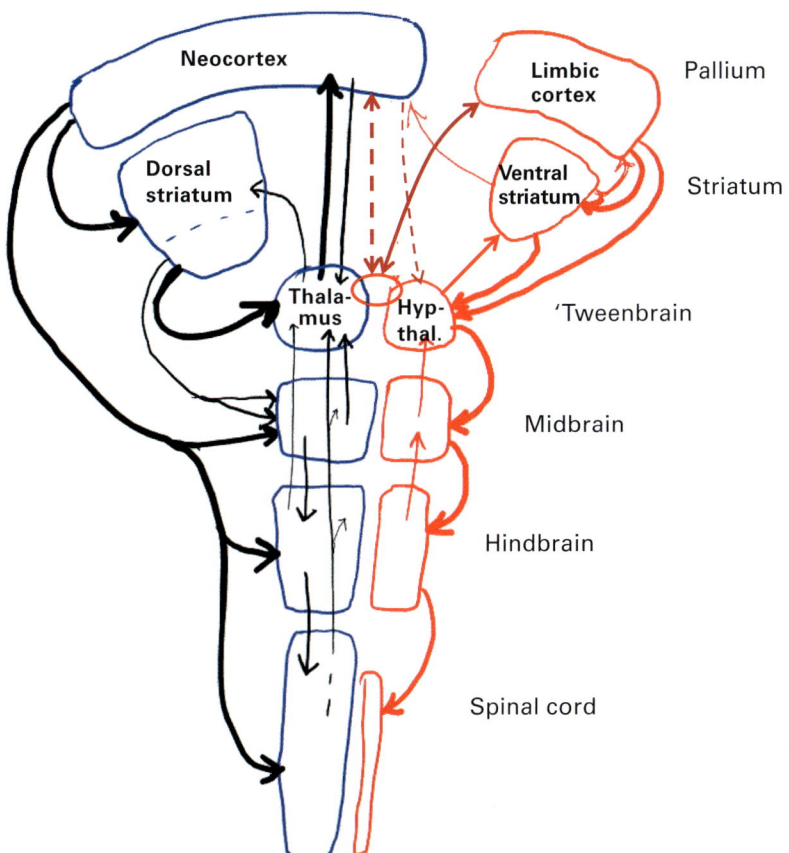

Figure 30.5
Diagram of major routes of information flow in and out of the endbrain of mammals. On the left side of this schematic are the structures related to the somatic systems (in blue color); on the right side are the structures of (or closely related to) the limbic system and the autonomic nervous system (in red color). Note especially the dorsal and ventral striatum where the links are believed to be plastic. Diffuse systems of axons (with very widespread connections), along with many others, are omitted. The diagram is a more detailed version of what was presented in figure 12.7.

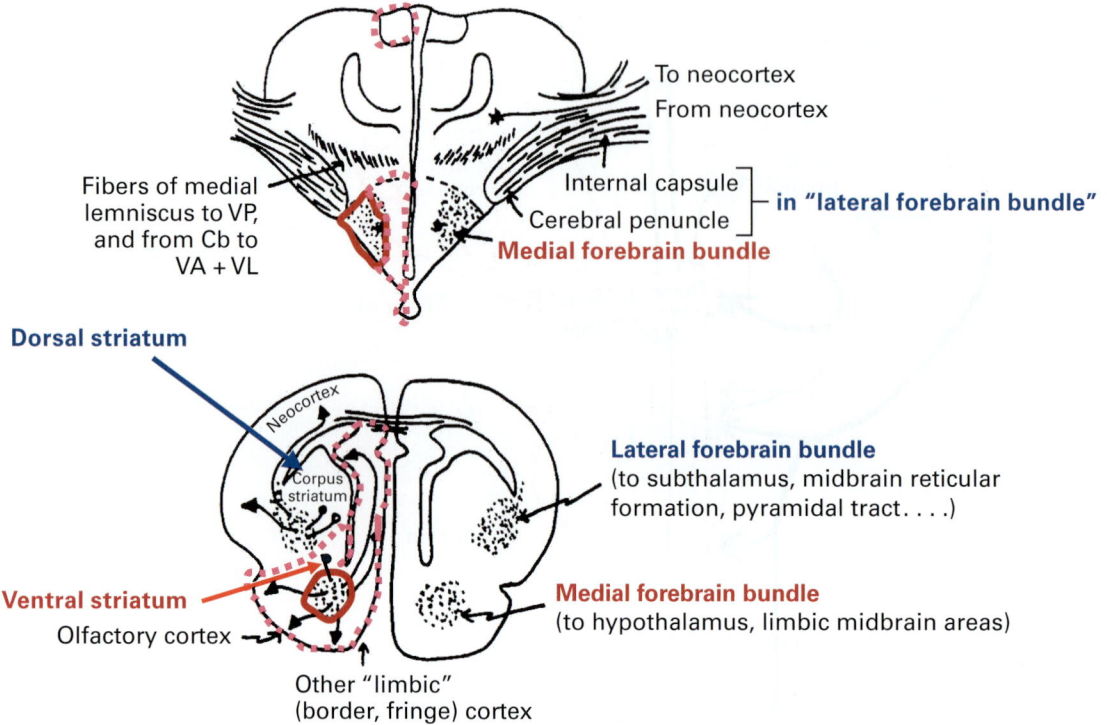

Figure 30.6
Sketches of frontal sections through the diencephalon (top) and telencephalon (bottom), similar to figures 12.4 and 12.5. Note that the dorsal striatum (together with its output structure, the globus pallidus) is related to the lateral forebrain bundle as is most of the neocortex. The ventral striatum (together with ventral pallidal structures) is related to the medial forebrain bundle as is the medial pallium. Cb, cerebellum; VA, thalamic nucleus ventralis anterior; VL, thalamic nucleus ventralis lateralis; VP, thalamic nucleus ventralis posterior.

Now, look at figure 30.8, where the same connections are illustrated in a somewhat more realistic schema. This illustration includes a depiction of the neocortical output coursing through the striatum in the internal capsule. It shows how the axons from the globus pallidus penetrate the bundle of corticofugal axons in order to reach the ventral thalamus (the ventral anterior and also the ventral lateral nuclei). This bundle of axons penetrates the cerebral peduncle (some of them going around its ventral edge) and, following a looping course around cells of the subthalamus, finally enters the thalamus. To get there, the axons first run caudally, then make the loop (around the zona incerta) and turn rostrally to reach their major destination in the anterior part of the ventral thalamic nucleus. Thus, the pathway has the shape of a handle, and for this it is named the *ansa lenticularis*—the handle of the lenticular nucleus. *Lenticular* means lens shaped, for the shape of the globus pallidus and putamen (lateral to the fibers of the internal capsule as in the figure). The details of the peculiar looping trajectory are not depicted in the figure.

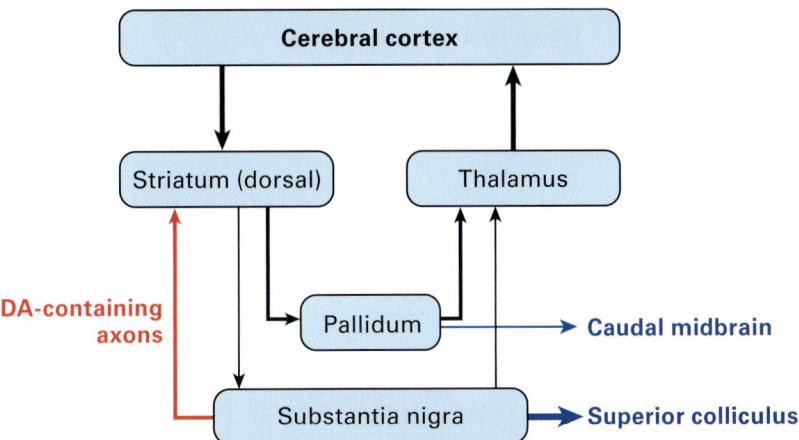

Figure 30.7
Simplified information-flow diagram showing the major connections among the cerebral neocortex, the dorsal striatum, the dorsal pallidum, the thalamus, and the substantia nigra. In addition, two important outputs from the nigra and the pallidum to the midbrain are indicated. The structure in the caudal midbrain is the midbrain locomotor area. Based partially on Brodal (2004).

In figure 30.8 there is also information about the physiology of these connections: many of them are inhibitory. In fact, you can see in the figure that there are inhibitory connections that inhibit neurons that themselves make inhibitory connections: a double inhibition in the striatal pathways. In thinking about how this system functions, we cannot assume that all long-axon connections are excitatory. These physiological facts help to explain some human behavioral disorders that arise from pathology of striatal components. These disorders include the uncontrollable writhing movements (choreiform or dance-like movements) of Huntington's chorea, and the sudden, violent flinging movements of the limbs in hemiballism. The tremors of Parkinson's disease cannot be explained simply in terms of reduced inhibition, but there is a loss of normal control of lower-level regulation of muscle length or tone.

Diaschisis after Loss of Inhibitory Connections

The disorders mentioned above are closely related to the diaschisis phenomena we discussed previously. Diaschisis was defined in chapter 7 as a depression of function in a structure that has suffered major loss of excitatory connections due to a lesion. However, when the brain damage causes a loss that is mainly of inhibitory rather than excitatory connections, the functional effect is not a depression but rather an overexcitation by the excitatory connections that are no longer kept in check by inhibition.

Thus, the opposite of the diaschisis described in the earlier chapter can occur. The functional problems caused by deafferentation are most likely to happen when the brain

572   Chapter 30

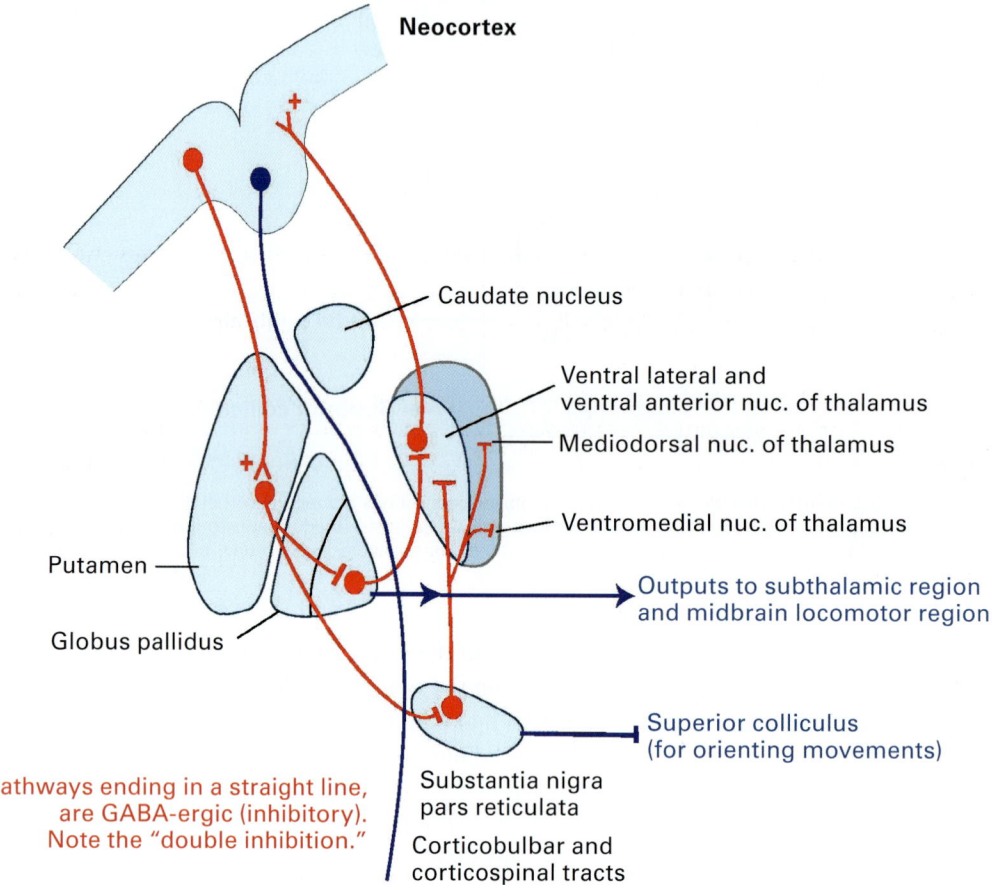

Figure 30.8
Another illustration of some major connections of the mammalian dorsal striatum. Pathways in red are inhibitory except where indicated by "+." Note the *double inhibition*. Striatal inputs from the neocortex are excitatory, but the pathway from dorsal striatum to globus pallidus is inhibitory, and the pathway from there to the thalamus is also inhibitory. Similarly, the routes involving the substantia nigra show an inhibition of an inhibitory input. The thalamocortical projection is excitatory. Also included are the ancient outputs to the midbrain locomotor region and to the superior colliculus. The pathway to the superior colliculus from the striatum also involves inhibition of an inhibitory pathway. Not shown are the dopamine pathways and the ventral striatum. Based partially on Brodal (2004).

damage involves a structure that has evolved so that it has become dominant over a function, because of the quantity of its axonal connections. In the case of the striatal structures, the function involves movements of the body used in learned habits, and these structures use inhibitory pathways in their major outputs. The result, a tonic overexcitation of the deafferented structures, causes chorea or ballismus; it may also contribute to abnormal tremors, depending on the structure with the major deafferentation.

You may recall the discussion in chapter 7 concerning recovery from diaschisis. In the case of depression of function, recovery means a gradual return of lost functions. Two kinds of changes have been discovered to underlie this recovery: collateral sprouting and denervation supersensitivity. There is evidence for both kinds of change in the dorsal striatum after partial denervation of dopamine-containing afferent axons. If it were not for these changes in response to the progressive loss of more and more axons coming from the substantia nigra, a person with Parkinson's disease would experience motor symptoms much earlier, before 80% of the nigral dopamine neurons were lost. However, it is also possible that too much sprouting or supersensitivity could result in other movement abnormalities.

The Striatal Satellites Add Complexity to the System

To understand the functions of the striatal system as well as the disorders of striatal pathologies, one must know more about the "satellites" of the corpus striatum. These include not only the substantia nigra illustrated in figure 30.8, but also the subthalamic nucleus (STN), a clearly delineated component of the subthalamus that is especially prominent in large mammals. Both of these satellite structures are illustrated in the network diagram of figure 30.9. Note the reciprocal connections of the STN with the globus pallidus (GP), and the projections of the STN to the nigra. Note also how the GP inhibits the STN neurons whereas the axons of STN cells form excitatory connections with the cells of both the GP and the nigra.

Because of the strength of its connections, a lesion of the subthalamic nucleus on one side in a human results in a release of normal restraints in the control of limb movements on the opposite side of the body, so the individual shows violent flinging movements when he or she tries to make a limb movement. The abnormal ballistic movements sometimes occur without any apparent intention.

There are two additional facts of particular importance to be gleaned from figure 30.9. First, note that the striatum (the putamen and the caudate nucleus) has separate projections to two distinct subdivisions of the substantia nigra. One is to the *pars compacta*, which contains the dopamine neurons that project back to the striatum, as shown by the gray arrow. The other is to the *pars reticulata*, which projects to the superior colliculus and to the locomotor region of the caudal midbrain, and also to specific thalamic nuclei. The two striatal projections have separable origins, as we shall discuss later.

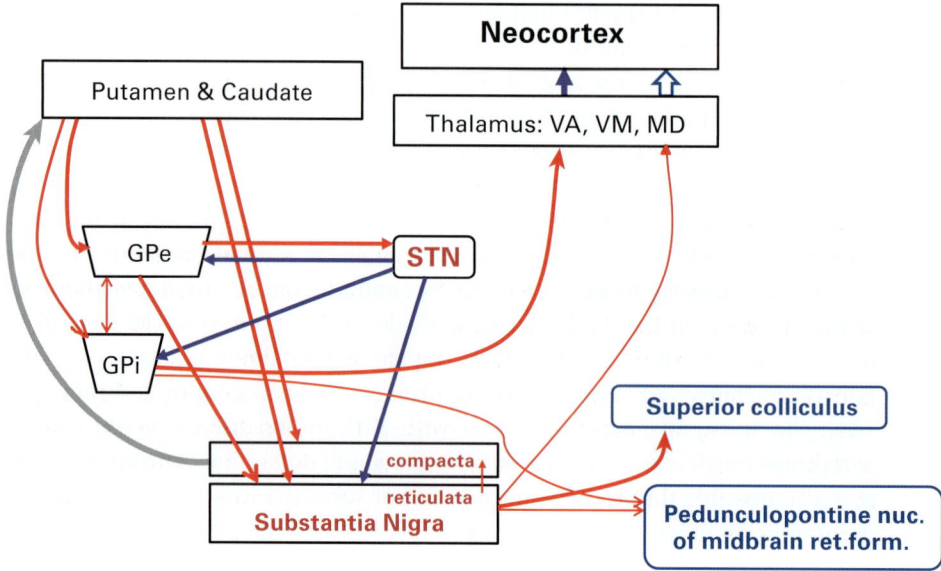

Figure 30.9
Explaining some movement disorders requires knowledge of striatal *satellites*. Both the substantia nigra and the subthalamic nucleus (STN, within the ventral thalamus, or subthalamus) are major satellite structures. The diagram shows the connections of these two structures with the dorsal striatum (caudate and putamen), the external and internal segments of the globus pallidus (GPe and GPi), and outputs to the thalamus and to the superior colliculus and the pedunculopontine nucleus of the midbrain locomotor area. Inhibitory connections in red (using GABA), excitatory in blue (using glutamate). The gray nigrostriatal connection uses dopamine as transmitter. MD, mediodorsal nucleus; VA, nucleus ventralis anterior; VM, nucleus ventralis medialis.

The other point for special notice in the figure returns to a point made above concerning the many inhibitory connections and the consequences of *double inhibition* in multisynaptic pathways. Note that there are direct and indirect routes from the dorsal striatum to the substantia nigra, pars reticulata. Following the direct route, there are two inhibitory connections in the pathway to motor control systems or to the neocortex. The inhibiting of an inhibitor should result in an increased excitation. The same is true for the pathway from striatum to the internal segment of the globus pallidus to the thalamic nucleus VA (nucleus ventralis anterior), which projects to motor cortex. Following the indirect pathways to the same structures, the consequences of striatal activity for the VA and for the midbrain tectum and the locomotor area should result in increased inhibition. We will return to these two routes to motor control systems later in the chapter.

Neocortex Dominates the Inputs to Dorsal Striatum of Large Mammals

Early in the evolution of the chordate line leading to the vertebrates, the dorsal striatum began to differentiate from the ventral striatum as connections with non-olfactory system axons formed. These axons came from the dorsal thalamus, and probably also the ventral thalamus (subthalamus). As such inputs also reached the pallium and the pallium also projected to the striatum, the pallial projections gained more and more importance as this region enlarged, especially the region that evolved into the neocortex of mammals. The projections to the striatum came from all parts of the cortex, both neocortex and limbic cortical areas. In present-day mammals, they are bilateral but are much stronger to the ipsilateral side. There is some topographic order in these projections, but it is not a simple point-to-point topography. Visual and auditory cortex, mainly the regions outside the primary sensory areas, project to the caudate nucleus's tail and middle regions, but not to the head of the caudate. The projections from central portions of the neocortical mantle, which are the motor cortex and the somatosensory cortex, go to the putamen, lateral to the fibers of the internal capsule in humans and many other large mammals.

The topographic distribution of axons is illustrated for the large primates in figure 30.10a. Not shown are complexities observed in axonal tracing studies. The functions of these complexities are not well understood. The multimodal association areas of the monkey project to longitudinally arranged regions that extend from the head of the caudate anteriorly and extend posteriorly throughout the portions known as the tail of the caudate (illustrated in the figure). The caudate tail can be followed to a position ventral to the internal capsule in its extension into the temporal lobe (see figure 30.12 later in text). The projections from posterior and anterior association areas are largely segregated, but they overlap with projections from unimodal areas.

Another interesting complexity of these projections is an interdigitation. For example, although there is some overlap between striatal afferents from the monkey's lateral prefrontal cortex and from posterior parietal cortex, there are larger regions where the two systems of axons terminate in separate patches immediately adjacent to each other.

The topography of cortex-to-striatum projections explains differences in functions of different parts of the dorsal striatum, because the functions depend on the inputs. Thus, damage to the putamen will cause problems with motor functions. Eye movements are affected by damage to the portions of the caudate head that receives inputs from the frontal eye fields, anterior to primary motor cortex, or inputs from parietal cortex—a region that is also important in eye movement control. Damage to various other portions of the caudate nucleus (particularly the large head of the nucleus), portions that receive inputs from prefrontal association cortex, results in problems with various cognitive functions.

Figure 30.10b illustrates findings on corticostriatal topography in the rodent brain. Note especially that the distinction between a dorsal and ventral striatum is more properly expressed as a separation of dorsolateral and ventromedial portions.

Topography in the Ventral Striatum

The details of topographic organization of ventral and ventromedial portions of the striatum and pallidum are less known. Larry Swanson has outlined three striatal–pallidal

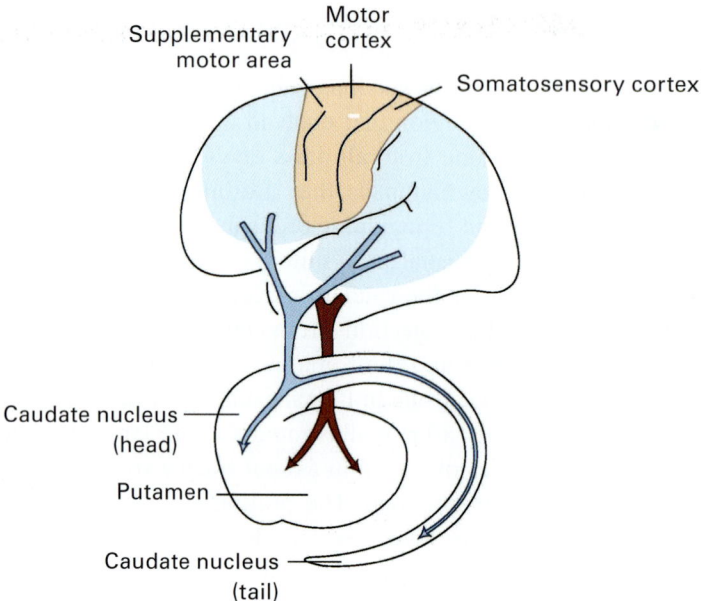

Figure 30.10a
Dominant inputs to the dorsal striatum in mammals come from the neocortex. Topography of corticostriatal projections: sensorimotor areas project to the putamen; prefrontal areas project to the head of the caudate nucleus; posterior neocortical areas project to the caudate—tail and the medial part of the head. Based on Brodal (2004).

divisions in addition to the dorsal striatum and pallidum. These three divisions are all part of what has been termed *ventral striatum* and *ventral pallidum* in this book (see figure 30.4). The three divisions are considerably interconnected, but the separations may prove very useful for interpreting functional studies. For reference, Swanson's groupings are shown in table 30.1.

As already indicated in figure 30.5, ventral striatal structures receive input from limbic cortical structures and other portions of the endbrain that are closely connected to the hypothalamus. Projections of some paralimbic cortical areas to ventromedial striatal areas are summarized in figure 30.10b. Projections from limbic areas also show some organization. For example, major inputs to the nucleus accumbens come from the amygdala and hippocampus, also from the dopamine-containing axons originating in the ventral tegmental area of the midbrain. The lateral septal nucleus receives major inputs from the fornix fibers originating in the hippocampal formation as well as from the hypothalamus. The portions of the amygdala that can be classified as striatal receive inputs from olfactory cortical portions of the temporal lobe and from other portions of the amygdala like the lateral nucleus—a nucleus that receives non-olfactory sensory inputs and inputs from association areas of neocortex.

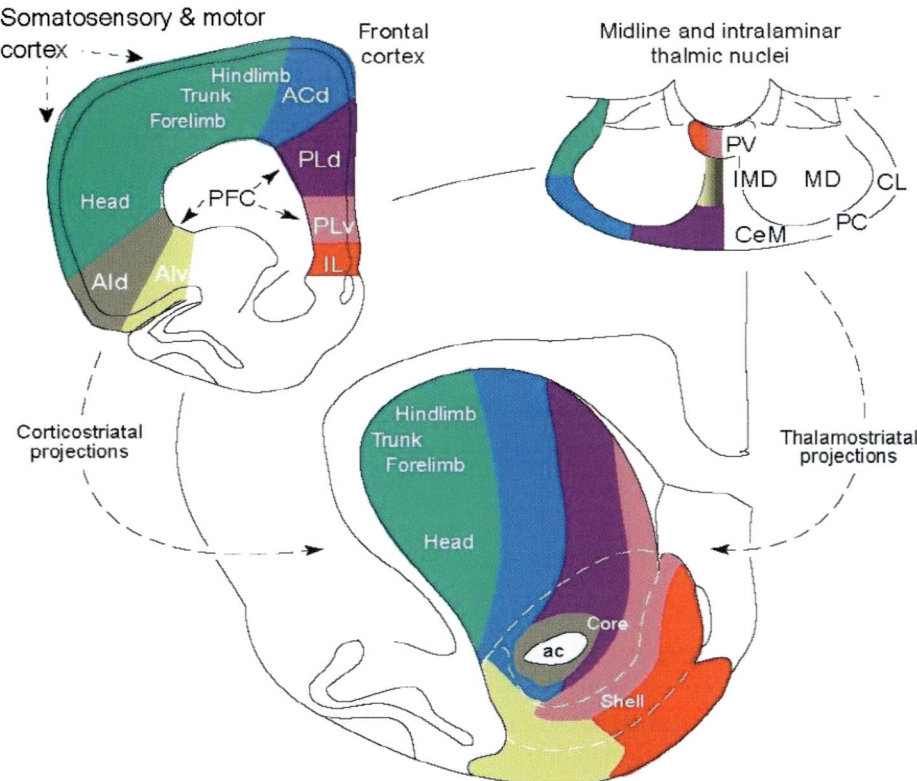

Figure 30.10b
Summary of findings on projections to the rodent striatum (lower central picture) from neocortex (upper left) and from paleothalamic cell groups (upper right). Such studies, as well as others, have indicated that the separation of dorsal and ventral striatum is not sharp but can be represented better as a gradient from dorsolateral to ventromedial. Not included here are results of studies that show projections from parts of the amygdala and from hippocampal formation to the ventral striatal structures. ac, anterior commissure; ACd, anterior cingulate cortex; AId, dorsal agranular insular cortex; AIv, ventral agranular insular cortex; CeM, central medial thalamic nucleus; CL, central lateral thalamic nucleus; IL, infralimbic cortex; IMD, intermediodorsal thalamic nucleus; MD, mediodorsal thalamic nucleus; PC, paracentral thalamic nucleus; PFC, prefrontal cortex; PLd, dorsal prelimbic cortex; PLv, ventral prelimbic cortex; PV, paraventricular thalamic nucleus. From Voorn et al. (2004).

Table 30.1
Four striatal–pallidal divisions

- **The dorsal striatum and pallidum:**
  Striatum: caudate-putamen
  Pallidum: external and internal segments of the globus pallidus (the latter is the entopeduncular nucleus in small rodents)
- **Ventral striatum and pallidum:**
  Striatum: nucleus accumbens, olfactory tubercle, fundus striati
  Pallidum: substantia innominata, magnocellular preoptic nucleus
- **Medial striatum and pallidum:**
  Striatum: lateral septal complex
  Pallidum: medial septal nucleus/nucleus of the diagonal band
- **Caudorostral striatum and pallidum:**
  Striatum: [baso-]medial and central amygdalar nuclei, anterior amygdalar area, intercalated amygdalar nuclei
  Pallidum: bed nucleus of the stria terminalis

Source: Based on Swanson (2003).

Complex Chemoarchitecture of the Striatum

To guide studies of striatal function, anatomists and physiologists have had to pay attention to much more than the excitatory or inhibitory nature of the axonal connections. Chemoarchitectural studies have revealed neurons and axons containing not only γ-aminobutyric acid (GABA) or acetylcholine (making inhibitory or excitatory connections, respectively), but in addition various peptides. More than 100 substances have been described in the striatum. Figure 30.11 illustrates three different, major cell types of the dorsal striatum plus dopamine axons coming in from the midbrain. Note that the acetylcholine-containing neurons are short-axon interneurons. Neurons with longer axons going to the external portion of the globus pallidus contain both GABA and enkephalin. Neurons with axons going to the internal segment of the globus pallidus and to the nigra contain both GABA and substance P.

When acetylcholine (ACh) is the neurotransmitter, as in the large interneurons of the striatum, acetylcholinesterase, the enzyme that breaks down ACh, is also found. Staining brain sections for acetylcholinesterase reveals an uneven distribution, a patchiness, indicating an organization into some kind of clustering. However, the meaning of the patchiness has been a puzzle. In the next section are some clues to this puzzle.

Compartmental Organization within the Striatal System

The acetylcholine-poor regions of the striatum have been dubbed *striasomes*. There are some tantalizing findings about the striasomes and the surrounding matrix of the striatum. One finding is that these two striatal compartments receive inputs from different

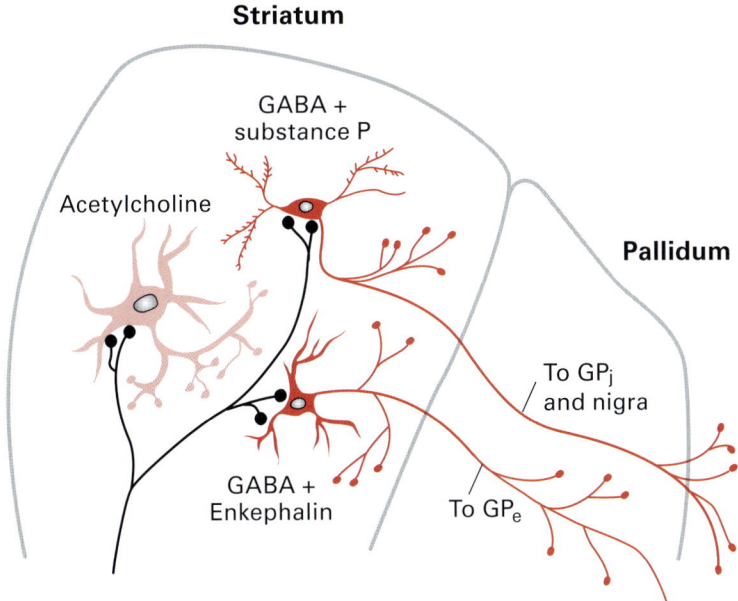

Figure 30.11
Drawing of three major cell types in the mammalian dorsal striatum. (Other cell types found there are not illustrated.) Most striatal efferents are GABAergic and are thus inhibitory. Large interneurons are cholinergic. GP_i, internal segment of the globus pallidus; GP_e, external segment of the globus pallidus. Based on Brodal (2004).

neocortical neurons. The striasomes receive projections from the deeper part of layer 5 and also from some layer 6 cells. The matrix receives inputs from a more superficial strata of layer 5 and also from some supragranular layer cells (above layer 4). These two compartments remain separate in the striatal projections to the substantia nigra. The striasomes project mainly to the *substantia nigra pars compacta*—the part containing dopamine neurons. Output neurons of the striatal matrix project to the *pars reticulata*.

We can speculate about what this means. Greater activity in the striasome compartment, influencing the dopamine-containing neurons, may indicate something worth reinforcing, like a sensorimotor sequence just carried out. Reinforcement, or reward, is commonly ascribed to ascending dopamine axons. Greater activity in the matrix compartment is more focused on influencing specific movements: orienting or locomotor movements or specific actions controlled by the motor cortical areas.

In this regard, it is interesting that striasomes located in the more ventral and medial portions of the striatum tend to receive mostly projections from limbic forebrain areas. The striasomes located more dorsally receive inputs from neocortex, but apparently of a different type than the inputs from neocortex to the matrix areas.

Finding the Corpus Striatum in Large Mammalian Brains

At this point, it is time to step back from the details and consider the whole brain (e.g., the brain of a human or a monkey, or a sheep or a cat). Assume that the brain has been hardened in a fixative and removed from the skull, as is the case if one is going to do a whole-brain dissection for purposes of study. In medical school, students may have an opportunity to participate in a human brain dissection. Outside of medical school, it is more likely that students would use the brain of a sheep for such study. Let's ask a few questions concerning the location of components of the corpus striatum.

The lenticular nucleus was mentioned earlier in this chapter. It is also known as the lentiform nucleus. Actually, it is not a single cell group at all, but includes the components of the corpus striatum and globus pallidus located lateral to the fibers of the internal capsule in a human brain. Its outermost portion is the putamen. Its inner portion is composed of the two segments of the globus pallidus. You can see these structures clearly in the fiber-stained section shown in figure 30.12. In a dissection, how could you reach the lenticular nucleus most easily? You would pry open the Sylvian fissure, exposing the normally hidden cortex of the insula. If you removed the insular cortex and scraped away the layer of deeper-lying cells called the claustrum, you would reach the outer boundary of the putamen. The putamen here is like the thick top layer of an ice-cream cone, covering the deeper-lying globus pallidus (the pale globe).

When inspecting the human brain section shown in figure 30.12, you may have noticed that the caudate nucleus is labeled in two locations on the left side. It is located next to the lateral ventricle, medial to the internal capsule fibers, and it is located also in a ventral position, below the putamen and the ventral portion of the internal capsule, in the temporal lobe. How is this possible? It is not difficult to understand. First, look at the pictures of four brains shown in figure 22.7a. The brain of each of the three larger animals has a temporal lobe, where the posterior cortex has not only protruded downward, but also this downward protrusion is pushed forward and upward (figure 22.7b). This results in the formation of a deep fissure separating the temporal lobe from the parietal and frontal cortex. As the temporal lobe forms in development, the posterior part of the internal capsule is pulled around with it, as shown in figure 22.8. As you look at this figure, imagine the caudate nucleus in this left hemisphere located just behind the fibers in the frontal region. Staying on that same side of the fibers and moving caudally, you come around into the temporal lobe, where you end up below the band of fibers. This is the position of the tail of the caudate in the temporal lobe. Its position there is an effect of the temporalization of the hemisphere.

Overview

Evolution has brought us far from the primitive origins of the striatum as a forebrain link between olfactory inputs and the output systems of the 'tweenbrain and midbrain—the

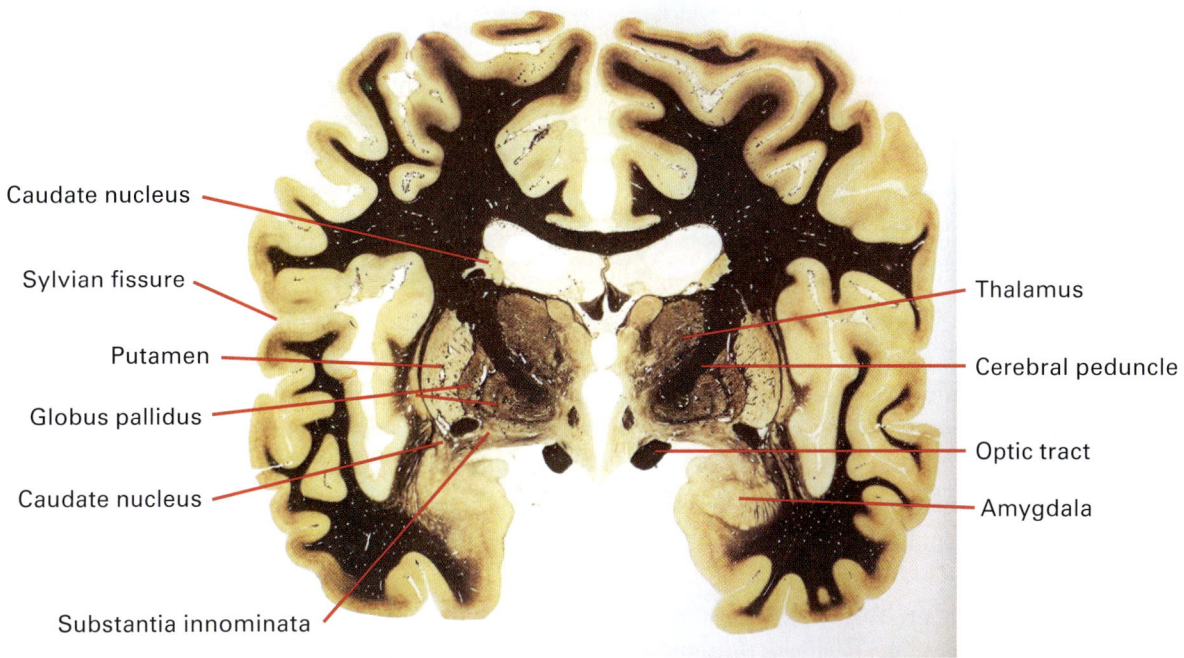

Figure 30.12
Myelin-stained frontal section of human forebrain showing, among other structures, the thalamus, corpus striatum, amygdala, and neocortex at the level of the insular cortex that overlies the lentiform nucleus (putamen and globus pallidus). From Nauta and Feirtag (1986).

basis of ancient learned sensorimotor habits. The input side has gone far beyond the invasion of non-olfactory modalities. As the neocortex has become the dominant source of inputs in the mammals, it has brought all the functions of neocortex into the striatal circuitry. On the output side, the early destinations probably led mainly to locomotor and orienting responses and shifts in instinctive action patterns, but subsequent evolution has added all the complex actions that neocortical motor areas can generate. The resulting habits are as diverse as neocortical functions are diverse: from simple responses of turning the eyes and head or of walking, and responses of the oral parts and of the limbs to useful objects, to habits of thought, traditions, social routines, and prejudices.

Moreover, in the mammals there are neocortical outputs that bypass the striatum. (There are equivalent structures and pathways in birds.) These outputs superimpose the effects of actions that originate in the cortex on the inherited action patterns of the brainstem and spinal cord and on the striatum-controlled learned habits. Especially in the animals with relatively the largest brains, these "voluntary actions" may sometimes altogether suppress those "lower" functions. In the final chapters of this book, we will focus on the neocortex.

Readings

Gerfen, C. R. (1992). The neostriatal mosaic: Multiple levels of compartmental organization. *Trends in Neuroscience 15*, 133–139.

Graybiel, A. M. (2008). Habits, rituals, and the evaluative brain. *Annual Review of Neuroscience, 31,* 359–387.

Haber, S. N., & Knutson, B. (2010). The reward circuit: Linking primate anatomy and human imaging. *Neuropsychopharmacology Reviews, 35,* 4–26.

McGeorge, A. J., & Faull, R. L. M. (1989). The organization of the projection from the cerebral cortex to the striatum in the rat. *Neuroscience 29,* 503–537.

Saint-Cyr, J. A., Ungerleider, L. G., & Desimone, R. (1990). Organization of visual cortical inputs to the striatum and subsequent outputs to the pallid-nigral complex in the monkey. *Journal of Comparative Neurology, 298,* 129–156.

Selemon, L. D., & Goldman-Rakic, P. S. (1985). Longitudinal topography and interdigitation of corticostriatal projections in the rhesus monkey. *Journal of Neuroscience 5,* 776–794.

Smeets, W. J. A. J., Marin, O., & Gonzalez, A. (2000). Evolution of the basal ganglia: New perspectives through a comparative approach. *Journal of Anatomy, 19,* 501–517.

Voorn, P., Vanderschuren, L. J. M. J., Groenewegen, H. J., Robbins, T. W., & Pennartz, C. M. A. (2004). Putting a spin on the dorsal–ventral divide of the striatum. *Trends in Neurosciences, 27,* 468–474.

For helpful simplifications of major connections of the corpus striatum, written primarily for medical students:

Brodal, P. (2004). *The central nervous system: Structure and function,* 3rd ed. New York: Oxford University Press.

See also chapter 24 readings.

31 Lost Dopamine Axons: Consequences and Remedies

In the previous chapter, pathologies of striatal structures came up. One of the diseases mentioned was Parkinson's disease. In this disease, the dopamine-producing neurons of the substantia nigra gradually degenerate. As they degenerate, more and more of the dopamine-containing axons that project to the striatum are lost. The remaining, intact axons appear to compensate sufficiently, because of collateral sprouting and denervation supersensitivity, to preserve striatal function—but only for a time. Once the loss of dopamine cells reaches a certain level, about 80%, symptoms of the disease start to become evident. These symptoms become progressively worse. A tremor of the hands, and sometimes of the mouth and chin and lower limbs, becomes more and more noticeable and bothersome. Walking, and movement in general, becomes affected more and more, becoming slower and slower. The gait becomes characteristically abnormal, and the facial muscles become rigid, the face expressionless. Cognitive functions also decline as mental abilities are slowed and dementia increases, especially in late stages of the disease. Obviously, patients suffering from this disease, and their families, become desperate for some kind of treatment.

Although other neuronal systems may play some role as well (e.g., norepinephrine neurons), research has focused on the role of the dopamine innervation of the striatum. Evidence from research with rats has indicated that replacing the dopamine innervation, even by transplanted neurons that are not in the normal midbrain location but are implanted directly into the dorsal striatum, is sufficient to reverse the movement deficits. The transplanted cells in this research were taken from the substantia nigra of living embryos.

Fetal Nigral Tissue Transplantation Initiated as a Therapy for Parkinson's Disease

The research findings led to treatments of human Parkinson's disease patients by implants of fetal nigral tissue taken from aborted fetuses. It was well known that tissue slabs taken from a fetal animal could be kept alive in tissue culture long enough for axons to grow from their neurons. Investigators concluded that if pieces of the substantia nigra could be taken from aborted human fetuses at certain stages of development and these pieces transplanted into a human corpus striatum, dopamine axons would grow out from the transplanted tissue, and these axons might form connections in the host brain, especially if the host striatum were lacking dopamine axons.

Procedures for doing this were worked out using rats. Small pieces of fetal nigra were taken up into a small pipette along with small amounts of fluid, and the pipette was inserted into the striatum of an adult rat that had suffered a lesion of the nigra, then the fetal tissue was injected. Using this procedure, studies were carried out to determine the optimal age of donor tissue for transplantation. Nigral tissue from human fetuses of various ages was injected into adult rat striatum. For example, tissue taken from a fetus aged 41 days postconception showed good survival, whereas tissue taken from a fetus aged 72 days postconception showed poor survival in the adult rat brain. Survival was judged using sections from the rat brains prepared after death of the animals and stained for tyrosine hydroxylase with an immunohistochemistry technique (figure 31.1).

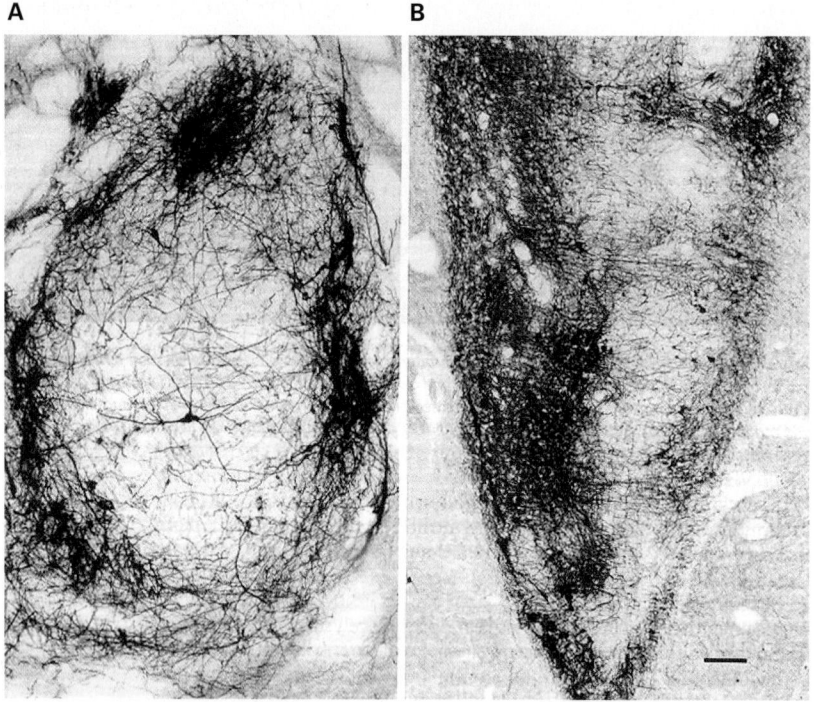

Figure 31.1
Grafts of human fetal substantia nigra tissue into the striatum of rat predict results in Parkinson's disease human patients. (A) Tyrosine hydroxylase–immunostained section from rat brain. The stained neurons are all from a transplant of human nigral tissue. (B) Similarly stained section from brain of a human patient. The patient had had a similar transplant. The dopamine neurons are similarly integrated into the striatal tissue in the two cases. (Tissue preservation, as expected, was better for the rat brain.) Scale bar, 100 micrometers (applies to both pictures). From Olanow, Kordower, and Freeman (1996).

This enzyme is always present in living dopamine-producing neurons. From such experiments, optimal ages of donor tissue were worked out. The experimental results with the rats were used to predict, successfully, the best fetal ages for survival in human patients.

The adult human striatum is very large and the fetal nigra is quite small. These facts made it necessary to use tissue from multiple fetuses for a single patient because of the limited distance of axonal growth into the adult tissue; research studies showed growth of no more than 2–3 millimeters. However, limited numbers of aborted fetuses were available at the best ages for use as donors. Therefore, with only a small number of transplants per patient, it was important to place the transplants in the host striatum locations where it was most likely to be beneficial for lessening the disturbed movements of the patient. Remember the discussion in the previous chapter about functional differences in different regions of the striatum. Striatal topography indicated that implants should be placed in the putamen in order to treat movement problems.

Initial Promise of Transplant Treatments

Many transplanted nigral neurons were able to survive and grow axons in the human host brains. This fact was confirmed by histological study of brain tissue taken from transplant recipients that died, but it was also assessed in living patients. This was done by injecting the compound fluorodopa, which was taken up by the dopamine neurons and could then be detected in the living cells by an imaging procedure, positron emission tomography (PET).

Behavioral improvement in the treated patients was variable and was never immediate. However, one would expect the recovery to be slow because of slow growth of the axons in the adult brain tissue (see chapter 13).

The variability of results led to doubts about the usefulness of the procedure. In 2003, results of an experimental study of human patients treated with fetal nigral transplants were published. A control group had been used for comparison. The findings were disappointing in that there was little or no certain effect of the treatment except in the Parkinson patients with milder symptoms. In more advanced Parkinson patients, defects besides the loss of dopamine neurons in the substantia nigra had to be considered.

Deep Brain Stimulation or Selective Surgical Lesions in Advanced Parkinson's Disease

The limited success of the transplants in patients with severe Parkinson's disease, together with the difficulties in obtaining large amounts of fetal nigral tissue at the best developmental stages, has led to more emphasis on other treatments. There was a change from a focus on undoing a major pathology to a focus on treating the motor symptoms. Deep brain stimulation became the most common method. Brain sites usually stimulated to alleviate the movement defects in Parkinson patients are the subthalamic nucleus and the globus pallidus. It is not difficult to find information on deep brain stimulation procedures and results on the Web.

Another procedure that has been used to alleviate debilitating symptoms suffered by Parkinson's disease patients involves making selective brain lesions. For example, a lesion of the internal segment of the globus pallidus, by destroying many inhibitory connections to the ventral anterior thalamic nucleus, can reduce an underactivity of the cells of that nucleus that contributes strongly to the severe reduction in movement initiation. This thalamic nucleus projects to the motor cortex.

Additional Alternative Treatments

Research studies using animals have been carried out using several different, additional procedures. Efforts have been made to produce pigs that are genetically altered to be more compatible with humans so that xenografts of fetal nigral tissue could become possible. Implants of non-neuronal cells have also been carried out with animals (e.g., fibroblasts have been genetically altered to produce dopamine for use in treating Parkinson's disease). Slow release of dopamine by implanted devices has also been tried.

But the most promising alternative to the fetal nigral tissue transplants has been the implantation of stem cells, ideally stem cells taken from the same patient's body, manipulated so they differentiate into dopamine-releasing cells. Research using adult rats has used stem cell lines that are maintained in culture and grown in sufficiently large numbers to use for implants into the brain. For example, when such stem cells were implanted through a canula in the basal forebrain, many were found to be surviving 4–9 months later (figure 31.2). Using antibody markers to identify cell types, the investigators found that some cells had differentiated into neurons (expressing neurofilament protein) and others had differentiated into astrocytes (expressing glial fibrillary acidic protein). Some of the implanted cells expressed proteins showing that they were cholinergic. Many large neurons in the normal basal forebrain are cholinergic. The findings showed that the stem cells appear to be able to pick up signals needed for differentiating into the cell types normally found in the region of the brain where they are placed.

Research on the use of stem cells in therapies for degenerative disorders will likely lead to new therapies for human patients with diseases of the striatal structures, as well as new therapies for a number of other diseases.

Lost Dopamine Axons: Consequences and Remedies

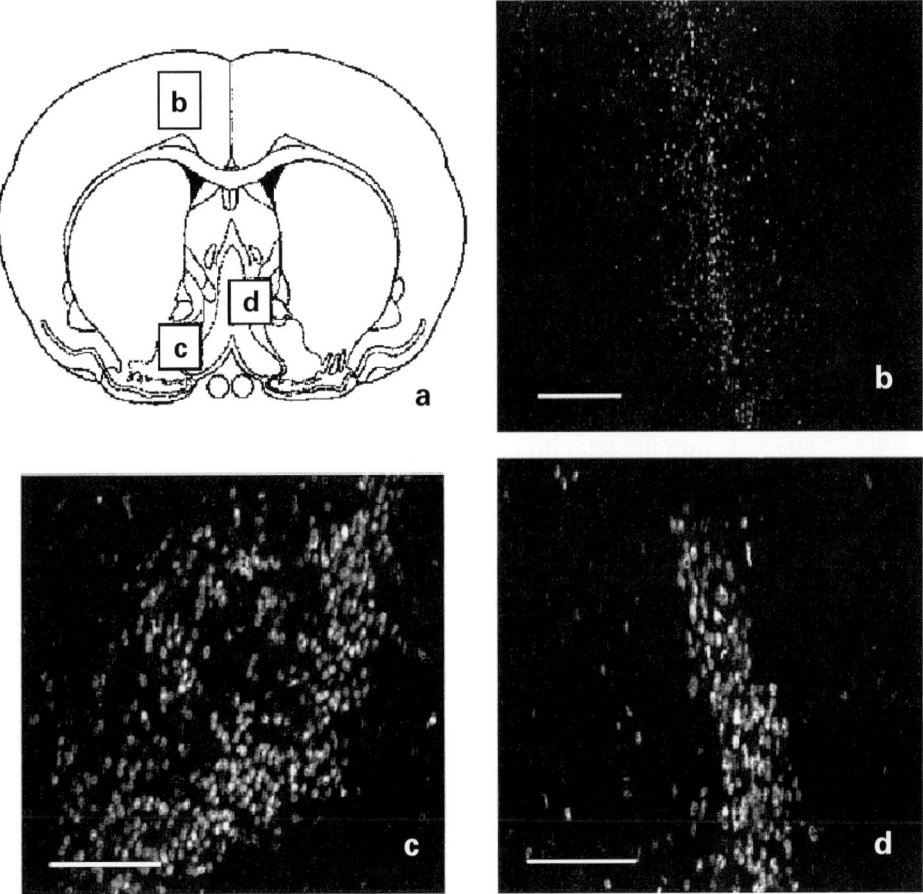

Figure 31.2
Implanted stem cells surviving after 4–9 months in adult rat brain. (a) Drawing of frontal section of rat brain from an atlas showing the positions of the photographs that follow. (b) Cells in neocortex labeled with bromodeoxyuridine (BrdU) found along a needle tract used for injecting the labeled stem cells 4 months earlier. (c) Labeled cells surviving in a 4-month-old graft in the horizontal limb of the diagonal band of Broca (in the basal forebrain). (d) Labeled cells surviving 9 months after being implanted near the ventral pallidum. The fluorescence images show binding by a specific antibody conjugated to fluorescein isothiocyanate (FITC). Scale bars: 500 µm for b, 400 µm for c and d. From Doering and Snyder (2000).

Readings

Cho, M. S., Lee, Y.-E., Kim, J. Y., Chung, S., Cho, Y. H., Kim, D.-S., & Kim, D.-W. (2008). Highly efficient and large-scale generation of functional dopamine neurons from human embryonic stem cells. *PNAS, 105,* 3392–3397.

Doering, L. C., & Snyder, E. Y. (2000). Cholinergic expression by a neural stem cell line grafted to the adult medial septum/ diagonal band complex. *Journal of Neuroscience Research, 61,* 597–604.

Olanow, C. W., Kordower, J. H., & Freeman, T. B. (1996). Fetal nigral transplantation as a therapy for Parkinson's disease. *Trends in Neuroscience, 19,* 102–109.

Wernig, M., Zhao, J.-P., Pruszak, J., Hedlund, E., Fu, D., Soldner, F., Broccoli, V., & Jaenisch, R. (2008). Neurons derived from reprogrammed fibroblasts functionally integrate into the fetal brain and improve symptoms of rats with Parkinson's disease. *PNAS, 105,* 5856–5861.

Little or no effect of transplants except in patients with milder symptoms:

Olanow, C. W., Goetz, C. J., Kordower, J. H., Stoessl, A. J., Sossi, V., & Freeman, T. B. (2003). A double-blind controlled trial of bilateral fetal nigral transplantation in Parkinson's disease. *Annals of Neurology, 54,* 403–414.

Intermission

Neurogenesis in Mature Brains

Discussions about stem cells, as in chapter 31, are likely to make us wonder about the possibility of generation of new neuronal cells in the adult brain of an animal or person. It was long a dogma in neuroscience that neurogenesis ceases very early in the life of a mammal. In the 1960s, a finding was reported that was so surprising to neuroscientists that most of them found it difficult to accept. The claim was that there was neurogenesis in specific regions of the mature rat brain. Injections of tritiated thymidine were used to label newly synthesized DNA and thereby mark newly generated cells. Labeled cells that appeared to be neurons were found among the granule cells in the olfactory bulbs, as described already in chapter 19. The neuroblasts migrate rostrally from the ventricular layer of the lateral ventricles. Newly generated neurons were also found in a different population of small neurons, the granules cells of the dentate gyrus of the hippocampus.

These findings were not confirmed for more than a decade, and then they were confirmed in new studies of rats. However, it was not until the 1980s that interest in the possibility of neurogenesis in adult animals was strongly rekindled. New research on mature songbirds was the trigger, with evidence that there are seasonal changes in the forebrain, particularly in male birds, with birth, migration, and differentiation of new neurons in the endbrain (see figure 27.6). The two groups of reports, on rats and on songbirds, could no longer be ignored, and similar investigations were undertaken with primates. As the claims of adult neurogenesis were verified in reports in the late 1990s of studies of several different primate species, interest in neurogenesis in the mature mammalian brain increased greatly. The finding of neurogenesis in the hippocampal dentate gyrus was confirmed in humans also. This was accomplished by histological studies of autopsy material from patients who had died of cancer. The growth of their tumors had been assessed using injections of bromodeoxyuridine (BrdU). BrdU is taken up by dividing cells and adds a detectable label to such cells (figure 31a.1).

After the resurgence of interest in adult neurogenesis, experiments with rats and mice have demonstrated that the rate of neurogenesis is not fixed. It can be stimulated by

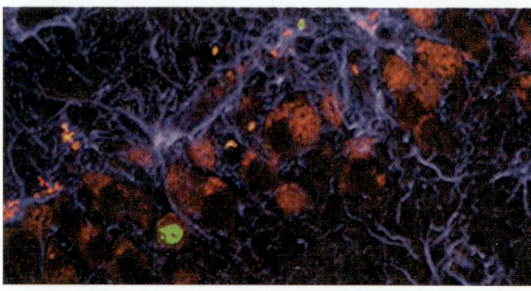
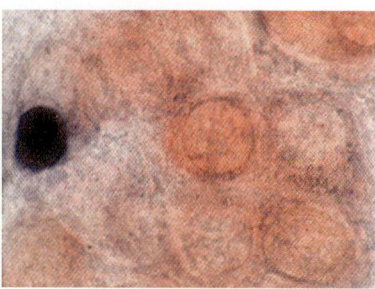

Figure 31a.1
Proof of the formation of new neurons in adult human brain. Tissue from the dentate gyrus of the hippocampal formation was taken from patients who had died of cancer. Bromodeoxyuridine had been injected previously in order to assess tumor growth. As the BrdU is retained only in cells that had incorporated it into their DNA during cell division, labeled neurons are cells that were born after the injection. Neurons in the photos are marked in red. Two different techniques are illustrated: (Left) BrdU indicated by green color. (Right) BrdU indicated by dark coloration. From Kempermann and Gage (1999).

keeping the animals in more complex environments, where they experience sensory enrichment and show more exploratory activity. Dentate gyrus neurogenesis in adult rats has been enhanced by training that requires spatial learning. The rate of new granule cell addition to the olfactory bulbs has been enhanced by the learning of complex odor discriminations.

In experimental studies of hippocampal long-term potentiation, there is evidence that the more recently generated neurons show enhanced plasticity. This parallels the greater plasticity of younger persons in their learning, as in the learning of language. If younger neurons are more plastic than older neurons, then we can see an adaptive advantage for adult neurogenesis and neuron turnover in a structure that is important throughout life in short-term storage. Rapid neurogenesis might be less important in sites of very-long-term memory storage where turnover of cells could destroy established engrams.

Experimental studies have produced evidence of some adult neurogenesis, less than in the hippocampus and olfactory bulb, in other brain regions. It has been found in neocortex, striatum, and amygdala, and also in the midbrain's substantia nigra, but the findings have been controversial and the nature of the newly generated cells is not always certain. Just how much neurogenesis occurs in these regions and how it varies with age may be determined in the future, and we can hope that the functions of new neurons in adult mammalian brains will become clarified.

Readings

Findings of neurogenesis in mature mammalian brains was first discovered at MIT by Joseph Altman and Gopal Das:

Altman, J. (1962). Are new neurons formed in the brains of adult mammals? *Science 135*, 1127–1128.

Altman, J. (1963). Autoradiographic investigation of cell proliferation in the brains of rats and cats. *The Anatomical Record 145*, 573–591.

Altman, J., & Das, G. (1965). Autoradiographic and histological evidence of postnatal hippocampal neurogenesis in rats. *Journal of Comparative Neurology, 124*, 319–336.

The findings of Altman and Das were finally verified for primates: See, for example, the review by Elizabeth Gould:

Gould, E. (2007). How widespread is adult neurogenesis in mammals? *Nature Reviews Neuroscience 8*, 481–488.

Later, the finding of adult neurogenesis was verified for the human dentate gyrus:

Eriksson, P. S., Perfilieva, E., Björk-Eriksson, T., Alborn, A.-M., Nordborg, C., Peterson, D. A., & Gage, F. H. (1998). Neurogenesis in the adult human hippocampus. *Nature Medicine, 4*, 1313–1317.

Evidence that more recently generated neurons show enhanced plasticity: reviewed by two investigators who have contributed in this area:

Ming, G. L., & Song, H. (2011). Adult neurogenesis in the mammalian brain: Significant answers and significant questions. *Neuron 70*, 687–702.

XII THE CROWN OF THE MAMMALIAN CNS: THE NEOCORTEX

XIV THE CROWN OF THE MAMMALIAN CNS: THE NEOCORTEX

32 Structural Origins of Object Cognition, Place Cognition, Dexterity, and Planning

In chapter 12, and again in chapter 24, we introduced the structures of the forebrain, including the neocortex. Now we want to consider how, within the endbrain's pallium, such a dominant expanse of multilayered neural tissue has evolved in the mammals. Figure 12.2 shows an interpretation of telencephalic evolution based on expression patterns of regulatory genes during development. Such studies have provided evidence that the neocortex in mammals arose from the same ancestral portions of the rostral neural tube as the dorsal cortex of amphibians and reptiles and the hyperpallium of birds. (Such studies have also provided evidence that the dorsal ventricular ridge structures of reptiles and birds, endbrain structures that receive non-olfactory sensory input from the thalamus, are related to the claustrum and portions of the amygdala in mammals.)

The recent data on gene expression are promising, but the story remains very incomplete. The full story of neocortical evolution is still unfolding and will have many complexities. Here we will focus mainly on the real drivers of evolution, the adaptive functions. Consideration of these functions, together with knowledge of brain structure in various species, has led to the ideas presented below. Although somewhat speculative, the ideas promote the acquisition of a useful picture of the functioning architecture of the brain.

The Primitive Olfactory Functions of Endbrain: Identification of Object and Place

Earlier in this book we described two major functions of the olfactory modality that have been of major importance in shaping the evolution of the endbrain. Using this sensory ability, animals have been able to identify and respond to objects and other animals, and they have been able to recognize places (in the sense of local regions or locales) in the environment and alter their behavior according to past experiences at those places. In many species, after lengthy periods of evolution, non-olfactory inputs have become dominant for these functions.

Place recognition, we will assume here, includes the rapid detection of the locale or the place of the organism within a region (space) of the animal's environment. A locale can be very small, but it is different from an object, which can move or be moved around from place to place. Place recognition is related to scene detection, which was discussed briefly in chapter 22 on the visual system. Prior to the evolution of such a function in the endbrain's visual system, it must have been an important function of the early olfactory system.

The endbrain functions of object perception and place perception are different from the midbrain functions of orienting to objects and animals or escaping from them. Without an endbrain, the midbrain tectum mediates innate reactions of orienting and escape. Object identity is represented in only a rudimentary way in this midbrain structure, as innate responses to salient visual features and movement. Place identity is only eye/head centered; it does not locate the animal with respect to the environment. Plasticity in the midbrain tectum is important but short-lived, consisting of sensitization (with temporary persistence of activation) and habituation. Similar statements can be made about other sensorimotor structures of the brainstem and spinal cord. The endbrain did not add only the olfactory sense early in its evolution. It also added new kinds of plasticity. A major addition was the alteration of connections resulting from feedback from the consequences of behavior. Effects that were experienced to be helpful or hurtful to the organism could alter connections that had been active just before the effects were experienced. In this way, a previously neutral object or place, signaled by its odor or by other inputs reaching the endbrain from the object or place, could acquire a positive or negative valence—a meaning for the animal.

Consider the first function, the detection of objects and animals and the ability to distinguish which were good and which were bad for the organism. Some inputs that had such effects without the need for much learning were from pheromones (substances produced by other members of the species that enabled, for example, identification of gender and sexual receptivity) detected by the vomeronasal system and from odorants detected by sniffing of animals and objects. Taste was also important, especially in initiating feedback after an action. Taste information reached the endbrain via projections ascending from the hindbrain and from the thalamus (chapter 18), whereas the rewarding effects of tastes activated ascending dopamine axons that originated in the ventral midbrain or in the hypothalamus. Later, the originally olfactory endbrain structures involved in this primitive object sense, the striatum and amygdala, were invaded also by visual, auditory, and somatosensory inputs. The ancient outputs went to hypothalamic and midbrain structures for motivational control and to the ventral striatum and pallidum for eliciting approach or avoidance or for modulation of orienting movements controlled by the midbrain.

The second function involved the learning of locations in the environment associated with good or bad experiences there. This knowledge enabled the prediction of odors encountered at different places during locomotion within that environment, knowledge that enabled a sensitivity to any changes. The importance of this ability led to the evolution of place cells in the medial pallium, with ancient inputs from olfactory bulbs and olfactory cortex, then non-olfactory inputs, especially visual inputs, that took over the major roles.

Outputs included connections to hypothalamic structures for motivational control and to ventral striatum for influencing locomotion and orienting *via* the midbrain.

The Advantages and Consequences of Non-olfactory Inputs to the Endbrain

For the endbrain functions of object cognition and place cognition, non-olfactory inputs brought major advantages. The distance senses of vision and audition in particular allowed more rapid assessments of the future, so better anticipatory actions were possible. (Initially in evolution, this probably meant only the immediate future.) The inputs to the forebrain from these distance senses and from somatosensory pathways reached the paleothalamus. There, reticular formation–like neurons had axons projecting to the corpus striatum and to the pallium. The evolutionary elaboration of these thalamic structures was not great in higher vertebrates because the striatal afferents from dorsal pallium became more important than the more direct inputs from the thalamus.

The dorsal cortex of the endbrain with little doubt was initially multimodal as in the lamprey. Early in evolution, this cortex was apparently what we would call in mammals limbic or paralimbic in its connections. This is what has been observed in neuroanatomical experiments with reptiles. Evolution of unimodal areas within this cortex brought advantages for sensory analysis, but other evolutionary changes were even more important. With the advent of mammals, cell migrations in some of the developing dorsal pallium formed additional, more superficial, layers so that in this cortex there were as many as five cellular layers rather than just one or two as in pallial regions in reptiles and amphibians. This evolution of a neocortex resulted in a clearer columnar organization with intracolumnar processing. New types of interneurons also appeared with migrations from the developing nonpallial regions (e.g., from the medial ganglionic eminence and the caudal ganglionic eminence).

It was the expansion of this new cortex (neocortex) that has been so characteristic of mammalian evolution. In the other major offshoot of the early reptiles, the birds, not only the nidopallium but also the hyperpallium evolved. Much of the Wulst of the hyperpallium functions like the primary visual neocortex of mammals. (In German, *Wulst* means "a bulge.") Into these avian endbrain regions, similar in many of their connections to mammalian neocortex, there was also a migration of new types of interneurons from striatal locations.

In Mammals, Why Did Neocortex Evolve Rather than Wulst?

John Allman has described the avian Wulst (see figure 22.2) as much more "efficient" than the mammalian neocortex. It is relatively much thicker with more layers of cells, supporting elaborate analysis of sensory input with less total length of axons in its interconnections. If this is the case, why has neocortex become so dominant in mammals?

Answers can be suggested. First, there was a kind of genetic efficiency: Single-gene alterations can lead to major changes in cortical area or in cortical thickness, simply by modifying the duration of symmetric or asymmetric cell division, respectively (see figure 8.7 and chapter 34). In addition, the neocortex has efficient local circuits in its columns of cells, and greater complexity of information analysis can be achieved without increasing cortical thickness by means of a multiplication of interconnected cortical areas. This occurs at the cost of adding to the total quantity of axons, but this cost can be minimized by the "small-world architecture" mentioned in the chapter 22 notes.

The Orderly Architecture of Neocortex Illustrated by the Human Brain

Figure 32.1A depicts part of a frontal section of human neocortex stained with three different cytoarchitectural methods. The left-most panel of figure 32.1A shows neurons as seen with the Golgi method. Only a small proportion of the cells is stained, but these cells are seen in silhouette with all or most of their processes. The dominant cell type is clearly the pyramidal cell, a cell with an apical dendrite that extends toward the pial surface plus many oblique dendrites and a skirt of dendrites around the base where an axon originates. The right-most panel of figure 32.1A shows results of staining with a Nissl method, revealing the cell bodies but not much more of the cells except a portion of the proximal portions of large dendrites. The middle panel is a drawing of the very different picture seen with the aid of a fiber stain.

Figure 32.1B is a more detailed composite of tracings of Golgi-stained neurons of the human neocortex by the Russian neuroanatomist Poliakov.

With each of the histological methods used for the illustrations, a layering (lamination) of the tissue is evident. The layering is often easiest to see in sections stained for fibers, but multiple layers (laminae) are easy to see with the other methods as well. It is traditional to specify six layers in all parts of the neocortex, although in many regions there are clear sublayers.

Another characteristic of neocortical tissue is a columnar arrangement. Narrow anatomical columns can be defined by the orientation of the apical dendrites of the large pyramidal cells of layer 5 or by the bundles of axons coursing from the white matter toward the brain surface, as indicated in the middle panel of figure 32.1A. However, columns of various widths have been defined by physiological characteristics of various types. For example, in the visual cortex there are columns of cells activated by stimulation of one eye in one type of column and by the other eye in adjacent columns. There are cells activated by contours of one particular orientation in one column and by a slightly different orientation in adjacent columns. The ocular dominance columns are much wider, each one encompassing multiple orientation columns.

Consistent patterns of connections have been found within cortical columns. This is illustrated in figure 32.2—a sketch that illustrates the information flow within a cortical

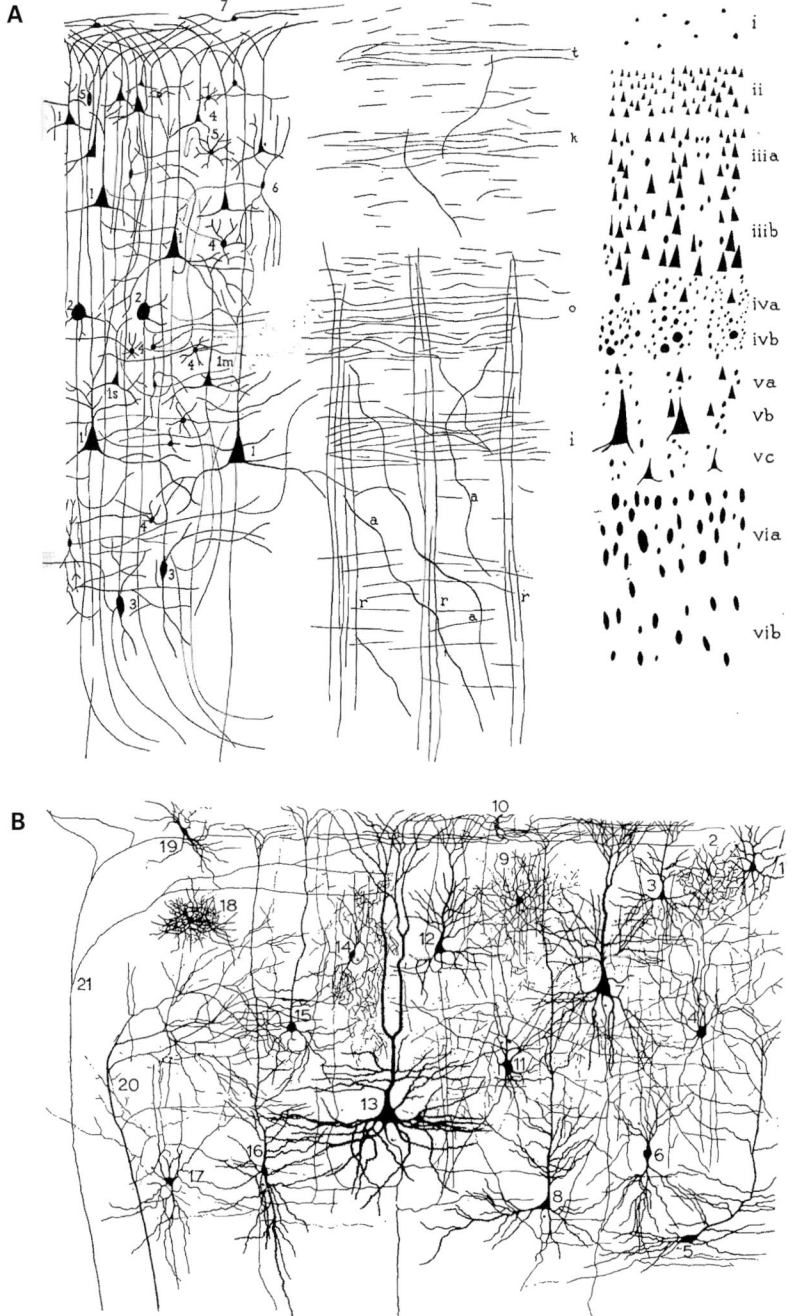

Figure 32.1
(A) Structure of the neocortex as seen in sections prepared with a Golgi method (left), Weigert stain for myelinated fibers (center), and a Nissl stain for cell bodies (right). The Roman numerals at the far right indicate layers. 1, pyramidal cell; 1m, medium pyramid; 1s, short pyramid; 2, star pyramid; 3, fusiform cell; 4, star cell; 5, spider cell; 6, double bush cell; 7, horizontal cell of Cajal; a, oblique fibers; r, radii; t, tangential fibers; k, stripe of Kaes–Bechterew; o, outer stripe of Baillarger; i, inner stripe of Baillarger. (B) "Chart illustrating the neuronal structure of the cerebral cortex (Poliakov, 1949). 1, 3, 5, 7, 8, 12, 13, 15, 16, Efferent (pyramidal and fusiform) neurons with a long axon; 2, 4, 6, 9, 10, 11, 14, 17, 18, 19, intermediate neurons (star cells) with a short axon; 20, 21, projectional and associative afferents of the cortex." Poliakov's paper in 1949 was focused on the human brain, so we can assume that this figure is based on Golgi staining of human brains. Panel (A) from von Bonin (1950); panel (B) from Poliakov (1968).

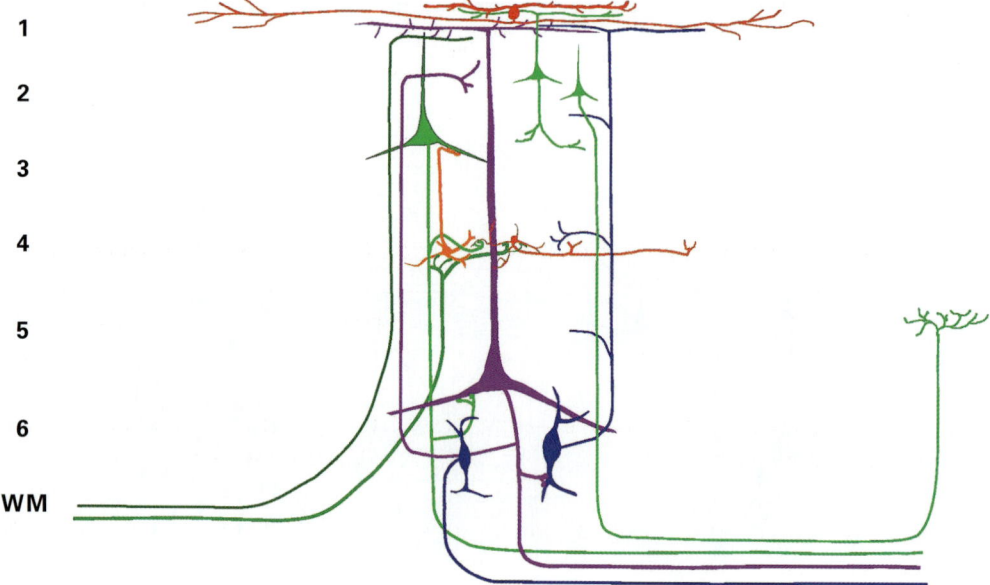

Figure 32.2
Sketch of three types of intracortical association axons: (1) axons that connect with other neurons within a single column, as from layer 4 stellates (orange) to layer 3 pyramidal cells (green), or from layer 6 fusiform cells to layer 4 and less strongly to layers 5 and 2–3 (blue cell on right); (2) axons that stay within a single layer and connect to other columns (orange horizontal axons); (3) axons that descend to the white matter where they can be followed to another part of the neocortex where they ascend and terminate (green axon from layer 2 pyramidal cell). Not illustrated is a third type of association connection formed by local collaterals of pyramidal cell axons; these collaterals can arborize in all layers, mostly in adjacent columns (see figure 33.2). WM, white matter.

column and from cells in a column to other regions of the brain. But how wide is a single column? This depends on the definition of a column and also on the region of the brain. In some regions, there may be more than one type of cortical column in the same region, superimposed on each other. An example was noted in the previous paragraph.

Axons within a cortical column arise not only from within the column but also from adjacent columns and from other regions of neocortex (figure 32.2). Some of them can be followed out of the cortex to subcortical regions (e.g., to the corpus striatum, to the thalamus, or to the midbrain or even farther). Axons that enter the white matter of the cortex and then find a connection in another region of the cortex are called U fibers, because from their origin they can be followed down, then laterally and then back up into another cortical column. Axons that enter the corpus callosum and form connections in the opposite hemisphere can be considered to be a particular type of U fiber. There are also shorter connections formed by tangential association fibers, where the axons travel tangentially within a cortical layer and never pass into the white matter.

Basic Sensorimotor Functions of Neocortex

The use of olfaction for discrimination of objects and for forming adaptive habits with that information remains a very useful ability in many animals, but survival and reproduction in a greater variety of habitats was vastly increased by the additions to this ability made possible by the neocortex. Neocortex made possible new discrimination abilities of imaging vision and of audition. Comparable new abilities were made possible by somatosensory neocortex, with active exploration by touch using whiskers and limbs. Spatial learning was greatly improved by these additions as well, with discrimination of landmarks in the environment. In the new cortex, there was an evolution of sensory analyzers for complex, high-resolution imaging (high sensory acuity) together with high-dexterity movement generation (motor acuity).

It appears that neocortex was more suited than striatum for evolution of higher acuity. It was more like the midbrain tectum: When the tectum increased sufficiently in surface area, it enabled a high acuity for various innate releasing mechanisms including orienting to novel stimuli or avoiding possible dangers that novel stimuli might represent. In the cerebral hemisphere, acuity is correlated with the area of the neocortex devoted to a given part of a sensory surface. For example, there is much more cortex representing the fovea than there is cortex representing the peripheral visual field. In somatosensory areas, there is much more cortex representing the fingertips than that representing the same amount of skin on the back of the hand. Motor acuity appears to be similarly related to neocortical area.

Functional Regions Based on the Early Evolution of Place Cognition and Object Cognition

The girdle of paralimbic areas at the fringes of the mammalian hemispheres—the areas, like the cingulate cortex and the parahippocampal areas, most closely connected to limbic system structures (see figure 26.9)—was divided into two functional regions by Marcel Mesulam: the olfactocentric and the hippocampocentric regions. His illustration of these two regions in the monkey is reproduced in figure 32.3. In the figure, the hippocampocentric regions are in light brown, and the olfactocentric regions are in darker brown. (Note that the hippocampocentric paralimbic regions follow the hippocampus as depicted for humans in figure 28.9.)

Related to this proposal is the hypothesis, presented with cytoarchitectural evidence and with some support from studies of transcortical connections, of a dual origin of neocortex by progressive differentiation from two margins of the hemisphere. The cortical areas nearest to these two margins appear to correspond to Mesulam's two paralimbic regions.

The hippocampocentric regions, closely connected to the hippocampus, receive inputs from all modalities. In mammals with relatively larger brains, this input comes primarily

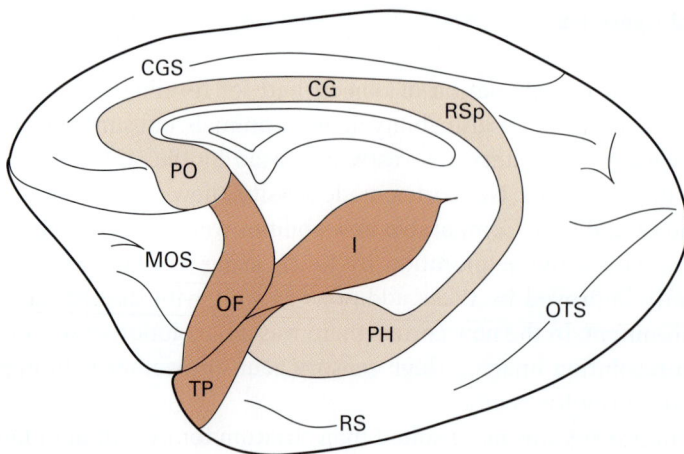

Figure 32.3
Schematic medial view of the rhesus monkey hemisphere showing the girdle of paralimbic cortical areas around the medial edge of the cortex. Mesulum distinguishes two divisions: hippocampocentric (light brown) and olfactocentric (darker brown). CG, cingulate cortex; CGS, cingulate sulcus; I, insula; MOS, medial orbitofrontal sulcus; OF, orbitofrontal cortex (caudal part); OTS, occipitotemporal sulcus; PH, parahippocampal cortex; PO, parolfactory cortex; RSp, retrosplenial cortex; TP, temporopolar cortex. Adapted from Mesulam (2000).

from association areas of the neocortex. Those areas are important in the retention of long-term memories (e.g., memories of places previously explored and memories of the landmarks at those places). With reciprocal connections between the neocortical association areas and paralimbic areas, which themselves are interconnected with the hippocampal formation, we can envision a system with activity that represents sensory inputs that are expected in the immediate future. Using the information on head movement and direction coming into the hemispheres by way of the anterior thalamic nuclei and the adjacent lateral-dorsal thalamic nucleus (LD) (as reviewed in chapters 22 and 28) and the information on the animal's location in the current environment coming from the hippocampus, the expectancy signals are constantly updated.

The neural representations of objects (i.e., the brain activities that are consistently correlated with object perception), independent of specific places, are also useful in increasing the ability to anticipate future inputs, especially in the immediate future. As the objects, living or nonliving, move around or are moved around, and as the animal moves itself around the objects, knowledge of the objects enables considerable predictability of inputs. The olfactocentric paralimbic cortical regions, much of it bordering the olfactory cortex and receiving olfactory and gustatory inputs from the thalamus, were no doubt involved in object sense early in evolution. These regions include the periamygdalar cortex of the temporal pole. However, the visual, somatosensory, and auditory inputs became more central to object perception, with formation of images retained in long-term memory. Object information from these modalities reaches the olfactocentric paralimbic regions in mammals via association areas of neocortex (e.g., the inferotemporal cortex).

The separation of hippocampocentric and olfactocentric paralimbic areas is not clear-cut in the anterior temporal region. The anterior parts of the hippocampus (ventral parts in small mammals) have many interconnections with olfactocentric structures.

Sensory Cortex That Led to the Evolution of Fine Motor Control

With the elaboration of somatosensory neocortex, mammals have evolved more acute touch senses. With outputs of this cortex to the corpus striatum and to the midbrain and subthalamus, more accurate orienting and detailed exploration of objects was enabled, with facilitation of the learning of movements. As in other sensory modalities, there was some multiplication of interconnected somatosensory cortical areas enabling more complex analysis of the information. Each of these areas had outputs to subcortical systems for eliciting movements, but one of those areas became particularly specialized for control of the motor system. In modern mammals, it is designated as the primary motor cortex.

In chapter 15, we described the motor cortex and noted the finding that in the opossum there is an "amalgam" of primary somatosensory and primary motor cortex (i.e., a very incomplete segregation). However, in the ventral thalamus, there is some segregation of the cell group that receives somatosensory input from the cell groups that receive cerebellar and striatal projections. Thus, an evolutionary parcellation may well have occurred first in the thalamus and later in the neocortex.

Anticipation as a Major Innovation of the Neocortex

The discussion above has included the idea that the object and place cognition of the neocortex enabled prediction of future inputs, especially the immediate future. The representations underlying this function can be conceived of as images—not only visual images but somatosensory and auditory images as well. These images depend on multiple levels of the neuronal pathways, from the receptor surfaces to central maps to the neural activities corresponding to objects and places. Imaging ability, dependent on activity in the association cortical areas and enabling the prediction of what is about to be perceived, facilitates the planning of movements. The animal can anticipate and respond in an adaptive manner to what is most likely about to occur.

The images of objects are constantly being updated by incoming sensory information and are organized in a spatial map of the environment. This organization is the role of the hippocampus, keeping track of the position of the animal in the current environment and constantly updating that position using information on head direction and locomotion (chapter 28). The multiple images of objects in the environment are represented by association cortical activity, and the activation of these images is constantly being updated by the information coming from the hippocampal formation by the connections we have already outlined.

Memory for the current environment is rapidly formed in the hippocampus and re-formed when the animal moves from one locale to another. With a move to another familiar environment, the hippocampus does not have to completely relearn the map: The new frame of reference, we can assume, is imposed on it by the long-term memories retained in the neocortex, especially in the parietal association cortex. Effects of damage to that region support this idea. Parietal damage in humans drastically degrades abilities that require spatial orientation with the aid of memory for specific places.

Thus, the posterior neocortex, including parietal, occipital, and temporal, is involved in perception and retention of images of objects, living and nonliving, and of the environmental spaces where these objects are encountered. The dynamics of this cortex include close interactions with hippocampus and other limbic system structures, with constant generation of expectancies as well as modulation by motivational states.

But what about the motor cortex and motor images? This leads to a discussion of planning and frontal neocortex.

Origins of Planning

The motor system also retains images—representations of movement patterns, including many that have been learned. Motor images involve retention of temporal patterns. The premotor cortex, when activated, controls organized pieces of movement. The representation of each of these movement patterns can be considered to be a kind of motor image. More complex motor images, representing longer movement sequences, are a kind of plan of the movements. The neocortex anterior to the primary motor cortex evolved for this kind of representation of plans. Thus, we can think of the learning of movement patterns retained in premotor and supplementary motor areas as similar to the images retained in posterior neocortical association areas. The anticipatory activity of posterior neocortex corresponds to the planning activity of the frontal areas.

Anterior to the premotor areas is the *prefrontal* cortex. Lateral prefrontal areas control anticipatory eye movements and other movements controlled by longer-range intentions. Orbital prefrontal areas, the cortex that in primates is above the eyes, control anticipatory and learned motivations.

Neocortex as Anticipator and Planner: Illustrative Evidence

Much of the evidence for the role of the neocortex in anticipation and planning is behavioral, but there is supportive data from studies of physiology and brain damage effects as well. Evidence that the brain of a human contains an active model of expected inputs—the images of objects and specific locations described above—was obtained in a series of experiments conducted in Moscow in the 1960s by E. Sokolov and his collaborators. The experiments demonstrated that novelty—unexpected sensory input—causes arousal and orienting responses with attraction of attention. The novelty could occur in an unexpected temporal pattern without other changes in the stimuli. It appears that whenever there is a clear mismatch between an expected input and the actual input, the orienting response is generated. When the inputs become familiar, there is a suppression of the arousal response.

Many of the experiments were conducted using auditory stimulation where the temporal pattern could easily be manipulated, but somatosensory and visual inputs could be used as well. Later, the Moscow lab used electrophysiological methods in animals in attempts to test the idea of an active central model (image). In America, one interesting experiment using electrical recording in the visual cortex of a cat yielded results that directly supported the hypothesis of a central model of expected input. Frank Morrell used a regularly flashing light as the stimulus, and he recorded visually evoked potentials from visual neocortex that kept time with the flashes. When the flashing was suddenly stopped, these cortical potentials continued for a brief period of time, but then they were abolished by the occurrence of an arousal response. (With the relative large electrode being used, the arousal response consisted of low-voltage fast activity as is seen using electroencephalographic recordings from the scalp in humans, reflecting activity in the underlying brain.)

The central models that are represented anterior to the motor cortex correspond to anticipation and planning of actions. These representations can control endogenous generation of actions—always under the influence of internal motivational states. This kind of action is what we call voluntary action. Damage to these cortical areas can cause reduction or abolition of voluntary action. For example, lesions of the frontal eye fields result in an inability voluntarily to move the eyes away from a moving visual stimulus that the eyes are following and an inability to initiate an eye movement in response to a verbal request.

Endbrain Evolution: Suggested Major Steps Reviewed and Extended

Before looking at neocortical organization in more detail, we will step back for a longer-term view, looking again at the outline of major evolutionary steps that are suggested by comparative anatomical studies and other relevant information (see figure 24.7). Added to the former outline are additional ideas about the forebrain focusing on neocortex and thalamus.

Stage 1 The earliest chordates, assuming that they resembled amphioxus (first introduced in chapter 3 and figure 3.3), had no cerebral hemispheres. The forebrain was little more than the forerunner of the pituitary and the hypothalamus and an epithalamus. An early input to this region was from chemoreceptors, conveyed into the forebrain by the olfactory nerve. Along with olfactory input there evolved endbrain structures connected to it. Thereafter, olfaction became the dominate exteroceptive input to the very early endbrain.

Stage 2 With an invasion of sensory inputs from non-olfactory modalities via midbrain and 'tweenbrain, the evolving endbrain structures became multimodal. These structures were the early pallial and subpallial structures. Both of these structures, pallium and subpallium, began to expand. Although they could function relatively independently, pallial projections to the striatum became increasingly important, especially at later stages.

Stage 3 The midbrain became the dominant source of non-olfactory inputs to the forebrain. This was true even for the visual system although the retina projects directly to diencephalic structures. The inputs reached the endbrain mostly via the evolving dorsal thalamus. Axons carrying somatosensory, visual, and auditory information contacted multimodal neurons of the paleothalamus. In modern mammals, such neurons are found in the intralaminar and midline nuclei and also in the posterior nuclear region of the lateral thalamus. Axons from the hypothalamus also reach some of these cell groups, particularly some of the midline nuclei. The hypothalamus also influences neurons that project from the midbrain into the thalamus.

Stage 4a Portions of the dorsal thalamus received axons from the more dorsal parts of the midbrain, the tectum. Portions of the tectum were primarily unimodal. Thus, the most superficial part of the rostral tectum was visual and the caudal tectum was auditory. Somatosensory and multimodal inputs dominated a deeper layer of the tectum. As the projections from unimodal neurons of these midbrain regions became denser in the thalamus, they segregated from each other, leading to the evolution of distinct cell groups in the thalamus. The axons from the different midbrain tectal regions followed separate routes into the thalamus, and their pattern appears to have influenced the segregation of modalities in the lateral part of the dorsal thalamus. This is illustrated for the hamster in figures 32.4 and 22.1.

Similarly, somatosensory axons reached the thalamus more ventrally and became segregated in the ventral thalamus. It was probably very early in vertebrate evolution that many of these axons reached the diencephalon from hindbrain or spinal cord.

In mammals, the striatum became increasingly dependent on the inputs from the pallium, which expanded more than the striatum in mammalian evolution. As we have noted earlier, in the striatum there was a parcellation of non-olfactory dorsal striatum from an olfactory ventral striatum. Similarly in the pallium, the dorsal pallium separated from the more ventral pallial areas. The evolutionary process of parcellation, beginning long before any neocortex evolved (long before the advent of mammals), separated the portions dominated by thalamic inputs from the portions dominated by olfactory bulb

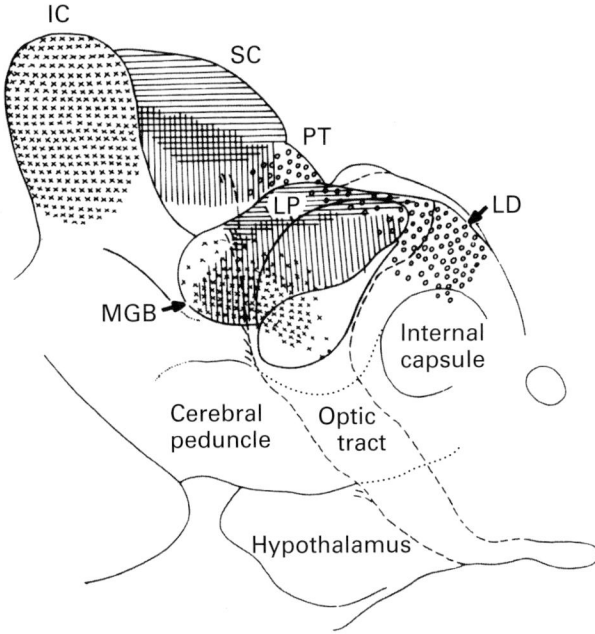

Figure 32.4
Side view of midbrain and 'tweenbrain of Syrian hamster showing where the different regions of the dorsal midbrain and pretectal area project axons to different regions of the thalamus. Structures are depicted as if the brain tissue were transparent except for the sources and terminations of the projections. The spatial arrangement of the midbrain and pretectal regions is matched by the arrangement of the terminal regions. IC, inferior colliculus; LD, lateral-dorsal nucleus; LP, lateral-posterior nucleus; MGB, medial geniculate body; PT, pretectal cell groups; SC, superior colliculus. Neuroanatomical reconstructions by author, from Schneider (1975).

inputs. The adaptive advantages of this parcellation were those of greater specialization of information processing.

With the separation of non-olfactory from olfactory pallium, the cell layers evolved as well. Exactly how and when this happened is a matter of speculation, but its great importance is abundantly clear. In part of the non-olfactory dorsal pallium, neuronal migrations formed new, more superficial layers. This *neocortex* is the hallmark of the mammalian endbrain, with five major cell layers. The new layers are numbered 2, 3, and 4. (Layer 1 contains only a few neurons, which are present very early in development; it is primarily a layer of dendritic processes and terminating axons and was present long before the new cell layers appeared.) The adaptive advantages of functions such as those noted above led to new connectional patterns with employment of new kinds of short-axon interneurons, mostly inhibitory neurons using the neurotransmitter GABA. These cortical interneurons have migrated from subcortical portions of the endbrain—mostly from the medial and the caudal ganglionic eminence and from the preoptic region (see figures 12.1, 12.10, and 12.11).

In humans and other large mammals, an additional source of neocortical neurons, very likely short-axon interneurons, has been found at early developmental stages: a transient region known as the subpial granular layer.

Next we turn again to the thalamus—the major source of inputs to the dorsal pallium: The posterior nuclear group and the ventral part of the lateral-posterior nucleus, abutting unimodal thalamic regions, remain multimodal in modern mammals. The proposal here is that multimodal convergence was the primitive state of much of the thalamus, long before the emergence of mammals or even reptiles. Then a parcellation occurred. Correlated changes occurred in the cortex (neocortex or its predecessor), the primary recipient of projections from the dorsal thalamus: Multimodal thalamus projects to multimodal association cortex. Unimodal thalamus projects to primary sensory areas and unimodal association cortex.

Findings of comparative neuroanatomical studies that fit the idea of thalamic parcellation in evolution are summarized in figure 32.5. Again, we can think of parcellation as adaptive because of the advantages of specialization in information processing. It came with a cost: the loss of integrative functions. Such integrative functions require the bringing together again of various streams of processed information for the purposes of controlling actions. The integration, of course, still occurs, but it is, so to speak, put off until later.

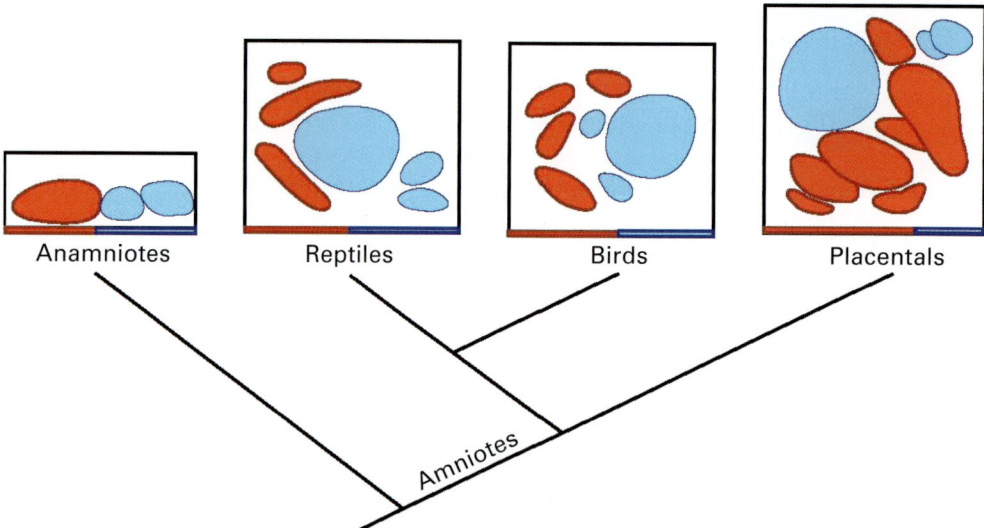

Figure 32.5
Schematic illustration of differences in segregation of distinct cell groups in the dorsal thalamus in major vertebrate groups. The cladogram shows the evolutionary descent of these groups: anamniotes and amniotes. Blue, regions dominated by inputs from midbrain tectum; red, regions dominated by other inputs (or by mixed inputs). Based on Striedter (2005) and Butler (1994).

Stage 4b Remember that the visual inputs entered the CNS at the diencephalic level. It is likely that a very early visual route to the endbrain did not have to go through the midbrain. Comparative neuroanatomical studies support this. However, such a direct visual route may not have been very well organized in its topography. It is possible that it was the midbrain tectal projection that first developed precise topography, as the first priority of imaging vision was probably orienting rather than precise object identification. It is interesting in this regard that the mammalian superior colliculus projects topographically to the lateral geniculate body. It seems possible that better topographic organization in the forerunners of this cell group was influenced by the tectal input.

It is also interesting that in experiments involving surgical ablation of the colliculus very early in life in a hamster, rat, or ferret, the retina develops a strong projection to the lateral posterior thalamus, adjacent to the lateral geniculate body (see figure 13.12). Thus, the specificity of the retinofugal axons, in these modern mammals, is compatible with synapse formation in thalamus as well as in tectum. Some retinofugal axon types, studied in the cat, are known to have terminal arbors in both the lateral geniculate and in the superior colliculus.

Expansions of Neocortical Areas

In the discussion of sensory and motor acuity above, it was noted that higher acuity is correlated with greater neocortical area. Expansions of individual cortical areas is one way that neocortex increased in size in evolution. Expansions also occurred by multiplication of the number of distinct representations of sensory surfaces. These processes have no doubt been of great importance in size increases in the cortex. However, such discussions usually occur with the assumption that unimodal cortex, the primary sensory areas, are also primary in evolutionary time. Then it is assumed that the association areas appeared later. We know that what we call the highest cognitive functions like human language, and the planning of actions well in advance of their execution, depend on multimodal association areas. It is common to assume that such areas evolved most recently. However, this may not be completely accurate.

A slightly different view of cortical evolution fits the findings of comparative anatomy better. Remember the stages of endbrain evolution outlined above. The earliest cortex, after what may have been a totally olfactory stage, was very likely multimodal, like the dorsal cortex of amphibians. Olfactory regions probably became segregated first. The dorsal non-olfactory cortex received multimodal inputs from the multimodal paleothalamic cell groups noted above. To be sure, at this early stage the cortex had not become a six-layered neocortex. The laminar structure of the cortex was evolving in parallel with the evolution of cortical areas.

As parcellation of thalamic cell groups occurred (figure 32.5) and predominantly unimodal regions emerged in the thalamus, unimodal cortical regions naturally appeared as well, with the functional advantages of specialization. Figure 22.11 depicts visual neocortical areas in a variety of mammals. Even in a primitive marsupial brain, there is a primary visual cortex, also a second visual area next to it. Such data can be used to argue that a geniculostriate projection arose in the earliest mammals or in mammal-like reptiles. Check figure 22.12 again: It shows a representation of "old" and "new" neocortical sensory and motor areas. What this picture fails to show is the reported existence of multimodal regions in between major sensory areas in hedgehog and other small mammals like the rat. (See also figure 34.11.) It also fails to indicate findings that there are multimodal inputs to primary sensory areas, for example, auditory inputs to primary visual cortex, even in the domestic cat. (Strong evidence on this has come more recently from electrophysiological studies of the prairie vole cortex.) These facts are important because they are clues to the evolutionary history of the tissue. Parcellation of unimodal areas from multimodal cortex, at least in some species, may be incomplete.

What, then, happened to the primitive multimodal cortex? The proposal here is that unimodal areas differentiated within it. These unimodal areas became the primary and secondary sensory cortical areas.

In the visual system, it is likely that at least two such very early unimodal areas evolved, one receiving the more direct visual input from the retinorecipient thalamic cell group (which became the lateral geniculate nucleus in mammals) and one receiving a projection from a tectorecipient portion of the lateral thalamic region (figure 22.1 and figure 32.4). As these figures indicate, it was likely that there was another such region, receiving lateral thalamic input from a pretecto-recipient portion of the lateral thalamus.

After the evolution of the unimodal sensory areas, and later the differentiation of a motor cortex from a somatosensory area, the neocortex expanded in two different ways. The first was the multiplication of unimodal association areas that received their input not only from thalamus but also via transcortical connections. Figure 6.4 illustrates this expansion of the visual areas in the primates. The second was a transformation of the residual multisensory cortex. Multisensory inputs reached this cortex not only from the thalamus, but increasingly via transcortical connections from the unimodal areas—not the primary sensory areas but the association areas around them. This resulted in an expansion of the multisensory cortical regions. They became far more sophisticated multimodal cortex than what had existed long before. This expansion has been the greatest in the large primates, especially in humans. This late expansion occurred with the evolution of higher cognitive abilities. We will discuss this further in the following chapters.

It is important to note here that there were energy constraints on expansions of the brain. The brain is a major user of the energy obtained from food, and acquiring food takes time. This places an upper limit on brain expansion. In the evolution of our species, there is evidence that the development of cooking abilities was a crucial breakthrough because it reduced the energy demands of the gut. As a result, more energy became available for the brain, and an important barrier to brain expansion was reduced.

Evolutionary Changes in Thalamocortical Axon Trajectories

Comparative neuroanatomical studies have given some support to the idea that ancestral brains have left clues in modern mammalian brains indicating that some pathways are more primitive and others are more recent in evolution. Structural clues can be found in the trajectories of thalamocortical axons. Figure 32.6 shows a sketch of a section through the endbrain of an amphibian—a frog—based on Golgi-stained sections, from the work of C. J. Herrick published in 1910. In the cortex, you can see the major characteristics of pyramidal neurons, with dendrites that radiate toward the surface from deeper-lying cell bodies. Axons from the lateral forebrain bundle, which includes axons from the thalamus, run tangentially through the cortex and are not concentrated in underlying white matter as is characteristic of the mammalian cortex. Such a morphology is also characteristic of the reptilian endbrain.

Figure 32.7 illustrates further the major differences between the mammalian neocortex and the reptilian or amphibian dorsal cortex.

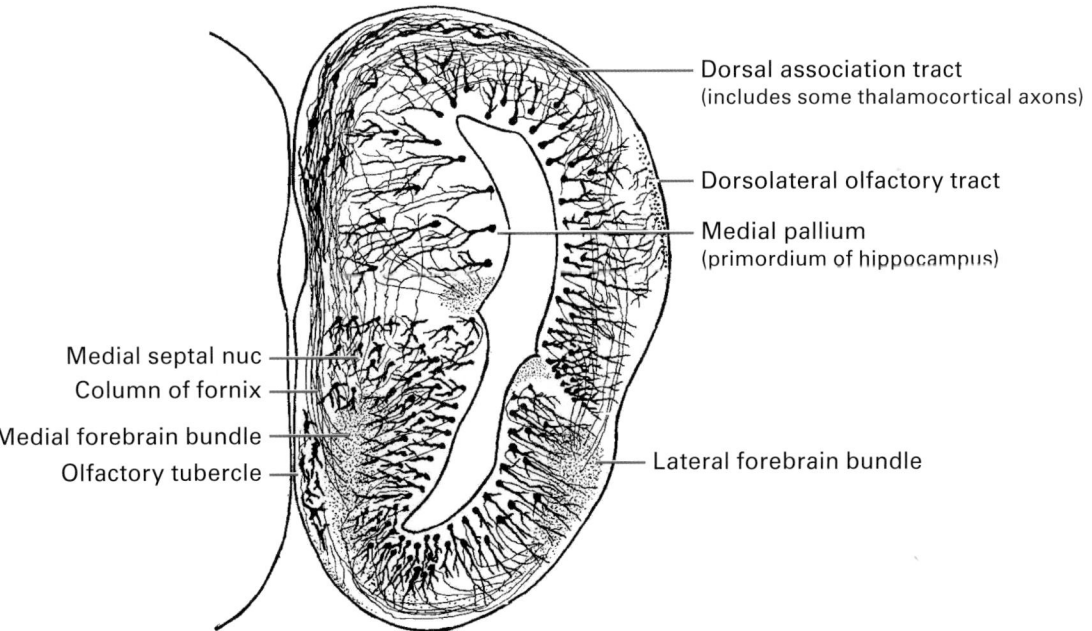

Figure 32.6
Drawing of a Golgi-stained section from the endbrain of a frog by C. Judson Herrick (1910). The cortex shows neurons with the major characteristics of pyramidal cells: dendrites that radiate toward the surface from deeper-lying cell bodies. Such a morphology is characteristic of the cortex of both amphibians and reptiles. The trajectory of axons coming from the thalamus and contacting dendrites of pallial neurons can be seen. The axons run tangentially through the cortex and are not concentrated in underlying white matter as is characteristic of the mammalian cortex. One can also see axons from dorsal pallial neurons to the medial pallium. From Herrick (1910).

(a) Amphibian dorsal pallium

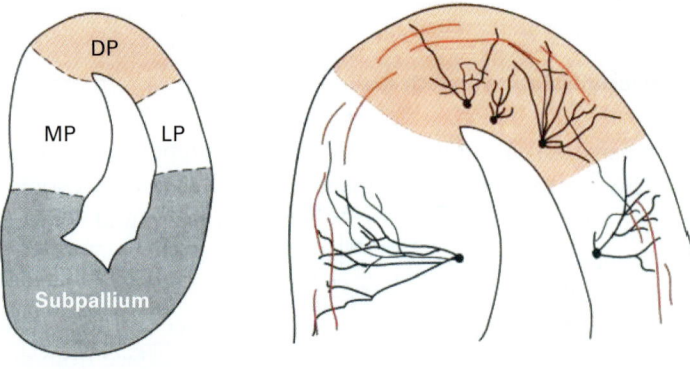

(b) Reptilian dorsal cortex

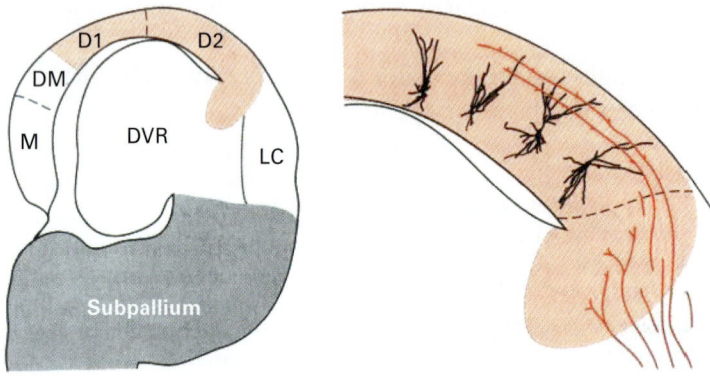

(c) Mammalian neocortex

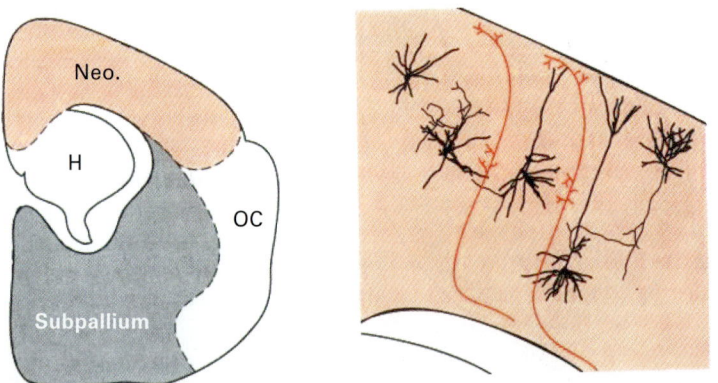

Figure 32.7
Drawings of frontal sections of the endbrain of (top) hedgehog, (middle) salamander, and (bottom) turtle with coloring of the neocortex of the hedgehog and related regions in the other two species (dorsal pallium in salamander, dorsal cortex in turtle). At the right are enlargements of the dorsal pallial region in each species showing typical neurons and the trajectories of axons entering the region from the thalamus. Note the difference between the mammal and the non-mammals in typical trajectories of axons from the thalamus (in red). D1, D2, dorsal cortex; DP, dorsal pallium; DM, dorsomedial cortex; DVR, dorsal ventricular ridge; H, hippocampus; LC, lateral cortex, LP, lateral pallium; M, medial cortex; MP, medial pallium; Neo, neocortex; OC, olfactory cortex. From Striedter (2005).

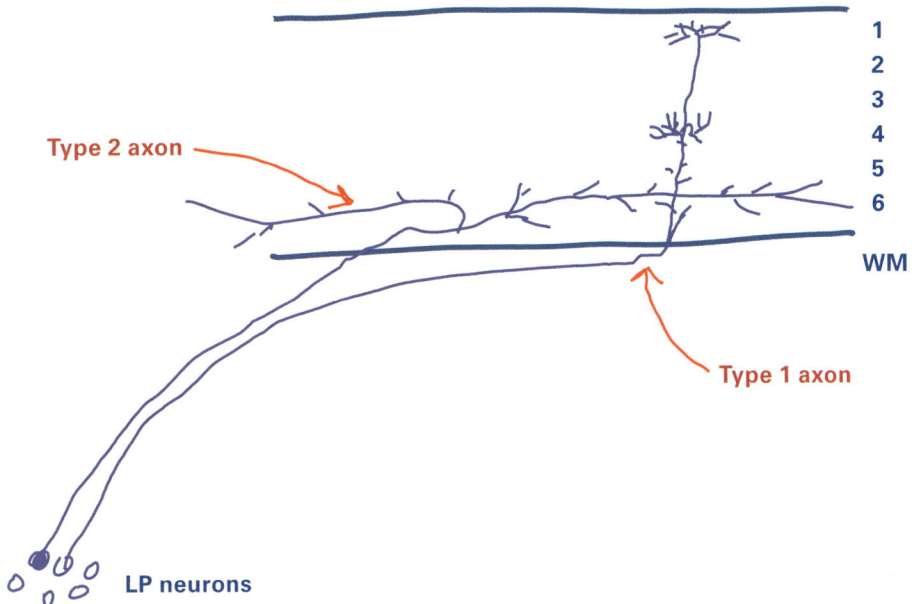

Figure 32.8
Trajectories of axons from the thalamic lateral posterior nucleus of the hamster to the posterior neocortex. There is evidence for at least two major types of axon. Type 1 axons are similar to the thalamocortical axons illustrated for mammals in figure 32.7. The terminal arbors are in cortex adjacent to the striate area. Type 2 axons are much more widespread in their distribution. The axon trajectories remind one somewhat of thalamocortical axons in reptiles and amphibians in that they course tangentially through the deeper cortical layers, mostly in layer 6, with evidence for terminations throughout posterior neocortex including primary visual, auditory, and somatosensory cortex and areas in between. Based on studies of Schneider using silver staining for degenerating axons (unpublished). Some type 2 axons may originate in the posterior nuclear group. Different types of thalamocortical axons, differing in how focal or widespread their terminations are, have been distinguished in a number of different studies of mammalian brains. LP, lateral-posterior nucleus; WM, white matter.

However, not all thalamocortical axons in mammals are so different from those axons in amphibians and reptiles. See figure 32.8 for a drawing of two types of axons from the lateral posterior nucleus (and posterior nucleus) of the hamster. Similar findings have been reported for the hedgehog. The axon labeled "type 1" has the morphology that is always described for mammalian thalamocortical axons. It follows the white matter to its terminal region, and then goes up along a column toward the surface with terminals in several layers, focused especially in layer 4 and in layer 1. We have also found some evidence for axons of the type labeled "type 2" in the figure: Such axons appear to follow a different trajectory, traveling tangentially through the deeper neocortical layers (especially layer 6) and forming scattered terminal arbors along their course. Such axons appear to distribute widely to posterior neocortical regions, not only in association areas but also in primary sensory areas. The trajectories of type 2 axons are somewhat reminiscent of thalamocortical axons in reptile and amphibian brains. They appear to arise mostly from the deeper portions of the lateral posterior nucleus and/or the posterior nucleus group.

The neocortical laminar terminations of many of the neurons of the intralaminar nuclei also indicate an early origin in thalamic evolution. The axons terminate mainly in layers 5 and 6—the first layers to develop. Details of the axon trajectories have not been described. As we have noted earlier, these cell groups project strongly to the striatum as well.

Readings

Extensive reviews of general neuroanatomy or comparative and evolutionary neuroanatomy:

> Brodal, P. (2004). *The central nervous system: Structure and function*, 3rd ed. New York: Oxford University Press.
>
> Butler, A. B., & Hodos. W. (2005). *Comparative vertebrate neuroanatomy: Evolution and adaptation*, 2nd ed. New York: John Wiley & Sons.
>
> Mesulam, M.-M. (2000). *Principles of behavioral and cognitive neurology*, chap. 1. New York: Oxford University Press.
>
> Striedter, G. (2005). *Principles of brain evolution*. Sunderland, MA: Sinauer Associates.

Migration of cells from ganglionic eminence to become interneurons in neocortex: summarized for mice with new data from Gord Fishell and co-workers, and updated for humans and other large-brained mammals by Rakic (below):

> Miyoshi, G., Hjerling-Leffler, J., Karayannis, T., Sousa, V. H., Butt, S. J. B., Battiste, J., & Fishell, G. (2010). *Journal of Neuroscience, 30*, 1582–1594.
>
> Rakic, P. (2009). Evolution of the neocortex: A perspective from developmental biology. *Nature Reviews Neuroscience, 10*, 724–735.

Results presented in the following paper indicate that the projections of the lizard dorsal cortex resemble those of the ventral part of the subiculum—part of the hippocampal formation—in mammalian brains. Thus, the dorsal cortex efferents in this lizard do not indicate a close similarity to the neocortex.

> Hoogland, P. V., & Vermeulen-Vanderzee, E. (1988). Efferent connections of the dorsal cortex of the lizard *Gekko gecko* studied with *Phaseolus vulgaris*-Leucoagglutinin. *Journal of Comparative Neurology, 285*, 289–303.

An impressive and comprehensive discussion of neocortical origins, with extensive references, can be found in the following article, although more recent findings and discussions can be found:

Aboitiz, F., Morales, D., & Montiel, J. (2003). The evolutionary origin of the mammalian isocortex: Towards an integrated developmental and functional approach. *Behavioral and Brain Sciences, 26,* 535–586.

For a more recent review, see the following paper:

Medina, L., & Abellán, A. (2009). Development and evolution of the pallium. *Seminars in Cell & Developmental Biology, 20,* 698–711.

Axons similar to the "type 2" axons in figure 32.8 are found in the hedgehog.

Gould, H. J. 3rd, Hall, W. C., & Ebner, F. F. (1978). Connections of the visual cortex in the hedgehog (*Paraechinus hypomelas*). *Journal of Comparative Neurology, 177,* 445–472.

Other publications that are particularly relevant to the chapter:

Butler, A. B. (1994). The evolution of the dorsal thalamus of jawed vertebrates, including mammals: Cladistic analysis and a new hypothesis. *Brain Research Reviews, 19,* 29–65.

Catania, K. C, Collins, C. E., & Kaas, J. H. (2000). Organization of sensory cortex in the East African hedgehog (*Atelerix albiventris*). *Journal of Comparative Neurology, 421,* 256–274.

Fernandez, A. S., Pieau, C., Reperant, J., Boncinelli, E.., & Wassef, M. (1998). Expression of the Emx-1 and Dlx-1 homeobox genes define three molecularly distinct domains in the telencephalon of mouse, chick, turtle and frog embryos: Implications for the evolution of telencephalic subdivisions in amniotes. *Development, 125,* 2099–2111.

Herkenham, M. (1980). Laminar organization of thalamic projections to the rat neocortex. *Science, 207,* 532–535.

Morrell, F. (1963). Information storage in nerve cells. In Fields, W. S., & Abbott, W., eds. *Information storage and neural control,* pp. 1–37. Springfield, IL: Charles C. Thomas. Republished in Pribram, K. H., ed. (1969). *Brain and behaviour 3: Memory mechanisms.* New York: Penguin.

Reiner, A. (1993). Neurotransmitter organization and connections of turtle cortex: Implications for the evolution of mammalian isocortex. *Cmparative Biochemistry and Physiology, 104A,* 735–748.

Rosa, M. G. P., & Krubitzer, L. A. (1999). The evolution of visual cortex: Where is V2? *Trends in Neuroscience, 22,* 242–248.

Sanides, F. (1969). Comparative architectonics of the neocortex of mammals and their evolutionary interpretation. *Annals of the New York Academy of Sciences, 167,* 404–423. (Special issue: Comparative and Evolutionary Aspects of the Vertebrate Central Nervous System.)

Sanides, F. (1970). Functional architecture of motor and sensory cortices in primates in the light of a new concept of neocortex evolution. In Noback, C. R., & Montagna, W, eds. *The primate brain* (pp. 137–208). New York: Appleton-Century-Crofts.

Schneider, G. E. (1973). Early lesions of superior colliculus: Factors affecting the formation of abnormal retinal projections. *Brain, Behavior and Evolution, 8,* 73–109.

Schneider, G. E. (1975). Two visuomotor systems in the Syrian hamster. In Ingle, D., & Sprague, J. M., eds. *Neurosciences Research Program bulletin, 13,* 255–257.

33 Basic Neocortical Organization: Cells, Modules, and Connections

In the previous chapter, we discussed the great specialty of the mammalian brain, the neocortex, from the viewpoint of evolution and the functions that drove that evolution. You saw pictures of the neocortical laminar pattern and were introduced to the shapes of some of its cells and trajectories of its axons.

Two Major Types of Neocortical Neuron

We have learned that the major gateway to the neocortex from subcortical regions is the thalamus. (For notes on the evolution of this region, see chapters 7 and 12.) Principal neurons of thalamic cell groups project axons to the various neocortical areas, and these axons are a major source of excitation of the most common class of neocortical neurons, the pyramidal cells (figure 32.1 and figure 33.1). (The pyramidal cells receive excitatory inputs not only directly from the thalamocortical axons, but also from intervening excitatory stellate cells of layer four.)

The second major group of neocortical neurons is composed of various types of short-axon interneurons, most of them stellate in shape. Figure 33.1 shows examples of specializations in this class of cells. These simplified drawings show several distinct patterns of local axonal distribution. For example, a cell may be activated in one column of cells and form inhibitory connections with the cell bodies of neurons in a number of adjacent columns.

Viewing the dendritic and axonal structure of neocortical neurons does not reveal the real diversity of interneuronal types. Studies of the neurotransmitters they use and the ultrastructure of the synapses they make indicate that there are both inhibitory and excitatory interneurons. In addition, various combinations of peptide neuromodulators are present.

The Major Morphological Characteristics of the Neocortex

To review the major structural features of neocortical tissue, look again at figure 32.1. Each of the three histological staining methods illustrated in the figure shows a clear laminar arrangement. Fiber stains indicate many more than the six layers usually named

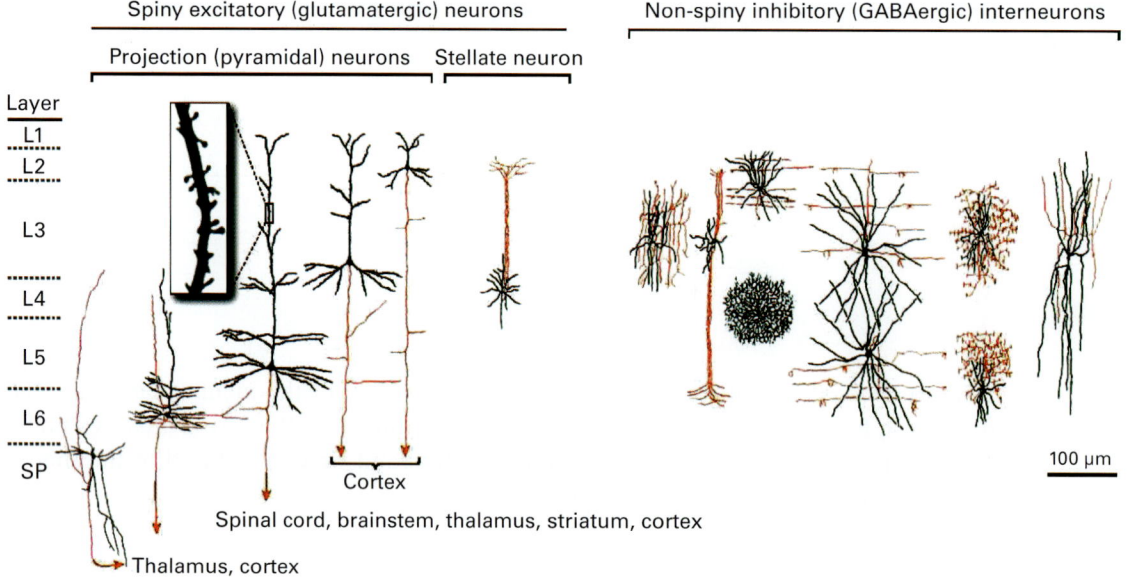

Figure 33.1
(Left) Simplified drawings of spiny, excitatory neurons (glutamatergic) of neocortex: pyramidal cells of layers 2, 3, 5, and 6, and stellate interneurons of layer 4. Note the enlargement of a segment of an apical dendrite showing its spines. Destinations of descending axons are named, and some axon collaterals in the cortex near the cells of origin are illustrated. (Right) Drawings of non-spiny inhibitory interneurons (GABAergic cells). SP, subplate. From Kwan, Šestan, and Anton (2012).

for neocortex, as they reveal a sublamination in some of the layers seen with cell-body stains.

Fiber pictures also reveal radial fascicles: small bundles of axons arranged perpendicular to the cortical layers (see figure 32.1a, middle picture). We have already learned about some of the axons contained in these bundles: axons carrying input information from the thalamus, axons carrying outputs to both subcortical structures and other areas of cortex, and short axons that interconnect cells in different cortical layers within a column (figure 32.2 and figure 33.1, right side). The outputs to subcortical structures go to the forebrain (corpus striatum, thalamus, subthalamus, epithalamus) and also, further caudally, at least as far as the midbrain and pons. (Cortical axons from somatosensory and motor cortex project further caudally, including to the spinal cord.)

Connections between neurons of nearby cortical columns are made not only by short-axon interneurons. They are also made by axon collaterals of the pyramidal cells with long axons carrying outputs to other brain regions (see figure 33.2).

Cell and fiber arrangements are sufficiently different from one region of neocortex to another to allow distinct boundaries to be drawn. As an example, the boundary in the human brain between human visual cortical areas 17 and 18 is illustrated in the

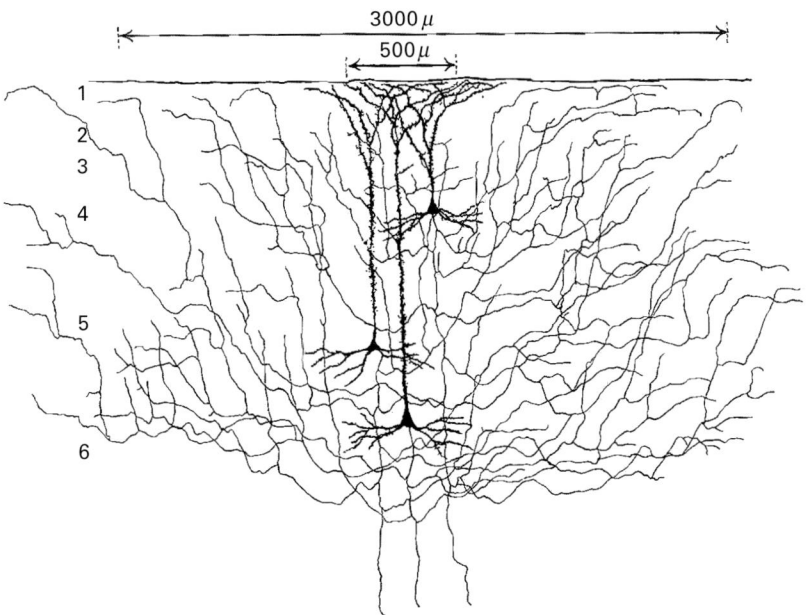

Figure 33.2
Drawings of three Golgi-stained neocortical pyramidal cells of a 60-day-old cat with tracings of their extensive axon collaterals. The dendrites of these cells spread over a column about 500 µm in diameter, whereas the axon collaterals spread over a column about 3,000 µm in diameter. From Scheibel and Scheibel (1970).

photograph of a Nissl-stained section in figure 33.3. Area 17 is seen in the left half of the picture, and area 18 is seen in the right half. Figure 33.4 shows an adjacent section stained for myelinated fibers. The distinct layer of fibers in the middle of layer 4 corresponds to a light band in the Nissl stain of the previous picture. This layer of tangential fibers can be seen even in human brain dissections and is named the *line of Gennari* for its discoverer, a medical student in Italy who first called attention to it in the eighteenth century.

Although many boundaries of neocortical areas are not as distinct as the area 17–18 boundary, Korbinian Brodmann was able to use Nissl-stained tissue to divide up the entire cortical mantle of the human neocortex and of other species as well. His map of the human cortex was published in 1909 (figure 33.5).

Modules of the Neocortex: Areas and Columns

The results of the cytoarchitectural studies of Brodmann and others in the late nineteenth and the early twentieth century provided neuroanatomical support for an increasing amount of evidence for functional specializations in localized regions of the endbrain. Many of the neocortical areas that he defined using morphological criteria are known to

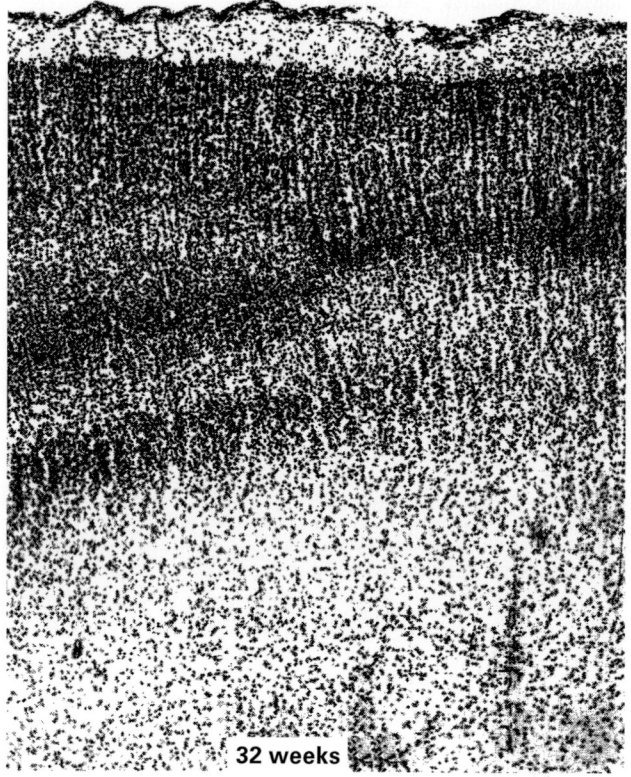

Figure 33.3
Area 17–18 border region: In this Nissl-stained section from the brain of a 32-week human fetus, area 17 occupies the left half of the photograph. Compare this figure to figure 2.3 showing the same region in an adult human neocortex. From Larroche (1996).

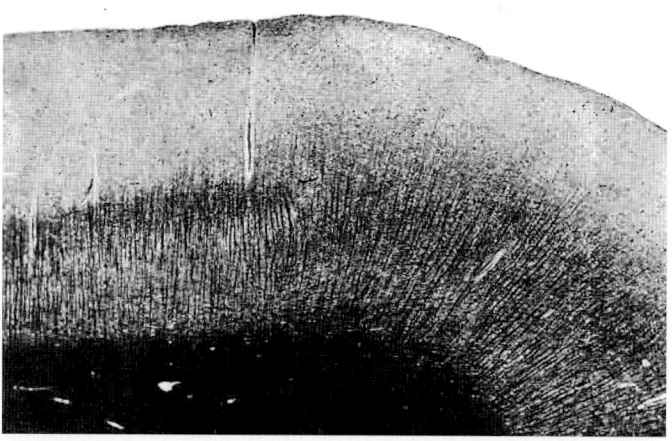

Figure 33.4
Myelin stain of a section of human occipital cortex similar to that of figure 33.3. Area 17, the primary visual cortex, occupies the left half of this photograph. From Le Gros Clark (1976).

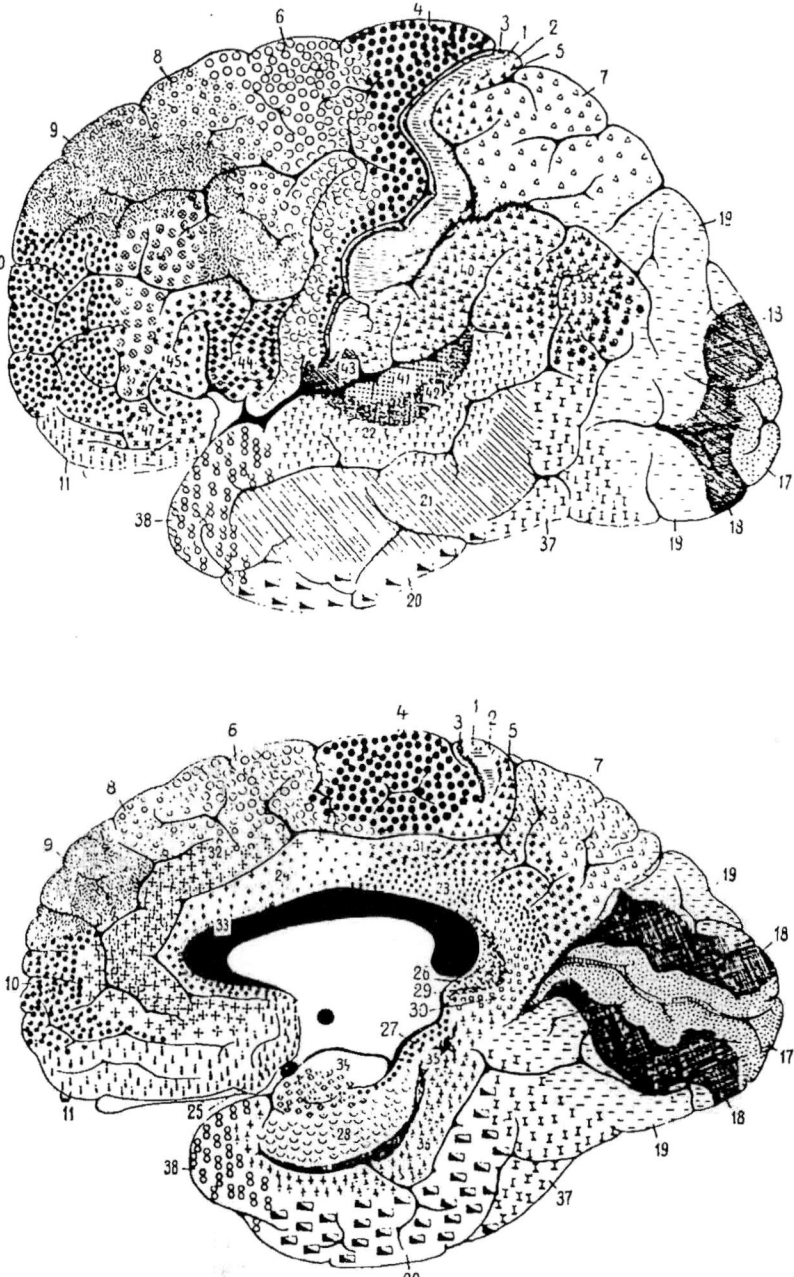

Figure 33.5
Brodmann's map of human neocortex, based on Nissl-stained sections. Areas distinguished by cytoarchitecture are filled with different tiny symbols. Small numbers designate each separate area. These numbers are still very commonly used in the literature. From Brodmann (1909).

have distinct functional correlates. The functions have been defined especially by electrical recording studies and by behavioral effects of localized cortical damage.

The subdividing of the neocortex into separate modules can be taken further. The different areas (e.g., the primary visual and primary somatosensory cortices) are each composed of many columns, as has been noted above and in the previous chapter. Functional correlates of some, but not all, neocortical columnar arrangements have been found.

The ocular dominance columns in the visual cortex have been clearly defined for large primates. As discussed in chapter 21, the right and left eyes of mammals send axons to separate layers of the lateral geniculate body of the thalamus: In the visual cortex of large primates, axons from right-eye layers terminate in columns separate from columns in which axons from left-eye layers terminate. Neuroanatomists have been able to follow the projections from one eye through the lateral geniculate body to neocortical area 17 by means of transneuronal transport of a labeling substance used for axon tracing (e.g., the tritium-labeled amino acid proline). When tangential sections are used to reconstruct this area of cortex in the rhesus monkey to reveal the pattern of terminations of the axons from a single eye as seen from the brain surface, a striped pattern is revealed (figure 33.6). The ocular dominance columns are distributed in a pattern resembling a zebra's stripes.

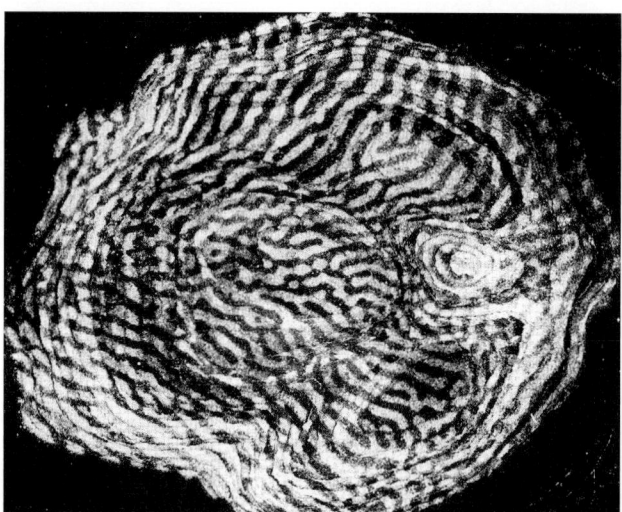

Figure 33.6
Ocular dominance stripes in layer 4 of macaque monkey area 17 (striate cortex, primary visual cortex). Sections were cut in a plane tangential to the cortical surface, processed by autoradiography in order to visualize the sites of the radioactive proline label that had been transported transneuronally from the ipsilateral eye over a period of 2 weeks. The radioactivity has exposed a photographic emulsion used to coat the sections that were kept in darkness during the exposure period. Silver grains in the developed emulsion are visualized as bright stripes by dark-field microscopy. The dark stripes are sites of termination of the opposite (contralateral) eye. The stripes are about 350 μm wide. From LeVay, Connolly, Houde, and Van Essen (1985).

A pattern of smaller columns has also been defined in these monkeys, as well as in other species like the cat. The smaller columns are defined by neuronal responses to small areas of the visual field where lines of a specific orientation are the best stimulus. The best orientation for stimulating the neurons in a column shifts systematically to a different orientation when one moves to adjacent cortical columns (see figure 34.10). In addition to preferences for one eye and for contours of a certain orientation, visual cortical neurons representing one small region of the visual field can have other preferences as well (e.g., for specific colors).

For some cortical regions, anatomical columns have been found without any specific physiological knowledge about functions. An example is the discovery of columns in the retrosplenial cortex of the monkey. This region is paralimbic cortex near the posterior end, or splenium, of the corpus callosum. A columnar pattern of terminations of axons from the prefrontal association neocortex has been observed there in axon-tracing studies (figure 33.7). The axon terminals were found in separated columns, with few terminals in between these columns. Projections to those in-between regions have been found to originate in parietal association cortex. Such projections to adjacent columns appears to be a means of bringing information together for integrative processing.

Cortical Types and Variations

Summarizing what we have discussed in previous chapters, the entire cortical mantle of the hemispheres has two major regions: the neocortex (also called the isocortex) and the other cortex—the *allocortex*. The allocortex includes the hippocampal cortex (sometimes called *archicortex*, with one or two layers of cells), and the olfactory cortex (piriform or *pre-piriform* cortex, sometimes called paleocortex). In addition, there are transitional types of cortex located between allocortex and neocortex. These include the paralimbic cortical areas outside the hippocampus medially and caudally (and temporally in mammals in which expansion of neocortex has led to temporalization of the posterior cortex) (see figure 32.3). Transitional cortex is also found laterally, between olfactory cortex and neocortex. These have also been called paralimbic areas.

Some anatomists have used the term *juxtallocortex* (next to the allocortex) for these paralimbic regions. Outside the hippocampal archicortex, these cortical areas have two to four discernible layers of cells plus the surface layer with few neurons. The simpler layering is in cortex closer to the hippocampus.

Within the expanse of neocortex there are regional differences—differences that have been used for cytoarchitectural mapping studies like those of Brodmann. The regions showing the greatest uniqueness are, at one extreme, the primary sensory areas with the most prominent layer 4—the layer of small stellate cells referred to as granular cells. Recall figure 33.3 illustrating the highly prominent granule cell layer, with sublayers, of human primary visual cortex. At the other extreme are the areas with no prominent

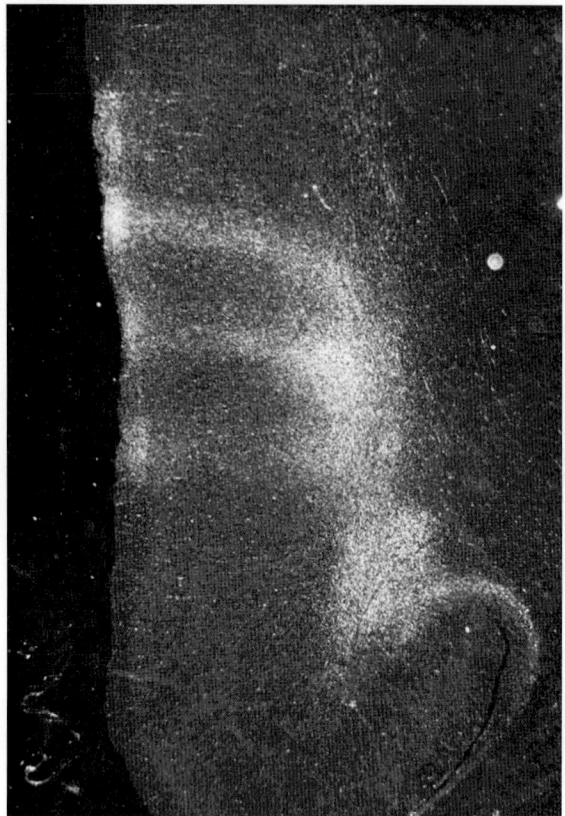

Figure 33.7
Projections to retrosplenial association neocortex of monkey from prefrontal association cortex at the frontal pole, labeled with radioactive amino acids and visualized with autoradiography (as in figure 33.6). The projection ends in columns, interleaved with a projection from parietal association cortex. From Goldman and Nauta (1977).

granular layer 4 at all. Agranular areas are the motor and premotor regions. The neuroanatomist von Economo in 1929 described these least typical neocortical areas as heterotypic cortex (sometimes called *ideotypic*). The remaining areas he called homotypic. He grouped homotypic areas into frontal type, parietal type, and polar type cortex according to major differences in the five cell layers (thickness, prominent cell types and densities, etc.) (figure 33.8).

Connections of the Neocortical Modules

The cortical columns described above and in the previous chapter (see figure 32.2) have **intracolumnar connections** that are similar throughout the cortical mantle. The schematic

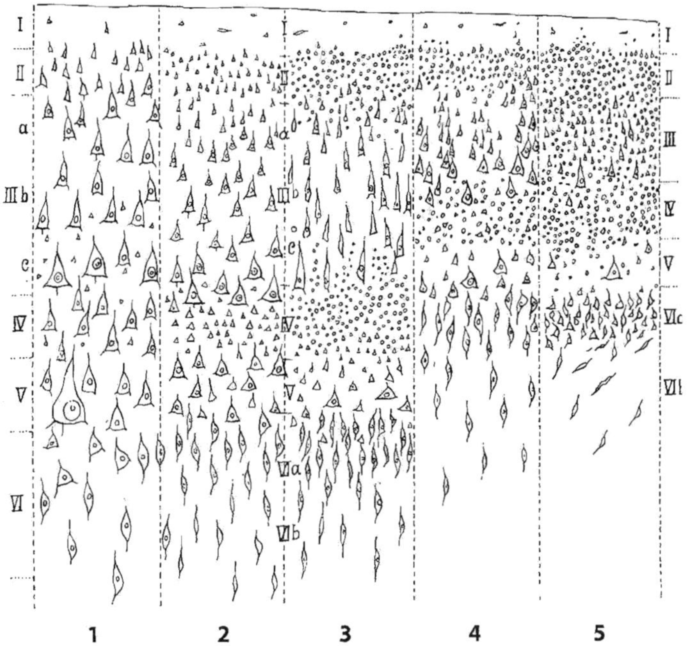

Figure 33.8
Major cytoarchitectural types of neocortex in the human brain, with each type numbered below the drawings.: from agranular cortex (motor cortex, type 1) at the left to the most granular cortex (primary sensory areas, type 5) at the right. In between are type 2, (pre)frontal type; type 3, parietal type; type 4, polar type. Type 5 is also called granulous or koniocortex. Neocortical layers are indicated by Roman numerals. From von Economo (1927).

diagram of figure 33.9 illustrates major intracolumnar connections by showing the course of axons from their cells of origin to the layer where those cells connect. These connections are made by collaterals of pyramidal cells (see figures 33.1 and 33.2) and by the axons of interneurons (see figure 33.1).

Although there may be some difference in details of these connections from area to area, the dominant theme is the similarity of the pattern. What most distinguishes one cortical area from another are the connections from outside the area. Figure 33.9 shows one type of connection from other brain regions; namely, the connections of **thalamocortical** axons. In many mammals, the thalamus projects mainly to layer 4 and the deeper part of layer 3, but there are connections to the other layers as well, particularly to layers 1 and 6 as illustrated. There are species differences in details (see, e.g., figure 32.8).

There are other cortical interconnections that are just as short as those within a column. These are the **local transverse connections** of axons that course within a layer from neurons in one column to neurons in nearby columns. When such connections are

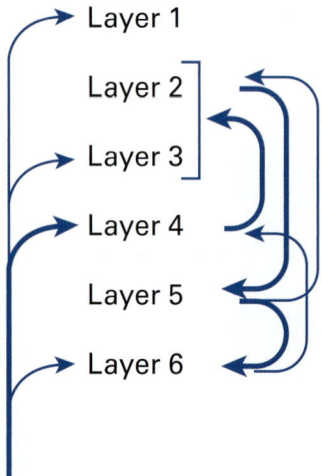

Figure 33.9
Schematic of intracolumnar neocortical connections made by short axons or axon collaterals from the neurons of the various layers. Also shown is the typical pattern of terminations of thalamocortical axons coming into a column from the white matter. Thicker lines indicate stronger connections. (There are many variations across species and within a species for different thalamic cell groups.) Outputs from such a column are not shown: from layer 6 to the thalamus, from layer 5 to other major subcortical targets (striatum, brainstem). Transcortical projections arise largely from layers 2 and 3.

inhibitory, they can result in a contrast enhancement similar to what has been demonstrated for the retina and the olfactory bulb.

Longer connections are formed by **transcortical U fibers**, fibers that take a U-shaped trajectory by going down into the cortical white matter below layer 6, then through the white matter to another cortical area, and then up toward the surface. Connections are made (most commonly) in layers 2 and 3; they usually originate also in these superficial layers. Connections with the opposite hemisphere are of this type.

Convergence of Inputs, Great and Small

All the types of connections summarized above converge on neurons in the same neocortical columns, and many of them on the very same pyramidal cells. We can get a sense of the amount of convergence from a fascinating study that was carried out by Brian Cragg using electron microscopic analysis of small pieces of the visual and motor cortex of the monkey (Brodmann's areas 17 and 4). Counts of all the neurons and all the synapses identified in these tissue samples yielded the following numbers: There were 4,000 synapses per neuron in the primary visual cortex and 60,000 synapses per neuron in the primary motor cortex.

Most of these synapses are on the pyramidal cell dendrites. Each axon connecting with these cells may form multiple synapses, but the number is not likely to be greater than a few hundred per axon. It is likely that the convergence of multiple transcortical inputs is very much larger for the cells of the motor area.

Expansions of the Neocortex Reviewed

This is a good time to review the nature of the great expansion of the neocortex described in the previous chapter, where an evolutionary progression is suggested. The resulting major types of neocortical territories are found in all modern mammals, but the segregation of these types is less in smaller mammals that more closely resemble early mammals (e.g., the hedgehog and opossum, also the Syrian hamster). The types are the primary sensory and the motor areas and the association areas—both unimodal and multimodal (heteromodal). Limbic cortical areas have also expanded.

As the cortical surface area has expanded, the need to fit this tissue into the skull case has led to the formation of gyri and sulci, and in large mammals also to large fissures. We have already discussed the formation of a temporal lobe and the Sylvian fissure that separates it from parietal and frontal cortex (see figures 22.7 and 22.8).

Axons to and from Neocortex

In chapter 12, we distinguished two major pathways of the forebrain—the lateral forebrain bundle and the medial forebrain bundle. The latter is related to the limbic system structures. Axons to and from neocortex are included within the lateral forebrain bundle. Review this by studying again figures 12.5 and 12.6.

Where in figure 12.6 would you look for the **thalamocortical axons**? They would be found only at the 'tweenbrain and endbrain levels, but are part of the illustrated bundle of axons only in the rostral-most section. They could be added to the longitudinal view, emerging from the thalamus, mostly in an anterolateral direction, and going through the corpus striatum, where they are part of the *internal capsule*. They enter the white matter of the cortex and follow their courses to their sites of termination.

Now look back at figure 22.8 where you can see the radiating pattern of axons going through and emerging from the internal capsule and coursing up to the surface gray matter—the neocortical layers of neurons. This radiating pattern has led to the name *corona radiata* for these axons. Included in these terms (internal capsule, corona radiata, and, of course, cortical white matter) are axons traveling in opposite directions: toward the cortex and away from the cortex (i.e., both the inputs and outputs of the neocortex).

In figure 22.8, you have to imagine the thalamus located just behind the axons in the position of the narrowing funnel. There the axons form a large bundle coursing over the lateral surface of the thalamus (figure 12.6). This bundle is called the **cerebral peduncle**. The peduncle, because it contains efferent axons as well as thalamocorticals, continues caudally and becomes the large band of axons at the base of the midbrain. We turn now to those efferent (output) axons.

When we restrict our attention to the output axons only, we can follow axons from the neocortex to all levels of the CNS. As it enlarged in evolution, the neocortex became more and more dominant in terms of numbers of connections with other structures.

Axons coming from the various portions of the neocortex can be followed caudally through the cortical white matter (corona radiata in large brains) into the internal capsule. Some of them terminate in the corpus striatum as noted previously in chapter 30. Large numbers terminate in the thalamus. These **corticothalamic** axons reciprocate the thalamocortical connections. Many efferent axons continue caudally in the cerebral peduncles, with some leaving these bundles to terminate in the pretectal region and midbrain. The **corticotectal** axons, together with those going to the pretectum, are so numerous that they have been called the *secondary optic radiations*.

The cerebral peduncles can be followed into the pons, where **corticopontine** axons terminate in the pontine gray matter, made up of cells that project their axons to the cerebellar cortex of the opposite side. Those that emerge from the caudal edge of the pons form bands of white matter, one on either side of the ventral midline, that constitute the **pyramidal tracts**. All these axonal tracts can be found easily in brain dissections.

Reaching the caudal end of the hindbrain (where there are many terminations), a large majority, or all, of these axons change course. They bend dorsally and decussate, entering the ventral portion of the dorsal columns in some species (like the rodents) and the lateral columns in other species (like the primates). These **corticospinal** axons terminate specifically in various parts of the spinal cord gray matter (see chapters 9 and 15), exactly where depending on their points of origin in the cerebral cortex.

Ascending Pathways to Neocortex Go Mainly through the Thalamus: The Basic Pattern

Except for the olfactory inputs, sensory pathways en route to the neocortex generally form connections in the thalamus. The olfactory inputs (chapter 19) pass from the primary sensory neurons to endbrain structures that are a kind of cortex, the olfactory bulbs, and the secondary sensory neurons of the bulbs project directly to the olfactory cortex—the piriform cortex. However, in order to reach the neocortex, the olfactory information must reach the dorsal thalamus. To do this, the olfactory cortex has projections to the medial part of the thalamic mediodorsal nucleus (MDm), which projects to an orbital part of the prefrontal cortex. It is interesting to note that this is the only part of the neocortex that projects directly to the hypothalamus. Thus, by this indirect route, olfactory information can affect motivational states and moods after being influenced by cognitive functions of the frontal neocortex.

Earlier in this book we have described the pathways whereby visual, auditory, and somatosensory information reaches neocortex after first reaching the thalamus. Thus, ascending **auditory** pathways go mainly through the medial geniculate body (MGB), which projects to auditory neocortical areas (chapter 23). **Visual** information reaches neocortex by several routes (chapter 22), the most direct one going through the lateral geniculate body (LGB) before reaching primary visual cortex. Other routes go by way of the lateral thalamus, for example, from the lateral posterior nucleus (LP), which receives input from the superficial layers of the superior colliculus; another route goes from the pretectal area to a more anterior part of the lateral nucleus called the lateral dorsal nucleus (LD). These portions of the lateral nucleus project to posterior association cortex dominated by visual inputs, outside the primary visual area.

Ascending **somatosensory** pathways reach the ventral posterior nucleus (VP, also called the ventrobasal nucleus), which projects to somatosensory areas of neocortex. The **taste** pathway goes through the medial-most part of the VP.

Activity crucial for **motor control** functions reaches the more anterior parts of the ventral nucleus. Thus, the cerebellum and the globus pallidus project to the ventral lateral and the ventral anterior nuclei (VL and VA), which project to motor and premotor neocortical areas.

The anterior nuclei of the thalamus (), receiving input from the mammillothalamic tract as described in chapters 26 and 28, project to the cingulate and retrosplenial cortical areas. These nuclei are three in number: the anterodorsal (AD), anteroventral (AV) and anteromedial (AM) nuclei.. Although cingulate areas are not in all respects neocortex (they are paralimbic areas that can be classified as juxtallocortex), they play a critical role in neocortical function because of strong reciprocal connections with multimodal association areas on the one hand and the hippocampal formation on the other.

Illustrations of Thalamocortical Connections: Human Brain

In order to visualize the organization of the thalamocortical projections, inspect figures 33.10, 33.11, and 33.12. The first of these drawings depicts the left hemisphere of a human brain with the territories receiving projections from the major thalamic cell groups distinguished. The thalamus is pictured in a simplified drawing to show the basic organization.

Note first the **central cortical areas**, the territory of the ventral thalamic nuclei. The most anterior part of the ventral nucleus (VA) projects most anteriorly in this central cortex, and the middle portion (VL) projects more caudally to the motor cortex, still anterior to the central sulcus. Behind the central sulcus is the projection area of the ventral posterior nucleus (VP). The lateral part of this nucleus (ventral posterolateral; VPL) projects more dorsally, representing the body surface below the head; the medial part (ventral posteromedial; VPM) projects more ventrally, representing the face and other parts of the head. The most medial portion is a small-celled territory receiving taste input; it projects to cortex hidden in the Sylvian fissure, the insula (island of hidden

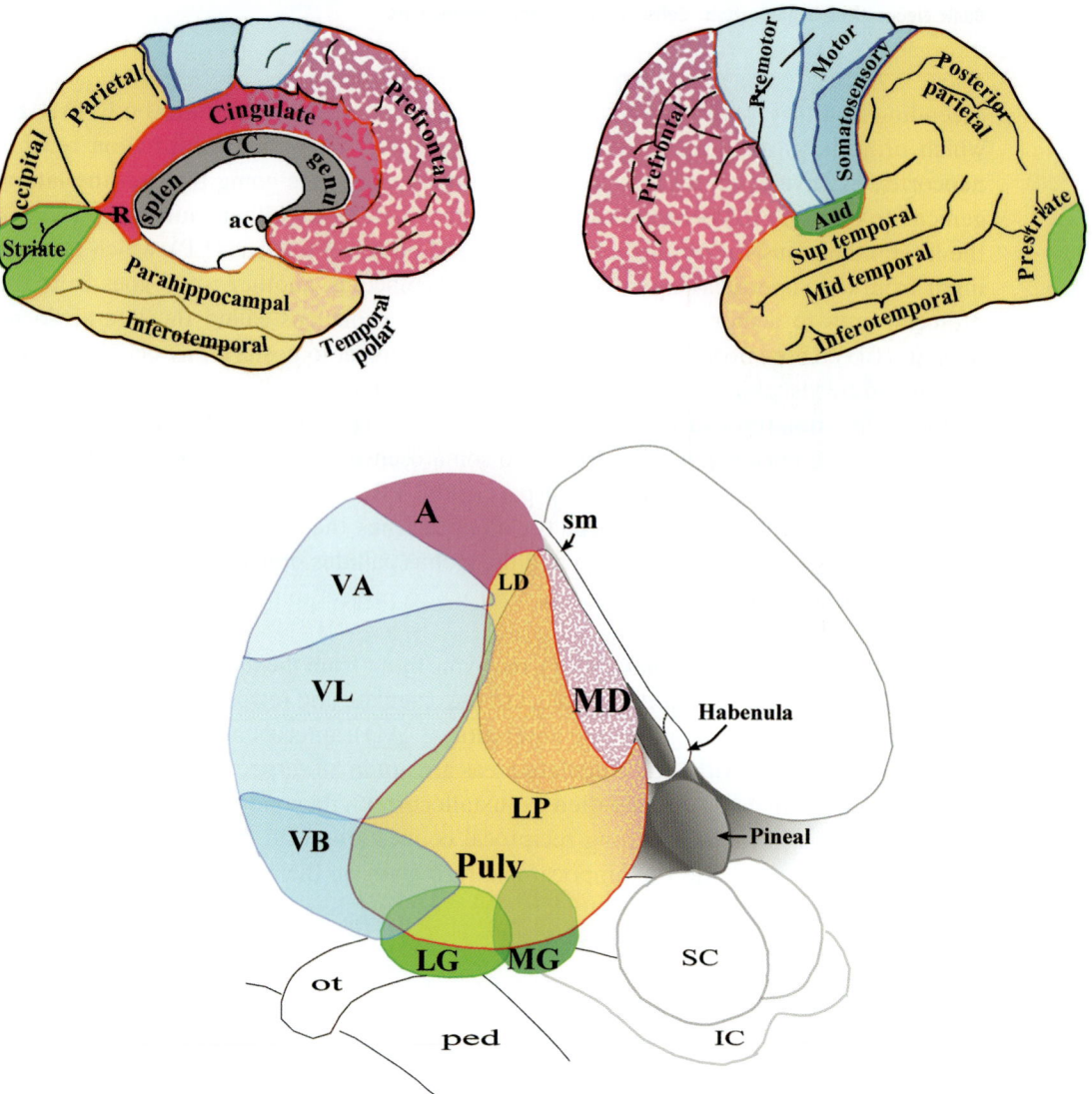

Figure 33.10
Drawings of primate thalamus (bottom) and left cerebral hemisphere of a human brain (top: medial view at left, lateral view at right). Color coding indicates thalamic cell groups and their major projection areas in the cortex. The intralaminar and midline nuclei of the thalamus are not indicated. A, anterior nuclei; ac, anterior commissure; Aud, primary auditory cortex; CC, corpus callosum; IC, inferior colliculus; LD, lateral dorsal nucleus; LG, lateral geniculate body; LP, lateral posterior nucleus; ot, optic tract; MD, mediodorsal nucleus; MG, medial geniculate body; Pulv, pulvinar nucleus; SC, superior colliculus; splen, splenium; VA, ventral anterior nucleus; VB, ventrobasal nucleus (ventral posterior nuclei); VL, ventral lateral nucleus. The posterior group, usually abbreviated as Po, is not indicated; its multisensory projections overlap with those of Pulv and LP. (Note: The medial parts of the pulvinar and the mediodorsal thalamic nuclei have projections that show considerable overlap. The overlap is greater than can easily be depicted; the figure indicates particularly strong overlap in the temporal pole and in ventral prefrontal areas. The MD and Pulv also overlap in projections to parietal areas caudal to primary somatosensory areas. Not indicated at all is overlap of projections of these nuclei with the VL projections. Also, LD projections have some overlap with those of the anterior nuclei. MD projections overlap with those of anterior nuclei in the rostral cingulate cortical area.) Additional structures: ped, cerebral peduncle; sm, stria medullaris.

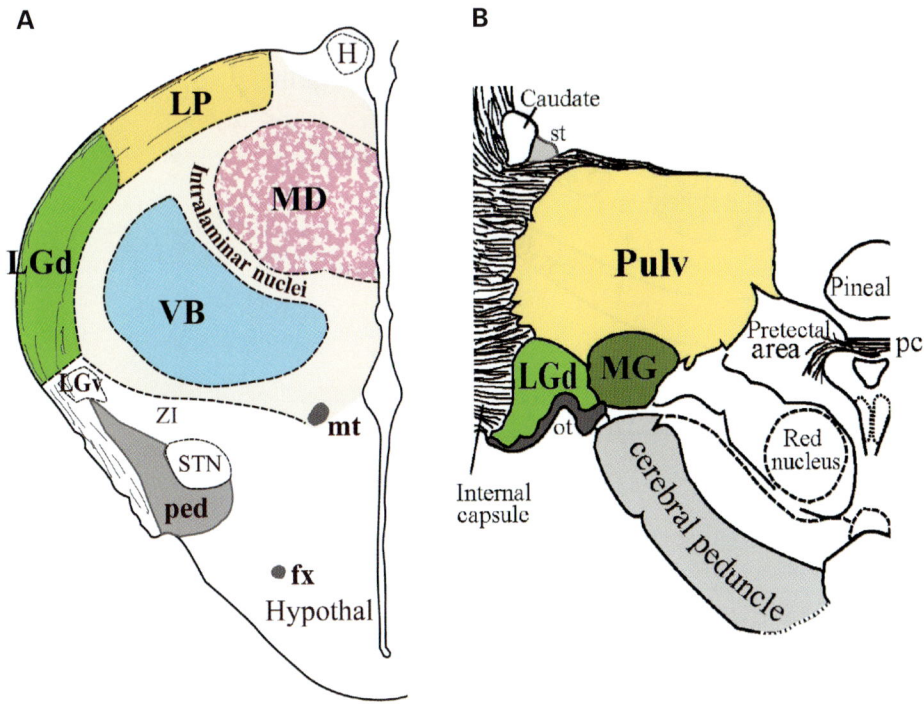

Figure 33.11
(A) The diencephalon in frontal section, sketched to resemble a section from an embryonic human brain (12 wk p.c.) or a mature brain of one of many small mammals. (B) The caudal thalamus of the adult human brain is illustrated in frontal section. (See figure 33.10.) fx, fornix; H, habenula; LGd, lateral geniculate body (nucleus)—dorsal part. LGv, lateral geniculate body, ventral part, LP, lateral posterior nucleus; Pulv, pulvinar; MD, mediodorsal nucleus; MG, medial geniculate body (nucleus); mt, mammillothalamic tract; ot, optic tract; pc, posterior commissure; st, stria terminalis; VB, ventrobasal nucleus: ZI, zona incerta of subthalamus.

cortex). This topographic organization, often pictured in textbooks as a little man in the brain (homunculus), is mirrored in the motor cortex just anterior to the primary somatosensory area.

Next note the more **anterior cortical areas**. This territory receives projections of the mediodorsal nucleus (MD) and is called the **prefrontal cortex** because it is not only in front of the central fissure but also in front of the motor areas. The prefrontal cortex is very large in the human brain, where it can be subdivided into a number of distinct areas. It is considerably smaller in small mammals, especially in those that most resemble primitive mammals. In these small animals, it is more difficult to subdivide the prefrontal cortex—there has undoubtedly been less specialization within the region in these animals—but in all mammals we can separate the territories of the medial and lateral divisions of the MD. It is the medial division (MDm) that receives inputs from the limbic

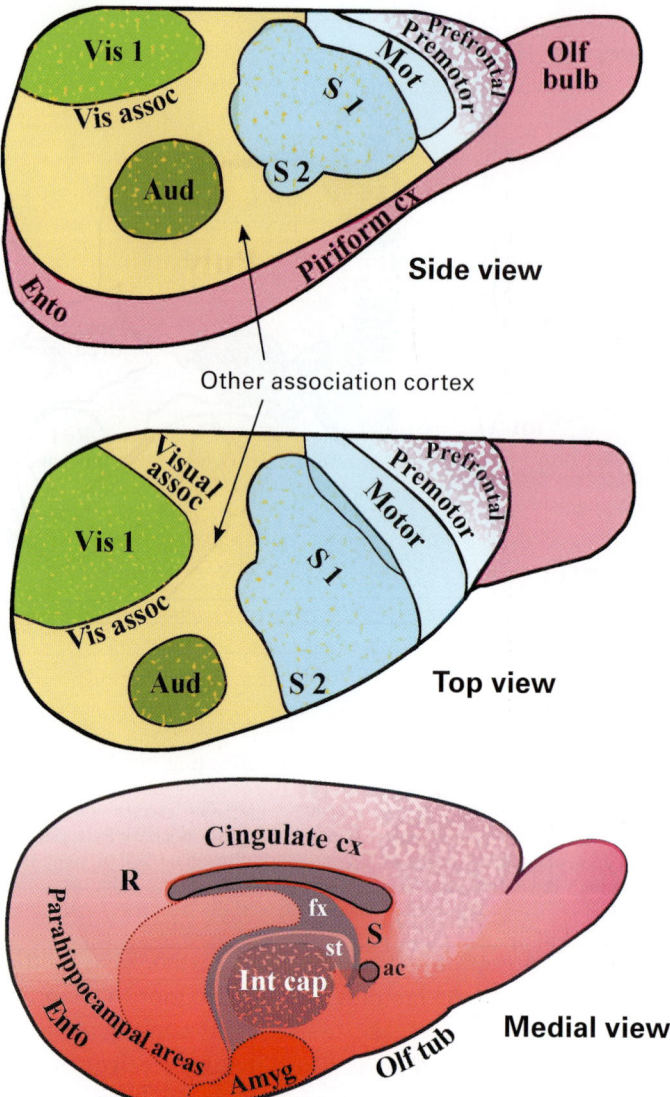

Figure 33.12
Sketch of the right hemisphere of a rat or hamster with approximate boundaries of primary sensory areas, primary motor area, association areas, and limbic cortical areas indicated. Color scheme matches that of figures 33.10 and 33.11. ac, anterior commissure; Amyg, amygdala; Aud, primary auditory cortex; cx, cortex; Ento, entorhinal cortex; fx, fornix; Int cap, internal capsule; Olf tub, olfactory tubercle (most ventral part of ventral striatum in the basal forebrain); Mot, primary motor cortex; R, retrosplenial cortex; S, septal area; S 1, primary somatosensory cortex; S 2, second somatosensory area; st, stria terminalis; Vis 1, visual area 1; Vis assoc, visual association areas. Other association areas include many multimodal neurons; the lateral nuclei of the thalamus project to these areas with some projections also to primary sensory areas. Prefrontal and premotor areas in some small mammals are not completely segregated, with overlapping inputs from MD and VA.

forebrain including the olfactory cortex. MDm projects to more ventral prefrontal cortex, the orbitofrontal cortex in large primates—the cortex located nearest to the bony orbits of the eyeballs. The lateral part of the MD projects more dorsally, especially to the lateral prefrontal cortex.

The third major territory, located posterior to the somatosensory cortex of the central territory, receives projections from the lateral nuclei of the thalamus. This territory of **posterior association cortex** is also greatly enlarged in the human brain, where a number of distinct subdivisions have been recognized. The greatest expansion has occurred in the projection area of the caudolateral portion of the lateral nucleus, termed the **pulvinar** nucleus in large mammals. (The name comes from the Latin word for a cushioned seat that in ceremonies was reserved for a visit from a god.)

In the dorsal thalamus, the mediodorsal nucleus is separated from the lateral nucleus and the ventral nucleus by a fiber layer—the internal medullary lamina—within which are the neurons of the **intralaminar nuclei** (see figure 33.11). These are continuous with cell groups of the thalamic **midline nuclei**.

Lateral to the ventral nuclei and more lateral than much of the lateral group of nuclei is another fiber layer (the external medullary lamina) that contains, for example, the corticotectal fibers and many thalamocortical fibers as well. Outside this fiber layer in the caudal thalamus we can find the external or **lateral geniculate** nucleus (LG or LGN), also called the lateral geniculate body (LGB) forming a knee-like bump on the thalamic surface. A second bump is located caudomedial to the LGN: the **medial geniculate** body (MG or MGB) or medial geniculate nucleus or nuclei (MGN). The MGB is separated from the ventral and the lateral nuclei of the thalamus by a region of multimodal neurons known as the posterior nuclear group (Po).

Besides the three major territories plus the geniculate bodies of the mammalian thalamus there remains one more major territory, separated from the others by a fiber lamina continuous with the internal medullary lamina. This is the territory of the **anterior nuclei**. The cells of the anterior nuclei send their axons anteriorly into the cingulum bundle located just dorsal to the fibers of the corpus callosum. The axons from the anterior nuclei project to the adjacent cingulate cortex plus the cortex just behind the callosum, the retrosplenial cortex. (The splenium is the caudal end of the corpus callosum.)

Illustrations of Thalamocortical Connections: Brain of Small Rodent or Insectivore

Now try to visualize the thalamocortical connections of the smaller brain (e.g., hamster or mouse) depicted in figures 33.11 and 33.12. Figure 33.11A shows a drawing of a frontal hemisection through the caudal diencephalon. The section is traced from an embryonic human, when the thalamus matches quite well the thalamus of a small mammal like a laboratory rodent. The cell groups of four cortical territories are distinguished. Next to this drawing there is a drawing of a portion of the caudal diencephalon of a human brain in order to illustrate the enlargement of the lateral nucleus, especially the pulvinar nucleus, in the caudal thalamus. In addition, this section shows the medial geniculate body.

The drawings of the 'tweenbrain do not show the more rostral level where the anterior nuclei are found.

Figure 33.12 shows surface views of the cerebral hemisphere of a small rodent (as seen from the lateral side, from the dorsal side, and also from the medial side) with territories of the thalamic regions of the previous figure indicated. Also shown is the territory of the anterior nuclei—the cingulate cortex—together with other limbic system areas, primarily the olfactory cortex plus hippocampal and parahippocampal regions.

Comparing the hemisphere of the small mammal with that of a human brain, the most conspicuous difference is the relatively much smaller size of the cortical territories of the mediodorsal nucleus and of the lateral nuclei (compared with the territories of the primary sensory and motor areas)—corresponding to the smaller thalamic nuclei.

These two neocortical regions are known to be critical for higher cognitive functions. They include unimodal association areas adjacent to the primary sensory and motor areas plus multimodal association areas. The functions of much of this cortex were very difficult to formulate or simply remained a mystery for a long time in the history of neuroscience, and some mysteries certainly remain.

However, neuroanatomical studies of axonal connections have provided a way to characterize the association areas. One approach would be to define them as the territories of the lateral thalamic nuclei and of the mediodorsal nuclei, leaving only the primary sensory and motor areas. Another approach, to which we now turn, is to define them in terms of their transcortical association connections.

Transcortical Association Fibers

Major transcortical association fiber bundles are very apparent in dissections of the fixed human brain (figure 33.13). Dissections do not reveal the exact connections formed by these bundles, but experimental neuroanatomical studies of the brains of monkeys have clarified this picture. The drawings of the dissected human brain show several fiber groups worth special attention at this point. First, note the relatively short U fibers that appear to connect the cortex of adjacent gyri. Such fibers are known to connect primary sensory and motor cortex with adjacent unimodal association cortex. They also connect nearby parts of association areas.

Longer transcortical connections, generally speaking, do not arise from the primary areas. Some of them arise from unimodal association cortex, and they are even more characteristic of multimodal association areas. The drawing in figure 33.14 summarizes results of a number of findings from axonal tracing studies of the rhesus monkey brain. Many of the very long connections are depicted; for example, from the posterior parietal and anterior occipital lobe to dorsal prefrontal regions, and from the temporal lobe cortex

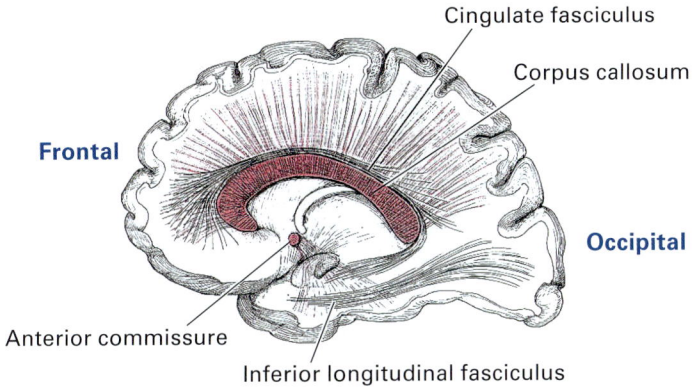
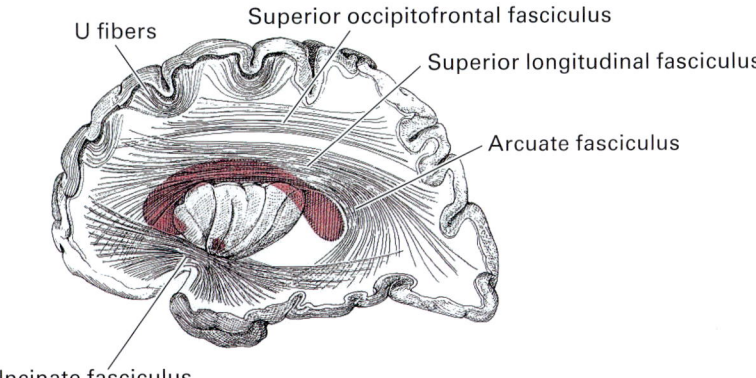

Figure 33.13
Drawings of two dissections of a human left cerebral hemisphere showing major transcortical fiber bundles. (Top) More medial levels are exposed. (Bottom) More lateral levels are exposed. Nearby neocortical areas are usually linked by U fibers. More distant association areas are linked by longer fibers that course mainly in the positions illustrated. From Nauta and Feirtag (1986).

to more ventral parts of the prefrontal region. Similar trajectories of transcortical fiber bundles in the human brain can be found in figure 33.13. Some of these fibers are critical for the functions of the language system. Transcortical connections of special importance for the production of speech have long been postulated to exist, and good evidence for their existence has been found using the method of diffusion tensor imaging for tract tracing, as illustrated in figure 33.15.

A useful way to group the various neocortical regions as a pathway from the external environment through the hemispheres of the brain to the body's internal environment was presented by the neurologist and neuroanatomist Marek-Marcel Mesulam. His figure was shown earlier in the book, in figure 26.9, and is well worth another look now. His groupings are, in order, the primary sensory and motor areas, the modality-specific

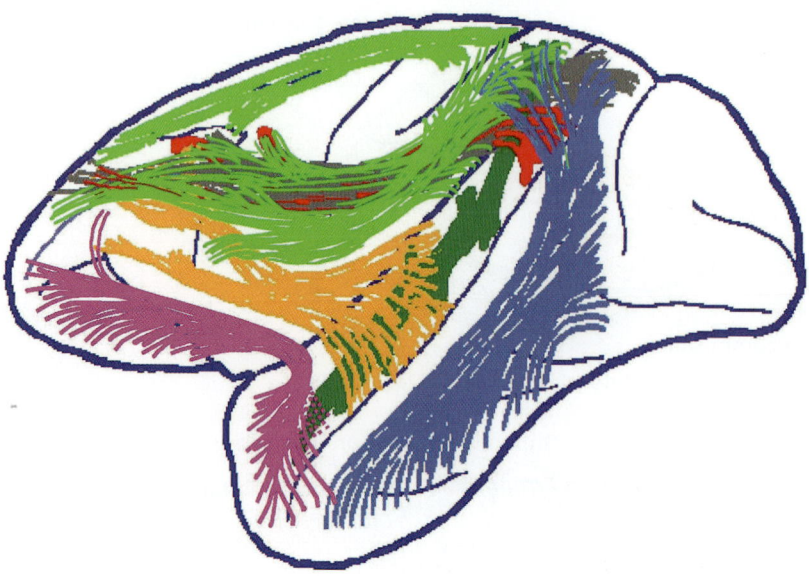

Arcuate fasciculus
Extreme capsule
Fronto-Occipital fasc.
Inferior longitudinal fasc.
Middle longitudinal fasciculus
Superior longitudinal fasciculus components 1, 2, 3
Uncinate fasciculus

Figure 33.14
Drawing of left hemisphere of rhesus monkey brain illustrating the course of major long association fiber pathways between association areas of neocortex. Seven groups of axons are named and drawn in distinct colors. These major groups of axons include fibers carrying information in both directions. The longer transcortical association fibers do not arise from or connect to the primary sensory and motor areas; large numbers pass to and from multimodal association areas, but some involve the unimodal association areas also. Based on Schmahmann and Pandya (2006), Petrides and Pandya (2007), and Schmahmann et al., (2007).

(unimodal) association areas, the high-order (heteromodal, or multimodal) association areas, the paralimbic areas, the limbic areas, and finally the hypothalamus, which most directly influences the internal environment. Each major group connects most strongly—via transcortical connections—with structures of the adjacent group although there are also some connections that pass over an adjacent group to the next group.

Figure 33.16 illustrates the actual locations of these groups of structures in the human hemisphere, recharacterized in terms of major functions. These functions are also indicated in the hemisphere of a small mammal like a hamster, mouse, or rat, and also in a sketch of the mammalian brainstem and spinal cord. *Sensory-perceptual* functions depend primarily on the primary sensory regions and the unimodal association areas, as well as on sensory structures of the brainstem and spinal cord. Lowest-level control of movements depends on motor and premotor neuronal pools in the brainstem and cord and on

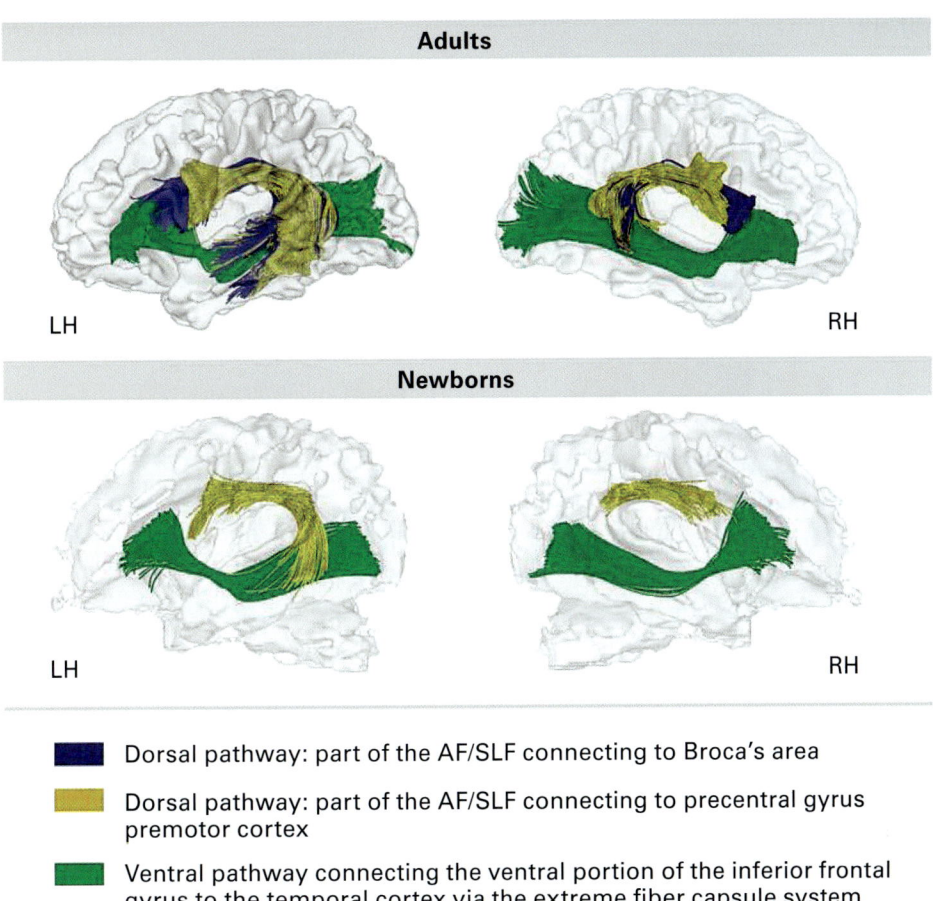

- Dorsal pathway: part of the AF/SLF connecting to Broca's area
- Dorsal pathway: part of the AF/SLF connecting to precentral gyrus premotor cortex
- Ventral pathway connecting the ventral portion of the inferior frontal gyrus to the temporal cortex via the extreme fiber capsule system

Figure 33.15
Transcortical fibers of the speech system as studied in living humans with the method of diffusion tensor imaging. The illustrations show reconstructions of the human cerebral hemispheres as seen from the left side (LH) or the right side (RH) of adults (top) and newborns (bottom). The fiber pathways depicted were those that were followed to and from Broca's area of the left prefrontal cortex (and the corresponding area on the right), and to and from the area of premotor cortex (located nearby) that is involved in speaking. Note that there are two major routes to these areas from association areas of the superior temporal lobe and caudally adjacent areas. One of the more dorsal pathways was not found in the newborns. LH, left hemisphere; RH, right hemisphere; AF, arcuate fasciculus; SLF, superior longitudinal fasciculus .From Perani et al. (2011).

638   Chapter 33

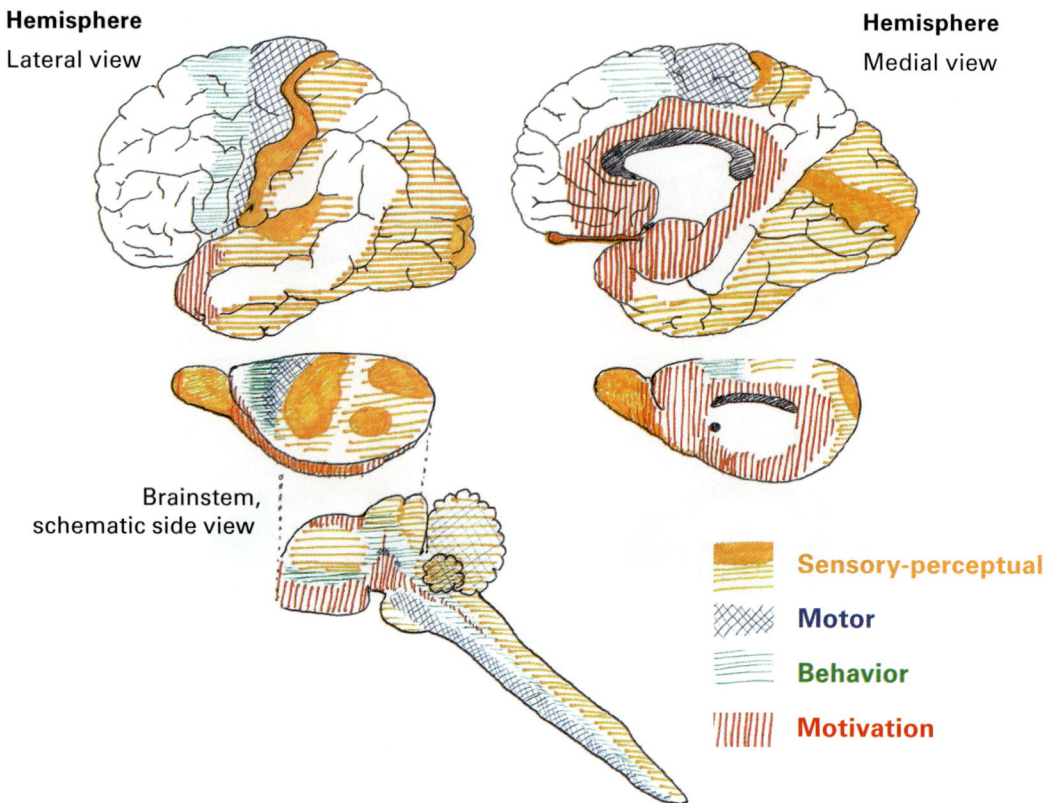

Figure 33.16
Major functional modules of human cerebral hemisphere: The CNS of a small mammal is also depicted. Different types of neocortical areas as distinguished by Mesulam (see figure 26.9) are recharacterized in terms of functions. Areas labeled "sensory" are the primary sensory and motor areas (Mesulam's ideotypic cortical areas). The "perceptual" areas are the modality-specific association areas of the posterior hemisphere (caudal to the central fissure). The primary "motor" cortex is labeled separately—the area most directly connected to premotor neuronal pools of the brainstem and spinal cord and to motor neurons directly. "Behavior" areas in the neocortex are Mesulam's motor association areas, known to be involved in the control of patterns of movement. Areas labeled "motivation" are the paralimbic areas of figure 26.9, areas that are closely connected to limbic system structures, including the subcortical structures most directly in control of motivational states. The white areas in neocortex correspond to the two large multimodal association areas of the human brain (prefrontal cortex anteriorly and posterior parietal and temporal areas posteriorly), known to be involved in complex cognitive functions as summarized in the text. Corresponding regions of the brain of a small mammal like the rat, mouse, or hamster are indicated below the human hemispheres. At the bottom of the figure, regions related to the major functions of the neocortical regions are sketched in a schematic parasagittal section of the rodent brainstem and spinal cord.

primary motor cortex. *Behavior*, meaning organized movements and sequences of movements, depends on premotor cortex and on the brainstem and spinal cord (and on the corpus striatum, not depicted in the figure). Motivational states depend on hypothalamus and limbic midbrain areas and on limbic endbrain structures.

Finally, we are left with the white areas, in the human brain occupying large regions of the frontal neocortex and of the posterior neocortex (in parietal and temporal lobes). These two regions are the multimodal association areas in the projection fields of the lateral thalamus (mostly the pulvinar nucleus) and of the mediodorsal nucleus. The multimodal areas cannot be characterized so easily in terms of sensory-perceptual, motor, behavioral, or motivational functions. In many years of neuroscientific research, they have been only partially characterized, and in recent years they have been subject to intensive studies with functional imaging methods. They include the areas underlying human language comprehension and speaking. They include the processing modules that underlie the abilities of large mammals, especially humans, to anticipate and to plan, to interpret and create.

Not shown in the neocortical maps we have been discussing are any neuroanatomical correlates of the functional differences between the two hemispheres in humans. Observations by clinical neurologists and by neuropsychologists have shown that humans generally have a left hemisphere dominance for verbal functions and right hemisphere dominance for some nonverbal functions. Neuroanatomical studies of human brains by Albert Galaburda in Boston provided some of the first evidence for striking size differences between the left and right hemispheres of these brains in the temporal lobe areas related to verbal abilities. Specific regions are much larger in the left hemisphere. There has been confirmation from more recent studies using magnetic resonance imaging. Some hemispheric differences have also been found in animal brains where the functional correlates are not clear.

Comparative Neocortical Anatomy from a Moscow Perspective

The great neuropsychologist Alexander Luria presented, in a neuropsychology book published in 1966, a summary of some of the neuroanatomical work of Poliakov in the mid-twentieth century. The Russian neuroanatomist used some terms not common in Western neuroscience, but some of the data he collected and figures he presented are unique. Poliakov distinguished between three major types of neocortex—which can be compared with the major subdivisions of Mesulam and others—see figure 26.9 and figure 33.16. His illustrations of the brains of six species including the human and other primates are shown in figure 33.17. Poliakov's "primary fields of the nuclear zones" are equivalent to the ideotypic cortex of Mesulam or the heterotypic cortex of von Economo. His secondary and tertiary fields are somewhat equivalent to Mesulam's unimodal and heteromodal (multimodal) association areas (except that the secondary fields corresponding to unimodal association areas are smaller).

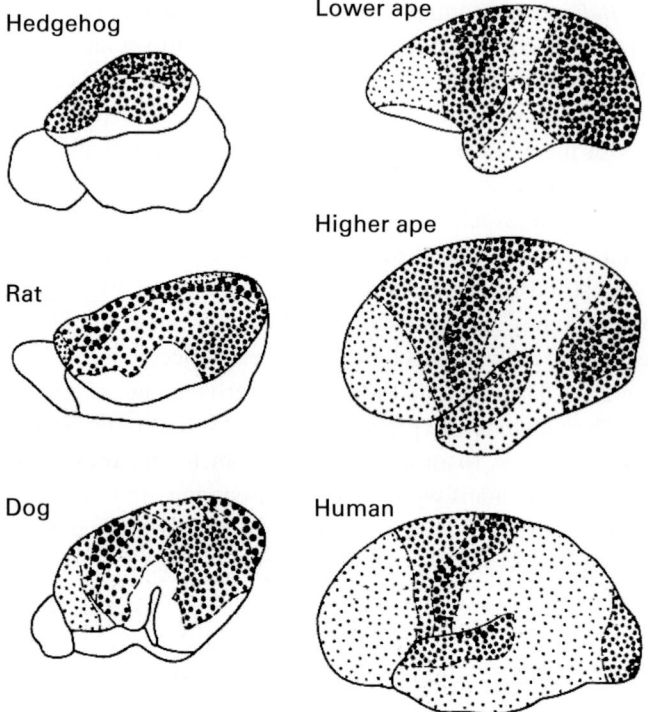

Figure 33.17
Poliakov, working in Russia during the middle part of the twentieth century, used cytoarchitectural methods to distinguish three major types of neocortex that have some correspondence to Mesulam's major neocortical types (see figure 26.9). In this figure, the large dots cover "primary" fields; middle-sized dots cover "secondary" fields; small dots cover tertiary fields—zones Poliakov considered to be of overlapping modalities. From Poliakov (1972).

Looking at the brains shown on the left side of figure 33.17 (dog, rat, and hedgehog), we see what looks like an absence or near absence of tertiary cortex, especially in the rat and hedgehog. A common interpretation is that there has been less evolution of higher functions in these species. Although this may be true, is it also true that there is really less, or no, multimodal cortex? I do not believe so. A likely interpretation is that in these species, there is less parcellation of unimodal areas out of a primordial multisensory cortex (discussed earlier, see chapter 32).

Studies of Development and Plasticity Lend Support to Classifications of Major Neocortical Types

The neocortical maps of Mesulam and of Poliakov showing three major types of neocortex have a fascinating resemblance to neocortical maps that distinguish areas where axons are early to myelinate from areas where myelin is acquired later and later (figure 33.18).

Basic Neocortical Organization: Cells, Modules, and Connections

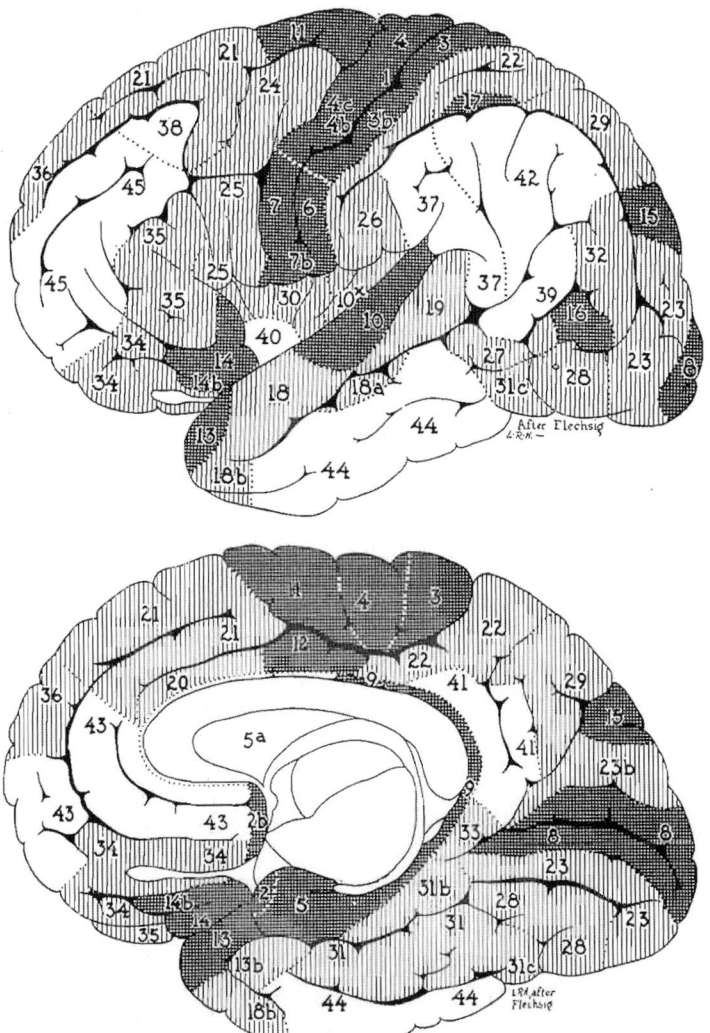

Figure 33.18
Human brain from lateral (top) and medial (bottom) sides, depicting the relative order of myelogenesis according to the work of Paul Flechsig (1920). In this redrawn illustration published by von Bonin, the areas earliest to myelinate are shown in a darker shade, the areas next to myelinate are shown in a lighter shade, and the areas last to myelinate are shown in white. The numbers show Flechsig's ranking of myelination order, from earliest to latest. The three groups depicted by the shading have arbitrary cutoffs. From von Bonin (1950).

The more specialized, primary sensory areas—the ideotypic areas—and their thalamocortical axons are the first to acquire myelin. Last to myelinate are the multimodal areas, which are also the areas that have undergone the most recent evolutionary expansions. Before myelin appears, axonal conduction is too slow to support much function; in fact, there is little indication of any function of cortical areas before myelination.

Corresponding to the findings on myelination, Poliakov presented data showing greater postnatal enlargement of selected areas underlying some of the most advanced cognitive functions in humans. He also showed greater postnatal thickening of neocortical layers 2 and 3, the *association layers*, in some of these areas.

Functional localization studies of neocortical modules have indicated that the multimodal association areas have a particular importance in various kinds of learning in mature animals and people. Correlated with this is an interesting neuroanatomical finding on the distribution of the growth-associated protein GAP-43, a molecule that is found in abundance in developing axons during their periods of greatest anatomical changes. Neve and Benowitz with co-workers at Harvard found that this molecule remains at high levels in adulthood in certain brain regions, most notably in the association layers (layers 2 and 3) of multimodal association cortex.

We will turn to the dynamics of neocortical development and plasticity in the next chapter.

Review of Ideas on Thalamic Evolution

The transcortical connections of association neocortex have become increasingly dominant in evolution of large brains, including the human. However, it appears that cortical connections from the thalamus, and from cortex back to thalamus, have remained important; thalamic organization has most likely played a crucial role in the shaping of the neocortex throughout its evolution. Here, we review two major ideas concerning thalamic evolution presented earlier in the book. These ideas have some support from findings of comparative anatomy supplemented by data on topographic arrangements and on development and plasticity of axonal systems.

Idea 1 As suggested in chapter 32, the earliest thalamus was multimodal and projected to primitive striatum and pallium. Within the primitive multimodal thalamus, unimodal regions evolved by functional segregation (parcellation) (see figure 32.5). Distinct territories corresponded to axonal entry positions, and this organization of axons is preserved in the thalamic outputs to the cortex. Some multimodal neurons remain.

Idea 2 Gating, or modulation of thalamic information destined for the endbrain, was important early in forebrain evolution. This resulted in the thalamus becoming a nearly obligatory way station for information passing to the endbrain. Modulatory projections came, and still come, from the hindbrain, the midbrain, and the hypothalamus. (In addition to a gating of information passing through the thalamus, these projections play critical roles in influencing the overall state of the brain—see chapter 17.) Many of these axons from hindbrain, midbrain, and hypothalamus have diffuse terminations in the thalamus and other parts of the 'tweenbrain (e.g., serotonin- and norepinephrine-containing axons and axons from the brainstem reticular formation using acetylcholine as the neurotransmitter). Many, including also dopamine-containing axons, in addition go directly to the endbrain.

Readings

Brodal, P. (2004). *The central nervous system: Structure and function*, 3rd ed. New York: Oxford University Press.

Catani, M., & Mesulam, M.-M. (2008). The arcuate fasciculus and the disconnection theme in language and aphasia: History and current state. *Cortex, 44*, 953–961.

Caviness, V. S., Jr., & Frost, D. O. (1980). Tangential neocortex organization of thalamic projections to the neocortex in the mouse. *Journal of Comparative Neurology, 194*, 335–367.

Cragg, B. G. (1967). The density of synapses and neurons in the motor and visual areas of the cerebral cortex. *Journal of Anatomy, 101*, 639–654.

Frost, D. O., & Caviness, V. S., Jr. (1980). Radial organization of thalamic projections to the neocortex in the mouse. *Journal of Comparative Neurology, 194*, 369–393.

Nauta & Feirtag (1986): see book listed in chapter 3.

Neve, R. L., Finch, E. A., Bird, E. D., & Benowitz, L. I. (1988). Growth-associated protein GAP-43 is expressed selectively in associative regions of the adult human brain. *PNAS, 85*, 3638–3642.

Nolte, J. *The human brain: An introduction to its functional anatomy* [various editions]. A good source of pictures of human brain.

Poliakov, G. I. (1966). Modern data on the structural organization of the cerebral cortex. In Luria, A. R., ed. *Higher cortical functions in man*, pp. 39–69. New York: Basic Books. (2nd ed., 1980: New York, Consultants Bureau.)

Sanides, F. (1970). Functional architecture of motor and sensory cortices in primates in the light of a new concept of neocortex evolution. In Noback, C. R., & Montagna, W., eds. *The primate brain*, pp. 137–208. New York: Appleton-Century-Crofts.

Weiller, C., Bormann, T., Saur, D., Musso, M., & Rijntjes, M. (2011). How the ventral pathway got lost—and what its recovery might mean. *Brain & Language, 118*, 29–39.

34 Structural Change in Development and in Maturity

In the previous chapter, we described the exquisite structure that unfolds at the rostral end of the mammalian neural tube. In its developed state, the neocortex governs many of the movements and expressions of the body and what we call our thoughts, both conscious and unconscious. How this most complex flower of evolution forms and blossoms in an individual is a study of gene expression with an orchestration of myriads of interacting cells and their processes. The basic factors at play are all read in a script of nature and in many interactive additions to that script by influences from the tissues and from external environments. We call this *nature and nurture*, and the nurture includes learning and memory of many sorts.

Its evolution took a very long time, but the simple evolutionary rules led inexorably from the neural tube of the earliest chordates to the elaborate CNS of the most advanced primates and other pinnacles of the evolutionary tree. It was inexorable, but it was also full of surprises, given the changing and variable environment of our planet over eons of geologic time since the appearance of replicating molecules and living cells.

We pictured the developing chordate neural tube early in the book (see chapters 8 and 13). Now we add to that story by describing cortical development with an emphasis on some of its unique aspects. We will also note how neuroscientists in recent decades have been exploring how certain aspects of that development continue in mature animals.

Neuronal Proliferation

In chapter 8, we discussed four major events in the development of the central nervous system:

1. Neurulation (differentiation of the CNS from a dorsal part of the ectoderm) with formation of the neural tube.
2. Proliferation of cells (by mitosis).
3. Migration.
4. Differentiation of the neurons, with growth of axons and dendrites.

Here we will focus on specific aspects of proliferation and migration in the neocortex. The illustrations of figure 34.1A show a small piece of neocortical tissue in frontal sections of the human neocortex at five successive embryonic stages. There are differences in neocortical embryogenesis of humans and other large primates compared with neocortical development in the small rodents commonly used in laboratory investigations. In the small brains, a subpial granular layer has not been found. (However, in rodents as well as in large mammals, such a layer is a major zone of cell proliferation in the cerebellum.) The additional mitotic zone seems to be necessary for producing the much larger numbers of cells present in larger brains. A number of quantitative differences are also found.

Figure 34.1B shows Nissl-stained sections of a small section through the visual cortex of postnatal hamsters. In these images, a subventricular zone is not evident, but the ventricular zone of mitotic cells is seen at both postnatal days 1 and 5 (P1 and P5). These images illustrate how different mammals can be born at very different stages of early development. A hamster at birth shows brain characteristics that resemble those of a human fetus at about 2.5 months after conception.

A sketch of some of the dynamics of cell proliferation was presented in figure 8.7. The figure depicts the movements of the cell body leading to mitosis. These movements are caused by changing positions of the nucleus within the elongated progenitor cell during the *cell cycle*, in the midst of which there is a synthesis of a copy of the genetic material (the cell's DNA). The division of the cell into two daughter cells is not always the same: there are two main types of cell division: symmetric and asymmetric. It is important to know the difference in order to understand neocortical development and evolution.

In symmetric cell division, each of the two daughter cells remains in a proliferative state. Thus, we can call them both stem cells. In asymmetric cell division, one daughter cell becomes postmitotic and migrates away from the ventricular layer, while the other one remains mitotic. The difference in the fate of the cells depends on exactly how the cell splits and the resulting distribution of certain intracellular proteins. Before the cell division, these proteins are not distributed evenly throughout the cell. A protein called numb is believed to be especially relevant to cell fate: It tends to collect on the side of the cell nearest the ventricle. If the cell divides along an axis passing through the region where numb is collected, then both daughter cells will contain similar amounts of it, and both of these cells will remain mitotic. In contrast, if the cell divides along an axis that separates the part containing numb from the rest, then the cell without the numb protein will become migratory.

Neocortical Expansion in Development and in Evolution

Consider next what the consequences would be if, during the period of neocortical development, the duration of symmetric cell division were increased. If it increased just enough for one additional division of all the cells in the ventricular zone, then the number of

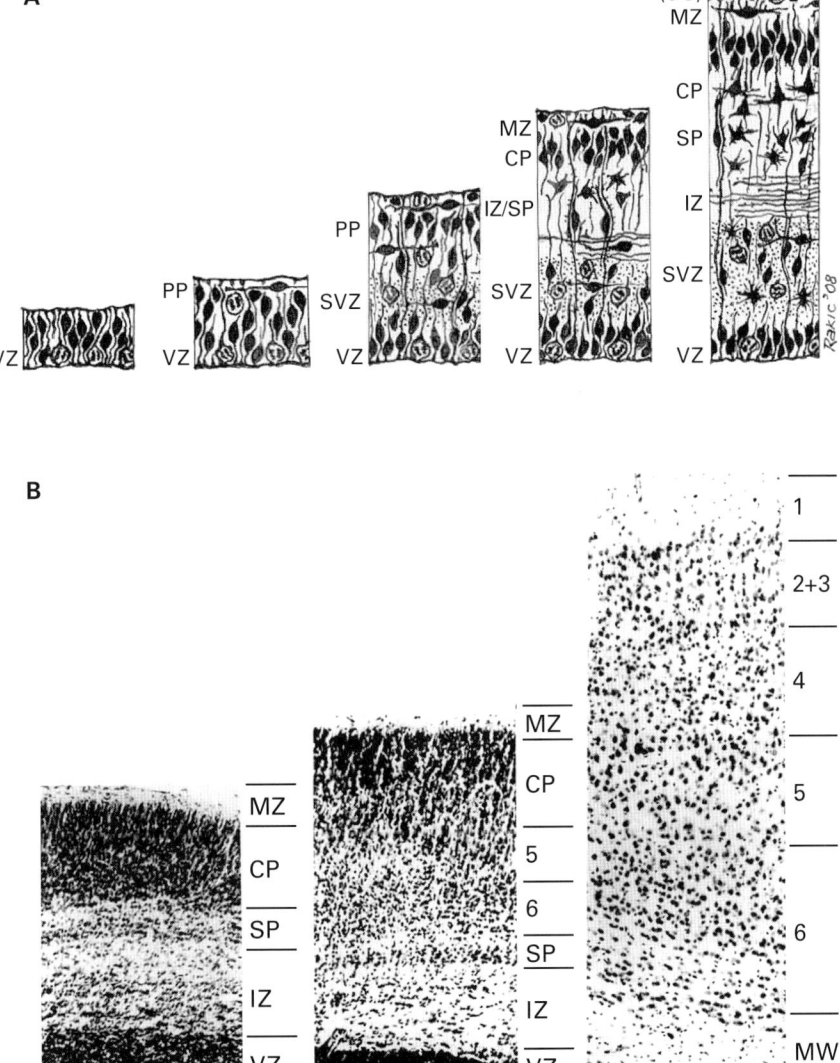

Figure 34.1
(A) Early stages in the development of the neocortex: Drawings by Pasko Rakic illustrate sections of a small piece of primate neocortex at five successive stages in embryonic development. CP, cortical plate; IZ, intermediate zone; MZ, marginal zone; PP, preplate; SP, subplate; SVZ, subventricular zone; VZ, ventricular zone; (SG), subpial granular layer, part of the marginal zone. (B) Postnatal differentiation of visual cortex (area 17) in Syrian hamster: In this cortex of a 1-day-old Syrian hamster (P1), the cortical plate has not yet begun to differentiate into distinct neocortical layers. By day 5, layers 5 and 6 can be distinguished from the cortical plate on the pial side. In the adult hamster, six layers can be distinguished. These layers are not as distinct in the hamster as they are in many other mammals. WM, white matter. Panel (A) from Bystron and Rakic (2008); panel (B) from Naegele, Jhaveri, and Schneider (1988).

neurons would be doubled and the area of the cortex would expand considerably. The consequences of an increase in the later period of asymmetric cell division would also increase the number of neurons, but if migrating cells from a single stem cell are mainly confined to the same column of cortex, then the area of cortex would not necessarily increase, but rather the thickness of the cortex would increase.

These considerations were reviewed by the neuroanatomist Pasco Rakic, who in 1995 published a hypothesis based on them (figure 34.2). He noted evidence that the period of neocortical neuron proliferation in the macaque monkey shifted from symmetric to predominantly asymmetric at about embryonic day 40. This is the first day of final mitoses of neuronal precursors that then migrate to the cortical plate. Such events of final mitosis, called the "birthday" of a cell, continue until about embryonic day 100.

By contrast, data on human brain development indicate that the period of asymmetric cell division begins 3 days later than in the monkey, and it continues about 20 days longer. The major consequences are twofold: In humans, a longer period of symmetric cell division results in a larger area of the cortical sheet, and the extended period of neuronal proliferation results in a thicker cortex with a much larger number of neocortical neurons.

In evolution, changes in neocortical size and thickness could result from mutations that altered the timing of symmetric and asymmetric cell division. If such changes resulted in adaptive changes in function (e.g., better sensory acuity resulting in better discrimination abilities), then the animal's likelihood of longer survival and greater reproduction would increase. The genetic change would be passed on in the gene pool of the population.

Changes in the size of neocortex are correlated with changes in the size of other brain structures, as pointed out in an interesting analysis by Finlay and Darlington (published in 1995), who used data from Heinz Stephan and his collaborators that had been published in the 1980s. Although the increases in neocortical size in the large primates are greater than in other parts of the brain, it is clear that variations in sizes of brain structures across species are not really independent, but show "linked regularities." They plotted brain structure sizes against total brain size, using log-log scales, for groups of simians, prosimians, insectivores, and bats (figure 34.3). When structure sizes are re-plotted, using a linear scale, against the logarithm of total brain size, the result shows the much greater increase in size of neocortex in the species with the larger brains (figure 34.4).

In addition, these same data on sizes of 11 brain structures were analyzed by a mathematical method called principal components analysis. The authors found that 99% of the variance could be accounted for by two factors. The first factor could account for more than 96% of total variance and was most highly correlated with neocortical size and least with olfactory bulb size. The second factor could account for 3% of total variance and was most highly correlated with olfactory bulb size, and next with size of limbic system structures. Thus, we can see the concerted evolution of the mammalian brain dominated by the neocortex, and also the effects of mosaic evolution of structures related to the olfactory system.

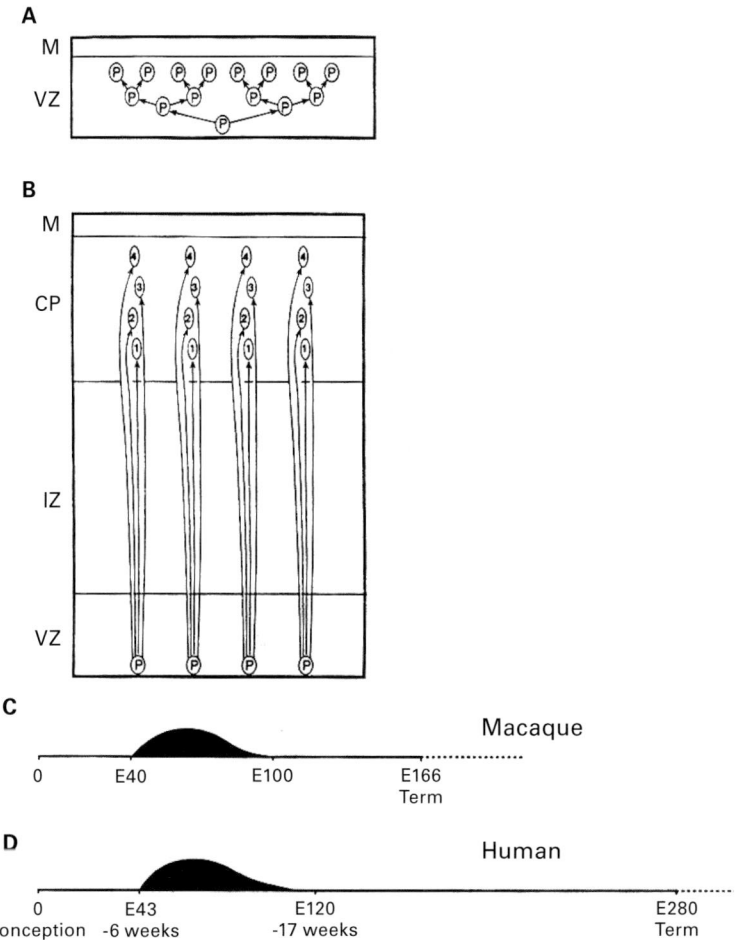

Figure 34.2
The relative durations of different modes of cell division determine cortical area and cortical thickness. (A) Symmetric cell division is represented; this mode dominates neurogenesis in the neocortex before embryonic day 40 (E40) in the rhesus macaque monkey. The more generations of cells produced, the greater the cortical area. (B) Asymmetric cell division is represented, where each cell division results in one cell that becomes postmitotic and migrates toward the pia. The first cells to migrate end up in the deepest layer of the cortical plate. The greater the number of asymmetric cell divisions, the thicker the cortex becomes. (C, D) Time periods and relative rates of neuron origin in macaque and human fetuses, respectively. Neuron origin is defined as the date of the last mitosis. This was determined in the macaque from labeling with [^{3}H]thymidine. In the human, it was estimated from measures of the number of mitotic figures detected histologically in the ventricular zone, DNA synthesis in slice preparations, and counts of migrating neurons in the intermediate zone. M, marginal layer; CP, cortical plate; IZ, intermediate zone; VZ, ventricular zone. From Rakic (1995).

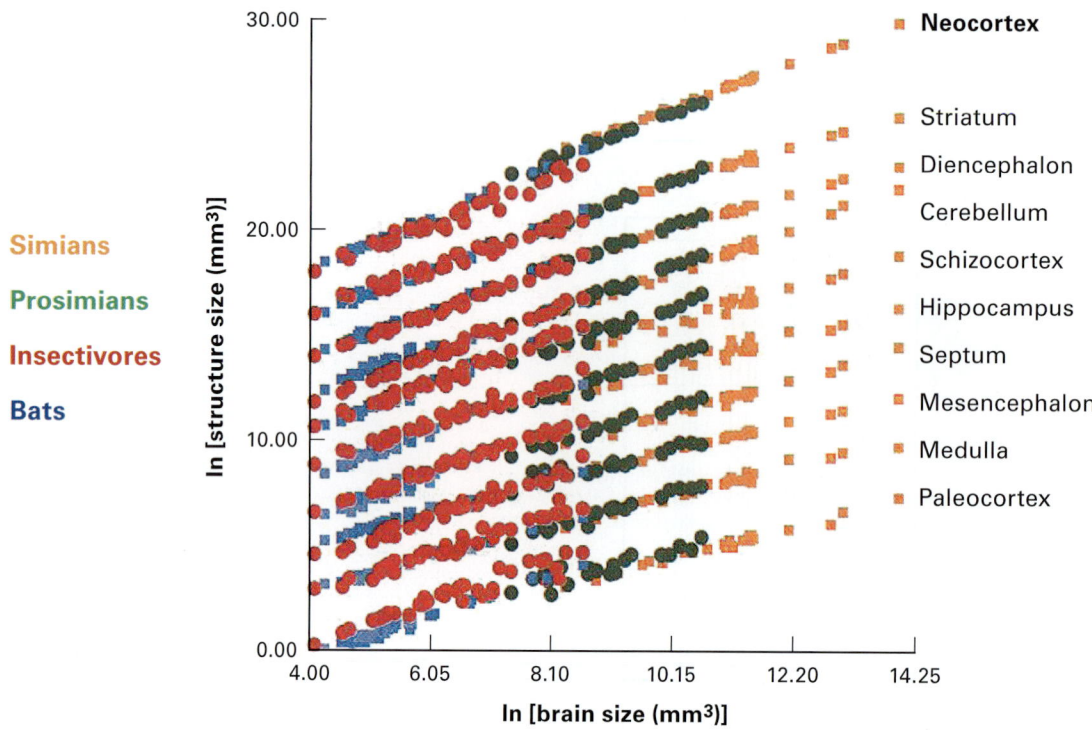

Figure 34.3
Size of brain subdivisions plotted as a function of total brain size for 131 species in four major groups of mammals (log-log scales). Arbitrary constants were added or subtracted for the sizes of each brain subdivision in order to get visual separation. The size of each structure is strongly related to total brain size across the different groups. However, note also that the slope of the increase with brain size for neocortex is greater than for other structures. Note also that certain structures are relatively larger for some groups, while other structures are relatively smaller: larger paleocortex and smaller neocortex for insectivores; larger mesencephalon and cerebellum for bats. For actual sizes, see figure 34.4. Data from Stephan et al. (1981, 1982, 1988); analysis by and figure from Finlay and Darlington (1995).

When Cortical Cells Are Born and How They Migrate

Neocortical neurons are born late in the period of brain development relative to the birth dates of other neurons of the CNS. This fits a rule described by the phrase "late equals large." Birth dates of neurons in the brain have been discovered using several different techniques. Prominent among these techniques are those that involve the labeling of cells that are synthesizing DNA by injection of embryos or pregnant female animals with tritiated thymidine (deoxythymidine, or thymine deoxyriboside, containing some ^{3}H as a label). Thymidine in the extracellular fluid, including the radioactive nucleosides, is taken up by the cells and becomes incorporated in the new genetic material. The tritium label is localized much later by autoradiography: the radioactivity exposes a photographic

Structural Change in Development and in Maturity

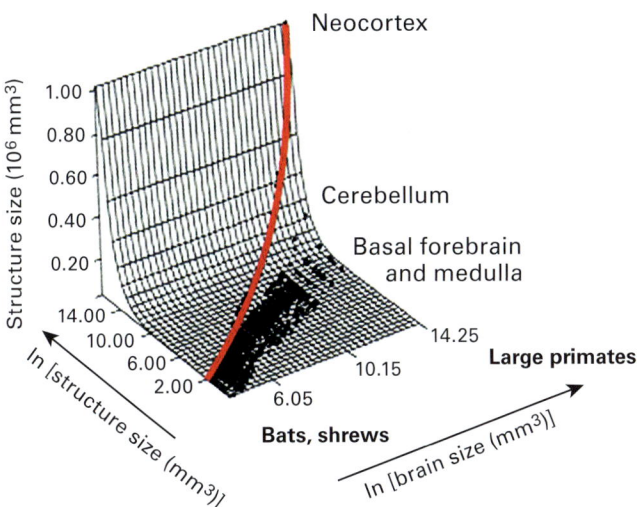

Figure 34.4
The same data used for figure 34.3 are re-plotted in a third dimension to show actual sizes of structures.

emulsion that is used to coat the thin brain sections (prepared in darkness). Such work was carried out initially in laboratory rodents. Results of a later study of neurogenesis in monkey neocortex are illustrated in figure 34.5.

The earlier injections result in the labeling of the neurons of the deepest cortical layers; cells that have reached the most superficial layers are born last. Thus, the later-born cells must migrate past the neurons that migrated earlier; therefore, the phenomenon is called an "inside-out" pattern of migration.

(Note: Not all the labeled neurons were born shortly after the injection because some radioactive cells continue to proliferate and thus contain successively more dilute label. This problem can be avoided by methods using retrovirus-mediated gene transfer to label dividing cells. Students can find information about this method on the Web.)

The increasing thickness of neocortex during development makes cell migration by a simple nuclear translocation mechanism alone more and more improbable. Thus, it is not so surprising that the migrating neuroblasts do not remain connected to both the pial surface and the ventricular surface except, perhaps, quite early when the cortex is very thin. Rakic and his collaborators obtained definitive evidence for guidance of the migrating neuronal cells by radial glia cells. Multiple postmitotic cells move along the same radial glial cell from the ventricular or subventricular zone toward the pial surface, reaching the cortical plate (figure 34.6). The radial glial cells remain attached to both limiting membranes during this entire period of migration.

Not all neocortical cells originate in this way. There are also longer migrations by movement of migrating neuroblasts from the ganglionic eminence (medial, caudal, lateral) (see

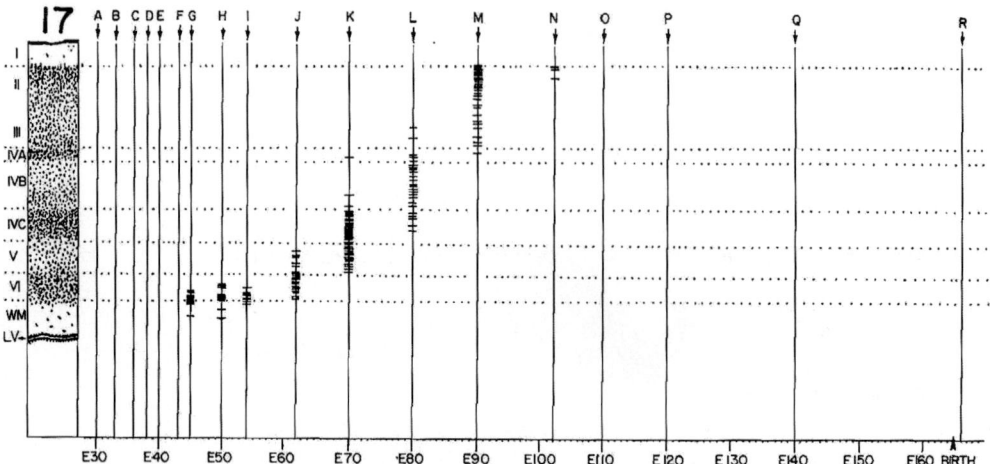

Figure 34.5
Birth dates of neurons in the various neocortical layers are displayed in this graphical representation of results of labeling of cells when they are synthesizing DNA. Injections of [^{3}H]thymidine were made on the embryonic days shown at the bottom. The locations of the cells after migrations were completed are indicated by the short horizontal lines. Positions of layers 1–6 above the white matter (WM) are shown at the left. LV, lateral ventricle (obliterated posterior horn). From Rakic (1981).

chapter 12 and figures 12.1, 12.10, and 12.11). This region of proliferating cells in the mammalian embryo is in the position of the later differentiating corpus striatum (more ventral portion); many striatal cells are also generated there.

How Do Distinct Cortical Areas Become Specified in Development?

The key issues in debates among neuroscientists concerning how the cortex becomes specified into different areas are common to many other areas of embryology. Debates concern genetic and epigenetic factors. With genetic determination, a map is laid down very early in the ventricular zone. This map has been termed a *protomap*. Epigenetic factors include the influence of cues from the local environment including afferent connections.

There are findings supporting both kinds of influence in cortical development. A debate pitting the two kinds of influence against each other has subsided with more and more findings of genetic differences between cortical areas. An early example was the finding of a membrane protein in limbic cortical areas. Once this protein appeared after embryonic day 12 in the rat, limbic cortex transplanted into the somatosensory neocortical region retained its fate as limbic cortex. Earlier transplants gave a different result, as these differentiated into somatosensory cortex. Similar results, without knowledge of genetic differences, were obtained with transplants of tissue from visual cortex into

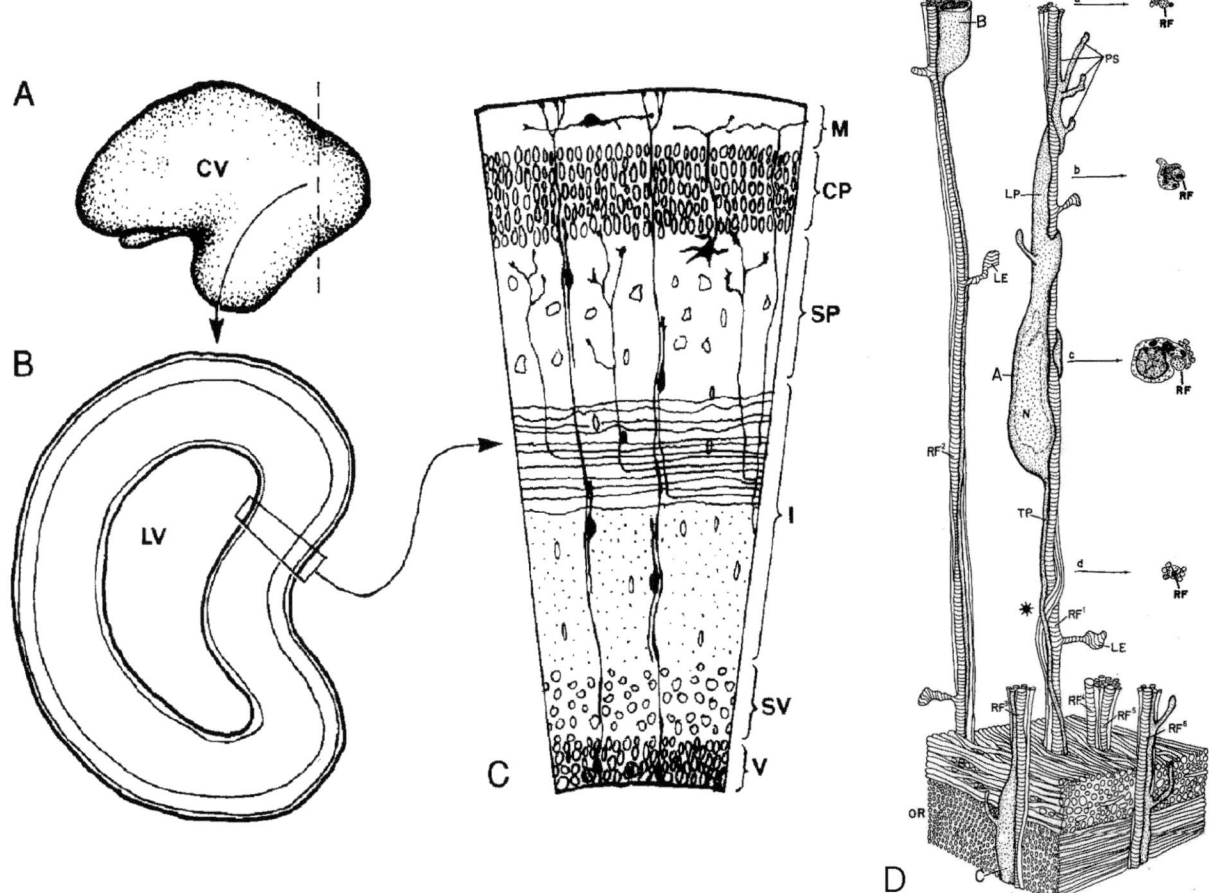

Figure 34.6
(A) Drawing of left cerebral hemisphere of 60- to 65-day-old monkey fetus. (B) Coronal section of the fetal hemisphere at the level shown in panel (A). (C) Histological section of the tissue reveals six embryonic layers: V, ventricular zone; SV, subventricular zone; I, intermediate zone; SP, subplate; CP, cortical plate; M, marginal zone. Radial glial cells serve as guides for migrating neurons. Early afferent axons extend up into the subplate. (D) Drawing of results from reconstructions from serial electron-microscopic sections showing parts of radial glial cells and three migrating neurons (*A, B, C*) attached to them. CV, cerebral vesicle; LE, lamellate expansions; LP leading processes; LV, lateral ventricle; N, nucleus; OR, optic radiation; PS pseudopodia; RF, radial fibers; TP, trailing process. Panels (A), (B), and (C) from Rakic (1995); panel (D) from Rakic (1972).

somatosensory cortex: specificity of the transplanted tissue depended on the age when the transplants were carried out.

The importance of intrinsic differences between regions of the developing cortex—a protomap—is undeniable. These differences appear very early in the embryo with developmental changes in gene expression patterns. However, the existence of a protomap does not deny a role for tissue interactions and influences of neuronal activity. We turn to activity effects below.

The Role of Activity in the Development of Neocortical Connections

The visual system has served as a good model for examining the role of activity in cortical development because of its well-studied organization and the accessibility of the eye. Projections in the visual system in many species become highly organized before birth. Because this is the case, it might seem that sensory experience could play no role in the development of the basic organization of the system. However, if we consider neuronal activity rather than sensory experience dependent on visual input from the outside environment, the question changes. In fact, activity in sensory pathways occurs before there is sensory input from outside the organism.

Figure 34.7 illustrates results of a study of spontaneous activity in a newborn ferret retina. The diagrams depict a wave of activity traveling across the retina, independent of any outside stimulus (confirmed also for the prenatal cat). Electrophysiological studies indicate that this kind of very early organized activity shapes the precision of termination of the retinofugal axons, visible in the terminal arbors. Such activity does not specify the polarity of the retinotopic maps in the thalamus and midbrain (see chapter 21), but it shapes details of the projections.

Neuronal activity also shapes the binocular responsiveness of neurons in the visual cortex of mammals that include the cat and the monkey. If one of a kitten's eyes is occluded shortly after birth, many cells in the primary visual cortex become less

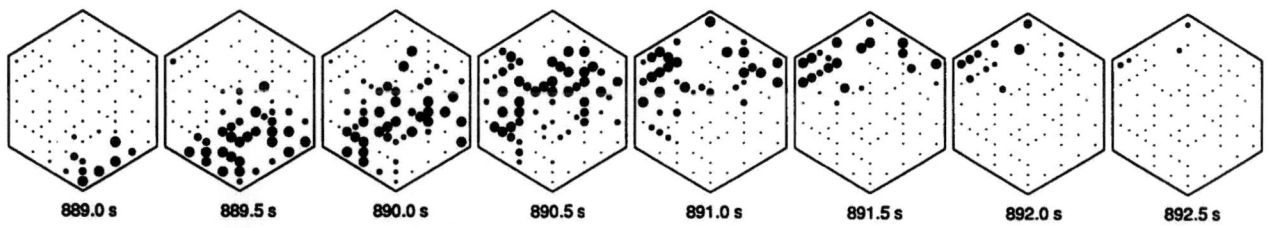

Figure 34.7
Simplified depictions of the retina of a kitten before eye opening, showing positions of retinal ganglion cells; cells spontaneously firing action potentials are shown as larger dots. The diagrams illustrate findings over a period of 3.5 seconds and show a wave of activity passing across the retina. From Meister, Wong, Baylor, and Shatz, (1991).

Structural Change in Development and in Maturity

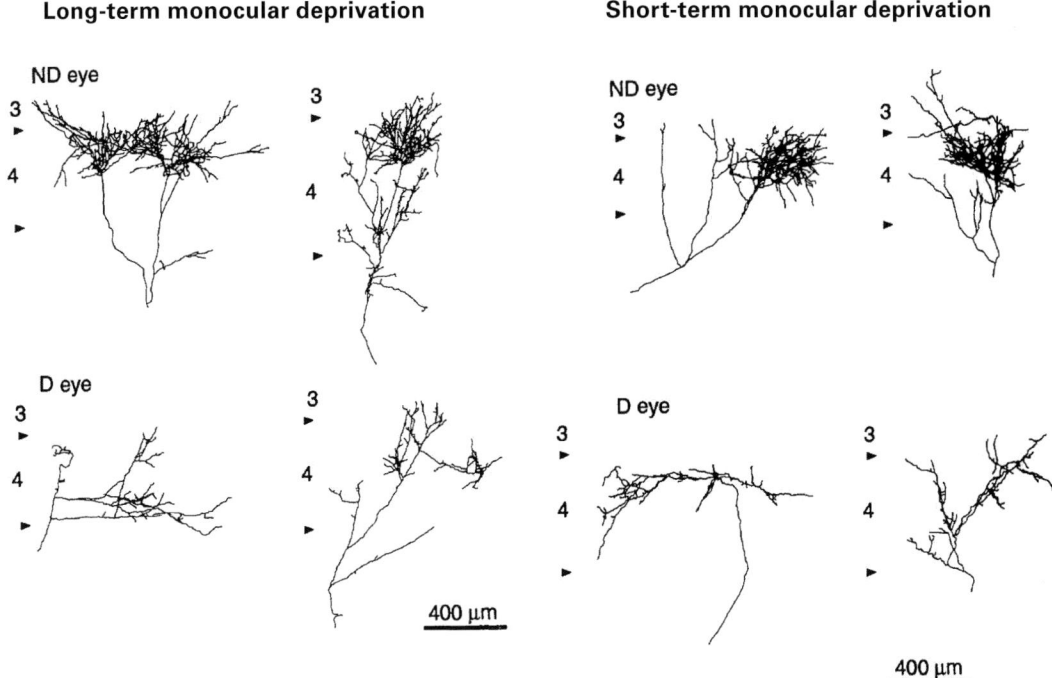

Figure 34.8
Arbors of thalamocortical axons in visual cortex: Morphological effects of short-term (6–7 days) and longer-term (33 days) monocular visual deprivation caused by lid suture in kittens during a critical period after eye opening. Axons connected (through the thalamus) with the deprived eye (D) are shown below axons connected with the open eye (ND). Positions of neocortical layers 3, 4, and 5 are indicated at the left of each drawing. From Antonini and Stryker (1993).

responsive to that eye and more responsive than usual to the nondeprived eye. In addition, there is a loss of binocular responsiveness of many neurons in that cortex. Examination of individual axons arborizing in layer 4, most of which come from the lateral geniculate body of the thalamus, has revealed that axons identified as getting their input from the open eye are much better developed than axons getting their input from the deprived eye (figure 34.8). This difference was found not only after a long period of visual deprivation (a month) but even after a shorter time (6–7 days). The thalamocortical axons in the developing animal apparently show surprisingly rapid morphological changes with large changes in neuronal activity.

Monocular deprivation is not the only possible cause of loss of binocularly responsive neurons during early development. In cases of strabismus, where the two eyes are not normally aligned (e.g., they appear to be crossed), neurons in the visual cortex become responsive to only one eye or the other, losing the normal binocular responsiveness. The axons receiving inputs from different eyes show a kind of competition in their formation of terminal end arbors (figure 34.9). (A normal pattern of termination of axons representing the two eyes in the visual cortex of a monkey was illustrated in figure 33.6.) Experiments demonstrating these effects of activity have been used to define a critical period during development when the changes occur. After this period, different for different species, such changes become much reduced or they no longer occur at all.

Apparent Shaping of Thalamocortical Projections by Abnormal Activity

In the discussions of the development of the central nervous system in chapter 13, we mentioned experiments that showed that the developing optic tract can be caused to terminate abnormally in the medial geniculate body, the major thalamic structure of the auditory system (see figure 13.13). Such findings raised an interesting question. How would the visually elicited activity in the abnormal pathway affect the developing organization of the visual pathway to the auditory cortex? Electrophysiological studies of ferrets with very early brain lesions were designed to answer this question. These studies were carried out by Mriganka Sur and his students at MIT. They showed that the organization of the projections resembled that of the visual cortex although their precision was somewhat reduced. Not only was a retinotopic order observed, but also the cells in the normally auditory region were found to have orientation selectivity with an arrangement resembling that found in a normal primary visual area (figure 34.10). Thus, it appears that the visually elicited activity reaching the auditory neocortex had a remarkable effect. It transformed the structures that normally process auditory information into processors of visual pattern information.

How Does Activity Affect Axons and Their Connections?

The cellular mechanisms that can account for plasticity of axons and axonal connections are explored in an area of neuroscience that has been expanding for several decades. It encompasses many studies of both developing and adult animals and includes the study of engram formation—the processes of brain alterations that underlie the formation of memories in all their varieties. This area of study was introduced in chapter 13.

Many investigations of changes underlying learning have focused on changes in the strength of synaptic connections and have often been guided by the proposals made by Donald Hebb in 1949. One major postulate can be summarized by the commonly repeated phrase, "cells that fire together wire together." This appears to happen commonly during development (e.g., in the refinement of topographic connections in the visual system). It has also been supported in studies of associative learning in adult animals. In these studies, differences in the roles of different glutamate receptors at synapses have been found. The N-methyl-d-aspartate (NMDA) receptor has been found to play an important role in the molecular-level dynamics of learning. Students can readily find material on the Web about this and many other aspects of learning and brain mechanisms.

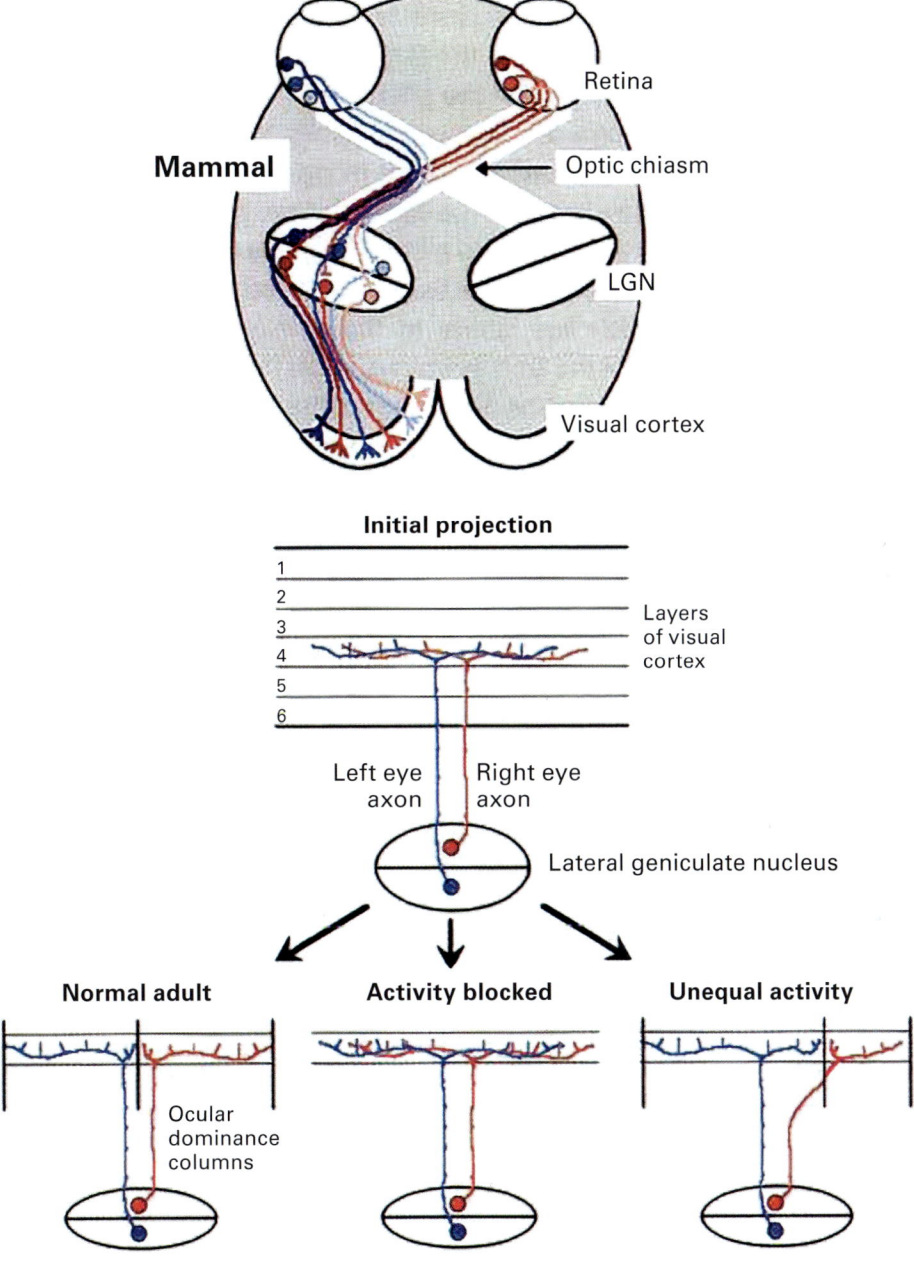

Figure 34.9
Summary drawing of the visual pathways from retina to visual cortex of a cat. Adult brain is depicted at the top, simplified so only two layers of the lateral geniculate body of the thalamus are shown. In the middle drawing, the initial projection is sketched, showing the overlap of axons connected with right and left eyes. At the bottom, the effects of neuronal activity during development are illustrated: In the normal adult (at left), terminal arbors of axons connected to right and left eyes are segregated into separate columns. If activity is blocked during development (center), the arbors retain their initial overlap. When activity is strongly biased toward one eye (right), the arbors of the more active eye occupy wider columns than those of the relatively deprived eye. LGN, lateral geniculate nucleus. From Goodman and Shatz (1993).

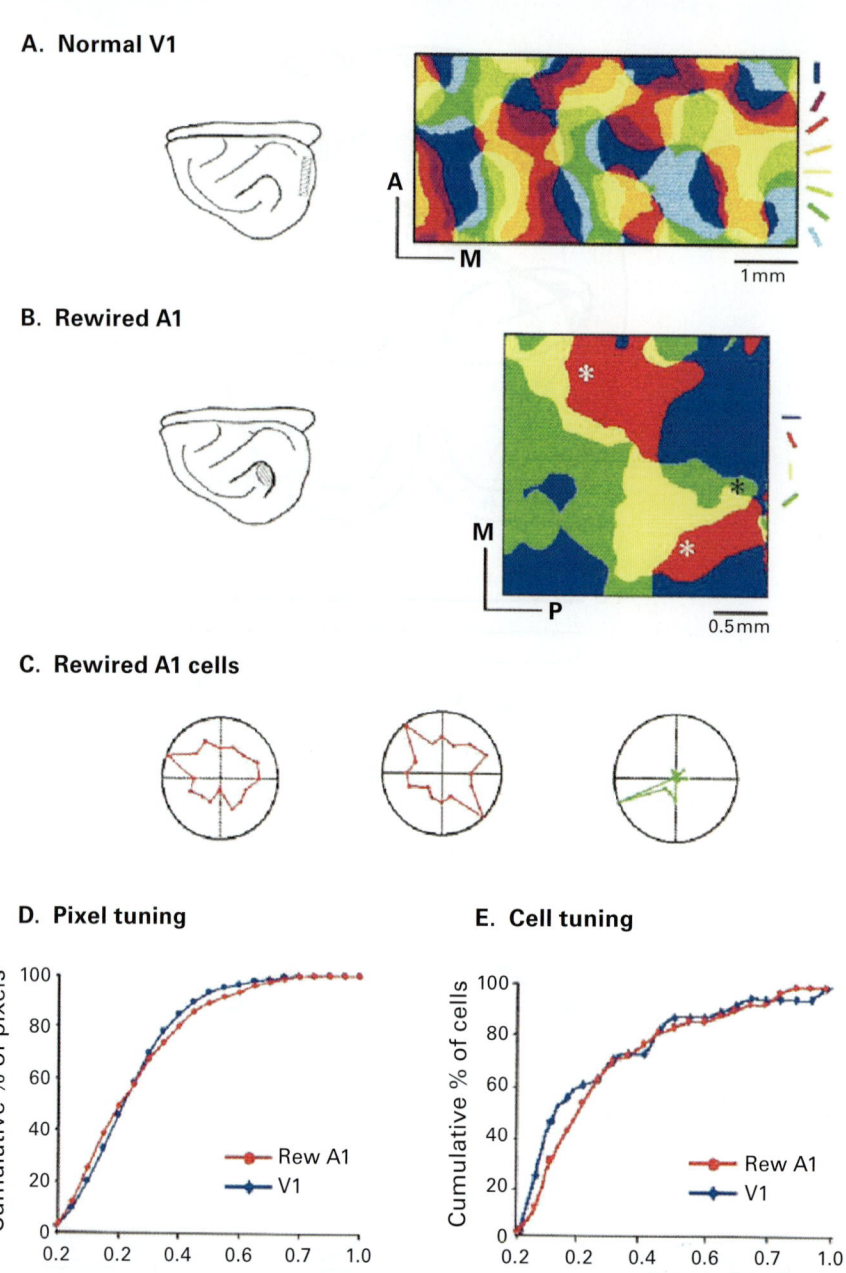

Figure 34.10
Data from studies of normal ferrets and ferrets with early midbrain lesions that caused an abnormal pathway to form from retina to medial geniculate body. The consequence of such early brain damage is visual activation of cells in neocortical areas normally activated by auditory inputs. (A) Normal visual cortex area V1 (area 17, the striate area). Colors represent activation by different line orientations as shown to the right of the figure. (B) Abnormal visually elicited activity pattern in rewired auditory cortical area A1, showing an organization that is courser than normal V1 but still recognizable as similar. (C) Graphical representation of orientation tuning found by recordings from three neurons of the rewired auditory cortex. (D, E) Two measures of "tuning" of orientation-selective cells comparing normal V1 and rewired A1. A, anterior; M, medial; P, posterior. From Sur, Angelucci, and Sharma (1999).

Structural Change in Development and in Maturity

Experimental studies that focused on molecular mechanisms of learning became more and more numerous, pioneered by the studies of simple forms of learning in the sea slug *Aplysia californica* and in the fruit fly. After the discovery of particular molecular changes underlying specific kinds of learning, morphological changes also began to be uncovered. The structural changes have included not only details of synaptic structure (e.g., the areas of membrane thickenings) but also changes in terminal arbors that indicate axonal sprouting. Thus, studies of plasticity during axonal development have shown some convergence with studies of learning in adults.

Altered Neocortical Structure after Early Blindness

Because structural plasticity seems greater during the developmental period, one would expect evidence of changes in the brains of animals or people who become totally deprived of one of their senses early in life. Early blinding has been the subject of several interesting investigations. Figure 34.11 illustrates results of studies of early blinding of

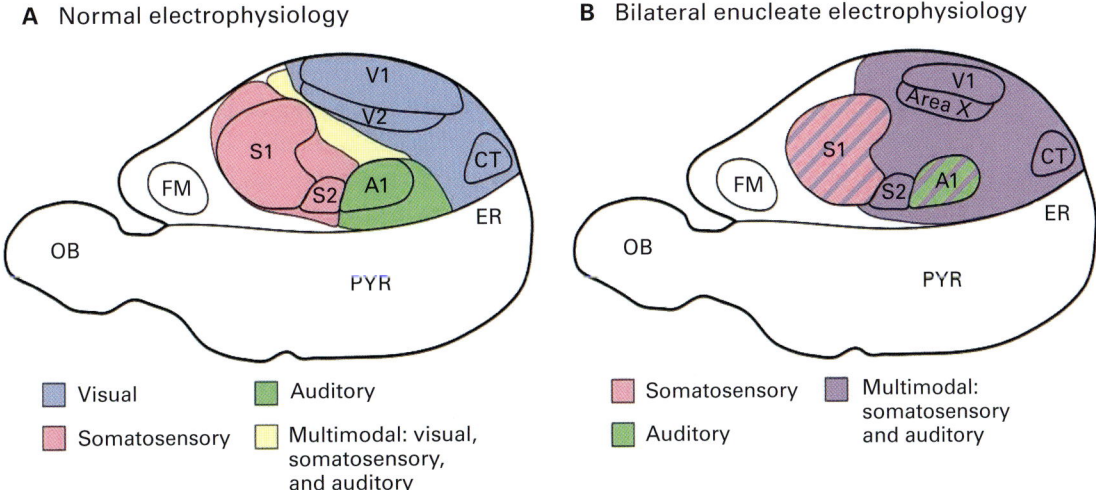

Figure 34.11
Effects of early bilateral eye removal in short-tailed opossums: Summary pictures of positions of visual, auditory, and somatosensory neocortical areas in a normal opossum (A) and an opossum blinded shortly after birth (B) when the optic tract is just beginning to form. Colors indicate sensory responsiveness of neurons according to electrophysiological studies. In early blinded opossums, the primary visual area (according to studies of cytoarchitecture and fiber architecture) is shrunken, and it and other visual areas contain cells that respond to auditory, somatic, or both auditory and somatic inputs. In addition, the primary auditory cortex was found to have cells that, unlike in the normal animals, responded also to somatic inputs, and similarly the primary somatosensory cortex had cells that responded also to auditory inputs. A1, auditory area; CT, caudotemporal area; ER, Entorhinal cortex; FM, frontal myelinated area; OB, olfactory bulb; PYR, pyriform cortex; S1, primary somatosensory area; S2, secondary somatosensory area; V1, primary visual area; V2, secondary visual area. From Karlen and Krubitzer (2009).

short-tailed opossums. In the blinded animals, cells in the cortex that can be structurally defined as area 17—the primary visual area in normal mammals—have been found to respond to auditory or somatosensory stimulation, unlike what has been found in non-blinded animals. Although this could indicate an invasion of axons from the other sensory systems into the normally visual area, another mechanism can also be postulated. Normally sparse connections from the other modalities, too sparse to activate cells, may have become stronger and therefore much more likely to activate the visual cortical neurons. This has been called an unmasking of *silent synapses*. In the study illustrated in the figure, some evidence was reported for anatomical changes in connections at both thalamic and neocortical levels; some of these connections are believed to be completely absent in normal animals.

In this regard, there has been evidence reported that both blind rats and blind humans use visual cortex in nonvisual functions (e.g., in Braille reading and in auditory localization).

Alteration of Receptive Fields by Repetitive Stimulation in *Adult* Somatosensory Cortex

The findings reviewed in the previous section are examples of brain plasticity during development, a period of time when we have come to expect more plasticity than in adults. In animals sufficiently young, even axonal regrowth occurs, regrowth that begins to fail when an animal reaches a certain age (see chapter 13). A number of experiments have shown distortions in topographic maps after early manipulations, but we had come to assume the relative rigidity of cortical organization in mature mammals. This picture became altered near the end of the twentieth century with reports of changes in the map of the body surface in primary somatosensory neocortex as a consequence of altered use of the hands in *adult* monkeys.

The experiments involved drastic changes in the normal timing of tactile stimulation of the hands because of a repetitive task that mature owl monkeys were required to learn and then repeatedly carry out. They were subjected to 4–6 weeks of training with several hundred trials per day. In this training, a monkey had to touch repeatedly a small bar using multiple digits simultaneously. The results revealed by microelectrode recordings in the hand representation of somatosensory cortex showed changes in the map with the appearance of much larger areas of multiple-digit receptive fields (figures 34.12 and 34.13).

What is changing in these monkeys that can account for the plasticity? The mechanism is not certain. Studies of the thalamus reveal no changes there. The speed at which cortical changes have been found indicates fairly rapid changes in synaptic connections or in synaptic strengths. Thus, it appears that new connections are grown rapidly or that previously inactive connections become active. The latter hypothesis, as noted previously when discussing visual cortex, is the silent-synapse hypothesis. There is electrophysiological evidence that such inactive connections do exist and can become active. The first

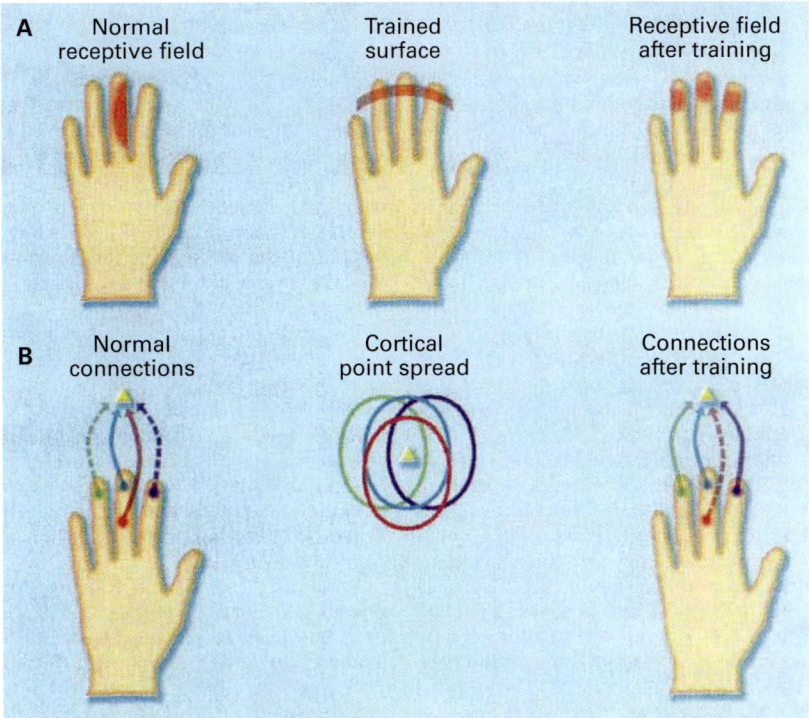

Figure 34.12
Repetitive digital stimulation in a mature monkey affects neocortical receptive fields. (A) Normally, single neurons in somatosensory area 3b respond to stimulation of part of a single digit (left-most drawing). If the monkey is trained to grasp and press a bar that stimulates repeatedly the outer part of three digits simultaneously (middle drawing), receptive fields of neurons appear that include the skin of all three digits (right-most drawing). (B) Postulated connections of single neurons of the type shown in (A): Effective connections are shown with solid lines, and ineffective (or "silent") connections are shown with broken lines. The left-most drawing shows the connections that explain receptive fields like the one depicted in (A) at the left. Postulated actual connections, both effective and silent ones, of the neuron are depicted diagrammatically in the center drawing: wide areas of skin have connections, some not strong, to a single cortical neuron. The connections as altered in relative strength after repetitive stimulation of three digits simultaneously are depicted at the right. From Sur (1995).

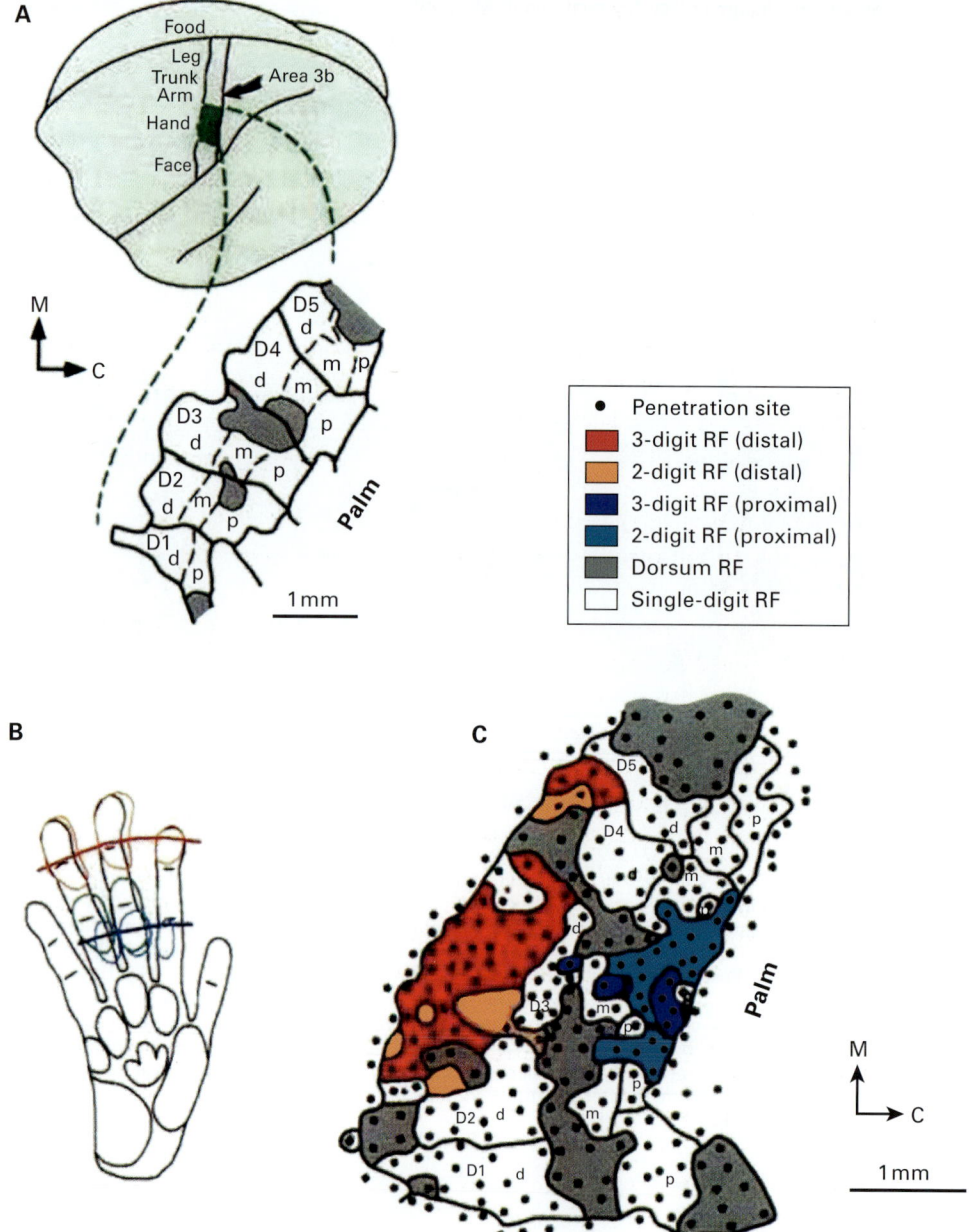

Figure 34.13
(A) Map of the representation of the hand surfaces in neocortical somatosensory area 3b of normal adult owl monkeys. In the representation of the digits (D1–D5), almost all receptive fields were on single digits; the darker gray shows areas representing dorsal surfaces; the remainder is for the ventral glabrous surfaces. (d, m, p: distal, middle, proximal finger segments.) (B) Three examples of multiple-digit receptive fields in monkeys trained as described in figure 34.12. The positions of the training bars touched by the hand during training are shown by the dark blue (distal) and dark red (proximal) lines. (C) A typical reorganized map of the representation of digits 2, 3, and 4, based on recordings at 300 locations in cortical layer 3. Dark red marks areas where three-digit receptive fields were found on distal digit segments; lighter red marks areas where two-digit fields were found. Dark blue marks areas where three-digit fields were found on proximal–middle digit segments; lighter blue shows areas where two-digit fields were found. White areas are where single-digit fields were found; gray areas represented dorsal surfaces. C, caudal; M, medial. From Wang, Merzenich, Sameshima, and Jenkins (1995).

hypothesis is unlikely because new axons would have to grow over distances that seem to be too great for the short time periods in the experiments.

Review: Where Are the Engrams?

In this book, we have noted that much habit learning, often called procedural memory formation, depends on connections in the corpus striatum (chapters 4, 7, 11, 12, and 30). What we can call emotional habits depend also on connections in the amygdala (chapter 29). Some kinds of motor learning have been found to depend on changes in the cerebellum (chapter 5); there is evidence that other motor learning may depend on changes in motor cortex (chapter 15).

Studies of the localization of synaptic changes that underlie the very simple kinds of nonassociative learning referred to as habituation and sensitization have found these simple engrams at various locations, including the spinal cord and the superior colliculus and elsewhere along sensory pathways, and in the hippocampus.

Neuroscientists have placed particularly strong emphasis on the hippocampus as a locus of learning—a place where engrams are found (i.e., the loci of changes underlying memories). This emphasis arose for good reasons. One reason was the dramatic finding that a person with bilateral hippocampal destruction is unable to form new long-term memories for events that occurred after the injury. Animal studies have shown that the hippocampus is critical for the learning of locations within a local environment (see chapter 28). A second reason is the anatomical arrangement of hippocampal circuitry. The circuit shown in figure 28.10 is found all along the long axis of Ammon's horn. Furthermore, long-term potentiation of the synaptic connections of this circuit has made the structure a favorite model for studies of learning. Slices of hippocampal tissue can be maintained in explant culture and used for electrophysiological stimulation and recording experiments that demonstrate the long-term potentiation.

There have been many findings of changes in the hippocampus correlated with memory formation. However, it is clear from neuropsychological studies of humans that the engrams for recalled long-term memories are not in that structure, but rather in the neocortex. The connections between hippocampus and neocortex, particularly the multimodal association areas, appear to play a crucial role (see figures 26.7 and 28.14). The question of where the changes in these neocortical areas may be happening was raised in chapter 13, where the context was axonal growth and plasticity. It is likely that changes occur in the small interneurons, primarily in the association layers—layers 2 and 3, as was suggested by Ramón y Cajal and later by Joseph Altman as well as others.

The short-axon interneurons, because of their small size and their short connections, are the most difficult neurons to study with neuroanatomical methods, and they are also not easy to record from. There is evidence that such neurons express the growth-associated protein GAP 43 in the adult mammal (see chapter 13). In human brains, this protein has been reported to be found at the highest levels in the association layers of the multimodal association cortex, a likely site of the engrams behind the recall of past events.

Recent studies (at MIT) have used techniques for imaging neurons in the live animal with a focus on neocortical interneurons of mice. It has been found that inhibitory short-axon neurons of various types (all using GABA as a neurotransmitter), located in a superficial zone of layers 2 and 3, show remodeling of the distal parts of dendritic branches. Structural changes were observed in dendritic processes observed over successive days and weeks. Much of the work has been focused on the visual cortex, but further studies have indicated that the superficial *dynamic zone* is very likely a general property of the neocortex.

Is Everything Plastic?

The learning and memory capacity of the neocortex and closely connected forebrain structures is certainly a great marvel. However, it would be wrong for us to conclude that everything is plastic. Remember that the major patterns and organization of CNS pathways and connections are inherited. They have resulted from a very long period of evolution and from genetically directed development. Everything is not connected to everything else! Connections matter. They determine function.

Conclusions about the Neocortex

The evolution of the neocortical mantle in the mammals added high acuity to visual and somatosensory perception and sophisticated discrimination of temporally patterned inputs in audition. The evolution of motor cortex added precision in control of movements. In this evolution, the role of learning became greatly expanded.

Its greatest innovations, I believe, were in the abilities to anticipate and to plan. Below, I expand this conception of the greatest innovations, most evident in our own species. Also, I note some of the great challenges for future research.

Posterior (Postcentral) Portions of the Hemisphere

High-resolution sensory information is relayed from neocortical analyzers to structures that generate both innate and learned habits (structures of the brainstem and of the basal ganglia). The posterior cortical areas have evolved mechanisms of object perception that enable anticipation of inputs in the immediate future. Major questions for future research include: How is object cognition accomplished? How are temporal patterns in auditory and somatosensory systems underlying object awareness analyzed and integrated? How do the objects represented in more than one modality form single representations?

Neocortex also represents spatial cognition with memory of object locations in the environment. How are entire scenes represented? How are representations of many different local environments retained and activated when needed? They appear to alter the activity of the hippocampus, within which cell activities represent different loci of the organism in the current environment.

Together, these abilities generate images that constitute an internal model used in the planning of actions. The transcortical connections between posterior and anterior association areas must be critical for such planning.

Rostral (Frontal) Portions of the Hemisphere

The central portions of the neocortex just rostral to the somatosensory areas constitute the motor and premotor cortical areas. Their activity results in high-resolution movement generation via direct projections to the brainstem and the spinal cord. Organized patterns of movement are generated that fit the images represented in the more caudal hemisphere. These patterns include temporal patterns. Much remains unknown concerning how cortical circuits generate those patterns.

Just rostral to the primary motor cortex, frontal cortex activity is involved in the planning of movements in the immediate future, with choices influenced by two kinds of connections: One is from connections with paralimbic cortex like the cingulate cortex and parolfactory regions. The other is a shaping by inputs anticipated by the images (models) of the posterior hemisphere. Here, we can envision the role of transcortical connections between posterior and anterior association areas.

The prefrontal areas, especially the more rostral areas, are also involved in the planning of actions of the more distant future. Such planning is of particular importance in highly social species like our own. The role of these areas in social interactions requires close integration with functions of the posterior association areas (like the areas important for face perception) and the amygdala—explaining some of the strong transcortical connections between anterior and posterior association cortex and also the amygdala–prefrontal interconnections. The importance and the complexity of anticipatory planning functions of the prefrontal cortex no doubt contributed to the evolutionary expansions of this region in the primates groups, for whom social life is so important. Much remains to be learned about how these functions are accomplished by the neuronal hardware.

Brain State Changes Can Alter the Processes of Both Posterior and Frontal Portions of the Neocortex

How changes in brain states (chapter 17) can alter the functioning of both posterior and anterior portions of the neocortex has not been well studied. Changes in state may alter the representation of temporal patterns—as has been demonstrated in studies of neural circuits of crustaceans, but more evidence is needed. Brain state changes must also alter the activation of specific images and movements (i.e., remembered objects and learned movements), but again, much more information is needed.

The Study of Brain Structure and Its Origins Lays the Groundwork for Understanding the Underpinnings of Behavior and Mental Abilities

Scientists who investigate the central nervous system have scarcely begun to explore some of the abilities just reviewed. Some of you, I hope, will be doing future explorations in this exciting area. As you proceed, I encourage you to remember the dictum attributed to the nineteenth century psychiatrist and neuroanatomist Bernhard von Gudden, who said (my translation), "First anatomy and then physiology, but if first physiology then not without anatomy." By physiology here, we can include all functional studies of the nervous system including all aspects of neuropsychology. Knowing about the underlying brain structure and its origins can greatly improve our understanding of animal and human behavior and its evolution. This behavior includes social behavior and all its ramifications and problems. The future of our species and others as well may depend on such understanding.

Readings

Chen, J. L., Flanders, G. H., Lee, W. C. A., Lin, W. C., & Nedivi, E. (2011). Inhibitory dendrite dynamics as a general feature of the adult cortical microcircuit. *Journal of Neuroscience, 31,* 12437–12443.

Cheung, A. F. P., Kondo, S., Abdel-Mannan, O., Chodroff, R. A., Sirey, T. M., Bluy, L. E., Webber, N., DeProto, J., Karlen, S. J., Krubitzer, L., Stolp, H. B., Saunders, N. R., & Molnar, Z. (2009). The subventricular zone is the developmental milestone of a 6-layered neocortex: Comparisons in metathcrian and eutherian mammals. *Cerebral Cortex, 20,* 1071–1081.

Hill, J., Inder, T., Neil, J., Dierker, D., Harwell, J., & Van Essen, D. (2010). Similar patterns of cortical expansion during human development and evolution. *PNAS, 107,* 13135–13140.

Kahn, D. M., & Krubitzer, L. (2002) Massive cross-modal cortical plasticity and the emergence of a new cortical area in developmentally blind mammals. *PNAS, 99,* 11429–11434.

Meister, M., Wong, R. O. L., Baylor, D. A., & Shatz, C. J. (1991). Synchronous bursts of action potentials in ganglion cells of the developing mammalian retina. *Science, 252,* 939–943.

Neve, R. I., Finch, E. A., Bird, E. D., & Benowitz, L. I. (1988). Growth-associated protein GAP-43 is expressed selectively in associative regions of the adult human brain. *PNAS, 85,* 3638–3642.

Rakic, P. (1995). A small step for the cell, a giant leap for mankind: A hypothesis of neocortical expansion during evolution. *Trends in Neurosciences, 18,* 383–388.

Rakic, P. (2009). Evolution of the neocortex: A perspective from developmental biology. *Nature Reviews Neuroscience 10,* 724–735.

Striedter, G. (2005): see book listed in chapter 4.

Sur, M., Angelucci, A., & Sharma, J. (1999). Rewiring cortex: The role of patterned activity development and plasticity of neocortical circuits. *Journal of Neurobiology, 41,* 33–43.

Wang, X., Merzenich, M. M., Sameshima, K., & Jenkins, W. M. (1995). Remodelling of hand representation in adult cortex determined by timing of tactile stimulation. *Nature, 378,* 71–75.

Data and analyses of sizes of brains and structures of the brain [see also the book by Georg Striedter (2005) noted above]:

Finlay, B. L., & Darlington, R. B. (1995). Linked regularities in the development and evolution of mammalian brains. *Science, 268,* 1578–1584.

Stephan, H., Frahm, H., & Baron, G. (1981). New and revised data on volumes of brain structures in insectivores and primates. *Folia Primatologica, 35,* 1–29.

Stephan, H., Baron, G., & Frahm, H. D. (1988). Comparative size of brain and brain components. In Steklis, H. D., & Erwin, J. eds. *Comparative primate biology.* Vol 4 (pp. 1–38). New York: Alan R. Liss.

The "rule" described by the phrase "late equals large": see two publications noted above; namely, Finlay and Darlington (1995) and a discussion in chapter 5 of Striedter (2005).

A revealing story about the difficulties many scientists had in accepting evidence for neurogenesis in adult neocortex:

Kaplan, M. S. (2001). Environment complexity stimulates visual cortex neurogenesis: Death of a dogma and a research career. *Trends in Neurosciences, 24,* 617–620.

Evidence that congenitally blind humans use visual cortical areas for auditory functions (i.e., they use regions that correspond to the visual cortex in nonblind persons):

Weeks, R., Horwitz, B., Aziz-Sultan, A., Tian, B., Wessinger, C. M., Cohen, L. G., & Rauschecker, J. P. (2000). A positron emission tomographic study of auditory localization in the congenitally blind. *Journal of Neuroscience, 20,* 2664–2672.

Evidence that stimulation of the basal nucleus of Meynart, the basal forebrain source of widespread cholinergic projections to neocortex, enhances neocortical plasticity and can improve auditory sensitivity in rats performing a behavioral task:

> Froemke, R. C., Carcea, I., Barker, A. J., Yuan, K., Seybold, B. A., Martins, A. R. O., & Schreiner, C. E. (2013). Long-term modification of cortical synapses improves sensory perception. *Nature Neuroscience, 16,* 79–88.

Von Gudden's sentence: "Zuerst also Anatomie und dann Physiologie, wenn aber erst Physiologie dann nicht ohne Anatomie."

Figure Credits

Chapter 1. Getting Ready for a Brain Structure Primer

1.1A & 1.2A Based on Butler, A. B., & Hodos, W. (2005). *Comparative vertebrate Neuroanatomy: Evolution and adaption*, 2nd ed. Hoboken, NJ: John Wiley & Sons. **1.4 (left)** From Herrick, C. J. (1922). *An introduction to neurology*, 3rd ed. Philadelphia and London: W. B. Saunders. **1.4 (right)** Photo courtesy of Prof. Robert Kretz at the Department of Medicine/Anatomy of the University of Fribourg, Switzerland. **1.7** Adapted from Ramón y Cajal, S. (1989). *Recollections of my life*. Translated by E. Horne Craigie. Cambridge, MA: The MIT Press. **1.8** From the Cajal Institute, Madrid. **1.11** From Becker, J. B., Breedlove, S. M., Crews, D., & McCarthy, M., eds. (2002). *Behavioral endocrinology*, 2nd ed. Cambridge, MA: The MIT Press. **1.12** Adapted from Breedlove, S. M., Rosenzweig, M. R., & Watson, N. V. (2007). *Biological psychology: An introduction to behavioral, cognitive, and clinical neuroscience*, 5th ed. Sunderland, MA: Sinauer Associates. **1.14** Adapted from Keeton, W. T. (1976). *Biological science*, 3rd ed. New York: W. W. Norton.

Chapter 2. Methods for Mapping Pathways and Interconnections That Enable the Integrative Activity of the CNS

2.1 Adapted from Breedlove, S. M., Rosenzweig, M. R., & Watson, N. V. (2007). *Biological psychology: An introduction to behavioral, cognitive, and clinical neuroscience*, 5th ed. Sunderland, MA: Sinauer Associates. **2.2** From Swanson, L. W. (1992). *Brain maps: Structure of the rat brain*. Amsterdam: Elsevier Science. **2.3** From Brodmann, K. (1999). *Localisation in the cerebral cortex*. Translated with editorial notes and an introduction by Laurence J. Garey. London: Imperial College Press. Original: Brodmann, K. (1909). *Vergleichende Lokalisationslehre der Grobhirnrinde*. Leipzig: Verlag von Johann Ambrosius Barth. **2.4A** From Nolte, J., & Angevine, J. B. (2007). *The human brain in photographs and diagrams*, 3rd ed. St. Louis: Mosby. **2.4B** From Jhaveri, S., Erzurumlu, R. S., Friedman, B. & Schneider, G. E. (1992). Oligodendrocytes and myelin formation along the optic tract of the developing hamster: An immunohistochemical study using the Rip antibody. *Glia 6*, 138–148. **2.5** Adapted from Flechsig, P. (1920). *Anatomie des menschlichen Gehims und Rückenmarks auf myelogenetischer Grundlage*. Leipzig: Georg Thieme. **2.6** Graybiel, A. M. (1978). A stereometric pattern of distribution of acetylthiocholinesterase in the deep layers of the superior colliculus. *Nature, 272*, 539–541. **2.7 (left)** From Karten, H. J., & Dubbeldam, J. L. (1973). The organization and projections of the paleostriatal complex in the pigeon (*Columba livia*). *Journal of Comparative Neurology, 148* (1), 61–89. **2.7 (right)** From MacLean, P. D. (1972). Cerebral evolution and emotional processes: New findings on the striatal complex. *Annals of the New York Academy of Sciences, 193*, 137–149. **2.8A** From Herkenhan, M., & Pert, C. B. (1982). Light microscopic localization of brain opiate receptors: A general autoradiographic method which preserves tissue quality. *Journal of Neuroscience, 2* (8), 1129–1149. **2.8B** From Lein, E. S., et al. (2007). Genome-wide atlas of gene expression in the adult mouse brain. *Nature, 445*, 168–176. **2.9** From Glaser, E. M., & Van der Loos, H. (1981). Analysis of thick brain sections by obverse-reverse computer microscopy: Application of a new, high clarity Golgi-Nissel stain. *Journal of Neuroscience Methods, 4*, 117–125. **2.10, 2.11, and 2.12** From Ramón y Cajal, S. (1995). *Histology of the nervous system of man and vertebrates*. Translated from the French by Neely Swanson and Larry W. Swanson. New York: Oxford University Press. **2.13** Photo courtesy of Haring Nauta. **2.20** Jellison, B. J., Field, A. S., Medow, J., Lazar, M.,

Salamat, M. S., & Alexander, A. L. (2004). Diffusion tensor imaging of cerebral white matter: A pictorial review of physics, fiber tract anatomy, and tumor imaging patterns. *American Journal of Neuroradiology, 25,* 356–369.

Chapter 3. Evolution of Multicellular Organisms with Neuron-Based Coordination

3.1A Mackie, G. O. (1970). Neuroid conduction and the evolution of conducting tissues. *Quarterly Review of Biology, 45* (4), 319–332. **3.1B** Based on Nauta, W. J. H., & Feirtag, M. (1986). *Fundamental neuroanatomy.* New York: W. H. Freeman. **3.3** Based on Striedter, G. F. (2005). *Principles of brain evolution.* Sunderland, MA: Sinauer Associates. **3.5** Based on Wicht, H., & Lacalli, T. C. (2005). The nervous system of amphioxus: structure, development, and evolutionary significance. *Canadian Journal of Zoology, 83,* 122–150.

Chapter 4. Expansions of the Neuronal Apparatus of Success

4.3, 4.4, 4.5, and 4.6 Based on Herrick, C. J. (1962) *Neurological foundations of animal behavior.* New York: Hafner Publishing Company. (Reprint of 1903 original.) **4.8B** From Schroeder, D. M. (1980). The telencephalon of teleosts. In Ebbesson, S. O. E., ed. *Comparative neurology of the telencephalon.* New York: Plenum Press. **4.11** Based on Striedter, G. F. (2005). *Principles of brain evolution.* Sunderland, MA: Sinauer Associates. Originally from Jerison, H. J. (1973). *Evolution of the brain and intelligence.* New York: Academic Press. **4.12** Based on Striedter, G. F. (2005). *Principles of brain evolution.* Sunderland, MA: Sinauer Associates. Data from several earlier studies.

Chapter 5. The Ancestors of Mammals: Sketch of a Pre-mammalian Brain

5.2 From Haeckel, E. H. P. A. (1897). *The evolution of man.* New York: D. Appleton & Company. Courtesy of the Wellcome Institute. **5.3** Procynosuchus and Cynognathus courtesy of Nobumichi Tamura. **5.8** From Herrick, C. J. (1922). *An introduction to neurology,* 3rd ed. Philadelphia and London: W. B. Saunders. **5.9** Based on Breedlove, S. M., Rosenzweig, M. R., & Watson, N. V. (2007). *Biological psychology: An introduction to behavioral, cognitive, and clinical neuroscience,* 5th ed. Sunderland, MA: Sinauer Associates. **5.11** Based on Striedter, G. F. (2005). *Principles of brain evolution.* Sunderland, MA: Sinauer Associates.

Chapter 6. Some Specializations Involving Head Receptors and Brain Expansions

6.1 Based on Friedman, M. A., & Hopkins, C. D. (1998). Neural substrates for species recognition in the time-coding electrosensory pathway of mormyrid electric fish. *Journal of Neuroscience, 18* (3), 1171–1185. **6.2** Based on Butler, A. B., & Hodos, W. (2005). *Comparative vertebrate neuroanatomy: Evolution and adaptation,* 2nd ed. New York: John Wiley & Sons. **6.3** Based on Striedter, G. F. (2005). *Principles of brain evolution.* Sunderland, MA: Sinauer Associates. **6.4** Based on Allman, J. M. (2000). *Evolving brains.* Scientific American Library. New York: W. H. Freeman. **6.5A** From Woolsey, T. A., & Van der Loos, H. (1970). The structural organization of layer IV in the somatosensory region (S I) of mouse cerebral cortex: The description of a cortical field composed of discrete cytoarchitectonic units. *Brain Research, 17* (2), 205–242. **6.5B** Photo courtesy of S. Jhaveri. **6.6** Adapted from Allman, J. M. (2000). *Evolving brains.* Scientific American Library. New York: W. H. Freeman. **6.7** Adapted from Ulinski, P. S. (1984). Thalamic projections to the somatosensory cortex of the echidna, Tachyglossus aculeatus. *Journal of Comparative Neurology, 229,* 153–170.

Chapter 8. The Neural Tube Forms in the Embryo, and CNS Development Begins

8.1 and 8.2 Based on Wolpert, L. (1991). *The triumph of the embryo.* New York: Oxford University Press. **8.4** Based on Brodal, P. (2004). *The central nervous system: Structure and function,* 3rd ed. New York: Oxford University Press. **8.5** Based on Breedlove, S. M., Rosenzweig, M. R., & Watson, N. V. (2007). *Biological psychology: An introduction to behavioral, cognitive, and clinical neuroscience,* 5th ed. Sunderland, MA: Sinauer Associates.

8.7 Based on McConnell, S. K. (1995). Constructing the cerebral cortex: Neurogenesis and fate determination. *Neuron, 15* (4), 761–768. **8.8, 8.9, and 8.11** From Ramón y Cajal, S. (1995). *Histology of the nervous system of man and vertebrates.* Translated from the French by N. Swanson & L. W. Swanson. New York and Oxford: Oxford University Press. **8.10** From Domesick, V. B., & Morest, D. K. (1977). Migration and differentiation of shepherd's crook cells in the optic tectum of the chick embryo. *Neuroscience, 2* (3), 477–491.

Chapter 9. The Lower Levels of Background Support: Spinal Cord and the Innervation of the Viscera

9.1 Based on Truex, R. C., & Carpenter, M. B. (1964). *Strong and Elwyn's human neuroanatomy*, 5th ed. Baltimore: Williams and Wilkins. **9.2 and 9.3** From Rexed, B. (1954). A cytoarchitectonic atlas of the spinal cord in the cat. *Journal of Comparative Neurology, 100* (2), 297–379. **9.4** From Gibson, S. J. (1981). The distribution of nine peptides in rat spinal cord with special emphasis on the substantia gelatinosa and on the area around the central canal (lamina X). *Journal of Comparative Neurology, 201* (1), 65–79. **9.5** Based on Per Brodal, (2004) The central nervous system: structure and functioning, 3rd ed. New York: Oxford University Press. **9.9** Based on Nauta, W. J. H., & Feirtag, M. (1986). *Fundamental neuroanatomy.* New York: W. H. Freeman. **9.11** Based on Brodal, P. (2004). *The central nervous system: Structure and function*, 3rd ed. New York: Oxford University Press, and others. **9.12** Based on Swanson, L. W. (2003). *Brain architecture: Understanding the basic plan*. New York: Oxford University Press.

Intermission: The Ventricular System, the Meninges, and the Glial Cells

9a.1 Based on Le Gros Clark, W. E. (1976). Central nervous system. In Hamilton, W. J., ed. *Textbook of human anatomy*, 2nd ed. New York: Macmillan. **9a.2** Based on Brodal, P. (1998). *The central nervous system: Structure and function*, 2nd ed. New York: Oxford University Press.

Chapter 10. Hindbrain Organization, Specializations, and Distortions

10.1, 10.9, and 10.11 Based on Nauta, W. J. H., & Feirtag, M. (1986). *Fundamental neuroanatomy.* New York: W. H. Freeman. **10.3 and 10.5** From Scheibel, M. E., & Scheibel, A. B. (1958). Structural substrates for integrative patterns in the brainstem reticular core. In Jasper, H. H. et al., eds. *Reticular formation of the brain*. Boston: Little Brown. **10.7** From Lumsden, A., & Krumlauf, R. (1996). Patterning the vertebrate neuraxis. *Science, 274,* 1109–1115. **10.8** From Wolpert, L., Jessell, T., Lawrence, P., Meyerowitz, E., Robertson, E., & Smith, J. (2007). *Principles of development*, 3rd ed. New York: Oxford University Press. Photographs by Alex P. Gould. **10.20** From Nolte, J., & Angevine, J. B., Jr. (1995). *The human brain: In photographs and diagrams.* St. Louis: Mosby.

Chapter 11. Why a Midbrain? Notes on Evolution, Structure, and Functions

11.4 Based on lectures by W. J. H. Nauta at MIT in his class "Outline of Neuroanatomy" in the late 1960s.

Chapter 12. Picturing the Forebrain with a Focus on Mammals

12.2 Based on Fernandez, A. S., Pieau, C., Reperant, J., Boncinelli, E., & Wassef, M. (1998). Expression of the Emx-1 and Dlx-1 homeobox genes define three molecularly distinct domains in the telencephalon of mouse, chick, turtle and frog embryos: Implications for the evolution of telencephalic subdivisions in amniotes. *Development, 125,* 2099–2111. **12.8** Based on Striedter, G. F. (2005). *Principles of brain evolution.* Sunderland, MA: Sinauer Associates. **12.9** From Reiner, A., Yamamoto, K., & Karten, H. J. (2005). Organization and evolution of the avian forebrain. *The Anatomical Record Part A, 287A,* 1080–1102. **12.10** Adapted from Welagen, J., & Anderson, S. (2011). Origins of neocortical interneurons in mice. *Developmental Neurobiology, 71,* 10–17. **12.11** From Rakic, P. (2009). Evolution of the neocortex: A perspective from developmental biology. *Nature Reviews Neuroscience 10,* 724–735.

Chapter 13. Growth of the Great Networks of Nervous Systems

13.1 From Ramón y Cajal, S. (1989). *Recollections of my life*. Cambridge, MA: The MIT Press. **13.2a** From Wessells, N. K., & Nuttall, R. P. (1978). Normal branching, induced branching, and steering of cultured parasympathetic motor neurons. *Experimental Cell Research, 115*, 111–122. **13.2b** From Bray, D., & Chapman, K. (1985). Analysis of microspike movements on the neuronal growth cone. *Journal of Neuroscience, 5* (12), 3204–3213. **13.3** Based on Lin, C.-H., Thompson, C. A., & Forscher, P. (1994). Cytoskeletal reorganization underlying growth cone motility. *Current Opinion in Neurobiology, 4* (5), 640–647. **13.4** From Levi-Montalcini, R. (1964). Growth control of nerve cells by a protein factor and its antiserum. *Science, 143*, 105–110. **13.5** From Bentley, D., & Caudy, M. (1983). Pioneer axons lose directed growth after selective killing of guidepost cells. *Nature, 304*, 62–65. **13.7** From Hibbard, E. (1965). Orientation and directed growth of Mauthner's cell axons from duplicated vestibular nerve roots. *Experimental Neurology, 13*, 289–301. **13.8** Based on Tessier-Lavigen, M., & Goodman, C. S. (1996). The molecular biology of axon guidance. *Science, 274* (5290) 1123–1133. **13.9a, 13.14, 13.15, and 13.16** Based on Jhaveri, S., Schneider, G. E., & Erzurumlu, R. S. (1991). Axonal plasticity in the context of development. In: Cronly-Dillon, J. R., ed. *Development and plasticity of the visual system* (Vision and Visual Dysfunction, vol. 11, pp. 232–256); Devor, M., & Schneider, G. E. (1975). Neuroanatomical plasticity: the principle of conservation of total axonal arborization. *Aspects of Neural Plasticity, 43*, 191–200. London: Macmillan Press. **13.9b** Courtesy of Michael P. Stryker, UCSF. **13.10** From O'Leary, D. D. M., Yates, P. A., & McLaughlin, T. (1999). Molecular development of sensory maps: Representing sights and smells in the brain. *Cell, 96* (2), 255–269. **13.11** Finlay, B. L., Schneps, S., & Schneider, G. E. (1979). Orderly compression of the retinotectal projection following partial tectal ablation in the newborn hamster. *Nature 280*, 153–155. **13.12** From Schneider, G. E. (1973). Early lesions of superior colliculus: Factors affecting the formation of abnormal retinal projections. *Brain, Behavior and Evolution, 8*, 73–109.

Chapter 14. Overview of Motor System Structure

14.1 Based on Orlovskii, G. N. (1970). On the connections of reticulospinal neurons with the "locomotor regions" of the brainstem. *Biofizika, 15*, 171–177. **14.2, 14.9, and 14.10** Based on Swanson, L. R. (2012). *Brain architecture: Understanding the basic plan,* 2nd ed. New York: Oxford University Press. **14.4** Based on Nauta, W. J. H., & Feirtag, M. (1986). *Fundamental neuroanatomy*. New York: W. H. Freeman. **14.5** From Goodlett, C. R. (2008). The cerebellum. In Conn, P. M., ed. *Neuroscience in medicine* (pp. 221–245). Totowa, NJ: Humana Press. **14.6** Based on Brodal, P. (2004). *The central nervous system: Structure and function,* 3rd ed. New York: Oxford University Press. **14.7** Based on Massion, J. (1988). Red nucleus: Past and future. *Behavioural Brain Research, 28*, 1–8. **14.8** Figure based on Striedter, G. F. (2005). *Principles of brain evolution*. Sunderland, MA: Sinauer Associates; and on three experimental studies by others in 1981 and 1982. **14.11** From Lawrence, D. G., & Kuypers, H. G. J. M. (1968). The functional organization of the motor system in the monkey II: The effects of lesions of the descending brain-stem pathways. *Brain, 91* (1), 15–36.

Chapter 15. Descending Pathways and Evolution

15.1 From Lawrence, D. G., & Kuypers, H. G. J. M. (1968). The functional organization of the motor system in the monkey II: The effects of lesions of the descending brain-stem pathways. *Brain, 91,* (1), 15–36. **15.5** Shoham, S., Halgren, E., Maynard, E. M., & Normann, R. A. (2001). Motor-cortical activity in tetraplegics. *Nature, 413*, 793. **15.6** Based on the following: Virginia opossum: Lende, R. A. (1963). Cerebral cortex: A sensorimotor amalgam in the marsupiala *Science, 141* (3582), 730–732. Brush-tailed possum: Haight, J. R., & Neylon, L. (1978). The organization of neocortical projections from the ventroposterior thalamic complex in the marsupial brush-tailed possum, *Trichosurus vulpecula*: A horseradish peroxidase study. *Journal of Anatomy, 126* (3) 459–485; and Haight, J. R., & Neylon, L. (1979). The organization of neocortical projections from the ventrolateral thalamic nucleus in the brush-tailed possum, *Trichosurus vulpecula*, and the problem of motor and sensory convergence within the mammalian brain. *Journal of Anatomy, 129* (4), 673–694. Rat: Donoghue, J. P., & Wise, S. P. (1982). The motor cortex of the rat: Cytoarchitecture and microstimulation mapping, *Journal of Comparative Neurology, 212* (1), 76–88. Galago: Kaas, J. H. (2004). Evolution of somatosensory and motor cortex in primates. *The Anatomical Record Part A, 281A* (1), 1148–1156. **15.7** Adapted from Nudo, R. J., & Masterton, R. B. (1990). Descending pathways to the spinal cord, III: Sites of origin of the corticospinal tract. *Journal of Comparative Neurology,*

296 (4), 559–583. **15.8** From Rathelot, J.-A., & Strick, P. L. (2009). Subdivisions of primary motor cortex based on cortico-motoneuronal cells. *PNAS, 106,* 918–923.

Chapter 16. The Temporal Patterns of Movements

16.1 Based on the description in Tinbergen, N. (1951). *The study of instinct.* Oxford: Clarendon Press.

Chapter 17. Widespread Changes in Brain State

17.2 (upper left) Levitt, P., & Moore, R. Y. (1978). Noradrenaline neuron innervation of the neocortex in the rat. *Brain Research, 139,* 219–231. **17.2 (lower)** Modified from Moore, R. Y., & Bloom, F. E. (1979). Central catecholamine neuron systems: Anatomy and physiology of the norepinephrine and epinephrine systems. *Annual Review of Neuroscience, 2,* 113–168. **17.3** Modified from Björklund, A., & Dunnett, S. B. (2007). Dopamine neuron systems in the brain: An update. *Trends in Neurosciences, 30,* 194–202. **17.4** From Heimer, L. (1995). *The human brain and spinal cord: Functional neuroanatomy and dissection guide,* 2nd ed. New York: Springer-Verlag. **17.5** From Törk, I. (1990). Anatomy of the serotonergic system. *Annals of the New York Academy of Sciences, 600,* 9–34. **17.6** From Baldo, B. A., Daniel, R. A., Berridge, C. W., & Kelley, A. E. (2003). Overlapping distributions of orexin/hypocretin- and dopamine-hydroxylase immunoreactive fibers in rat brain regions mediating arousal, motivation, and stress. *Journal of Comparative Neurology, 464,* 220–237.

Chapter 18. Taste

18.1 and 18.2 Based on Zigmond, M. J., Bloom, F. E., Lands, S. C., Roberts, J. L., & Squire, L. R. (1999). *Fundamental neuroscience.* New York: Academic Press.

Chapter 19. Olfaction

19.1 Based on Herrick, C. J. (1933). The functions of the olfactory parts of the cerebral cortex. *PNAS, 19,* 7–14. **19.5** Based on Polenova, O. A., & Vesselkin, N. P. (1993). Olfactory and nonolfactory projections in the river lamprey (*Lampetra fluviatilis*) telencephalon. *Journal für Hirnforschung, 34,* (2), 261–279. **19.6** From Wicht, H., & Northcutt, R. G. (1993). Secondary olfactory projections and pallial topography in the pacific hagfish, *Eptatretus stouti*. *Journal of Comparative Neurology, 337,* 529–542. **19.7.** From Striedter, G. F. (2005). *Principles of brain evolution.* Sunderland, MA: Sinauer Associates. **19.8** From Heimer, L. (1995) *The human brain and spinal cord: Functional neuroanatomy and dissection guide,* 2nd ed. New York: Springer-Verlag. **19.9** Based on Parker, C. A. (1906). *A guide to diseases of the nose and throat and their treatment.* London: Edward Arnold; and on other sources including studies of nasal air flow. **19.10** Based on Heimer, L. (1995). *The human brain and spinal cord: Functional neuroanatomy and dissection guide,* 2nd ed. New York: Springer-Verlag. **19.11, 19.12, 19.13 and 19.15** From Ramón y Cajal, S. (1995). *Histology of the nervous system of man and vertebrates.* Translated from the French by N. Swanson & L. W. Swanson. New York: Oxford: Oxford University Press. **19.14** From Haberly, L. B. (2001). Parallel-distributed processing in olfactory cortex: New insights from morphological and physiological analysis of neuronal circuitry. *Chemical Senses, 26* (5), 551–576. **19.16** From Mori, K., Nagao, H., & Yoshihara, Y. (1999). The olfactory bulb: Coding and processing of odor molecule information. *Science, 286,* 711–715.

Chapter 20. Visual Systems: Origins and Functions

20.1 From Ling, C., Jhaveri, S. & Schneider, G. E. (1998). Target-specific morphology of retinal axon arbors in the adult hamster. *Visual Neuroscience, 15,* 559–579. **20.5** Based on Finlay, B. L., Schneps, S. E., Wilson, K. G., & Schneider, G. E. (1978). Topography of visual and somatosensory projections to the superior colliculus of the golden hamster. *Brain Research, 142,* 223–235; and on Frost, D. O., & Schneider, G. E (1979). Plasticity of retinofugal projections after partial lesions of the retina in newborn Syrian hamsters. *Journal of Comparative Neurology, 185,* 517–567; using results also from S. Jhaveri. **20.6** Deschênes, M., Bourassa, J., & Parent, A. (1996).

Striatal and cortical projections of single neurons from the central lateral thalamic nucleus in the rat. *Neuroscience, 72,* 679–687.

Chapter 21. Visual Systems: The Retinal Projections

21.1b and 21.4c From Le Gros Clark, W. E. (1932). A morphological study of the lateral geniculate body. *British Journal of Ophthalmology, 16* (5), 264–284. **21.4a** From Hubel, D. H. (1988). *Eye, brain, and vision.* Scientific American Library. New York: Freeman. **21.4b** Courtesy of Dr. Bjorn Merker of Sweden. **21.11** Reconstruction by K. Hsiao and G. E. Schneider. **21.12** Butler, A. B., & Northcutt, R. G. (1992). Retinal projections in the bowfin, *Amia calva*: Cytoarchitectonic and experimental analysis. *Brain, Behavior and Evolution, 39,* 169–194. Color added by G. Schneider. **21.13** Based on Butler, A. B., & Hodos, W. (2005). *Comparative vertebrate neuroanatomy: Evolution and adaptation,* 2nd ed. New York: John Wiley & Sons. **21.14** From Parent, A., Dube, L., Braford, M. R., Jr., & Northcutt, R. G. (1978). The organization of monoamine-containing neurons in the brain of the sunfish (*Lepomis gibbosus*) as revealed by fluorescence microscopy. *Journal of Comparative Neurology, 182,* 495–516. **21.15** From Butler, A. B., & Hodos, W. (2005). *Comparative vertebrate neuroanatomy: Evolution and adaptation,* 2nd ed. New York: John Wiley & Sons. Their figure based on: Vanegas, H., Laufer, M., & Amat, J. (1974). The optic tectum of perciform teleost I. general configuration and cytoarchitecture. *Journal of Comparative Neurology, 154* (1) 43–60. **21.16a** Foster, R. E., & Hall, W. C. (1975). The connections and laminar organization of the optic tectum in a reptile (*Iguana iguana*). *Journal of Comparative Neurology, 163* (4), 397–425. **21.16b** Szekely, G., Setalo, G., & Lazar, G. (1973). Fine structure of the frog's optic tectum: Optic fibre termination layers. *Journal für Hirnforschung, 14,* 189–225. **21.17** Schneider, G. E. (1973). Early lesions of superior colliculus: Factors affecting the formation of abnormal retinal projections. *Brain, Behavior and Evolution, 8,* 73–109.

Chapter 22. The Visual Endbrain Structures

22.1 Drawings based on neuroanatomical tracing studies by the author; ideas from developmental studies, and from S. O. E. Ebbesson. **22.3** From Balaban, C. D., & Ulinski, P. S. (1981). Organization of thalamic afferents to anterior dorsal ventricular ridge in turtle. I. projections of thalamic nuclei. *Journal of Comparative Neurology, 200* (1), 95–129. Colors added using findings of Heller, S. B., & Ulinksy, P. S. (1987). Morphology of geniculocortical axons in turtles of the genera Pseudemys and Chrysemys. *Anatomy and Embryology, 175* (4), 505–515. **22.4** From Hubel, D. H. (1988) *Eye, brain, and vision.* Scientific American Library. New York: Freeman. **22.5** From Flechsig, P. (1920). *Antomie des menschlichen gehirns und rückenmarls auf mylogenetischer grundsalge, Vol. 1.* Leipzig: Georg Thieme. **22.6** Larroche, J.-C. (1966). The development of the central nervous system during intrauterine life. In Falkner, F., ed. *Human development.* Philadelphia: W. B. Saunders. **22.7a and 22.8** Based on Nauta, W. J. H & Feirtag, M. (1986). *Fundamental neuroanatomy.* New York: W. H. Freeman. **22.9** From Allman, J. M., & Kaas, J. H. (1976). Representation of the visual field on the medial wall of occipital-parietal cortex in the owl monkey. *Science, 191* (4227), 572–575. **22.10** From Wang, Q., & Burkhalter, A. (2007). Area map of mouse visual cortex. *Journal of Comparative Neurology, 502* (3), 339–357. **22.11** From Rosa, M. G. P., & Krubitzer, L. A. The evolution of visual cortex: Where is V2? *Trends in Neuroscience, 22* (6), 242–248. **22.12** Based on: Sereno, M., & Allman, J. (1991). Cortical visual areas in mammals. In Leventhal, A. G., ed. *The neural basis of visual function* (pp. 160–172). London: Macmillan. Stepniewska, I., Preuss, T. M., & Kaas, J. H. (1993). Architectonics, somatotopic organization, and ipsilateral cortical connections of the primary motor area (M1) of owl monkeys. *Journal of Comparative Neurology, 330* (2), 238–271. Catania, K. C., Collins, C. E., & Kaas, J. H. (2000). Organization of sensory cortex in the East African hedgehog (*Atelerix albiventris*). *Journal of Comparative Neurology, 421*(2), 256–274. **22.13 and 22.14** Based on Schmahmann, J. D., & Pandya, D. N. (2006). *Fiber pathway of the brain.* New York: Oxford University Press.

Chapter 23. Auditory Systems

23.3 From Frost, S. B., & Masterton, B. (1992). Origin of auditory cortex. In Webster, D. B., Fay, R. R., & Popper, A. N., eds. *The evolutionary biology of hearing* (pp. 655–671). New York: Springer-Verlag. **23.4** From Lindsay, P. H., & Norman, D. A. (1977) *Human information processing: An introduction to psychology* (2d ed.). New York: Academic Press. **23.5** From Allman, J. A. (2000). *Evolving brains.* Scientific American Library. New York: Freeman. **23.6** From Dooling, R. J. (1980). Behavior and psychophysics of hearing in birds. In Popper, A. N., &

Fay, R. B., eds. *Comparative studies of hearing in vertebrates* (pp. 261–288). New York: Springer-Verlag. **23.7 (top)** From von Békésy, G. (1957). The ear. *Scientific American, 197,* (2), 66–79. Courtesy of the estate of Bunji Tagawa. **23.7 (bottom)** Davis, H. (2012). Peripheral coding of auditory information. In: Rosenblith, W. A., ed. *Sensory communication* (pp. 110–141). Cambridge, MA: The MIT Press. Originally published in 1961, reprinted in 2012. **23.8** From Lindsay, P. H., & Norman, D. A. (1977) *Human information processing: An introduction to psychology* (2d ed.). New York: Academic Press. **23.9** Adapted from von Békésy, G. (1949). On the resonance curve and the decay period at various points on the cochlear partition. *Journal of the Acoustical Society of America, 21,* (3), 245–254. Frequency designation changed from cycles per second to hertz (Hz). **23.13** From Evans, E. F. (1968). Cortical representation. In De Reuck, A. V. S., & Knight, J., eds. *Hearing mechanisms in vertebrates; a Ciba Foundation symposium.* Boston: Little, Brown. **23.14** From Lee, C. C., & Winer, J. A. (2008). Connections of cat auditory cortex: I. Thalamocortical system. *Journal of Comparative Neurology, 507,* (6), 1879–1900. **23.15** From Whitfield, I. C., & Evans, E. F. (1965). Responses of auditory cortical neurons to stimuli of changing frequency. *Journal of Neurophysiology, 28,* 655–672. **23.16** Based on Whitfield, I. C. (1969). Response of the auditory nervous system to simple time-dependent acoustic stimuli. *Annals of the New York Academy of Sciences, 156,* 671–677. **23.17** From Zatorre, R. J., & Belin, P. (2001). Spectral and temporal processing in human auditory cortex. *Cerebral Cortex, 11* (10), 946–953. **23.18** From Rauschecker, J. P., & Scott, S. K. (2009). Maps and streams in the auditory cortex: Nonhuman primates illuminate human speech processing. *Nature Neuroscience 12,* 718–724. **23.19** Bird: Karten, H. J. (1968). The ascending auditory pathway in the pigeon (*Columba livia*) II. Telencephalic projections of the nucleus ovoidalis thalami. *Brain Research, 11,* 134–153. Reptile (turtle): Balaban, C. D., & Ulinski, P. S. (1981). Organization of thalamic afferents to anterior dorsal ventricular ridge in turtles. I. Projections of thalamic nuclei. *Journal of Comparative Neurology, 200,* 95–129. **23.20** From Karten, H. J., & Hodos, W. (1967). *A stereotaxic atlas of the brain of the pigeon,* Columba livia. Baltimore: Johns Hopkins University Press.

Chapter 24. Forebrain Origins: From Primitive Appendage to Modern Dominance

24.1 From Northcutt, R. G. (2002). Understanding the vertebrate brain evolution. *Integrative and Comparative Biology, 42* (4), 743–756. **24.2** Based on: (a) Striedter, G. F. (2005). *Principles of brain evolution.* Sunderland, MA: Sinauer Associates. (b) Nieuwenhuys, R., Donkelaar, H. J. T., & Nicholson, C. (1998). *The central nervous system of vertebrates.* New York: Springer. With additional information from Romer, A. S., & Parsons, T. S. (1977). *The vertebrate body,* 5th ed. Philadelphia: Saunders. **24.3** Based on: Striedter, G. F. (2005). *Principles of brain evolution.* Sunderland, MA: Sinauer Associates. Nieuwenhuys, R., Donkelaar, H. J. T., & Nicholson, C. (1998). *The central nervous system of vertebrates.* New York: Springer. With additional information from Papez, J. W. (1929). *Comparative neurology: A manual and text for the study of the nervous system of vertebrates.* New York: Thomas Y. Crowell. **24.4** Based on Striedter, G. F. (2005). *Principles of brain evolution.* Sunderland, MA: Sinauer Associates. **24.5** From Sherry, D., & Duff, S. (1996). Behavioural and neural bases of orientation in food-storing birds. *Journal of Experimental Biology, 199,* 165–172.

Chapter 25. Regulating the Internal Milieu and the Basic Instincts

25.1 Based on Mesulam, M.-M. (2000). *Principles of behavioral and cognitive neurology,* 2nd ed. New York: Oxford University Press. **25.2** Based on Brownstein, M. J., Russell, J. T., & Gainer, H. (1980). Synthesis, transport, and release of posterior pituitary hormones. *Science, 207,* (4429) 373–378. **25.3** Based on Chi, C. C., & Flynn, J. P. (1971). Neural pathways associated with hypothalamically elicited attack behavior in cats. *Science, 171,* 703–706.

Chapter 26. Core Pathways of the Limbic System, with Memory for Meaningful Places

26.1 From Le Gros Clark, W. E. (1936). The topography and homologies if the hypothalamic nuclei in man. *Journal of Anatomy, 70* (Pt 2), 203–214. **26.3.** based on Johnson, A. K., & Gross, P. M. (1993). Sensory circumventricular organs and brain homeostatic pathways. *FASEB Journal, 7* (8), 678–686. **26.4** From Broadwell, R. D., Balin, B. J., Charlton, H. M., & Salcman, M. (1987). Angioarchitecture of the CNS, pituitary gland, and intracerebral grafts revealed with peroxidase cytochemistry. *Journal of Comparative Neurology, 260* (1), 47–62. **26.9** Adapted from Mesulam, M.-M. (2000). *Principles of behavioral and cognitive neurology,* 2nd ed. New York: Oxford University Press.

Chapter 27. Hormones and the Shaping of Brain Structures

27.1, 27.2, and table 27.1 Adapted from Swaab, D. F., & Hofman, M. A. (1995). Sexual differentiation of the human hypothalamus in relation to gender and sexual orientation. *Trends in Neuroscience, 18* (6), 264–270. **27.3** From Nottebohm, F., & Arnold, A. P. (1976). Sexual dimorphism in vocal control areas of the songbird brain. *Science, 194,* 211–213. **27.4 and 27.5** Based on Butler, A. B., & Hodos, W. (1996). *Comparative vertebrate neuroanatomy: Evolution and adaption.* New York: Wiley-Liss. **27.6** From Nottebohm, F. (1989). From bird song to neurogenesis. *Scientific American, 260,* 74–79. Courtesy of Patricia J. Wynne.

Chapter 28. The Medial Pallium Becomes the Hippocampus

28.1 Figure from Matthew Wilson, MIT; originally from Wilson, M. A., & McNaughton, B. L. (1993). Dynamics of the hippocampal ensemble code for space. *Science, 261,* 1055–1059. **28.3** Adapted from Northcutt, R. G., Reiner, A., & Karten, H. J. (1988). Immunohistochemical study of the telencephalon of the spiny dogfish, *Squalus acanthias. Journal of Comparative Neurology, 277,* 250–267. **28.4** Adapted from Reiner, A., & Northcutt, R. G. (1987). An immunohistochemical study of the telencephalon of the African lungfish, *Protopterus annectens. Journal of Comparative Neurology, 256,* 463–481. **28.5** Adapted from Wilczynski, W., & Northcutt, G. (1983). Connections of the bullfrog striatum: Efferent projections. *Journal of Comparative Neurology, 214,* 333–343. **28.6** From Oswaldo-Cruz, E., & Rocha-Miranda, C. E. (1968). *The brain of the opossum* (Didelphis marsupialis). Instituto de Biofisica, Universidade Federal do Rio de Janeiro. **28.7** From Paxinos, G., & Watson, C. (1982). *The rat brain in stereotaxic coordinates.* New York: Academic Press. **28.8** Rat sketches from Witter, M. P., Wouterlood, F. G., Naber, P. A., & Van Haeften, T. (2000). Anatomical organization of the parahippocampal-hippocampal network. *Annals of the New York Academy of Sciences, 911* (The Parahippocampal Region: Implications for Neurological and Psychiatric Diseases), 1–24. **28.9 and 28.11** Based on Brodal, P. (2004). *The central nervous system: Structure and function,* 3rd ed. New York: Oxford University Press. **28.10a** From Ramón y Cajal, S. (1995). *Histology of the nervous system of man and vertebrates.* Translated from the French by N. Swanson & L. W. Swanson. New York and Oxford: Oxford University Press. **28.10b** From van Strien, N. M., Cappaert, N. L. M., & Witter, M. P. (2009). The anatomy of memory: An interactive overview of the parahippocampal—hippocampal network. *Nature Reviews Neuroscience 10,* 272–28. **28.12** Based on Lund, R. D. (1978). *Development and plasticity of the brain: An introduction.* New York: Oxford University Press. **28.14** From Hasselmo, M. E. (1999). Neuromodulation: Acetylcholine and memory consolidation. *Trends in Cognitive Sciences, 3,* 351–359.

Chapter 29. The Limbic Striatum and Its Outpost in the Temporal Lobe

29.1 Modification of figure 26.9, adapted from Mesulam, M.-M. (2000). Patterns in behavioral neuroanatomy: Association areas, the limbic system, and hemispheric specialization. In Mesulam, M.-M., ed. *Principles of behavioral and cognitive neurology,* 2nd ed. New York: Oxford University Press. **29.2** Based on Lammers, H. J. (1972). The neural connections of the amygdaloid complex in mammals. In Eleftheriou, B., ed. *The neurobiology of the amygdalia; the proceedings of a Symposium on the Neurobiology of the Amygdala, Bar Harbor, Maine, June 6–17, 1971.* New York: Plenum Press. **29.3** Modification of figure 2.2, adapted from Swanson, L. W. (1992). *Brain maps: Structure of the rat brain.* Amsterdam: Elsevier Science. **29.6** Gluhbegovic, N., & Williams, T. H. (1980). *The human brain: A photographic guide.* New York: Harper and Row. **29.7** Based on descriptions by Downer, J. D. C. (1961). Changes in visual gnostic function and emotional behavior following unilateral temporal lobe damage in the "split-brain" monkey. *Nature, 191,* 50–51. **29.8** From Ledoux, J. E. (1994). Emotion, memory and the brain. *Scientific American, 270* (6), 50–57. Figure drawn by Roberto Osti. **29.11 (top)** Hyde, T. M., & Weinberger, D. K. (1990). The brain in schizophrenia. *Seminars in Neurology, 10,* 276–286. **29.11 (bottom)** Shenton, M. E., Kikinis, R., Jolesz, F. A., Pollak, S. D., LeMay, M., Wible, C. G., & McCarley, R. W. (1992). Abnormalities of the left temporal lobe and thought disorder in schizophrenia. *New England Journal of Medicine, 327,* 604–612.

Chapter 30. The Major Subpallial Structure of the Endbrain

Table 30.1 Based on Swanson, L. W. (2003). *Brain architecture: Understanding the basic plan.* New York: Oxford University Press. **30.2, 30.7, 30.8, 30.10a, and 30.11** Based on Brodal, P. (2004). *The central nervous system:*

Structure and function, 3rd ed. New York: Oxford University Press. **30.3** Deschênes, M., Bourassa, J., & Parent, A. (1996). Striatal and cortical projections of single neurons from the central lateral thalamic nucleus in the rat. *Neuroscience, 72,* 679–687. **30.10b** From Voorn, P., Louk Vanderschuren, L. J. M. J., Groenewegen, H. J., Robbins, T. W., & Pennartz, C. M. A. (2004). Putting a spin on the dorsal-ventral divide of the striatum. *Trends in Neurosciences, 2,* 468–474. **30.12** From Nauta, W. J. H., & Feirtag, M. (1986). *Fundamental neuroanatomy.* New York: W. H. Freeman.

Chapter 31. Lost Dopamine Axons: Consequences and Remedies

31.1 From Olanow, C. W., Kordower, J. H., & Freeman, T. B. (1996). Fetal nigral transplantation as a therapy for Parkinson's disease. *Trends in Neuroscience, 19,* 102–109. **31.2** From Doering, L. C., & Snyder, E. Y. (2000). Cholinergic expression by a neural stem cell line grafted to the adult medial septum/diagonal band complex. *Journal of Neuroscience Research, 61,* 597–604.

Intermission: Neurogenesis in Mature Brains

31a.1 Kempermann, G., & Gage, F. H. New nerve cells for the adult brain. *Scientific American,* May 1999. Photographs by Linda Kitabayashi / Salk Institute.

Chapter 32. Structural Origins of Object Cognition, Place Cognition, Dexterity, and Planning

32.1a From von Bonin, G. (1950). *Essay on the cerebral cortex.* Springfield, IL: Charles C Thomas. **32.1b** From Poliakov, G. L. (1968). On the structural mechanisms of functional states in neurons of the cerebral cortex. In Asratyan, E. A., ed. *Brain reflexes* (Progress in Brain Research, Vol. 22, pp. 98–106). Amsterdam: Elsevier. **32.3** Adapted from Mesulam, M.-M. (2000). *Principles of behavioral and cognitive neurology,* chap. 1. Oxford: Oxford University Press. **32.4** From Schneider, G. E. (1975). Two visuomotor systems in the Syrian hamster. In Ingle, D., & Sprague, J. M., eds. *Neurosciences Research Program bulletin, 13,* 255–257. **32.5** Based on: Striedter, G. F. (2005). *Principles of brain evolution.* Sunderland, MA: Sinauer Associates; and Butler, A. B. (1994). The evolution of the dorsal thalamus of jawed vertebrates, including mammals: Cladistic analysis and a new hypothesis. *Brain Research Reviews, 19,* 29–65. **32.6** From Herrick, C. J. (1910). The morphology of the forebrain in amphibia and reptilian. *Journal of Comparative Neurology and Psychology, 20* (5), 413–547. Labels altered. **32.7** From Striedter, G. F. (2005). *Principles of brain evolution.* Sunderland, MA: Sinauer Associates.

Chapter 33. Basic Neocortical Organization: Cells, Modules, and Connections

33.1 From Kwan, K. Y., Šestan, N., & Anton, E. S. (2012). Transcriptional co-regulation of neuronal migration and laminar identity in the neocortex. *Development, 139,* 1535–1546. **33.2** From Scheibel, M. E., & Scheibel, A. B. (1970). Elementary processes in selected thalamic and cortical subsystems—the structural substrates. In Schmitt, F. O. ed. *The neurosciences: Second study program* (1970). New York: The Rockefeller University Press. **33.3** From Larroche, J.-C. (1966). The development of the central nervous system during intrauterine life. In Falkner, F., ed. *Human development* (pp. 257–276). Philadelphia: W. B. Saunders Company. **33.4** From Le Gros Clark, W. E. (1976). Central nervous system. In: Hamilton, W. J., ed. *Textbook of human anatomy,* 2nd ed. London: Macmillan Press. **33.5** From Brodmann, K. (1868–1918). Brodmann's localisation in the cerebral cortex. Translated with editorial notes and an introduction by Laurence J. Garey. London: Imperial College Press. **33.6** From LeVay, S., Connolly, M., Houde, J., & Van Essen, D. C. (1985). The complete pattern of ocular dominance stripes in the striate cortex and visual field of the macaque monkey. *Journal of Neuroscience, 5* (2), 486–501. **33.7** From Goldman, P. S., & Nauta, W. J. H. (1977). Columnar distribution of cortico-cortical fibers in the frontal association, limbic, and motor cortex of the developing rhesus monkey. *Brain Research, 122,* 393–414. **33.8** [**Credit line to come.**] **33.13** From Nauta, W. J. H., & Feirtag, M. (1986). *Fundamental neuroanatomy.* New York: W. H. Freeman. **33.14** Based on: Schmahmann, J., & Pandya, D. N. (2006). Fiber pathways of the brain. New York: Oxford University Press; Petrides, M., & Pandya, D. N. (2007). Efferent association pathways from the rostral prefrontal cortex in the macaque monkey. *Journal of Neuroscience, 27,* 11573–11586; Schmahmann, J. D., Pandya,

D. N., Wang, R., Dai, G., D'Arceuil, H. E., de Crespigny, A. J., & Wedeen, V. J. (2007). Association fibre pathways of the brain: Parallel observations from diffusion spectrum imaging and autoradiography. *Brain, 130,* 630–653. **33.15** From Perani, D., Saccuman, M. C., Scifo, P., Anwander, A., Spada, D., Baldoli, C. & Friederici, A. D. (2011). Neural language networks at birth. *PNAS, 108,* (38) 16056–16061. **33.17** From Poliakov, G. I. (1972). *Neuron structure of the brain.* Cambridge, MA: Harvard University Press. **33.18** From von Bonin, G. (1950). *Essay on the cerebral cortex.* Springfield, IL: Charles C Thomas.

Chapter 34. Structural Change in Development and in Maturity

34.1a From Bystron, I., Blakemore, C., & Rakic, P. (2008). Development of the human cerebral cortex: Boulder Committee revisited. *Nature Reviews Neuroscience, 9,* 110–122. **34.1b** From Naegele, J. R., Jhaveri, S., & Schneider, G. E. (1988). Sharpening of topographical projections and maturation of geniculocortical axon arbors in the hamster. *Journal of Comparative Neurology, 277* (4), 593–607. **34.2 and 34.6 a, b, and c** From Rakic, P. (1995). A small step for the cell, a giant leap for mankind: A hypothesis of neocortical expansion during evolution. *Trends in Neurosciences, 18* (9), 383–388. **34.3 and 34.4** Data from: Stephan, H., Frahm, H., & Baron, G. (1981). New and revised data on volumes of brain structures in insectivores and primates. *Folia Primatologica, 35,* 1–29;. Analysis by and figure from Finlay, B. L., & Darlinton, R. B. (1995). Linked regularities in the development and evolution of mammalian brains. *Science, 268,* 1578–1584. **34.5** From Rakic, P. (1981). Development events leading to lamina and areal organization of the neocortex. In Schmitt, F. O., ed. *The organization of the cerebral cortex: Proceedings of a neurosciences research program colloquium.* Cambridge, MA: The MIT Press. **34.6d** From Rakic, P. (1972). Mode of cell migration to the superficial layers of fetal monkey neocortex. *Journal of Comparative Neurology, 145* (1), 61–84. **34.7** From Meister, M., Wong, R. O. L., Baylor, D. A., & Shatz, C. J. (1991). Synchronous bursts of action potentials in ganglion cells of the developing mammalian retina. *Science, 252,* 939–943. **34.8** From Antonini, A., & Stryker, M. P. (1993). Rapid remodeling of axonal arbors in the visual cortex. *Science, 260,* 1819–1821. **34.9** From Goodman, C. S., & Shatz, C. J. (1993). Development mechanisms that generate precise patterns of neuronal connectivity. *Cell, 72* (Suppl.), 77–98. **34.10** From Sur, M., Angelucci, A., & Sharma, J. (1999). Rewiring cortex: The role of patterned activity development and plasticity of neocortical circuits. *Journal of Neurobiology, 41* (1), 33–43. **34.11** From Karlen, S. J., & Krubitzer, L. (2009). Effects of bilateral enucleation on the size of visual and nonvisual areas of the brain. *Cerebral Cortex 19* (6), 1360–1371. **34.12** From Sur, M. (1995). Maps of time and space. *Nature, 378,* 13–13. **34.13** From Wang, X., Merzenich, M. M., Sameshima, K., & Jenkins, W. M. (1995). Remodelling of hand representation in adult cortex determined by timing of tactile stimulation. *Nature, 378,* 71–75.

Index

5-hydroxytryptamine. *See* serotonin
γ-aminobutyric acid. *See* GABA

A1 (auditory cortical area 1), 409, 435–436, 442, 658–659
abducens nerve (cranial nerve 6), 192, 194, 272
Acceleransstoff, 21. *See also* norepinephrine (noradrenaline)
accelerator nerve, 20–21, 171
accessory olfactory bulb, 335, 341, 348, 451, 540
accessory optic tract (system, nuclei), 360–361, 377–378, 386, 388–389, 391
accumbens, 495, 543, 551–552, 576, 578
acetylcholine, 21, 25, 32, 167, 171, 173, 312, 318–320, 491, 528–529, 532, 549, 552, 578–579, 643
acetylcholinesterase, 25, 32, 34–35, 578
action patterns (includes fixed action patterns), 56, 63–64, 69–70, 80, 119, 128, 134, 164, 183–184, 188, 205, 208, 265, 267, 269, 281, 290, 300–303, 305–306, 419, 473–474, 478, 581
action potential, 13–19, 21–24, 26–27, 42, 121–122, 156–157, 168, 255, 268, 303–305, 423, 429–430, 433, 470, 513, 654
activity effects in development, 255–256, 654–660
  motor, 120, 134, 157, 601, 609
  sensory, 75–76, 80, 111–112, 114, 120, 134, 157, 364, 367–368, 407, 450, 601, 609, 648, 664
adaptive functions, 3, 12, 25, 72, 74, 77, 81, 128, 183, 208, 249, 355–356, 364, 366, 369, 410, 414, 421, 456–457, 547, 562, 601–603, 607–608
adaptive habits, 207, 331, 562, 601
adenohypophysis, 471–472. *See also* pituitary
adhesion
  changes in, 139–140
  differential, 224
adrenaline (epinephrine), 170–171, 471
adrenal medulla, adrenal gland, 141, 143, 170–171, 471–472
affective associations, 413, 556. *See also* affective tags

affective tags, 470, 475, 539, 546, 556
agonistic behavior (aggressive, defensive behavior), 20, 70, 102, 119, 124, 170, 184, 196, 212, 267. 281, 419, 421, 450, 467, 473–474, 477–479, 497–498, 547, 556
agranular cortical areas, 406, 414, 577, 624, 625
alar plate, 142, 144, 181–183, 188, 191–192, 197–198, 201, 270, 432
alimentary tract, 139, 173
Allman, John, 116, 320, 350, 369, 405, 597
allocentric direction, 491, 515–517
allocortex, 496, 513, 538, 623
Altman, Joseph, 256, 350, 353, 590, 591, 663
Alzheimer's disease, 312, 549
Ammon's horn, 520, 522–523, 527, 663
amniotes, 104, 131, 462, 608
amphibian, 6, 14–15, 72, 78, 89, 91–92, 94, 131, 161, 219, 242–243, 258, 267, 313, 333–335, 337, 339, 361, 366, 385, 419, 424–425, 462, 472, 525, 564, 595, 597, 609, 611–614
  hindbrain, 242–243
amphioxus (branchiostoma), 60–63, 65–67, 69, 103–104, 124, 205, 320, 356, 369, 452–453, 467, 605
  forebrain, 61–62, 69, 124, 205, 356, 369, 467, 605
amygdala, 126, 206, 219, 224, 229–230, 232, 315, 319, 329, 330–331, 335, 340, 348, 352, 366–367, 369, 383, 410, 416, 418, 422, 429–430, 443–445, 456–457, 460, 470, 496, 506, 515, 521, 529, 537–552, 554–557, 562–564, 567, 576, 578, 581, 595, 632, 663, 665
  basolateral nuclei, 540–541
  central nucleus, 506, 540–542, 545, 551, 562–563
  cortical nucleus, 340, 540–542, 551
  corticomedial nuclei, 540–541
  lateral nucleus, 366, 369, 418, 422, 521, 539–541, 545–546
  lesion, 547–548, 554–555
  striatal, 460
amygdala-prefrontal interconnections, 554, 665
amygdalectomized monkey, 547–548

amygdalo-hippocampal area, 551
anencephalic infants, 476
ansa lenticularis, 570
anterior association cortex (areas), 494, 575, 665
anterior cingulate cortex, 296, 577
anterior commissure, 349, 378, 381, 383, 484, 488, 495, 505, 522, 543–544, 577, 630, 632
anterior hypothalamic region, 206, 357, 472–473, 485, 492, 505, 506, 551
anterior nuclei of thalamus, 486, 489, 491–494, 515, 602, 629, 630, 633–634
anterior perforated space, 344
anterior pituitary, 93, 127, 129, 472. *See also* adenohypophysis
anterior thalamic radiation, 489
anterodorsal thalamic nucleus, 391, 629
anterograde degeneration, 42, 44, 258
anterograde transport, 25, 45, 47, 406
anteromedial nucleus of thalamus (AM), 629
anteroventral nucleus of thalamus (AV), 629
anticipation of inputs (stimuli), 81, 106, 134, 136, 296, 352, 413, 442, 450, 596, 602–604, 664, 665
anticipation of location, 492–493, 515–516, 534
antidiuretic hormone (ADH, vasopressin), 471–472
antidromic stimulation and recording, 42
Apatosaurus. See Brontosaurus
ape, 112, 276, 291, 329, 374, 381, 385, 407, 450
apical dendrite, 598, 618
Aplysia (sea slug), 26–27, 304, 659
apoptosis, 257–258
appetitive behavior (actions), 268, 474–476
approach movements (actions, behavior, reaction), 55, 64, 72, 74, 207–208, 211, 217, 265, 268–269, 271–272, 275, 331, 349, 351–352, 355, 364, 367, 409, 410, 411, 413, 435, 452, 457, 460, 475, 547, 551, 596
aqueduct of Sylvius, 176, 210, 420
arachnoid membrane, 176–177
arborization
 axonal, 16, 38, 249, 255, 389, 396, 564, 600, 655
 dendritic, 345, 389
 mode of growth, 247–249, 255, 257–258
arches of fornication, 529
archicortex, 520, 623
arcopallium (avian), 219, 443–444, 506–509
arcuate fasciculus, 635–636, 637, 643
area 17, 32, 366, 368, 395, 398, 399, 411, 414, 619, 620, 622, 647, 658, 660. *See also* primary visual cortex
area 18, 32, 399, 618–619. *See also* prestriate areas; juxtastriate cortical areas
area 18b, 406, 414
area L pallii of the nidopallium (birds), 443. *See also* field L
arousal
 limbic, 469, 470
 response, 469, 605
 somatic, nonlimbic, 184, 311, 315, 318, 468–469, 470, 473, 480, 562–563, 605

sympathetic, 222, 469, 471, 498, 547
systems, 321, 469
ascending auditory pathways, 417–418, 421, 428, 431–432, 435, 443, 629
ascending dopamine axons, 315, 462, 579, 596
ascending lemniscal channels, pathways, projections, 82, 103, 125, 155, 158, 161, 198, 212, 214, 224, 328, 330, 364, 395, 418, 421, 428–429, 431, 435, 479, 628–629
ascending pathways, taste, pain, visceral, 214, 224, 328, 329, 470, 596
ascending reticular activating system (ARAS), 311, 316, 469, 480
ascending somatosensory pathways (channels), 102–103, 131, 158, 161, 198, 214, 629
ascending visual pathways, 364, 395, 429
association areas of neocortex, 81, 293, 491–492, 494, 496, 517, 523, 525, 532, 538, 545, 576, 602–604, 609, 613, 627, 632, 634, 642
 anterior (frontal), 352, 494, 575, 630, 632, 665
 auditory, 442
 motor, 632, 638
 multimodal (heteromodal), 115, 257, 296, 407, 442, 494, 496, 538, 575, 608–610, 627, 629, 634, 636, 638–639, 642, 663–664
 parietal, 412, 414, 517, 604, 623–624, 630
 posterior (somatosensory, visual), 402, 575, 604, 629–630, 632–633, 638, 665
 prefrontal, 415, 604, 623–624, 630, 632
 retrosplenial, 624, 630
 unimodal (modality-specific), 407, 496, 538–539, 608, 634, 639
 visual, 111, 411–413, 415, 539, 545, 630, 632, 636
association connections of cortex
 hippocampal, 527, 530–531
 intracortical, 600
 long transcortical, 600, 634–637
association layers of neocortex (layers 2 and 3), 642, 663–664
association pathways (bundles) of neocortex, 634–637
associative learning, 257, 421, 430, 460, 528, 538, 656
astrocytes, 150, 177, 586
asymmetric cell division, 145, 598, 646, 648, 649
auditory areas, 111, 428–430, 432, 435–438, 440, 630, 632, 650
auditory cortex, 115, 409, 564, 575, 630, 632, 650, 656, 658–659
auditory images, 603
auditory input, 75, 207, 268, 273, 506, 508, 545, 547, 602, 610, 658–659
auditory pathways, 110, 210, 229, 329, 379, 629
auditory tectum, 210, 395. *See also* colliculus, inferior
auditory thalamus, cell group, 368, 421–422, 443, 445, 457. *See also* geniculate body, medial
auditory systems, pathways, 417–446

auditory-vestibular nerve (cranial nerve 8), 96, 190, 194, 270–274, 380, 417–418, 421, 423, 425–426, 428–429, 433
autonomic activity, functions, 102, 118, 153–174, 547
autonomic cranial nerves, 192
autonomic ganglia, 59, 93, 97, 142, 153–174
autonomic nervous system (ANS), 59, 93, 97, 124, 133, 142, 153–174, 166–173, 214, 267, 279, 326, 356, 449, 467, 469, 471, 497, 541, 537, 547, 569
 and chemoreceptors, 326, 153–174
 control of inputs to, 102, 124–125, 133, 212–222, 267–278, 356, 449, 457, 467, 470, 497–498, 537, 539, 541, 547, 551
 definition and summary, 166–173
 development, 142
 involuntary and voluntary, 497
 nature of innervation, 153–174
 and sympathetic arousal, 469, 471
autoradiography, techniques, applications, 44, 623, 624, 650
avian brains (bird CNS), 3, 35, 67, 78–79, 81–83, 89–90, 96, 103–105, 119, 131, 207, 212, 219, 227, 229–230, 232–233, 275, 361–362, 364, 367, 370, 395, 397, 424–425, 431, 443–445, 450, 452, 456, 458, 472, 506–510, 581, 589, 595, 597, 608
avoidance (behavior, reactions, responses), 55, 64, 72, 74, 77, 107, 110, 196, 207–208, 211, 217, 265–266, 268–269, 271–272, 274, 326, 331, 349, 351–352, 355, 358–359, 364, 367–368, 409–411, 413, 418–419, 421, 424, 435, 452, 457, 518, 547, 551, 596, 601
axial muscles, 164, 280, 283–289
axo-axonal synapses, 22–24
axo-dendritic synapses, 22–24
axon
 collaterals, 38, 351, 527, 531, 618, 619, 626
 competition, 79, 85, 246, 254–256, 351, 396, 655
 end arbors (telodendria), 16, 37, 42, 45, 98, 157, 235–236, 247, 249, 256–258, 260, 306, 343, 345, 351, 374, 385, 396, 529, 566, 656
 end-arbor formation, 235–261
 growth, 235–261
 growth cone, 149, 151, 235–261
 growth vigor, 246, 254–256, 258–259
 plasticity, 235, 249, 251–252, 254–255, 257, 259, 261, 529–531, 535, 642, 656, 659–660, 663–664
 regeneration (regrowth), 248–249, 251–252, 258–261, 290, 351
 sprouting (collateral sprouting), 122, 247–248, 252–257, 261, 306, 351, 529–31, 534, 555, 573, 583, 659
 trajectory, 240, 241, 260, 410, 432–433, 542–543, 570, 611–613, 626
axo-somatic synapse, 22, 24
azimuthal location (position), 432–434

Baer, Karl Ernst von, 89–91
Bard, P., 117–118, 124
barrel fields, 111–113

basal forebrain, 81, 126, 181, 217, 224, 312, 239, 351, 352, 356, 410, 488, 506, 517, 586, 587, 632, 651, 668
 structures, 537–558
basal ganglia, 340, 421, 430, 460, 462, 463, 539, 547, 664
basal nucleus of Meynart, 549, 552, 668
basal plate (region, cells of), 142, 181, 182–183, 188, 192, 201, 266
basolateral nuclei of the amygdala, 540–541. *See also* amygdala
bat, 84, 110, 116, 418, 442, 648, 650–651
Bcl-2, 258, 261
bed nucleus of the stria terminalis, 316, 329, 488–489, 506, 542–544, 551–552, 578
Beer, Randall, 475
behavioral pattern controller, 269–270, 475
behavioral state, 206, 315–316, 318, 320–321, 364, 498
Bell-Magendie law (law of roots), 60, 182
Benowitz, Larry, 642
bichir, 452, 454
bilateral somatosensory projections, early evolution, 102, 195, 196
bilateral visual projections, early evolution, 195–196, 207, 359
binocular responsiveness (of neurons), 654–655
biological clock, 27, 124, 206, 304–305, 356, 467, 473, 499, 504
biological drive, 474
bipolar cell, 14–15, 96, 346, 349, 428
birds. *See* avian brain
birthday (of a cell), 468
blastula, 90, 139–140
blind animals, 659–660
blind humans, blindness in humans, 404, 660, 667
blood-brain barrier, 326, 485–487
Blumberg, Bruce, 475, 481
body size, 55, 64, 92, 105, 452
Bonhoeffer, Friedrich, 249–250
Bosma, J., 302
Bottjer, Sarah, 508
bouton, 14, 22, 44, 430
bowfin, 385
brachium conjunctivum, 214, 266, 329
brachium of inferior colliculus, 210, 378, 429, 495
Braille reading, 660
brain
 expansion (*see* expansions of brain in evolution)
 size, weight, 81–84, 452, 455, 648, 650–651
 state, 311–321, 531–532, 665
brainstem reticular formation, 165, 195, 305, 312, 365, 643
brainstem trigeminal nuclei (complex), 109, 193, 198, 323
branchial arches, 183, 186–188
branchial motor column, 187–188, 193
Branchiostoma. *See* amphioxus
BrdU, 587, 589–590

682   Index

Broca, Pierre Paul, 468, 537
Broca's area, 444, 506, 637
Brodmann, Korbinian, 32, 395, 619, 621, 623, 626
bromodeoxyuridine (BrdU), 587, 589–590
Brontosaurus (*Apatosaurus*), 156
brush-tailed possum, 292–293
buffalofish, 70, 72–73, 201, 327
bullfrog, 438, 520–521
Butler, Ann B., 6–7, 204, 281–282, 370, 608, 668

CA fields of hippocampus (*cornu Ammonis*), 520
Cajal. *See* Ramón y Cajal, Santiago
cardiac ganglion, 171, 173
Carlsson, A., 313
cartilaginous fish, 83, 104–105, 450, 456, 462
cascade processing, 305, 364
cat
  behavior and brain, 117–118, 121, 476–477
  CNS, 154–155, 158, 249, 255–267, 277, 281, 367, 373, 400, 436–439, 445, 470, 473, 477, 605, 609–610, 619, 623, 654, 657
  fetal, 37, 39, 654
catecholamine axons, 313, 554–555
catfish, 71, 73, 327
caudal, definition, 5–6
caudal ganglionic eminence, 230, 233, 597, 607
caudate nucleus, 422, 520, 546, 549, 562–563, 565, 568, 573–576, 580–581
caudate-putamen, 319, 365–366, 578
caudate tail, 575–576, 580
cell
  cycle, 145, 646
  death, 9, 41, 246, 257–258
  division, 145–146, 590, 598, 664, 648–649
  growth, 9, 139
  migration (*see* migration)
  movement, 139, 146 (*see also* migration)
  types, 105, 147–148, 343, 350, 390–391, 481, 533, 578–579, 586, 598, 624
cell-cell contact, 244
cell-free layers, 385
center of vital functions, 183
central gray area (CGA, also called periaqueductal gray), 176, 185, 208, 210, 212, 214–215, 220, 222, 224, 267, 281, 313, 318, 331, 419–420, 469–470, 478, 495, 498, 537, 568
central model, 605
central nervous system (CNS), definition, 3–5, 8, 58
central nucleus of the amygdala, 506, 540–542, 545, 551, 562–563
central pattern controllers, 269, 468, 474
central pattern generators, 269–270, 278–279, 282, 289, 291, 302, 306–307
central pattern initiators, 269–270, 475
central programs, 65, 302
central sulcus, 295, 629
cerebellar channel (of sensory information), 102–103, 105, 158, 162–163, 198, 428–429

cerebellar cortex, 105, 201, 203, 271, 273–274, 628
cerebellar outputs, 105, 276, 329, 470
cerebellar white matter, 32
cerebellum, 34, 70–73, 75, 81, 102–106, 158, 162–165, 184, 192, 201, 214, 270–271, 273–276, 285, 329, 361, 398, 451–452, 454, 462, 467,629, 663
  cognitive functions, 106
  and electroreception, 108–109
  enlargement, 106, 203, 236, 381, 646
cerebral aqueduct. *See* aqueduct of Sylvius
cerebral hemisphere
  definition, 8–9, 77–78
cerebral peduncle, 31, 33, 50, 210, 213–214, 223–225, 372, 379, 384, 398, 404, 477, 495, 581, 607, 627–628, 630
cerebral ventricles, 8, 150, 175–177, 193, 198, 210, 229, 350, 353, 388, 420, 486–487, 505, 510, 522, 551, 553, 653
cerebrospinal fluid (CSF), 175–176, 311
cervical flexure, 181–182
CGA. *See* central gray area
chaining of reflexes or stimulus-elicited action patterns, 300–302
changes in adhesion, 139–140
channels of conduction. *See* conduction, channels of
chemical guidance, 243–246, 249–250
chemoaffinity, 246, 260
chemoarchitecture, 32, 34–36, 578
chemoreceptor, 19, 22, 325–327, 332, 605
chemorepulsion, 244–247
chemosensory, 68, 325–326, 452
chemotropism, 244. *See also* chemical guidance
chick, chicken, 89, 91, 146–149, 219, 250, 258, 432–433, 445
cholinergic axons, systems, 312, 469, 579, 586, 668
chordate animals, evolution, 55, 60–68, 70, 74, 76–78, 82, 92–93, 103–104, 124, 126, 139, 156, 181, 183, 195–197, 205–207, 217, 235, 267–269, 271, 320, 323, 331, 333, 336–337, 355–356, 359, 452–453, 456, 461, 467, 517, 561–562, 568, 574, 605, 645
chorea, choreiform movements, 571, 573
choroid plexus, 175–176, 486
ciliary ganglion, 168–169
cingulate cortex, 296, 315, 415, 459, 486, 490–492, 494, 496, 515, 520, 522–523, 525, 538, 577, 601–602, 629, 633–634, 665
cingulum bundle, 527, 633
circadian rhythm, activity rhythm, 27, 124, 304–305, 307, 356, 369. *See also* biological clock
circuit of Papez, 486–487, 489, 494, 510, 515, 523, 528, 537
circulating hormones, 467, 471–472, 485, 501
cirumvallate papillae, 327
Clarke, Jacob A. L., 163
Clarke's column, 163
claustrum, 219, 232, 546, 580, 595
Cnidaria, 12, 26, 56–57, 326
coatimundi, 112, 114

cochlear duct, 426–427
cochlear nucleus (nuclei), 96, 194, 272, 380, 418, 426, 428–433, 435–436
  dorsal, 428–429, 431–432
  ventral, 426, 428–429, 431–432, 435
cognitive disorders, 502, 551, 575, 583, 586
cognitive function(s), 96, 194, 272, 380, 418, 426, 428–433, 435–436. *See also* higher cognitive functions
  abilities, processes, activities, 40, 49, 64, 81, 106, 184, 331, 412–414, 497, 533, 537, 545, 551, 575, 583, 609–610, 628, 634, 638, 642
collateral connections. *See* axon collaterals
collateral sprouting. *See* axonal sprouting
colliculus
  inferior, 11, 253, 266, 269, 281, 315, 377, 378–381, 418–419, 428–432, 442, 444, 523–524, 607, 630
  superior, anterior, 11, 33–34, 44, 47, 49, 75, 313, 160, 208, 210–213, 215–216, 244, 251, 253, 266, 282, 285, 315, 359–363, 365, 367–369, 372–375, 377–381, 384–385, 387–391, 394, 398, 409, 416, 431–432, 434–435, 462, 524–524, 562–563, 571–574, 607, 609, 629–630, 663
columns
  in central gray area, 281, 498
  motor neuron groups in hindbrain, 188, 191–194, 198, 272, 330
  in neocortex, 248–249, 443, 597–598, 600, 613, 617–619, 622–626, 648, 657
  secondary sensory neuron groups in hindbrain, 188, 191, 194, 198, 272, 325–327, 330
competition among axons for termination sites (areas, space), 79, 85, 256, 351, 396, 655–656
competitive growth vigor, 254–255
compression of retinotectal map, 251. *See also* plasticity of brain map
computational method, 49, 51–52
computational models, 305–306, 475
concerted evolution, 212, 339–340, 648
conduction
  axonal, non-decremental, 13, 16–17, 19, 42, 57, 96, 156–157, 30, 433–434, 642
  channels of, 12, 94, 97–98, 101, 160, 198, 417, 429 25, 103, 105, 153, 168, 160–162, 197–198, 200, 210, 374, 417, 428–429
  delays, 105, 300, 433
  graded (in dendrites, cell bodies), 16, 57
  neuroid or myoid, 56–57
  primitive cellular mechanism, 12
connectome, connectomics, 49
conservation of terminal quantity, 256–257, 351
consummatory actions, behavior, patterns, 265, 275, 450, 474–476, 562
contact attraction, 245
contact inhibition of extension, retraction reaction, 245, 261
contact repulsion, 245

contractile cells, 12, 14, 19, 20, 56–58, 61, 70, 98, 139–141, 167, 237
contrast enhancement, 626
convergence
  of axons on other neurons, 350, 363, 428–429, 433–434, 564, 608, 626, 627, 659
  multisensory, 79, 363, 428, 608, 626–627, 659, 610
cooking abilities in human evolution, 610
cornu Ammonis (CA), the hippocampus, 520, 523, 526–527
coronal sections (frontal sections, transverse), 5, 7–8, 30–31, 33, 95, 333
corona radiata, 50, 404, 627–628
corpus callosum, 492, 504, 522, 525–527, 529, 547–548, 600, 630, 623, 635
corpus striatum, 35, 72, 74, 78–81, 93–94, 103, 123–124, 126–129, 133, 160, 164, 208, 211, 217, 218–219, 222–224,226, 228, 229, 232, 268–271, 275, 296, 313, 315, 317, 328, 330–331, 339–340, 351, 352, 364–366, 367, 379,383, 394, 398, 404, 407, 409–410, 418, 420–421, 430, 444, 456–457, 459–463, 467, 470, 475, 490, 509, 517–518, 520, 537–588, 590, 596–597, 600, 603, 606, 614, 618, 642, 650, 652, 663
cortex. *See also* pallium
  definition and selected photographs, 31–32, 34, 36, 78–79, 82, 290, 334, 399, 401–404, 437, 460, 468, 522–523, 599–600, 611–612, 618, 620–621, 625–626, 630, 632, 638
  development, 235–259, 645–660
  neocortex, 595–668
cortical nucleus of the amygdala, 340, 540, 542, 551
cortical plate, 647–649, 651, 653
cortical vision, 368, 385, 393–415
corticoid, 496, 538
corticomedial nuclei of the amygdala, 540–541
corticopontine pathways and connections, 201, 210–213, 273, 628
corticortropin-releasing hormone, 318, 484
corticospinal axons, 121–123, 131–133, 134–135, 158, 164–165, 193, 203, 210–213, 224–225, 283–286, 289–291, 293, 295, 567, 572, 628
corticospinal diaschisis, 122–123, 135, 289
corticotectal axons, 628, 633
corticothalamic axons, 618, 628
corticotrophic hormone, adrenal, 472
courtship behavior, movements, 267, 281, 301, 306, 474
Cragg, Brian, 626
cranial nerves and columns of neurons in hindbrain, 188–192
cranio-sacral system (parasympathetic), 169–170
cribriform plate, 341, 433, 349
critical period, 255, 655–656
crossed descending pathway, 164, 207–208, 211, 274, 621–363. *See also* decussation
crossed projections, evolution of, 75–77, 194, 197
cyclic behavior, 80, 206, 268, 473
cynodonts, 90, 92–93, 97, 106

cytoarchitecture, 30–31, 219, 266, 335, 341, 541
  amygdala, 539–540
  cortex, 32, 399, 450, 464, 598, 601, 619, 621, 623, 625, 640, 659
  hypothalamus, 483–484
  spinal cord, 156–158

Dahlström, A, 313, 315
Darlington, R. B., 648, 650, 667
Darwin, Charles, 63, 89–91
Das, Gopal, 350, 353, 590–591
daughter cells, 145, 510, 646
Deacon, Terrence, 297
Deacon's rule, 293, 297
deafferentation, deafferentation depression, 120, 122, 571, 573
decelerator nerve, 20–21, 171
decerebration, 117–120, 135, 450, 474, 476
decussation, 75, 102–103, 105, 133–134, 162, 164, 194–197, 200, 214, 224, 242, 264, 287, 314, 330, 359, 628
deep brain stimulation, 585–586
deep nuclei of the cerebellum, 203, 296
defensive behaviors, 102, 119, 212, 267, 281, 467, 477, 498, 547
Deiter's nucleus (lateral vestibular nucleus), 270
déjà vu experience, 546
dementia, 312, 583
dendrites, definition and functions, 13–16, 21, 37, 186
dendro-axonal synapses, 22–24
dendro-dendritic synapses, 22, 24
denervation supersensitivity, 212, 573, 583
dentate gyrus of the hippocampus, 315, 317, 251, 490–491, 522–527, 530–532, 589–591
depolarization, 13–16, 19, 21–22, 27, 304, 423
deprivation, visual, 655, 657, 659
dermatome, 92, 99, 101
Descartes, René, 39, 300
dexterity, 81, 112, 134, 275, 290–291 293, 295–296, 298, 601
diabetes insipidus, 472
diagonal band of Broca, 521, 551–552, 578, 587
diaschisis, 119–123, 135, 287, 289, 450, 571, 573
diencephalon
  basic components, 124–125, 175
  cell groups summary, 629–634
  defined, 8–9, 31
  locomotor regions, 208, 211, 266, 269–271
  neuromeres, 227, 372–373, 391, 392
  state changing systems, 316–319
differential adhesion, 224
differentiation of developing neurons, 142, 149, 150, 232, 235–260, 501–510; 589, 645, 647
diffuse connections (projections), 306, 311, 313, 315, 320, 328, 330, 364–365, 395, 434, 494, 547, 469, 643
diffusion tensor imaging, 48, 50, 635, 637
dilation of pupils, 168, 170

disconnection syndromes, 252, 267, 281, 289, 473–474, 575, 480, 497, 547–548
distal muscle innervation, control, 211, 275, 278, 280, 283–287, 289
dog, 19, 92, 119, 291, 338, 400, 640
dolphin and marine mammals, 67, 84, 110, 511–512
dopamine (DA), 313, 315, 318, 320, 330–331, 462, 509, 534, 554–555, 562, 564–565, 568, 572–574, 576–579, 583–587, 596, 643
dorsal cochlear nucleus, 428–429, 431–432
dorsal column, dorsal column nuclei, 60, 131–133, 154, 158, 160–162, 165, 185, 193, 323, 628
dorsal cortex (pallium), 78–79, 93–94, 117, 127–129, 218, 222, 228–229, 232, 334–335, 337–339, 393, 366, 393, 397–398, 409, 450, 456, 459, 462, 525, 564, 595, 597, 608–609, 611–612, 614
dorsal horn (of spinal cord), 38, 96, 103, 142, 155, 157, 160, 162, 183, 246–247, 316, 485, 497
dorsal longitudinal fasciculus, 214
dorsal nerve cord, 60–61
dorsal pallidum, 568, 571
dorsal psalterium (harp of David), 529
dorsal-root ganglion and cells (DRG cells), 13–15, 40, 56, 100, 142–144, 149, 171–172, 237–240, 246, 255
dorsal roots, inputs, 38, 60, 96–98, 158–163, 172, 270, 323
dorsal spinocerebellar tract, 163
dorsal striatum, 79, 222, 224, 226, 365, 418, 430, 457, 539, 549, 561–588, 606
dorsal terminal nucleus of the accessory optic tract, 378
dorsal thalamus, defined, 206, 220, 227–228, 232, 328, 372–373
dorsal ventricular ridge (DVR), 78–79, 82, 219, 229–230, 232, 395, 397–398, 443, 664, 595, 612
dorsolateral placodes, 417
Doty, R. W., 302
double inhibition, 571–572, 574
Downer J.D.C., 548
DRG cell. *See* dorsal-root ganglion and cells
drive and reward axons, 479
drives, 473–476, 478–479
drug addiction, 551
dura mater, 177
dynamic zone (in neocortex), 257, 261, 664
dynorphin, 318, 484

earthworm, 14–15
eating or feeding reflexes, 28, 118–119, 476
Ebbesson, Sven O. E., 85
echidna, 115
echolocation, 110, 116, 417, 442
Economo, Constantin von, 624–625, 639
ectoderm, 11, 140–142, 144, 146, 150, 235, 417, 645
ectostriatum, 229
ectothermic animals, 472

Edinger, Ludwig, 168–169, 194, 272, 450
efferent axons, 241, 579, 599, 614, 627
  definition, 628
efficient local circuits, 207, 394
  connections, 597–598
egocentric direction (in object localization), 411–412
electric eel, 107, 452
electric fishes, 105, 452, 454. *See also* mormyrid electric fish
electroreception, electrosensory inputs and receptors, 69–70, 107–108, 115–116, 201, 214, 418
electrosensory lateral line lobes. *See* lateral line lobes
Elliott, H. Chandler, 62, 67, 84
Ellis-Behnke, Rutledge, 259, 261
elongation mode of axonal growth, 247, 255
elongation protein, 258
embryonic human brain, 5–9, 143, 188, 203, 231, 371–372, 381, 401–402, 494, 564, 631, 633, 646, 648–649
embryonic stages, 89, 91, 94, 139, 147–148, 191–192, 197, 203, 235, 240, 249, 257–258, 381, 384, 400, 494, 585, 608, 646–647
emotional habits, 546, 556, 663
emotional-motivational disorders, 502, 551, 553, 556
encephalin, 578–579
encephalization of function, 119
encoding of odor qualities, 347, 350
endbrain
  definition, 8–9, 62
  structures, 126, 130, 222, 229, 333, 350, 393–416, 451, 460, 470, 513–668
endbulb of Held, 429–430, 433
end buttons (boutons), 14, 22
endocrine system, 20, 64, 69, 78, 124, 167–168, 206, 278–279, 356, 449, 452, 457, 470–471, 486, 511, 517, 551
endocytosis, 23, 25
endoderm, endodermal tissue, 11, 144, 150
endogenous activity, 12, 25–27, 40, 59, 124, 173, 206, 268, 299, 302, 304–305, 356, 654
endogenous clock. *See* biological clock
endothermic animals, 472
energy constraints (on brain expansion), 610
engram, 257, 590, 656, 663–664
enteric nervous system, 168, 171–173, 279
entopallium, 395, 397
entorhinal cortex, 316, 335, 340–341, 348, 352, 458, 490–491, 522–525, 527, 529–530
entrainment of daily rhythm, 27, 124, 206, 304–305, 356, 369, 467, 473
ependyma, ependymal cells, 148, 150, 175–177
ephrin ligands and receptors, 246, 249–250
epigenetic factors in brain development, 652
epilepsy, 552–554
epinephrine (adrenaline), 170–171, 471
epiphysial nerve, 449. *See also* pineal eye
episodic behavior, 473

episodic memory, 483, 528, 532
epithalamus, epithalamic afferents, 74, 81, 124–125, 130, 168, 206, 211, 215, 220, 222, 224, 227, 275, 352, 356, 360–361, 371–373, 460, 499, 517–518, 531, 605, 618
epithelium, epithelial cells, 14–15, 19, 56–58, 60, 96, 145–148, 190, 192, 217, 341, 343, 345–350, 456
EPSP (excitatory postsynaptic potential), 21, 23, 303, 430
escape behavior (movements), 64, 68, 70, 75, 77–78, 119, 157, 196–197, 207–208, 211, 216, 242, 265, 268, 272–274, 31, 355, 358–361, 363, 369, 393, 417–419, 431, 445, 473, 547, 596
estrogen, 501, 509
excitatory postsynaptic potential. *See* EPSP
exocytosis, 23
expansions of brain in evolution, 68–72, 74–78, 80–82, 105–106, 107–116, 128–129, 134, 201–213, 217, 268, 275, 325, 328, 335, 351, 367–369, 375, 393, 400, 403, 407, 410, 421, 442, 450, 452–459, 461–462, 517–518, 537, 561–562, 564, 587, 609–610, 623, 627, 633, 642, 646–650, 665
expectancies, expectancy signals, 296, 602, 604–605. *See also* novelty responses
explant culture, 146, 240, 664
external medullary lamina, 633
extracellular matrix (ECM), 245
extrapyramidal motor system, 564, 567
extrastriate cortex areas, 360, 394, 397
extreme capsule, 522, 636
exuberant projections, 247–249
eyeblink reflex, response, 118–119, 184, 193, 197, 199
eye spot of Amphioxus, 61, 356

face area in neocortex, 200, 410
facial lobe, 71, 73
facial motor nucleus, 187, 192–193, 197, 199–200, 419
facial muscles, 184, 187–188, 300, 583
facial nerve (cranial nerve 7), 71, 73, 187, 190, 192–194, 199, 272, 320–321, 327
Falck, B., 313
FAP. *See* fixed action patterns
fasciculus, fasciculi (definition, formation), 60, 245, 247
fastigial nucleus, 271, 285–286
fear conditioning, learning, 420–421, 430, 547–549
feedback circuitry, control circuits, 23, 74, 128, 207, 299, 302–305, 460, 484–485, 562, 568, 596
Feirtag, Michael, 55, 106, 178
Fentriss, John, 302
ferret, 252, 373, 428, 609, 654, 656, 658
fiber architecture, 30–31, 266, 406, 659
fiber stain, 33–34, 157, 203, 212, 385, 390, 400, 580–581, 598–599, 617, 620
field L, 443–444, 506, 508–509
fight or flight, 171, 547

filopodium, filopodia, 140–141, 236–237, 239–240, 249, 257
fimbria of the fornix, 527
Finger, Thomas, 326
Finlay, Barbara, 392, 648, 650, 667
fish, 6, 14–17, 27,56–57, 70–73, 76, 78, 83, 89, 91, 103–108, 156, 166, 183, 186, 201, 212, 227–228, 233, 249, 251, 258, 275, 301–302, 306, 313, 320, 325–327, 331, 336–337, 339, 352, 363, 367, 385–389, 418, 450, 452–453, 456–457, 462, 475, 477, 518–519
fixation of brain tissue, 30, 41, 44, 379, 503, 580, 634
fixed action patterns (FAPs), 80, 119, 134, 164, 183–184, 188, 205, 208, 267, 269, 281, 290, 300–302, 305–306, 419, 473, 478
fixed motor patterns, 136, 164–165, 301, 306, 474
Flanagan, John, 249–250
flexor motor neurons, flexion response of limb, 98, 119, 159–160, 222
flexures of developing CNS, 181–183, 218, 221, 225, 372, 381
flocculus, 271, 273–274
floor plate, 142, 144, 147, 149, 153, 160, 183, 246–247
fluorescent marker, molecules, tracer, 47–49, 113, 237–238, 242, 257, 313, 406, 587
fluorodopa, 585
Flynn, John, 125, 281, 477–478
foraging behavior, 26, 55, 64, 205–206, 211, 265–266, 268, 300, 419, 473
foramen of Magendie, 176
foramina of Luschke, 176
forebrain
  bundles, medial and lateral, 222–225, 398, 476, 479, 483, 498–499, 528, 568, 570, 611, 627
  definition and components, 8–9, 61, 68–69, 93, 95, 130
  evolution (ideas), 90, 194, 219, 227–232, 268, 337, 449–464, 605–614, 642–643
  expansions, 68, 72–74, 77–84, 111–115, 129, 201–213, 217, 268, 275, 328, 335, 351, 367–369, 375, 393–400, 403, 407, 410, 421, 450, 452–44, 456–463, 517–520, 537, 561–562, 564, 597, 609–610, 623, 627, 633, 642, 646–650, 665
  removal effects, 117–123, 135, 450, 474
  vesicle, 217, 333, 467
fornix, fornix bundle, fornix column (fibers of postcommisural fornix), 477, 488, 491–493, 495, 517, 523–525, 527–529, 541–542, 576, 611, 631–632
Fourier analysis, 27, 304
fovea (and sensory acuity), 601
frame of reference, 493, 515, 517
frequency coding of sound stimuli, 425–428, 436, 440
frequency modulation of sound, cortical responses, 438–440
Fritsch, G., 291
frog, 20, 39–40, 104, 121, 249, 275, 291, 296, 361, 364, 376, 385, 390–391, 438, 520–521, 611
frontal eye fields, 296, 410–413, 575, 605

frontal section (transverse, coronal) defined and illustrated, 5, 7–8, 30–31, 33, 95, 333
fronto-occipital fasciculus, 636
fruit fly (*Drosophila* species), 187, 659
Fuxe, K., 313, 315

GABA (gamma-amino butyric acid), 230, 257, 318, 563, 572, 574, 578–579, 607, 618, 664
Galaburda, Albert, 639
galvanotropism, 244
ganglion, ganglia, ganglion cell (of peripheral nervous system, definition), 13, 58–60
ganglionic eminence, 218–219, 230–231, 597, 607, 614, 651
ganglionic motor neuron (with postganglionic axon), 97, 167,169, 171, 278–279, 313
gap junctions, 21, 56
GAP-43, 257, 261, 642, 663
gastrulation, 139–141, 235
gating, 24, 124, 222, 494, 642–643
general purpose movements, 265–266, 269
geniculate body or nucleus
  lateral, 41, 45–46, 49, 253, 273, 359–361, 363, 366–368, 371–375, 377, 379, 381, 384, 387–389, 391, 394–398, 413, 495, 609–610, 622, 630–631, 633, 655, 657
  medial, 247, 252–253, 260, 369, 379, 418, 422, 428–432, 442–443, 545, 547, 607, 629–631, 633–634, 656, 658
geniculostriate axons, 397, 399–400, 610
Gennari, 619. *See also* line of Gennari
genu of the corpus callosum, 630
girdle muscles, innervation, 280, 283
glial cells, glia, 13, 17, 20, 31, 33, 146, 150–151, 156, 173, 175–177, 244–245, 259, 350, 353, 485, 586, 651, 653
globus pallidus (GP), 269, 319, 365, 566, 568, 570, 572–574, 578–581, 586, 629
glomerulus, 343–350
glossopharyngeal nerve (cranial nerve 9), 190, 192, 194, 272, 327
glutamate, glutamate receptors, 282, 574, 618, 656
goldfish, 249, 251
Golgi, Camillo, 35, 37
Golgi method, stain, 35, 37–41, 45, 146–149, 185, 187, 229, 235–239, 249, 300, 341, 344–347, 385, 389–390, 564, 598–599, 611, 619
gonadal hormones, testosterone and estrogen, 501, 503, 508–520
granular cells, granule cells (in adult brain), 256–257, 345–346, 348, 350, 527, 530–531, 589–590, 608, 623–625
grasping movements, 81, 208, 210–211, 265, 257, 275, 293, 367, 410, 459, 462
grasshopper leg, axon guidance, 240–242, 260
gray matter
  definition of, 41
  illustrations, 379; 382–383, 441, 546, 553

Gregory, R. L., 9, 28–29
growth-associated protein. *See* GAP-43
growth cones, 149, 151, 235–240, 244–245, 247, 249, 254, 259
growth factors, 239–240, 245–246, 254–255, 257, 509
growth vigor, 246, 254–256, 258–259
Gudden, Bernhard von, 666, 668
guidepost cells, 241–242
guinea pig, 426
gustatory system and taste inputs, 19, 68–73, 126, 192, 201, 207, 222, 224, 268, 325–331, 339, 341, 462, 545, 596, 602, 629
gyrus, gyri (definition), 400, 459

habenula, habenular nuclei, 211, 215, 224, 372, 495, 499, 543, 631
habit formation, learning, 80, 124, 128, 224, 296, 313, 328, 331, 352, 366, 407–410, 420, 450, 457, 459–460, 475, 490, 517, 528, 546, 556, 562, 573, 581, 601, 663–664
habits of thought and prejudices, 581
habituation, 438, 470, 596, 663
Haeckel, Ernst, 89, 91
hagfish, 17, 103–104, 106, 156, 275, 320, 326, 337, 339, 452–453, 456–457
hamartoma, 554
Hamburger, Victor, 239–240
hamster, 3, 7, 11, 31, 33, 43–48, 134, 210, 212, 213, 216, 237–238, 242, 247, 249, 251–253, 260, 290, 295–296, 300, 327–328, 335–336, 338, 357, 363, 366, 368, 371, 373–375, 377–384, 387–388, 390, 392, 394, 396–397, 400, 410, 416, 419–420, 432, 435, 457, 468, 483, 485, 490, 520, 524–525, 535, 544, 564, 606–607, 609, 613, 616, 627, 632–633, 636, 638, 646–647
harp of David, 529
Harrison, Ross Granille, 235, 259
head direction, head-direction cells (HD cells), 491–493, 515–515, 525, 533, 603
heart innervation, control, 20–21, 64, 102, 165, 170–171, 173, 184, 305, 497–498
Hebb, Donald O., 257, 656
hedgehogs, 84, 294, 340, 407–409, 422, 459–460, 545, 610, 612–613, 615, 627, 640
Heimer, Lennart, 43, 316, 342, 344, 353, 549, 557
hemiballism, 571
hemispheric dominance, 444, 506
Herrick, Charles Judson, 78, 611
Herrick, Clarence Luther, 70–71, 611
heterotypic cortex (ideotypic cortex), 624, 638–639, 642
Hibbard, Emerson, 242–243
higher cognitive functions (higher function, cognitive function), 81, 106, 184, 331, 497, 533, 537, 545, 575, 583, 609–610, 628, 634, 638, 640, 642
higher vocal center (HVC in birds), 443, 506–509
Hillarp, N. A., 313

hindbrain
 definition and illustrations, 8–9, 68, 70–73, 93
 reticular formation, 184–185, 188, 211, 266, 283–286, 311–312, 331, 361
 segmentation, rhombomeres, 187, 189, 227–228, 372–373
hippocampal association fibers, 527, 530–532
hippocampal formation, 46, 126, 268, 315, 317, 334–335, 366, 412, 414, 416, 442, 457, 470, 486–487, 490–492, 513–535, 542, 545, 576–577, 590, 602, 604, 614, 629. *See also* medial pallium
hippocampal gyrus, 415, 522
hippocampal place cells, 513–517, 533, 596
hippocampal rudiment, 334, 457, 522, 525
hippocampocentric regions, areas, 601–603
hippocampus, 31, 79, 94, 126, 207, 257, 315–316, 329, 334–336, 340, 351–352, 402–403, 414, 456–458, 488, 490–493, 496, 499, 513–535, 538, 541, 543, 553, 556, 564, 567, 576, 589–590, 601–604, 611–612, 623, 650, 663, 665. *See also* hippocampal formation
hippocampus circuitry (intrahippocampal), 490, 527, 530, 534, 663
histamine, 318, 494
histological staining methods, 11, 30–39, 41–49, 51, 60, 78, 112–113, 154–155, 157–158, 188, 190, 203, 227, 376–377, 390, 399–400, 420, 581, 584, 599, 617, 620
Hitzig, E., 291
hodology (definition), 126
Hodos, William, 6–7, 204, 281–282, 370
homeobox genes (hox genes), 187–188, 190, 218–219, 615
homeostatic control circuits, functions, 257, 304–305, 472–473, 480
homeostatic regulation of terminal quantity, 254–257
homologous structures, 62, 94, 131, 229–230, 368, 395, 407–408, 431
homotypic cortex, 496, 538, 624
horizontal section, 5, 7, 31, 34, 36, 95, 176, 280, 399–400, 520, 523–524
hormonal influences on brain, 474, 476, 501–512
horseradish peroxidase (HRP), 44–45, 47–48, 248–249, 487
Horton, Jonathan, 246
hox genes (homeobox genes), 187–188, 190, 218–219, 615
Hubel, David, 305, 307, 392, 399
human cortex, map, 252, 290, 334, 403, 576, 619, 621, 623, 630, 637–641
hummingbirds, 362, 370
hunger motivation, 118, 364, 476–477, 479, 497, 502
Huntington's chorea. *See* chorea
hydra, 12, 26, 28, 56–57, 59, 65, 326
hyperpallium (birds), 131, 219, 229, 397, 595, 597
hyperstriatum, 229
hypocretin (orexin), 318–319, 484, 494
hypoglossal nerve (cranial nerve 12), 194, 272, 327, 508

hypophysial artery, 471
hypothalamus
   cell groups, 268, 318, 356, 369, 471, 484–487, 503–504, 541
   functions, 467–558
   gating functions, 124, 222, 494, 642–643
   locomotor area, 208, 211, 266, 270, 331, 352
   outputs, 315, 319, 358, 457, 461, 471, 477, 490–491, 495–496, 517–518, 523, 538, 561, 569, 561. *See also* medial forebrain bundle
   stimulation, 125, 266, 269, 281, 471, 473–474, 476–479, 481

ibex, 110
ideotypic cortex. *See* heterotypic cortex
idiodendritic, 185–186
Iguana, 385, 390
immunohistochemistry, 31–33, 36, 45–46, 548
impedance matching in auditory system, 423–424
implicit learning, 124, 128, 490, 528, 663
incus, 423–424, 427
indirect vocal control pathway (birds). *See* vocal control systems, birds
inferior colliculus (IC). *See* colliculus, inferior
inferior longitudinal fasciculus, 635–636
inferior olive, 187, 201, 276
inferotemporal cortex, 410, 412–413, 442, 538–539, 603
infrared sensors, 19, 107, 109
inhibition of axon growth, 245–247, 259–260, 657
inhibitory cell types, 105, 578–579, 618
inhibitory postsynaptic potential (IPSP), 21, 23
inhibitory short-axon neurons, 230–231, 438, 607, 664, 618
innate movement patterns. *See* fixed action patterns; fixed motor patterns
innate releasing mechanisms, 364, 601
insectivores, 340, 361, 375, 408, 418, 459–460, 633, 448, 650
inside-out pattern of migration, 651–652
instinctive behaviors. *See* innate movement patterns
insula, insular cortex 437, 496, 538, 545–546, 577, 580–581, 602
intensity of sound, coding, 423
internal capsule, 50, 132, 223–225, 381, 383, 398, 404, 430, 477–488, 522, 540, 542–543, 546, 570, 575, 580, 627–628, 632
internal medullary lamina (with the intralaminar nuclei), 631, 633
internal milieu, 69, 120, 208, 214, 467, 472–473, 496, 510, 538
internal model, 135–136, 665
interneurons
   definition, 59
   types in neocortex, 597, 599–600, 607–608, 614, 617–618, 626, 663–664
interpeduncular fossa, 344
intersegmental reflexes, 98, 158, 163–164

interventricular foramen, 176
intracolumnar connections (in neocortex), 597, 624–626
intrahippocampal connections. *See* hippocampus circuitry
intralaminar thalamic nuclei, cell groups, 215, 304, 312, 316, 320, 328, 365–366, 418, 421, 429–430, 459, 470, 477–478, 494, 563–566, 606, 614, 630, 633
intrathalamic axons, 222, 494
involuntary actions, 697
ion channels, 15–18
iris, reflex action and innervation, 168–170
irritability, a primitive cellular mechanism, 12–13, 16, 19
isocortex, 496, 538, 623
isodendritic morphology, 185, 186
isthmus, 228, 372–373

Jacobson's organ, 341, 335, 449, 540, 596. *See also* vomeronasal organ
jawless vertebrate, 83, 103, 156, 232, 275, 336–337, 339, 452–453, 456
Jessell, Tom, 246
juxtallocortex, 623, 629
juxtastriate cortical areas, 394

kangaroo rat, 419
Karten, Harvey, 35, 229–230, 397, 443–444, 518
Kuypers, Hans G. J. M., 278, 280, 282, 284, 286–288, 296

lamellipodia, 237, 239
laminar focalization of terminal arbors, 248
lamination, laminar patterns, 32, 155–156, 248, 345–347, 367–368, 371, 374, 376–377, 385–387, 389–390, 399, 598–599, 609, 611–612, 614, 617–618, 620, 625, 611–612, 618
lamprey (sea lamprey), 17, 103–104, 106, 156, 227–228, 232, 275, 281–282, 320, 336–339, 352, 452–453, 456–457, 464, 597
language system, abilities, 115, 184, 417, 438, 443–445, 506–507, 590, 609, 635, 637, 639
Lashley, Karl, 39–40, 42, 302
late equals large, 560, 667
lateral apertures. *See* foramina of Luschke
lateral column of spinal cord, 60, 154, 156, 160, 163, 165, 170, 247, 628
lateral dorsal nucleus of thalamus (LD, laterodorsal nucleus), 366, 391, 394, 410, 414–415, 491, 493, 515, 517, 563, 602, 607, 629–630
lateral forebrain bundle, 222–225, 398, 498, 568, 570, 611, 627
lateral geniculate body (nucleus, nuclei) (LGB, LGN, LG). *See* geniculate body or nucleus
lateral horn of spinal cord, 154, 156, 163, 166, 171, 316, 358
lateral hypothalamic area, cell groups, 125, 224, 267, 318–319, 476, 478, 481, 483, 494, 562, 568

lateral lemniscus, 210, 266, 380, 418, 429–431
lateral line lobes, 108
lateral line nerves, 69, 107–108, 190, 418
lateral line receptors, lateral line system, 69, 107–108, 418
lateral olfactory tract, 335, 343–346, 349, 351, 379, 539–540
lateral pallium, 219, 228, 232, 234, 338–339, 518–519, 521, 612
lateral posterior nucleus (LP), 253, 359–360, 366, 369, 371–372, 374–375, 377, 381, 394–395, 397, 416, 435, 495, 607–609, 613–614, 629–631
lateral prefrontal cortex, 296, 435, 442, 575, 633
lateral septal nucleus, 576, 578
lateral tegmental pathway, auditory system 428–429, 431
lateral terminal nucleus of the accessory optic tract, 378
lateral ventricles, 175–176, 350, 486, 522, 542, 553, 580, 589, 652–653
lateral ventricular angle, 218, 229–230
law of roots. See Bell-Magendie law
Lawrence, D. G., 278, 280, 284, 286–288
Le Gros Clark, W. E., 176, 372, 483–484, 620
learned actions, movements (motor learning), 134, 290, 603–604, 663, 666
learned affects, 413, 470, 475, 539, 546, 556–557
learned fears, 420–421, 430, 547–549
learning of locations. See place learning
lemniscal pathways, somatosensory, 101–103, 106, 128, 131–133, 158, 160–162, 193, 195, 198–200, 210–211, 214, 220, 223, 227, 323, 495, 570
lenticular nucleus (lentiform nucleus), 570, 580–581
Levi Montalcini, Rita, 239–240, 259
Levinthal, Cyrus, 51
Leyhausen, Paul, 124, 136
Lichtman, Jeff, 244
light detection, detectors, 125, 217, 355–358, 369, 420, 452, 456, 467
limbic endbrain structures, 483–499, 513–558
limbic midbrain areas, 210, 215, 223–224, 331, 469–470, 478, 480, 493, 498, 537, 542, 545, 568, 570, 639
limbic striatum, 537–557
limbic system arousal, 469–470, 480, 483
limbic system, defined, 81, 94, 125–127, 214, 220, 222–224
limb movement, control, 211, 265, 267, 270, 275, 278, 280, 283–284, 290–291, 296, 302, 573
line of Gennari, 619–620
Lissauer's zone, 162–163
local reflex, 40, 58, 97–98, 158–159
locomotion, control of, 19, 26, 59, 72, 74, 212, 266, 268–272, 274, 281–282, 289, 291, 302, 304, 306–307, 331, 352, 356, 360–361, 462, 473, 475, 467, 517, 581, 597, 571–572, 574, 579, 561, 568
locomotor areas, 207, 211–212, 266, 269, 271, 281, 331, 352, 360–361, 462, 561, 571, 574

locus coeruleus, 313–314, 320, 554–555
Loewi, Otto, 20–21, 171
long-term memory, 120, 590, 527–528, 531, 533
long-term potentiation, 527, 529, 590, 663
Lorenz, Konrad, 268, 281, 301–302, 306–307
lungfish, 104, 462, 518–519
Luria, Alexander, 639

Macht, M. B., 117–118, 124
Mackie, George, 56–57, 65
MacLean, Paul, 35, 468, 537
macrosmatic mammals, 325, 338
magnetic resonance imaging (MRI), 48, 289–290, 135, 438, 512, 553, 639
magnocellular preoptic nucleus, 551, 578
Magoun, Horace Winchell, 469
male-female brain differences. See sex differences in brains
malleus, 423–424, 427
mammillary bodies, nuclei, 316, 318, 344, 471, 484–489, 491–492, 494–495, 544, 629, 631
mammillothalamic tract, 372, 477, 484, 486, 488–489, 491–492, 494–495, 544, 629, 631
mandibular branch of trigeminal nerve, 101, 109
Mangold, Hilde, 150
map plasticity in midbrain. See plasticity in brain map
Marchi method, 42
matrix layer (ventricular layer of developing CNS), 145, 150, 217–218, 229, 325, 350, 353, 539, 589, 646
matrix of adult striatum, 578–579
Mauthner cells, 242–243, 419
maxillary branch of trigeminal nerve, 101, 109
mechanosensory lateral line nerves, 418
medial brainstem pathways, 165, 283, 286–288
medial forebrain bundle. See forebrain bundles, medial and lateral
medial ganglionic eminence, 218, 230–231, 597
medial geniculate body. See geniculate body or nucleus
medial geniculate nucleus or nuclei (MGN). See geniculate body or nucleus
medial lemniscus, 131–133, 158, 199, 210–211, 214, 223, 227, 317, 323, 495
medial longitudinal fasciculus (MLF), 214, 270, 271
medial pallium, 94, 126, 207, 228, 232, 268, 334–335, 339, 366, 414, 453, 456–457, 459, 487, 513, 517–523, 525, 528, 567, 570, 596, 611–612
medial terminal nucleus of the accessory optic tract, 360, 378
median aperture. See foramen of Magendie
median eminence, 471–472, 485–487
mediodorsal nucleus of thalamus (MD), 115, 185, 350–352, 372, 477, 488–489, 494–495, 543–544, 563, 572, 574, 577, 628, 630–634, 639
medulla oblongata, 11, 70–71, 182, 185, 329, 340, 380, 451
medulla spinalis, 93, 127, 129, 182

melanin concentrating hormone, 318, 484
melatonin, 206, 358
memory consolidation, 531–532, 535
meninges, meningeal layers, 175–177
mental disorders, 502, 551–556
mesencephalic flexure, 181–182, 372, 381
mesoderm, mesodermal tissues, 11, 140, 144, 150, 186
mesopallium (birds), 506, 508
Mesulam, Marek-Marcel, 468, 494, 538, 557, 601–602, 635, 638–640, 643
metazoan, 12, 67, 356
Meyer's loop, 404
Meynert, Theodor Hermann, 312
microsmatic mammals, 325, 338
midbrain locomotor area (MLA). *See* locomotor areas
midbrain reticular formation, 160, 165, 208, 215, 223, 266, 311, 318, 470, 478, 570
midbrain tectum, 33, 75–76, 78, 82, 103, 109, 119, 131, 194, 197, 207–208, 211–213, 216, 220, 232, 244, 247, 249, 250, 266–268, 273–274, 282, 285–286, 311, 315, 259–368, 371, 382, 384–385, 387, 395, 407, 409, 412, 434, 435, 450, 452, 457, 562, 565, 574, 596, 601, 606, 608–609. *See also* colliculus, superior
 map, 207, 249, 251, 258, 273, 363–364, 385, 434–435
midbrain vision, 368, 385
middle ear, 423–425
 bones, ossicles, 423–425
middle longitudinal fasciculus, 636
midline nuclei of thalamus, 185, 215, 320, 478, 606, 630, 633
migration, developing neurons, 20, 79, 141–151, 153, 201, 203, 218, 229–231, 235, 237, 244, 325, 350, 353, 389, 432, 508, 510, 539, 589, 597, 607, 614, 645–646, 648–653
mitosis, mitotic cells, 143, 145, 235, 432, 508, 510, 645–646, 648–649
mitral cells, 96, 339–341, 343–350
modes of axon growth, 247–249, 255, 260
modulatory projections, 24, 80, 102, 124–125, 165, 173, 184, 208, 217, 222, 268, 270, 291, 316, 331, 364, 394, 374, 478–479, 528, 532, 596, 604, 642
modules of the neocortex, 415, 468, 617, 619–626, 638–639, 642
molecular layer of cortex, 344, 346, 530
mollusk, 14–15, 57
Monakow, C. von, 122, 135
monkeys (rhesus macaque, owl, squirrel, spider), 35, 42, 84, 111–112, 255, 276, 278, 283–289, 291, 293–295, 329, 340, 368, 374, 377, 381, 385, 392, 395, 399, 403, 405, 407–411, 415–416, 435, 438, 440, 487, 490, 545, 547–548, 557, 580, 601–602, 622–624, 626, 634, 636, 648–649, 651, 653–654, 656, 660–662
monoamine-containing axon systems, 312–316, 320, 554. *See also specific monoamines*
monocular deprivation, 255, 655–657
mooneye (fresh water mooneye), 70–72, 75
mormyrid electric fish, 105–106, 108, 201, 452, 454

Morrell, Frank, 605
morula, 139–140, 235
Moruzzi, Giuseppe, 469
mosaic evolution, 108, 110, 199, 201, 212, 325, 339–340, 648
mossy fibers, 527
motivational states, control, systems, 55, 59, 64, 75–76, 78, 80–81, 124–125, 164, 170, 205–206, 208, 265, 267–268, 296, 318, 331, 351, 356, 410, 449, 467–468, 473–479, 493, 497–498, 517, 537, 542, 546, 551, 604–605, 628, 638–639
motivational functions of cortex, 638, 468. *See also* limbic endbrain structures; orbital prefrontal areas
motivational modulators. *See* modulatory projections
motivation-initiated behavior, 26, 59, 120, 268, 302–303, 605
motor acuity, 120, 134, 601, 609. *See also* dexterity
motor areas, 291–296, 466, 638
motor control systems, 123, 165, 265–307
motor cortex. *See* motor areas
motor endplates (muscle endplates), 166
motor functions of cortex, 265–296, 466, 638
motor images, 604
motor learning, 290, 663, 666
motor neuron
 cell groups, brainstem, 188, 191, 193–194, 198–199, 210, 272, 280
 defined, 13–14, 38, 59
 locations, 38–40, 156, 159, 167, 169, 183, 188, 191–194, 198–199, 272, 280, 284, 295
motor pattern generators, 269–270, 279, 282, 289, 291, 299–307
motor system hierarchy, 269, 281, 468, 474
motor Wulst (birds), 131
mouse (mice), 3, 7, 11, 27, 36, 84, 111–113, 190, 201, 210, 219, 230–231, 239, 257, 259, 302, 336, 344, 346, 350, 373–374, 377, 384, 388, 394, 400, 405–406, 408, 468, 477, 483, 485, 490, 509, 535, 564, 589, 614, 633, 636, 638, 643, 664
multimodal convergence. *See* convergence, multisensory
multimodal cortex in evolution, 79, 608, 610, 659
multimodal inputs in primary sensory areas of neocortex, 610, 659
multimodal thalamus, 608, 642, 565–566. *See also* intralaminar thalamic nuclei; midline nuclei of thalamus
multimodal (heteromodal) association areas, neocortex. *See* association areas of neocortex, multimodal (heteromodal)
multiple-digit receptive fields, somatosensory cortex, 660, 662
multistage lesions, 487–498
muscle cells, 13–14, 18–20, 56–59, 62, 93, 98, 166–167, 172, 305
mutations, 63, 256, 318, 648
myelin, 13, 17, 31, 33–34, 41–42, 141, 143, 154, 156–158, 203, 379, 381, 390, 399–400, 522, 581, 599, 619–620,

640–642, 659. *See also* histological staining methods
myelination of primary sensory cortex, 399–400, 641
myoid conduction, 56
myotome, 98

nasal placode, 190
nature and nurture, 645
Nauta, Walle J.H., 42–44, 55–57, 78, 92, 106, 120, 163, 173, 178, 182, 184, 186, 191, 194, 204, 215, 227, 229, 233, 253, 272, 278, 320, 378, 402, 404, 415, 469, 477, 480, 497–499, 529, 537, 545, 549, 581, 624, 635
Nauta methods. *See* silver stains
neencephalon, 450
negative affects, feelings, 410, 436, 469, 497, 556. *See also* fear conditioning; central gray area
neocortex, 595–667
  association areas (*see* association areas)
  cells, 618 (*see also* pyramidal cells; stellate cells; granular cells)
  dual origin hypothesis, 601–602 (*see also* Sanides, Friedrich)
  embryogenesis, 645–660
  expansions, 81–83, 111–115, 128–130, 134, 201, 213, 275, 328, 335, 367–369, 393, 400, 403, 407, 410, 421, 450, 459, 461–462, 564, 697, 609–610, 623, 627, 633, 642, 646–650, 665
  maps, 111, 252, 297, 364, 404–405, 407, 619, 621, 623, 639–640, 652, 654, 660, 662
  origins, 79, 81, 128–134, 217–220, 227–232, 459–463
  outputs, 106, 132, 164, 223, 225, 270 284–285, 288, 296, 407, 413, 459, 565–572, 576, 581, 600, 603, 618, 627–628
  removal effects, 117–123, 134–135
  size, 201, 212, 293–294, 296, 451–452, 490, 504, 609–610, 634, 639–640, 648–651, 667
  specializations (including bird homologs), 111–115, 256, 293, 367–369, 410, 415, 417, 442–445, 496, 557, 610, 619, 631, 635, 637, 642, 665
  structural types, 600, 623–626, 627, 638–642, 596, 538, 602
  white matter, 32, 41, 224–225, 398, 400, 539, 546, 598, 600, 611, 613, 616–628, 647, 652
neolemniscus pathways, 128, 131–133
neopallium. *See* neocortex
neostriatum, 222, 444. *See also* dorsal striatum
neothalamus, 269–270, 366, 547
nerve, definition, 13, 60
nerve growth factor (NGF), 239–240, 245–246, 254–255, 259–260
netrin molecules, 245–247
neural crest cells, 141–144, 150, 171, 190
neural groove, 142–144
neural plate, 140, 142, 144, 146, 235
neural tube, 8–9, 60–64, 67–69, 77, 79, 94–95, 103, 124, 139–147, 150, 153, 167, 175–176, 181–183, 198, 206, 208, 217–218, 220, 235, 268, 283, 325, 356,

364, 372–373, 3381, 456–457, 467, 476, 595, 645
neuroblasts, 149, 201, 203, 236, 241, 539, 589, 651
neuroendocrine cells, 167–168, 278–279, 486, 471. *See also* pituitary; posterior pituitary
neurogenesis, 151, 353, 508, 510, 589–591, 649, 651, 667
neurohormones. *See* neuroendocrine cells; oxytocin; vasopressin; antidiuretic hormone
neurohypophysis (posterior pituitary), 93, 118, 125, 127, 129, 471–472, 485, 498
neuroid conduction, 56
neuromeres, 227–228, 232, 372–373, 391–392
neuromodulators, 21, 299, 532, 617
neuronal activity effects in development, 255–256, 654–660
neuron-target dependence, 258
neuropil segments in hindbrain, 186–187
neurosecretory cells, 456. *See also* neurohyophysis; neurohormones
neurotransmitter, 14, 19, 21–22
neurotrophins, 240, 245–246, 258, 509
neurula, neurulation, 139–142, 150, 645
Neve, Rachel L., 642
NGF. *See* nerve growth factor
nidopallium (birds), 229, 397, 443–444, 506, 508–509, 597
nigral dopamine neurons, 313, 315, 573, 579, 584–585
nigral tissue transplants, 583–588
nigrotectal projection (nigra to superior colliculus), 563, 571–572, 574
nigrothalamic axons, 562–563, 571, 574
Nissl stain, method, 30–31, 36–37, 112–113, 154–155, 287, 377, 385, 388, 390, 395, 398–399, 420, 506–507, 522–523, 540–541, 598–599, 619–621, 646
NMDA (N-methyl-D-aspartic acid), 565
node of Ranvier, 19, 157
nodulus of cerebellum, 271, 273–274
Nogo, growth inhibitory protein, 245, 260
noise aversion, 419–420
nonassociative learning, 663. *See also* habituation
non-limbic arousal. *See* arousal
nonspecific widespread modulation, 184
non-taste chemoreceptor, 326
noradrenaline. *See* norepinephrine
norepinephrine (noradrenaline), 21, 167, 171, 255, 313–314, 318–320, 358, 532, 554, 583, 643
Northcutt, R. Glenn, 339, 386, 388, 450–451, 518–519, 521
nose brain, 126, 333, 487, 539
notochord, 60–62, 90, 140–142, 144, 235
Nottebohm, Fernando, 506–507, 510–511
novelty (sensory) responses, 196, 207–208, 211, 272, 312–313, 355, 363, 368, 419, 421, 438, 469–470, 556, 601, 605
nuclei of the lateral lemniscus, 210, 418, 429, 431
nuclear translocation (type of cell migration), 145–146, 148, 151, 651

nucleus accumbens, 495, 543, 551–552, 576, 578
nucleus ambiguus. 192–194, 200, 272, 280, 329, 331
nucleus basalis of Meynert (basal nucleus of Meynert), 312, 544, 549, 552, 668
nucleus bulbocavernosis, 505
nucleus centralis lateralis, 365, 566
nucleus cuneatus, 132, 162, 193
nucleus gracilis, 132, 162, 185, 193
nucleus laminaris (of bird auditory system), 433, 445
nucleus magnocellularis (of chicken auditory system), 433, 445
nucleus of Edinger-Westphal, 168–169, 194, 272
nucleus of the solitary tract. *See* solitary tract, nucleus of
nucleus of the bulbocavernosis (Onuf's nucleus in humans), 505
nucleus of the lateral olfactory tract, 539–540
nucleus of the optic tract (a superficial pretectal nucleus), 374, 378, 384, 389, 391
nucleus ovoidalis (birds), 443
nucleus raphé magnus, 315–316, 498
nucleus raphé obscurus, 316
nucleus robustus (in avian brain), 443, 506–509
nucleus ruber. *See* red nucleus
numb (protein), 646

obex, 176, 181, 326
object identification, representation, 64, 80, 126, 364, 412, 457, 595–597, 609
object identity (perception of), 126, 206, 355, 411, 527, 596
object location (perception of), 350, 410–411, 413, 665
oblique section, 5, 7
obsessive-compulsive behavior, 551
occipital visual areas, 41, 111, 229, 367, 393–416, 610, 629–632, 656, 659–660
ocular dominance columns, 260, 598, 622
oculomotor nerve (cranial nerve 3), 100, 192, 194, 272
oculomotor nuclei, 192, 194, 210, 214, 266, 270–272, 280
olfactocentric paralimbic cortical regions, 601–603. *See also* Sanides, Friedrich; Mesulam, Marek-Marcel
olfactory bulb, 71, 74, 76, 78, 93–94, 96, 100, 108, 126–127, 129–130, 181, 222, 228, 259, 271, 325, 333, 335–350, 352, 379, 382–383, 385, 403, 451, 453–454, 456–457, 460, 519, 525, 537, 540–542, 549, 561, 589–590, 596, 606, 626, 628, 648, 659
  projections, 333–354
  size, 338–340, 648
olfactory cortex (piriform cortex), 31, 93–94, 112, 127–129, 229–230, 325, 333, 340–341, 344, 346, 349, 352, 379, 456, 490, 494, 516, 522, 539, 541, 570, 596, 602, 609, 623, 628, 633–634. *See also* lateral pallium; piriform cortex
olfactory epithelium, receptor neurons, 96, 190, 192, 341, 342, 345–347, 349–350, 456

olfactory glomeruli. *See* glomerulus
olfactory-guided habits, 74, 206–207, 352, 460
olfactory inputs, 72, 117, 128, 205, 207, 222, 269, 331, 333, 337–338, 351, 456–457, 459, 461–462, 467, 490, 545, 547, 561–562, 564, 580, 606, 628
olfactory nerve (cranial nerve 1), 62, 96, 192, 351, 606
olfactory nerve fibers (olfactory filaments), 96, 341, 605–606
olfactory peduncle, 344
olfactory systems and pathways, 74, 78, 126–127, 217, 261, 269, 325, 333–353, 394, 402, 493, 537, 553, 556, 574, 596, 648
olfactory tract. *See* lateral olfactory tract
olfactory tubercle, 222, 335, 340–341, 344, 348, 456, 540, 544, 549, 551, 561, 578, 611, 632
ontogeny and phylogeny, 89–91
Onuf's nucleus, 505
ophthalmic branch of trigeminal nerve, 101, 109
opossums, 277, 333–334, 336, 442, 459–460, 522, 545, 603, 627, 659–660
optic chiasm, 206, 253, 335, 357, 361, 371, 373–375, 379, 471, 483, 485, 505, 540, 548, 657
optic lobe, 75, 77, 212, 367–368, 384. *See also* optic tectum; colliculus, superior
optic nerve (cranial nerve 2), 100, 108, 168, 192, 194, 249, 251, 272, 344, 358, 360, 381, 387, 449–450, 484, 522
optic radiations, 50, 397–400, 404, 628, 653
optic tectum, 41, 51, 76, 147, 165, 208, 212, 251, 274, 360, 364, 367–368, 372, 384–390, 393–395, 397, 451, 562. *See also* colliculus, superior
  lamination, 385–390
optic thalamus, 371
optic tract
  defined, 360, 372, 374, 378–379
  distortions in large primates, 374–375
  embryonic, 381–384
  terminations, 355–392
optical image formation, 207
optogenetics, 478
optokinetic nystagmus, 362
orbital prefrontal areas (orbitofrontal cortex), 344, 348, 350, 413, 494, 496, 516–517, 538–539, 602, 604, 628, 633
orexin (hypocretin), 318–319, 484, 494
organ of Corti, 426, 433
orientation columns, visual cortex, 598
orienting movements, responses, 55, 64, 68–69, 75, 78, 80, 109, 119, 134, 197, 205, 207–208, 211, 216, 249, 265, 267, 271–272, 274–275, 282, 287, 330–331, 355, 361–364, 367–368, 393, 407, 409, 413, 431–432, 435, 457, 459, 462, 469, 517, 561–561, 568, 572, 579, 581, 596–598, 601, 603–605, 609
oval window, 425, 427
owl, 434
owl monkey, 111, 405, 407, 409, 660, 662
oxytocin, 471–472

pacemaker loci, 27, 304, 306
pain pathways, 162, 174, 198, 212, 222
paleocortex, 385, 623, 650
paleolemniscus, 103, 162
paleothalamic cell groups, paleothalamus, 269, 319, 330, 394, 461–462, 545, 547, 577, 597, 609. *See also* intralaminar thalamic nuclei
pallidum (including ventral pallidum), 218, 228, 232, 269, 365, 394, 565–566, 568, 570–576, 578–581, 586–587, 596, 629
pallium, definition, 77–79, 81, 94
Papez, James, 487, 537
Papez's circuit. *See* circuit of Papez
parabrachial nucleus, 329–331, 345
parahippocampal cortical areas, 338, 391, 394, 410, 414–415, 458, 496, 517, 523, 525, 527, 535, 538, 601–602, 634
parahippocampal gyrus, 410, 414–415, 523
paralimbic cortical areas, regions, 459, 486, 490, 492, 494, 496, 523, 527, 538, 576, 597, 601–603, 623, 629, 636, 638, 665
parasagittal section, 5, 7, 185, 187, 266, 327–329, 341, 343
parasubiculum, 522, 524, 527
parasympathetic functions, 168, 170, 171, 192, 471, 498
parasympathetic nervous system, innervation, 166–171, 173, 188, 192–193, 279, 330, 471, 498
paraventricular nucleus of hypothalamus (PVH), 356, 358, 472, 484–485, 505
paravertebral ganglia, 10, 142, 144, 169, 171–172, 358
parcellation of brain structures, 78–79, 277, 293, 295–296, 459, 462, 603, 606–610, 640, 642
parental behavior, 70, 351, 474, 502
parietal association cortex, 111, 115, 298, 410–412, 414–415, 435, 442, 444 445, 517, 575, 580, 604, 623–625, 630, 634, 638–639
parietal damage, 604
parietal eye (pineal eye), 61, 124–125, 206, 365, 358
parietal lobe, 403, 412, 442
Parker, George, 56, 56, 343
Parkinson's disease, 313, 571, 573, 583–586
 therapy, 583–588
parolfactory regions of cortex, 459, 602, 665
pathologies of striatal structures, 561–587
pattern identification, auditory, 431, 435, 438
Pavlov, Ivan, 39, 300, 306
PCBs (polychlorinated biphenols), 501, 512
peduncle. *See* cerebral peduncle
peptide neuromodulators. *See* neuromodulators
perforant path, 525–527
periamygdalar cortex, 340–341, 602
peripheral ganglia. *See* autonomic, cardiac, ciliary, and dorsal root ganglia; ganglion, definition; paravertebral ganglia; ganglionic motor neuron; prevertebral ganglion; superior cervical ganglion
peripheral innervation, 19, 40, 58–59, 70, 73, 93, 97–98, 101, 155, 158, 167–172, 187, 192, 200, 252, 258, 280, 284, 601, 661. *See also* peripheral ganglia
peripheral nerve, 59–60, 62, 68, 141, 252, 259

peripheral nervous system (PNS), 19, 58–59, 60, 140–142, 153, 157, 240, 279
PET scan (positron emission tomography), 441, 585
pheromones, 341, 348, 353, 596. *See also* Jacobson's organ; accessory olfactory bulb
pia mater, pial cells, 147, 150, 177, 649, 485
pig, 586
pigeon, 35, 117, 119, 135, 229, 443–444, 450
pigmented cells, 144, 211, 313, 456, 467. *See also* eye spot of Amphioxus
piloerection, 118, 472. *See also* temperature regulation
pineal eye. *See* parietal eye
pineal gland, organ, 61, 125, 206, 356, 358, 449, 453, 484, 486
pioneering axon, 240, 242
piriform cortex, 333–335, 338, 340–341, 344, 346, 349, 352, 357, 379, 496, 538–540, 542, 561, 623, 628
pit viper. *See* vipers
pituitary, 61, 69, 94, 118, 124–125, 206, 217, 279, 356, 379, 449–451, 456, 467, 471–472, 474, 483, 484–485, 498, 605. *See also* adenohypophysis; neurohypophysis
place cells, 513–517, 533, 596
place learning, place memory, and recognition, 126, 128, 207, 268, 349, 352, 364, 367, 412, 414, 457–459, 473, 492–493, 513–518, 528, 533, 535, 546, 556, 595–597, 601–604
placodes, 190, 417–418
planning of movements, actions, 81, 545, 603, 605, 609, 665
plastic links, connections, 72–74, 77, 79, 124, 126, 136, 207, 352, 407, 421, 457, 460–462, 517–518, 539, 546, 556, 562, 569
plastic song, 507
plasticity in brain map, 207–208, 249–252, 258, 435, 660–663
platypus (duck-billed platypus), 108, 115
plexi of enteric nervous system, 173
Poliakov, G. I., 598–599, 639–640, 642
polychorinated biphenyls (PCBs). *See* PCBs
polysynaptic pathways of limbic system, 220, 485, 497, 569
pons, pontine gray, pontine cells, 106, 132–133, 187, 201, 203, 210, 266, 273, 224–225, 360, 398, 618, 628
pontine flexure, 181–183
pontine gray cells, 133, 201, 203, 224, 266, 273, 398, 404, 628
pontine taste area, 329
Poo, Mu-ming, 247
Porifera, 326
positron emission tomography (PET). *See* PET scan
posterior hypothalamus, 473, 484–485
posterior neocortical association areas, 402, 629, 633, 665, 394, 404, 604, 613, 639
posterior nuclear group (Po), 418, 429–431, 608, 613–614, 630, 633
posterior parietal cortex, 410–412, 414–415, 435, 442, 575, 634, 638

posterior pituitary. *See* neurohypophysis
posterior tuberculum, 313, 386, 462
post-orbital bar, 424
postsubiculum, 414, 523, 525
post-traumatic stress disorder (PTSD), 556
potassium channels, potassium ions, 14–15, 17–18
pre-cerebellar cell groups, 201. *See also* inferior olive; pontine gray cells
precommisural fornix, 524
precursors of ventral striatum, 217
prediction of inputs. *See* anticipation of inputs (stimuli)
prefrontal association cortex. *See* association areas of neocortex, prefrontal; lateral prefrontal cortex; orbital prefrontal areas; ventral prefrontal areas
preganglionic motor neurons
  parasympathetic system, 166–167, 169–171, 188, 279, 330
  sympathetic system, 154, 156, 166–167, 169–171, 279, 316, 358
prehensile tail, 112
prejudices, learning of, 581
premotor cortex, 407, 409–411, 604, 624, 629, 632, 637, 639, 665
preoptic area, 230, 357, 484–485, 503–503, 521–522, 551, 578, 607. See also magnocellular preoptic nucleus
pre-piriform cortex, 623. *See also* piriform cortex
pre-rubral field, 275
prestriate areas, 410–412, 415. *See also* juxtastriate cortical areas
presubiculum, 522–524, 527
presynaptic facilitation, 23–24
presynaptic inhibition, 23–24
pretectal cell groups, area, 168, 206, 268, 274–275, 360–367, 370–372, 374, 377, 380–382, 384, 386, 388–391, 394–398, 410, 414–415, 491, 495, 515, 525, 523, 535, 607, 610, 628–629
pretectal functions, 168, 268, 274, 361–362, 364, 367, 370, 389–380, 410, 414–415, 515, 525
prevertebral ganglion, 142, 144, 170–172
primary auditory area (A1), 409, 429, 432, 435–438, 442–443, 445, 630, 632, 658–659
primary brain vesicles, 61, 68–69, 81, 150, 178, 181, 205, 351
primary motor cortex. *See* motor cortex
primary sensory areas, myelination, 399–400, 641
primary sensory neurons, 12–15, 19, 58–59, 93, 96, 108, 131–132, 142, 155, 160–162, 190, 199–200, 240, 271, 274, 326–327, 341, 346–349, 417–418, 428–430, 628
primary somatosensory neurons in evolution, 15, 57
primary visual cortex (striate area in neocortex), 32, 111, 134, 249, 336, 368, 375, 393–395, 397, 399–400, 404–411, 414–415, 442, 525, 535, 597, 610, 613, 620, 622–623, 626, 629, 654, 656, 658–660. *See also* area 17

primitive cellular mechanisms, 12, 25
principal components analysis, 648
procedural learning, memories (implicit learning), 124, 128, 490, 528, 663. *See also* habit formation, learning
Proescholdt, Hilde, 141, 150
proliferation of cells in development, 60, 143–145, 150, 153, 175, 183, 201–203, 217–218, 230, 235, 645–652
propriospinal axons, connections, 98, 102, 161, 163–164, 269, 291, 305
prosencephalon, 8–9, 95, 220–221, 227, 339. *See also* endbrain; limbic endbrain; telencephalon
prosimians, 340, 408, 648, 650
protomap in developing cortex, 652, 654
protozoa, 12, 20, 25, 356
proximal muscles, 158, 283. *See also* axial muscles
pruning effect, 255–257, 554–555
pruning in normal development. *See* self-pruning
pruning of catecholamine axons, 554–555
pseudostratified epithelium, 145
pulvinar nucleus of thalamus, 375, 377, 395, 387, 630–631, 633–634, 639
Purves, Dale, 244, 261
putamen, 319, 365, 422, 520, 522, 546, 549, 562–563, 565–566, 568, 570, 572–576, 578, 580–581, 585
pyramidal cells, neurons, 131–133, 231, 291, 291–295, 312, 346–348, 542, 527, 598–600, 611, 617–619, 625–625
pyramidal motor system, 546, 567
pyramidal tract, 131–134, 186–187, 223–225, 286–288, 404, 567, 570, 628
pyramidal tract lesion, 286–290
pyramidotomy, 287, 289–290

quadriplegic, tetraplegic, 289–290

rabbit, 84, 91, 294, 335, 340–341, 344, 347, 419, 434, 450, 497–498, 539–540
raccoon, 81, 112, 114, 201, 294
Rademacher, G. G. J., 119
radial fascicles (in neocortex), 599–600, 618
radial glial cells, radial glia, 146, 150–151
radial glial guidance of migrating cells, 146, 151, 651, 653
Rakic, Pasko, 146, 231, 614, 647–649, 651–653
Ramón y Cajal, Santiago, 14–16, 22, 35, 37, 38–40, 51, 146–147, 149–150, 235–237, 256, 259, 300, 341, 345–347, 349, 367, 527, 539, 663
Randall, Walter, 475, 481
raphé nuclei, 315–317, 498
rat, Norway rat, 11, 30–31, 34, 36–38, 42, 112–113, 118, 121, 153, 157, 165, 185, 187, 210, 212–213, 232–233, 237, 259, 267, 277, 281, 291–293, 313–315, 319–320, 328–329, 331, 335–336, 338, 348, 350, 353, 365–366, 368–369, 371, 373–374, 381, 384, 388, 392, 394, 400, 402, 407, 409, 468, 470, 477, 483, 485, 490, 513–516, 520, 523–524, 530, 533–535, 540–541, 549,

557,563, 584, 587, 589, 609–610, 632, 636, 638, 640, 652
rattlesnake, 109, 419, 445
ray-finned fish, 83, 104, 385,454, 462
rays (sharks, skates, and rays), 465, 462, 518
receptor cells, receptor neurons, 19,27, 40, 68–71, 73, 107–109, 115, 165, 195, 201, 217, 325–327, 341, 347–348, 350, 368, 417–418, 456, 606
reciprocal inhibition, 159
reciprocal synapses, 23–24
re-crossing of the midline, 207
red nucleus (nucleus ruber), 165, 210–212, 214–215, 266–267, 275–277, 284, 286–287, 495. See also rubrospinal tract
 magnocellular part, 275–277
 parvocellular part, 275–277
reflex chaining, 300–302
reflex channel, 40, 97–98, 158, 198, 428
reflex deglutition (swallowing), 184, 187–188, 193, 200, 302, 331, 450, 474, 476
reflex model of behavior, 26, 39–40, 300–301
reflex spasticity, 123, 289
refractory period, 479
regeneration in CNS, 248–249, 251–252, 258–259, 261, 290, 351
reinforcement learning, 528
reptile, 14–15, 72, 89–90, 106, 219, 333–334, 385, 390, 397, 424, 443, 452, 614
reticular activating system. See ascending reticular activating system
reticular formation, definition, 69
reticular nucleus of thalamus, 365, 398, 422, 566
reticulospinal axons, pathways, 164–165, 172, 193, 211, 282
retinal ganglion cells, 47–49, 192, 257–258, 357, 364, 391, 397, 654
retinal projections, retinofugal axon projections, 355–391
retinal receptor cells, 19
retinocollicular map, retinotectal map. See midbrain tectal map
retinogeniculostriate pathway (retinothalamocortical), illustration, 360, 397
retinotopic map formation and its plasticity, 247, 249, 252
retrograde degeneration, atrophy, 41–42, 287
retrograde transport, 25, 45, 47, 277, 293–294
retrosplenial cortex, 315, 406, 414–415, 492, 495, 515, 522–523, 524, 527, 602, 23–624, 629–630, 632–633
retrovirus-mediated gene transfer, 651
reverberating circuits, 303, 305
reward systems, 80, 126, 210–211, 222, 224, 328–329, 331, 420, 461–462, 479, 498, 562, 579, 598
reward-driven learning, 80. See also reward systems
Rexed, Bror, 156–158
Rexed's layers in spinal cord, 155–159
rhinal fissure, 319, 335, 340, 522, 540–541
rhinencephalon, 126, 333, 487, 537

rhombic lip region, 201–203
rhombomeres, 187–189, 227–228, 372–373
rhythmic potentials, 26–27, 299, 304–305
rodents, 7, 30,111–113, 119, 188, 212–213, 329, 336, 340, 364, 375, 381, 384, 400, 408, 414, 435, 460, 471, 477, 483, 485, 487–488, 494, 528, 534, 540, 564, 567, 575, 577–578, 628, 633–634, 646, 651. See also mouse, rat, hamster, squirrel
roof plate, 94, 142, 144, 147, 153, 176, 181, 183, 201, 203
round window, 423, 427–428
rubrospinal tract, 164–165, 210–212, 267, 275–277, 282, 286–287

sacral spinal cord, 10, 99–101, 153–154, 166, 169–170, 505
sagittal (or parasagittal) section, 5, 7, 185, 187, 266, 274, 314, 327–329, 341, 346–347, 357, 483–484, 486, 638
salivatory nuclei, 192, 194, 200, 272, 330
Sanides, Friedrich, 464
satellite oligodendrocytes, 177
satellite structures of the striatum, 573–574. See also subthalamic nucleus; substantia nigra
scenes, scene perception, 134, 136, 407, 414, 596, 665
Schaffer collaterals, 527
schizophrenia, 502, 551, 553–555, 558
Schwann cells, 17, 141, 143, 156, 159
sea lamprey. See lamprey
secondary auditory areas, 437–438
secondary optic radiations, 398, 628
secondary sensory cell groups, cells, neurons, 13, 58–59, 70, 72–73, 93, 96, 102–103, 108–109, 131, 155, 158, 163, 182–183, 186, 188, 191–200, 270, 272, 325–330, 338–339, 346, 348, 418, 428–429, 431, 433–434, 456, 628
secretory cells (not including neurotransmitter release), 20, 26, 142, 170, 175, 206, 239, 245–246, 311, 333, 456, 467, 472. See also endocrine system; neuroendocrine cells, adrenal, pituitary
segmentation of brain structures, 61, 98, 101, 143, 154, 186–189, 227, 232, 372–373. See also neuromeres; homobox genes; diencephalic neuromeres
segments of neuropil in hindbrain, 186–187
self-assembling proteins, 259, 261
self-pruning, 257
semaphorins, 245–247
sensitization (a type of nonassociative learning), 596, 663
sensorimotor amalgam hypothesis, 291–292, 603
sensory acuity. See acuity, sensory
sensory analyzers, 70, 75, 123, 126
sensory channels. See conduction, channels of
sensory-perceptual functions, neocortical localization, 636, 638–639
sensory placodes, 190, 417–419
septal area (septal nuclei), 230, 470, 488–489, 491–492, 518, 522, 524–525, 528–529, 538, 542, 544, 550–552, 555, 564, 576, 578, 611, 632

septohippocampal connections, 528–529
serial synapses, 24
serotonin system (cells, projections), 315–318, 320, 498, 532, 554, 643
sex differences in brains, sexual differentiation, 501–512
sexual behavior, mating, 64, 70, 72, 74, 102, 118, 120, 165, 168, 212, 301, 341, 351, 450, 474, 498, 506, 540, 696
sexual hormones, steroids. See estrogen; gonadal hormones
sexually dimorphic nucleus, hypothalamus, 501–511
sexual orientation, 501–511
sham rage, 118
shark, 456, 462, 518
shepherd's crook cell, 147–149
Sherrington, Charles Scott, 14, 42, 124, 173, 467
short-axon interneurons (mostly inhibitory), 96–97, 230–231, 252, 257, 261, 438, 607–608, 617–618, 663–664
silent synapses, 660–661
silver stains, Nauta method, 42, 43–44, 78, 227, 253, 377–378, 613, 477
simians, 340, 408, 648, 650
skates, 462
small-world architecture, networks, 416, 598
smooth muscle, 58–59, 93, 97, 166–167, 169, 171–172, 279, 358
snake, 109, 116, 275, 282, 419
sodium channels, ions, 14–15, 17–18
sodium pump, 17, 27
Sokolov, Eugene N., 605
solitary chemosensory cells, 326
solitary tract, nucleus of, 192, 194, 272, 316, 325–330, 484
soma, defined, 14, 16, 22, 237
somatic motor system, 98, 112, 165, 188, 191–194, 214, 220, 272, 278, 280, 283, 569. See pyramidal motor system; pyramidal tract
somatosensory areas of neocortex, 37, 105, 108, 112–116, 131–133, 193, 197, 200, 252, 275, 291–293, 321, 323, 399–400, 406–407, 409, 462, 564, 567, 575, 601–603, 610, 618, 622, 628–632, 652, 654, 659–663
somatosensory inputs, 68, 70, 75, 78, 97, 103, 109, 112, 131, 142, 162–163, 191, 195, 193–194, 197–199, 208, 222, 267–268, 272, 281, 293, 323, 359, 387, 419, 421, 457, 473, 545, 596, 603
somatosensory map in midbrain tectum, 273, 435
somatosensory map in neocortex (a homunculus), 113, 629–631
somatosensory pathways, 113, 629–631. See leminiscal pathways, somotosensory; medial lemniscus; neolemniscus pathways; paleolemniscus, trigeminal lemniscus; spinoreticular
somatosensory Wulst (birds), 113
somites, 143, 186
somitomeres, branchial arches, 183, 186–188
song patterns, birdsong, 443–444, 506–509

songbirds, 443–444, 458, 506–507, 589
sonic hedgehog protein, 142
sound localization, 417, 421, 431, 434–435
spasticity. See reflex spasticity
spatial cognition, 665. See also parietal damage; hippocampal place cells; circuit of Papez; spatial learning, memory
spatial learning, memory, 490, 513, 545, 590, 601
spatial localization. See allocentric direction; egocentric direction; sound localization; place cells; place learning, place memory and recognition
spatial map of the environment, 126, 603
spatial summation, 121, 433
Speidel, 236, 259
Spemann, Hans, 141, 150
Sperry, Roger, 246, 249
spider monkey, 112
spinal accessory nerve (cranial nerve 11), 192, 280
spinal cord development, 139–152
spinal cord, general, 139–174
spinal enlargements, 156, 158, 165, 211, 277–278, 280, 284–285
spinal nerves, 10, 96, 98–101, 109, 171–172, 186, 188, 323. See also Bell-Magendie law
spinal reflexes, 40, 58, 64, 97–98, 121–123, 121–122, 158–160, 163, 165, 270, 289, 291, 362, 450, 474, 450, 505
spinal shock, 121
spinocerebellar tracts, 102–103, 163
spinoreticular pathway, tract, 102–103, 131, 158, 160–162, 195, 323, 359
spinothalamic tract, 102–103, 105, 128, 130–131, 158, 160–162, 193, 199, 210, 214, 246, 323
spinothalamic tract decussation
  development, 246–247
  evolution, 75, 194–197
splenium of corpus callosum, 492, 522, 623, 630, 633
split-brain surgery, 547–548
sponges, sponge cells, 12, 20–21, 25, 56, 326
spontaneous activity, prenatal, effects on development, 654
spontaneous activity. See endogeneous activity
spread of effect (in spinal cord with summation effects), 98
sprouting, 122, 247–248, 252–257, 261, 306, 351, 529–31, 534, 555, 573, 583, 659
squirrel, 212–213, 294, 368, 384, 408
S-R models, 26, 39–40, 59, 299–302
stable song (birdsong), 507
stages of nervous system evolution, suggested, 55–57, 181, 196, 355, 364, 396, 461–462, 517, 562, 568, 605–606, 609
staining, histological, 11, 30–35, 37–39, 41–45, 29–51, 60, 78, 112–113, 157, 190, 203, 227, 390, 400, 420, 508, 581, 584, 598, 617, 619–620
startle reflex, 300

stellate cells, 37, 600, 617–618, 623. *See also* Golgi method; granular cells, granule cells
stem cells, 586–589, 646, 648
Stephan, Heinz, 339–340, 648, 650
stereotropism, 244
strabismus, 655
striatal amygdala, 460
striatal outputs, 457, 459, 462, 409, 561–581
striatonigral tract, 563, 571–572, 574, 579
stretch reflex pathway, monosynaptic, 40, 158–159, 246
stria medularis, 224
striasomes, 578–579
striatal compartments, 578–579
striatal components, 561–581. *See also* dorsal striatum; corpus striatum; limbic striatum; matrix of adult striatum; ventral striatum
striatal connections, pathways, 537–581. *See also* ventral striatum
striatal satellites. *See* satellite structures of the striatum
striate area in neocortex. *See* primary visual cortex (V1, striate cortex)
striated muscle, 13, 40, 58, 93, 97–98, 166, 172, 279
stria terminalis, 314, 316, 319, 329, 383, 488–489, 506, 523, 541–544, 551–552, 578, 631–632
striatum. *See* corpus striatum
Striedter, George F., 61, 83–84, 104, 110, 228, 277, 297, 452–455, 608, 667
Strumwasser, Felix, 26–27
subarachnoid space, 176–177
subcortical proliferative zones. *See* ganglionic eminence; caudal ganglionic eminence; medial ganglionic eminence
subiculum, 490, 522–524, 527, 529, 614
submucosal plexus. *See* enteric nervous system
subpial granular layer, 608, 646, 647
substance P, 578
substantia innominata (SI), 488–489, 496, 538, 544, 550–552, 578, 581
substantia nigra, 211, 313, 315, 462, 562–565, 568, 571–574, 578–579, 583–586, 590
substantia nigra, pars compacta, 565, 573–574
substantia nigra, pars reticulata, 572–574
subthalamic locomotor area, 561
subthalamic nucleus (STN), 573–574, 586. *See also* satellite structures of the striatum
subthalamus, 124–125, 133, 160, 185, 206, 214–215, 200, 223, 268–269, 271, 273, 275, 318–319, 359–361, 367, 371–372, 394, 396, 409, 457, 459, 462, 470, 476–477, 495, 517, 543, 561, 566, 570, 572–575, 586, 603, 618, 631
subthalamus, defined, 125, 215, 220, 372. *See also* diencephalic neuromeres
subventricular zone, 231, 353, 646–647
sulcus limitans, 142, 150, 183, 188
sunfish, 385
superior cervical ganglion, 10, 100, 168, 358

superior colliculi. *See* colliculus, superior
superior longitudinal fasciculus, 635–637
superior olive, medial and lateral, 430–434
supernumerary limb, 258
supplementary motor area, 576, 604
suprachiasmatic nucleus (SCN), 206, 216, 356–360, 373–374, 471, 473, 499, 504–505
supragranular layers (cell layers 2–3 of neocortex), 36, 257, 579, 600, 618, 626, 642, 663–664. *See also* association layers of neocortex
supraoptic nucleus, 357, 471, 484–485, 505
suprasegmental pathways, defined, 98
Sur, Mriganka, 656, 658, 661
swallowing reflex. *See* reflex deglutition
Swanson, Larry, 31, 168, 269–270, 279–281, 320, 474–475, 541, 575, 578
Sylvian fissure, 334, 403, 546, 580–581, 627, 629
symmetric cell division (mitosis), 145, 598, 646, 648–649
sympathetic arousal. *See* arousal, sympathetic
sympathetic nervous system, 142–144, 154, 156, 166–172, 192, 222, 240, 246, 259, 279, 313, 316, 358, 469, 471, 498, 547
functions, 166–171
sympathetic preganglionic motor neurons. *See* preganglionic motor neurons, sympathetic system
synapses, introduced and defined, 14, 16, 20–25
synaptic change, 257, 290, 306, 312, 513, 663. *See also* long-term potentiation; plastic connections, links
synthetic screen characters, 475, 481
Syrian hamster, 33, 44, 247, 290, 300, 327, 357, 363, 368, 378, 392, 419, 457, 524, 607, 627, 647. *See also* hamster
syrinx, vocal output structure in birds, 444, 506, 508–509
system-wide modulations. *See* nonspecific widespread modulation

tail of the caudate, 539, 575–576, 580
tangential association fibers, neocortex. *See* association connections
taste buds, 192, 326–328
taste nucleus (gustatory nucleus), 192, 328–331
taste perception, 69–73, 80, 126, 207, 222, 224, 325–331, 341, 545, 562, 596, 629
taste receptors, 71, 73, 201, 326–327, 330, 341
taste system, inputs, pathways. *See* gustatory system
tectospinal tract, pathway, 164–165, 193, 197, 208, 210–211, 266, 274, 283, 286, 360–361, 363, 390
tectum of midbrain. *See* colliculus
tegmental nuclei (of Gudden), 491–492, 515
telencephalon (endbrain), defined, 8–9, 93–95, 219, 222
teleosts, 76, 104, 108, 212, 275, 367, 385, 388–389, 456
temperature regulation, control, 56, 69, 85, 102, 118, 162, 173–174, 205, 246, 472–473, 484–485, 497
temporalization of the posterior cortex, 400, 402–404, 522, 580, 623

temporal lobe, 334, 340–341, 344, 393, 399–400, 402–404, 410, 415, 422, 430, 438, 440, 442, 444, 520, 522, 524–525, 537, 539, 545–547, 551, 555, 558, 575–576, 580, 627, 634, 637, 639
temporal lobe epilepsy, seizures, 551, 554
temporal neocortical areas, 111, 229, 334, 393, 399–400, 402, 410, 412–413, 415–416, 422, 430, 432, 435, 437–438, 440–442, 520, 538–539, 545, 603–604, 630, 634, 637–639, 659
temporal patterns of movements, 299–307, 604, 665
temporal summation, 98, 300
terminal nerve, 62, 449, 452
testosterone (male hormone). *See* gonadal hormones
tetraplegic, 289–290
tetrapods, 105, 450, 525, 564
thalamic cell groups and their projections, summary, 627–634
thalamic evolution, 124, 366–367, 395–396, 494, 606–610, 642–643, 614, 642–643
thalamic parcellation, 395–396, 603, 608–610, 642
thalamocortical axon trajectories, 611–614, 489
thalamocortical pathways that myelinate early, 34, 399–400
thalamocortical projections summary. *See* thalamic cell groups and their projections, summary
thalamus (dorsal thalamus) defined, 125, 131, 130. *See* diencephalic neuromeres
Thauer, Rudolf, 497–499
third ventricle, 175–176, 372, 388, 484, 486, 505, 553
third visual pathway, transcortical, 412–415
thoracolumbar system. *See* sympathetic nervous system
three motor systems, 278–279
three-spined stickleback fish, 301, 306
three transcortical visual pathways, 409–415
thyroid hormone and thyroid stimulating hormone, 472, 501, 511
Tinbergen, Niko, 301–302, 306
tongue innervation, 188, 192–193, 327
tonotopic maps, 436–438, 440
Toran-Allerand, C.D., 509
torus semicircularis, 388, 418
tracing methods, axonal tract tracing, 42–52, 78, 227, 247,253, 622
transcortical association fibers, pathways. *See* association connections
transitional cortex, 623. *See* paralimbic cortical areas; olfactocentric cortical areas; hippocampocentric areas
transneuronal transport (in axon tracing), 622
transplants, transplanted cells, 242–244, 258–259, 583–586, 587–588, 652, 654
transport, axonal
  anterograde, 25, 45, 47, 406
  retrograde, 25, 45, 47, 277, 293–294
transverse section (frontal, coronal sections), 5,7, 30, 33, 62, 333–334
trapezoid body, 429–432, 434

tree shrew (*Tupaia*), 212–213, 294, 361, 368, 375, 377, 384, 408, 422, 545
trigeminal lemniscus, 193, 199–200
trigeminal nerve (cranial nerve 5), 101, 109, 112, 162, 193–194, 197–200, 267, 272, 275, 280, 284, 323, 327
  mandibular branch, 101, 109
  maxillary branch, 101, 109
  ophthalmic branch, 101, 109
trigeminal nuclei, brainstem trigeminal nuclei, 109, 193, 198, 323
trigeminoreticular pathways, 194,–196, 199, 359
tritiated thymidine, 508, 510, 589, 649–650, 652
trochlear nerve (cranial nerve 4), 192, 194, 272
trophic factors, effects, 239–240, 246, 257–258, 472
tropic effects, 246–247
trout, 452
tuberomammillary nucleus, 318
tufted cells, 294, 343–345, 348–350
turnover of neuronal cells in adults, 350–351, 590
turtle, 104, 219, 232, 395, 398, 425, 612, 615
'tweenbrain (diencephalon), defined, 8, 9, 31. *See also* epithalamus; thalamus; subthalamus; hypothalamic cell groups
two visual pathways. *See* three transcortical visual pathways
tyrosine hydroxylase, 584

U fibers, transcortical, 50, 600, 626, 634–635
umami (specific taste sense), 326
uncinate fasciculus, 635–636
uncus, 344, 525
unimodal association areas of cortex. *See* association areas of neocortex, unimodal (modality specific)
uvula, 271

vagal lobe, 70–73, 201, 327
vagus nerve (10th cranial nerve), 21, 70, 72, 169–171, 173, 187–188, 190, 192, 194, 272, 280, 327, 476, 484
Vagusstoff, 21
valvula of cerebellum, 108, 201, 288
vasopressin. *See* antidiuretic hormone (ADH)
vasopressin-containing subnucleus of SCN, 504–505
vegetative nervous system (autonomic nervous system), 59. *See* autonomic nervous system
venous sinus (next to pituitary), 471
ventral anterior thalamic nucleus (VA), 223, 570, 572, 574, 586, 629–630, 632
ventral cochlear nucleus, 426, 428–429, 431–432, 435
ventral column, spinal cord, defined, 60, 154, 157, 160, 165
ventral horn, of spinal cord, 38, 158, 160, 183, 266, 270, 278, 280, 193, 316, 505
ventral lateral nucleus of thalamus (VL), 223, 292, 570, 629–630
ventral pallidum, 232, 269, 568,570, 575–578, 587
ventral pallium, 79, 218, 228, 232, 338, 539, 570, 576, 587, 606

ventral posterior nucleus (VP and its subdivision), 160, 193, 200, 214, 222–223, 227, 292, 329–330, 570, 629–630
ventral posterolateral nucleus of thalamus (VPL). *See* ventral posterior nucleus; ventrobasal nucleus
ventral posteromedial nucleus (VPM). *See* ventral posterior nucleus; ventrobasal nucleus
ventral prefrontal areas, 348, 410, 412–413, 442, 494, 516, 539, 546, 554, 557, 604, 628, 630, 633, 635, 637. *See* orbital prefrontal areas; association areas of neocortex
ventral roots of spinal nerves, 60, 98, 149, 150, 171, 182, 194, 272. *See also* Bell-Magendie law
ventral striatum (ventromedial striatum), 126, 206, 208, 217, 222, 224, 268–269, 331, 340, 351–352, 457, 460–461, 517–518, 521, 537, 541, 548–549, 551, 556–557, 561–562, 564, 567–570, 572, 574–578, 596, 597, 606, 632
ventral tegmental area (VTA), 210, 214–215, 220, 222, 224, 267, 313, 315, 318, 328–331, 462, 469, 470, 485, 495, 498, 537, 554–55, 562, 568, 576
ventral thalamic nuclei, 630, sometimes called ventral thalamus (of adult brain): VA, VL and VB
ventral thalamus of embryo (subthalamus), 220, 273, 275, 313, 371–373, 394, 462, 575
ventralizing factor (in spinal development), 142, 151. *See* sonic hedgehog protein
ventricular layer, zone, 145, 150, 217–218, 229, 325, 350, 353, 539, 589, 646
ventrobasal nucleus of thalamus: VB, same as VP and its subdivision, 132, 200, 227, 247, 252, 260, 629–630
ventromedial nucleus of hypothalamus (VMH), 31, 319, 357, 477, 484–485, 541–542, 544
ventromedial striatum. *See* ventral striatum
ventromedial thalamic nucleus (VM), 316, 320, 562–563, 572
ventroposteromedial nucleus (VPM). *See* ventral posteromedial nucleus
vermis of cerebellum, 273
vestibular canals, 192, 427
vestibular system, inputs, nerves, nuclei, 14, 68–70, 75, 77, 96, 165, 184, 192–194, 214, 269–274, 285–286, 362, 389, 417, 425, 470, 492, 515–516, 565
vestibulocerebellum, 271, 273–274
vestibulospinal tract, 164–165, 270–271, 274, 283, 286, 288
visceral motor column, 188, 191, 194, 272, 330
visceral nervous system. *See* autonomic nervous system
Vilain, Eric, 509, 512
vipers, 109, 419
Virginia opossum, 291–294, 522
visceral arches, 186. *See also* branchial arches
visceral inputs, 62, 191–194, 205, 210, 214, 220, 267, 272, 325–327, 330, 470 484
visceral sensory column. *See* visceral inputs
Visser, J.A, 119

visual areas, 355–415. *See also* occipital visual areas; neocortex, maps
visual cortical areas, older and newer, 393, 407–409
visual deprivation. *See* deprivation, visual; monocular deprivation
visual grasp reflex, 363
visual images, 76, 81, 134, 136, 207, 352, 355, 358–359, 361, 364, 366–367, 603–605, 666
visual neocortex, 393–415. *See also* occipital visual areas; neocortex, maps
visual object identity. *See* object identification, representation; object identity (perception of)
visual pathways, multiple routes to endbrain, 365, 394–398
visual system, systems. *See* ascending visual pathways; association areas of neocortex, visual; bilateral visual projections, early evolution; occipital visual areas; orientation columns, visual cortex; primary visual cortex; three transcortical visual pathways; visual neocortex
visuomotor pathways through superior colliculus, 212, 409
visuotopic map in the midbrain, 207. *See also* midbrain tectum, map; retinocollicular map, retinotectal map
vocal control systems, birds, 443–444, 506, 508–509
voltage-gated ion channels, 15–18
voluntary actions, 290, 296, 410, 413, 497, 545–546, 567, 581, 605
vomeronasal organ. *See* Jacobson's organ
vomeronasal system in brain. *See* accessory olfactory bulb
vomiting, vomiting reflex, 184, 326, 486
von Baer, Karl Ernst, 89–91
von Economo, Constantin, 624–625, 639
von Gudden, Bernhard, 666, 668
von Monakow, C., 122, 135

Ware, Randle, 51
West African hedgehog, 459–460
whisker fields. *See* barrel fields
widely projecting axon systems, 311–320. *See also* diffuse connections (projections)
Wiesel, Torsten, 305
withdrawal reflex, flexion reflex, 98, 159–160
Wolpert, Lewis, 139
Woods, J.W., 118–119, 135, 511
word deafness, 440
working memory and prefrontal cortex, 115, 410
Wulst, 82, 131, 230, 397, 597

xenografts, 586

zebra finch, 506–507
zona incerta, 185, 319, 360, 394, 477, 570, 631